# Engineering Materials and Processes Desk Reference

# Note from the Publisher

This book has been compiled using extracts from the following books within the range of Materials and Process Engineering books in the Elsevier collection:

Furlani, E.P. (2001) Permanent Magnet and Electromechanical Devices, 9780122699511

Ashby, M. (2005) Materials Selection in Mechanical Design 9780750661683

Smallman, R. E. and Ngan A.H.W. (2007) Physical Metallurgy and Advanced Materials, 9780750669061

Asthana, R. et al (2006) Materials Processing and Manufacturing Science 9780750677165

Messler R.W. (2004) Joining of Materials and Structures, 9780750677578

Ashby, M. and Jones, D. H. (2005) Engineering Materials 2 9780750663816

Crawford, R.J. (1998) Plastics Engineering, 9780750637640

Mills, N. (2005) Plastics 9780750651486

The extracts have been taken directly from the above source books, with some small editorial changes. These changes have entailed the re-numbering of Sections and Figures. In view of the breadth of content and style of the source books, there is some overlap and repetition of material between chapters and significant differences in style, but these features have been left in order to retain the flavour and readability of the individual chapters.

### End of chapter questions

Within the book, several chapters end with a set of questions; please note that these questions are for reference only. Solutions are not always provided for these questions.

### Units of measure

Units are provided in either SI or IP units. A conversion table for these units is provided at the front of the book.

### Upgrade to an Electronic Version

An electronic version of the Desk reference, the *Engineering Materials and Processes e-Mega Reference*, 9781856175876

- A fully searchable Mega Reference eBook, providing all the essential material needed by Engineering Materials and Processes Engineers on a day-to-day basis.
- Fundamentals, key techniques, engineering best practice and rules-of-thumb at one quick click of a button
- Over 1,500 pages of reference material, including over 1,000 pages not included in the print edition

Go to http://www.elsevierdirect.com/9781856175869 and click on **Ebook Available**

# Engineering Materials and Processes Desk Reference

Amsterdam · Boston · Heidelberg · London · New York · Oxford
Paris · San Diego · San Francisco · Sydney · Tokyo
Butterworth-Heinemann is an imprint of Elsevier

Butterworth-Heinemann is an imprint of Elsevier
Linacre House, Jordan Hill, Oxford OX2 8DP, UK
30 Corporate Drive, Suite 400, Burlington, MA 01803, USA

First edition 2009

Copyright © 2009 Elsevier Inc. All rights reserved

No part of this publication may be reproduced, stored in a retrieval system or transmitted in any form
or by any means electronic, mechanical, photocopying, recording or otherwise without the prior written
permission of the publisher

Permissions may be sought directly from Elsevier's Science & Technology Rights Department in Oxford,
UK: phone (+44) (0) 1865 843830; fax (+44) (0) 1865 853333; email: permissions@elsevier.com.
Alternatively visit the Science and Technology website at www.elsevierdirect.com/rights for further information

Notice
No responsibility is assumed by the publisher for any injury and/or damage to persons or property
as a matter of products liability, negligence or otherwise, or from any use or operation of any methods,
products, instructions or ideas contained in the material herein. Because of rapid advances in the
medical sciences, in particular, independent verification of diagnoses and drug dosages should be made

**British Library Cataloguing in Publication Data**
A catalogue record for this book is available from the British Library

**Library of Congress Cataloging-in-Publication Data**
A catalog record for this book is availabe from the Library of Congress

ISBN: 978-1-85-617586-9

For information on all Butterworth-Heinemann publications
visit our web site at elsevierdirect.com

Printed and bound in the United States of America

09 10 11 11 10 9 8 7 6 5 4 3 2 1

Working together to grow
libraries in developing countries

www.elsevier.com | www.bookaid.org | www.sabre.org

ELSEVIER   BOOK AID International   Sabre Foundation

# Contents

Author Biographies ............................................................. vii

Section 1  INTRODUCTION ........................................................ 1
    1.1  Introduction to Engineering materials ................................... 3
    1.2  Science of materials behavior .......................................... 13

Section 2  MATERIALS SELECTION ................................................ 49
    2.1  Materials selection .................................................... 51

Section 3  PROCESSES AND PROCESS SELECTION .................................... 67
    3.1  Processes and process selection ........................................ 69

Section 4  METALS ............................................................. 97
    4.1  Introduction to Metals ................................................. 99
    4.2  Metal structures ...................................................... 107
    4.3  Equilibrium constitution and phase diagrams ........................... 115
    4.4  Physical properties of metals ......................................... 121
    4.5  Mechanical properties of metals ....................................... 157

Section 5  PRODUCTION, FORMING AND JOINING OF METALS ......................... 223
    5.1  Production, forming and joining of metals ............................. 225

Section 6  LIGHT ALLOYS ...................................................... 235
    6.1  Light alloys .......................................................... 237

Section 7  PLASTICS .......................................................... 245
    7.1  Introduction to plastics .............................................. 247
    7.2  General properties of plastics ........................................ 261
    7.3  Processing of plastics ................................................ 287

Section 8  CERAMICS AND GLASSES .............................................. 341
    8.1  Ceramics and glasses .................................................. 343

Section 9  COMPOSITE MATERIALS ............................................... 349
    9.1  Composite materials ................................................... 351

Section 10  MAGNETIC MATERIALS ............................................... 411
    10.1  Magnetic materials .................................................. 413

| | | |
|---|---|---|
| Section 11 | NANOMATERIALS | 447 |
| **11.1** | **Nanomaterials** | **449** |
| Section 12 | JOINING MATERIALS | 495 |
| **12.2** | **Joining materials** | **497** |
| | Index | 525 |

# Author Biographies

**Professor Michael Ashby** is Royal Society Research Professor at Cambridge Engineering Design Centre. He has been associated with the Engineering Design Centre since its inception, as one of the three Principal Investigators. He previously held the post of Professor of Applied Physics in the Division of Engineering and Applied Physics at Harvard University. He is a member of the Royal Society, the Royal Academy of Engineering and the U.S. National Academy of Engineering. He was also the Editor of *Acta Metallurgica* and is now Editor of *Progress in Materials Science*.

**Dr. Rajiv Asthana** is a Professor of Engineering and Technology at the University of Wisconsin-Stout. He is the author or co-author of three books and 132 refereed journal and conference publications, and book chapters. He has served on several committees of the American Society for Materials and has been a Visiting Scientist at NASA Glenn Research Center, and a scientist with the Council of Scientific & Industrial Research (India), among other institutions. He has received multiple awards, including the ASM-IIM Lectureship of American Society for Materials, and the U.S. National Academy of Sciences/NRC COBASE Research Award.

**Professor Roy Crawford** is Vice-Chancellor and President of the University of Waikato, New Zealand. His previous positions include Director of the School of Mechanical and Process Engineering and the Polymer Processing Research Centre at Queens University Belfast and Professor of Mechanical Engineering at the University of Auckland. He has given keynote lectures, courses and seminars all over the world and has published eight books and over 350 research papers. He is a Fellow of a number of professional and academic organisations and was elected to the Association of Rotational Moulders Hall of Fame.

**Dr. Edward P. Furlani** is currently a senior scientist in the research laboratories in the Eastman Kodak Company. He is also Research Professor in the Institute for Lasers, Photonics and Biophotonics at the University at Buffalo. Dr. Furlani has extensive experience in the area of applied magnetics. He has authored over 60 publications in scientific journals and holds over 140 US patents.

**Dr Robert Messler**, after 16 years in the materials industry, served as Technical Director and Associate Director of the Center for Manufacturing Productivity at Rensselaer. He joined the faculty as Associate Professor and Director of the Materials Joining Laboratory, and served as Associate Dean for Academic & Student Affairs for the School. He has authored four technical books in welding and joining, and over 140 papers in diverse areas of materials engineering. He has received numerous departmental, School, Institute, and national awards.

**Dr. Nigel Mills** was Reader in Polymer Engineering in the Metallurgy and Materials Department, earning honorary status after retirement, and chairman of the British Standards committee for motorcycle helmets. He previously worked for ICI Petrochemical and Polymer Laboratory in Runcorn. He has published many papers on foam and polymer properties and applications and authored the *Polymer Foams Handbook*.

**Professor Ray Smallman**, is Emeritus Professor of Metallurgy and Materials Science and honorary senior research fellow at the University of Birmingham. He spent his early career with the Atomic Energy Research Establishment in Harwell, UK. He is now President of the Federation of European Materials Societies and was prominent in its development. He has served on many committees and councils of various associations and received the *Acta Materialia* Gold Medal for his ability and leadership in materials science.

**Dr. Ashok V. Kumar** is Associate Professor in the Department of Mechanical Engineering at the University of Florida. His main research focus is the broad area of computational methods and design optimization.

**Dr. Narendra B. Dahotre** is a Professor with joint appointment with Oak Ridge National Laboratory and Department of Materials Science and Engineering of the University of Tennessee-Knoxville. He is also a senior faculty member of the Center for Laser Applications at the University of Tennessee Space Institute-Tullahoma. He is author of two technical books and editor/co-editor of 14 books. He is author of over 125 reviewed technical journal articles.

**Dr. D.R.H. Jones** is Emeritus Professor in the Mechanics and Materials Division at the University of Cambridge, UK

**Professor A.H.W. Ngan** is a member of the Mechanical Engineering Department at the University of Hong Kong. In 2007, he was awarded the Rosehain Medal and Prize by the Institute of Materials, Minerals and Mining, UK, and in 2008, he was conferred a higher doctorate (DSc) by the University of Birmingham.

# Section One

## Introduction

# Chapter 1.1

# Introduction to Engineering materials

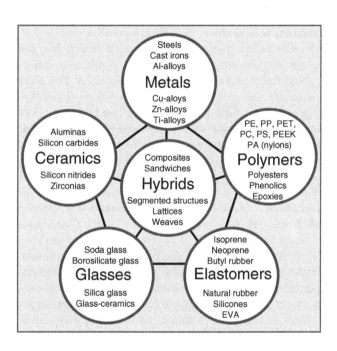

## 1.1.1 Introduction and synopsis

Materials, one might say, are the food of design. This chapter presents the menu: the full shopping list of materials. A successful product—one that performs well, is good value for money and gives pleasure to the user—uses the best materials for the job, and fully exploits their potential and characteristics. Brings out their flavor, so to speak.

The families of materials—metals, polymers, ceramics, and so forth—are introduced in Section 1.1.2. But it is not, in the end, a *material* that we seek; it is a certain *profile of properties*—the one that best meets the needs of the design. The properties, important in thermo-mechanical design, are defined briefly in Section 1.1.3. It makes boring reading. The reader confident in the definitions of moduli, strengths, damping capacities, thermal and electrical conductivities and the like, may wish to skip this, using it for reference, when needed, for the precise meaning and units of the data in the Property Charts that come later. Do not, however, skip Sections 1.1.2—it sets up the classification structure that is used throughout the book. The chapter ends, in the usual way, with a summary.

## 1.1.2 The families of engineering materials

It is helpful to classify the materials of engineering into the six broad families shown in Figure 1.1-1: metals, polymers, elastomers, ceramics, glasses, and hybrids. The members of a family have certain features in common: similar properties, similar processing routes, and, often, similar applications.

*Metals* have relatively high moduli. Most, when pure, are soft and easily deformed. They can be made strong by alloying and by mechanical and heat treatment, but they remain ductile, allowing them to be formed by deformation processes. Certain high-strength alloys (spring steel, for instance) have ductilities as low as 1 percent, but even this is enough to ensure that the material yields before it fractures and that fracture, when it occurs, is of a tough, ductile type. Partly because of their ductility, metals are prey to fatigue and of all the classes of material, they are the least resistant to corrosion.

*Engineering Materials and Processes Desk Reference*; ISBN: 9781856175869
Copyright © 2009 Elsevier Ltd; All rights of reproduction, in any form, reserved.

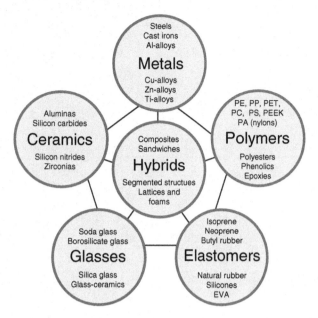

**Figure 1.1-1** The menu of engineering materials. The basic families of metals, ceramics, glasses, polymers, and elastomers can be combined in various geometries to create hybrids.

*Ceramics* too, have high moduli, but, unlike metals, they are brittle. Their "strength" in tension means the brittle fracture strength; in compression it is the brittle crushing strength, which is about 15 times larger. And because ceramics have no ductility, they have a low tolerance for stress concentrations (like holes or cracks) or for high-contact stresses (at clamping points, for instance). Ductile materials accommodate stress concentrations by deforming in a way that redistributes the load more evenly, and because of this, they can be used under static loads within a small margin of their yield strength. Ceramics cannot. Brittle materials always have a wide scatter in strength and the strength itself depends on the volume of material under load and the time for which it is applied. So ceramics are not as easy to design with as metals. Despite this, they have attractive features. They are stiff, hard, and abrasion-resistant (hence their use for bearings and cutting tools); they retain their strength to high temperatures; and they resist corrosion well.

*Glasses* are non-crystalline ("amorphous") solids. The commonest are the soda-lime and boro-silicate glasses familiar as bottles and ovenware, but there are many more. Metals, too, can be made non-crystalline by cooling them sufficiently quickly. The lack of crystal structure suppresses plasticity, so, like ceramics, glasses are hard, brittle and vulnerable to stress concentrations.

*Polymers* are at the other end of the spectrum. They have moduli that are low, roughly 50 times less than those of metals, but they can be strong—nearly as strong as metals. A consequence of this is that elastic deflections can be large. They creep, even at room temperature, meaning that a polymer component under load may, with time, acquire a permanent set. And their properties depend on temperature so that a polymer that is tough and flexible at 20°C may be brittle at the 4°C of a household refrigerator, yet creep rapidly at the 100°C of boiling water. Few have useful strength above 200°C. If these aspects are allowed for in the design, the advantages of polymers can be exploited. And there are many. When combinations of properties, such as strength-per-unit-weight, are important, polymers are as good as metals. They are easy to shape: complicated parts performing several functions can be molded from a polymer in a single operation. The large elastic deflections allow the design of polymer components that snap together, making assembly fast and cheap. And by accurately sizing the mold and pre-coloring the polymer, no finishing operations are needed. Polymers are corrosion resistant and have low coefficients of friction. Good design exploits these properties.

*Elastomers* are long-chain polymers above their glass-transition temperature, $T_g$. The covalent bonds that link the units of the polymer chain remain intact, but the weaker Van der Waals and hydrogen bonds that, below $T_g$, bind the chains to each other, have melted. This gives elastomers unique property profiles: Young's moduli as low as $10^{-3}$ GPa ($10^5$ time less than that typical of metals) that increase with temperature (all other solids show a decrease), and enormous elastic extension. Their properties differ so much from those of other solids that special tests have evolved to characterize them. This creates a problem: if we wish to select materials by prescribing a desired attribute profile (as we do later in this book), then a prerequisite is a set of attributes common to all materials. To overcome this, we settle on a common set for use in the first stage of design, estimating approximate values for anomalies like elastomers. Specialized attributes, representative of one family only, are stored separately; they are for use in the later stages.

*Hybrids* are combinations of two or more materials in a pre-determined configuration and scale. They combine the attractive properties of the other families of materials while avoiding some of their drawbacks. Their design is the subject of Chapters 13 and 14. The family of hybrids includes fiber and particulate composites, sandwich structures, lattice structures, foams, cables, and laminates. And almost all the materials of nature—wood, bone, skin, leaf—are hybrids. Fiber-reinforced composites are, of course, the most familiar. Most of those at present available to the engineer have a polymer matrix reinforced by fibers of glass, carbon or Kevlar (an aramid). They are light, stiff and strong, and they can be tough. They, and other hybrids using a polymer as one component, cannot be used above 250°C because the polymer softens, but at room temperature their performance can be outstanding. Hybrid components are

# Introduction to Engineering materials CHAPTER 1.1

**Table 1.1-1** Basic design-limiting material properties and their usual SI units*

| Class | Property | Symbol and units | |
|---|---|---|---|
| General | Density | $\rho$ | (kg/m$^3$ or Mg/m$^3$) |
| | Price | $C_m$ | ($/kg) |
| Mechanical | Elastic moduli (Young's, shear, bulk) | $E, G, K$ | (GPa) |
| | Yield strength | $\sigma_y$ | (MPa) |
| | Ultimate strength | $\sigma_u$ | (MPa) |
| | Compressive strength | $\sigma_c$ | (MPa) |
| | Failure strength | $\sigma_f$ | (MPa) |
| | Hardness | $H$ | (Vickers) |
| | Elongation | $\varepsilon$ | (—) |
| | Fatigue endurance limit | $\sigma_e$ | (MPa) |
| | Fracture toughness | $K_{1C}$ | (MPa.m$^{1/2}$) |
| | Toughness | $G_{1C}$ | (kJ/m$^2$) |
| | Loss coefficient (damping capacity) | $\eta$ | (—) |
| Thermal | Melting point | $T_m$ | (C or K) |
| | Glass temperature | $T_g$ | (C or K) |
| | Maximum service temperature | $T_{max}$ | (C or K) |
| | Minimum service temperature | $T_{max}$ | (C or K) |
| | Thermal conductivity | $\lambda$ | (W/m.K) |
| | Specific heat | $C_p$ | (J/kg.K) |
| | Thermal expansion coefficient | $\alpha$ | (K$^{-1}$) |
| | Thermal shock resistance | $\Delta T_s$ | (C or K) |
| Electrical | Electrical resistivity | $\rho_e$ | ($\Omega$.m or $\mu\Omega$.cm) |
| | Dielectric constant | $\varepsilon_d$ | (—) |
| | Breakdown potential | $V_b$ | (10$^6$ V/m) |
| | Power factor | $P$ | (—) |
| Optical | Optical, transparent, translucent, opaque | Yes/No | |
| | Refractive index | n | (—) |
| Eco-properties | Energy/kg to extract material | $E_f$ | (MJ/kg) |
| | $CO_2$/kg to extract material | $CO_2$ | (kg/kg) |
| Environmental resistance | Oxidation rates | Very low, low, average, high, very high | |
| | Corrosion rates | | |
| | Wear rate constant | $K_A$ | MPa$^{-1}$ |

* Conversion factors to imperial and cgs units appear inside the back and front covers of this book.

expensive and they are relatively difficult to form and join. So despite their attractive properties the designer will use them only when the added performance justifies the added cost. Today's growing emphasis on high performance and fuel efficiency provides increasing drivers for their use.

## 1.1.3 The definitions of material properties

Each material can be thought of as having a set of attributes: its properties. It is not a material, *per se*, that the designer seeks; it is a specific combination of these attributes: a *property-profile*. The material name is the identifier for a particular property-profile.

The properties themselves are standard: density, modulus, strength, toughness, thermal and electrical conductivities, and so on (Tables 1.1-1). For completeness and precision, they are defined, with their limits, in this section. If you think you know how properties are defined, you might jump to Section 1.1.5, returning to this section only if need arises.

## General properties

The *density* (units: kg/m$^3$) is the mass per unit volume. We measure it today as Archimedes did: by weighing in air and in a fluid of known density.

The *price*, $C_m$ (units: $/kg), of materials spans a wide range. Some cost as little as $0.2/kg, others as much as $1000/kg. Prices, of course, fluctuate, and they depend on the quantity you want and on your status as a "preferred customer" or otherwise. Despite this uncertainty, it is useful to have an approximate price, useful in the early stages of selection.

## Mechanical properties

The *elastic modulus* (units: GPa or GN/m$^2$) is defined as the slope of the linear-elastic part of the stress–strain curve (Figure 1.1-2). Young's modulus, $E$, describes response to tensile or compressive loading, the shear modulus, $G$, describes shear loading and the bulk modulus, $K$, hydrostatic pressure. Poisson's ratio, $\nu$, is dimensionless: it is the negative of the ratio of the lateral strain, $\varepsilon_2$, to the axial strain, $\varepsilon_1$, in axial loading:

$$\nu = -\frac{\varepsilon_2}{\varepsilon_1}$$

In reality, moduli measured as slopes of stress–strain curves are inaccurate, often low by a factor of 2 or more, because of contributions to the strain from anelasticity, creep and other factors. Accurate moduli are measured dynamically: by exciting the natural vibrations of a beam or wire, or by measuring the velocity of sound waves in the material.

In an isotropic material, the moduli are related in the following ways:

$$E = \frac{3G}{1 + G/3K}; \quad G = \frac{E}{2(1+\nu)}; \quad K = \frac{E}{3(1-2\nu)}$$
(1.1.1)

Commonly $\nu \approx 1/3$ when

$$G \approx \frac{3}{8}E \text{ and } K \approx E \qquad (1.1.2a)$$

Elastomers are exceptional. For these $\nu \approx 1/2$ when

$$G \approx \frac{1}{3}E \text{ and } K \gg E \qquad (1.1.2b)$$

Data sources like those described in Chapter 15 list values for all four moduli. In this book we examine data for $E$; approximate values for the others can be derived from equation (1.1.2) when needed.

The *strength* $\sigma_f$, of a solid (units: MPa or MN/m$^2$) requires careful definition. For metals, we identify $\sigma_f$ with the 0.2 percent offset yield strength $\sigma_y$ (Figure 1.1-2), that is, the stress at which the stress–strain curve for axial loading deviates by a strain of 0.2 percent from the linear-elastic line. It is the same in tension and compression. For polymers, $\sigma_f$ is identified as the stress at which the stress–strain curve becomes markedly non-linear: typically, a strain of 1 percent (Figure 1.1-3). This may be caused by shear-yielding: the irreversible slipping of molecular chains; or it may be caused by crazing: the formation of low density, crack-like volumes that scatter light, making the polymer look white. Polymers are a little stronger ($\approx 20$ percent) in compression than in tension. Strength, for ceramics and glasses, depends strongly on the mode of loading (Figure 1.1-4). In tension, "strength" means the fracture strength, $\sigma_t$. In compression it means the crushing strength $\sigma_c$, which is much larger; typically

$$\sigma_c = 10 \text{ to } 15 \, \sigma_t \qquad (1.1.3)$$

When the material is difficult to grip (as is a ceramic), its strength can be measured in bending. The *modulus of rupture* or MoR (units: MPa) is the maximum surface stress in a bent beam at the instant of failure (Figure 1.1-5).

One might expect this to be the same as the strength measured in tension, but for ceramics it is larger (by

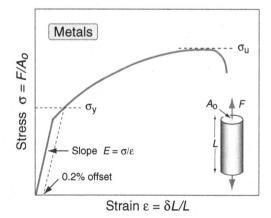

**Figure 1.1-2** The stress–strain curve for a metal, showing the modulus, $E$, the 0.2 percent yield strength, $\sigma_Y$, and the ultimate strength, $\sigma_u$.

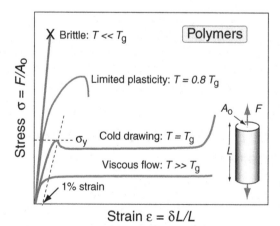

**Figure 1.1-3** Stress–strain curves for a polymer, below, at and above its glass transition temperature, $T_g$.

# Introduction to Engineering materials  CHAPTER 1.1

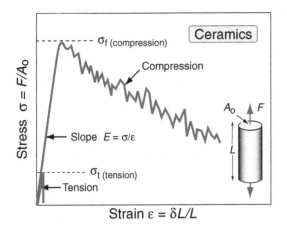

**Figure 1.1-4** Stress–strain curves for a ceramic in tension and in compression. The compressive strength $\sigma_c$ is 10 to 15 times greater than the tensile strength $\sigma_t$.

a factor of about 1.3) because the volume subjected to this maximum stress is small and the probability of a large flaw lying in it is small also; in simple tension all flaws see the maximum stress.

The strength of a composite is best defined by a set deviation from linear-elastic behavior: 0.5 percent is sometimes taken. Composites that contain fibers (and this includes natural composites like wood) are a little weaker (up to 30 percent) in compression than tension because fibers buckle. In subsequent chapters, $\sigma_f$ for composites means the tensile strength.

Strength, then, depends on material class and on mode of loading. Other modes of loading are possible: shear, for instance. Yield under multi-axial loads is related to that in simple tension by a yield function. For metals, the Von Mises' yield function is a good description:

$$(\sigma_1 - \sigma_2)^2 + (\sigma_2 - \sigma_3)^2 + (\sigma_3 - \sigma_1)^2 = 2\sigma_f^2 \quad (1.1.4)$$

where $\sigma_1$, $\sigma_2$, and $\sigma_3$ are the principal stresses, positive when tensile; $\sigma_1$, by convention, is the largest or most positive, $\sigma_3$ the smallest or least. For polymers the yield function is modified to include the effect of pressure:

$$(\sigma_1 - \sigma_2)^2 + (\sigma_2 - \sigma_3)^2 + (\sigma_3 - \sigma_1)^2$$
$$= 2\sigma_f^2\left(1 + \frac{\beta p}{K}\right)^2 \quad (1.1.5)$$

where $K$ is the bulk modulus of the polymer, $\beta \approx 2$ is a numerical coefficient that characterizes the pressure dependence of the flow strength and the pressure $p$ is defined by

$$p = -\frac{1}{3}(\sigma_1 + \sigma_2 + \sigma_3)$$

For ceramics, a Coulomb flow law is used:

$$\sigma_1 - B\sigma_2 = C \quad (1.1.6)$$

where $B$ and $C$ are constants.

The *ultimate (tensile) strength*, $\sigma_u$ (units: MPa), is the nominal stress at which a round bar of the material, loaded in tension, separates (see Figure 1.1-2). For brittle solids—ceramics, glasses, and brittle polymers—it is the same as the failure strength in tension. For metals, ductile polymers and most composites, it is larger than the strength, $\sigma_f$, by a factor of between 1.1 and 3 because of work hardening or (in the case of composites) load transfer to the reinforcement.

Cyclic loading not only dissipates energy; it can also cause a crack to nucleate and grow, culminating in fatigue failure. For many materials there exists a fatigue or *endurance limit*, $\sigma_e$ (units: MPa), illustrated by the $\Delta\sigma - N_f$ curve of Figure 1.1-6. It is the stress amplitude

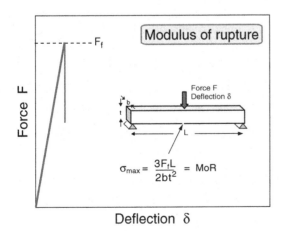

**Figure 1.1-5** The MoR is the surface stress at failure in bending. It is equal to, or slightly larger than the failure stress in tension.

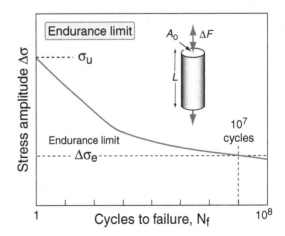

**Figure 1.1-6** The endurance limit, $\Delta\sigma_e$, is the cyclic stress that causes failure in $N_f = 10^7$ cycles.

$\Delta\sigma$ below which fracture does not occur, or occurs only after a very large number ($N_f > 10^7$) of cycles.

The *hardness*, $H$, of a material is a crude measure of its strength. It is measured by pressing a pointed diamond or hardened steel ball into the surface of the material (Figure 1.1-7). The hardness is defined as the indenter force divided by the projected area of the indent. It is related to the quantity we have defined as $\sigma_f$ by

$$H \approx 3\sigma_f \tag{1.1.7}$$

and this, in the SI system, has units of MPa. Hardness is most usually reported in other units, the commonest of which is the Vickers $H_v$ scale with units of kg/mm². It is related to $H$ in the units used here by

$$H_v = \frac{H}{10}$$

The *toughness*, $G_{1C}$, (units: kJ/m²), and the *fracture toughness*, $K_{1C}$, (units: MPa.m$^{1/2}$ or MN/m$^{1/2}$), measure the resistance of a material to the propagation of a crack. The fracture toughness is measured by loading a sample containing a deliberately-introduced crack of length $2c$ (Figure 1.1-8), recording the tensile stress $\sigma_c$ at which the crack propagates. The quantity $K_{1C}$ is then calculated from

$$K_{1C} = Y\sigma_c\sqrt{\pi c} \tag{1.1.8}$$

and the toughness from

$$G_{1C} = \frac{K_{1C}^2}{E(1+\nu)} \tag{1.1.9}$$

where $Y$ is a geometric factor, near unity, that depends on details of the sample geometry, $E$ is Young's modulus and

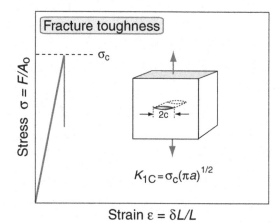

**Figure 1.1-8** The fracture toughness, $K_{1c}$, measures the resistance to the propagation of a crack. The failure strength of a brittle solid containing a crack of length $2c$ is $K_{1C} = Y(\sigma_c/\sqrt{\pi c})$ where $Y$ is a constant near unity.

$\nu$ is Poisson's ratio. Measured in this way $K_{1C}$ and $G_{1C}$ have well-defined values for brittle materials (ceramics, glasses, and many polymers). In ductile materials a plastic zone develops at the crack tip, introducing new features into the way in which cracks propagate that necessitate more involved characterization. Values for $K_{1C}$ and $G_{1C}$ are, nonetheless, cited, and are useful as a way of ranking materials.

The *loss-coefficient*, $\eta$ (a dimensionless quantity), measures the degree to which a material dissipates vibrational energy (Figure 1.1-9). If a material is loaded elastically to a stress, $\sigma_{max}$, it stores an elastic energy

$$U = \int_0^{\sigma_{max}} \sigma d\varepsilon \approx \frac{1}{2}\frac{\sigma_{max}^2}{E}$$

per unit volume. If it is loaded and then unloaded, it dissipates an energy

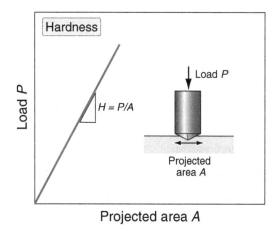

**Figure 1.1-7** Hardness is measured as the load $P$ divided by the projected area of contact, A, when a diamond-shaped indenter is forced into the surface.

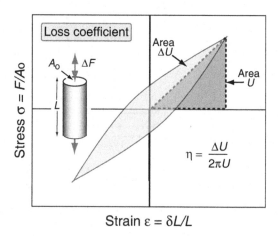

**Figure 1.1-9** The loss coefficient $\eta$ measures the fractional energy dissipated in a stress–strain cycle.

$$\Delta U = \oint \sigma \, d\varepsilon$$

The loss coefficient is

$$\eta = \frac{\Delta U}{2\pi U} \qquad (1.1.10)$$

The value of $\eta$ usually depends on the time-scale or frequency of cycling.

Other measures of damping include the *specific damping capacity*, $D = \Delta U/U$, the *log decrement*, $\Delta$ (the log of the ratio of successive amplitudes of natural vibrations), the *phase-lag*, $\delta$, between stress and strain, and the "*Q*"-*factor* or *resonance factor*, $Q$. When damping is small ($\eta < 0.01$) these measures are related by

$$\eta = \frac{D}{2\pi} = \frac{\Delta}{\pi} = \tan\delta = \frac{1}{Q} \qquad (1.1.11)$$

but when damping is large, they are no longer equivalent.

## Thermal properties

Two temperatures, the *melting temperature*, $T_m$, and the *glass temperature*, $T_g$ (units for both: K or C) are fundamental because they relate directly to the strength of the bonds in the solid. Crystalline solids have a sharp melting point, $T_m$. Non-crystalline solids do not; the temperature $T_g$ characterizes the transition from true solid to very viscous liquid. It is helpful, in engineering design, to define two further temperatures: the *maximum* and *minimum service temperatures* $T_{max}$ and $T_{min}$ (both: K or C). The first tells us the highest temperature at which the material can reasonably be used without oxidation, chemical change, or excessive creep becoming a problem. The second is the temperature below which the material becomes brittle or otherwise unsafe to use.

The rate at which heat is conducted through a solid at steady state (meaning that the temperature profile does not change with time) is measured by the *thermal conductivity*, $\lambda$ (units: W/m.K). Figure 1.1-10 shows how it is measured: by recording the heat flux $q$ (W/m$^2$) flowing through the material from a surface at higher temperature $T_1$ to a lower one at $T_2$ separated by a distance $X$. The conductivity is calculated from Fourier's law:

$$q = -\lambda \frac{dT}{dX} = \lambda \frac{(T_1 - T_2)}{X} \qquad (1.1.12)$$

The measurement is not, in practice, easy (particularly for materials with low conductivities), but reliable data are now generally available.

When heat flow is transient, the flux depends instead on the *thermal diffusivity*, $a$ (units: m$^2$/s), defined by

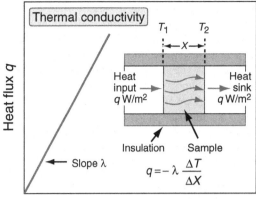

**Figure 1.1-10** The thermal conductivity $\lambda$ measures the flux of heat driven by a temperature gradient $dT/dX$.

$$a = \frac{\lambda}{\rho C_p} \qquad (1.1.13)$$

where $\rho$ is the density and $C_p$ is the *specific heat at constant pressure* (units: J/kg.K). The thermal diffusivity can be measured directly by measuring the decay of a temperature pulse when a heat source, applied to the material, is switched off; or it can be calculated from $\lambda$, via equation (1.1.13). This requires values for $C_p$. It is measured by the technique of calorimetry, which is also the standard way of measuring the glass temperature $T_g$.

Most materials expand when they are heated (Figure 1.1-11). The thermal strain per degree of temperature change is measured by the *linear thermal-expansion coefficient*, $\alpha$ (units: K$^{-1}$ or, more conveniently, as "microstrain/C" or $10^{-6}$C$^{-1}$). If the material is thermally isotropic, the volume expansion, per degree, is $3\alpha$. If it is anisotropic, two or more coefficients are required, and the volume expansion becomes the sum of the principal thermal strains.

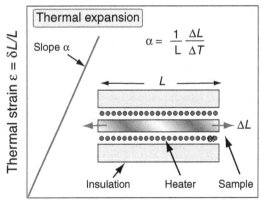

**Figure 1.1-11** The linear-thermal expansion coefficient $\alpha$ measures the change in length, per unit length, when the sample is heated.

The *thermal shock resistance* $\Delta T_s$ (units: K or C) is the maximum temperature difference through which a material can be quenched suddenly without damage. It, and the *creep resistance*, are important in high-temperature design. Creep is the slow, time-dependent deformation that occurs when materials are loaded above about $\frac{1}{3}T_m$ or $\frac{2}{3}T_g$. Design against creep is a specialized subject. Here we rely instead on avoiding the use of a material above its maximum service temperature, $T_{max}$, or, for polymers, its "heat deflection temperature".

## Electrical properties

The *electrical resistivity*, $\rho_e$ (SI units $\Omega$.m, but commonly reported in units of $\mu\Omega$.cm) is the resistance of a unit cube with unit potential difference between a pair of it faces. It is measured in the way shown in Figure 1.1-12. It has an immense range, from a little more than $10^{-8}$ in units of $\Omega$.m (equal to 1 $\mu\Omega$.cm) for good conductors to more than $10^{16}$ $\Omega$.m ($10^{24}$ $\mu\Omega$.cm) for the best insulators. The electrical conductivity is simply the reciprocal of the resisitivity.

When an insulator is placed in an electric field, it becomes polarized and charges appear on its surfaces that tend to screen the interior from the electric field. The tendency to polarize is measured by the *dielectric constant*, $\varepsilon_d$ (a dimensionless quantity). Its value for free space and, for practical purposes, for most gasses, is 1. Most insulators have values between 2 and 30, though low-density foams approach the value 1 because they are largely air.

The *breakdown potential* (units: MV/m) is the electrical potential gradient at which an insulator breaks down and a damaging surge of current flows through it. It is measured by increasing, at a uniform rate, a 60 Hz alternating potential applied across the faces of a plate of the material until breakdown occurs.

Polarization in an electric field involves the motion of charge particles (electrons, ions, or molecules that carry a dipole moment). In an oscillating field, the charged particles are driven between two alternative configurations. This charge-motion corresponds to an electric current that—if there were no losses—would be 90° out of phase with the voltage. In real dielectrics, the motion of the charged particles dissipates energy and the current leads the voltage by something less that 90°; the loss angle $\theta$ is the deviation. The loss tangent is the tangent of this angle. The *power factor* (dimensionless) is the sine of the loss angle, and measures the fraction of the energy stored in the dielectric at peak voltage that is dissipated in a cycle; when small, it is equal to the loss tangent. The *loss factor* is the loss tangent times the dielectric constant.

## Optical properties

All materials allow some passage of light, although for metals it is exceedingly small. The speed of light when in the material, $v$, is always less than that in vacuum, $c$. A consequence is that a beam of light striking the surface of such a material at an angle $\alpha$, the angle of incidence, enters the material at an angle $\beta$, the angle of refraction. The *refractive index*, $n$ (dimensionless), is

$$n = \frac{c}{v} = \frac{\sin\alpha}{\sin\beta} \qquad (1.1.14)$$

It is related to the dielectric constant, $\varepsilon_d$, by

$$n \approx \sqrt{\varepsilon_d}$$

It depends on wavelength. The denser the material, and the higher its dielectric constant, the larger is the refractive index. When $n = 1$, the entire incident intensity enters the material, but when $n > 1$, some is reflected. If the surface is smooth and polished, it is reflected as a beam; if rough, it is scattered. The percentage reflected, $R$, is related to the refractive index by

$$R = \left(\frac{n-1}{n+1}\right)^2 \times 100 \qquad (1.1.15)$$

As $n$ increases, the value of $R$ tends to 100 percent.

## Eco properties

The *contained* or *production energy* (units MJ/kg) is the energy required to extract 1 kg of a material from its ores and feedstocks. The associated $CO_2$ production (units: kg/kg) is the mass of carbon dioxide released into the atmosphere during the production of 1 kg of material. These and other eco-attributes are the subject of Chapter 16.

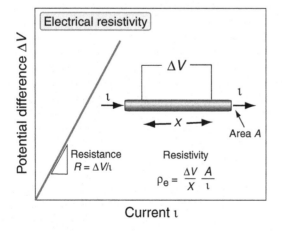

**Figure 1.1-12** Electrical resistivity is measured as the potential gradient $\Delta V/X$ divided by the current density, $\iota/A$.

## Environmental resistance

Some material attributes are difficult to quantify, particularly those that involve the interaction of the material within the environments in which it must operate. Environmental resistance is conventionally characterized on a discrete 5-point scale: very good, good, average, poor, very poor. "Very good" means that the material is highly resistant to the environment, "very poor" that it is completely non-resistant or unstable. The categorization is designed to help with initial screening; supporting information should always be sought if environmental attack is a concern. Ways of doing this are described later.

Wear, like the other interactions, is a multi-body problem. None-the-less it can, to a degree, be quantified. When solids slide (Figure 1.1-13) the volume of material lost from one surface, per unit distance slid, is called the wear rate, $W$.

The wear resistance of the surface is characterized by the *Archard wear constant*, $K_A$ (units: $MPa^{-1}$), defined by the equation

$$\frac{W}{A} = K_A P \qquad (1.1.16)$$

where $A$ is the area of the surface and $P$ the normal force pressing them together. Approximate data for $K_A$ appear in Chapter 4, but must be interpreted as the property of the sliding couple, not of just one member of it.

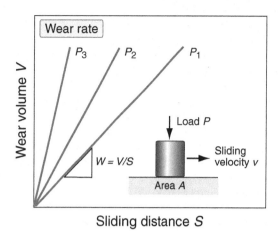

**Figure 1.1-13** Wear is the loss of material from surfaces when they slide. The wear resistance is measured by the Archard wear constant $K_A$.

### 1.1.4 Summary and conclusions

There are six important families of materials for mechanical design: metals, ceramics, glasses, polymers, elastomers, and hybrids that combine the properties of two or more of the others. Within a family there is certain common ground: ceramics as a family are hard, brittle, and corrosion resistant; metals are ductile, tough, and good thermal and electrical conductors; polymers are light, easily shaped, and electrical insulators, and so on—that is what makes the classification useful. But in design we wish to escape from the constraints of family, and think, instead, of the material name as an identifier for a certain property-profile—one that will, in later chapters, be compared with an "ideal" profile suggested by the design, guiding our choice. To that end, the properties important in thermo-mechanical design were defined in this chapter. In Chapter 4 we develop a way of displaying these properties so as to maximize the freedom of choice.

### 1.1.5 Further reading

Definitions of material properties can be found in numerous general texts on engineering materials, among them those listed here.

Ashby, M.F. and Jones, D.R.H. (1996). *Engineering Materials 1, and Introduction to their Properties and Applications*, 2nd edition, Pergamon Press, Oxford, U.K. ISBN 0-7506-3081-7.

ASM Engineered Materials Handbook (2004). "Testing and characterisation of polymeric materials", ASM International, Metals Park, OH, USA *(An on-line, subscription-based resource, detailing testing procedures for polymers.)*

ASM Handbooks Volume 8 (2004) "Mechanical testing and evaluation". ASM International, Metals Park, Ohio, USA *(An on-line, subscription-based resource, detailing testing procedures for metals and ceramics.).*

ASTM Standards (1988) Vol. 08.01 and 08.02 Plastics; (1989) Vol. 04.02 Concrete; (1990) Vols. 01.01 to 01.05 Steels; Vol. 0201 Copper alloys; Vol. 02.03 Aluminum alloys; Vol. 02.04 Non-ferrous alloys; Vol. 02.05 Coatings; Vol. 03.01 Metals at high and low temperatures; Vol. 04.09 Wood; Vols 09.01 and 09.02 Rubber, American Society for Testing Materials, 1916 Race Street, Philadelphia, PA, USA. ISBN 0-8031-1581-4. *(The ASTM set standards for materials testing.)*

Callister, W.D. (2003) *Materials Science and Engineering, an Introduction*, 6th edition. John Wiley, New York, USA. ISBN 0-471-13576-3. *(A well-respected materials text, now in its 6th edition, widely used for materials teaching in North America.)*

Charles, J.A. Crane, F.A.A. and Furness, J.A.G. (1997) *Selection and Use of Engineering Materials*, 3rd edition. Butterworth-Heinemann, Oxford, UK. ISBN 0-7506-3277-1. *(A materials-science approach to the selection of materials.)*

Dieter, G.E. (1991). *Engineering Design, a Materials and Processing Approach*, 2nd edition. McGraw-Hill, New York, USA. ISBN 0-07-100829-2. *(A well-balanced and respected text focussing on the place of materials and processing in technical design.)*

Farag, M.M. (1989) *Selection of Materials and Manufacturing Processes for Engineering Design*. Prentice-Hall, Englewood Cliffs, NJ, USA. ISBN 0-13-575192-6. *(A materials-science approach to the selection of materials.)*

# Chapter 1.2

# Science of materials behavior

## Introduction

Innovative materials and processes to produce them are enabling technologies. Materials that are multifunctional, smart, and possess physical and engineering properties superior to the existing materials are constantly needed for continued technical advances in a variety of fields. In modern times, the development, processing, and characterization of new materials have been greatly aided by novel approaches to materials design and synthesis that are based on a fundamental and unified understanding of the processing-structure-properties-performance relationships for a wide range of materials.

The subject matter of materials and manufacturing processes is very broad, and integrates the understanding derived from the study of materials science and engineering, process engineering, physical sciences, and the applied knowledge about practical manufacturing technologies. There are several complementary ways to approach this subject matter, and one that this book follows is from the viewpoint of materials science and engineering, which is the study of structure, processing, and properties, and their interrelationship. Perhaps more than any other technical discipline (with the exception perhaps of computer science and engineering), the discipline of materials science and engineering (MSE) builds a bridge between scientific theory and engineering practice. This is clearly reflected in its widely accepted title; we generally talk of mechanical engineering and electrical engineering rather than mechanical science and engineering, or electrical science and engineering! MSE has intellectual roots in physical sciences, but ultimately it represents the marriage of the "pure" and the "applied" and of the "fundamental" and the "practical."

In this chapter, we shall briefly review some of the foundational topics in materials science and engineering in order to develop a better understanding of the topics related to manufacturing processes that are covered in later chapters. We shall, however, first present some examples of innovations in materials and processes—taken from a National Academies Report—that touch upon our everyday lives. These examples also highlight how premeditated design based on the scientific method has led to technical innovations (with the exception perhaps of the tungsten filament for which the technological advance preceded a scientific understanding of the materials behavior).

## Process innovation as driver of technological growth

### Single-crystal turbine blades

Turbine blades for gas turbine engines are made out of Ni-base high-temperature superalloys that retain their strength even at 90% of their melting temperature. This has permitted an increase in the fuel inlet temperatures and increased engine efficiency (the efficiency increases about 1% for every 12°F increment in the fuel inlet temperature). However, even super-alloys become susceptible to creep and failure at high fuel-combustion temperatures under the centrifugal force generated by a rotational speed of 25,000 revolutions per minute. Early approaches succeeded in strengthening the superalloys by adding C, B, and Zr to the superalloys. These elements segregate at and strengthen the grain boundaries, thus providing resistance to creep and

*Engineering Materials and Processes Desk Reference*; ISBN: 9781856175869
Copyright © 2009 Elsevier Ltd; All rights of reproduction, in any form, reserved.

fracture. Unfortunately, these additives also lower the melting temperature of the superalloy.

In the 1960s, the problem was addressed from a different angle. It was demonstrated that eliminating grain boundaries that were oriented perpendicular to the centrifugal stress could increase the blade's service life. This is because such grain boundaries experience greater stress for deformation and fracture than boundaries oriented parallel to the blade axis. During casting of the blades, directional solidification was initiated with the help of a chill, which led to large, columnar grains oriented parallel to the blade axis (i.e., direction of the centrifugal stress). The method increased the high-temperature strength of superalloy turbine blades by several hundred percent.

A breakthrough in further enhancing the high-temperature strength was subsequently achieved by eliminating all grain boundaries, resulting in the growth of the entire blade as a single crystal. Single crystals of semiconducting materials (Si and Ge) had already been grown using special techniques (crystal pulling, floating-zone directional solidification, etc., see Chapter 2). The key innovation in the growth of single-crystal turbine blades centered around extremely slow directional cooling and design of a "crystal selector," a pigtail-shaped tortuous opening at the base of the casting mold that would annihilate all but one grain. This single grain would then grow into the liquid alloy when the mold was slowly withdrawn out of the hot zone of the furnace. Since 1982, single-crystal turbine blades have become a standard element in the hot zone of gas turbine engines.

## Copper interconnects for microelectronic packages

Faster and more efficient microcircuits require an increasing number of transistors to be interconnected on chips. At first, Al metal proved convenient as the interconnect material for microelectronic packages, although in terms of electrical resistivity Cu was known to be far superior (with about 40% less resistance than Al). Tiny Cu micro wires could also withstand higher current densities so they could be packed closer together for increased chip efficiency and miniaturization. However, Cu had a major drawback over Al; Cu readily diffuses into silicon wafer. In addition, depositing and patterning Cu microcircuitry proved more difficult than Al. With continued push toward miniaturization, the limitations of Al relative to its resistivity and current density became more pronounced. Research on depositing Cu interconnects continued, and around 1997, a viable technology for Cu interconnects was unveiled that relied on the development of a reliable diffusion barrier for Cu.

## Tungsten filament for light bulbs

The development of tungsten filament wire for use in incandescent lamps is a well-known example of process innovation that drove major technological advance. More than a century ago, carbon filaments were used in light bulbs. However, carbon filaments were fragile, brittle, short-lived, and reacted with the residual gas in the bulb, leading to soot deposition and diminished lumens. Tungsten was known to provide better light output than carbon, and its extremely high melting point and tendency to retain strength at high temperatures suggested longer filament life. However, making tungsten into a filament (to increase its light-emitting surface area) was nearly impossible because of its extremely poor ductility. William Coolidge at General Electric was able to make long W filaments by heating the metal ingot and pulling the hot metal piece through a series of wire-drawing dies. The W metal that was used for wire drawing was first obtained by reducing tungsten oxide to tungsten metal in a clay crucible. Interestingly, W metal obtained via reduction in other (non-clay) crucibles was brittle and not amenable to wire drawing. Only around the 1960s did researchers find the scientific reason for this anomaly. Potassium from the clay crucibles had dissolved into the metal during the reduction process and made the metal ductile. Potassium then turned into tiny bubbles during high-temperature processing. These bubbles elongated into tubes during wire drawing. After annealing, the tubes pinched off into a series of tiny bubbles that anchored the tungsten grain boundaries (whose movement would otherwise cause filament failure).

## Tailor-welded blanks

In the past, structural automobile body parts were made by cutting steel sheets into starting shapes or blanks. These steel sheets had a specific thickness, protective coating, and metallurgical structure required for the application. The blanks were then stamped into the three-dimensional forms of the finished body parts. Areas such as side panels and wheel housings required selective reinforcement with heavier steel for safety or to withstand stresses. These composite assemblies were made by first making individual parts and then welding these parts together into finished assemblies.

A manufacturing innovation of the 1980s, called tailor-welded blanks (TWBs), considerably simplified auto body assembly. The key was the incorporation of the heterogeneous material properties needed for auto parts into a single blank that could be formed into the finished shape with a single set of forming dies. The blanks were tailor-made by laser-welding flat steel sheets with different thicknesses, strengths, and coatings. As laser welding was already a mature technology, TWBs saw rapid and wide industrial acceptance.

As the preceding example of tailor-welded blanks shows, most applications of engineering materials require welding (or fastening and adhesive bonding) of materials into parts, devices, or structural elements, and these into assemblies, packages, or structural systems. In some applications, however, material synthesis and fabrication of the device or component may occur concurrently and seamlessly so that the boundaries between materials and devices based on them can no longer be distinguished as separate entities. An example is the junction between negative-type and positive-type extrinsic semiconductors; junctions between these semiconductors for use in transistors are synthesized at the same time as the semiconducting materials themselves (see Chapter 7 for a discussion of semiconducting materials and devices). Thus, innovations in materials and processes also bring about evolutionary changes in prevailing manufacturing paradigms.

## Atomic bonding in materials

The origin of the physical and mechanical behaviors of materials can be traced to the interatomic forces in solids. The two fundamental forces between atoms are the attractive forces due to the specific type of chemical bonding in a particular solid, and the repulsive forces due to overlapping of the outer electron shells of neighboring atoms. The magnitude of these forces decreases as the separation between atoms increases. The net force between atoms is the sum of the attractive and repulsive forces, and varies with the distance between atoms as shown in Figure 1.2-1. The net force approaches zero at a distance (typically, a few angstroms) where these two forces exactly balance each other, and a mechanical equilibrium is reached. Because the net force, $F_N$, is related to the total energy, $E$, by

$$E = \int_{\infty}^{r} F_N . dr,$$

where $r$ is the distance, the net energy is a minimum at the equilibrium separation, $r_0$, between atoms. In other words, $(dE/dr) = 0$ at $r = r_0$. The binding energy between atoms is the net energy corresponding to this equilibrium separation.

In a real solid composed of a large number of mutually interacting atoms, the estimation of binding energies from interatomic forces becomes complex, although in principle a binding energy can be specified for the atoms of any real solid. Properties such as thermal expansion, stiffness, and melting temperatures are derived from the shape of the interaction energy curve (Figure 1.2-1), the magnitude of minimum (binding) energy, and the nature of chemical bond between atoms of the solid.

Atoms in solids exhibit three types of chemical bonds: ionic, covalent, and metallic. Ionic bonds are universally found in solids that are made of a metallic and a nonmetallic element (e.g., NaCl, $Al_2O_3$). In ionic bonding, the outer (valence) electrons from the metallic element are transferred to the nonmetallic element, causing a positive charge on the atoms of the former and a negative charge on the atoms of the latter. These ionized atoms develop electrostatic (Coulombic) forces of attraction and repulsion between them, which decrease with increasing separation between atoms. Ceramic materials exhibit predominantly ionic bonding.

In covalent solids, the electrons between neighboring atoms are shared, and the shared electron belongs to both atoms. Molecules such as $Cl_2$, $F_2$, HF, and polymeric materials are covalently bonded. Covalent bonds are directional in the sense that the bond forms only between atoms that share an electron. In contrast, ionic bonds are nondirectional; that is, the bond strength is same in all

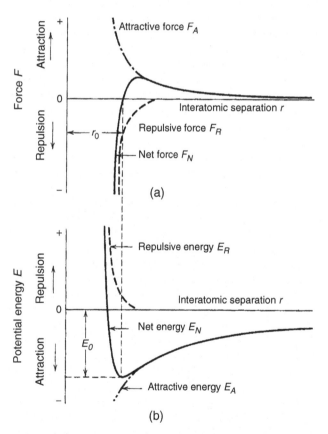

**Figure 1.2-1** *(a) Variation of the repulsive force, attractive force, and net force between two isolated atoms as a function of the interatomic separation, and (b) variation of the repulsive, attractive, and net potential energies as a function of the interatomic separation.* (W. D. Callister, Jr., Materials Science & Engineering: An Introduction, 5th ed., Wiley, New York, 2000, p. 19).

directions around an ion. Many solids are partially ionic and partially covalent; the larger the separation of two elements in the periodic table (i.e., the greater the difference in the electronegativity of the two elements), the greater will be the degree of ionic bonding in compounds of the two elements.

The third type of primary bond, metallic bond, occurs in metals and alloys, and is due to the freely drifting valence electrons that are shared by all the positively charged ions in the metal. Thus, an "electron cloud" or "electron sea" permeates the entire metal, and provides shielding against mutual repulsion between the positive ions. The electron cloud also acts as a "binder" to hold the ions together in the solid via electrostatic attractive forces. Table 1.2-1 shows the relationship between the binding energy and melting temperatures of some solids; solids of high bond energy exhibit high melting points.

In addition to the primary or chemical bonds, secondary bonds exist between atoms and influence properties such as surface energies. Secondary bonds are weaker than primary bonds and have energies of a few tens of kJ/mol as opposed to a few hundred kJ/mol or higher for the primary bonds. Nevertheless, secondary bonds are ubiquitous; they are present between all atoms and molecules, but their presence can be masked by the stronger chemical bonds. The genesis of secondary bonds lies in the transient and permanent dipole moments of atoms or molecules. The constant thermal vibration of an atom causes transient (short-lived) distortion of an otherwise spatially symmetric electron distribution

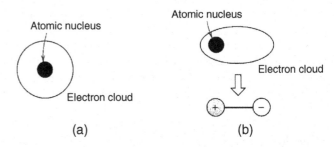

**Figure 1.2-2** Schematic representation of (a) an electrically symmetric atom, and (b) an induced atomic dipole due to a net shift in the centers of the positive and negative charges. (W. D. Callister, Jr., Materials Science & Engineering: An Introduction, 5th ed., Wiley, New York, 2000, p. 25).

around the nucleus, leading to a separation of the centers of positive and negative charges (Figure 1.2-2). This results in a transient induced dipole that induces a dipole in a neighboring atom by disturbing its charge symmetry, and so forth. The resulting electrostatic forces fluctuate with time. Weak secondary forces that have their origin in such induced atomic dipoles are called van der Waals bonds. Certain molecules, called polar molecules, possess a permanent dipole moment because of an asymmetric charge distribution in their atoms and molecules. Such molecules can induce dipoles in neighboring nonpolar molecules, causing an attractive force or bond to develop between the molecules.

## Crystal structure

Upon slow cooling, the disordered structure of a liquid transforms into an ordered structure characteristic of crystalline solids. In crystalline solids, the atoms are arranged in three-dimensional periodic arrays over large distances; these arrays could be relatively simple as in common metals or extremely complex as in polymeric materials. Certain materials, however, do not exhibit the long-range atomic order characteristic of crystalline solids. Such materials form either a completely amorphous or partially crystalline structure when they are cooled from the liquid state. Many complex polymers comprised of long-chain molecules show only partial crystallinity under normal cooling because of the entanglement of chain segments that create pockets of atomic disorder. Similarly, most metals and alloys that would normally crystallize under slow cooling may exhibit an amorphous or glassy structure under ultra-fast cooling conditions, which restrict atomic diffusion needed to form a periodic atomic array, thus causing the random structure of the liquid to be "frozen" in the solid state. Our current knowledge of the atomic arrangements in solids is largely derived from the use of x-rays as a tool to probe the crystal structure. The periodically arranged

| Table 1.2-1 Melting temperatures and bond energies | | |
|---|---|---|
| **Material** | **$T_m$, °K** | **Bond Energy, $kJ \cdot mol^{-1}$** |
| NaCl | 1074 | 640 |
| MgO | 3073 | 1000 |
| Si | 1683 | 450 |
| Al | 933 | 324 |
| Fe | 1811 | 406 |
| W | 3683 | 849 |
| Ar | 84 | 7.7 |
| $Cl_2$ | 172 | 31 |
| $NH_3$ | 195 | 35 |
| $H_2O$ | 273 | 51 |

Source: Adapted from W. D. Callister, Jr., Materials Science & Engineering: An Introduction, 5th ed., Wiley, New York, 2000.

atoms scatter the x-rays of a wavelength comparable to the spacing between atoms, and give rise to the phenomena of diffraction or specific phase relationships between scattered waves.

Many physical attributes of crystalline solids are determined by the type of geometric arrangement of their constituent atoms. A useful model to visualize the atomic packing in crystalline solids is to first liken each atom as a "hard sphere," and then identify the smallest repeating cluster of atoms (unit cell) that could be stacked in three dimensions to generate the long-range atomic order. For each metallic element, the hard spheres (with a characteristic atomic radius) represent a positive ion in a sea of electrons. Many metallic elements crystallize in one of three basic geometric forms or crystal structure: face-centered cubic (FCC) (Figure 1.2-3), body-centered cubic (BCC) (Figure 1.2-4), and hexagonal close-packed (HCP) (Figure 1.2-5).

Many features of atomic packing depend on the crystal directions and crystal planes. For example, the packing density of atoms (or atomic density) and interatomic voids depend upon crystal directions and crystal planes. The packing density and the void content in the crystal structure influence the alloying behavior, diffusion processes, plastic deformation, and various other material properties. Thus, knowledge about the crystal structure is important in understanding the materials behavior.

Crystallographic directions are specified in terms of a line between two points in the unit cell, and denoted as [uvw] where u, v, and w are the projections along the x, y, and z axes, respectively (with reference to the origin of the coordinate system, conveniently located at a corner in the unit cell). Crystal planes or atomic planes are specified relative to the unit cell as (hkl) and for HCP (Figure 1.2-5) as (hklm). The intercepts made by a crystal plane along the x, y, and z axes are written in terms of the intercepts within the unit cell (i.e., normalized with respect to the length of the sides of the cube to obtain integral values for h, k, and l), and

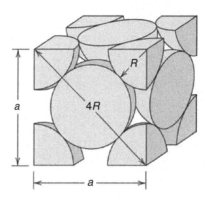

**Figure 1.2-3** *A hard-sphere unit cell representation of the face centered cubic (FCC) crystal structure. R is the atomic radius, and a is the side of the unit cell.* (W. D. Callister, Jr., Materials Science & Engineering: An Introduction, 5th ed., Wiley, New York, 2000, p. 32).

the reciprocal of these intercepts is then written in a reduced form (i.e., in terms of the smallest integers). Figure 1.2-6 illustrate some examples of crystal planes and crystal directions.

A large number of physical and mechanical properties of materials depend on the crystal-lographic orientation along which the property is measured, and significant differences can occur along different directions. For example, the modulus of elasticity of metals is orientation-dependent. The modulus of Fe along [100], [110], and [111] directions is 125 GPa, 211 GPa, and 273 GPa, respectively. Similarly, the modulus of Cu along [100], [110], and [111] directions has been measured to be 67 GPa, 130 GPa, and 191 GPa, respectively.

## Defects in crystalline solids

Perfect long-range atomic order does not exist in crystalline solids even in most carefully prepared materials. Various types of imperfections or irregularities exist

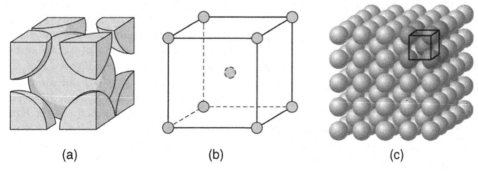

**Figure 1.2-4** *(a) A hard-sphere unit cell representation of the body-centered cubic (BCC) crystal structure, (b) a reduced-sphere unit cell, and (c) an aggregate of many atoms with the BCC arrangement.* (W. D. Callister, Jr., Materials Science & Engineering: An Introduction, 5th ed., Wiley, New York, 2000, p. 34).

# CHAPTER 1.2  Science of materials behavior

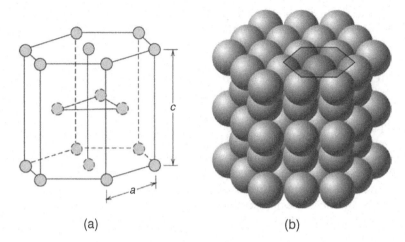

**Figure 1.2-5** *(a) A reduced-sphere unit cell representation of the hexagonal close-packed (HCP) crystal structure, and (b) an aggregate of many atoms with the HCP arrangement.* (W. D. Callister, Jr., Materials Science & Engineering: An Introduction, 5th ed., Wiley, New York, 2000, p. 35).

in the atomic arrangement in all solid materials. The most common imperfections in crystalline solids include vacancies, interstitials, solute (or impurity) atoms, dislocations, grain boundaries, and surfaces and interfaces. Many physical and mechanical properties of crystalline solids are determined by the nature of the defects, and their distribution and concentration in the material.

Vacancies or vacant atomic sites (Figure 1.2-7) exist in solids at all temperatures; their concentration increases exponentially with temperature according to the Boltzmann distribution function, $N = N_0 \exp(-Q/RT)$, where $N$ is the equilibrium number of vacancies per unit volume of the material at an absolute temperature, $T$, $N_0$ is the total number of atomic sites per unit volume (which depends upon the crystal structure), $Q$ is the activation energy to form the vacancy (related to the energy barrier that must be surmounted to dislodge an atom from its normal position and create a vacancy), and $k$ is the Boltzmann's constant ($13.81 \times 10^{-24}$ J/K). Vacancies increase the disorder (entropy) in the crystal, thus making their presence a thermo-dynamic necessity. Unlike a vacancy, an interstitial defect forms when either a host atom or an impurity atom resides in a preexisting void in the crystal lattice. Figure 1.2-7 shows the formation of a self-interstitial when a host atom is dislodged from its normal site and forced into the (smaller) void between atoms.

Most practical engineering materials are alloys rather than pure elemental solids. Alloys are solid solutions and form when an impurity atom either substitutes a host atom (substitutional solid solution) or enters the interstices of the parent lattice (interstitial solid solution). Figure 1.2-8 shows substitutional and interstitial impurity atoms in a crystal lattice. Solute atoms in a host crystal lattice can also form a compound (e.g.,

intermetallic compounds in which the different types of atoms are combined in a fixed or nearly fixed proportion). High solubility of impurity atoms in substitutional solid solutions is favored when a set of criteria, called the Hume-Rothery rules, are satisfied. These rules stipulate that solid solutions form when one or more of the following criteria are met: (1) the difference in the atomic radii of the two atom types should be less than 15%, (2) crystal structure of the two metals should be the same, (3) the electronegativity difference between the two atom types should be small (metals widely separated in the periodic table, i.e., those exhibiting large electronegativity difference, are more likely to form an intermetallic compound rather than a solid solution), and (4) a metal of higher valence will dissolve more readily in the host metal than a metal of valence lower than the host metal. A classic substitutional solid solution is Cu-Ni, which exhibits complete solubility. In contrast to substitutional solid solutions, in an interstitial solid solution, the need for impurity atoms to fit in the interstices of the host lattice limits the solubility (usually <10%). In iron-carbon alloys, the much smaller carbon atoms occupy the interstitial positions in the iron lattice and form an interstitial solid solution.

Dislocations are linear defects in crystals that form in a region where a plane of atoms terminates abruptly in the lattice, as shown in Figure 1.2-9. This figure shows an edge dislocation where an atomic plane is shown missing in the bottom half of the crystal. Due to the disturbance in the periodicity of the lattice near a dislocation line, there is some distortion (stress) around the atomic planes, which in turn influences the physical and mechanical behaviors of the material. A screw dislocation forms when a shear stress causes the atomic planes across a region within the crystal to be shifted one atomic spacing relative to the other planes (Figure 1.2-10). In

Science of materials behavior | CHAPTER 1.2

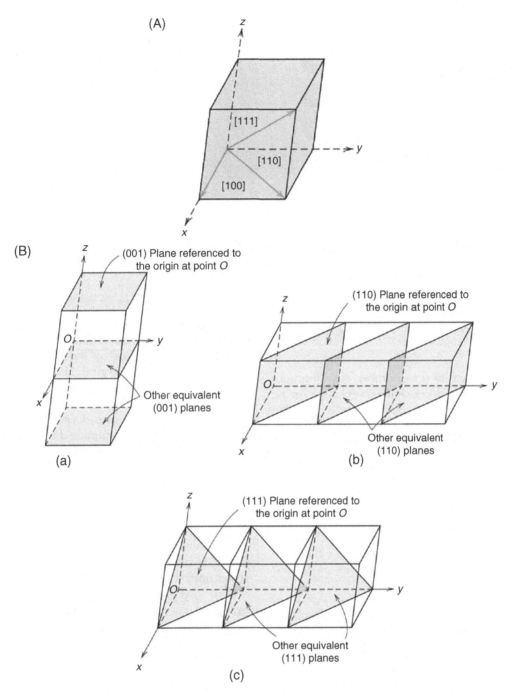

**Figure 1.2-6** *(A) The [100], [110], and [111] crystallographic directions in a unit cell. (W. D. Callister, Jr., Materials Science & Engineering: An Introduction, 5th ed., Wiley, New York, 2000, p. 41) (B) Representation of a series each of (a) (001), (b) (110), and (c) (111) crystal-lographic planes. (W. D. Callister, Jr., Materials Science & Engineering: An Introduction, 5th ed., Wiley, New York, 2000, p. 44).*

reality, mixed dislocations (comprised of edge and screw components) are more common than either pure edge or screw dislocations. Transmission electron microscopy (TEM) techniques permit visual observation of the dislocations. Figure 1.2-11 shows TEM photomicrographs of dislocation lines in a deformed intermetallic compound, NiAl, alloyed with a small amount of chromium; the dislocation lines are tangled and pinned by secondary phases. All crystalline materials contain dislocations, and it is virtually impossible to produce a dislocation-free crystal even under the most stringent processing conditions. Plastic deformation, phase transformations (e.g., solidification), thermal stresses, and irradiation increase the concentration of dislocations in the solid.

# CHAPTER 1.2  Science of materials behavior

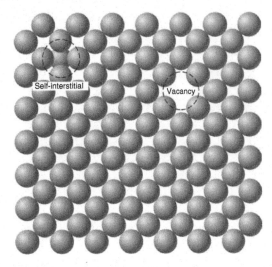

**Figure 1.2-7** *Schematic illustration of two types of point defects in crystalline solids: a vacancy and a self-interstitial.* (W. G. Moffat, G. W. Pearsall, and J. Wulff, Properties of Materials, vol. 1, Structure, Wiley, 1964, p. 77). Reprinted with permission from Janet M. Moffat, 7300 Don Diego NE, Albuquerque, NM.

The dislocation density (i.e., number of dislocations per unit area) in real crystalline solids is very large, on the order of $10^{12}$ dislocations per square meter in annealed metals, and $10^{15}$ to 10 in cold-worked metals. Dislocations can be mobile under an applied stress, and hindrances to dislocation motion such as grain boundaries, secondary precipitates, inclusions, and other dislocations that form tangles and impede one another's motion lead to strengthening. Dislocations move on slip planes along certain preferred (close-packed) directions under stress, and emerge at the crystal surface in the form of a step. Such a process of dislocation exhaustion should in fact promote the solid's progression toward

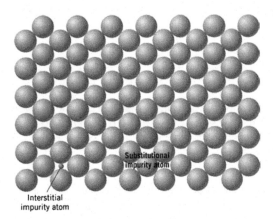

**Figure 1.2-8** *Schematic illustration of an interstitial and a substitutional impurity atom in crystalline solids.* (W. G. Moffat, G. W. Pearsall, and J. Wulff, Properties of Materials, vol. 1, Structure, Wiley, 1964, p. 77). Reprinted with permission from Janet M. Moffat, 7300 Don Diego NE, Albuquerque, NM.

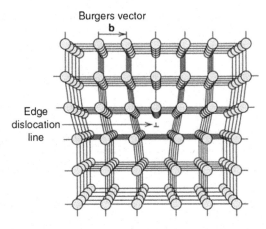

**Figure 1.2-9** *The atom positions around an edge dislocation; extra half-plane of atoms shown in perspective.* (A. G. Guy, Essentials of Materials Science, McGraw-Hill, New York, 1976, p. 153).

crystallographic perfection through migration of dislocations toward the surface and their elimination from the crystal lattice by formation of a step on an external surface of the crystal. However, this process of dislocation exhaustion is more than compensated by the generation of new dislocations during deformation. An important mechanism of dislocation generation, called the Frank-Reed source, mimics the evolution of a soap bubble at the end of a capillary under air pressure. A dislocation line D-D′, with ends pinned by solute atoms (or immovable points of intersection with other dislocations), steadily grows under an applied stress, $\tau$, as shown in Figure 1.2-12, until it bends into a semicircle. Beyond this stage, the dislocation continues to grow at a decreasing stress, a closed dislocation loop forms by joining at points 6-6′ and 7-7′, and the dislocation loop grows until it reaches the solid's surface where it forms a step. In the process, an internal dislocation, D-D′, is generated at which the preceding mechanism repeats itself, thus generating new dislocation loops and increasing the overall dislocation density in the material.

Surfaces and interfaces represent discontinuity in the ordered arrangement of atoms and are also crystal defects. The free surface of a solid is associated with an excess energy due to the unsaturated atomic bonds at the surface. This excess energy is the surface energy (or surface tension) of the solid and is a measure of the driving force needed to minimize the free surface of the solid (in reality, a solid's free surface is actually an interface between the solid and the surrounding atmosphere or vapor). Grain boundaries in crystalline solids separate grains or regions having different crystallographic orientations. Both grain boundaries and interfaces have some atomic mismatch (lattice strain) because of the different crystallographic orientations of the neighboring regions and are, therefore, associated

# Science of materials behavior CHAPTER 1.2

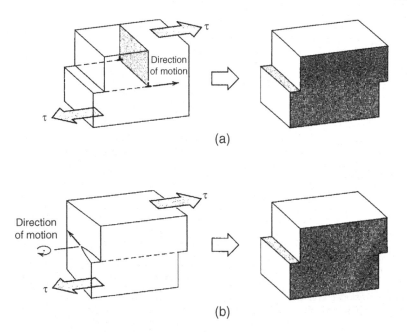

**Figure 1.2-10** *The formation of a step on the surface of a crystal by the motion of (a) an edge dislocation, and (b) a screw dislocation. For the edge dislocation, the dislocation line moves in the direction of the applied shear stress, and for the screw dislocation, the dislocation line moves perpendicular to the direction of the shear stress.* (H. W. Hayden, W. G. Moffat, and J. Wulff, The Structure and Properties of Materials, vol. 3, Mechanical Behavior, Wiley, New York, 1965, p. 70). Reprinted with permission from Janet M. Moffat, 7300 Don Diego NE, Albuquerque, NM.

with an excess energy. Small- or low-angle grain boundaries form when the mismatch is small and can be accommodated by an array of dislocations. A special type of grain boundary, called a twin boundary, forms when the atoms across the boundary are located to form a mirror image of the other side. Twin boundaries form across definite crystallographic planes when either shear forces are applied or the material is annealed following plastic deformation. Figure 1.2-13 shows a schematic illustration of low-angle boundaries and twin boundaries.

## Annealing

Annealing is a heat treatment that is applied to cold-worked metals to allow the structure and properties of the pre-cold-worked state to be regained. This occurs

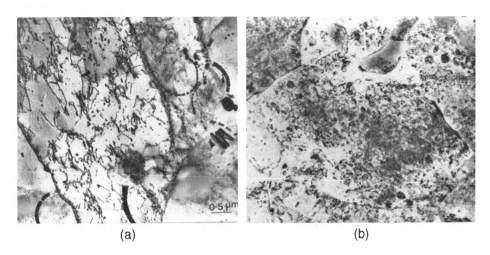

**Figure 1.2-11** *(a) Transmission electron micrograph of an extruded Ni-48.3Al-1W alloy showing dislocation networks, and (b) transmission electron micrograph of an extruded Ni-43Al-9.7Cr alloy showing dislocation networks.* (R. Tiwari, S. N. Tewari, R. Asthana and A. Garg, J. Materials Science, 30, 1995, 4861–70).

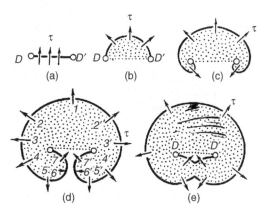

**Figure 1.2-12** Schematic illustration of the operation of a Frank-Reed dislocation source. (a) Initial position of the dislocation line D-D', (b) bending of the dislocation line under an applied stress, (c) and (d) continuing development of a symmetric dislocation loop, (e) formation of external dislocation loop spreading across the crystal and of internal dislocation D-D' returning to the original position. (G. I. Epifanov, Solid State Physics, Mir Publishers, Moscow, 1979, p. 68, English translation).

through the temperature-sensitive processes of recovery, recrystallization, and grain growth. During recovery, some of the physical properties (thermal and electrical conductivities) are recovered, although mechanical properties do not revert to the pre-cold-worked state. No observable microstructural changes occur during recovery. During the next stage of recrystallization, strain-free grains nucleate within the cold-worked material and slowly consume the entire cold-worked structure. Complete restoration of mechanical properties to their pre-cold-worked state occurs during recrystallization. Figure 1.2-14 shows the metallurgical structure of a cold-worked Sn-Pb alloy that was recrystallized at room temperature.

Highly cold-worked metals recrystallize faster, and the recrystallization temperature decreases with increasing degree of cold-working. Pure metals recrystallize faster than alloys, and alloying raises the recrystallization temperature. The recrystallization temperature, $T_{cr}$, is defined as the temperature at which the cold-worked structure fully recrystallizes in 1 h. For pure metals, $T_{cr} \sim 0.3T_m$ ($T_m$ is the absolute melting temperature), but for alloys, $T_{cr} \sim 0.7T_m$. High-melting-point metals have a high $T_{cr}$. Many deformation processes use hot-working to shape parts; hot-working is done above $T_{cr}$. Note that room temperature deformation of Pb, Sn, and Sn-Pb alloys ($T_{cr} \sim -4°C$) is actually hot-working, whereas the deformation of W($T_{cr} \sim 1200°C$) at 1000°C is cold-working.

The new stress-free recrystallized grains continue to grow if a high temperature is maintained for a long period. This is because there is a distribution of grain sizes in the recrystallized material, and the need to decrease the grain boundary area provides the driving force for grain growth (or grain coarsening). Grain coarsening occurs by competitive dissolution of small grains and growth of larger grains in the distribution through mass transport via atomic diffusion. Grain coarsening in many polycrystalline metallic and ceramic materials follows the relationship: $d^n - d_0^n = Kt$, where $d$ is the initial grain diameter (at $t = 0$, i.e., at the onset of grain growth), and $K$ and $n$ are time-independent constants, with $n$ being greater or equal to 2. The coefficient $K$ is estimated from the curve-fitting of experimental $d$ versus $t$ data.

## Diffusion in crystalline solids

Diffusion in crystalline solids involves movement of atoms in steps within a crystal lattice. It could involve either one type of atoms (e.g., self-diffusion of like atoms in a pure metal) or different types of atoms (e.g., interdiffusion or impurity diffusion). The process of diffusion requires breaking of existing bonds by an atom, its migration to a vacant site in the lattice via an atomic jump process, and formation of chemical bonds with its new neighbors. Small atoms (such as C, N, etc.) diffuse in a crystal via interstitial positions (interstitial diffusion), whereas larger substitutional impurity atoms diffuse by jumping into vacancies whose concentration exponentially increases with temperature. This makes diffusion easier at high temperatures. In addition, at high temperatures atoms have high thermal (vibrational) energy, which also facilitates their migration.

Atomic diffusion at a constant temperature causes concentration variations with time and position. The concentration of diffusing atoms is specified by the diffusion flux, $J$, which is defined as the mass, $M$ (or concentration, $C$), diffusing per unit time across a plane of area, $A$, normal to the diffusion direction. The diffusion flux, $J = (1/A)(dM/dt)$, where $t$ is the time. If a steady-state is reached, then $J$ becomes independent of time, and a linear concentration gradient of diffusing atoms is attained as shown for the case of gaseous diffusion across a thin metal foil in Figure 1.2-15. The steady-state diffusion process is described by Fick's first equation, which in one dimension reads, $J = -D(dC/dx)$, where $C$ is the concentration and $D$ is the diffusion coefficient. $D$ represents the mobility of the diffusing atoms and has the dimensions of (length)$^2$/time. If the diffusion flux and the concentration gradient at a point change with time, then the diffusion is non-steady (Figure 1.2-16), and the concentration, $C$, is related to the position, $x$,

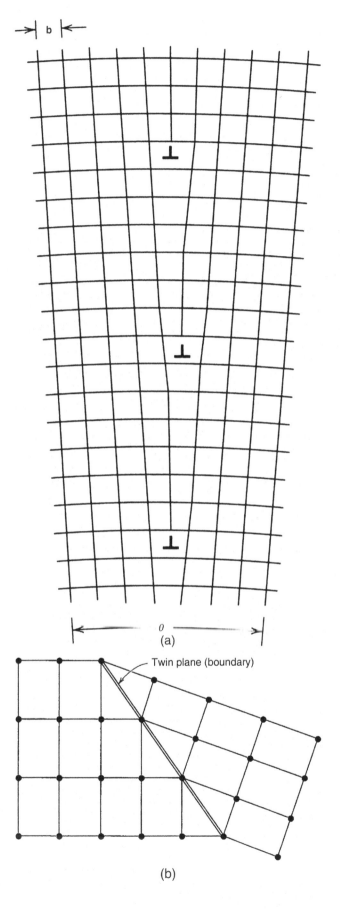

and time, $t$, by Fick's second equation, whose one-dimensional form at a constant temperature is

$$\frac{\partial C}{\partial t} = \frac{\partial}{\partial x}\left(D\frac{\partial C}{\partial x}\right). \tag{1.2.1}$$

If $D$ is constant, then it can be taken out of the partial derivative in Equation 1.2.1, thus yielding the simpler equation $\frac{\partial C}{\partial t} = D\frac{\partial^2 C}{\partial x^2}$. In many situations, however, $D$ is a strong function of concentration, C, and Equation 1.2.1 must be solved. Fick's equations are analogous to the Fourier equations for heat conduction through solids (with the diffusion coefficient replaced with the thermal diffusivity, and C replaced with temperature, $T$). Many mathematical solutions to the above equation have been derived for boundary conditions that are encountered in a variety of physical processes. An important practical situation involving doping of semiconducting materials to control their electronic properties involves diffusion of impurity atoms, and the applicable solution to the Fick's equation is discussed in Chapter 7.

The diffusion coefficient, $D$, is very sensitive to temperature, and for atomic diffusion in solids, $D = D_0 \exp(-Q/RT)$, where $Q$ is the activation energy for diffusion, $R$ is the gas constant, and $D_0$ is a pre-exponential term, called the frequency factor, which depends on the atomic vibration frequency and the crystal structure. The activation energy, $Q$, represents the energy consumed in distorting the local lattice to permit an atomic jump. Table 1.2-2 summarizes the values of $Q$ and $D_0$ for several materials, and Figures 1.2-17 and 1.2-18 show the dependence of the diffusion coefficient on temperature for some common elements. By taking the natural logarithm of the preceding expression for $D$, a linear relationship between $\ln D$ and inverse absolute temperature, $T$, is obtained, which has the form, $\ln D = \ln D_0 - (Q/R)(1/T)$. Thus, by plotting experimentally measured $D$ values at different temperatures, it is possible to obtain the activation energy for diffusion and the frequency factor.

Figure 1.2-18a shows the diffusion coefficient versus $1/T$ plots for various solutes in common metals and in the semiconducting element Si. This figure shows that at a given temperature, Group III elements (Al, Ga, B, In) and Group V elements (P, As, Sb) diffuse faster than the Si atoms in Si crystals. As the slopes of lines for Group III and Group V solutes in Si in Figure 1.2-18a are roughly

**Figure 1.2-13** (a) A low-angle grain boundary formed by the arrangement of edge dislocations. This type of low-angle boundary is called a tilt boundary. (b) A twin boundary and the adjacent atom positions. The atoms on one side of the boundary are mirror images of the atoms on the other side. (W. D. Callister, Jr., Materials Science & Engineering: An Introduction, 5th ed., Wiley, New York, 2000, p. 80).

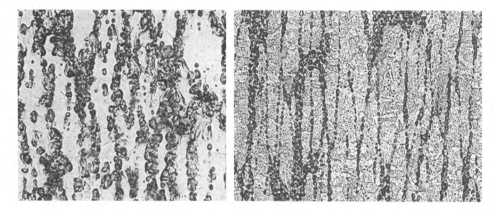

**Figure 1.2-14** *Photomicrographs showing recrystallized grains in Sn-Pb alloys with two different volume fractions of the eutectic, which were mechanically deformed (swaged) and then recrystallized at room temperature. (R. Asthana, unpublished research, NASA Glenn Research Center, Cleveland, OH, 1995).*

the same, the activation energies for diffusion are approximately identical. Figure 1.2-18a also shows that solutes such as Ni, Cu, Li, and Fe diffuse much faster in Si (and have lower activation energies) than other common solutes.

There are several basic differences in atomic diffusion in semiconductors such as Si and Ge, and in common metals such as Ni and Cu. Silicon and germanium form a diamond cubic crystal lattice characteristic of diamond (see Chapter 3), which is more open than the crystal lattice of close-packed metals. Second, the energy to form vacancies in Si and Ge is higher relative to the thermal energy at the melting point (i.e., $Q/kT_m$ is large, where $T_m$ is the melting point), and the energy to form vacancies and self-interstitials are more nearly equal in Si and Ge than in metals. Third, atoms occupy interstitial positions much more often in Si and Ge than in metals;

therefore, interstitials play a more important role in self-diffusion in semiconductors. In addition, the presence or absence of bonding between solute atoms and Si or Ge determines the mobility of solute atoms. For example, oxygen atom is small and occupies interstitial sites in Si but bonds with Si and diffuses with a relatively high activation energy (2 eV), whereas the larger Ni and Cu atoms that form no bonds with Si move with a lower activation energy (0.5 eV). Another difference between metals and semiconductors is related to the difference in the dislocation density. In a carefully grown Ge or Si crystal, the dislocation density may average $10^4$ per m$^2$ or less, whereas in metals, the dislocation density is high, $10^9$ per m$^2$ even in well-annealed condition. Thus, in a metal, the distance a vacancy must diffuse to find a dislocation is much shorter than in a semiconductor (dislocations pin vacancies). As a result,

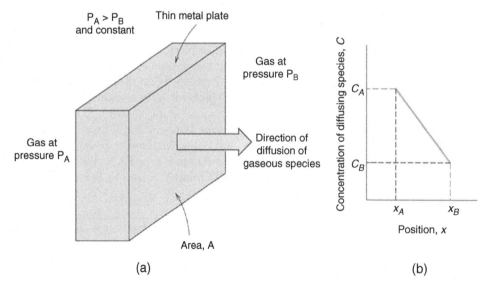

**Figure 1.2-15** *(a) Steady-state diffusion of a gas across a thin plate in a pressure gradient, and (b) a linear concentration profile for the diffusion situation in (a). (W. D. Callister, Jr., Materials Science & Engineering: An Introduction, 5th ed., Wiley, New York, 2000, p. 96).*

# Science of materials behavior CHAPTER 1.2

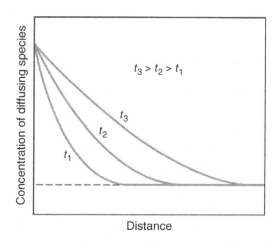

**Figure 1.2-16** *Time modulation of the concentration distribution of a diffusing species.* (W. D. Callister, Jr., Materials Science & Engineering: An Introduction, 5th ed., Wiley, New York, 2000, p. 98).

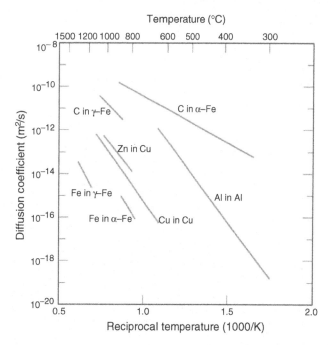

**Figure 1.2-17** *Plot of the logarithm of the diffusion coefficient versus the reciprocal of absolute temperature for C and Fe in $\alpha$- and $\gamma$-Fe, Zn in Cu, Al in Al, and Cu in Cu.* (W. D. Callister, Jr., Materials Science & Engineering: An Introduction, 5th ed., Wiley, New York, 2000, p. 103).

whereas in a metal the equilibrium concentration of vacancies is maintained throughout the crystal, in Si or Ge, the vacancy concentration may deviate from the equilibrium value over a large fraction of the crystal.

The self-diffusion in Si and Ge at their respective melting points is orders of magnitude slower than that in metals at their melting points, as seen from Figure 1.2-18b. This difference between metals and semiconductors increases yet more at lower temperatures due to the relatively larger activation energies for Si and Ge. Experiments show that self-diffusion in Ge and Si is dominated by vacancy motion at low temperatures, but at high temperatures it is dominated by the motion of interstitials.

In the manufacture of Si-based devices, it is customary to oxidize the surface at intermediate temperatures to form an insulating silica layer. The growth of this layer speeds up the diffusion of Group III elements (B, Al, Ga, and P), and it slows down the diffusion of Group V elements (Sb and As). The diffusion of Group V solutes in Si depends on their atomic radius. Phosphorus (the smallest in the group) diffuses primarily by an interstitial mechanism, whereas the largest (Sb) diffuses by a vacancy mechanism. The role of diffusion in the manufacture of silicon-based devices is discussed in greater depth in Chapter 7.

## Mechanical behavior

For many crystalline solids, mechanical deformation is elastic at low applied normal stresses and follows Hooke's law, according to which $\sigma = E\varepsilon$, where $\sigma$ is the applied stress, $\varepsilon$ is the elastic strain, and $E$ is the modulus of elasticity or Young's modulus, which represents the

**Table 1.2-2** Diffusion data for metals and semimetals

| Diffusing Atom | Host material | $D_0$, m²·s⁻¹ | Q, kJ·mol⁻¹ |
|---|---|---|---|
| Fe | $\alpha$-Fe | $2.8 \times 10^{-4}$ | 251 |
| Fe | $\gamma$-Fe | $5.0 \times 10^{-5}$ | 284 |
| C | $\gamma$-Fe | $2.3 \times 10^{-5}$ | 148 |
| Cu | Cu | $7.8 \times 10^{-5}$ | 211 |
| Zn | Cu | $2.4 \times 10^{-5}$ | 189 |
| Cu | Al | $6.5 \times 10^{-5}$ | 136 |
| Cu | Ni | $2.7 \times 10^{-5}$ | 256 |
| W | W | $1.88 \times 10^{-4}$ | 586 |
| Al | Al | $4.7 \times 10^{-6}$ | 123 |
| Si | Si | $20 \times 10^{-4}$ | 424 |
| Ge | Ge | $25 \times 10^{-4}$ | 318 |
| Cr | Cr | $9.7 \times 10^{-2}$ | 435 |

Note: $D_0$, frequency factor; $Q$, activation energy for diffusion.

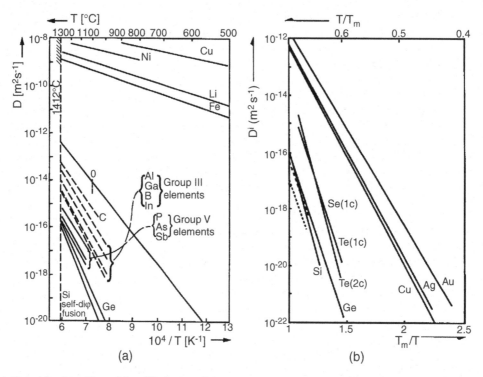

**Figure 1.2-18** *(a) Plot of the logarithm of the diffusion coefficient versus the reciprocal of absolute temperature for various solutes in silicon. (P. G. Shewmon, Diffusion in Solids, 2nd ed., The Minerals, Metals & Materials Society, Warrendale, PA, 1989, p. 175). (b) Plot of the logarithm of the diffusion coefficient versus the reciprocal of the absolute homologous temperature ($T_m/T$) for Si, Ge, and noble metals. (P. G. Shewmon, Diffusion in Solids, 2nd ed., 1989, p. 175). Reprinted with permission from The Minerals, Metals & Materials Society, Warrendale, PA (www.tms.org).*

stiffness of the material. The elastic strain is completely recovered on removal of the stress, and there is no permanent deformation. For certain materials (many polymers and concrete), the stress-strain relationship is not linear, and it is a standard practice to characterize the stiffness or modulus of such materials either at a particular value of the strain or as an average over a range of strain values.

The origin of elastic modulus or stiffness lies in the strength of the interatomic bonds. As per the potential energy curve between neighboring atoms in a solid shown in Figure 1.2-1, the equilibrium separation is $r_0$ and the energy is $E(r_0)$. If an external stress increases the distance between the two atoms by a small amount, "$x$," then the potential energy increases to $E(r)$ where $r = r_0 + x$. The work done for displacement through "$x$" is therefore, $E(x) = E(r) - E(r_0)$. Expanding $E(r)$ into a Taylor series in terms of "$x$" yields

$$E(r) = \left(\frac{\partial E}{\partial r}\right)_0 \cdot x + \frac{1}{2} \cdot \left(\frac{\partial^2 E}{\partial r^2}\right)_0 \cdot x^2 + \frac{1}{6} \cdot \left(\frac{\partial^3 E}{\partial r^3}\right)_0 \cdot x^3 + \ldots \quad (1.2.2)$$

If all the terms higher than the quadratic are neglected due to the small value of the slope ($dE/dr$) near $r_0$, then we obtain

$$E(x) = \frac{1}{2}\left(\frac{\partial^2 E}{\partial r^2}\right)_0 \cdot x^2 = 0.5\beta x^2, \quad (1.2.3)$$

where $\beta$ specifies the strength of the interatomic bond. The force needed to displace the atoms through "$x$" is therefore $F = -dE(x)/dx = -\beta x$, i.e., the force to increase the separation between neighboring atoms is directly proportional to the displacement. This relationship may be considered as a microscopic analogue of Hooke's law; summing up the force between all atom pairs and adding all the increments in atomic displacements, one would arrive at an equation substantially similar to Hooke's law. Figure 1.2-19 shows the interatomic force versus distance curves for strongly bonded and weakly bonded solids. The larger slope of the strongly bonded solid indicates the greater stiffness (higher modulus) of this material.

In a manner similar to the case of normal stress considered above, the deformation of solids under low shear and torsional stresses is also elastic; for example, shear strain, $\gamma$, is proportional to shear stress, $\tau$, so that $\tau = G\gamma$, where $G$ is the shear modulus.

The initial portion of a typical tensile stress–strain curve for a metal is shown in Figure 1.2-20. The initial linear regime is the elastic behavior described by Hooke's law just discussed. At a certain stress value (elastic limit), a transition occurs to a nonlinear behavior and the onset

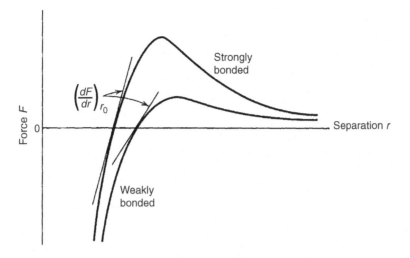

**Figure 1.2-19** *Interatomic force versus separation plot for strongly bonded and weakly bonded atom pairs. The slope of the force – separation curve is proportional to the elastic modulus.* (W. D. Callister, Jr., Materials Science & Engineering: An Introduction, 5th ed., Wiley, New York, 2000, p. 120).

of plastic (permanent) deformation. The transition stress is the yield strength of the material, often not precise or distinct on the curve, as in the example of Figure 1.2-20. Often the stress rises to a peak value (upper yield point), and then drops to a lower value (lower yield point) about which the stress fluctuates before rising again, this time in a nonlinear fashion. The yield strength in such a case is taken as the stress corresponding to 0.2% offset strain. A straight line is drawn parallel to the linear portion of the stress–strain curve starting at a strain of 0.002, and the intersection of the line with the stress–strain curve gives the yield strength. After the onset of plastic deformation, the stress needed to further deform the material continues to increase, reaches a maximum (tensile strength, point M in Figure 1.2-21), and then continuously decreases until the material fractures at the fracture stress or breaking stress. The deformation beyond the tensile stress is confined to a small region of the sample or "neck," which is a region of highly localized deformation. Fracture occurs at the necked region at the breaking stress (Figure 1.2-21).

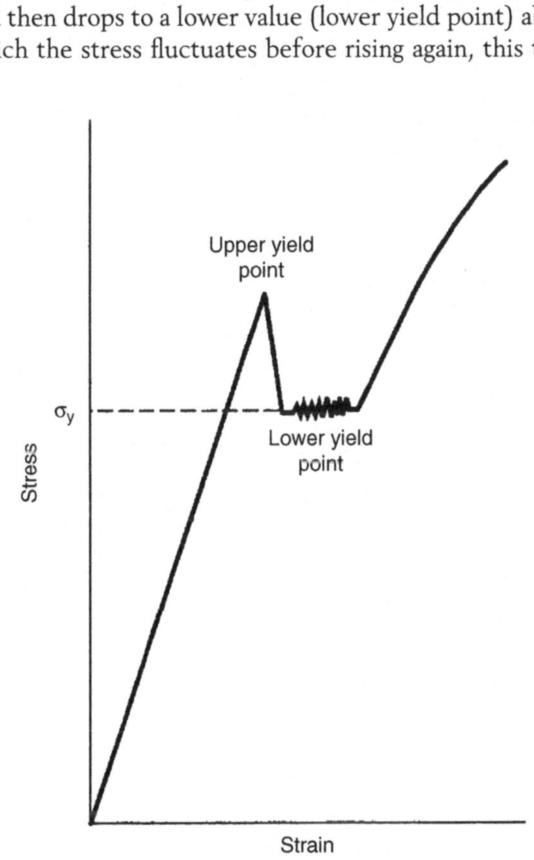

Two important measures of the energy absorbed by a material during deformation are resilience and toughness. The resilience of a material is the energy absorbed during elastic deformation. It is estimated from the area under the linear portion of the stress–strain diagram under uniaxial tension, i.e.,

$$U = \int_0^{\varepsilon_y} \sigma \, d\varepsilon, \tag{1.2.4}$$

where $U$ is the resilience and $\varepsilon_y$ is the yield strain. If the material obeys Hooke's law, the expression for resilience becomes

$$U = \frac{1}{2}\sigma_y \varepsilon_y = \frac{\sigma_y^2}{2E}, \tag{1.2.5}$$

**Figure 1.2-20** *The yield point phenomenon in steels that results in an upper yield point and a lower yield point.* (W. D. Callister, Jr., Materials Science & Engineering: An Introduction, 5th ed., Wiley, New York, 2000, p. 124).

which shows that materials with high yield strength and low elastic modulus will be highly resilient. Toughness of a material is the total energy absorbed by a material until

# CHAPTER 1.2   Science of materials behavior

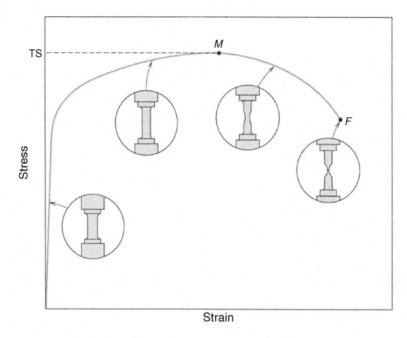

**Figure 1.2-21** *Typical engineering stress–strain behavior to fracture, point F. The tensile strength, TS, is indicated by point M. The circular inserts represent the geometry of the deformed specimen at various points along the curve.* (W. D. Callister, Jr., Materials Science & Engineering: An Introduction, 5th ed., Wiley, New York, 2000, p. 126).

it fractures, and is estimated from the area under the entire stress versus strain curve of a material.

A distinction is usually made between engineering stress and strain, and true stress and strain. Engineering stress is obtained by dividing the applied load by the original cross-sectional area of the test specimen. As the load-bearing area continuously decreases and specimen length continuously increases during tensile deformation, a more logical approach would be to define the true stress and true strain in terms of the instantaneous area and length. True stress is defined as $\sigma_t = F/A_i$, and true strain as $\varepsilon_t = \ln(l_i/l_0)$, where $A_i$, $l_i$, and $l_0$ are the instantaneous area, instantaneous length, and initial length, respectively. Figure 1.2-22 shows a schematic plot of true stress versus true strain superimposed on the corresponding plot of engineering stress versus engineering strain. Because the total volume of the material is conserved during loading, $A_0 l_0 = A_i l_i$, where $A_0$ is the original cross-section of the sample. This relationship allows one to relate the true stress and true strain to engineering stress and engineering strain, and it can be readily shown that $\sigma_t = \sigma(1 + \varepsilon)$ and $\varepsilon_t = \ln(1 + \varepsilon)$, where the engineering stress, $\sigma$, and engineering strain, $\varepsilon$, are given from $\sigma = F/A_0$, and $\varepsilon = (l_i - l_0)/l_0$. The preceding equations for true stress and true strain are, however, valid only up to the onset of necking.

For a large number of metals and alloys, the region between the inception of plastic deformation (yield point) and the onset of necking (tensile strength) on a true stress–true strain diagram can be described by a power law relationship of the form $\sigma_t = K \varepsilon_t^n$, where $K$ and $n$ are material-specific constants, and $n$ is called the strain-hardening exponent. The strain-hardening exponent, $n$, is a measure of the ability of a metal to work harden (i.e., strengthen through plastic deformation). A large value of $n$ indicates that strain hardening will be large for a given amount of plastic deformation. Thus, annealed copper ($n = 0.54$) will strain-harden more than annealed Al ($n = 0.20$) as would 70/30 brass ($n = 0.49$) compared to 4340 alloy steel ($n = 0.15$). Table 1.2-3 gives the values of $n$ and $K$ for common metals and alloys.

The rate of deformation, or strain rate, is a parameter of importance to the mechanical behavior of metals. High

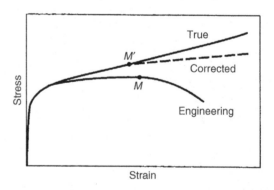

**Figure 1.2-22** *A comparison of the engineering stress–strain and true stress-strain behaviors. Necking begins at point M on the engineering curve, which corresponds to point M' on the true curve. The corrected true stress–strain curve takes into account the complex stress state within the neck region.* (W. D. Callister, Jr., Materials Science & Engineering: An Introduction, 5th ed., Wiley, New York, 2000, p. 132).

**Table 1.2-3** Values of exponents *n* and *K* in the flow stress equation, $\sigma_T = K\epsilon_T^n$, for different alloys

| Alloy | n | K, MPa |
|---|---|---|
| Low-C steel (annealed) | 0.26 | 530 |
| 4340 alloy steel (annealed) | 0.15 | 640 |
| 304 stainless steel (annealed) | 0.45 | 1275 |
| 2024 Al alloy (annealed) | 0.16 | 690 |
| Brass (70Cu-30Zn, annealed) | 0.49 | 895 |

strain rate increases the flow stress of the metal and the temperature of the deformed metal because adiabatic conditions could exist due to rapid loading. A general relationship between flow stress, $\sigma$, and strain rate ($\dot{\epsilon}$) at constant temperature and strain is $\sigma = C(\dot{\epsilon})^m$, where $m$ is the strain-rate sensitivity of the metal. Metals with a high strain-rate sensitivity ($0.3 < m < 1.0$) do not exhibit localized necking. Superplastic alloys (capable of 100–1000% elongation) have a high value of strain-rate sensitivity. This behavior occurs in materials with a very fine grain size ($\sim 1$ μm) and at temperatures of about 0.4 $T_m$, where $T_m$ is the absolute melting point of the metal. The major advantage of superplastic materials is the ease of shaping a part because of the very low flow stress required for deformation at low strain rates. Superplasticity is lost at strain rates above a critical value ($0.01\ s^{-1}$).

# Strengthening of metals

The fundamental approach to strengthening metals is to devise methods that increase the resistance to the motion of dislocations responsible for the plastic deformation. Reducing the grain size by rapid cooling of a casting, or by adding inoculants or "seed" crystals to a solidifying alloy to promote nucleation and grain refinement (see Chapter 2) is one method to strengthen metals. This is because fine-grained materials have a large-grain boundary area, and the motion of dislocation is hindered by grain boundaries because of crystallographic misorientation between grains, and because of the discontinuity of slip planes between neighboring grains. The strengthening by grain size reduction leads to an increase in the yield strength of the material. The grain size dependence of the yield strength, $\sigma_0$, of monolithic crystalline solids is described by the Hall-Petch equation according to which, $\sigma_0 = \sigma_i + k \cdot D^{-1/2}$, where $\sigma_i$ is a friction stress that opposes dislocation motion, $k$ is a material constant, and $D$ is the average grain diameter. Thus, a plot of the yield strength as a function of the inverse square root of the average grain diameter will yield a straight line. Figure 1.2-23 shows an example of

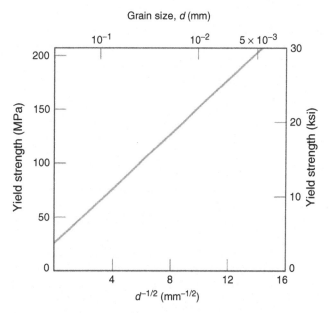

**Figure 1.2-23** *The influence of grain size on the yield strength of a 70Cu-30Zn brass alloy. The theoretical relationship expressing the influence of grain size on yield strength is the Hall-Petch equation.* (Adapted from H. Suzuki, The Relation Between the Structure and Mechanical Properties of Metals, vol. II, National Physical Laboratory Symposium No. 15, 1963, p. 524). Reprinted with permission from National Physical Laboratory, Tedclington, Middlesex, U.K.

the effect of grain size on the yield strength of a brass alloy, which is consistent with the Hall-Petch equation. The Hall-Petch equation correctly describes the grain size dependence of yield stress for micrometer size grains but fails at grain diameters smaller than about 10 nm. This is because the equation was derived for relatively large dislocation pile-ups at grain boundaries, with about 50 dislocations in the pile-up. Clearly, at nanometric grain sizes, the pile-ups would contain fewer dislocations, thus violating the key assumption of the Hall-Petch equation.

The properties and behavior of materials change, often dramatically and in surprising ways, when the grain size approaches nanoscale, with the lower limit approaching the size of several crystal units. At nanometric grain sizes the fraction of the disordered interfacial grain boundary area becomes very large, and the grain diameter approaches some characteristic physical length such as the size of the Frank-Reed loop for dislocation slip. Furthermore, interfacial defects such as grain boundaries, triple points, and solute atoms segregated at an interface begin to increasingly influence the physical and mechanical behaviors. In addition, interfacial defects from processing such as micropores, and minute quantities of oxide contaminants and residual binders mask the grain size effects in nanomaterials, making it very difficult to isolate the true effect of grain-size reduction in such materials.

The other methods to strengthen metals include alloying, work hardening, and precipitation hardening. Pure metals are almost always soft, and alloying strengthens a metal. This is because the solute atoms (interstitial or substitutional) locally distort the crystal lattice, and generate stresses that interact with the stress field around dislocations, which hinder the dislocation motion. Impurity atoms segregate around the dislocation line in configurations that lower the total energy. This causes the dislocation to experience a drag due to the impurity "atmosphere" surrounding it, which must be carried along if the dislocation must move to cause deformation. Besides alloying, plastic deformation can also strengthen metals, a process called work-hardening or strain-hardening. Plastic deformation rapidly increases the density of dislocations, whose stress fields inhibit the motion of other dislocations. Frequently, dislocation tangles form, which provide considerable resistance to continued plastic deformation. The strain-hardening exponent, $n$, in equation $\sigma_t = K\varepsilon_t^n$ introduced in the preceding section is a measure of the ability of a metal to work harden; metals with a large value of $n$ strain harden more than metals with a low value of $n$ for a given amount of plastic deformation or degree of cold work. Certain alloys can be strengthened via a special heat treatment that creates finely dispersed hard second-phase particles, which resist the motion of dislocations. This is discussed in the section on heat treatment. Finally, hard second-phase particles and fibers may be added to metals from outside for strengthening by creating a composite material; this is discussed in Chapter 6.

# Fracture mechanics

The theoretical fracture strength of brittle materials, calculated on the basis of atomic bonding considerations, is very high, typically on the order of $E/10$, where $E$ is the modulus of elasticity. For example, the theoretical strengths of aluminum oxide ($Al_2O_3$) and zirconia ($ZrO_2$) are approximately 39.3 GPa and 20.5 GPa, respectively ($E_{Al2O3} = 393$ GPa, and $E_{ZrO2} = 205$ GPa). The actual strengths are in the range 0.80–1.50 GPa for zirconia and 0.28–0.70 GPa for alumina. The actual fracture strength of most brittle materials is 10 to 1000 times lower than their theoretical strength. The discrepancy arises because of minute flaws (cracks) that are universally present at the surface and in the interior of all solids. These cracks locally amplify or concentrate the stress, especially at the tip of the crack. The actual stress experienced by the solid near the crack tip could exceed the fracture strength of the material even when the applied stress is only a fraction of the fracture strength. For a long, penny-shaped crack of length $2a$, with a tip of radius $\rho_t$, the distribution of stress in the material is schematically shown in Figure 1.2-24. The maximum stress, $\sigma_m$, occurs at the crack tip, and is given from

$$\sigma_m = 2\sigma_0 \left(\frac{a}{\rho_t}\right)^{1/2} \qquad 1.2.6$$

where $\sigma_0$ is the applied stress, and the stress ratio ($\sigma_m/\sigma_0$) is called the stress concentration factor, $K$. Thus, long

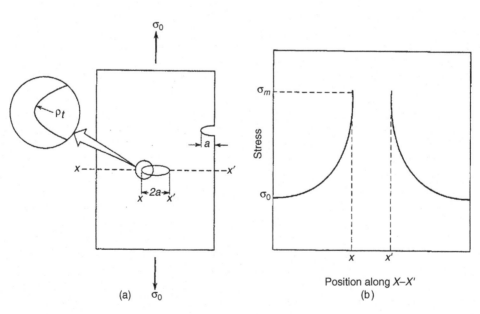

**Figure 1.2-24** (a) The geometry of surface and internal cracks, and (b) schematic stress profile along the line X–X' in (a), demonstrating stress amplification at the crack tip. (W. D. Callister, Jr., Materials Science & Engineering: An Introduction, 5th ed., Wiley, New York, 2000, p. 191).

cracks with sharp tips will raise the stress more than small cracks with large tip radii. For fracture to occur, a crack must extend through the solid, a process that creates additional free surface within the material. The elastic strain energy released on crack propagation is consumed in creating the new surface that has an excess energy (surface energy) associated with it. The critical stress, $\sigma_c$ (i.e., the fracture strength), required to propagate a crack of length $2a$ through a solid of elastic modulus, $E$, and surface energy, $\gamma_{sv}$, is given from the Griffith equation, applicable to brittle solids, which is

$$\sigma_c = \left(\frac{2E\gamma_{sv}}{\pi a}\right)^{1/2}. \quad (1.2.7)$$

In the Griffith equation, the parameter $a$ is the depth of a surface crack, or half-length of an internal crack. For metals (and some polymers), fracture is accompanied by plastic deformation, so that a portion of the stored mechanical energy is used up in the plastic deformation accompanying crack propagation. For such materials, the Griffith equation has been modified to account for the material plasticity in terms of a plastic energy, $\gamma_p$, so that

$$\sigma_c = \left[\frac{2E(\gamma_{sv} + \gamma_p)}{\pi a}\right]^{1/2}. \quad (1.2.8)$$

A useful parameter to characterize a material's resistance to fracture is the fracture toughness of the material. Fracture toughness is most commonly specified for the tensile fracture mode, or Mode I, in which a preexisting crack extends under a normal tensile load. Two other modes of fracture are encountered, albeit to a lesser degree, in solids. These are shear failure and failure by tearing (Figure 1.2-25). Mode I fracture toughness defined for the tensile fracture mode is also called the plain-strain fracture toughness, $K_{IC}$, and is given from

$$K_{IC} = Y\sigma\sqrt{\pi a}, \quad (1.2.9)$$

where $Y$ is a geometric factor. Table 1.2-4 gives the $K_{IC}$ values of several materials. Brittle materials have a low $K_{IC}$, whereas ductile materials have a high $K_{IC}$. The mode I toughness, $K_{IC}$, depends also on the microstructure, temperature, and strain rate. Generally, fine-grained materials exhibit a high $K_{IC}$, and high temperatures and low strain rates increase the $K_{IC}$.

The engineering value of $K_{IC}$ is in calculating the critical crack size that would cause catastrophic failure in a component under a given applied stress, $\sigma$. This critical size is given from

$$a_c = \frac{1}{\pi}\left(\frac{K_{IC}}{Y\sigma}\right)^2. \quad (1.2.10)$$

Alternatively, a maximum stress, $\sigma$, can be specified for a given material containing cracks of known size such that the material will not fail. This stress is

**Table 1.2-4** Mode I fracture toughness of selected metallic, polymeric and ceramic materials

| Material | $K_{IC}$ (MPa·m$^{1/2}$) |
|---|---|
| Cu alloys | 30–120 |
| Ni alloys | 100–150 |
| Ti alloys | 50–100 |
| Steels | 80–170 |
| Al alloys | 5–70 |
| Polyethylene | 1–5 |
| Polypropylene | 3–4 |
| Polycarbonate | 1–2.5 |
| Nylons | 3–5 |
| GFRP | 20–60 |
| alumina | 3–5 |
| $Si_3N_4$ | 4–5 |
| MgO | 3 |
| SiC | 3 |
| $ZrO_2$ (PSZ) | 8–13 |
| Soda glass ($Na_2O$-$SiO_2$) | 0.7–0.8 |
| Concrete | 0.2–1.4 |

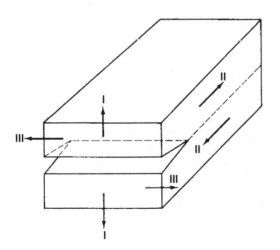

**Figure 1.2-25** The three modes of crack displacement: Mode I (crack opening or tensile mode), Mode II (sliding mode), and Mode III (tearing mode).

$$\sigma = \frac{K_{IC}}{Y\sqrt{(\pi a)}}. \qquad (1.2.11)$$

Sophisticated tests and analytical methods have been developed to measure the $K_{IC}$ of materials in various geometrical configurations. A qualitative assessment of the relative fracture toughness of different materials, especially at high strain rates, is readily obtained from a simple test, called the Charpy V-notch test. The test utilizes a standard specimen with a V-notch machined on one of its faces. A pendulum weight is allowed to fall from a fixed height and strike the sample securely positioned on a fixture to expose its back-face (i.e., the face opposite the V-notch) to the striking pendulum. The energy needed to fracture the specimen is obtained from the difference in the height of the swinging pendulum before and after the impact. A semiquantitative assessment of toughness can also be obtained from the measurement of total crack length in a material when a microscopic indentation is made on its surface under a fixed load. The plastic yielding and stress accommodation in a ductile material will prevent cracking, whereas fine, hairline cracks will emanate from the indentation in a brittle material (Figure 1.2-26). Examination of the fracture surface of a material can also reveal whether the failure is predominantly brittle or ductile. The scanning electron micrographs in Fig. 1.2-27a show that the intermetallic compound, NiAl, fractures in a predominantly intergranular manner, whereas the same material with tungsten alloying exhibits transgranular fracture (Fig. 1.2-27b), which is indicative of a predominantly ductile mode of fracture in this material.

## Fatigue

Failure caused by a fluctuating stress (periodic or random) acting on a material that is smaller in magnitude than the material's tensile or yield strength under static load, is called the fatigue failure. Parameters used to characterize fatigue include: stress amplitude, $\sigma_a$, mean stress, $\sigma_m$, and stress range, $\sigma_r$. These parameters are defined as follows: $\sigma_a = \frac{\sigma_{max} - \sigma_{min}}{2}$, $\sigma_m = \frac{\sigma_{max} + \sigma_{min}}{2}$, and $\sigma_r = \sigma_{max} - \sigma_{min}$. Fatigue behavior is most conveniently characterized by subjecting a specimen to cyclic tensile and compressive stresses until the specimen fails. The results (Figure 1.2-28) of a series of such tests depict the number of cycles to failure, $N$, as a function of the applied stress, $S$ (usually expressed as stress amplitude, $\sigma_a$). The general response of most materials is represented by one of the two S–N curves shown in Figure 1.2-28. Some iron and titanium alloys attain a limiting stress, called the fatigue limit or endurance limit (Fig. 1.2-28a), below which they survive an infinite number of stress cycles. In contrast, many nonferrous alloys do not exhibit a fatigue limit, and the number of cycles to failure, $N$, continuously decreases with increasing stress, $S$ (Fig. 1.2-28b). For these materials, fatigue strength is specified as the value of stress at which the material will fail after a specified number of cycles, $N$ (usually, $10^7$). Alternatively, fatigue life of a material can be defined as the number of cycles to failure at a specified stress level. Fatigue life is influenced by a large number of variables. For example, minute surface or bulk imperfections (e.g., cracks, microporosity) in the material,

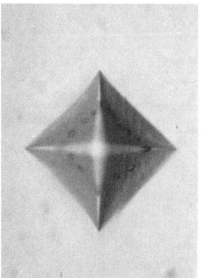 

**Figure 1.2-26** *Qualitative assessment of fracture toughness from the material's response to indentation. Cracks emanate from the indentation area in a brittle material, whereas no cracks are seen in a tough material.* (R. Asthana, R. Tiwari and S. N. Tewari, Materials Science & Engineering, A336, 2002, 99–109).

Science of materials behavior  CHAPTER 1.2

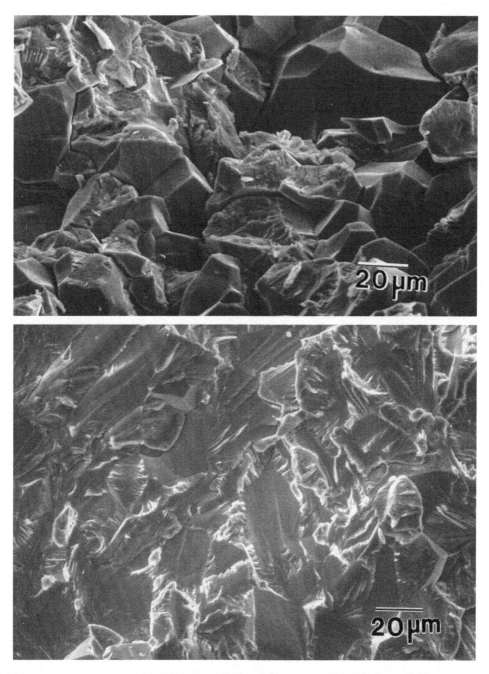

**Figure 1.2-27** (a) Scanning electron micrograph (SEM) view of the fracture surface of an extruded Ni-46Al alloy, compression tested at 300 K, showing intergranular fracture. (R. Tiwari, S. N. Tewari, R. Asthana and A. Garg, Materials Science & Engineering, A 192/193, 1995, 356–363). (b) SEM view of the fracture surface of a directionally solidified and compression deformed Ni-48.3Al-1W alloy showing evidence of transgranular fracture. (R. Asthana, R. Tiwari and S. N. Tewari, Materials Science & Engineering, A336, 2002, 99–109).

design features that could raise the stress locally, surface treatments, and second-phase particles all influence the fatigue life.

In reality, the $S$–$N$ curves of the type shown in Figure 1.2-28 are characterized by a marked scatter in the data, usually caused due to the sensitivity of fatigue to metallurgical structure, sample preparation, processing history, and the test conditions. As a result, for design purpose, fatigue data are statistically analyzed and presented as probability of failure curves. Fatigue failures occur in three distinct steps: crack initiation, crack propagation, and a rather catastrophic final failure. Introducing compressive stresses in the surface of the part (through surface hardening such as carburizing or shot peening) improves the fatigue life because the compressive stresses counter the tensile stresses during service.

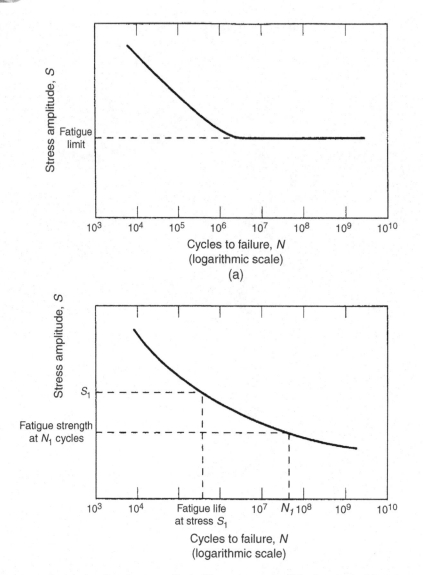

**Figure 1.2-28** *Fatigue response depicted as the stress amplitude (S) versus number (N) of cycles to failure curves. In (a) the fatigue limit or endurance limit is reached at a specific N value, and in (b) the material does not reach a fatigue limit.* (W. D. Callister, Jr., Materials Science & Engineering: An Introduction, 5th ed., Wiley, New York, 2000, p. 212).

## Creep

Creep is the time-dependent deformation at elevated temperatures under constant load or constant stress. Creep deformation occurs in all type of materials, including metals, alloys, ceramics, plastics, rubbers, and composites. The deformation strain versus time behavior for high-temperature creep under constant load (Figure 1.2-29) is characterized by three distinct regions, following an initial instantaneous (elastic) deformation region. These three creep regimes are (1) primary creep with a continuously decreasing creep rate; (2) secondary or steady-state creep, during which the creep rate is constant; and (3) tertiary creep, during which creep rate is accelerated, leading to eventual material failure. In primary creep, the decrease in the creep rate is caused by the material's strain hardening. In the steady-state regime, usually the longest in duration, equilibrium is attained between the competing processes of strain hardening and recovery (softening). The creep failure toward the end of the third stage is caused by the structural and chemical changes such as de-cohesion at the grain boundaries, and nucleation and growth of internal cracks and voids. From the material and component design perspectives, the steady-state creep rate ($d\varepsilon_s/dt$) is of considerable importance. The steady-state creep rate depends on the magnitude of applied stress, $\sigma$, and temperature, $T$, according to

$$\frac{d\varepsilon_s}{dt} = K'\sigma^n \exp\left(-\frac{Q}{RT}\right). \quad (1.2.12)$$

where $Q$ is the activation energy for creep, $K'$ and $n$ are empirical constants, and $R$ is the gas constant. The

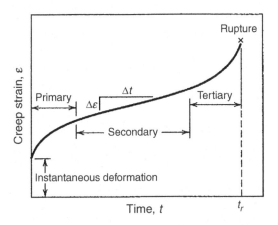

**Figure 1.2-29** *Schematic creep curve showing strain versus time at constant stress and temperature. The minimum creep rate is the slope of the linear region in the secondary creep regime. (W. D. Callister, Jr., Materials Science & Engineering: An Introduction, 5th ed., Wiley, New York, 2000, p. 226).*

exponent $n$ depends on the dominant mechanisms of creep under a given set of experimental conditions, which include vacancy diffusion in a stress field, grain boundary migration, dislocation motion, grain boundary sliding, and others. The materials used in modern aircraft engines provide a classic example of how high-temperature creep resistance is enhanced through microstructure design in Ni-base alloys by dispersing nanometer-size oxide particles; this is discussed in Chapter 6.

## Deformation processing

A large number of manufacturing processes employ solid-state deformation of hot or cold metals and alloys to shape parts. Hot-working is the mechanical shaping operation that is performed at temperatures high enough to cause the processes of recovery and recrystallization to keep pace with the work hardening due to deformation. In contrast, cold-working is performed below the recrystallization temperature of the metal; however, both hot- and cold-working are done at high strain rates. Cold-working strain hardens a metal, and excessive deformation without intermediate annealing causes fracture. It is a standard practice to anneal the metal between multiple passes during cold-working, to soften the metal and prevent fracture. In hot-working, the work hardening and distorted grain structure produced by mechanical deformation are rapidly erased, and new stress-free equiaxed grains recrystallize. In contrast, during cold-working the material hardens and its flow stress increases with continued deformation. In this section, we focus on the metallurgical changes that occur during hot working.

Hot-working erases compositional inhomogeneity of the cast structure (e.g., coring); creates a uniform, equiaxed grain structure; and welds any residual porosity. However, some structural inhomogeneity may be caused in hot-working because the component surface experiences greater deformation than the interior, resulting in finer recrystallized grains on the surface. Also, because the interior cools slower than the surface, some grain coarsening is possible in the interior, leading to low strength. In addition, surface contamination from scale formation and chemical reactions with gases in the furnace atmosphere, and the disintegration and entrapment of the scale in the surface and subsurface layers, can cause surface defects, poor finish, and part embrittlement. Hot-worked steels can decarburize at the surface, leading to strength loss and poor surface finish. Tighter dimensional tolerances are possible with cold-working than hot-working because of the absence of thermal expansion and contraction in the former.

In practice, hot-working is done in a temperature range bound by an upper and a lower limit. The upper temperature is limited by the melting temperature, oxidation kinetics, and the melting temperatures of grain boundary secondary phases (e.g., low-melting eutectics) that segregate at the grain boundaries during primary fabrication (e.g., casting). The melting of even a small amount of a low-melting-point secondary phase residing at the grain boundaries could cause material failure (hot-shortness). The lower hot-working temperature depends on the rate of recrystallization and the residence time at temperature. Because the recrystallization temperature decreases with increasing deformation, a heavily deformed metal will usually require a lower recrystallization temperature.

Temperature gradients develop in the workpiece during hot-working. This is because of the chilling action of the die. Because the flow stress is strongly temperature-dependent, a chilled region produces local hardening, and a non-deformable zone in the vicinity of a hotter, softer region. This could lead to the development of shear bands, and the localization of flow into these bands, resulting in very high shear strains and shear fracture.

## Heat treatment

Heat treatment is done to transform the metallurgical structure in alloys to develop the desired properties. Iron-carbon alloys serve as a classic example of how thermal treatments can be designed to transform the metallurgical structure for enhanced properties. Heat treatment practices make use of the thermodynamic data provided by the Fe-C (more often the Fe-Fe$_3$C) phase diagram, and the experimental measurements of the transformation kinetics. Many (but not all) phase transformations involve atomic diffusion and occur in a manner somewhat similar to the crystallization of a liquid via nucleation and growth (discussed in Chapter 2). In this section, we briefly review the basic elements of practical heat treatment approaches that are used to design the structure and properties of alloys.

Iron-carbon alloys are the most widely used structural materials, and the phase transformations in these alloys are of considerable importance in heat treatment. The Fe-Fe$_3$C diagram (Figure 1.2-30) exhibits three important phase reactions: (1) a eutectoid reaction at 727°C, consisting of decomposition of austenite ($\gamma$-phase, FCC) into ferrite ($\alpha$-Fe, BCC) and cementite, i.e., $\gamma \leftrightarrow \alpha + $ Fe$_3$C; (2) a peritectic reaction at 1493°C, involving reaction of the BCC $\delta$-phase and the liquid to form austenite ($\delta + L \leftrightarrow \gamma$); and (3) a eutectic reaction at 1147°C, resulting in the formation of austenite and cementite (L $\leftrightarrow \gamma + $ Fe$_3$C). Figure 1.2-31 depicts the development of microstructure in three representative steels at different stages of cooling: a hypoeutectoid steel, a steel of eutectoid composition (0.76% C), and a hypereutectoid steel.

Consider a eutectoid steel that is cooled from a high temperature where it is fully austenitic to a temperature below 727°C, and isothermally held for different times at the temperature. The isothermal hold will transform the austenite into pearlite, which is a mixture of ferrite ($\alpha$-Fe) and cementite (iron carbide, Fe$_3$C). Experiments show that transformation (conversion) kinetics at a fixed temperature exhibit an S-shaped curve (Figure 1.2-32) that is mathematically described by the Avrami equation, $y = 1 - \exp(-kt^n)$, where $y$ is the fraction transformed, $k$ is a temperature-sensitive rate constant, and $t$ is transformation time. The Avrami equation applies to a variety of diffusion-driven phase transformations, including recrystallization of cold-worked metals and to many nonmetallurgical phenomena whose kinetics display a sigmoid behavior of Fig. 1.2-32. Usually the isothermal transformation data of Figure 1.2-32 is organized as temperature-versus-time map (with time on a logarithmic scale) to delineate the fraction (0 to 100%) transformed. The approach to construct an isothermal transformation map is shown in Figure 1.2-33. A complete isothermal-transformation (I-T) diagram for a eutectoid steel (0.77% C) is shown in Figure 1.2-34. The different iron-carbon phases (e.g., coarse and fine pearlite, upper and lower bainite, etc.) that form under different isothermal conditions are also displayed on this diagram. The physical appearance of some of the phases as revealed under a microscope is shown in the photomicrographs of Figure 1.2-35.

In practice, steels are rapidly cooled in a continuous fashion from the fully austenitic state to the ambient temperature, rather than held isothermally to trigger the

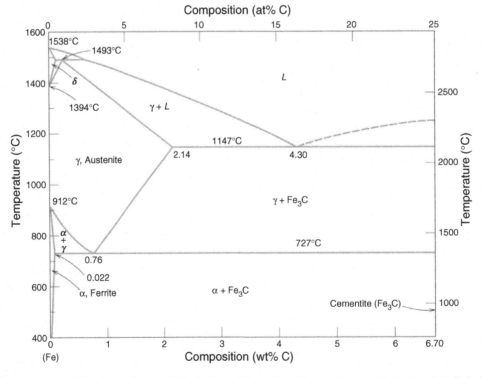

**Figure 1.2-30** *The iron-iron carbide phase diagram.* (Adapted from Binary Alloy Phase Diagrams, 2nd ed., vol. 1, T. B. Massalski, editor-in-chief, 1990). Reprinted with permission from ASM International, Materials Park, OH (www.asminternational.org).

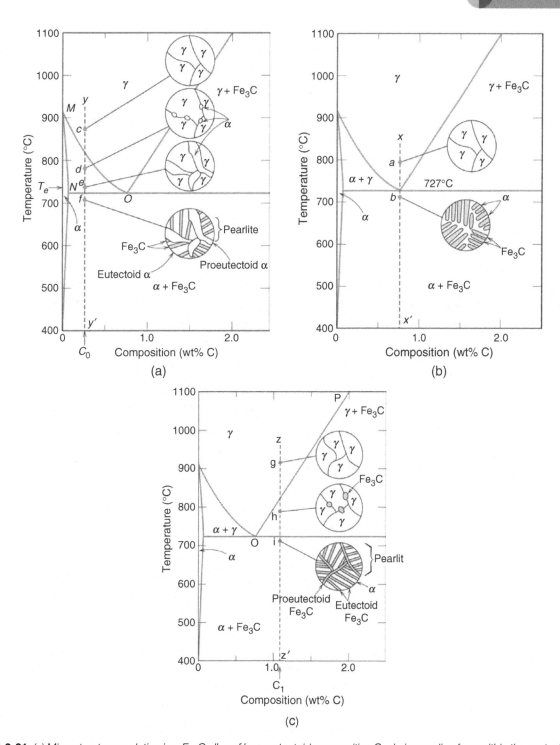

**Figure 1.2-31** (a) Microstructure evolution in a Fe-C alloy of hypoeutectoid composition $C_0$ during cooling from within the austenite region to below the eutectoid temperature of 727°C (the alloy contains less than 0.76 wt% C) (W. D. Callister, Jr., Materials Science & Engineering: An Introduction, 5th ed., Wiley, New York, 2000, p. 279). (b) Microstructure evolution in a Fe-C alloy of eutectoid composition (0.76% C) during cooling from within the austenite region to below 727°C (W. D. Callister, Jr., Materials Science & Engineering: An Introduction, 5th ed., Wiley, New York, 2000, p. 277). (c) Microstructure evolution in a Fe-C alloy of hypereutectoid composition $C_1$ during cooling from within the austenite region to below the eutectoid temperature of 727°C (the alloy contains more than 0.76 wt% C) (W. D. Callister, Jr., Materials Science & Engineering: An Introduction, 5th ed., Wiley, New York, 2000, p. 282).

transformation. In continuous cooling of steel, the time required for a transformation to start and end is delayed, and the isothermal transformation curves all shift to longer times. Figure 1.2-36 shows a continuous cooling transformation (CCT) diagram for a eutectoid steel with lines representing different cooling rates superimposed

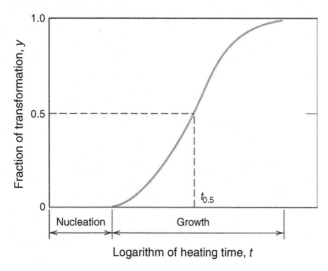

**Figure 1.2-32** Plot showing fraction reacted (y) versus the logarithm of time (t) typical of many isothermal solid-state transformations. The mathematical relationship between y and t is expressed by the Avrami equation, $y = 1 - \exp(-kt^n)$, where k and n are time-independent constants for a particular reaction. (W. D. Callister, Jr., Materials Science & Engineering: An Introduction, 5th ed., Wiley, New York, 2000, p. 296).

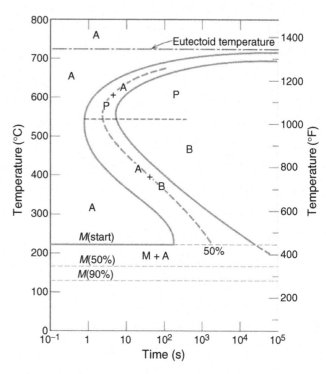

**Figure 1.2-34** The complete I–T diagram for a Fe-C alloy of eutectoid composition. A, austenite; B, bainite; M, martensite; P, pearlite. (Adapted from H. Boyer, editor, Atlas of Isothermal Transformation and Cooling Transformation Diagrams, 1977, p. 28). Reprinted with permission from ASM International, Materials Park, OH (www.asminternational.org).

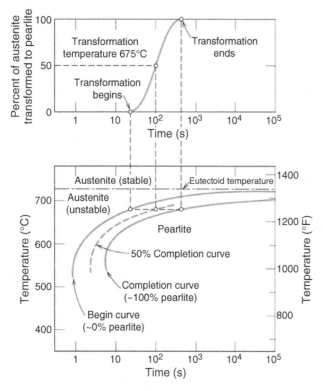

**Figure 1.2-33** Procedure to develop an isothermal transformation (I-T) diagram (bottom) from an experimental Avrami (fraction transformed versus log time) curve. (Adapted from H. Boyer, ed., Atlas of Isothermal Transformation and Cooling Transformation Diagrams, 1977, p. 369). Reprinted with permission from ASM International, Materials Park, OH (www.asminternational.org).

on the diagram (the constant cooling rate lines appear as curves rather than straight lines because time is plotted on a logarithmic scale).

Heat treated steels harden because of the formation of a hard and brittle phase, martensite, which forms only on rapid cooling below a certain temperature, called the martensite start temperature, $M_s$, which is shown in Figure 1.2-36. Martensite is a non-equilibrium phase and does not appear on the equilibrium Fe-Fe$_3$C phase diagram. Martensite has a body-centered tetragonal (BCT) structure (Figure 1.2-37), and a density lower than that of the FCC austenite. Thus, there is volume expansion on martensite formation that could cause thermal stresses and cracks in rapidly cooled parts. The situation is exacerbated for relatively large parts in which the center cools more slowly than the surface. Special surface-hardening treatments for steels such as nitriding do not, however, require martensite formation; instead, hard nitride compounds from the alloying elements in steel, and the solid solution hardening due to nitrogen dissolution, result in a hard and wear-resistant exterior in nitrided steels.

Steels hardened using martensitic transformation are frequently given a secondary treatment called tempering to impart some toughness and ductility. Hardened steels

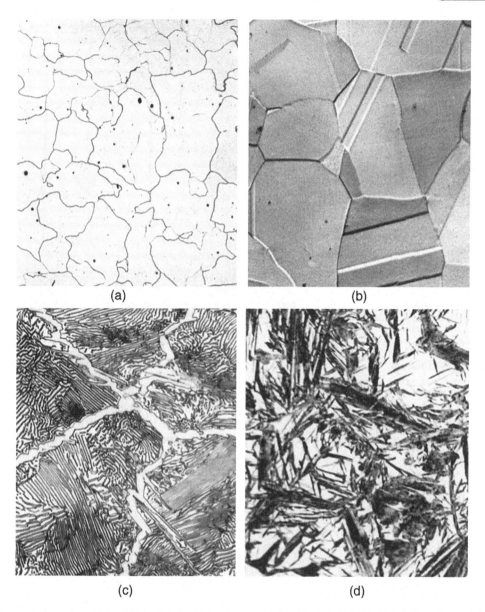

**Figure 1.2-35** Photomicrographs of (a) α-Fe (ferrite), and (b) γ-Fe (austenite). (Copyright 1971, United States Steel Corporation). (c) Photomicrograph of a 1-4wt% C steel having a microstructure consisting of a white pro-eutectoid cementite network surrounding the pearlite colonies (Copyright 1971, United States Steel Corporation). (d) Photomicrograph showing plate martensite. The needle-shaped dark regions are martensite, and the lighter regions are untransformed austenite (Copyright 1971, United States Steel Corporation). Reprinted with permission from U.S. Steel Corporation, Pittsburgh, PA.

are tempered by heating and isothermal holding at temperatures in the range 250°C to 650°C; this allows for diffusional processes to form tempered martensite, which is composed of stable ferrite and cementite phases. Figure 1.2-38 shows the effect of tempering temperature and time on the hardness of a heat-treated eutectoid steel.

The ability of a steel to form martensite for a given quenching treatment is represented by the "hardenability" of the steel. A standardized test procedure to characterize the hardenability of steels of different compositions is the Jominy end-quench test (ASTM Standard A 255). A cylindrical test specimen (25.4 mm diameter and 100 mm length) is austenitized at a specified temperature for a fixed time, whereafter it is quickly mounted on a fixture and its one end is cooled with a water jet. After cooling, shallow flats are ground along the specimen length along which the hardness is measured, to develop a hardness distribution curve (hardenability curve). Because a distribution of cooling rates is achieved along the length of the specimen (whose one end is cooled), the hardness variation along

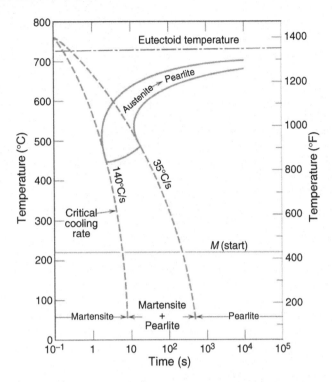

**Figure 1.2-36** Continuous cooling transformation (CCT) diagram for a eutectoid steel (0.76wt% C) and superimposed cooling curves, demonstrating the dependence of the final microstructure on the transformations that occur during cooling. (W. D. Callister, Jr., Materials Science & Engineering: An Introduction, 5th ed., Wiley, New York, 2000, p. 313).

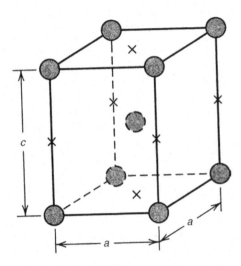

**Figure 1.2-37** The body-centered tetragonal (BCT) structure of martensite. (W. D. Callister, Jr., Materials Science & Engineering: An Introduction, 5th ed., Wiley, New York, 2000, p. 306).

quenched end, cooling rate progressively decreases, thus allowing more time for diffusion of carbon to form softer phases (e.g., pearlite). A steel with a high hardenability will exhibit high hardness to relatively large distances from the quenched end. The hardenability of plain carbon steels increases with increasing carbon content. Likewise, alloying elements such as Ni, Cr, and Mo in alloy steels improve the hardenability by delaying the diffusion-limited formation of pearlite and bainite, thus aiding martensite formation. Figure 1.2-39 shows a schematic hardenability curve obtained from the Jominy end-quench test.

the specimen length becomes a measure of cooling rate. The maximum hardness occurs at the quenched end where nearly 100% martensite forms. Away from the

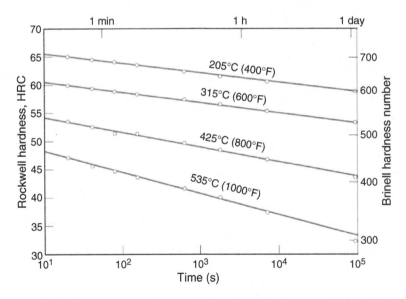

**Figure 1.2-38** Hardness versus tempering time for a water-quenched eutectoid plain carbon (1080) steel. (Adapted from E.C. Bain, Functions of the Alloying Elements in Steels, 1939, p. 233). Reprinted with permission from ASM International, Materials Park, OH (www.asminternational.org).

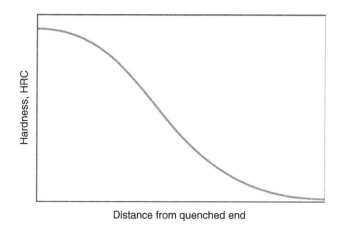

**Figure 1.2-39** *Typical hardenability curve showing Rockwell C hardness as a function of the distance from the quenched end in a Jominy end-quench test cooling. (W. D. Callister, Jr., Materials Science & Engineering: An Introduction, 5th ed., Wiley, New York, 2000, p. 333).*

## Precipitation hardening

Some nonferrous alloys (e.g., Al alloys) can be strengthened through a special thermal treatment that forms a distribution of very fine and hard second-phase precipitates that act as barriers to dislocation motion. These alloys generally exhibit high solubility of the solute in the parent matrix, and a rather rapid change in the solute concentration with temperature. The alloy is first solutionized (or homogenized) by heating it to a temperature $T_0$ within the single-phase ($\alpha$) region of the phase diagram (Figure 1.2-40) followed by quenching to near room temperature in the two-phase ($\alpha + \beta$) field. Quenching preserves the single-phase structure, and the

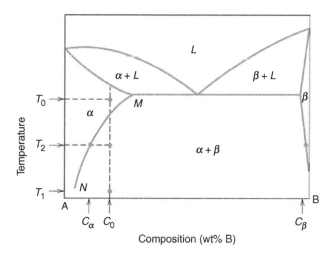

**Figure 1.2-40** *Hypothetical phase diagram of a precipitation hardening alloy of composition $C_0$. (W. D. Callister, Jr., Materials Science & Engineering: An Introduction, 5th ed., Wiley, New York, 2000, p. 342).*

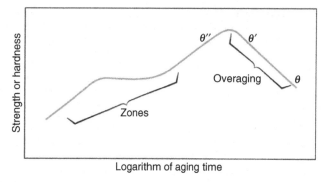

**Figure 1.2-41** *Schematic diagram showing strength and hardness as a function of the logarithm of aging time at constant temperature during the precipitation hardening treatment. (W. D. Callister, Jr., Materials Science & Engineering: An Introduction, 5th ed., Wiley, New York, 2000, p. 343).*

alloy becomes thermodynamically unstable in the ($\alpha + \beta$) field because the $\alpha$-phase becomes supersaturated with the solute. The atomic rearrangement via diffusion is sluggish because of the low temperatures after quenching. The supersaturated $\alpha$-phase solid solution is then heated to a temperature $T_2$ in the ($\alpha + \beta$) field to enhance the atomic diffusion and allow finely dispersed $\beta$ particles of composition, $C_\beta$, to form during isothermal hold, a treatment called "aging." An optimum aging time at a fixed temperature is required for optimum hardening. Overaging (prolonged aging) weakens the alloy through the coarsening (competitive dissolution) of fine precipitates in a distribution of coarse and fine precipitates. Higher temperatures enhance the aging kinetics, thus allowing maximum hardening to be achieved in a shorter time. The most widely studied precipitation-hardening alloy is Al-4%Cu, in which the $\alpha$-phase is a substitutional solid solution of Cu in Al, and the second phase ($\theta$-phase) is an intermetallic compound, $CuAl_2$. During aging of this alloy, intermediate non-equilibrium phases (denoted by $\theta''$ and $\theta'$) form prior to the formation of the equilibrium $\theta$-phase; the maximum hardness is associated with the $\theta''$ phase, whereas overaging is caused by the continued precipitate coarsening and formation of $\theta'$ and $\theta$ phases. A schematic hardness (or strength) versus aging time profile after precipitation hardening for this alloy is shown in Figure 1.2-41.

## Thermal properties

Solid materials experience an increase in temperature upon heating due to energy absorption. The heat capacity, C, of a material specifies the quantity of heat energy, Q, absorbed to result in a unit temperature rise, i.e., $C = QdQ/dT$. The quantity of heat absorbed by a unit mass to produce a unit increase in the solid's temperature is the specific heat, c, of the material. The heat capacity is expressed in J/mol·K or cal/mol·K, and the specific heat in J/kg·K or cal/g·K. At a fixed temperature, the atoms in a solid vibrate at high frequencies with a certain amplitude; the vibrations between atoms are coupled because of interatomic forces. These coupled vibrations produce elastic waves of various frequency distributions that propagate through the solid at the speed of sound. Wave propagation is possible only at certain allowed energy values or quantum of vibration energy, the "phonon" (which is analogous to a photon of electromagnetic radiation). Thus, the thermal energy of a solid can be represented in terms of the energy distribution of phonons.

For many solids, the heat capacity at constant volume, $C_v$, increases with increasing temperature according to $C_v = AT^3$, where $A$ is a constant. Above a critical temperature, called the Debye temperature, $\theta_D$, $C_v$ levels off to a value $\sim 3/R$ (where $R$ is gas constant, 8.314 J/mol·K), or $\sim 25$ J/mol·K. Thus, the amount of energy required to produce a unit temperature change becomes constant at $T > \theta_D$.

The coefficient of linear thermal expansion (CTE), $\alpha$, of a solid represents the change in length, $\Delta l$, per unit length, $l_0$, of a solid when its temperature changes by an amount, $\Delta T$, due to heating or cooling, i.e., $\alpha = \frac{\Delta l}{(l_0 \cdot \Delta T)}$. The CTE represents the thermal strain per unit change in the temperature. The change in the volume, $\Delta V$, of a solid on heating or cooling is related to the coefficient of volumetric expansion, $\alpha_v$, by $\alpha_v = \frac{\Delta V}{(V_0 \cdot \Delta T)}$, where $V_0$ is the original volume. For uniform expansion along all directions upon heating, $\alpha_v = 3\alpha$; in general, however, $\alpha_v$ is anisotropic. Table 1.2-5 presents the average CTE data on some solid materials.

The genesis of thermal expansion of solids lies in the shape of the potential energy curve for atoms in solids (Figure 1.2-42). At 0°K, the equilibrium separation

**Table 1.2-5** Coefficient of thermal expansion (CTE), thermal conductivity ($k$), and specific heat of selected materials

| Material | CTE, $10^{-6} \times C$ | $k$, W/m·K | $C_p$, J/Kg·K |
| --- | --- | --- | --- |
| Al | 23.6 | 247 | 900 |
| Cu | 17.0 | 398 | 386 |
| Au | 14.2 | 315 | 128 |
| Fe | 11.8 | 80 | 448 |
| 316 Stainless steel | 16.0 | 15.9 | 502 |
| Invar (64Fe-36Ni) | 1.6 | 10.0 | 500 |
| $Al_2O_3$ | 7.6 | 39 | 775 |
| $SiO_2$ | 0.4 | 1.4 | 740 |
| Pyrex (borosilicate glass) | 3.3 | 1.4 | 850 |
| Mullite ($3Al_2O_3 \cdot 2SiO_2$) | 5.3 | 5.9 | – |
| BeO | 9.0 | 219 | – |
| $ZrO_2$(PSZ) | 10.0 | 2.0–3.3 | 481 |
| TiC | 7.4 | 25 | – |
| High-density polyethylene | 106–198 | 0.46–0.50 | 1850 |
| Polypropylene | 145–180 | 0.12 | 1925 |
| Polystyrene | 90–150 | 0.13 | 1170 |
| Teflon (polytetrafluoroethylene) | 126–216 | 0.25 | 1050 |

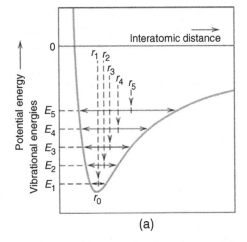
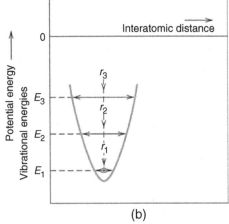

**Figure 1.2-42** (a) *Plot of potential energy versus interatomic distance, demonstrating the increase in interatomic separation with rising temperature. On heating, the interatomic separation increases from $r_0$ to $r_1$ to $r_2$ and so on. (b) For a symmetric potential energy-interatomic separation curve, there is no net increase in interatomic separation with rising temperature ($r_1 = r_2 = r_3$, etc.), and therefore, no net expansion. (Adapted from R. M. Rose, L. A. Shepard, and J. Wulff, The Structure and Properties of Materials, vol. 4, Electronic Properties, 1966, Wiley, New York.)*

between atoms is $r_0$; as the solid's temperature is successively raised to $T_1, T_2 \ldots T_n$, the potential energy rises to $E_1, E_2, \ldots E_n$, and the mean position of an atom changes from $r_0$ to $r_1, r_2 \ldots r_n$. Because the shape of the potential energy trough in Figure 1.2-42 is nonsymmetric, the mean positions $r_1, r_2 \ldots r_n$ represent a net shift in the mean position of the atoms, i.e., a net expansion of the solid (for a symmetric potential energy trough, an increase in the temperature will lead to a greater vibrational amplitude but no increase in the mean atomic positions, i.e., zero expansion). For solids with strong atomic bonding, the mean positions will shift by only a small amount for a given temperature rise; as a result, such solids will exhibit a small CTE. For example, Al with a bond energy of 324 kJ·mol$^{-1}$ has a CTE of $23.6 \times 10^{-6}$ °K$^{-1}$, whereas W with a bond energy of 849 kJ·mol$^{-1}$ has a CTE of $4.5 \times 10^{-6}$ °K$^{-1}$.

Thermal conductivity, $k$, represents the ability of a solid to spatially distribute the thermal energy, i.e., $k$ characterizes the rate at which energy is transported down a thermal gradient, $dT/dx$, between a region at a high temperature and a region at a low temperature within a solid. Experimental measurements of $k$ for selected materials are summarized in Table 1.2-5. The thermal conductivity, $k$, is given from $k = -q/(dT/dx)$, where $q$ is the heat flux (energy transferred per unit area per unit time) assumed to be constant, and the negative sign indicates that heat flows from the high- to low-temperature region. If $q$ is not constant (nonsteady thermal conduction), then the temperature changes with time, $t$, according to the Fourier equation,

$$\frac{\delta T}{\delta t} = \left(\frac{k}{c\rho}\right)\frac{\delta^2 T}{\delta x^2}, \quad (1.2.13)$$

where $\rho$ is the solid's density, and the ratio, $\frac{k}{c\rho}$ is called the thermal diffusivity, $\alpha$, of the solid. By analogy to the process of mass transport by diffusion in solids, to which Fick's equation (i.e., Equation 1.2.1) applies, (with C as the concentration), it is seen that $D$ is analogous to thermal diffusivity, $\alpha$ (both $\alpha$ and $D$ have the SI units, m$^2 \cdot$ s$^{-1}$).

Thermal conduction in solids occurs via energy transmission through lattice waves (phonons) and also via free electrons, so that the conductivity, $k = k_l + k_e$, where the subscripts "l" and "e" denote the lattice and electronic contributions to conductivity, respectively. The lattice contribution is due to the migration of phonons (and transfer of thermal energy) from the high- to low-temperature regions. The electronic contribution is due to the migration of high-energy free electrons from the high-temperature regions to the low-temperature regions, where some of the kinetic energy of free electrons is transferred to atoms. However, electron collisions with crystal defects rapidly dissipate this energy. In very pure metals, free electron energy is not rapidly dissipated through collisions with defects, because of fewer crystal defects; as a result, the high-energy electrons can travel larger distances before their energy is transferred to atoms. In addition, a large number of free electrons exist in metals, which also aids heat conduction. Because electrons contribute to both thermal and electrical conduction in pure metals, the thermal and electrical conductivities are related. The fundamental law describing this relationship is called the Wiedemann-Franz law, according to which $L = k/\sigma T$, where $\sigma$ is the electrical conductivity, and $L$ is a constant, with an approximate value of $2.44 \times 10^{-8}$ $\Omega \cdot W/K^2$. Alloying reduces both the thermal and electrical conductivity of metals because of the increased scattering of electron energy by the impurity atoms.

# Electrical properties

The conduction of electricity in solids follows a fundamental law, called Ohm's law, according to which $V = IR$, where $V$ is the voltage (in volts), $I$ is the electrical current (in amps or coulombs per s), and $R$ is the electrical resistance of the material (in Ohms), respectively. The resistance, $R$, is influenced by the specimen geometry. In contrast, the electrical resistivity, $\rho$, is independent of specimen geometry and is related to the electrical resistance, $R$, by $\rho = RA/l$, where $A$ is the cross-sectional area of the specimen perpendicular to the direction of current, and $l$ is the distance between the points over which the voltage is applied. The units of $\rho$ are Ohm·meters ($\Omega \cdot$m). The electrical conductivity, $\sigma$, is the inverse of the resistivity, i.e., $\sigma = \rho^{-1}$, and the units of $\sigma$ are $(\Omega \cdot$m$)^{-1}$, or mho·m$^{-1}$. Solids exhibit an extremely wide range of $\sigma$ (Table 1.2-6), with $\sigma$ values varying from $10^7 (\Omega \cdot$m$)^{-1}$ for conductive metals to $10^{-20} (\Omega \cdot$m$)^{-1}$ for insulators.

Electronic conduction is due to motion of charged particles (electrons and positively and negatively charged ions). However, not all electrons in atoms of a solid contribute to electronic conduction, and the number of electrons available to participate in conduction depends on the electron energy levels. Electrons in each atom may exist in certain allowed shells (1, 2, 3, and so on) and subshells (s, p, d, and f), with the electrons filling the states having the lowest energies. The electron configuration of an isolated atom (or an atom far removed from its neighbors in a solid) begin to be perturbed by the forces due to electrons and positive nuclei of its neighbors when the isolated atom is brought within relatively short distances of its neighbors. This leads to a splitting of closely spaced electron states in the solid's atoms, and the formation of electron energy bands. The energy gap

Table 1.2-6 Representative room-temperature electrical conductivity ($\sigma$) of metals, alloys, and semiconductors

| Material | $\sigma(\Omega\text{-m})^{-1}$ |
|---|---|
| Al | $3.8 \times 10^7$ |
| Cu | $6.0 \times 10^7$ |
| Ag | $6.8 \times 10^7$ |
| Fe | $1.0 \times 10^7$ |
| Stainless steel | $0.2 \times 10^7$ |
| W | $1.8 \times 10^7$ |
| Nichrome (80%Ni-20%Cr) | $0.09 \times 10^7$ |
| 60%Pb-40%Sn | $0.66 \times 10^7$ |
| Si | $4 \times 10^{-4}$ |
| Ge | 2.2 |
| GaAs | $1 \times 10^{-6}$ |

between bands is large, and electron occupancy of the band gap region is normally prohibited. The electrons in the outermost energy band participate in the electron conduction process. The energy of the highest filled electron state at 0°K is called the Fermi energy, $E_f$. For metals such as Cu, the outermost energy band is only partially filled at 0°K, i.e., there are available electron states above and adjacent to filled states, within the same band. For metals such as Mg, an empty outer band overlaps a filled band. For both these types of metals, even small electrical excitation could cause electrons to be promoted to lower-level empty states (above $E_f$) within the same energy band; these "free" electrons will then contribute to electrical conduction. For insulators and semiconductors, one energy band (called the valence band) is completely filled with electrons but is separated by an energy gap from an empty band (conduction band). For insulating materials, the band gap is large ($E_g > 2$ eV), and for semiconducting materials, the band gap is smaller ($E_g < 2$ eV); the Fermi energy, $E_f$, lies near the center of the band gap. For electrical conduction to occur in such materials, electrons must be promoted (usually, by thermal or light energy) across the band gap into low-lying empty states of the conduction band. For materials with large $E_g$, only a few electrons can be promoted into the conduction band at a given temperature; increasing the temperature increases the number of thermally excited electrons into the conduction band, and the conductivity increases. Ionic insulators and strongly covalent insulators have tightly bound valence electrons, which makes it difficult for thermal excitation to increase the conductivity. In contrast, semiconductors are mainly covalently bonded but have relatively weak bonding. Thermal excitation of valence electrons into the conduction band across the relatively small $E_g$ is, therefore, possible. As a result the conductivity rises faster with increasing temperature than for insulators. The electrical and electronic properties of semiconducting materials are discussed in Chapter 7.

Upon application of an electrical field to a solid, electrons are accelerated (in a direction opposite to the field), and the electrical current should increase with time as more and more electrons are excited. However, a saturation value of current is reached almost instantaneously upon the application of a field, because of frictional resistance offered to electron motion by the universally present crystal imperfections such as impurity atoms (substitutional and interstitial), dislocations, grain boundaries etc. The electron energy loss due to scattering by such defects leads to a drift velocity, $v_d$, which is the average electron velocity in the direction of the applied field. The drift velocity, $v_d = \mu_e E$, where $E$ is the electrical field strength, and $\mu_e$ is called the electron mobility (m$^2$/V·s). Clearly, both the electron mobility and the number of electrons will influence the solid's conductivity. The electrical conductivity, $\sigma$, increases with both the number of free electrons, and the electron mobility. The conductivity $\sigma = ne\mu_e$, where $n$ is the number per unit volume of free or conduction electrons, and $e$ is the elementary charge ($1.6 \times 10^{-19}$ C).

In the case of metals, the electrical resistivity, $\rho$, is influenced by the temperature, impurities (alloying), and plastic deformation. The temperature dependence of $\rho$ is given from the linear relationship, $\rho = \rho_0 + aT$, where $\rho_0$ and $a$ are material constants. Thus increasing the temperature increases the thermal vibrations and lattice defects (e.g., vacancies), both of which scatter electron energy, thereby increasing the resistivity or decreasing the conductivity. The increased crystal defect content because of plastic deformation also increases the energy scattering centers in the crystal lattice, thus adversely affecting the conductivity. Alloying usually decreases the resistivity. The impurity content, C, of a single impurity in a metal decreases the resistivity according to $\rho = AC(1 - C)$, where $A$ is constant, which is independent of the impurity content.

The electrical conductivity of a class of materials comprised of Si, Ge, GaAs, InSb, CdS, and others is extremely sensitive to minute concentration of impurities. These materials are called semiconducting material (see Chapter 7). In intrinsic semiconductors (e.g., Si and Ge), the electrical behavior is governed by the electronic structure of the pure elemental material, and in extrinsic semiconductors, the small concentration of the impurity (solute) atoms determines the electrical behavior. Both silicon and germanium are covalently bonded elements, with a small $E_g$ (1.1 eV for Si and 0.7 for Ge); in addition, compounds such as gallium arsenide (GaAs), indium antimonide (InSb), cadmium sulfide (CdS), and zinc telluride (ZnTe) also exhibit intrinsic semiconduction. The wider the separation of the two elements forming the compound semiconductor in the periodic table, the stronger will be their bonding, and larger will be the $E_g$, making the compound more insulating than semiconducting.

In all semiconducting materials, every electron excited into the conduction band leaves behind a missing electron or "hole" in the valence band. Under an electric field, the motion of the electrons within the valence band is aided by the motion in an opposite direction of the electron hole, which is considered as a positively charged electron. The electrical conductivity of a semiconductor is, therefore, sum of the contributions made by electrons and holes, i.e.,

$$\sigma = ne\mu_e + pe\mu_h, \quad (1.2.14)$$

where $n$ and $p$ are the number of electrons and holes per unit volume, respectively, and $\mu_e$ and $\mu_h$, are the electron and hole mobility, respectively. Because in intrinsic semiconductors every electron elevated to conduction band leaves behind a hole in the valence band, the numbers of electrons, n, and holes, p, are identical (n = p) and, therefore,

$$\sigma = ne(\mu_e + \mu_h). \quad (1.2.15)$$

The electron and hole mobilities of common intrinsic semiconductors are in the range 0.01-8 m$^2$/V·s, and $\mu_h < \mu_e$. Both electrons and holes are scattered by lattice defects.

In the case of extrinsic semiconductors, the electrical behavior is dominated by the impurity atoms, and extremely minute impurity levels (e.g., one impurity atom in every 1000 billion atoms) can influence the electrical behavior. The intrinsic semiconductor, silicon, has four valence electrons, and it can be made into an extrinsic semiconductor by adding an element that has five valence electrons per atom. For example, P, As, or Sb from Group VA of the periodic table can be added to pure Si to leave one electron that is loosely bound to the impurity atom, and can take part in electronic conduction. Because of the extra negative electron, this type of material is called an n-type extrinsic semiconductor, with the electrons constituting the majority carriers and holes (positive charge) constituting the minority carriers. If Si is doped with a small amount of a Group IIIA element such as Al, B, or Ga, that have three valence electrons per atom, then each Si atom becomes deficient in one negative charge in so far as electron bonding is concerned. This electron deficiency is likened to availability of a positive hole, loosely bound to the impurity atom, and capable of taking part in conduction. Such materials containing an excess of positive charge are called p-type extrinsic semiconductors. Devices based on n- and p-type semiconductors are discussed in Chapter 7.

## Dielectric and magnetic properties

Dielectric properties are exhibited by materials that are electrical insulators in which the centers of positive and negative charges do not coincide. These materials consist of electrical dipoles, which interact with an external electric field such as in a capacitor. The dielectric properties are in fact best understood in terms of the behavior of an electrical capacitor. In a parallel plate capacitor (plate separation, $l$, and area, $A$) the capacitance, C, is related to the charge stored on either plate, Q, by Q = CV, where V is the voltage applied across the capacitor. With a vacuum between the plates, the capacitance is given from C = $\varepsilon_0 A/l$, where $\varepsilon_0$ is the permittivity of vacuum ($\varepsilon_0 = 8.85 \times 10^{-12}$ Farads/m, and Farad = Coulombs per volt). With a dielectric material of permittivity $\varepsilon$ residing in the gap between the plates, the capacitance, C, becomes, C = $\varepsilon A/l$, and $\varepsilon > \varepsilon_0$. The relative permittivity, $\varepsilon_r = \varepsilon/\varepsilon_0$ is called the material's dielectric constant, which is always greater than unity. In an electric field, the dipoles in the dielectric material experience a force that aligns (polarizes) them along the direction of electric field. As a result of polarization, the surface charge density on the plates of the capacitor's changes. The surface charge density D (in coulombs/m$^2$) on a capacitor with a dielectric material between the plates is defined from, D = $\varepsilon$E, where E is the electric field strength. Alternatively, the surface charge density, D = $\varepsilon_0 E + P$, where P is the polarization defined as the increase in charge density relative to vacuum in the presence of a dielectric material in the capacitor.

The magnetic properties of solids have their origin in the magnetic moments of individual electrons. In essence, each electron is a tiny magnet whose magnetic properties originate from the orbital motion of the electron around the positive nucleus, and the spinning of each electron about its own axis. Each electron in an atom has a spin magnetic moment of absolute magnitude, $\mu$, and an orbital magnetic moment of $m\mu$, where $\mu$ is called the Bohr magneton, and has a value of $9.27 \times 10^{-24}$ A·m$^2$, and $m$ is termed the magnetic quantum number of the electron. The net magnetic moment of an atom is the vector sum of magnetic moments of all electrons constituting the atom. Elements such as He, Ne, Ar, etc., have completely filled electron shells and cannot be permanently magnetized, because the orbital and spin moments of all electrons cancel out.

Different materials exhibit different types of magnetic behaviors, and some materials exhibit more than one type of magnetic behavior. Diamagnetic materials exhibit very weak magnetic properties when subjected to an external field, which perturbs the orbital motion of the electrons. Actually, all materials are diamagnetic, but the effect is usually too weak to be observed especially when the other types of magnetic behavior are stronger. In solids in which the electron spin or orbital magnetic moments do not completely cancel out, a net permanent dipole moment exists, giving rise to paramagnetic behavior. These randomly oriented dipoles rotate to align themselves when subjected to an external field, thus enhancing the field. The magnetic behavior is characterized in terms of the magnetic susceptibility, $\chi_m$, which for a paramagnetic material, is a positive number in the range, $10^{-5}$ to $10^{-2}$. The magnetic susceptibility, $\chi_m$, is defined from $\chi_m = \mu_r - 1$, and $\mu_r$ is the permeability ($\mu$) of the solid relative to that of the vacuum ($\mu_0$), i.e., $\mu_r = \mu/\mu_0$. The permeability, $\mu$, characterizes the magnetic induction (or the internal field strength), $B$, in a solid when subjected to an external field of strength $H$, and $\mu = B/H$.

Ferromagnetic materials such as Fe, Co, and Ni possess a permanent magnetic moment in the absence of an applied field. These materials are strongly magnetic, with very large positive ($\sim 10^6$) susceptibilities. In these materials, the electron spin moments do not cancel out, and the mutual interactions of these moments cause the dipoles to align themselves even in the absence of an external field. These materials reach a maximum possible magnetization (saturation magnetization) when all the magnetic dipoles are aligned with an external field. Many ceramics also exhibit ferromagnetic behavior; examples include cubic ferrites (e.g., $Fe_3O_4$), hexagonal ferrites (e.g., $BaFe_{12}O_{19}$), and garnets (e.g., $Y_3Fe_5O_{12}$).

The behavior called antiferromagnetism arises from the antiparallel alignment of electronic spin moments of neighboring atoms or ions in a solid. The oxide ceramic manganese oxide (MnO) is antiferromagnetic by virtue of an antiparallel alignment of $Mn^{2+}$ ions in the crystal lattice. However, because the opposing magnetic moments cancel out, there is no net magnetic moment in the solid. Finally, ferrimagnetic behavior arises from the incomplete cancellation of the electron spin moments, but the saturation magnetization for these materials is less than that of ferromagnetic materials.

The magnetic behavior is sensitive to temperature changes because enhanced thermal vibration of atoms perturbs the alignment of magnetic moments in a crystal. At the absolute zero temperature ($0°K$), the thermal vibrations are minimum and the saturation magnetization is a maximum for ferro- and ferrimagnetic materials. The saturation magnetization decreases with increasing temperature, and rapidly drops to zero at a critical temperature called the Curie temperature, $T_C$, because of complete destruction of coupling of spin moments. The value of the Curie temperature varies with the metal, but it is typically a few hundred degrees. At $T < T_C$, a ferromagnetic material consists of small regions or domains within each of which all magnetic dipole moments are aligned along the same direction; several domains could form within each grain of a polycrystalline solid. The adjacent domains are separated by domain walls or boundaries, across which the orientation of the dipole moments changes gradually. In a manner similar to the ferromagnetic behavior, antiferromagnetic behavior also diminishes with increasing temperature, and completely disappears at a temperature called the Neel temperature, $T_N$. Depending on their intrinsic magnetic property, different magnetic materials become paramagnetic above either $T_C$ or $T_N$.

The internal field strength of materials is characterized in terms of the magnetic induction, $B$. For most materials, the magnetic induction, $B$, is not a simple function of the magnetic field strength, $H$, and the $B-H$ plot (i.e., magnetization curve) usually forms a loop whose area represents the loss of magnetic energy per unit volume of the material during each magnetization-demagnetization cycle (Figure 1.2-43). For soft magnetic

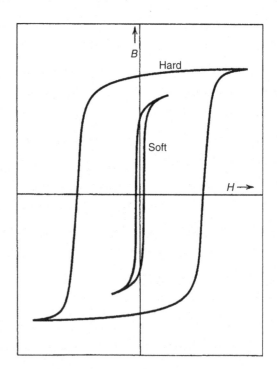

**Figure 1.2-43** Schematic magnetization curve for soft and hard magnetic materials. (From K. M. Ralls, T. H. Courtney and J. Wulff, Introduction to Materials Science & Engineering, Wiley, New York, 1976).

materials (99.95% pure Fe, Fe-3Si alloy, etc.), the area under the loop (also called the hysteresis curve) is small and the saturation magnetization is reached at small applied field strength. For a given material composition, structural defects such as voids and nonmagnetic inclusions limit the magnetic behavior of soft magnets. The hysteresis losses manifest themselves in heat generation. Another source of heating is the small-scale or eddy currents induced by an external magnetic field in the material (this is minimized by alloying to raise the electrical resistance of the material). Hard magnetic materials such as magnetic steels, Al-Ni-Co alloys, and Cu-Ni-Fe alloys are characterized by a higher hysteresis energy loss than soft magnetic materials. The relative ease with which magnetic boundaries move determines the hysteresis behavior, and microstructural features such as fine second-phase particles that inhibit the magnetic wall movement will lead to large energy losses. Judicious selection of alloying additions and heat treatment practices that lead to fine precipitates (e.g., WC and $Cr_3C_2$ in steels containing W and Cr) increase the resistance to the domain wall movement.

## Optical properties

The interaction of electromagnetic (EM) radiation (including visible radiation) with solids gives rise to physical processes such as absorption, scattering, and transmission. The range of wavelength of visible light is 0.4 μm to 0.7 μm, with each wavelength corresponding to a specific spectral color. The energy corresponding to each wavelength is expressed in terms of the energy, $E$, of a discrete packet or quantum (photon) from the Planck's equation, $E = h\nu = hc/\lambda$, where $h$ is Planck's constant ($h = 6.63 \times 10^{-34}$ J·s), and $\nu$ and $\lambda$ are the frequency and wavelength of the EM radiation ($c = \nu\lambda$), and $c$ is the velocity of light in vacuum. Thus, EM radiation of short wavelength has higher energy than radiation of large wavelength. When EM radiation crosses an interface between two different phases, the flux or intensity, $I_0$ (in units of W/m$^2$), of the incident EM radiation is partitioned between the intensities of the transmitted, reflected, and absorbed radiation. From the conservation of energy, $I_0 = I_T + I_R + I_A$, or $1 = (I_T/I_0 + I_R/I_0 + I_A/I_0)$. Materials with high transmissivity (i.e., large $I_T/I_0$) are transparent, materials that are impervious to visible light are opaque, and materials that cause internal scattering of light are translucent. Internal scattering of radiation occurs because of factors such as grain boundaries in polycrystalline materials (different refractive indices of grains having different crystallographic orientations), finely dispersed second-phase particles of different index of refraction, fabrication-related defects such as dispersed porosity, and in the case of polymers, the degree of crystallinity. In metals, the incident radiation is absorbed in a thin surface layer (typically less than 0.1 μm), which excites the electrons into unoccupied energy states. The reflectivity of the metals is a consequence of the reemitted radiation from transition of an electron from a higher energy level to a lower energy level, resulting in the emission of a photon. The surface color of a metal is determined by the wavelength of radiation that is reflected and not absorbed. For example, a bright, shiny metal surface will emit the photons in the same number and of same frequency as the incident light. Nonmetallic materials exhibit refraction and transmission in addition to absorption and reflection.

The velocity, $v$, of light in a transparent material is different from the velocity in vacuum, $c$, and the ratio $c/v = n$ is the refractive index of the material; the slower the velocity $v$ in the medium, greater is the $n$. For nonmagnetic materials, $n \approx \sqrt{\varepsilon_r}$, where $\varepsilon_r$ is the dielectric constant of the transparent material. The refraction phenomenon (i.e., bending of EM radiation due to velocity change at an interface) is related to electronic polarization, and large ions or atoms lead to greater polarization and larger $n$. For example, addition of large lead or barium ions to common soda-lime glass appreciably increases the $n$.

The reflectivity ($I_R/I_0$) represents the fraction of incident light scattered or reflected at an interface between media having different refractive indices. Generally, the higher the refractive index of a material, the greater is its reflectivity. Light reflection from lenses and other optical devices is minimized by applying a very thin coating of a dielectric material such as magnesium fluoride ($MgF_2$) to the reflecting surface.

The light absorption in nonmetals occurs by electronic polarization and by electron transfer between conduction band and valence band. A photon of energy $E = hc/\lambda$ may be absorbed by a material if it can excite an electron in the nearly filled valence band to the conduction band, i.e., if $E > E_g$, where $E_g$ is the energy of the gap between the valence band and conduction band. Taking $c = 3 \times 10^8$ m/s, and $h = 4.13 \times 10^{-15}$ eV·s, and considering the minimum and maximum $\lambda$ values in the visible spectrum to be 0.4 μm and 0.7 μm, respectively, the maximum and minimum values of band gap energy, $E_g$, for which absorption of visible light is possible, are 3.1 eV and 1.8 eV, respectively. Thus, nonmetallic materials with $E_g > 3.1$ eV will not absorb visible light, and will appear as transparent and colorless. In contrast, materials with $E_g < 1.8$ eV will absorb all visible radiation for electronic transitions from the valence band to conduction band, and will appear colored. Different materials could become opaque to some type of EM radiation depending on the wavelength of radiation and the band gap, $E_g$, of the material.

The intensity, $I'_T$, of radiation transmitted through a material depends also on the distance, $x$, traveled through the medium; longer distances cause a greater decrease in the intensity, which is given from $I'_T = I'_0 e^{-\beta x}$, where $I'_0$ is intensity of incident radiation, and $\beta$ is called the absorption coefficient of the material (a high $\beta$ indicates that the material is highly absorptive). When light incident on one face of a transparent solid of thickness, $L$, travels through the solid, the intensity of radiation transmitted from the opposite face is, $I_T = I_0(1 - R)^2 e^{-\beta x}$, where R is the reflectance ($I_R/I_0$). Thus, energy losses due to both absorption and reflection determine the intensity of transmitted radiation.

The color of solids arises out of the phenomena of absorption and transmission of radiation of specific wavelengths. As mentioned earlier, a colorless material such as single-crystal sapphire will uniformly absorb radiation of all wavelengths. In contrast, the semiconducting material cadmium sulfide ($E_g = 2.4$ eV) appears orange-yellow because it absorbs visible light photons of energy greater than 2.4 eV, which include blue and violet portions, the total energy range for visible light photons being 1.8 to 3.1 eV. Impurity atoms or ions influence the phenomenon of color. Thus, even though single-crystal sapphire is colorless, ruby is bright red in color because in the latter $Cr^{3+}$ ions (from $Cr_2O_3$ added to sapphire) substitute $Al^{3+}$ ions of $Al_2O_3$ crystal, and introduce impurity energy levels within the band gap of sapphire. This permits electronic transitions in multiple steps. Some of the light absorbed by ruby that contributes to the electron excitation from the valence band to the conduction band is reemitted when electrons transition between impurity levels within the band gap. Colored glasses incorporate different ions for color effects, such as $Cu^{2+}$ for blue-green and $Co^{2+}$ for blue-violet.

# References

Ashby M.F., and Jones D.R.H. *Engineering Materials—An Introduction to their Properties and Applications*, 2$^{nd}$ ed. Butterworth-Heinemann, Boston, 1996.

Askeland D.R., and Phule P.P. *The Science & Engineering of Materials*, 4$^{th}$ ed. Thompson Brooks/Cole, Pacific Grove, CA, 2003.

Budinski K. *Engineering Materials: Properties and Selection*. Prentice Hall, Englewood Cliffs, NJ, 1979.

Cahn R.W. *Coming of Materials Science*. Elsevier, New York, 2001.

Callister, W.D. Jr. *Materials Science & Engineering: An Introduction*, 5$^{th}$ ed. Wiley, New York, 2000.

Darken L.S., and Gurry W.R. *Physical Chemistry of Metals*. McGraw-Hill, New York, 1953.

DeGarmo E.P., Black J.T., Kohser R.A., and Klanecki B.E. *Materials and Processes in Manufacturing*, 9$^{th}$ ed. Wiley, New York, 2003.

Dieter G.E. *Mechanical Metallurgy*, 3$^{rd}$ ed. McGraw-Hill, New York, 1986.

Epifanov G.I. *Solid State Physics*. Mir Publishers, Moscow, 1974.

Flinn R.A., and Trojan P.K. *Engineering Materials and Their Applications*, 4$^{th}$ ed. Wiley, New York, 1990.

Gaskell D.R. *Introduction to the Thermodynamics of Materials*, 3$^{rd}$ ed. Taylor & Francis, Washington D.C., 1995

Ghosh, A., and A.K. Mallik. *Manufacturing Science*. New York: Wiley.

Jastrzebski, Z.D. *The Nature and Properties of Engineering Materials*. New York: Wiley.

John, V. *Introduction to Engineering Materials*. New York: Industrial Press, Inc.

Kalpakjian S., and Schmid S.R. *Manufacturing Engineering & Technology*. Prentice Hall, Englewood Cliffs, NJ, 2001.

Kittle C. *Introduction to Solid State Physics*, 6$^{th}$ ed. Wiley, New York, 1986.

Meyers M.A., and Chawla K.K. *Mechanical Metallurgy, Principles and Applications*. Prentice Hall, Englewood Cliffs, NJ, 1984.

Mitchell B.S. *An Introduction to Materials Engineering and Science for Chemical and Materials Engineers*. Wiley Interscience, New York, 2004.

National Materials Advisory Board, National Research Council. *Materials Science & Engineering: Forging Stronger Links to Users*. NMAB-492, Washington, D.C.: National Academy Press.

Shackelford J.F. *Introduction to Materials Science for Engineers*, 6$^{th}$ ed. Prentice Hall, Englewood Cliffs, NJ, 2005.

Shewmon P.G. *Diffusion in Solids*, 2$^{nd}$ ed. The Minerals, Metals & Materials Society (TMS), Warrendale, PA, 1989.

Swalin R.A. *Thermodynamics of Solids*. Wiley, New York, 1972.

Trivedi R. *Materials in Art and Technology*. Taylor Knowlton, Ames, 1998.

# Section Two

## Materials selection

# Chapter 2.1

# Materials selection

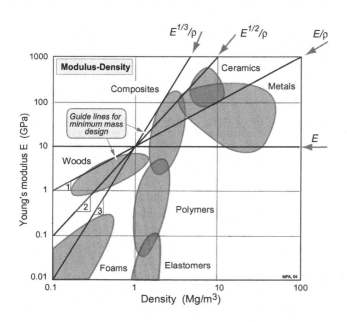

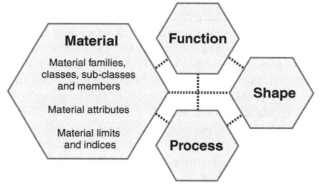

**Figure 2.1-1** Material selection is determined by function. Shape sometimes influences the selection. This chapter and the next deal with materials selection when this is independent of shape.

## 2.1.1 Introduction and synopsis

This chapter sets out the basic procedure for selection, establishing the link between material and function (Figure 2.1-1). A material has *attributes*: its density, strength, cost, resistance to corrosion, and so forth. A design demands a certain profile of these: a low density, a high strength, a modest cost and resistance to sea water, perhaps. It is important to start with the full menu of materials in mind; failure to do so may mean a missed opportunity. If an innovative choice is to be made, it must be identified early in the design process. Later, too many decisions have been taken and commitments made to allow radical change: it is now or never. The task, restated in two lines, is that of

**(1)** identifying the desired attribute profile and then

**(2)** comparing it with those of real engineering materials to find the best match.

The first step in tackling it is that of *translation*, examining the design requirements to identify the constraints that they impose on material choice. The immensely wide choice is narrowed, first, by *screening-out* the materials that cannot meet the constraints. Further narrowing is achieved by *ranking* the candidates by their ability to maximize performance. Criteria for screening and ranking are derived from the design requirements for a component by an analysis of *function, constraints, objectives*, and *free variables*. This chapter explains how to do it.

The materials property charts introduced in Chapter 4 are designed for use with these criteria. Property

*Engineering Materials and Processes Desk Reference*; ISBN: 9781856175869
Copyright © 2009 Elsevier Ltd; All rights of reproduction, in any form, reserved.

constraints and material indices can be plotted onto them, isolating the subset of materials that are the best choice for the design. The whole procedure can be implemented in software as a design tool, allowing computer-aided selection. The procedure is fast, and makes for lateral thinking. Examples of the method are given in Chapter 6.

## 2.1.2 The selection strategy

### Material attributes

Figure 2.1-2 illustrates how the kingdom of materials is divided into families, classes, sub-classes, and members. Each member is characterized by a set of *attributes*: its properties. As an example, the materials kingdom contains the family "metals", which in turn contains the class "aluminum alloys", the subclass "6000 series" and finally the particular member "Alloy 6061". It, and every other member of the kingdom, is characterized by a set of attributes that include its mechanical, thermal, electrical, optical, and chemical properties, its processing characteristics, its cost and availability, and the environmental consequences of its use. We call this its *property-profile*. Selection involves seeking the best match between the property-profiles of the materials in the kingdom and that required by the design.

There are four main steps, which we here call *translation, screening, ranking, and supporting information* (Figure 2.1-3). The steps can be likened to those in selecting a candidate for a job. The job is first analyzed and advertised, identifying essential skills and experience required of the candidate ("translation"). Some of these are simple go/no go criteria like the requirement that the applicant "must have a valid driving license", or "a degree in computer science", eliminating anyone who does not ("screening"). Others imply a criterion of excellence,

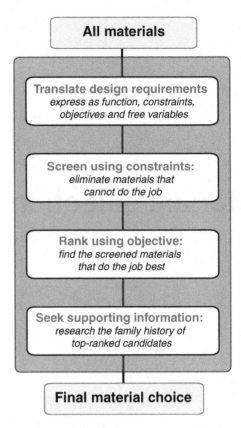

**Figure 2.1-3** The strategy for material selection. The four main steps—translation, screening, ranking, and supporting information—are shown here.

such as "typing speed and accuracy are priorities", or "preference will be given to candidates with a substantial publication list", implying that applicants will be ranked by these criteria ("ranking"). Finally references and interviews are sought for the top ranked candidates, building a file of supporting information—an opportunity to probe deeply into character and potential.

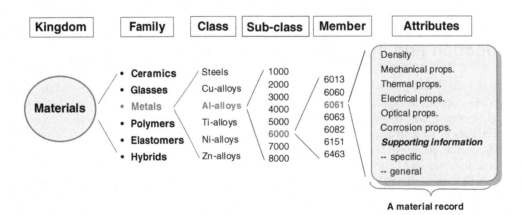

**Figure 2.1-2** The taxonomy of the kingdom of materials and their attributes. Computer-based selection software stores data in a hierarchical structure like this.

## Translation

How are the design requirements for a component (defining what it must do) translated into a prescription for a material? Any engineering component has one or more *functions*: to support a load, to contain a pressure, to transmit heat, and so forth. This must be achieved subject to *constraints*: that certain dimensions are fixed, that the component must carry the design loads or pressures without failure, that it insulates or conducts, that it can function in a certain range of temperature and in a given environment, and many more. In designing the component, the designer has an *objective*: to make it as cheap as possible, perhaps, or as light, or as safe, or perhaps some combination of these. Certain parameters can be adjusted in order to optimize the objective—the designer is free to vary dimensions that have not been constrained by design requirements and, most importantly, free to choose the material for the component. We refer to these as *free variables*. Function and constraints, objective and free variables (Table 2.1-1) define the boundary conditions for selecting a material and—in the case of load-bearing components—a shape for its cross-section. The first step in relating design requirements to material properties is a clear statement of function, constraints, objective, and free variables.

## Screening: attribute limits

Unbiased selection requires that all materials are considered to be candidates until shown to be otherwise, using the steps in the boxes below "translate" in Figure 2.1-3. The first of these, *screening*, eliminates candidates that cannot do the job at all because one or more of their attributes lies outside the limits set by the constraints. As examples, the requirement that "the component must function in boiling water", or that "the component must be transparent" imposes obvious limits on the attributes of *maximum service temperature* and *optical transparency* that successful candidates must meet. We refer to these as *attribute limits*.

## Ranking: material indices

Attribute limits do not, however, help with ordering the candidates that remain. To do this we need optimization criteria. They are found in the material indices, developed below, which measure how well a candidate that has passed the screening step can do the job. Performance is sometimes limited by a single property, sometimes by a combination of them. Thus the best

**Table 2.1-1** Function, constraints, objectives and free variables

| | |
|---|---|
| Function | What does component do? |
| Constraints* | What non-negotiable conditions must be met? What negotiable but desirable conditions …? |
| Objective | What is to be maximized or minimized? |
| Free variables | What parameters of the problem is the designer free to change? |

\* It is sometimes useful to distinguish between "hard" and "soft" constraints. Stiffness and strength might be absolute requirements (hard constraints); cost might be negotiable (a soft constraint).

materials for buoyancy are those with the lowest density, $\rho$; those best for thermal insulation the ones with the smallest values of the thermal conductivity, $\lambda$. Here maximizing or minimizing a single property maximizes performance. But—as we shall see–the best materials for a light stiff tie-rod are those with the greatest value of the specific stiffness, $E/\rho$, where E is Young's modulus. The best materials for a spring are those with the greatest value of $\sigma_f^2/E$ where $\sigma_f$ is the failure stress. The property or property-group that maximizes performance for a given design is called its *material index*. There are many such indices, each associated with maximizing some aspect of performance.[1] They provide *criteria of excellence* that allow ranking of materials by their ability to perform well in the given application.

To summarize: screening isolate candidates that are capable of doing the job; ranking identifies those among them that can do the job best.

## Supporting information

The outcome of the steps so far is a ranked short-list of candidates that meet the constraints and that maximize or minimize the criterion of excellence, whichever is required. You could just choose the top-ranked candidate, but what bad secrets might it hide? What are its strengths and weaknesses? Does it have a good reputation? What, in a word, is its credit-rating? To proceed further we seek a detailed profile of each: its *supporting information* (Figure 2.1-3, bottom).

Supporting information differs greatly from the structured property data used for screening. Typically, it is descriptive, graphical or pictorial: case studies of previous uses of the material, details of its corrosion behavior in particular environments, information of availability and pricing, experience of its environmental impact. Such

---

[1] Maximizing performance often means minimizing something: cost is the obvious example; mass, in transport systems, is another. A low-cost or light component, here, improves performance.

information is found in handbooks, suppliers' data sheets, CD-based data sources and the world-wide web. Supporting information helps narrow the short-list to a final choice, allowing a definitive match to be made between design requirements and material attributes.

Why are all these steps necessary? Without screening and ranking, the candidate-pool is enormous and the volume of supporting information overwhelming. Dipping into it, hoping to stumble on a good material, gets you nowhere. But once a small number of potential candidates have been identified by the screening–ranking steps, detailed supporting information can be sought for these few alone, and the task becomes viable.

## Local conditions

The final choice between competing candidates will, often, depend on local conditions: on in-house expertise or equipment, on the availability of local suppliers, and so forth. A systematic procedure cannot help here—the decision must instead be based on local knowledge. This does not mean that the result of the systematic procedure is irrelevant. It is always important to know which material is best, even if, for local reasons, you decide not to use it.

We will explore supporting information more fully in other chapters. Here we focus on the derivation of property limits and indices.

## 2.1.3 Attribute limits and material indices

Constraints set property limits. Objectives define material indices, for which we seek extreme values. When the objective in not coupled to a constraint, the material index is a simple material property. When, instead, they are coupled, the index becomes a group of properties like those cited above. Both are explained below. We start with two simple examples of the first—uncoupled objectives.

### Heat sinks for hot microchips

A microchip may only consume milliwatts, but the power is dissipated in a tiny volume. The power is low but the *power-density* is high. As chips shrink and clock-speeds grow, heating becomes a problem. The Pentium chip of today's PCs already reaches 85°C, requiring forced cooling. Multiple-chip modules (MCMs) pack as many as 130 chips on to a single substrate. Heating is kept under control by attaching the chip to a heat sink (Figure 2.1-4), taking pains to ensure good thermal contact between the chip and the sink. The heat sink now becomes a critical component, limiting further development of the electronics. How can its performance be maximized?

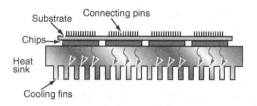

**Figure 2.1-4** A heat sink for power micro-electronics. The material must insulate electrically, but conduct heat as well as possible.

To prevent electrical coupling and stray capacitance between chip and heat sink, the heat sink must be a good electrical insulator, meaning a resistivity, $\rho_e > 10^{19}$ $\mu\Omega.cm$. But to drain heat away from the chip as fast as possible, it must also have the highest possible thermal conductivity, $\lambda$. The translation step is summarized in Table 2.1-2, where we assume that all dimensions are constrained by other aspects of the design.

To explain: resistivity is treated as a *constraint*, a go/no go criterion. Materials that fail to qualify as "good insulator", or have a resistivity greater than the value listed in the table, are screened out. The thermal conductivity is treated as an *objective*: of the materials that meet the constraint, we seek those with the largest values of $\lambda$ and rank them by this—it becomes the material index for the design. If we assume that all dimensions are fixed by the design, there remains only one *free variable* in seeking to maximize heat-flow: the choice of material. The procedure, then, is to *screen* on resistivity, then *rank* on conductivity.

The steps can be implemented using the $\lambda$—$\rho_e$ chart, reproduced as Figure 2.1-5. Draw a vertical line at $\rho_e = 1019$ $\mu\Omega.cm$, then pick off the materials that lie above this line, and have the highest $\lambda$. The result: aluminum nitride, AlN, or alumina, $Al_2O_3$. The final step is to seek supporting information for these two materials. A web-search on "aluminum nitride" leads immediately to detailed data-sheets with the information we seek.

### Materials for overhead transmission lines

Electrical power, today, is generated centrally and distributed by overhead or underground cables. Buried

**Table 2.1-2** Function, constraints, objective, and free variables for the heat sink

| | |
|---|---|
| Function | Heat sink |
| Constraints | • Material must be "good insulator", or $\rho_e > 10^{19}$ $\mu\Omega.cm$<br>• All dimensions are specified |
| Objective | Maximize thermal conductivity, $\lambda$ |
| Free variables | Choice of material |

# Materials selection    CHAPTER 2.1

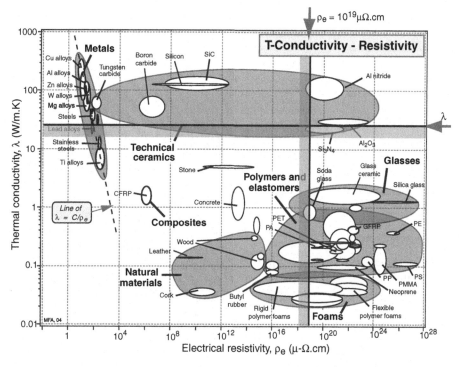

**Figure 2.1-5** The $\lambda - \rho_e$ chart of Figure 4.10 with the attribute limit $\rho_e > 10^{19}$ $\mu\Omega$.cm and the index $\lambda$ plotted on it. The selection is refined by raising the position of the $\lambda$ selection line.

lines are costly so cheaper overhead transmission (Figure 2.1-6) is widely used. A large span is desirable because the towers are expensive, but so too is a low electrical resistance to minimize power losses. The span of cable between two towers must support the tension needed to limit its sag and to tolerate wind and ice loads. Consider the simple case in which the tower spacing $L$ is fixed at a distance that requires a cable with a strength $\sigma_f$ of at least 80 MPa (a constraint). The objective then becomes that of minimizing resistive losses, and that means seeking materials with the lowest possible resistivity, $\rho_e$, defining the material index for the problem. The translation step is summarized in Table 2.1-3.

The prescription, then, is to *screen* on strength and *rank* on resistivity. There is no $\sigma_f - \rho_e$ chart in Chapter 4 (though it is easy to make one using the software described in Section 2.1.5). Instead we use the $\lambda - \rho_e$ chart of Figure 4.10 to identify materials with the lowest resistivity (Cu and Al alloys) and then check, using the $\sigma_f - \rho$ chart of Figure 4.4 that the strength meets the constraint listed in the table. Both do (try it!).

The two examples have been greatly simplified—reality is more complex than this. We will return to both again later. The aim here is simply to introduce the disciplined way of approaching a selection problem by identifying its key features: function, constraints, objective, and free variables. Now for some slightly more complex examples.

## Material indices when objectives are coupled to constraints

Think for a moment of the simplest of mechanical components, helped by Figure 2.1-7. The loading on a component can generally be decomposed into

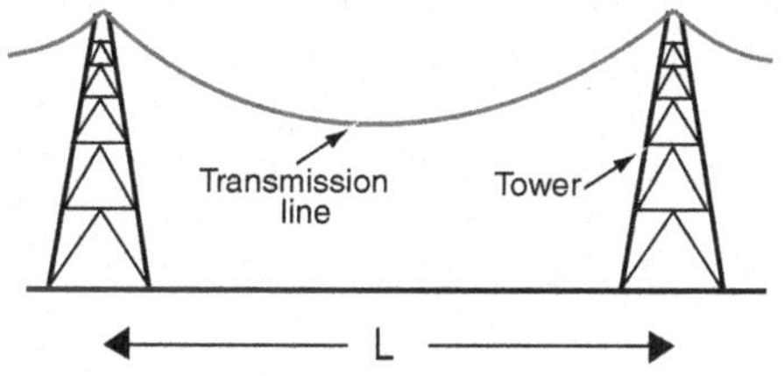

**Figure 2.1-6** A transmission line. The cable must be strong enough to carry its supporting tension, together with wind and ice loads. But it must also conduct electricity as well as possible.

| Table 2.1-3 Function, constraints, objective, and free variables for the transmission line | |
|---|---|
| Function | Long span transmission line |
| Constraints | • Span $L$ is specified <br> • Material must be strength $\sigma_f > 80$ MPa |
| Objective | Minimize electrical resistivity $\rho_e$ |
| Free variables | Choice of material |

# CHAPTER 2.1　Materials selection

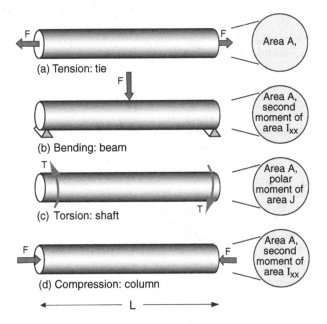

**Figure 2.1-7** A cylindrical tie-rod loaded (a) in tension, (b) in bending, (c) in torsion and (d) axially, as a column. The best choice of materials depends on the mode of loading and on the design goal; it is found by deriving the appropriate material index.

some combination of axial tension, bending, torsion, and compression. Almost always, one mode dominates. So common is this that the functional name given to the component describes the way it is loaded: *ties* carry tensile loads; *beams* carry bending moments; *shafts* carry torques; and *columns* carry compressive axial loads. The words "tie", "beam", "shaft", and "column" each imply a function. Many simple engineering functions can be described by single words or short phrases, saving the need to explain the function in detail. Here we explore property limits and material indices for some of these.

*Material index for a light, strong tie-rod*
A design calls for a cylindrical tie-rod of specified length $L$ to carry a tensile force $F$ without failure; it is to be of minimum mass, as in the uppermost sketch in Figure 2.1-7. The length $L$ is specified but the cross-section area $A$ is not. Here, "maximizing performance" means "minimizing the mass while still carrying the load $F$ safely". The design requirements, translated, are listed in Table 2.1-4.

We first seek an equation describing the quantity to be maximized or minimized. Here it is the mass $m$ of the tie, and it is a minimum that we seek. This equation, called *the objective function*, is

**Table 2.1-4** Design requirements for the light tie

| Function | Tie rod |
|---|---|
| Constraints | • Length $L$ is specified<br>• Tie must support axial tensile load $F$ without failing |
| Objective | Minimize the mass m of the tie |
| Free variables | • Cross-section area, $A$<br>• Choice of material |

$$m = AL\rho \qquad (2.1.1)$$

where $A$ is the area of the cross-section and $\rho$ is the density of the material of which it is made. The length $L$ and force $F$ are specified and are therefore fixed; the cross-section $A$, is free. We can reduce the mass by reducing the cross-section, but there is a constraint: the section-area $A$ must be sufficient to carry the tensile load $F$, requiring that

$$\frac{F}{A} \leq \sigma_f \qquad (2.1.2)$$

where $\sigma_f$ is the failure strength. Eliminating $A$ between these two equations give

$$m \geq (F)(L)\left(\frac{\rho}{\sigma_f}\right) \qquad (2.1.3)$$

Note the form of this result. The first bracket contains the specified load $F$. The second bracket contains the specified geometry (the length $L$ of the tie). The last bracket contains the material properties. The lightest tie that will carry $F$ safely[2] is that made of the material with the smallest value of $\rho/\sigma_f$. We could define this as the material index of the problem, seeking a minimum, but it is more usual, when dealing with specific properties, to express them in a form for which a maximum is sought. We therefore invert the material properties in equation (2.1.3) and define the material index $M$, as

$$M = \frac{\sigma_f}{\rho} \qquad (2.1.4)$$

The lightest tie-rod that will safely carry the load $F$ without failing is that with the largest value of this index, the "specific strength", plotted in the chart of Figure 4.6. A similar calculation for a light *stiff* tie (one for which the stiffness $S$ rather than the strength $\sigma_f$ is specified) leads to the index

---

[2] In reality a safety factor, $S_f$, is always included in such a calculation, such that equation (2.1.2) becomes $F/A = \sigma_f/S_f$. If the same safety factor is applied to each material, its value does not influence the choice. We omit it here for simplicity.

$$M = \frac{E}{\rho} \qquad (2.1.5)$$

where $E$ is Young's modulus. This time the index is the "specific stiffness", also shown in Figure 4.6. The material group (rather than just a single property) appears as the index in both cases because minimizing the mass $m$—the objective—was coupled to one of the constraints, that of carrying the load F without failing or deflecting too much.

That was easy. Now for a slightly more difficult (and important) one.

*Material index for a light, stiff beam*

The mode of loading that most commonly dominates in engineering is not tension, but bending—think of floor joists, of wing spars, of golf-club shafts. Consider, then, a light beam of square section $b \times b$ and length $L$ loaded in bending. It must meet a constraint on its stiffness $S$, meaning that it must not deflect more than $\delta$ under a load $F$ (Figure 2.1-8). Table 2.1-5 translates the design requirements.

Appendix A of this book catalogues useful solutions to a range of standard problems. The stiffness of beams is one of these. Turning to Section A3 we find an equation for the stiffness $S$ of an elastic beam. The constraint requires that $S = F/\delta$ be greater than this:

$$S = \frac{F}{\delta} \geq \frac{C_1 E I}{L^3} \qquad (2.1.6)$$

where $E$ is Young's modulus, $C_1$ is a constant that depends on the distribution of load and $I$ is the second moment of the area of the section, which, for a beam of square section ("Useful Solutions", Appendix A, Section A.2), is

$$I = \frac{b^4}{12} = \frac{A^2}{12} \qquad (2.1.7)$$

The stiffness $S$ and the length $L$ are specified; the section area $A$ is free. We can reduce the mass of the beam by reducing $A$, but only so far that the stiffness constraint is still met. Using these two equations to eliminate $A$ in equation (2.1.1) for the mass gives

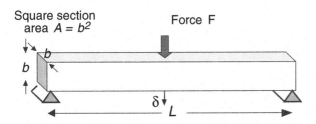

**Figure 2.1-8** A beam of square section, loaded in bending. Its stiffness is $S = F/\delta$ where $F$ is the load and $\delta$ is the deflection.

**Table 2.1-5** Design requirements for the light stiff beam

| Function | Beam |
|---|---|
| Constraints | • Length $L$ is specified<br>• Beam must support a bending load $F$ without deflecting too much, meaning that the bending stiffness $S$ is specified |
| Objective | Minimize the mass of the beam |
| Free variables | • Cross-section area, $A$<br>• Choice of material |

$$m \geq \left(\frac{12S}{C_1 L}\right)^{1/2} (L^3) \left(\frac{\rho}{E^{1/2}}\right) \qquad (2.1.8)$$

The brackets are ordered as before: functional requirement, geometry and material. The best materials for a light, stiff beam are those with the smallest values of $\rho/E^{1/2}$. As before, we will invert this, seeking instead large values of the material index

$$M = \frac{E^{1/2}}{\rho} \qquad (2.1.9)$$

In deriving the index, we have assumed that the section of the beam remained square so that both edges changed in length when $A$ changed. If one of the two dimensions is held fixed, the index changes. A panel is a flat plate with a given length $L$ and width $W$; the only free variable (apart from material) is the thickness $t$. For this the index becomes (via an identical derivation)

$$M = \frac{E^{1/3}}{\rho} \qquad (2.1.10)$$

Note the procedure. The length of the rod or beam is specified but we are free to choose the section area $A$. The objective is to minimize its mass, $m$. We write an equation for $m$: it is the objective function. But there is a constraint: the rod must carry the load $F$ without yielding in tension (in the first example) or bending too much (in the second). Use this to eliminate the free variable $A$ and read off the combination of properties, $M$, to be maximized. It sounds easy, and it is so long as you are clear from the start what the constraints are, what you are trying to maximize or minimize, which parameters are specified and which are free.

## Deriving indices—how to do it

This is a good moment to describe the method in more general terms. *Structural elements* are components that perform a physical function: they carry loads, transmit heat, store energy, and so on: in short, they satisfy

# CHAPTER 2.1  Materials selection

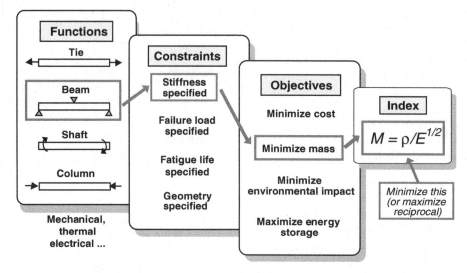

**Figure 2.1-9** The specification of function, objective, and constraint leads to a materials index. The combination in the highlighted boxes leads to the index $E^{1/2}/\rho$.

*functional requirements*. The functional requirements are specified by the design: a tie must carry a specified tensile load; a spring must provide a given restoring force or store a given energy, a heat exchanger must transmit heat a given heat flux, and so on.

The performance of a structural element is determined by three things: the functional requirements, the geometry and the properties of the material of which it is made. The performance $P$ of the element is described by an equation of the form

$$P = [(\text{Functional requirements}, F),$$
$$(\text{Geometric parameters}, G),$$
$$(\text{Material properties}, M)]$$

or

$$P = f(F, G, M) \qquad (2.1.11)$$

where $P$, the *performance metric*, describes some aspect of the performance of the component: its mass, or volume, or cost, or life for example; and "$f$" means "a function of". *Optimum design* is the selection of the material and geometry that maximize or minimize $P$, according to its desirability or otherwise.

The three groups of parameters in equation (2.1.11) are said to be *separable* when the equation can be written

$$P = f_1(F) \cdot f_2(G) \cdot f_3(M) \qquad (2.1.12)$$

where $f_1$, $f_2$, and $f_3$ are separate functions that are simply multiplied together. When the groups are separable, as they frequently are, the optimum choice of material becomes independent of the details of the design; it is the same for all geometries, $G$, and for all values of the function requirement, $F$. Then the optimum subset of materials can be identified without solving the complete design problem, or even knowing all the details of $F$ and $G$. This enables enormous simplification: the performance for *all F and G* is maximized by maximizing $f_3(M)$, which is called the material efficiency coefficient, or *material index* for short. The remaining bit, $f_1(F) \cdot f_2(G)$, is related to *the structural efficiency coefficient*, or *structural index*. We do not need it now, but will examine it briefly in Section 2.1.7.

Each combination of function, objective and constraint leads to a material index (Figure 2.1-9); the index is characteristic of the combination, and thus of the function the component performs. The method is general, and, in later chapters, is applied to a wide range of problems. Table 2.1-6 gives examples of indices and the design problems that they characterize. A fuller catalogue of indices is given in Appendix B. New problems throw up new indices, as the case studies of the next chapter will show.

## 2.1.4 The selection procedure

We can now assemble the four steps into a systematic procedure.

### Translation

Table 2.1-7 says it all. Simplified: identify the material attributes that are constrained by the design, decide what you will use as a criterion of excellence (to be minimized

# Materials selection CHAPTER 2.1

| Table 2.1-6 Examples of material indices | |
|---|---|
| **Function, objective, and constraints** | **Index** |
| Tie, minimum weight, stiffness prescribed | $\dfrac{E}{\rho}$ |
| Beam, minimum weight, stiffness prescribed | $\dfrac{E^{1/2}}{\rho}$ |
| Beam, minimum weight, strength prescribed | $\dfrac{\sigma_y^{2/3}}{\rho}$ |
| Beam, minimum cost, stiffness prescribed | $\dfrac{E^{1/2}}{C_m \rho}$ |
| Beam, minimum cost, strength prescribed | $\dfrac{\sigma_y^{2/3}}{C_m \rho}$ |
| Column, minimum cost, buckling load prescribed | $\dfrac{E^{1/2}}{C_m \rho}$ |
| Spring, minimum weight for given energy storage | $\dfrac{\sigma_y^2}{E\rho}$ |
| Thermal insulation, minimum cost, heat flux prescribed | $\dfrac{1}{\lambda C_p \rho}$ |
| Electromagnet, maximum field, temperature rise prescribed | $\dfrac{C_p \rho}{\rho_e}$ |

$\rho$ = density; $E$ = Young's modulus; $\sigma_y$ = elastic limit; $C_m$ = cost/kg; $\lambda$ = thermal conductivity; $\rho_e$ = electrical resistivity; $C_p$ = specific heat.

| Table 2.1-7 Translation | | |
|---|---|---|
| **Step** | **Action** | |
| 1 | Define the design requirements: | |
| | (a) | Function: what does the component do? |
| | (b) | Constraints: essential requirements that must be met: stiffness, strength, corrosion resistance, forming characteristics, … |
| | (c) | Objective: what is to be maximized or minimized? |
| | (d) | Free variables: what are the unconstrained variables of the problem? |
| 2 | List the constraints (no yield; no fracture; no buckling, etc.) and develop an equation for them if necessary | |
| 3 | Develop an equation for the objective in terms of the functional requirements, the geometry and the material properties (the objective function) | |
| 4 | Identify the free (unspecified) variables | |
| 5 | Substitute for the free variables from the constraint equations into the obejective function | |
| 6 | Group the variables into three groups: functional requirements, $F$, geometry, $G$, and material properties, $M$, thus<br>Performance metric $P \leq f_1(F) \cdot f_2(G) \cdot f_3(M)$<br>or<br>Performance metric $P \geq f_1(F) \cdot f_2(G) \cdot f_3(M)$ | |
| 7 | Read off the material index, expressed as a quantity $M$, that optimizes the performance metric $P$. $M$ is the criterion of excellence. | |

or maximized), substitute for any free variables using one of the constraints, and read off the combination of material properties that optimize the criterion of excellence.

## Screening: Applying attribute limits

Any design imposes certain non-negotiable demands ("constraints") on the material of which it is made. We have explained how these are translated into attribute limits. Attribute limits plot as horizontal or vertical lines on material selection charts, illustrated in Figure 2.1-10. It shows a schematic $E - \rho$ chart, in the manner of Chapter 4. We suppose that the design imposes limits on these of $E > 10$ GPa and $\rho < 3$ Mg/m$^3$, shown on the figure. The optimizing search is restricted to the window boxed by the limits, labeled "Search region". Less quantifiable properties such as corrosion resistance, wear resistance or formability can all appear as primary limits, which take the form

$$A > A^*$$

or

$$A < A^* \quad (2.1.13)$$

where $A$ is an attribute (service temperature, for instance) and $A^*$ is a critical value of that attribute, set by

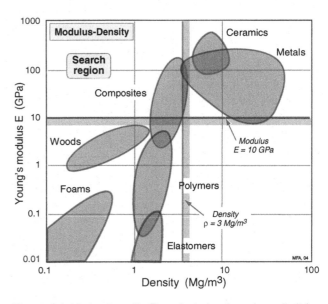

**Figure 2.1-10** A schematic $E-\rho$ chart showing a lower limit for $E$ and an upper one for $\rho$.

the design, that must be exceeded, or (in the case of corrosion rate) must *not* be exceeded.

One should not be too hasty in applying attribute limits; it may be possible to engineer a route around them. A component that gets too hot can be cooled; one that corrodes can be coated with a protective film. Many designers apply attribute limits for fracture toughness, $K_{1C}$ and ductility $\varepsilon_f$ insisting on materials with, as rules of thumb, $K_{1C} > 15$ Mpa.m$^{1/2}$ and $\varepsilon_f > 2\%$ in order to guarantee adequate tolerance to stress concentrations. By doing this they eliminate materials that the more innovative designer is able to use to good purpose (the limits just cited for $K_{1C}$ and $\varepsilon_f$ eliminate all polymers and all ceramics, a rash step too early in the design). At this stage, keep as many options open as possible.

## Ranking: indices on charts

The next step is to seek, from the subset of materials that meet the property limits, those that maximize the performance of the component. We will use the design of light, stiff components as an example; the other material indices are used in a similar way.

Figure 2.1-11 shows, as before, modulus $E$, plotted against density $\rho$, on log scales. The material indices $E/\rho$, $E^{1/2}/\rho$, and $E^{1/3}/\rho$ can be plotted onto the figure. The condition

$$\frac{E}{\rho} = C$$

or, taking logs,

$$\text{Log}(E) = \text{Log}(\rho) + \text{Log}(C) \quad (2.1.14)$$

is a family of straight parallel lines of slope 1 on a plot of Log($E$) against Log($\rho$) each line corresponds to a value of the constant C. The condition

$$\frac{E^{1/2}}{\rho} = C \quad (2.1.15)$$

or, taking logs again,

$$\text{Log}(E) = 2\text{Log}(\rho) + 2\text{Log}(C) \quad (2.1.16)$$

gives another set, this time with a slope of 2; and

$$\frac{E^{1/3}}{\rho} = C \quad (2.1.17)$$

gives yet another set, with slope 3. We shall refer to these lines as *selection guidelines*. They give the slope of the family of parallel lines belonging to that index. Where appropriate the charts of Chapter 4 show the slopes of guidelines like these.

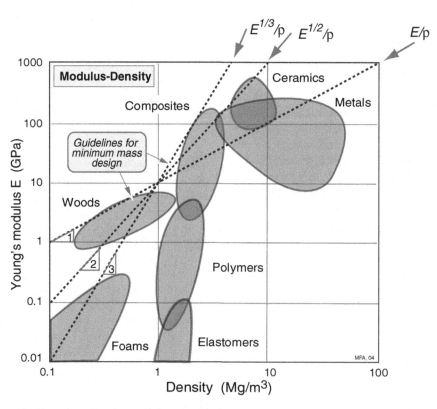

**Figure 2.1-11** A schematic $E$–$\rho$ chart showing guidelines for the three material indices for stiff, lightweight design.

# Materials selection CHAPTER 2.1

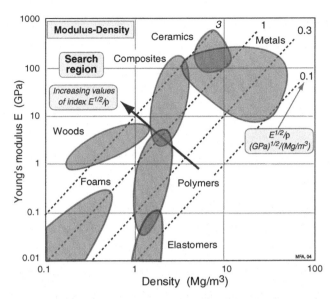

**Figure 2.1-12** A schematic $E - \rho$ chart showing a grid of lines for the material index $M = E^{1/2}/\rho$. The units are $(GPa)^{1/2}/(Mg/m^3)$.

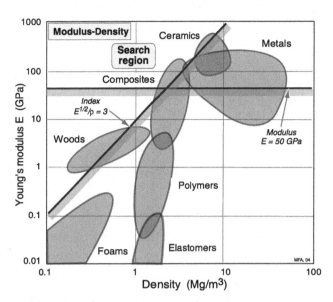

**Figure 2.1-13** A selection based on the index $M = E^{1/2}/\rho$ together with the property limit $E > 50$ GPa. The shaded band with slope 2 has been positioned to isolate a subset of materials with high $E^{1/2}/\rho$; the horizontal one lies at $E = 50$ GPa. The materials contained in the search region become the candidates for the next stage of the selection process.

It is now easy to read off the subset materials that optimally maximize performance for each loading geometry. All the materials that lie on a line of constant $E^{1/2}/\rho$ perform equally well as a light, stiff beam; those above the line are better, those below, worse. Figure 2.1-12 shows a grid of lines corresponding to values of $E^{1/2}/\rho$ from 0.1 to 3 in units of $GPa^{1/2}/(Mg/m^3)$. A material with $M = 1$ in these units gives a beam that has one tenth the weight of one with $M = 0.1$. The subset of materials with particularly good values of the index is identified by picking a line that isolates a search area containing a reasonably small number of candidates, as shown schematically in Figure 2.1-13 as a diagonal selection line. Attribute limits can be added, narrowing the search window: that corresponding to $E > 50$ GPa is shown as a horizontal line. The short-list of candidate materials is expanded or contracted by moving the index line.

## Supporting information

We now have a ranked short-list of potential candidate materials. The last step is to explore their character in depth. The list of constraints usually contains some that cannot be expressed as simple attribute limits. Many of these relate to the behavior of the material in a given environment, or when in contact with another material, or to aspects of the ways in which the material can be shaped, joined, or finished. Such information can be found in handbooks, manufacturers' data-sheets, or on the internet. And then—it is to be anticipated—there are the constraints that have been overlooked simply because they were not seen as such. Confidence is built by seeking design guidelines, case studies or failure analyses that document each candidate, building a dossier of its strengths, its weaknesses, and ways in which these can be overcome. All of these come under the heading of supporting information. Finding it is the subject of Chapter 15.

The selection procedure is extended in Chapters 9 and 11 to deal with multiple constraints and objectives and to include section shape. Before moving on to these, it is a good idea to consolidate the ideas so far by applying them to a number of case studies. They follow in Chapter 6.

## 2.1.5 Computer-aided selection

The charts of Chapter 4 give an overview, but the number of materials that can be shown on any one of them is obviously limited. Selection using them is practical when there are very few constraints, as the examples of Section 2.1.3 showed, but when there are many—as there usually are—checking that a given material meets them all is cumbersome. Both problems are overcome by computer implementation of the method.

The CES material and process selection software[3] is an example of such an implementation. A database

---

[3] Granta Design Ltd, Cambridge, UK (www.grantadesign.com).

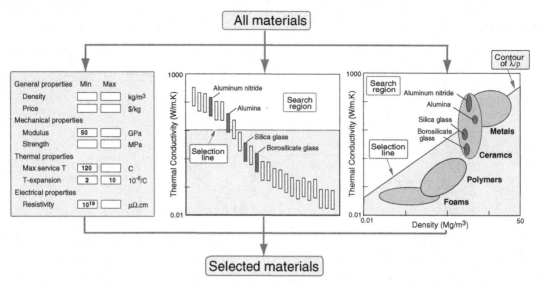

**Figure 2.1-14** Computer-aided selection using the CES software. The schematic shows the three types of selection window. They can be used in any order and any combination. The selection engine isolates the subset of material that pass all the selection stages.

contains records for materials, organized in the hierarchical manner shown in Figure 2.1-2. Each record contains structured property-data for a material, each stored as a range spanning the typical (or, often, the permitted) range of values of that property. It also contains limited unstructured data in the form of text, images, and references to sources of information about the material. The data are interrogated by a search engine that offers search interfaces shown schematically in Figure 2.1-14. On the left is a simple query interface for screening on single properties. The desired upper or lower limits for constrained attributes are entered; the search engine rejects all materials with attributes that lie outside the limits. In the center is shown a second way of interrogating the data: a bar chart like that shown earlier as Figure 4.1. It and the bubble chart shown on the right are the ways both of applying constraints and of ranking. Used for ranking, a selection line or box is super-imposed on the charts with edges that lie at the constrained values of the property (bar chart) or properties (bubble chart), eliminating the material in the shaded areas, and leaving the materials that meet all the constraints. If instead, ranking is sought (having already applied all necessary constraints) the line or box is positioned so that a few—say, three—materials are left in the selected area; these are the top ranked candidates.

The figure illustrates an elaboration of the heat sink example given earlier, in which we now add more constraints (Table 2.1-8). The requirements are as before, plus the requirement that the modulus be greater than 50 GPa, that the expansion coefficient $\alpha$, lies between 2 and $10 \times 10^{-6}/°C$ and that the maximum service temperature exceeds 120°C. All are applied as property limits on the left-hand window, implementing a screening stage.

**Table 2.1-8** Function, expanded constraints, objective, and free variable for the heat sink

| Function | Heat sink |
|---|---|
| Constraints | • Material must be "good insulator", or $\rho_e > 10^{19}$ $\mu\Omega.cm$<br>• Modulus $E > 50$ Gpa<br>• Maximum service temperature $T_{max} < 120°C$<br>• Expansion coefficient $2 \times 10^{-6} < \alpha < 10 \times 10^{-6}/°C$<br>• All dimensions are specified |
| Objective | Maximize thermal conductivity, $\lambda$ or conductivity per unit mass $\lambda/\rho$ |
| Free variables | Choice of material |

**Table 2.1-9** The selection

| Material |
|---|
| Diamond |
| Beryllia (Grade 99) |
| Beryllia (Grade B995) |
| Beryllia (Grade BZ) |
| Aluminum nitride (fully dense) |
| Aluminum nitride (97 percent dense) |

**Table 2.1-10** Part of a record for aluminum nitride, showing structured and unstructured data, references and the web-search facility

## Aluminum Nitride

### General properties
| | |
|---|---|
| Density | 3.26–3.33 Mg/m$^3$ |
| Price | *70–95 $/kg |

### Mechanical properties
| | |
|---|---|
| Young's M modulus | 302–348 GPa |
| Hardness—Vickers | 990–1260 HV |
| Compressive strength | 1970–2700 MPa |
| Fracture toughness | 2.5–3.4 MPa.m$^{1/2}$ |

### Thermal properties
| | |
|---|---|
| Thermal conductivity | 80–200 W/m.K |
| Thermal expansion | 4.9–6.2 µstrain/K |
| Max. service temperature | *1027–1727 °C |

### Electrical properties
| | |
|---|---|
| Resistivity | 1e18–1e21 µΩ.cm |
| Dielectric constant | 8.3–9.3 |

### Supporting information

**Design guidelines.** Aluminum nitride (AlN) has an unusual combination of properties: it is an electrical insulator, but an excellent conductor of heat. This is just what is wanted for substrates for high-powered electronics; the substrate must insulate yet conduct the heat out of the microchips. This, and its high strength, chemical stability, and low expansion give it a special role as a heat sinks for power electronics. Aluminum nitride starts as a powder, is pressed (with a polymer binder) to the desired shape, then fired at a high temperature, burning off the binder and causing the powder to sinter.

**Technical notes.** Aluminum nitride is particularly unusual for its high thermal conductivity combined with a high electrical resistance, low dielectric constant, good corrosion, and thermal shock resistance.

**Typical uses.** Substrates for microcircuits, chip carriers, heat sinks, electronic components; windows, heaters, chucks, clamp rings, gas distribution plates.

### References

*Handbook of Ceramics, Glasses and Diamonds*, (2001) Harper, C.A. editor, McGraw-Hill, New York, NY, USA. ISBN 0-07-026712-X. (A comprehensive compilation of data and design guidelines.)
*Handbook of structural ceramics*, editor: M.M. Schwartz, McGraw-Hill, New York, USA (1992)
Morrell, R. *Handbook of properties of technical & engineering ceramics*, Parts I and II, National Physical Laboratory, Her Majesty's Stationery Office, London, UK (1985)

---

Ranking on thermal conductivity is shown in the central window. Materials that fail the screening stage on the left are grayed-out; those that pass remain colored. The selection line has been positioned so that two classes of material lie in the search region. The top-ranked candidate is aluminum nitride, the second is alumina. If, for some reason, the mass of the heat sink was also important, it might instead be desired to rank using material index $\lambda/\rho$, where $\rho$ is the density. Then the window on the right, showing a $\lambda$–$\rho$ chart, allows selection by $\lambda/\rho$, plotted as diagonal contour on the schematic. The materials furthest above the line are the best choice. Once again, AlN wins.

The software contains not one, but two databases. The first of these contains the 68 material classes shown in the charts of Chapter 4—indeed all these charts were made using the software. They are chosen because they are those most widely used; between them they account for 98 percent of material usage. This database allows a first look at a problem, but it is inadequate for a fuller exploration. The second database is much larger—it contains data for over 3000 materials. By changing the database, the selection criteria already entered are applied instead to the much larger population. Doing this (and ranking on $\lambda$ as in the central window) gives the top rank candidates listed in Table 2.1-9, listed in order of decreasing $\lambda$. Diamond is outstanding but is probably impracticable for reasons of cost; and compounds of beryllium (beryllia is beryllium oxide) are toxic and for this reason perhaps undesirable. That leaves us with aluminum nitride, our earlier choice. Part of a record for one grade of aluminum nitride is shown in Table 2.1-10. The upper part lists structured data (there is more, but it's not relevant in this example). The lower part gives the limited unstructured data provided by the record itself, and references to sources that are linked to the record in which more supporting information can be found. The search engine has a further feature, represented by the button labeled "search web" next to the material name at the top. Activating it sends the material name as a string to a web search engine, delivering supporting information available there.

Examples of the use of the software appear later in the book.

## 2.1.6 The structural index

Books on optimal design of structures (e.g. Shanley, 1960) make the point that the efficiency of material usage in mechanically loaded components depends on the product of three factors: the material index, as defined here; a factor describing section shape, the subject of our Chapter 11; and a *structural index*,[4] which contains elements of the $G$ and $F$ of equation (2.1.12).

The subjects of this book—material and process selection—focuses on the material index and on shape; but we should examine the structural index briefly, partly to make the connection with the classical theory of optimal design, and partly because it becomes useful (even to us) when structures are scaled in size.

In design for minimum mass [equations (2.1.3) and (2.1.8)], a measure of the efficiency of the design is given by the quantity $m/L^3$. Equation (2.1.3), for instance, can be written

$$\frac{m}{L^3} \geq \left(\frac{F}{L^2}\right)\left(\frac{\rho}{\sigma_f}\right) \qquad (2.1.18)$$

and equation (2.1.8) becomes

$$\frac{m}{L^3} \geq \left(\frac{12}{C_1}\right)^{1/2} \left(\frac{S}{L}\right)^{1/2} \left(\frac{\rho}{E^{1/2}}\right) \qquad (2.1.19)$$

This $m/L^3$ has the dimensions of density; the lower this pseudo-density the lighter is the structure for a given scale, and thus the greater is the structural efficiency. The first bracketed term on the right of the equation is merely a constant. The last is the material index. The middle one, $F/L^2$ for strength-limited design and $S/L$ for stiffness limited design, is called the *structural index*. It has the dimensions of stress; it is a measure of the intensity of loading. Design proportions that are optimal, minimizing material usage, are optimal for structures of any size provided they all have the same structural index. The performance equation (2.1.8), was written in a way that isolated the structural index.

The structural index for a component of minimum cost is the same as that for one of minimum mass; it is $F/L^2$ again for strength limited design, $S/L$ when it is stiffness. For beams or columns of minimum mass, cost, or energy content, it is the same. For panels (dimensions $L \times W$) loaded in bending or such that they buckle it is $FW/L^3$ and $SW^2/L^3$ where $L$ and $W$ are the (fixed) dimensions of the panel.

### 2.1.7 Summary and conclusions

Material selection is tackled in four steps.

- *Translation*—reinterpreting the design requirements in terms of function, constraints, objectives, and free variables.
- *Screening*—deriving attribute limits from the constraints and applying these to isolate a subset of viable materials.
- *Ranking*—ordering the viable candidates by the value of a material index, the criterion of excellence that maximizes or minimizes some measure of performance.
- Seeking supporting information for the top-ranked candidates, exploring aspects of their past history, their established uses, their behavior in relevant environments, their availability and more until a sufficiently detailed picture is built up so that a final choice can be made.

Hard-copy material charts allow a first go at the task, and have the merit of maintaining breadth of vision: all material classes are in the frame, so to speak. But materials have many properties, and the number of combinations of these appearing in indices is very much larger. It is impractical to print charts for all of them. Even if you did, their resolution is limited. Both problems are overcome by computer implementation, allowing freedom to explore the whole kingdom of materials and also providing detail when required.

## Further reading

The books listed below discuss optimization methods and their application in materials engineering. None contain the approach developed here.
Dieter, G.E. (1991). *Engineering Design, a Materials and Processing Approach*, 2nd edition. McGraw-Hill, New York, USA. ISBN 0-07-100829-2. (*A well-balanced and respected text focusing on the place of materials and processing in technical design.*)
Gordon, J.E. (1976). *The New Science of Strong Materials, or why you don't Fall Through the Floor*, 2nd edition. Penguin Books, Harmondsworth, UK. ISBN 0-1402-0920-7 (*This very readable book presents ideas about plasticity and fracture, and ways of designing materials to prevent them.*)
Gordon, J.E. (1978). *Structures, or why Things don't Fall Down*. Penguin Books, Harmondsworth, UK. ISBN 0-1402-1961-7. (*A companion to the other book

---

[4] Also called the "structural loading coefficient", the "strain number" or the "strain index".

by Gordon (above), this time introducing structural design.)

Shanley, F.R. (1960). *Weight-Strength Analysis of Aircraft Structures*, 2nd edition. Dover Publications, Inc, New York, USA. Library of Congress Number 60-50107. *(A remarkable text, no longer in print, on the design of light-weight structures.)*

Arora, J.S. (1989). *Introduction to Optimum Design*. McGraw-Hill, New York, USA. ISBN 0-07-002460-X. *(An introduction to the terminology and methods of optimization theory.)*

# Section Three

**Processes and process selection**

# Chapter 3.1

# Processes and process selection

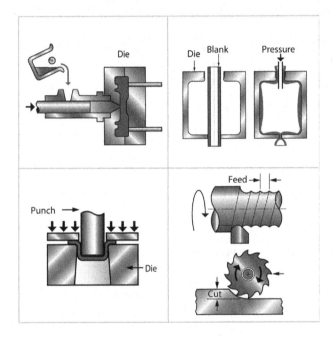

## 3.1.1 Introduction and synopsis

A *process* is a method of shaping, joining, or finishing a material. *Sand casting*, *injection molding*, *fusion welding*, and *electro-polishing* are all processes; there are hundreds of them. It is important to choose the right process-route at an early stage in the design before the cost-penalty of making changes becomes large. The choice, for a given component, depends on the material of which it is to be made, on its size, shape and precision, and on how many are to be made—in short, on the *design requirements*. A change in design requirements may demand a change in process route.

Each process is characterized by a set of *attributes:* the materials it can handle, the shapes it can make and their precision, complexity, and size. The intimate details of processes make tedious reading, but have to be faced: we describe them briefly in Section 3.1.3, using process selection charts to capture their attributes.

*Process selection*—finding the best match between process attributes and design requirements—is the subject of Sections 3.1.4 and 3.1.5. In using the methods developed there, one should not forget that material, shape, and processing interact (Figure 3.1-1). Material properties and shape limit the choice of process: ductile materials can be forged, rolled, and drawn; those that are brittle must be shaped in other ways. Materials that melt at modest temperatures to low-viscosity liquids can be

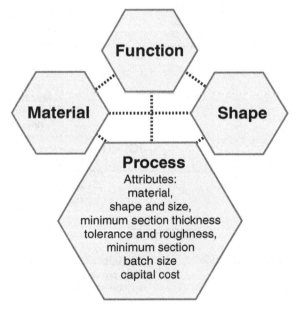

**Figure 3.1-1** Processing selection depends on material and shape. The "process attributes" are used as criteria for selection.

*Engineering Materials and Processes Desk Reference*; ISBN: 9781856175869
Copyright © 2009 Elsevier Ltd; All rights of reproduction, in any form, reserved.

cast; those that do not have to be processed by other routes. Shape, too, can influence the choice of process. Slender shapes can be made easily by rolling or drawing but not by casting. Hollow shapes cannot be made by forging, but they can by casting or molding. Conversely, processing affects properties. Rolling and forging change the hardness and texture of metals, and align the inclusions they contain, enhancing strength, and ductility. Composites only acquire their properties during processing; before, they are just a soup of polymer and a sheaf of fibers.

Like the other aspects of design, process selection is an iterative procedure. The first iteration gives one or more possible processes-routes. The design must then be rethought to adapt it, as far as possible, to ease of manufacture by the most promising route. The final choice is based on a comparison of *process-cost*, requiring the use of cost models developed in Section 3.1.6, and on *supporting information*: case histories, documented experience and examples of process-routes used for related products (Section 3.1.7). Supporting information helps in another way: that of dealing with the coupling between process and material properties. Processes influence properties, sometimes in a desired way (e.g. heat treatment) sometimes not (uncontrolled casting defects, for instance). This coupling cannot be described by simple processes attributes, but requires empirical characterization or process modeling.

The chapter ends, as always, with a summary and annotated recommendations for further reading.

## 3.1.2 Classifying processes

Manufacturing processes can be classified under the headings shown in Figure 3.1-2. *Primary processes* create *shapes*. The first row lists seven primary forming processes: casting, molding, deformation, powder methods, methods for forming composites, special methods, and rapid prototyping. *Secondary processes* modify shapes or properties; here they are shown as "machining", which adds features to an already shaped body, and "heat treatment", which enhances surface or bulk properties. Below these comes *joining*, and, finally, *finishing*.

The merit of Figure 3.1-2 is as a flow chart: a progression through a manufacturing route. It should not be treated too literally: the order of the steps can be varied to suit the needs of the design. The point it makes is that there are three broad process families: those of shaping, joining, and finishing. The attributes of one family differ so greatly from those of another that, in assembling and structuring data for them, they must be treated separately.

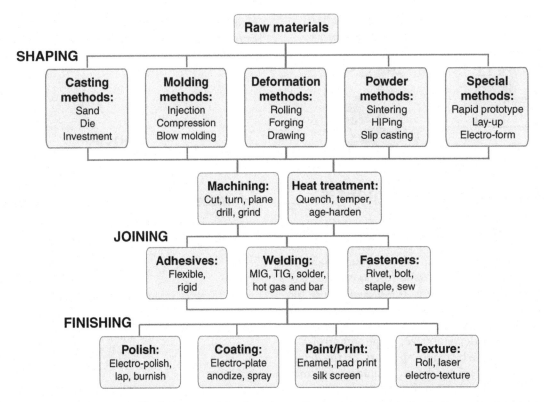

**Figure 3.1-2** The classes of process. The first row contains the primary shaping processes; below lie the secondary processes of machining and heat treatment, followed by the families of joining and finishing processes.

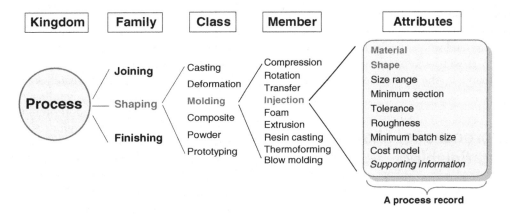

**Figure 3.1-3** The taxonomy of the kingdom of process with part of the *shaping* family expanded. Each member is characterized by a set of attributes. Process selection involves matching these to the requirements of the design.

To structure processes attributes for screening, we need a hierarchical classification of the processes families themselves, in the spirit of that used in Chapter 5 for materials. This gives each process a place, enabling the development of a computer-based tool. Figure 3.1-3 shows part of it. The process kingdom has three families: shaping, joining, and finishing. In this figure, the shaping family is expanded to show classes: casting, deformation, molding, etc. One of these—molding—is again expanded to show its members: rotation molding, blow molding, injection molding, and so forth. Each of these have certain attributes: the materials it can handle, the shapes it can make, their size, precision, and an optimum batch size (the number of units that it can make economically). This is the information that you would find in a record for a shaping-process in a selection database.

The other two families are partly expanded in Figure 3.1-4. There are three broad joining classes: adhesives, welding, and fasteners. In this figure one of them—welding—is expanded to show its members. As before each member has attributes. The first is the material or materials that the process can join. After that the attribute-list differs from that for shaping. Here the geometry of the joint and the way it will be loaded are important, as are requirements that the joint can, or cannot, be disassembled, be watertight, be electrically conducting and the like.

The lower part of the figure expands the family of finishing. Some of the classes it contains are shown; one—coating—is expanded to show some of its members. As with joining, the material to be coated is an important attribute but the others again differ. Most

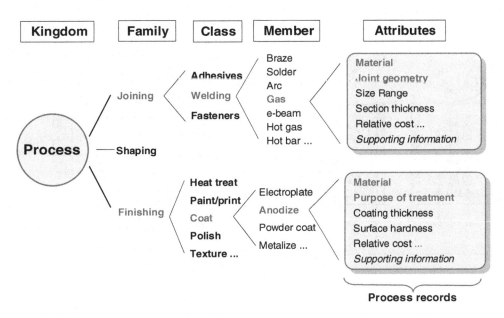

**Figure 3.1-4** The taxonomy of the process kingdom again, with the families of *joining* and *finishing* partly expanded.

important is the purpose of the treatment, followed by properties of the coating itself.

With this background we can embark on our lightning tour of processes. It will be kept as concise as possible; details can be found in the numerous books listed in Section 3.1.9.

### 3.1.3 The processes: shaping, joining, and finishing

#### Shaping processes

In *casting* (Figure 3.1-5), a liquid is poured or forced into a mold where it solidifies by cooling. Casting is distinguished from molding, which comes next, by the low viscosity of the liquid: it fills the mold by flow under its own weight (as in gravity sand and investment casting) or under a modest pressure (as in die casting and pressure sand casting). Sand molds for one-off castings are cheap; metal dies for die-casting large batches can be expensive. Between these extremes lie a number of other casting methods: shell, investment, plaster-mold and so forth.

Cast shapes must be designed for easy flow of liquid to all parts of the mold, and for progressive solidification that does not trap pockets of liquid in a solid shell, giving shrinkage cavities. Whenever possible, section thicknesses are made uniform (the thickness of adjoining sections should not differ by more than a factor of 2). The shape is designed so that the pattern and the finished casting can be removed from the mold. Keyed-in shapes are avoided because they lead to "hot tearing" (a tensile

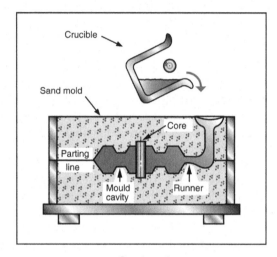

Sand casting

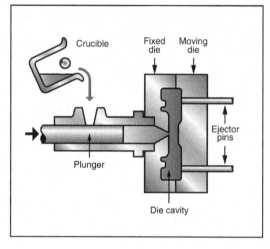

Die casting

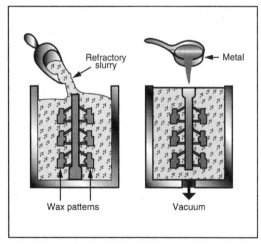

Investment casting

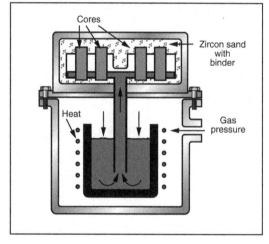

Low pressure casting

**Figure 3.1-5** Casting processes. In *sand casting*, liquid metal is poured into a split sand mold. In *die casting*, liquid is forced under pressure into a metal mold. In *investment casting*, a wax pattern is embedded in refractory, melted out, and the cavity filled with metal. In *pressure casting*, a die is filled from below, giving control of atmosphere and of the flow of metal into the die.

creep-fracture) as the solid cools and shrinks. The tolerance and surface finish of a casting vary from poor for sand-casting to excellent for precision die-castings; they are quantified in Section 3.1-5.

When metal is poured into a mold, the flow is turbulent, trapping surface oxide and debris within the casting, giving casting defects. These are avoided by filling the mold from below in such a way that flow is laminar, driven by a vacuum or gas pressure as shown in Figure 3.1-4.

## Molding

(Figure 3.1-6). Molding is casting, adapted to materials that are very viscous when molten, particularly thermoplastics and glasses. The hot, viscous fluid is pressed or injected into a die under considerable pressure, where it cools and solidifies. The die must withstand repeated application of pressure, temperature and the wear involved in separating and removing the part, and therefore is expensive. Elaborate shapes can be molded, but at the penalty of complexity in die shape and in the way it separates to allow removal. The molds for thermoforming, by contrast, are cheap. Variants of the process use gas pressure or vacuum to mold form a heated polymer sheet onto a single-part mold. Blow-molding, too, uses a gas pressure to expand a polymer or glass blank into a split outer-die. It is a rapid, low-cost process well suited for mass-production of cheap parts like milk bottles. Polymers, like metals, can be extruded; virtually all rods, tubes and other prismatic sections are made in this way.

## Deformation processing

(Figure 3.1-7). This process can be hot, warm or cold—cold, that is, relative to the melting point $T_m$ of the material being processed. Extrusion, hot forging and hot

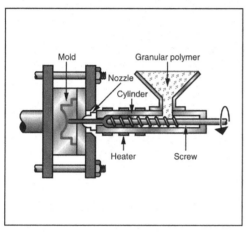

Injection-molding

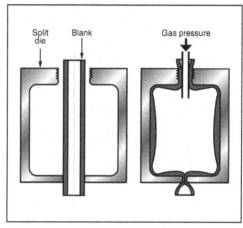

Blow-molding

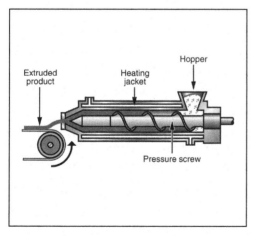

Polymer extrusion

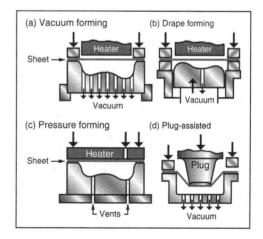

Thermo-forming

**Figure 3.1-6** Molding processes. In *injection-molding*, a granular polymer (or filled polymer) is heated, compressed and sheared by a screw feeder, forcing it into the mold cavity. In *blow-molding*, a tubular blank of hot polymer or glass is expanded by gas pressure against the inner wall of a split die. In *polymer extrusion*, shaped sections are formed by extrusion through a shaped die. In *thermoforming*, a sheet of thermoplastic is heated and deformed into a female die by vacuum or gas pressure.

73

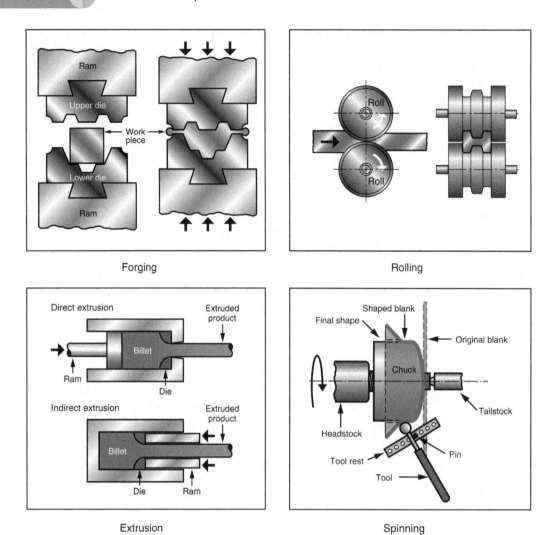

**Figure 3.1-7** Deformation processes. In *forging*, a slug of metal is shaped between two dies held in the jaws of a press. In *rolling*, a billet or bar is reduced in section by compressive deformation between the rolls. In *extrusion*, metal is forced to flow through a die aperture to give a continuous prismatic shape. All three process can be hot ($T > 0.85T_m$), warm ($0.55T_m < T < 0.85T_m$) or cold ($T < 0.35T_m$). In *spinning*, a spinning disk of ductile metal is shaped over a wooden pattern by repeated sweeps of the smooth, rounded, tool.

rolling ($T > 0.55T_m$) have much in common with molding, though the material is a true solid not a viscous liquid. The high temperature lowers the yield strength and allows simultaneous recrystallization, both of which lower the forming pressures. Warm working ($0.35T_m < T < 0.55T_m$) allows recovery but not recrystallization. Cold forging, rolling, and drawing ($T < 0.35T_m$) exploit work hardening to increase the strength of the final product, but at the penalty of higher forming pressures.

Forged parts are designed to avoid rapid changes in thickness and sharp radii of curvature since both require large local strains that can cause the material to tear or to fold back on itself ("lapping"). Hot forging of metals allows larger changes of shape but generally gives a poor surface and tolerance because of oxidation and warpage. Cold forging gives greater precision and finish, but forging pressures are higher and the deformations are limited by work hardening.

### Powder methods

(Figure 3.1-8). These methods create the shape by pressing and then sintering fine particles of the material. The powder can be cold-pressed and then sintered (heated at up to $0.8\, T_m$ to give bonding); it can be pressed in a heated die ("die-pressing"); or, contained in a thin preform, it can be heated under a hydrostatic pressure ("hot isostatic pressing" or "HIPing"). Metals that are too high-melting to cast and too strong to deform, can be made (by chemical methods) into powders and then shaped in this way. But the processes are not limited to "difficult" materials; almost any material can be shaped by subjecting it, as a powder, to pressure and heat.

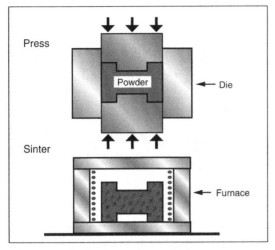
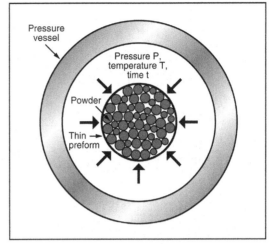
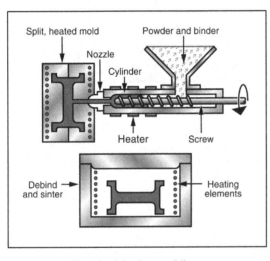
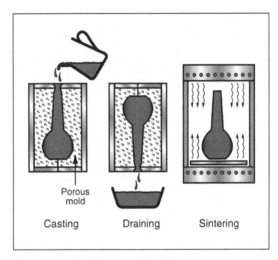

**Figure 3.1-8** Powder processing. In *die-pressing and sintering* the powder is compacted in a die, often with a binder, and the green compact is then fired to give a more or less dense product. In *hot isostatic pressing*, powder in a thin, shaped, shell or pre-form is heated and compressed by an external gas pressure. In *powder injection molding*, powder and binder are forced into a die to give a green blank that is then fired. In *slip casting*, a water-based powder slurry is poured into a porous plaster mold that absorbs the water, leaving a powder shell that is subsequently fired.

Powder processing is most widely used for small metallic parts like gears and bearings for cars and appliances. It is economic in its use of material, it allows parts to be fabricated from materials that cannot be cast, deformed or machined, and it can give a product that requires little or no finishing. Since pressure is not transmitted uniformly through a bed of powder, the length of a die-pressed powder part should not exceed 2.5 times its diameter. Sections must be near-uniform because the powder will not flow easily around corners. And the shape must be simple and easily extracted from the die.

Ceramics, difficult to cast and impossible to deform, are routinely shaped by powder methods. In slip casting, a water-based powder slurry is poured into a plaster mold. The mold wall absorbs water, leaving a semi-dry skin of slurry over its inner wall. The remaining liquid is drained out, and the dried slurry shell is fired to give a ceramic body. In powder injection molding (the way spark-plug insulators are made) a ceramic powder in a polymer binder is molded in the conventional way; the molded part is fired, burning off the binder and sintering the powder.

## Composite fabrication methods

(Figure 3.1-9). These make polymer–matrix composites reinforced with continuous or chopped fibers. Large components are fabricated by filament winding or by laying-up pre-impregnated mats of carbon, glass or Kevlar fiber ("pre-preg") to the required thickness,

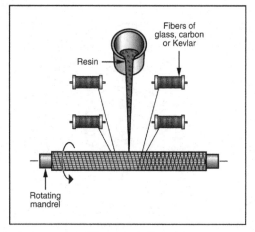
Filament winding

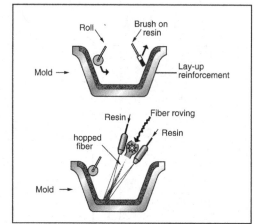
Roll and spray lay-up

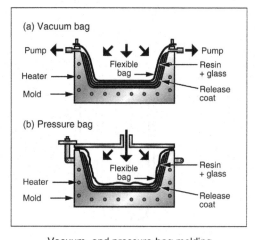
Vacuum- and pressure-bag molding

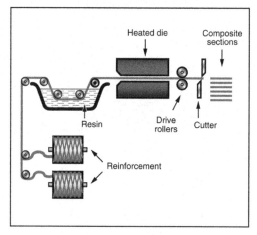
Pultrusion

**Figure 3.1-9** Composite forming methods. In *filament winding*, fibers of glass, Kevlar or carbon are wound onto a former and impregnated with a resin-hardener mix. In *roll* and spray *lay-up*, fiber reinforcement is laid up in a mold onto which the resin-hardener mix is rolled or sprayed. In *vacuum-* and *pressure-bag molding*, laid-up fiber reinforcement, impregnated with resin-hardener mix, is compressed and heated to cause polymerization. In *pultrusion*, fibers are fed through a resin bath into a heated die to form continuous prismatic sections.

pressing and curing. Parts of the process can be automated, but it remains a slow manufacturing route; and, if the component is a critical one, extensive ultrasonic testing may be necessary to confirm its integrity. Higher integrity is given by vacuum- or pressure-bag molding, which squeezes bubbles out of the matrix before it polymerizes. Lay-up methods are best suited to a small number of high-performance, tailor-made, components. More routine components (car bumpers, tennis racquets) are made from chopped-fiber composites by pressing and heating a "dough" of resin containing the fibers, known as bulk molding compound (BMC) or sheet molding compound (SMC), in a mold, or by injection molding a rather more fluid mixture into a die. The flow pattern is critical in aligning the fibers, so that the designer must work closely with the manufacturer to exploit the composite properties fully.

### Rapid prototyping systems

(RPS—Figure 3.1-10). The RPS allow single examples of complex shapes to be made from numerical data generated by CAD solid-modeling software. The motive may be that of visualization: the aesthetics of an object may be evident only when viewed as a prototype. It may be that of pattern-making: the prototype becomes the master from which molds for conventional processing, such as casting, can be made or—in complex assemblies—it may be that of validating intricate geometry, ensuring that parts fit, can be assembled, and are accessible. All RPS can create shapes of great complexity with internal cavities, overhangs and transverse features, though the precision, at present, is limited to ±0.3 mm at best.

All RP methods build shapes layer-by-layer, rather like three-dimensional (3D) printing, and are slow

## Processes and process selection CHAPTER 3.1

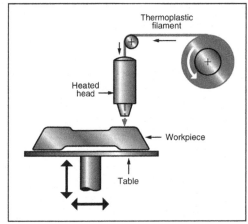
Deposition modeling

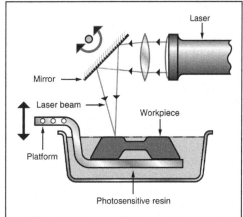
Stereo-lithography, SLA

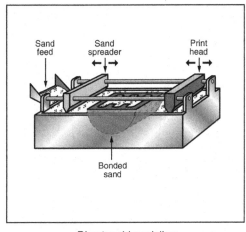
Direct mold modeling

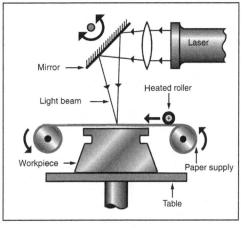
Laminated object manufacture, LOM

**Figure 3.1-10** Rapid prototyping. In *deposition modeling* and *ballistic particle manufacture* (BPM), a solid body is created by the layer-by-layer deposition of polymer droplets. In *stereo-lithography* (SLA), a solid shape is created layer-by-layer by laser-induced polymerization of a resin. In *direct mold modeling*, a sand mold is built up layer-by-layer by selective spraying of a binder from a scanning print-head. In *laminated object manufacture* (LOM), a solid body is created from layers of paper, cut by a scanning laser beam and bonded with a heat-sensitive polymer.

(typically 1–40 h per unit). There are at least six broad classes of RPS:

**(i)** The shape is built up from a thermoplastic fed to a single scanning head that extrudes it like a thin layer of toothpaste ("fused deposition modeling" or FDM), exudes it as tiny droplets ("ballistic particle manufacture", BPM), or ejects it in a patterned array like a bubble-jet printer ("3D printing").

**(ii)** Scanned-laser induced polymerization of a photo-sensitive monomer ("stereo-lithography" or SLA). After each scan, the work piece is incrementally lowered, allowing fresh monomer to cover the surface. Selected laser sintering (SLS) uses similar laser-based technology to sinter polymeric powders to give a final product. Systems that extend this to the sintering of metals are under development.

**(iii)** Scanned laser cutting of bondable paper elements. Each paper-thin layer is cut by a laser beam and heat bonded to the one below.

**(iv)** Screen-based technology like that used to produce microcircuits ("solid ground curing" or SGC). A succession of screens admits UV light to polymerize a photo-sensitive monomer, building shapes layer by layer.

**(v)** SLS allows components to be fabricated directly in thermoplastic, metal or ceramic. A laser, as in SLA, scans a bed of particles, sintering a thin surface layer where the beam strikes. A new layer of particles is swept across the surface and the laser-sintering step is repeated, building up a 3-dimensional body.

**(vi)** Bonded sand molding offers the ability to make large complex metal parts easily. Here a multi-jet print-head squirts a binder onto a bed of loose casting sand, building up the mold shape much as selected laser sintering does, but more quickly. When complete the mold is lifted from the remaining loose sand and used in a conventional casting process.

To be useful, the prototypes made by RPS are used as masters for silicone molding, allowing a number of replicas to be cast using high-temperature resins or metals.

## Machining

(Figure 3.1-11). Almost all engineering components, whether made of metal, polymer, or ceramic, are subjected to some kind of machining during manufacture. To make this possible they should be designed to make gripping and jigging easy, and to keep the symmetry high: symmetric shapes need fewer operations. Metals differ greatly in their *machinability*, a measure of the ease of chip formation, the ability to give a smooth surface, and the ability to give economical tool life (evaluated in a standard test). Poor machinability means higher cost.

Most polymers are molded to a final shape. When necessary they can be machined but their low moduli mean that they deflect elastically during the machining operation, limiting the tolerance. Ceramics and glasses can be ground and lapped to high tolerance and finish (think of the mirrors of telescopes). There are many "special" machining techniques with particular applications; they include electro-discharge machining (EDM), ultrasonic cutting, chemical milling, cutting by water-jets, sand-jets, electron and laser beams.

Sheet metal forming involves punching, bending and stretching. Holes cannot be punched to a diameter less

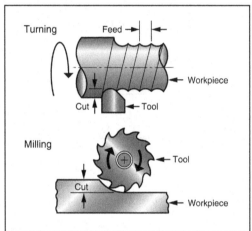
Turning and milling

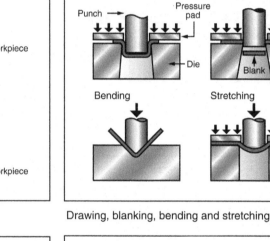
Drawing, blanking, bending and stretching

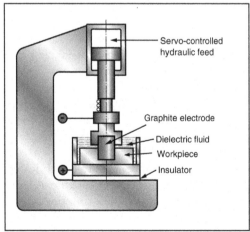
Electro-discharge machining

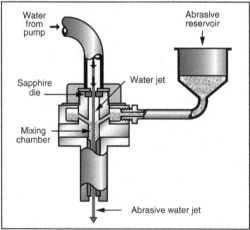
Water-jet cutting

**Figure 3.1-11** Machining operations. In *turning* and *milling*, the sharp, hardened tip of a tool cuts a chip from the workpiece surface. In *drawing, blanking, bending* and *stretching*, a sheet is shaped and cut to give flat and dished shapes. In *electro-discharge machining*, electric discharge between a graphite electrode and the workpiece, submerged in a dielectric such as paraffin, erodes the workpiece to the desired shape. In *water-jet cutting*, an abrasive entrained in a high speed water-jet erodes the material in its path.

than the thickness of the sheet. The minimum radius to which a sheet can be bent, its *formability*, is sometimes expressed in multiples of the sheet thickness *t*: a value of 1 is good; one of 4 is average. Radii are best made as large as possible, and never less than *t*. The formability also determines the amount the sheet can be stretched or drawn without necking and failing. The *forming limit diagram* gives more precise information: it shows the combination of principal strains in the plane of the sheet that will cause failure. The part is designed so that the strains do not exceed this limit.

Machining is often a secondary operations applied to castings, moldings or powder products to increase finish and tolerance. Higher finish and tolerance means higher cost; over-specifying either is a mistake.

## Joining processes

### Joining

(Figure 3.1-12). Joining is made possible by a number of techniques. Almost any material can be joined with adhesives, though ensuring a sound, durable bond can be difficult. Bolting, riveting, stapling, and snap fitting are commonly used to join polymers and metals, and have the feature that they can be disassembled if need be. Welding, the largest class of joining processes, is widely used to bond metals and polymers; specialized techniques have evolved to deal with each class. Ceramics can be diffusion-bonded to themselves, to glasses and to metals. Friction welding and friction-stir welding rely on the heat and deformation generated by friction to create a bond.

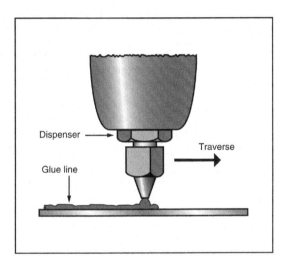

Adhesives

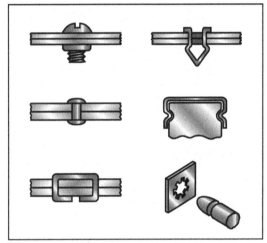

Fasteners

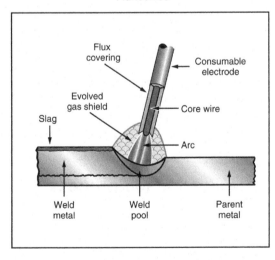

Welding: manual metal arc

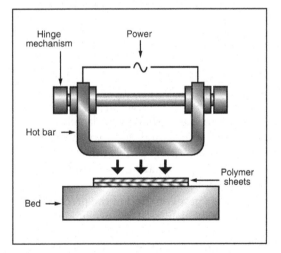

Welding: hot bar polymer welding

**Figure 3.1-12** Joining operations. In *adhesive bonding*, a film of adhesive is applied to one surface, which is then pressed onto the mating one. *Fastening* is achieved by bolting, riveting, stapling, push-through snap fastener, push-on snap fastener or rod-to-sheet snap fastener. In *metal fusion-welding*, metal is melted, and more added from a filler rod, to give a bond or coating. In *thermoplastic polymer welding*, heat is applied to the polymer components, which are simultaneously pressed together to form a bond.

If components are to be welded, the material of which they are made must be characterized by a high *weldability*. Like *machinability*, it measures a combination of basic properties. A low thermal conductivity allows welding with a low rate of heat input but can lead to greater distortion on cooling. Low thermal expansion gives small thermal strains with less risk of distortion. A solid solution is better than an age-hardened alloy because, in the heat-affected zone on either side of the weld, over-ageing and softening can occur.

Welding always leaves internal stresses that are roughly equal to the yield strength of the parent material. They can be relaxed by heat treatment but this is expensive, so it is better to minimize their effect by good design. To achieve this, parts to be welded are made of equal thickness whenever possible, the welds are located where stress or deflection is least critical, and the total number of welds is minimized.

The large-volume use of fasteners is costly because it is difficult to automate; welding, crimping or the use of adhesives can be more economical. Design for assembly (DFA) provides a check-list to guide minimizing assembly time.

## Finishing processes

Finishing describes treatments applied to the surface of the component or assembly. Some aim to improve mechanical and other engineering properties, others to enhance appearance.

### Finishing treatments to improve engineering properties

(Figure 3.1-13). Grinding, lapping, and polishing increase precision and smoothness, particularly important for bearing surfaces. Electro-plating deposits a thin metal layer onto the surface of a component to give resistance to corrosion and abrasion. Plating and painting are both made easier by a simple part shape with largely convex surfaces: channels, crevices, and slots are difficult to reach. Anodizing, phosphating and chromating create a thin layer of oxide, phosphate or chromate on the surface, imparting corrosion resistance.

Heat treatment is a necessary part of the processing of many materials. Age-hardening alloys of aluminum, titanium and nickel derive their strength from a precipitate produced by a controlled heat treatment: quenching from a high temperature followed by ageing at a lower one. The hardness and toughness of steels is controlled in a similar way: by quenching from the "austenitizing" temperature (about 800°C) and tempering. The treatment can be applied to the entire component, as in bulk carburizing, or just to a surface layer, as in flame hardening, induction hardening and laser surface hardening.

Quenching is a savage procedure; thermal contraction can produce stresses large enough to distort or crack the component. The stresses are caused by a non-uniform temperature distribution, and this, in turn, is related to the geometry of the component. To avoid damaging stresses, the section should be as uniform as possible, and nowhere so large that the quench-rate falls below the critical value required for successful heat treatment. Stress concentrations should be avoided: they are the source of quench cracks. Materials that have been molded or deformed may contain internal stresses that can be removed, at least partially, by stress-relief anneals—another sort of heat treatment.

### Finishing treatments that enhance aesthetics

(Figure 3.1-14). The processes just described can be used to enhance the visual and tactile attributes of a material: electroplating and anodizing are examples. There are many more, of which painting is the most widely used. Organic-solvent based paints give durable coatings with high finish, but the solvent poses environmental problems. Water-based paints overcome these, but dry more slowly and the resulting paint film is less perfect. In polymer powder-coating and polymer powder-spraying a film of thermoplastic—nylon, polypropylene or polyethylene—is deposited on the surface, giving a protective layer that can be brightly colored. In screen printing an oil-based ink is squeegeed through a mesh on which a blocking-film holds back the ink where it is not wanted; full color printing requires the successive use of up to four screens. Screen printing is widely used to print designs onto flat surfaces. Curved surfaces require the use of pad printing, in which a pattern, etched onto a metal "cliche", is inked and picked up on a soft rubber pad. The pad is pressed onto the product, depositing the pattern on its surface; the compliant rubber conforms to the curvature of the surface.

Enough of the processes themselves; for more detail the reader will have to consult Further reading, Section 3.1.9.

## 3.1.4 Systematic process selection

### The strategy

The strategy for selecting processes parallels that for materials. The starting point is that all processes are considered as possible candidates until shown to be otherwise. Figure 3.1-15 shows the now-familiar steps: *translation, screening, ranking* and search for *supporting information*. In translation, the design requirements

# Processes and process selection  CHAPTER 3.1

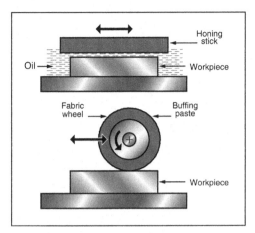

Mechanical polishing

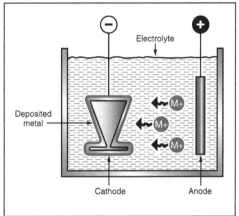

Electro-plating

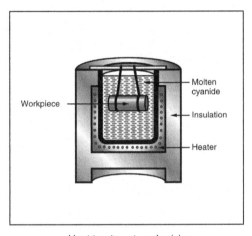

Heat treatment: carburizing

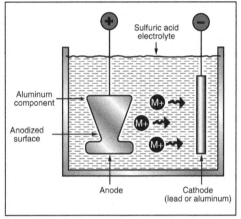

Anodizing

**Figure 3.1-13** Finishing processes to protect and enhance properties. In *mechanical polishing*, the roughness of a surface is reduced, and its precision increased, by material removal using finely ground abrasives. In *electro-plating*, metal is plated onto a conducting workpiece by electro-deposition in a plating bath. In *heat treatment*, a surface layer of the workpiece is hardened, and made more corrosion resistant, by the inward diffusion of carbon, nitrogen, phosphorous or aluminum from a powder bed or molten bath. In *anodizing*, a surface oxide layer is built up on the workpiece (which must be aluminum, magnesium, titanium, or zinc) by a potential gradient in an oxidizing bath.

are expressed as constraints on material, shape, size, tolerance, roughness, and other process-related parameters. It is helpful to list these in the way suggested by Table 3.1-1. The constraints are used to screen out processes that are incapable of meeting them, using process selection diagrams (or their computer-based equivalents) shown in a moment. The surviving processes are then ranked by economic measures, also detailed below. Finally, the top-ranked candidates are explored in depth, seeking as much supporting information as possible to enable a final choice. Chapter 8 gives examples.

## Selection charts

As already explained, each process is characterized by a set of attributes. These are conveniently displayed as simple matrices and bar charts. They provide the selection tools we need for screening. The hard-copy versions, shown here, are necessarily simplified, showing only a limited number of processes and attributes. Computer implementation, the subject of Section 3.1.6, allows exploration of a much larger number of both.

### The process-material matrix

(Figure 3.1-16). A given process can shape, or join, or finish some materials but not others. The matrix shows the links between material and process classes. Processes that cannot shape the material of choice are non-starters.

### The process-shape matrix

(Figure 3.1-17). Shape is the most difficult attribute to characterize. Many processes involve rotation or

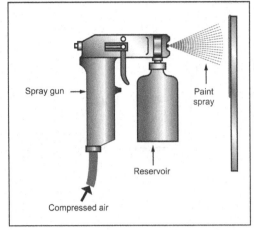 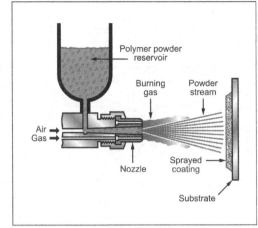
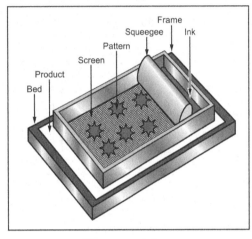 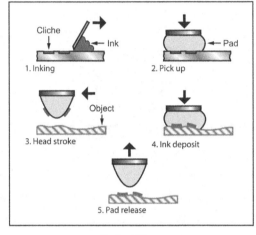

**Figure 3.1-14** Finishing processes to enhance appearance. In *paint spraying*, a pigment in an organic or water-based solvent is sprayed onto the surface to be decorated. In *polymer powder-spraying* a layer of thermoplastic is deposited on the surface by direct spraying in a gas flame, or by immersing the hot workpiece in a bed of powder. In *silk-screen printing*, ink is wiped onto the surface through a screen onto which a blocking-pattern has been deposited, allowing ink to pass in selected areas only. In *pad printing*, an inked pattern is picked up on a rubber pad and applied to the surface, which can be curved or irregular.

translation of a tool or of the material, directing our thinking towards axial symmetry, translational symmetry, uniformity of section and such like. Turning creates axisymmetric (or circular) shapes; extrusion, drawing and rolling make prismatic shapes, both circular and non-circular. Sheet-forming processes make flat shapes (stamping) or dished shapes (drawing). Certain processes can make 3D shapes, and among these some can make hollow shapes whereas others cannot. Figure 3.1-18 illustrates this classification scheme.

The process-shape matrix displays the links between the two. If the process cannot make the shape we want, it may be possible to combine it with a secondary process to give a process-chain that adds the additional features: casting followed by machining is an obvious example. But remember: every additional process step adds cost.

### The mass bar-chart

(Figure 3.1-19). The bar-chart—laid on its side to make labeling easier—shows the typical mass-range of components that each process can make. It is one of four, allowing application of constraints on size (measured by mass), section thickness, tolerance and surface roughness. Large components can be built up by joining smaller ones. For this reason the ranges associated with joining are shown in the lower part of the figure. In applying a constraint on mass, we seek single shaping-processes or shaping-joining combinations capable of making it, rejecting those that cannot.

The processes themselves are grouped by the material classes they can treat, allowing discrimination by both material and shape.

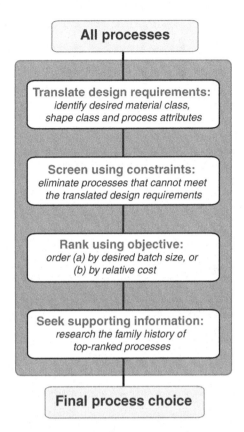

**Figure 3.1-15** A flow chart of the procedure for process selection. It parallels that for material selection.

## The section thickness bar-chart

(Figure 3.1-20). Surface tension and heat-flow limit the minimum section and slenderness of gravity cast shapes. The range can be extended by applying a pressure as in centrifugal casting and pressured die casting, or by preheating the mold. But there remain definite lower limits for the section thickness achievable by casting. Deformation processes—cold, warm, and hot—cover a wider range. Limits on forging-pressures set a lower limit on thickness and slenderness, but it is not nearly as severe as in casting. Machining creates slender shapes by removing unwanted material. Powder-forming methods are more limited in the section thicknesses they can create, but they can be used for ceramics and very hard metals that cannot be shaped in other ways. Polymer-forming methods—injection molding, pressing, blow-molding, etc.—share this regime. Special techniques, which include electro-forming, plasma-spraying and various vapor-deposition methods, allow very slender shapes.

The bar-chart of Figure 3.1-20 allows selection to meet constraints on section thickness.

## The tolerance and surface-roughness bar-charts

(Figures 3.1-21, 3.1-22 and Table 3.1-2). No process can shape a part *exactly* to a specified dimension. Some deviation $\Delta x$ from a desired dimension $x$ is permitted; it is referred to as the *tolerance*, $T$, and is specified as $x = 100 \pm 0.1$ mm, or as $x = 50^{+0.01}_{-0.001}$ mm. Closely related to this is the *surface roughness*, $R$, measured by the root-mean-square amplitude of the irregularities on the surface. It is specified as $R < 100$ μm (the rough surface of a sand casting) or $R < 100$ μm (a highly polished surface).

Manufacturing processes vary in the levels of tolerance and roughness they can achieve economically. Achievable tolerances and roughness are shown in Figures 3.1-21 and 3.1-22. The tolerance $T$ is obviously greater than $2R$; indeed, since $R$ is the root-mean-square roughness, the peak roughness is more like $5R$. Real processes give tolerances that range from about $10R$ to $1000R$. Sand casting gives rough surfaces; casting into metal dies gives a better finish. Molded polymers inherit the finish of the molds and thus can be very smooth, but tolerances better than $\pm 0.2$ mm are seldom possible because internal stresses left by molding cause distortion and because polymers creep in service. Machining, capable of high-dimensional accuracy and surface finish, is commonly used after casting or deformation processing to bring the tolerance or finish up to the desired level. Metals and ceramics can be surface-ground and lapped to a high tolerance and smoothness: a large telescope reflector has a tolerance approaching 5 μm over a dimension of a meter or more, and a roughness of about 1/100 of this. But such precision and finish are expensive: processing costs increase almost exponentially as the requirements for tolerance and surface finish are made more severe. It is an expensive mistake to over-specify precision.

## Use of hard-copy process selection charts

The charts presented here provide an overview: an initial at-a-glance graphical comparison of the capabilities of various process classes. In a given selection exercise they are not all equally useful: sometimes one is discriminating, another not—it depends on the design requirements. They should not be used blindly, but used to give guidance in selection and engender a feel for the capabilities and

| Table 3.1-1 Translation of process requirements | |
|---|---|
| Function | What must the process do? (Shape? Join? Finish?) |
| Constraints | What material, shape, size, precision, etc. must it provide? |
| Objective | What is to be maximized or minimized? (Cost? Time? Quality?) |
| Free variables | • Choice of process<br>• Process chain options |

# CHAPTER 3.1 Processes and process selection

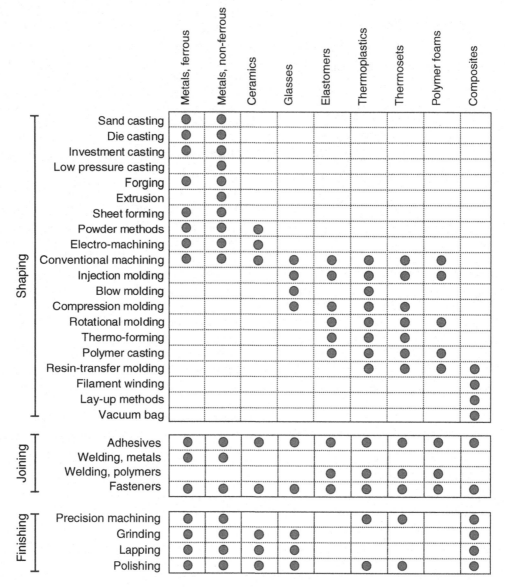

**Figure 3.1-16** The process—material matrix. A dot indicates that the pair are compatible.

limitations of various process types, remembering that some attributes (precision, for instance) can be added later by using secondary processes. That is as far as one can go with hard-copy charts. The number of processes that can be presented on hard-copy charts is obviously limited and the resolution is poor. But the procedure lends itself well to computer implementation, overcoming these deficiencies. It is described in Section 3.1.6.

The next step is to rank the survivors by economic criteria. To do this we need to examine process cost.

## 3.1.5 Ranking: process cost

Part of the cost of a component is that of the material of which it is made. The rest is the cost of manufacture, that is, of forming it to a shape, and of joining and finishing. Before turning to details, there are four common-sense rules for minimizing cost that the designer should bear in mind. They are these.

### Keep things standard

If someone already makes the part you want, it will almost certainly be cheaper to buy it than to make it. If nobody does, then it is cheaper to design it to be made from standard stock (sheet, rod, tube) than from non-standard shapes or from special castings or forgings. Try to use standard materials, and as few of them as possible: it reduces inventory costs and the range of tooling the manufacturer needs, and it can help with recycling.

# Processes and process selection — CHAPTER 3.1

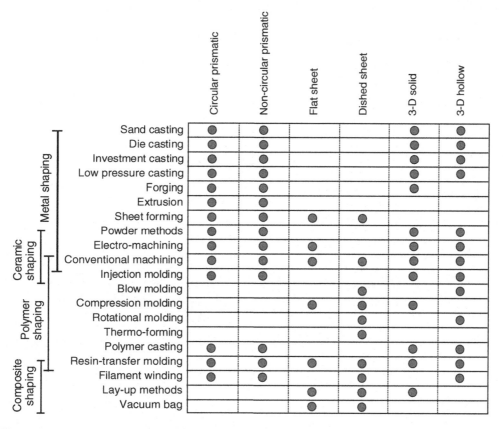

**Figure 3.1-17** The process–shape matrix. Information about material compatibility is included at the extreme left.

## Keep things simple

If a part has to be machined, it will have to be clamped; the cost increases with the number of times it will have to be re-jigged or re-oriented, specially if special tools are necessary. If a part is to be welded or brazed, the welder must be able to reach it with his torch and still see what he is doing. If it is to be cast or molded or forged, it should be remembered that high (and expensive) pressures are required to make fluids flow into narrow channels, and that re-entrant shapes greatly complicate mold and die design. Think of making the part yourself: will it be awkward? Could slight re-design make it less awkward?

## Make the parts easy to assemble

Assembly takes time, and time is money. If the overhead rate is a mere $60 per hour, every minute of assembly time adds another $1 to the cost. Design for assembly (DFA) addresses this problem with a set of common-sense criteria and rules. Briefly, there are three:

- minimize part count,
- design parts to be self-aligning on assembly,

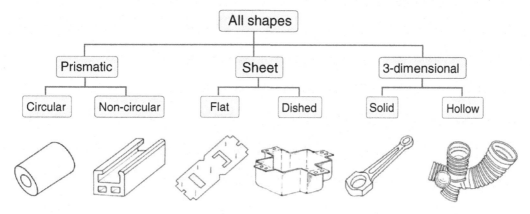

**Figure 3.1-18** The shape classification. More complex schemes are possible, but none are wholly satisfactory.

85

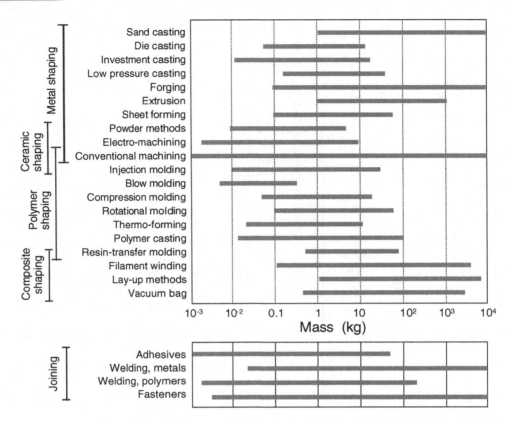

**Figure 3.1-19** The process—mass-range chart. The inclusion of joining allows simple process chains to be explored.

- use joining methods that are fast. Snap-fits and spot welds are faster than threaded fasteners or, usually, adhesives.

**Do not specify more performance than is needed**

Performance must be paid for. High strength metals are more heavily alloyed with expensive additions; high

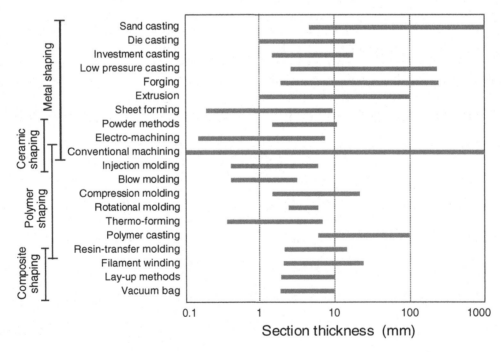

**Figure 3.1-20** The process—section thickness chart.

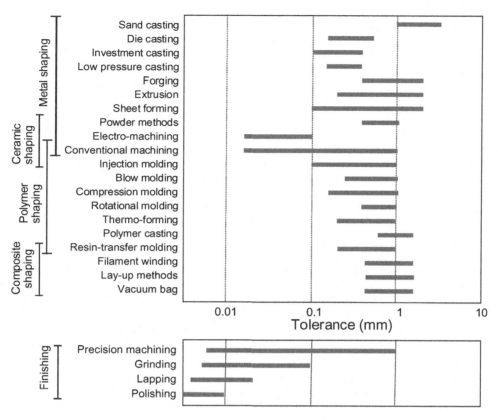

**Figure 3.1-21** The process–tolerance chart. The inclusion of finishing processes allows simple process chains to be explored.

performance polymers are chemically more complex; high performance ceramics require greater quality control in their manufacture. All of these increase material costs. In addition, high strength materials are hard to fabricate. The forming pressures (whether for a metal or a polymer) are higher; tool wear is greater; ductility is usually less so that deformation processing can be difficult or impossible. This can mean that new processing routes must be used: investment casting or powder forming instead of conventional casting and mechanical working; more expensive molding equipment operating at higher temperatures and pressures, and so on. The better performance of the high strength material must be paid for, not only in greater material cost but also in the higher cost of processing. Finally, there are the questions of tolerance and roughness. Cost rises exponentially with precision and surface finish. It is an expensive mistake to specify tighter tolerance or smoother surfaces than are necessary. The message is clear. Performance costs money. Do not over-specify it.

To make further progress, we must examine the contributions to process costs, and their origins.

### Economic criteria for selection

If you have to sharpen a pencil, you can do it with a knife. If, instead, you had to sharpen a thousand pencils, it would pay to buy an electric sharpener. And if you had to sharpen a million, you might wish to equip yourself with an automatic feeding, gripping, and sharpening system. To cope with pencils of different length and diameter, you could go further and devise a microprocessor-controlled system with sensors to measure pencil dimensions, sharpening pressure and so on—an

**Table 3.1-2** Levels of finish

| Finish (μm) | Process | Typical application |
|---|---|---|
| $R = 0.01$ | Lapping | Mirrors |
| $R = 0.1$ | Precision grind or lap | High quality bearings |
| $R = 0.2-0.5$ | Precision grinding | Cylinders, pistons, cams, bearings |
| $R = 0.5-2$ | Precision machining | Gears, ordinary machine parts |
| $R = 2-10$ | Machining | Light-loaded bearings. Non-critical components |
| $R = 3-100$ | Unfinished castings | Non-bearing surfaces |

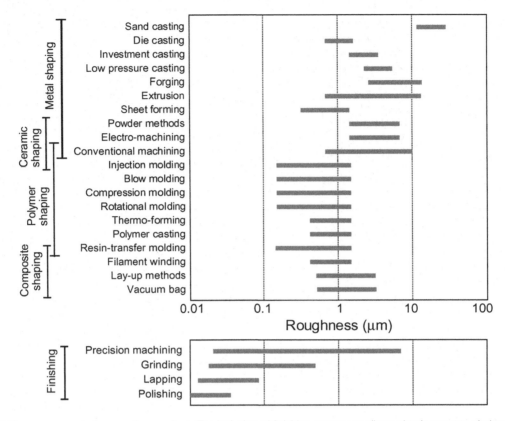

**Figure 3.1-22** The process–surface roughness chart. The inclusion of finishing processes allows simple process chains to be explored.

"intelligent" system that can recognize and adapt to pencil size. The choice of process, then, depends on the number of pencils you wish to sharpen, that is, on the *batch size*. The best choice is that one that costs least per pencil sharpened.

Figure 3.1-23 is a schematic of how the cost of sharpening a pencil might vary with batch size. A knife does not cost much but it is slow, so the labor cost is high. The other processes involve progressively greater capital investment but do the job more quickly, reducing labor costs. The balance between capital cost and rate gives the shape of the curves. In this figure the best choice is the lowest curve—a knife for up to 100 pencils; an electric sharpener for $10^2$ to $10^4$, an automatic system for $10^4$ to $10^6$, and so on.

## Economic batch size

Process cost depends on a large number of independent variables, not all within the control of the modeler. Cost modeling is described in the next section, but—given the disheartening implications of the last sentence—it is comforting to have an alternative, if approximate, way out. The influence of many of the inputs to the cost of a process are captured by a single attribute: the *economic batch size*; those for the processes described in this chapter are shown in Figure 3.1-24. A process with an economic batch size with the range $B_1$–$B_2$ is one that is found by experience to be competitive in cost when the output lies in that range, just as the electric sharpener was economic in the range $10^2$ to $10^4$. The economic batch size is commonly cited for processes. The easy way to introduce economy into the selection is to rank

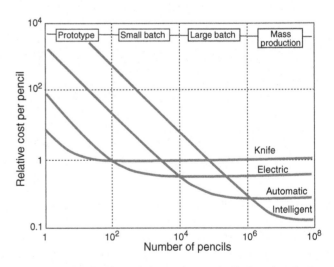

**Figure 3.1-23** The cost of sharpening a pencil plotted against batch size for four processes. The curves all have the form of equation (3.1.5).

# Processes and process selection CHAPTER 3.1

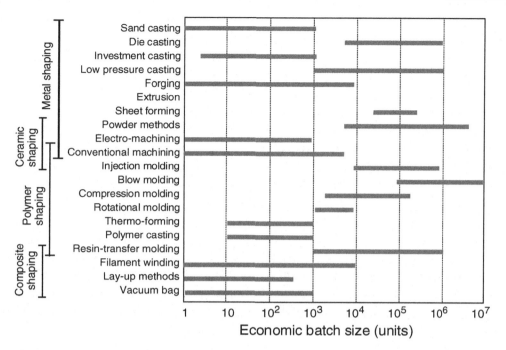

**Figure 3.1-24** The economic batch-size chart.

candidate processes by economic batch size and retain those that are economic in the range you want. But do not harbor false illusions: many variables cannot be rolled into one without loss of discrimination. A cost model gives deeper insight.

## Cost modeling

The manufacture of a component consumes resources (Figure 3.1-25), each of which has an associated cost. The final cost is the sum of those of the resources it consumes. They are detailed in Table 3.1-3. Thus the cost of producing a component of mass $m$ entails the cost $C_m$ ($/kg) of the materials and feed-stocks from which it is made. It involves the cost of dedicated tooling, $C_t$ ($), and that of the capital equipment, $C_c$ ($), in which the tooling will be used. It requires time, chargeable at an overhead rate $\dot{C}_{oh}$ (thus with units of $/h), in which we include the cost of labor, administration and general plant costs. It requires energy, which is sometimes charged against a process-step if it is very energy intensive but more usually is treated as part of the overhead and lumped into $\dot{C}_{oh}$, as we shall do here. Finally there is the cost of information, meaning that of research and development, royalty or licence fees; this, too, we view as a cost per unit time and lump it into the overhead.

Think now of the manufacture of a component (the "unit of output") weighing $m$ kg, made of a material costing $C_m$ $/kg. The first contribution to the unit cost is that of the material $mC_m$ magnified by the factor $1/(1-f)$ where $f$ is the scrap fraction—the fraction of the starting material that ends up as sprues, risers, turnings, rejects or waste:

$$C_1 = \frac{mC_m}{(1-f)} \qquad 3.1.1$$

The cost $C_t$ of a set of tooling—dies, molds, fixtures, and jigs—is what is called a *dedicated cost:* one that must be wholly assigned to the production run of this single component. It is written off against the numerical size $n$ of the production run. Tooling wears out. If the run is a long one, replacement will be necessary. Thus tooling cost per unit takes the form

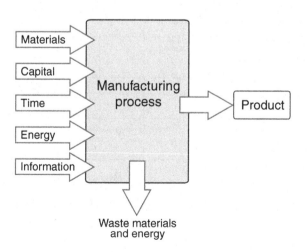

**Figure 3.1-25** The inputs to a cost model.

$$C_2 = \frac{C_t}{n}\left\{\text{Int}\left(\frac{n}{n_t} + 0.51\right)\right\} \quad 3.1.2$$

where $n_t$ is the number of units that a set of tooling can make before it has to be replaced, and 'Int' is the integer function. The term in curly brackets simply increments the tooling cost by that of one tool-set every time $n$ exceeds $n_t$.

The capital cost of equipment, $C_c$, by contrast, is rarely dedicated. A given piece of equipment—for example a powder press—can be used to make many different components by installing different die-sets or tooling. It is usual to convert the capital cost of *non-dedicated* equipment, and the cost of borrowing the capital itself, into an overhead by dividing it by a capital write-off time, $t_{wo}$ (5 years, say), over which it is to be recovered. The quantity $C_c/t_{wo}$ is then a cost per hour—provided the equipment is used continuously. That is rarely the case, so the term is modified by dividing it by a load factor, $L$—the fraction of time for which the equipment is productive. The cost per unit is then this hourly cost divided by the rate $\dot{n}$ at which units are produced:

$$C_3 = \frac{1}{\dot{n}}\left(\frac{C_c}{Lt_{wo}}\right) \quad 3.1.3$$

Finally there is the overhead rate $\dot{C}_{oh}$. It becomes a cost per unit when divided by the production rate $\dot{n}$ units per hour:

$$C_4 = \frac{\dot{C}_{oh}}{\dot{n}} \quad 3.1.4$$

The total shaping cost per part, $C_s$, is the sum of these four terms, taking the form:

$$C_s = \frac{mC_m}{(1-f)} + \frac{C_t}{n}\left\{\text{Int}\left(\frac{n}{n_t}+0.51\right)\right\} + \frac{1}{\dot{n}}\left(\frac{C_c}{Lt_{wo}}+\dot{C}_{oh}\right) \quad 3.1.5$$

The equation says: the cost has three essential contributions—a material cost per unit of production that is independent of batch size and rate, a dedicated cost per unit of production that varies as the reciprocal of the production volume $(1/n)$, and a gross overhead per unit of production that varies as the reciprocal of the production rate $(1/\dot{n})$. The equation describes a set of curves, one for each process. Each has the shape of the pencil-sharpening curves of Figure 3.1-23.

Figure 3.1-26 illustrates a second example: the manufacture of an aluminum con-rod by three alternative processes: sand casting, die casting and low pressure casting. At small batch sizes the unit cost is dominated by the "fixed" costs of tooling [the second term on the right of equation (3.1.5)]. As the batch size $n$ increases, the contribution of this to the unit cost falls (provided, of course, that the tooling has a life that is greater than $n$)

**Table 3.1-3** Symbols definitions and units

| Resource | Symbol | Unit |
|---|---|---|
| **Materials** | | |
| Including consumables | $C_m$ | \$/kg |
| **Capital** | | |
| Cost of tooling | $C_t$ | \$ |
| Cost of equipment | $C_c$ | \$/h |
| **Time** | | |
| Overhead rate, including labor, administration, rent, … | $\dot{C}_{oh}$ | \$/h |
| **Energy** | | |
| Cost of energy | $C_e$ | \$/h |
| **Information** | | |
| R & D or royalty payments | $C_i$ | \$/year |

until it flattens out at a value that is dominated by the "variable" costs of material, labor and other overheads. Competing processes usually differ in tooling cost $C_t$ and production rate $\dot{h}$, causing their $C-n$ curves to intersect, as shown in the schematic. Sand casting equipment is cheap but the process is slow. Die-casting equipment costs much more but it is also much faster. Mold costs for low pressure casting are greater than for sand casting, and the process is a little slower. Data for the terms in equation (3.1.5) are listed in Table 3.1-4. They show that the capital cost of the die-casting equipment is much greater than that for sand casting, but it is also much faster. The material cost, the labor cost per hour and the capital write-off time are, of course, the same for all. Figure 3.1-26 is a plot of equation (3.1.5), using these

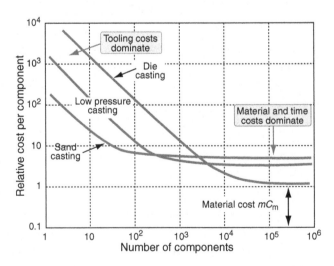

**Figure 3.1-26** The relative cost of casting as a function of the number of components to be cast.

**Table 3.1-4** Data for the cost equation

| Relative cost* | Sand casting | Die casting | Low pressure casting | Comment |
|---|---|---|---|---|
| Material, $mC_m/(1-f)$ | 1 | 1 | 1 | |
| Basic overhead, $\dot{C}_{oh}$ (per hour) | 10 | 10 | 10 | Process |
| Capital write-off time, $t_{wo}$ (yrs) | 5 | 5 | 5 | independent |
| Load factor | 0.5 | 0.5 | 0.5 | parameters |
| Dedicated tool cost, $C_t$ | 210 | 16,000 | 2000 | |
| Capital cost, $C_c$ | 800 | 30,000 | 8000 | Process |
| Batch rate, $\dot{n}$ (per hour) | 3 | 100 | 10 | dependent |
| Tool life, $n_t$ (number of units) | 200,000 | 1,000,000 | 500,000 | parameters |

* All costs normalized to the material cost.

data, all of which are normalized to the material cost. The curves for sand and die casting intersect at a batch size of 3000: below this, sand casting is the most economical process; above, it is die casting. Low pressure casting is more expensive for all batch sizes, but the higher quality casting it allows may be worth the extra. Note that, for small batches, the component cost is dominated by that of the tooling — the material cost hardly matters. But as the batch size grows, the contribution of the second term in the cost equation diminishes; and if the process is fast, the cost falls until it is typically about twice that of the material of which the component is made.

### Technical cost modeling

Equation (3.1.5) is the first step in modeling cost. Greater predictive power is possible with technical cost models that exploit understanding of the way in which the design, the process and cost interact. The capital cost of equipment depends on size and degree of automation. Tooling cost and production rate depend on complexity. These and many other dependencies can be captured in theoretical or empirical formulae or look-up tables that can be built into the cost model, giving more resolution in ranking competing processes. For more advanced analyses the reader is referred to the literature listed at the end of this chapter.

## 3.1.6 Computer-aided process selection

### Computer-aided screening

#### Shaping

If process attributes are stored in a database with an appropriate user-interface, selection charts can be created and selection boxes manipulated with much greater freedom. The CES platform, mentioned earlier, is an example of such a system. The way it works is described here; examples of its use are given in Chapter 8. The database contains records, each describing the attributes of a single process. Table 3.1-5 shows part of a typical record: it is that for injection molding. A schematic indicates how the process works; it is supported by a short description. This is followed by a listing of attributes: the shapes it can make, the attributes relating to shape and physical characteristics, and those that describe economic parameters; a brief description of typical uses, references and notes. The numeric attributes are stored as ranges, indicating the range of capability of the process. Each record is linked to records for the materials with which it is compatible, allowing choice of material to be used as a screening criterion, like the material matrix of Figure 3.1-16 but with greater resolution.

The starting point for selection is the idea that all processes are potential candidates until screened out because they fail to meet design requirements. A short list of candidates is extracted in two steps: screening to eliminate processes that cannot meet the design specification, and ranking to order the survivors by economic criteria. A typical three stage selection takes the form shown in Figure 3.1-27. It shows, on the left, a screening on material, implemented by selecting the material of choice from the hierarchy described in Chapter 5 (Figure 3.1-2). To this is added a limit stage in which desired limits on numeric attributes are entered in a dialog box, and the required shape class is selected. Alternatively, bar-charts like those of Figures 3.1-19–3.1-22 can be created, selecting the desired range of the attribute with a selection box like that on the right of Figure 3.1-27. The bar-chart shown here is for economic batch size, allowing approximate economic ranking. The

**Table 3.1-5** Part of a record for a process

### Injection molding

**The process.** No other process has changed product design more than injection molding. Injection molded products appear in every sector of product design: consumer products, business, industrial, computers, communication, medical and research products, toys, cosmetic packaging, and sports equipment. The most common equipment for molding thermoplastics is the reciprocating screw machine, shown schematically in the figure. Polymer granules are fed into a spiral press where they mix and soften to a dough-like consistency that can be forced through one or more channels ("sprues") into the die. The polymer solidifies under pressure and the component is then ejected.

Thermoplastics, thermosets, and elastomers can all be injection molded. Co-injection allows molding of components with different materials, colors and features. Injection foam molding allows economical production of large molded components by using inert gas or chemical blowing agents to make components that have a solid skin and a cellular inner structure.

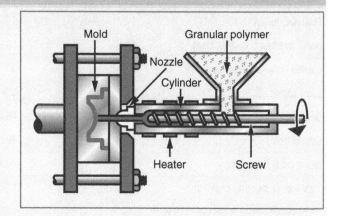

### Physical attributes

| | |
|---|---|
| Mass range | 0.01–25 kg |
| Range of section thickness | 0.4–6.3 mm |
| Tolerance | 0.2–1 mm |
| Roughness | 0.2–1.6 μm |
| Surface roughness (A = v. smooth) | A |

### Economic attributes

| | |
|---|---|
| Economic batch size (units) | 1E4–1E6 |
| Relative tooling cost | very high |
| Relative equipment cost | high |
| Labor intensity | low |

### Cost modeling

| | |
|---|---|
| Relative cost index (per unit) | 15.65–47.02 |
| Capital cost | * 3.28E4–7.38E5 USD |
| Material utilization fraction | * 0.6–0.9 |
| Production rate (units) | * 60–1000/h |
| Tooling cost | * 3280–3.28E4 USD |
| Tool life (units) | * 1E4–1E6 |

### Shape

| | |
|---|---|
| Circular prismatic | True |
| Non-circular prismatic | True |
| Solid 3D | True |
| Hollow 3D | True |

### Supporting Information

**Design guidelines.** Injection molding is the best way to mass-produce small, precise, polymer components with complex shapes. The surface finish is good; texture and pattern can be easily altered in the tool, and fine detail reproduces well. Decorative labels can be molded onto the surface of the component (see in-mold decoration). The only finishing operation is the removal of the sprue.

**Technical notes.** Most thermoplastics can be injection molded, although those with high melting temperatures (e.g. PTFE) are difficult. Thermoplastic based composites (short fiber and articulate filled) can be processed providing the filler-loading is not too large. Large changes in section area are not recommended. Small re-entrant angles and complex shapes are possible, though some features (e.g. undercuts, screw threads, inserts) may result in increased tooling costs.

**Typical uses.** Extremely varied. Housings, containers, covers, knobs, tool handles, plumbing fittings, lenses, etc.

# Processes and process selection  CHAPTER 3.1

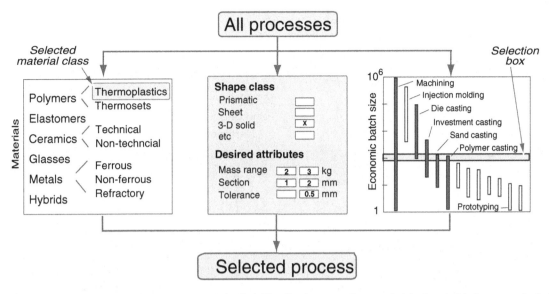

**Figure 3.1-27** The steps in computer-based process selection. The first imposes the constraint of material, the second of shape and numeric attributes, and the third allows ranking by economic batch size.

system lists the processes that pass all the stages. It overcomes the obvious limitation of the hard-copy charts in allowing selection over a much larger number of processes and attributes.

The cost model is implemented in the CES system. The records contain approximate data for the ranges of capital and tooling costs ($C_c$ and $C_t$) and for the rate of production ($\dot{n}$). Equation (3.1.5) contains other parameters not listed in the record because they are not attributes of the process itself but depend on the design, or the material, or the economics (and thus the location) of the plant in which the processing will be done. The user of the cost model must provide this information, conveniently entered through a dialog box like that of Figure 3.1-28.

One type of output is shown in Figure 3.1-29. The user-defined parameters are listed on the figure. The shaded band brackets a range of costs. The lower edge of the band uses the lower limits of the ranges for the input parameters — it characterizes simple parts requiring only a small machine and an inexpensive mold. The upper edge uses the upper limits of the ranges; it describes large complex parts requiring both a larger machine and a more complex mold. Plots of this sort allow two processes to be compared and highlight cost drivers, but they do not easily allow a ranking of a large population of competing processes. This is achieved by plotting unit cost for each member of the population for a chosen batch size. The output is shown as a bar chart in Figure 3.1-30. The software evaluates equation (3.1.5) for each member of the population and orders them by the mean value of the cost suggesting those that are the most economic. As explained earlier, the ranking is based on very approximate data; but since the most expensive processes in the figure are over 100 times more costly than the cheapest, an error of a factor of 2 in the inputs changes the ranking only slightly.

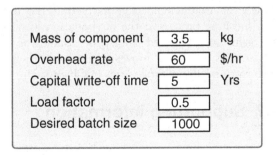

**Figure 3.1-28** A dialog box that allows the user-defined conditions to be entered into the cost model.

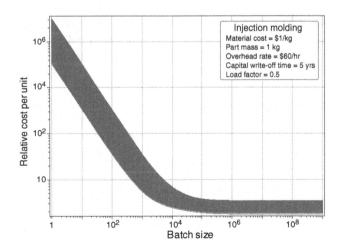

**Figure 3.1-29** The output of a computer-based cost model for a single process, here injection molding.

93

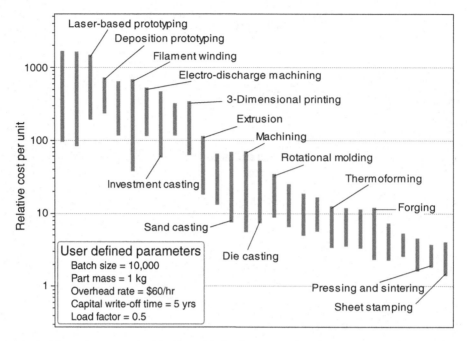

**Figure 3.1-30** The output of the same computer-based model allowing comparison between competing processes to shape a given component.

### Joining

The dominant constraints in selecting a joining process are usually set by the material or materials to be joined and by the geometry of the joint itself (butt, sleeve, lap, etc.). When these and secondary constraints (e.g. requirements that the joint be water-tight, or demountable, or electrical conducting) are met, the relative cost becomes the discriminator.

Detailed studies of assembly costs—particularly those involving fasteners—establish that it is a significant cost driver. The key to low-cost assembly is to make it fast. Design for assembly (DFA) addresses this issue. The method has three steps. The first is an examination of the design of the product, questioning whether each individual part is necessary. The second is to estimate the assembly time $t_a$ by summing the times required for each step of assembly, using historical data that relate the time for a single step to the features of the joint — the nature of the fastener, the ease of access, the requirement for precise alignment and such like. At the same time an "ideal" assembly time $(t_a)_{ideal}$ is calculated by assigning three seconds (an empirical minimum) to each step of assembly. In the third step, this ideal assembly time is divided by the estimated time $t_a$ to give a DFA index, expressed as a percentage:

$$\text{DFA index} = \frac{(t_a)_{ideal}}{t_a} \times 100 \qquad 3.1.6$$

This is a measure of assembly efficiency. It can be seen as a tool for motivating redesign: a low value of the index (10%, for instance) suggests that there is room for major reductions in $t_a$ with associated savings in cost.

### Finishing

A surface treatment imparts properties to a surface that it previously lacked (dimensional precision, smoothness, corrosion resistance, hardness, surface texture, etc.). The dominant constraints in selecting a treatment are the surface functions that are sought and the material to which they are to be applied. Once these and secondary constraints (the ability to treat curved surfaces or hollow shapes, for instance) are met, relative cost again becomes a discriminator and again it is time that is the enemy. Organic solvent-based paints dry more quickly than those that are water-based, electro-less plating is faster that electro-plating, and laser surface-hardening can be quicker than more conventional heat-treatments. Economics recommend the selection of the faster process. But minimizing time is not the only discriminator. Increasingly, environmental concerns restrict the use of certain finishing processes: the air pollution associated with organic solvent-based paints, for instance, is sufficiently serious to prompt moves to phase out their use. We return to environmental issues in Chapter 16.

### 3.1.7 Supporting information

Systematic screening and ranking based on attributes common to all processes in a given family yields a short list of candidates. We now need supporting information

— details, case studies, experience, warnings, anything that helps form a final judgement. The CES database contains some. But where do you go from there?

Start with texts and handbooks — they don't help with systematic selection, but they *are* good on supporting information. They and other sources are listed in Section 3.1.9 and in Chapter 15. The handbook of Bralla and the ASM Handbook series are particularly helpful, though increasingly dated. Next look at the data sheets and design manuals available from the makers and suppliers of process equipment, and, often, from material suppliers. Leading suppliers exhibit at major conferences and exhibitions and an increasing number have helpful web sites. Specialized software allowing greater discrimination within a narrow process domain is still largely a research topic, but will in time become available.

## 3.1.8 Summary and conclusions

A wide range of shaping, joining, and finishing processes is available to the design engineer. Each has certain characteristics, which, taken together, suit it to the processing of certain materials in certain shapes, but disqualify it for others. Faced with the choice, the designer has, in the past, relied on locally-available expertise or on common practice. Neither of these lead to innovation, nor are they well matched to current design methods. The structured, systematic approach of this chapter provides a way forward. It ensures that potentially interesting processes are not overlooked, and guides the user quickly to processes capable of achieving the desired requirements.

The method parallels that for selection of material, using process selection matrices and charts to implement the procedure. A component design dictates a certain, known, combination of process attributes. These design requirements are plotted onto the charts, identifying a subset of possible processes. The method lends itself to computer implementation, allowing selection from a large portfolio of processes by screening on attributes and ranking by economic criteria.

There is, of course, much more to process selection than this. It is to be seen, rather, as a first systematic step, replacing a total reliance on local experience and past practice. The narrowing of choice helps considerably: it is now much easier to identify the right source for more expert knowledge and to ask of it the right questions. But the final choice still depends on local economic and organizational factors that can only be decided on a case-by-case basis.

## 3.1.9 Further reading

ASM Handbook Series (1971–2004), "Volume 4: Heat treatment; Volume 5: Surface engineering; Volume 6: Welding, brazing and soldering; Volume 7: Powder metal technologies; Volume 14: Forming and forging; Volume 15: Casting; Volume 16: Machining", ASM International, Metals Park, OH, USA. (*A comprehensive set of handbooks on processing, occasionally updated, and now available on-line at www. asminternational.org/hbk/index.jsp*)

Bralla, J.G. (1998) *Design for Manufacturability Handbook*, 2nd edition, McGraw-Hill, New York, USA. ISBN 0-07-007139-X. (*Turgid reading, but a rich mine of information about manufacturing processes.*).

Bréchet, Y., Bassetti, D., Landru, D. and Salvo, L. (2001) Challenges in materials and process selection. *Prog. Mater. Sci* **46**, 407–428 (*An exploration of knowledge-based methods for capturing material and process attributes.*).

Campbell, J. (1991). *Casting*, Butterworth-Heinemann, Oxford, UK. ISBN 0-7506-1696-2. (*The fundamental science and technology of casting processes.*).

Clark, J.P. and Field, F.R., III (1997) Techno-economic issues in materials selection. In: *ASM Metals Handbook*, Vol 20. American Society for Metals, Metals Park, OH, USA. (*A paper outlining the principles of technical cost-modeling and its use in the automobile industry.*).

Dieter, G.E. (1991) *Engineering Design, a Materials and Processing Approach*, 2nd edition. McGraw-Hill, New York, USA. ISBN 0-07-100829-2. (*A well-balanced and respected text focusing on the place of materials and processing in technical design.*)

Esawi, A. and Ashby, M.F. (1998) Computer-based selection of manufacturing processes: methods, software and case studies. *Proc. Inst. Mech. Eng.* **212**, 595–610 (*A paper describing the development and use of the CES database for process selection.*).

Esawi, A. and Ashby, M.F. (2003) Cost estimates to guide pre-selection of processes. *Mater. Design* **24**, 605–616 (*A paper developing the cost-estimation methods used in this chapter.*).

Grainger, S. and Blunt, J. (1998). *Engineering Coatings, Design and Application*, Abington Publishing, Abington, Cambridge, UK. ISBN 1-85573-369-2. (*A handbook of surface treatment processes to improve surface durability—generally meaning surface hardness.*).

Houldcroft, P. (1990). *Which Process?*, Abington Publishing, Abington, Cambridge, UK. ISBN 1-85573-008-1. (*The title of this useful book is misleading — it deals only with a subset of joining process: the welding of steels. But here it is good, matching the process to the design requirements.*).

Kalpakjian, S. and Schmid, S.R. (2003) *Manufacturing Processes for Engineering Materials*, 4th edition. Prentice Hall Pearson Education, Inc, New Jersey, USA. ISBN 0-13-040871-9 (*A comprehensive and widely used text on material processing.*).

Lascoe, O.D. (1988) *Handbook of Fabrication Processes*. ASM International, Metals Park, Columbus, OH, USA. ISBN 0-87170-302-5. (*A reference source for fabrication processes.*).

Shercliff, H.R. and Lovatt, A.M. (2001) Selection of manufacturing processes in design and the role of process modelling. *Prog. Mater. Sci.* **46**, 429–459 (*An introduction to ways of dealing with the coupling between process and material attributes.*)

Swift, K.G. and Booker, J.D. (1997) *Process Selection, from Design to Manufacture.* Arnold, London, UK. ISBN 0-340-69249-9. (*Details of 48 processes in a standard format, structured to guide process selection.*)

Wise, R.J. (1999) *Thermal Welding of Polymers*. Abington Publishing, Abington, Cambridge, UK. ISBN 1-85573-495-8. (*An introduction to the thermal welding of thermoplastics.*)

# Section Four

## Metals

# Chapter 4.1

# Introduction to metals

## Introduction

This first group of chapters looks at metals. There are so many different metals – literally hundreds of them – that it is impossible to remember them all. It isn't necessary – nearly all have evolved from a few "generic" metals and are simply tuned-up modifications of the basic recipes. If you know about the generic metals, you know most of what you need.

This chapter introduces the generic metals. But rather than bore you with a catalogue we introduce them through three real engineering examples. They allow us not only to find examples of the uses of the main generic metals but also to introduce the all-important business of how the characteristics of each metal determine how it is used in practice.

## Metals for a model traction engine

Model-making has become big business. The testing of scale models provides a cheap way of getting critical design information for things from Olympic yacht hulls to tidal barrages. Architects sell their newest creations with the help of miniature versions correct to the nearest door-handle and garden shrub. And in an age of increasing leisure time, many people find an outlet for their energies in making models – perhaps putting together a miniature aircraft from a kit of plastic parts or, at the other extreme, building a fully working model of a steam engine from the basic raw materials in their own "garden-shed" machine shop.

Figure 4.1-1 shows a model of a nineteenth-century steam traction engine built in a home workshop from plans published in a well-known modellers' magazine. Everything works just as it did in the original – the boiler even burns the same type of coal to raise steam – and the model is capable of towing an automobile! But what interests us here is the large range of metals that were used in its construction, and the way in which their selection was dictated by the requirements of design. We begin by looking at metals based on *iron* (*ferrous*) metals). Table 4.1-1 lists the generic iron-based metals.

How are these metals used in the traction engine? The design loads in components like the wheels and frames are sufficiently low that *mild steel*, with a yield strength $\sigma_y$ of around 220 MPa, is more than strong enough. It is also easy to cut, bend or machine to shape. And last, but not least, it is cheap.

The stresses in the machinery – like the gear-wheel teeth or the drive shafts – are a good deal higher, and these parts are made from either *medium-carbon*, *high-carbon* or *low-alloy* steels to give extra strength. However, there are a few components where even the strength of high-carbon steels as delivered "off the shelf" ($\sigma_y \approx 400$ MPa) is not enough. We can see a good example in the mechanical lubricator, shown in Fig. 4.1-2, which is essentially a high-pressure oil metering pump. This is driven by a ratchet and pawl. These have sharp teeth which would quickly wear if they were made of a soft alloy. But how do we raise the hardness above that of ordinary high-carbon steel? Well, medium- and high-carbon steels can be hardened to give a yield strength of up to 1000 MPa by heating them to bright red heat and then quenching them into cold water. Although the quench makes the hardened steel brittle, we can make it tough again (though still hard) by *tempering* it – a process that involves heating the steel again, but to a much lower temperature. And so the

## CHAPTER 4.1  Introduction to metals

Figure 4.1-1 A fully working model, one-sixth full size, of a steam traction engine of the type used on many farms a hundred years ago. The model can pull an automobile on a few litres of water and a handful of coal. But it is also a nice example of materials selection and design.

Figure 4.1-2 A close-up of the mechanical lubricator on the traction engine. Unless the bore of the steam cylinder is kept oiled it will become worn and scored. The lubricator pumps small metered quantities of steam oil into the cylinder to stop this happening. The drive is taken from the piston rod by the ratchet and pawl arrangement.

ratchet and pawls are made from a quenched and tempered high-carbon steel.

*Stainless steel* is used in several places. Figure 4.1-3 shows the fire grate - the metal bars which carry the burning coals inside the firebox. When the engine is working hard the coal is white hot; then, both oxidation and creep are problems. Mild steel bars can burn out in a season, but stainless steel bars last indefinitely.

Finally, what about *cast iron*? Although this is rather brittle, it is fine for low-stressed components like the cylinder block. In fact, because cast iron has a lot of

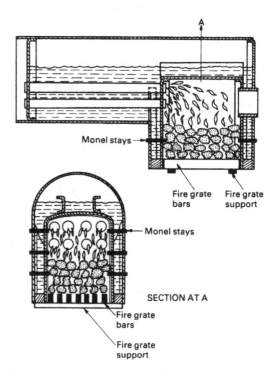

Figure 4.1-3 The fire grate, which carries the white-hot fire inside the firebox, must resist oxidation and creep. Stainless steel is best for this application. Note also the threaded monel stays which hold the firebox sides together against the internal pressure of the steam.

Table 4.1-1 Generic iron-based metals

| Metal | Typical composition (wt%) | Typical uses |
|---|---|---|
| Low-carbon ("mild") steel | Fe + 0.04 to 0.3 C (+ ≈0.8 Mn) | Low-stress uses. General constructional steel, suitable for welding. |
| Medium-carbon steel | Fe + 0.3 to 0.7C (+ ≈0.8 Mn) | Medium-stress uses: machinery parts – nuts and bolts, shafts, gears. |
| High-carbon steel | Fe + 0.7 to 1.7 C (+ ≈0.8 Mn) | High-stress uses: springs, cutting tools, dies. |
| Low-alloy steel | Fe + 0.2 C 0.8 Mn 1 Cr 2 Ni | High-stress uses: pressure vessels, aircraft parts. |
| High-alloy ("stainless") steel | Fe + 0.1 C 0.5 Mn 18 Cr 8 Ni | High-temperature or anti-corrosion uses: chemical or steam plants. |
| Cast iron | Fe + 1.8 to 4 C (+ ≈0.8 Mn 2 Si) | Low-stress uses: cylinder blocks, drain pipes. |

| Table 4.1-2 Generic copper-based metals | | |
|---|---|---|
| Metal | Typical composition (wt%) | Typical uses |
| Copper | 100 Cu | Ductile, corrosion resistant and a good electrical conductor: water pipes, electrical wiring. |
| Brass | Zn | Stronger than copper, machinable, reasonable corrosion resistance: water fittings, screws, electrical components. |
| Bronze | Cu + 10–30 Sn | Good corrosion resistance: bearings, ships' propellers, bells. |
| Cupronickel | Cu + 30 Ni | Good corrosion resistance, coinage. |

**Figure 4.1-4** Miniature boiler fittings made from brass: a water-level gauge, a steam valve, a pressure gauge, and a feed-water injector. Brass is so easy to machine that it is good for intricate parts like these.

carbon it has several advantages over mild steel. Complicated components like the cylinder block are best produced by casting. Now cast iron melts much more easily than steel (adding carbon reduces the melting point in just the way that adding anti-freeze works with water) and this makes the pouring of the castings much easier. During casting, the carbon can be made to separate out as tiny particles of graphite, distributed throughout the iron, which make an ideal boundary lubricant. Cylinders and pistons made from cast iron wear very well; look inside the cylinders of your car engine next time the head has to come off, and you will be amazed by the polished, almost glazed look of the bores – and this after perhaps $10^8$ piston strokes.

These, then, are the basic classes of ferrous alloys. Their compositions and uses are summarised in Table 4.1-1, and you will learn more about them in Chapters 11 and 12, but let us now look at the other generic alloy groups.

An important group of alloys are those based on copper (Table 4.1-2).

The most notable part of the traction engine made from copper is the boiler and its firetubes (see Fig. 4.1-1). In full size this would have been made from mild steel, and the use of copper in the model is a nice example of how the choice of material can depend on the *scale* of the structure. The boiler plates of the full-size engine are about 10 mm thick, of which perhaps only 6 mm is needed to stand the load from the pressurised steam safely – the other 4 mm is an allowance for corrosion. Although a model steel boiler would stand the pressure with plates only 1 mm thick, it would still need the same corrosion allowance of 4 mm, totalling 5 mm altogether. This would mean a very heavy boiler, and a lot of water space would be taken up by thick plates and firetubes.

Because copper hardly corrodes in clean water, this is the obvious material to use. Although weaker than steel, copper plates 2.5 mm thick are strong enough to resist the working pressure, and there is no need for a separate corrosion allowance. Of course, copper is expensive – it would be prohibitive in full size - but this is balanced by its ductility (it is very easy to bend and flange to shape) and by its high thermal conductivity (which means that the boiler steams very freely).

*Brass* is stronger than copper, is much easier to machine, and is fairly corrosion-proof (although it can "dezincify" in water after a long time). A good example of its use in the engine is for steam valves and other boiler fittings (see Fig. 4.1-4). These are intricate, and must be easy to machine; dezincification is a long-term possibility, so occasional inspection is needed. Alternatively, corrosion can be avoided altogether by using the more expensive *bronzes*, although some are hard to machine.

*Nickel* and its alloys form another important class of non-ferrous metals (Table 4.1-3). The superb creep resistance of the nickel-based superalloys is a key factor in

| Table 4.1-3 Generic nickel-based metals | | |
|---|---|---|
| Metals | Typical composition (wt%) | Typical uses |
| Monels | Ni + 30 Cu 1Fe 1Mn | Strong, corrosion resistant: heat-exchanger tubes. |
| Superalloys | Ni + 30 Cr 30 Fe 0.5 Ti 0.5 Al | Creep and oxidation resistant: furnace parts. |
| | Ni + 10 Co 10 W 9 Cr 5 A 12 Ti | Highly creep resistant: turbine blades and discs. |

# CHAPTER 4.1  Introduction to metals

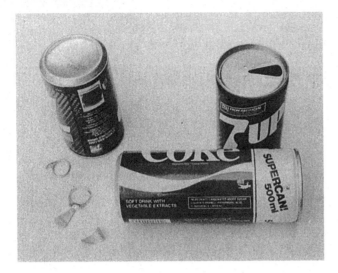

**Figure 4.1-5** The aluminium drink can is an innovative product. The body is made from a single slug of a 3000 series aluminium alloy. The can top is a separate pressing which is fastened to the body by a rolled seam once the can has been filled. There are limits to one-piece construction.

**Table 4.1-4** Generic aluminium-based metals

| Metal | Typical composition (wt%) | Typical uses |
|---|---|---|
| 1000 Series unalloyed Al | >99 Al | Weak but ductile and a good electrical conductor: power transmission lines, cooking foil. |
| 2000 Series major additive Cu | Al + 4 Cu + Mg, Si, Mn | Strong age-hardening alloy: aircraft skins, spars, forgings, rivets. |
| 3000 Series major additive Mn | Al + 1 Mn | Moderate strength, ductile, excellent corrosion resistance: roofing sheet, cooking pans, drinks can bodies. |
| 5000 Series major additive Mg | Al + 3 Mg 0.5 Mn | Strong work-hardening weldable plate: pressure vessels, ship superstructures. |
| 6000 Series major additives Mg + Si | Al + 0.5 Mg 0.5 Si | Moderate-strength age-hardening alloy: anodised extruded sections, e.g. window frames. |
| 7000 Series major additives Zn + Mg | Al + 6 Zn + Mg, Cu, Mn | Strong age-hardening alloy: aircraft forgings, spars, lightweight railway carriage shells. |
| Casting alloys | Al + 11 Si | Sand and die castings. |
| Aluminium–lithium alloys | Al + 3 Li | Low density and good strength: aircraft skins and spars. |

designing the modern gas-turbine aero-engine. But nickel alloys even appear in a model steam engine. The flat plates in the firebox must be stayed together to resist the internal steam pressure (see Fig. 4.1-3). Some model-builders make these stays from pieces of monel rod because it is much stronger than copper, takes threads much better and is very corrosion resistant.

## Metals for drinks cans

Few people would think that the humble drink can (Fig. 4.1-5) was anything special. But to a materials engineer it is high technology. Look at the requirements. As far as possible we want to avoid seams. The can must not leak, should use as little metal as possible and be recyclable. We have to choose a metal that is ductile to the point that it can be drawn into a single-piece can body from one small slug of metal. It must not corrode in beer or coke and, of course, it must be non-toxic. And it must be light and must cost almost nothing.

*Aluminium-based* metals are the obvious choice[*] (Table 4.1-4) – they are light, corrosion resistant and non-toxic. But it took several years to develop the process for forming the can and the alloy to go with it. The end product is a big advance from the days when drinks only came in glass bottles, and has created a new market for aluminium (now threatened, as we shall see in Chapter 21, by polymers). Because aluminium is lighter than most other metals it is also the obvious choice for transportation: aircraft, high-speed trains, cars, even. Most of the alloys listed in Table 4.1-4 are designed with these uses in mind. We will discuss the origin of their strength, and their uses, in more detail in Chapter 10.

## Metals for artificial hip joints

As a last example we turn to the world of medicine. Osteo-arthritis is an illness that affects many people as they get older. The disease affects the joints between different bones in the body and makes it hard – and

---

[*] One thinks of aluminium as a cheap material - aluminium spoons are so cheap that they are thrown away. It was not always so. Napoleon had a set of cutlery specially made from the then-new material. It cost him more than a set of solid silver.

# Introduction to metals CHAPTER 4.1

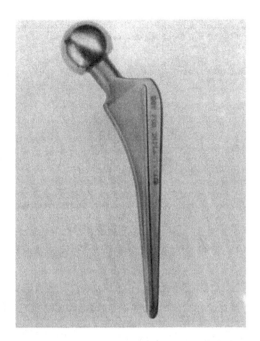

**Figure 4.1-6** The titanium alloy implant for a replacement hip joint. The long shank is glued into the top of the femur. The spherical head engages in a high-density polythene socket which is glued into the pelvic socket.

painful – to move them. The problem is caused by small lumps of bone which grow on the rubbing surfaces of the joints and which prevent them sliding properly. The problem can only be cured by removing the bad joints and putting artificial joints in their place. The first recorded hip-joint replacement was done as far back as 1897 – when it must have been a pretty hazardous business – but the operation is now a routine piece of orthopaedic surgery. In fact half a million hip joints are replaced world-wide every year.

Figure 4.1-6 shows the implant for a replacement hip joint. In the operation, the head of the femur is cut off and the soft marrow is taken out to make a hole down the centre of the bone. Into the hole is glued a long metal shank which carries the artificial head. This fits into a high-density polythene socket which in turn is glued into the old bone socket. The requirements of the implant are stringent. It has to take large loads without bending. Every time the joint is used ($\approx 10^6$ times a year) the load on it fluctuates, giving us a high-cycle fatigue problem as well. Body fluids are as corrosive as sea water, so we must design against corrosion, stress corrosion and corrosion fatigue. The metal must be bio-compatible. And, ideally, it should be light as well.

The materials that best meet these tough requirements are based on *titanium*. The $\alpha$–$\beta$ alloy shown in Table 4.1-5 is as strong as a hardened and tempered high-carbon steel, is more corrosion resistant in body fluids than stainless steel, but is only half the weight. A disadvantage is that its modulus is only half that of steels, so that it tends to be "whippy" under load. But this can be overcome by using slightly stiffer sections. The same alloy is used in aircraft, both in the airframes and in the compressor stages of the gas turbines which drive them.

## Data for metals

When you select a metal for any design application you need *data* for the properties. Table 4.1-6 gives you *approximate* property data for the main generic metals, useful for the first phase of a design project. When you have narrowed down your choice you should turn to the more exhaustive data compilations given in Appendix 3. Finally, before making final design decisions you should get detailed material specifications from the supplier who will provide the materials you intend to use. And if the component is a critical one (meaning that its failure could precipitate a catastrophe) you should arrange to test it yourself.

There are, of course, many more metals available than those listed here. It is useful to know that some properties depend very little on microstructure: the density, modulus, thermal expansion and specific heat of *any* steel are pretty close to those listed in the table. (Look at the table and you will see that the variations in these properties are seldom more than ±5%.) These are the *"structure-insensitive"* properties. Other properties, though, vary greatly with the heat treatment and mechanical treatment, and the detailed alloy composition. These are the *"structure-sensitive"* properties: yield and tensile strength, ductility, fracture toughness, and creep and fatigue strength. They cannot be guessed from data for other alloys, even when the composition is almost the same. For these it is *essential* to consult manufacturers' data sheets listing the properties of the alloy you intend to use, with the same mechanical and heat treatment.

| Table 4.1-5 Generic titanium-based metals | | |
|---|---|---|
| **Metal** | **Typical composition (wt%)** | **Typical uses** |
| $\alpha$-$\beta$ titanium alloy | Ti–6 Al4 V | Light, very strong, excellent corrosion resistance, high melting point, good creep resistance. The alloy workhouse: turbofans, air frames, chemical plant, surgical implants. |

# CHAPTER 4.1 Introduction to metals

Table 4.1-6 Properties of the generic metals

| Metal | Cost (UK£ (US$) tonne$^{-1}$) | Density (Mg m$^{-3}$) | Young's modulus (GPa) | Yield strength (MPa) | Tensile strength (MPa) | Ductility | Fracture toughness (MPa m$^{1/2}$) | Melting Temperature (K) | Specific heat (J kg$^{-1}$ K$^{-1}$) | Thermal conductivity (W m$^{-1}$ K$^{-1}$) | Thermal expansion coefficient (MK$^{-1}$) |
|---|---|---|---|---|---|---|---|---|---|---|---|
| Iron | 100 (140) | 7.9 | 211 | 50 | 200 | 0.3 | 80 | 1809 | 456 | 78 | 12 |
| Mild steel | 200–230 (260–300) | 7.9 | 210 | 220 | 430 | 0.21 | 140 | 1765 | 482 | 60 | 12 |
| High-carbon steel | 150 (200) | 7.8 | 210 | 350–1600 | 650–2000 | 0.1–0.2 | 20–50 | 1570 | 460 | 40 | 12 |
| Low-alloy steels | 180–250 (230–330) | 7.8 | 203 | 290–1600 | 420–2000 | 0.1–0.2 | 50–170 | 1750 | 460 | 40 | 12 |
| High-alloy steels | 1100–1400 (1400–1800) | 7.8 | 215 | 170–1600 | 460–1700 | 0.1–0.5 | 50–170 | 1680 | 500 | 12–30 | 10–18 |
| Cast irons | 120 (160) | 7.4 | 152 | 50–400 | 10–800 | 0–0.18 | 6–20 | 1403 | | | |
| Copper | 1020 (1330) | 8.9 | 130 | 75 | 220 | 0.5–0.9 | >100 | 1356 | 385 | 397 | 17 |
| Brasses | 750–1060 (980–1380) | 8.4 | 105 | 200 | 350 | 0.5 | 30–100 | 1190 | | 121 | 20 |
| Bronzes | 1500 (2000) | 8.4 | 120 | 200 | 350 | 0.5 | 30–100 | 1120 | | 85 | 19 |
| Nickel | 3200 (4200) | 8.9 | 214 | 60 | 300 | 0.4 | >100 | 1728 | 450 | 89 | 13 |
| Monels | 3000 (3900) | 8.9 | 185 | 340 | 680 | 0.5 | >100 | 1600 | 420 | 22 | 14 |
| Superalloys | 5000 (6500) | 7.9 | 214 | 800 | 1300 | 0.2 | >100 | 1550 | 450 | 11 | 12 |
| Aluminium | 910 (1180) | 2.7 | 71 | 25–125 | 75–135 | 0.1–0.5 | 45 | 933 | 917 | 240 | 24 |
| 1000 Series | 910 (1180) | 2.7 | 71 | 28–165 | 75–180 | 0.1–0.45 | 45 | 915 | | | 24 |
| 2000 Series | 1100 (1430) | 2.8 | 71 | 200–500 | 300–600 | 0.1–0.25 | 10–50 | 860 | | 180 | 24 |
| 5000 Series | 1000 (1300) | 2.7 | 71 | 40–300 | 120–430 | 0.1–0.35 | 30–40 | 890 | | 130 | 22 |
| 7000 Series | 1100 (1430) | 2.8 | 71 | 350–600 | 500–670 | 0.1–0.17 | 20–70 | 890 | | 150 | 24 |
| Casting alloys | 1100 (1430) | 2.7 | 71 | 65–350 | 130–400 | 0.01–0.15 | 5–30 | 860 | | 140 | 20 |
| Titanium | 4630 (6020) | 4.5 | 120 | 170 | 240 | 0.25 | | 1940 | 530 | 22 | 9 |
| Ti–6 A14 V | 5780 (7510) | 4.4 | 115 | 800–900 | 900–1000 | 0.1–0.2 | 50–80 | 1920 | 610 | 6 | 8 |
| Zinc | 330 (430) | 7.1 | 105 | | 120 | 0.4 | | 693 | 390 | 120 | 31 |
| Lead—tin solder | 2000 (2600) | 9.4 | 40 | | | | | 456 | | | |
| Diecasting alloy | 800 (1040) | 6.7 | 105 | | 280–330 | 0.07–0.15 | | 650 | 420 | 110 | 27 |

## Examples

**1.1** Explain what is meant by the following terms:

(a) structure-sensitive property;

(b) structure-insensitive property.

List five different structure-sensitive properties.
List four different structure-insensitive properties.

## Answers

Structure-sensitive properties: yield strength, hardness, tensile strength, ductility, fracture toughness, fatigue strength, creep strength, corrosion resistance, wear resistance, thermal conductivity, electrical conductivity. Structure-insensitive properties: elastic moduli, Poisson's ratio, density, thermal expansion coefficient, specific heat.

**1.2** What are the five main generic classes of metals? For each generic class:

(a) give one example of a specific component made from that class;

(b) indicate why that class was selected for the component.

# Chapter 4.2

# Metal structures

## Introduction

At the end of Chapter 1 we noted that structure-sensitive properties like strength, ductility or toughness depend critically on things like the composition of the metal and on whether it has been heated, quenched or cold formed. Alloying or heat treating work by controlling the *structure* of the metal. Table 4.2-1 shows the large range over which a material has structure. The bracketed subset in the table can be controlled to give a wide choice of structure-sensitive properties.

## Crystal and glass structures

We begin by looking at the smallest scale of controllable structural feature – the way in which the atoms in the metals are packed together to give either a crystalline or a glassy (amorphous) structure. Table 4.2-2 lists the crystal structures of the pure metals at room temperature. In nearly every case the metal atoms pack into the simple crystal structures of face-centred cubic (f.c.c), body-centred cubic (b.c.c.) or close-packed hexagonal (c.p.h.).

Metal atoms tend to behave like miniature ball-bearings and tend to pack together as tightly as possible. F.c.c. and c.p.h. give the highest possible packing density, with 74% of the volume of the metal taken up by the atomic spheres. However, in some metals, like iron or chromium, the metallic bond has some directionality and this makes the atoms pack into the more open b.c.c. structure with a packing density of 68%.

Some metals have more than one crystal structure. The most important examples are iron and titanium. As Figure 4.2-1 shows, iron changes from b.c.c. to f.c.c. at 914°C but goes back to b.c.c. at 1391°C; and titanium changes from c.p.h. to b.c.c. at 882°C. This multiplicity of crystal structures is called *polymorphism*. But it is

| Table 4.2-1 | | |
|---|---|---|
| **Structural feature** | **Typical scale (m)** | |
| Nuclear structure | $10^{-15}$ | |
| Structure of atom | $10^{-10}$ | |
| Crystal or glass structure | $10^{-9}$ | |
| Structures of solutions and compounds | $10^{-9}$ | Range that can be controlled to alter properties |
| Structures of grain and phase boundaries | $10^{-8}$ | |
| Shapes of grains and phases | $10^{-7}$ to $10^{-3}$ | |
| Aggregates of grains | $10^{-5}$ to $10^{-2}$ | |
| Engineering structures | $10^{-3}$ to $10^{3}$ | |

*Engineering Materials and Processes Desk Reference*; ISBN: 9781856175869
Copyright © 2009 Elsevier Ltd; All rights of reproduction, in any form, reserved.

Table 4.2-2 Crystal structures of pure metals at room temperature

| Pure metal | Structure | Unit cell dimensions (nm) | |
|---|---|---|---|
| | | a | c |
| Aluminium | f.c.c. | 0.405 | |
| Beryllium | c.p.h. | 0.229 | 0.358 |
| Cadmium | c.p.h. | 0.298 | 0.562 |
| Chromium | b.c.c. | 0.289 | |
| Cobalt | c.p.h. | 0.251 | 0.409 |
| Copper | f.c.c. | 0.362 | |
| Gold | f.c.c. | 0.408 | |
| Hafnium | c.p.h. | 0.320 | 0.506 |
| Indium | Face-centred tetragonal | | |
| Iridium | f.c.c. | 0.384 | |
| Iron | b.c.c. | 0.287 | |
| Lanthanum | c.p.h. | 0.376 | 0.606 |
| Lead | f.c.c. | 0.495 | |
| Magnesium | c.p.h. | 0.321 | 0.521 |
| Manganese | Cubic | 0.891 | |
| Molybdenum | b.c.c. | 0.315 | |
| Nickel | f.c.c. | 0.352 | |
| Niobium | b.c.c. | 0.330 | |
| Palladium | f.c.c. | 0.389 | |
| Platinum | f.c.c. | 0.392 | |
| Rhodium | f.c.c. | 0.380 | |
| Silver | f.c.c. | 0.409 | |
| Tantalum | b.c.c. | 0.331 | |
| Thallium | c.p.h. | 0.346 | 0.553 |
| Tin | Body-centred tetragonal | | |
| Titanium | c.p.h. | 0.295 | 0.468 |
| Tungsten | b.c.c. | 0.317 | |
| Vanadium | b.c.c. | 0.303 | |
| Yttrium | c.p.h. | 0.365 | 0.573 |
| Zinc | c.p.h. | 0.267 | 0.495 |
| Zirconium | c.p.h. | 0.323 | 0.515 |

obviously out of the question to try to control crystal structure simply by changing the temperature (iron is useless as a structural material well below 914°C). Polymorphism can, however, be brought about at room temperature by alloying. Indeed, many stainless steels are f.c.c. rather than b.c.c. and, especially at low temperatures, have much better ductility and toughness than ordinary carbon steels.

This is why stainless steel is so good for cryogenic work: the fast fracture of a steel vacuum flask containing liquid nitrogen would be embarrassing, to say the least, but stainless steel is essential for the vacuum jackets needed to cool the latest superconducting magnets down to liquid helium temperatures, or for storing liquid hydrogen or oxygen.

If molten metals (or, more usually, alloys) are cooled very fast – faster than about $10^6 K\ s^{-1}$ – there is no time for the randomly arranged atoms in the liquid to switch into the orderly arrangement of a solid crystal. Instead, a *glassy* or *amorphous* solid is produced which has essentially a "frozen-in" liquid structure. This structure – which is termed *dense random packing (drp)* – can be modelled very well by pouring ball-bearings into a glass jar and shaking them down to maximise the packing density. It is interesting to see that, although this structure is disordered, it has well-defined characteristics. For example, the packing density is always 64%, which is why corn was always sold in bushels (1 bushel = 8 UK gallons): provided the corn was always shaken down well in the sack a bushel always gave $0.64 \times 8 = 5.12$ gallons of corn material! It has only recently become practicable to make glassy metals in quantity but, because their structure is so different from that of "normal" metals, they have some very unusual and exciting properties.

## Structures of solutions and compounds

As you can see from the tables in Chapter 1, few metals are used in their pure state – they nearly always have other elements added to them which turn them into *alloys* and give them better mechanical properties. The alloying elements will always dissolve in the basic metal to form *solid solution*, although the solubility can vary between <0.01% and 100% depending on the combinations of elements we choose. As examples, the iron in a carbon steel can only dissolve 0.007% carbon at room temperature; the copper in brass can dissolve more than 30% zinc; and the copper–nickel system – the basis of the monels and the cupronickels – has complete solid solubility.

There are two basic classes of solid solution. In the first, small atoms (like carbon, boron and most gases) fit between the larger metal atoms to give *interstitial solid solutions* (Figure 4.2-2a). Although this interstitial

# Metal structures  CHAPTER 4.2

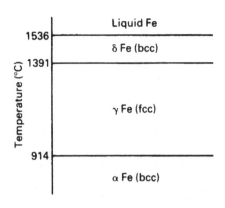
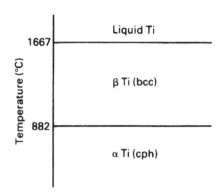

**Figure 4.2-1** Some metals have more than one crystal structure. The most important examples of this *polymorphism* are in iron and titanium.

solubility is usually limited to a few per cent it can have a large effect on properties. Indeed, as we shall see later, interstitial solutions of carbon in iron are largely responsible for the enormous range of strengths that we can get from carbon steels. It is much more common, though, for the dissolved atoms to have a similar size to those of the host metal. Then the dissolved atoms simply replace some of the host atoms to give a *substitutional solid solution* (Fig. 4.2-2b). Brass and cupronickel are good examples of the large solubilities that this atomic substitution can give.

Solutions normally tend to be *random* so that one cannot predict which of the sites will be occupied by which atoms (Fig. 4.2.2c). But if A atoms prefer to have A neighbours, or B atoms prefer B neighbours, the solution can *cluster* (Fig. 4.2.2d); and when A atoms prefer B neighbours the solution can *order* (Fig. 4.2-2e).

Many alloys contain more of the alloying elements than the host metal can dissolve. Then the surplus must separate out to give regions that have a high concentration of the alloying element. In a few alloys these regions consist of a solid solution based on the *alloying* element. (The lead–tin alloy system, on which most soft solders are based, Table 1.6, is a nice example of this – the lead can only dissolve 2% tin at room temperature and any surplus tin will separate out as regions of *tin* containing 0.3% dissolved lead.) In most alloy systems, however, the surplus atoms of the alloying element separate out as *chemical compounds*. An important example of this is in the aluminium-copper system (the basis of the 2000 series alloys, Table 1.4) where surplus copper separates out as the compound $CuAl_2$. $CuAl_2$ is hard and is not easily cut by dislocations. And when it is finely dispersed throughout the alloy it can give *very* big increases in strength. Other important compounds are $Ni_3Al$, $Ni_3Ti$, $Mo_2C$ and $TaC$ (in super-alloys) and $Fe_3C$ (in carbon steels). Figure 4.2-3 shows the crystal structure of $CuAl_2$. As with most compounds, it is quite complicated.

**Figure 4.2-2** Solid-solution structures. In *interstitial* solutions small atoms fit into the spaces between large atoms. In *substitutional* solutions similarly sized atoms replace one another. If A–A, A–B and B–B bonds have the same strength then this replacement is *random*. But unequal bond strengths can give *clustering* or *ordering*.

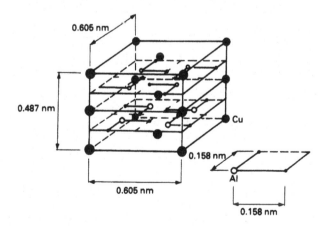

**Figure 4.2-3** The crystal structure of the "intermetallic" compound $CuAl_2$. The structures of compounds are usually more complicated than those of pure metals.

109

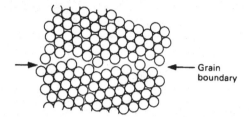

**Figure 4.2-4** The structure of a typical grain boundary. In order to "bridge the gap" between two crystals of different orientation the atoms in the grain boundary have to be packed in a less ordered way. The packing density in the boundary is then as low as 50%.

## Phases

The things that we have been talking about so far – metal crystals, amorphous metals, solid solutions, and solid compounds – are all *phases*. A phase is a region of material that has uniform physical and chemical properties. Water is a phase – any one drop of water is the same as the next. Ice is another phase – one splinter of ice is the same as any other. But the mixture of ice and water in your glass at dinner is not a single phase because its properties vary as you move from water to ice. Ice + water is a *two-phase* mixture.

## Grain and phase boundaries

A pure metal, or a solid solution, is single-phase. It is certainly possible to make single crystals of metals or alloys but it is difficult and the expense is only worth it for high-technology applications such as single-crystal turbine blades or single-crystal silicon for microchips. Normally, any single-phase metal is *polycrystalline* – it is made up of millions of small crystals, or *grains*, "stuck" together by *grain boundaries* (Fig. 4.2-4). Because of their unusual structure, grain boundaries have special properties of their own. First, the lower bond density in the boundary is associated with a boundary surface-energy: typically 0.5 Joules per square metre of boundary area (0.5 J m$^{-2}$). Secondly, the more open structure of the boundary can give much faster diffusion in the boundary plane than in the crystal on either side. And finally, the extra space makes it easier for outsized impurity atoms to dissolve in the boundary. These atoms tend to *segregate* to the boundaries, sometimes very strongly. Then an *average* impurity concentration of a few parts per million can give a *local* concentration of 10% in the boundary with very damaging effects on the fracture toughness.*

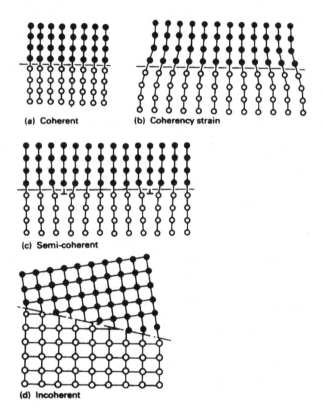

**Figure 4.2-5** Structures of interphase boundaries.

As we have already seen, when an alloy contains more of the alloying element than the host metal can dissolve, it will split up into *two* phases. The two phases are "stuck" together by *interphase boundaries* which, again, have special properties of their own. We look first at two phases which have different chemical compositions but the same crystal structure (Fig. 4.2-5a). Provided they are oriented in the right way, the crystals can be made to match up at the boundary. Then, although there is a sharp change in chemical composition, there is no structural change, and the energy of this *coherent* boundary is low (typically 0.05 J m$^{-2}$). If the two crystals have slightly different lattice spacings, the boundary is still coherent but has some strain (and more energy) associated with it (Fig. 4.2-5b). The strain obviously gets bigger as the boundary grows sideways: full coherency is usually possible only with small second-phase particles. As the particle grows, the strain builds up until it is relieved by the injection of dislocations to give a *semi-coherent* boundary (Fig. 4.2-5c). Often the two phases which meet at the boundary are large, and differ in both chemical composition *and* crystal structure. Then the boundary between them is *incoherent*; it is like a grain

---

* Henry Bessemer, the great Victorian ironmaster and the first person to mass-produce mild steel, was nearly bankrupted by this. When he changed his suppliers of iron ore, his steel began to crack in service. The new ore contained phosphorus, which we now know segregates badly to grain boundaries. Modern steels must contain less than ≈ 0.05% phosphorus as a result.

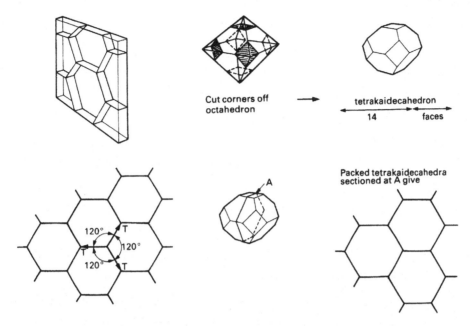

**Figure 4.2-6** (a) The surface energy of a "two-dimensional" array of soap bubbles is minimised if the soap films straighten out. Where films meet the forces of surface tension must balance. This can only happen if films meet in "120° three-somes". (b) In a three-dimensional polycrystal the grain boundary energy is minimised if the boundaries flatten out. These flats must meet in 120° three-somes to balance the grain boundary tensions. If we fill space with equally sized tetrakaidecahedra we will satisfy these conditions. Grains in polycrystals therefore tend to be shaped like tetrakaidecahedra when the grain-boundary energy is the dominating influence.

boundary across which there is also a change in chemical composition (Fig. 4.2-5d). Such a phase boundary has a high energy – comparable with that of a grain boundary – and around $0.5\ \mathrm{J\ m^{-2}}$.

## Shapes of grains and phases

Grains come in all shapes and sizes, and both shape and size can have a big effect on the properties of the polycrystalline metal (a good example is mild steel – its strength can be *doubled* by a ten-times decrease in grain size). Grain shape is strongly affected by the way in which the metal is processed. Rolling or forging, for instance, can give stretched-out (or "textured") grains; and in casting the solidifying grains are often elongated in the direction of the easiest heat loss. But if there are no external effects like these, then the energy of the grain boundaries is the important thing. This can be illustrated very nicely by looking at a "two-dimensional" array of soap bubbles in a thin glass cell. The soap film minimises its overall energy by straightening out; and at the corners of the bubbles the films meet at angles of 120° to balance the surface tensions (Fig. 4.2-6a). of course a polycrystalline metal is three-dimensional, but the same principles apply: the grain boundaries try to form themselves into flat planes, and these planes always try to meet at 120°. A grain shape does indeed exist which not only satisfies these conditions but also packs together to fill space. It has 14 faces, and is therefore called a tetrakaidecahedron (Fig. 4.2-6b). This shape is remarkable, not only for the properties just given, but because it appears in almost every physical science (the shape of cells in plants, of bubbles in foams, of grains in metals and of Dirichlet cells in solid-state physics).*

If the metal consists of *two* phases then we can get more shapes. The simplest is when a single-crystal particle of one phase forms inside a grain of another phase. Then, if the energy of the interphase boundary is *isotropic* (the same for all orientations), the second-phase particle will try to be *spherical* in order to minimise the interphase boundary energy (Fig. 4.2-7a). Naturally, if coherency is possible along some planes, but not along others, the particle will tend to grow as a *plate*, extensive along the low-energy planes but narrow along the high-energy ones (Fig. 4.2-7b). Phase shapes get more complicated when interphase boundaries and grain boundaries meet. Figure 4.2-7(c) shows the shape of a second-phase particle that has formed at a grain boundary. The particle is shaped by two spherical caps which meet the grain boundary at an angle $\theta$. This angle is set by the balance of boundary tensions

---

* For a long time it was thought that soap foams, grains in metals and so on were icosahedra. It took Lord Kelvin (of the degree K) to get it right.

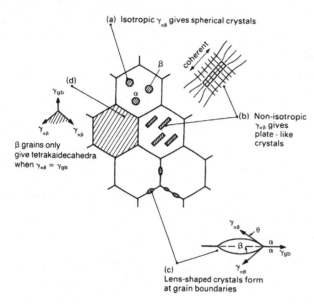

**Figure 4.2-7** Many metals are made up of *two* phases. This figure shows some of the shapes that they can have when boundary energies dominate. To keep things simple we have sectioned the tetrakaidecahedral grains in the way that we did in Fig. 4.2-6(b). Note that Greek letters are often used to indicate phases. We have called the major phase α and the second phase β. But γ is the symbol for the energy (or tension) of grain boundaries ($\gamma_{gb}$) and interphase interfaces ($\gamma_{\alpha\beta}$).

$$2\gamma_{\alpha\beta}\cos\theta = \gamma_{gb} \quad (4.2.1)$$

where $\gamma_{\alpha\beta}$ is the tension (or energy) of the interphase boundary and $\gamma_{gb}$ is the grain boundary tension (or energy).

In some alloys, $\gamma_{\alpha\beta}$ can be $\leqslant \gamma_{gb}/2$ in which case $\theta = 0$. The second phase will then spread along the boundary as a thin layer of β. This "wetting" of the grain boundary can be a great nuisance - if the phase is brittle then cracks can spread along the grain boundaries until the metal falls apart completely. A favourite scientific party trick is to put some aluminium sheet in a dish of molten gallium and watch the individual grains of aluminium come apart as the gallium whizzes down the boundary.

The second phase can, of course, form complete grains (Fig. 4.2-7d). But only if $\gamma_{\alpha\beta}$ and $\gamma_{gb}$ are similar will the phases have tetrakaidecahedral shapes where they come together. In general, $\gamma_{\alpha\beta}$ and $\gamma_{gb}$ may be quite different and the grains then have more complicated shapes.

## Summary: constitution and structure

The structure of a metal is defined by two things. The first is the *constitution*:

**(a)** The overall composition – the elements (or *components*) that the metal contains and the relative weights of each of them.

**(b)** The number of phases, and their relative weights.

**(c)** The composition of each phase.

The second is the geometric information about *shape and size*:

**(d)** The shape of each phase.

**(e)** The sizes and spacings of the phases.

Armed with this information, we are in a strong position to re-examine the mechanical properties, and explain the great differences in strength, or toughness, or corrosion resistance between alloys. But where does this information come from? The *constitution* of an alloy is summarised by its phase diagram – the subject of the next chapter. The *shape and size* are more difficult, since they depend on the details of how the alloy was made. But, as we shall see from later chapters, a fascinating range of microscopic processes operates when metals are cast, or worked or heat-treated into finished products; and by understanding these, shape and size can, to a large extent, be predicted.

## Examples

**4.2.1** Describe, in a few words, with an example or sketch where appropriate, what is meant by each of the following:
  **(a)** polymorphism;
  **(b)** dense random packing;
  **(c)** an interstitial solid solution;
  **(d)** a substitutional solid solution;
  **(e)** clustering in solid solutions;
  **(f)** ordering in solid solutions;
  **(g)** an intermetallic compound;
  **(h)** a phase in a metal;
  **(i)** a grain boundary;
  **(j)** an interphase boundary;
  **(k)** a coherent interphase boundary;
  **(l)** a semi-coherent interphase boundary;
  **(m)** an incoherent interphase boundary;
  **(n)** the constitution of a metal;
  **(o)** a component in a metal.

**4.2.2** Why do impurity atoms segregate to grain boundaries?

**4.2.3** A large furnace flue operating at 440°C was made from a steel containing 0.10% phosphorus as an impurity. After two years in service, specimens were removed from the flue and tested for fracture toughness. The value of $K_c$ was 30 MPam$^{1/2}$, compared to 100 MPam$^{1/2}$ for new steel. Because of this, the flue had to be scrapped for safety reasons. Explain this dramatic drop in toughness.

**4.2.4** Indicate the shapes that the following adopt *when boundary energies dominate*:

(a) a polycrystalline pure metal (isotropic $\gamma_{gb}$);

(b) an intermetallic precipitate inside a grain (isotropic $\gamma_{\alpha\beta}$);

(c) an intermetallic precipitate at a grain boundary ($\gamma_{\alpha\beta} > \gamma_{gb}/2$);

(d) an intermetallic precipitate at a grain boundary ($\gamma_{\alpha\beta} < \gamma_{gb}/2$).

# Chapter 4.3

# Equilibrium constitution and phase diagrams

## Introduction

Whenever you have to report on the structure of an alloy – because it is a possible design choice, or because it has mysteriously failed in service – the first thing you should do is reach for its *phase diagram*. It tells you what, at equilibrium, the *constitution* of the alloy should be. The real constitution may not be the equilibrium one, but the equilibrium constitution gives a base line from which other non-equilibrium constitutions can be inferred.

Using phase diagrams is like reading a map. We can explain how they work, but you will not feel confident until you have used them. Hands-on experience is essential. So, although this chapter introduces you to phase diagrams, it is important for you to work through the "Teaching Yourself Phase Diagrams" section at the end of the book. This includes many short examples which give you direct experience of using the diagrams. The whole thing will only take you about four hours and we have tried to make it interesting, even entertaining. But first, a reminder of some essential definitions.

## Definitions

An alloy is a metal made by taking a pure metal and adding other elements (the "alloying elements") to it. Examples are brass ($Cu + Zn$) and monel ($Ni + Cu$).

The *components* of an alloy are the elements which make it up. In brass, the components are copper and zinc. In monel they are nickel and copper. The components are given the atomic symbols, e.g. Cu, Zn or Ni, Cu.

An *alloy system* is all the alloys you can make with a given set of components: "the Cu-Zn system" describes all the alloys you can make from copper and zinc. A *binary* alloy has two components; a *ternary* alloy has three.

A *phase* is a region of material that has uniform physical and chemical properties. Phases are often given Greek symbols, like $\alpha$ or $\beta$ But when a phase consists of a solid solution of an alloying element in a host metal, a clearer symbol can be used. As an example, the phases in the lead-tin system may be symbolised as (Pb) – for the solution of tin in *lead*, and (Sn) – for the solution of lead in *tin*.

The *composition* of an alloy, or of a phase in an alloy, is usually measured in weight %, and is given the symbol $W$. Thus, in an imaginary A–B alloy system:

$$W_A = \frac{\text{wt of A}}{\text{wt of A + wt of B}} \times 100\%, \quad (4.3.1)$$

$$W_B = \frac{\text{wt of B}}{\text{wt of A + wt of B}} \times 100\%, \quad (4.3.2)$$

and

$$W_A + W_B = 100\%. \quad (4.3.3)$$

Sometimes it is helpful to define the atom (or mol)%, given by

$$X_A = \frac{\text{atoms of A}}{\text{atoms of A + atoms of B}} \times 100\% \quad (4.3.4)$$

and so on.

The *constitution* of an alloy is described by

(a) the overall composition;
(b) the number of phases;

(c) the composition of each phase;

(d) the proportion by weight of each phase.

An alloy has its *equilibrium constitution* when there is no further tendency for the constitution to change with time.

The equilibrium diagram or *phase diagram* summarises the equilibrium constitution of the alloy system.

## The lead-tin phase diagram

And now for a real phase diagram. We have chosen the lead–tin diagram (Fig. 4.3-1) as our example because it is pretty straightforward and we already know a bit about it. Indeed, if you have soldered electronic components together or used soldered pipe fittings in your hot-water layout, you will already have had some direct experience of this system.

As in all binary phase diagrams, we have plotted the composition of the alloy on the horizontal scale (in weight %), and the temperature on the vertical scale. The diagram is simply a two-dimensional map (drawn up from experimental data on the lead-tin system) which shows us where the various phases are in composition–temperature space. But how do we use the diagram in practice? As a first example, take an alloy of overall composition 50 wt% lead at 170°C. The *constitution point* (Fig. 4.3-2a) lies inside a two-phase field. So, *at equilibrium*, the alloy must be a two-phase mixture: it must consist of "lumps" of (Sn) and (Pb) stuck together.

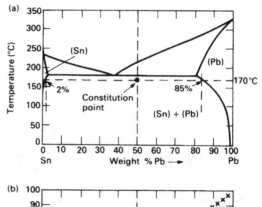

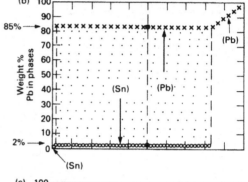

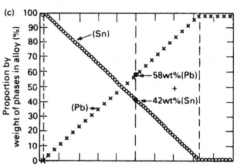

**Figure 4.3-2** (a) A 50-50 lead-tin alloy at 170°C has a constitution point that puts it in the (Sn) + (Pb) two-phase field. The compositions of the (Sn) and (Pb) phases in the two-phase mixture are 2 wt% lead and 85 wt% lead. Remember that, in any overall composition, or in any phase, wt% tin + wt% lead = 100%. So the compositions of the (Sn) and (Pb) phases could just as well have been written as 98 wt% tin and 15 wt% tin. (b) This diagram only duplicates information that is already contained in the phase diagram, but it helps to emphasise how the compositions of the phases depend on the overall composition of the alloy. (c) The 50-50 alloy at 170°C consists of 58 wt% of the (Pb) phase and 42 wt% of the (Sn) phase. The straight-line relations in the diagram are a simple consequence of the following requirements: (i) mass (Pb) phase + mass (Sn) phase = mass alloy; (ii) mass lead in (Pb) + mass lead in (Sn) = mass lead in alloy; (iii) mass tin in (Pb) + mass tin in (Sn) = mass tin in alloy.

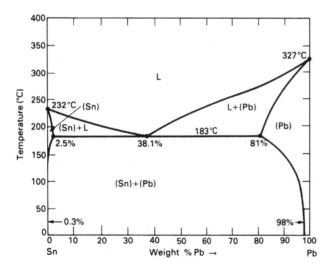

**Figure 4.3-1** The phase diagram for the lead-tin alloy system. There are three phases: L – a liquid solution of lead and tin; (Pb) – a solid solution of tin in lead; and (Sn) – a solid solution of lead in tin. The diagram is divided up into *six fields* – three of them are single-phase, and three are two-phase.

More than this, the diagram tells us (Fig. 4.3-2b) that the (Sn) phase in our mixture contains 2% lead dissolved in it (it is 98% tin) and the (Pb) phase is 85% lead (it has 15% tin dissolved in it). And finally the diagram tells us (Fig. 4.3-2c) that the mixture contains 58% by weight of the (Pb) phase. To summarise, then, with the help of the phase diagram, we now know what the equilibrium *constitution* of our alloy is – we know:

(a) the overall composition (50 wt% lead + 50 wt% tin),
(b) the number of phases (two),
(c) the composition of each phase (2 wt% lead, 85 wt% lead),
(d) the proportion of each phase (58 wt% (Pb), 42 wt% (Sn)).

What we *don't* know is how the lumps of (Sn) and (Pb) are sized or shaped. And we can only find *that* out by cutting the alloy open and looking at it with a microscope.*

Now let's try a few other alloy compositions at 170°C. Using Figs 4.3-2(b) and 4.3-2(c) you should be able to convince yourself that the following equilibrium constitutions are consistent.

(a) 25 wt% lead + 75 wt% tin,
(b) two phases,
(c) 2 wt% lead, 85 wt% lead,
(d) 30 wt% (Pb), 70 wt% (Sn).

(a) 75 wt% lead + 25 wt% tin,
(b) two phases,
(c) 2 wt% lead, 85 wt% lead,
(d) 87 wt% (Pb), 13 wt% (Sn).

(a) 85 wt% lead + 15 wt% tin,
(b) one phase (just),
(c) 85 wt% lead,
(d) 100 wt% (Pb).

(a) 95 wt% lead + 5 wt% tin,
(b) one phase,
(c) 95 wt% lead,
(d) 100 wt% (Pb).

(a) 2 wt% lead + 98 wt% tin,
(b) one phase (just),
(c) 2 wt% lead,
(d) 100 wt% (Sn).

(a) 1 wt% lead + 99 wt% tin,
(b) one phase,
(c) 1 wt% lead,
(d) 100 wt% (Sn).

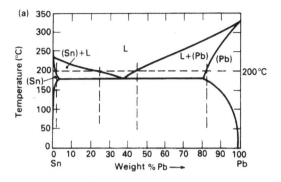

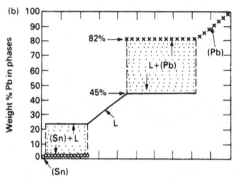

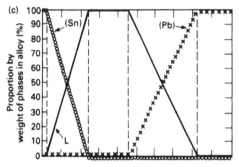

**Figure 4.3-3** Diagrams showing how you can find the equilibrium constitution of any lead-tin alloy at 200°C. Once you have had a little practice you will be able to write down constitutions directly from the phase diagram without bothering about diagrams like (b) or (c).

For our second example, we look at alloys at 200°C. We can use exactly the same method, as Fig. 4.3-3 shows. A typical constitution at 200°C would be:

(a) 50 wt% lead + 50 wt% tin,
(b) two phases,
(c) 45 wt% lead, 82 wt% lead,
(d) 87 wt% (L), 13 wt% (Pb),

and you should have no problem in writing down many others.

---

* A whole science, called metallography, is devoted to this. The oldest method is to cut the alloy in half, polish the cut faces, etch them in acid to colour the phases differently, and look at them in the light microscope. But you don't even need a microscope to see some grains. Look at any galvanised steel fire-escape or cast brass door knob and you will see the grains, etched by acid rain or the salts from people's hands.

# CHAPTER 4.3
## Equilibrium constitution and phase diagrams

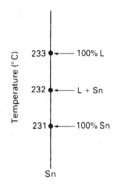

**Figure 4.3-4** At 232°C, the melting point of pure tin, we have a L + Sn two-phase mixture. But, without more information, we can't say what the relative weights of L and Sn *are*.

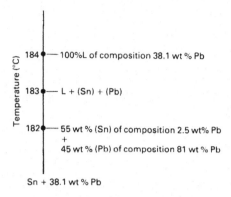

**Figure 4.3-5** At 183°C we have a three-phase *mixture* of L + (Sn) + (Pb). Their relative weights can't be found from the phase diagram.

## Incompletely defined constitutions

There *are* places in the phase diagram where we *can't* write out the full constitution. To start with, let's look at pure tin. At 233°C we have single-phase liquid tin (Fig. 4.3-4). At 231°C we have single-phase solid tin. At 232°C, the melting point of pure tin, we can either have solid tin about to melt, or liquid tin about to solidify, or a mixture of both. If we started with solid tin about to melt we could, of course, supply latent heat of melting at 232°C and get some liquid tin as a result. But the phase diagram knows nothing about external factors like this. Quite simply, the constitution of pure tin at 232°C is incompletely defined because we cannot write down the relative weights of the phases. And the same is, of course, true for pure *lead* at 327°C.

The other place where the constitution is not fully defined is where there is a horizontal line on the phase diagram. The lead–tin diagram has one line like this – it runs across the diagram at 183°C and connects (Sn) of 2.5 wt% lead, L of 38.1% lead and (Pb) of 81% lead. Just above 183°C an alloy of tin + 38.1% lead is single-phase liquid (Fig. 4.3-5). Just below 183°C it is two-phase, (Sn) + (Pb). At 183°C we have a *three-phase mixture* of L + (Sn) + (Pb) but we can't of course say from the phase diagram what the relative weights of the three phases are.

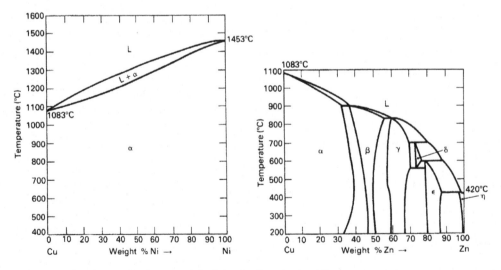

**Figure 4.3-6 (a)** The copper–nickel diagram is a good deal simpler than the lead-tin one, largely because copper and nickel are completely soluble in one another in the solid state. **(b)** The copper-zinc diagram is much more involved than the lead-tin one, largely because there are extra *(intermediate)* phases in between the end *(terminal)* phases. However, it is still an assembly of single-phase and two-phase fields.

# Equilibrium constitution and phase diagrams — CHAPTER 4.3

| Table 4.3-1 | | | |
|---|---|---|---|
| Feature | Cu–Ni (Fig. 3.6a) | Pb–Sn (Fig. 3.1) | Cu–Zn (Fig. 3.6b) |
| Melting point | Two: Cu, Ni | Two: Pb, Sn | Two: Cu, Zn |
| Three-phase horizontals | None | One: L + (Sn) + (Pb) | Six: $\alpha + \beta + L$<br>$\beta + \gamma + L$<br>$\gamma + \delta + L$<br>$\gamma + \delta + \varepsilon$<br>$\delta + \varepsilon + L$<br>$\varepsilon + \eta + L$ |
| Single-phase fields | Two: L, $\alpha$ | Three: L, (Pb), (Sn) | Seven: L, $\alpha, \beta, \gamma, \delta, \varepsilon, \eta$ |
| Two-phase fields | One: L + $\alpha$ | Three: L + (Pb), L + (Sn), (Sn) + (Pb) | Twelve: L + $\alpha$, L + $\beta$, L + $\gamma$, L + $\delta$, L + $\varepsilon$, L + $\eta$, $\alpha + \beta$, $\beta + \gamma$, $\gamma + \delta$, $\gamma + \varepsilon$, $\delta + \varepsilon$, $\varepsilon + \eta$ |

## Other phase diagrams

Phase diagrams have been measured for almost any alloy system you are likely to meet: copper–nickel, copper–zinc, gold–platinum, or even water–antifreeze. Some diagrams, like copper–nickel (Fig. 4.3-6a), are simple; others, like copper–zinc (Fig. 4.3-6b) are fairly involved; and a few are positively horrendous! But, as Table 4.3-1 shows, the only real difference between one phase diagram and the next is one of degree.

So that you know where to find the phase diagrams you need we have listed published sources of phase diagrams in Appendix 3. The determination of a typical phase diagram used to provide enough work to keep a doctoral student busy for several years. And yet the most comprehensive of the references runs to over a thousand different phase diagrams!

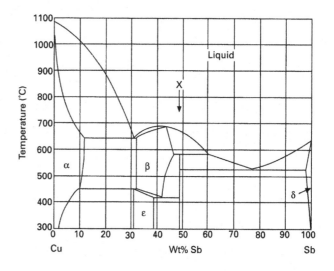

## Examples

Do not attempt examples 3.1 to 3.5 until you have worked through the Teaching Yourself Phase Diagrams section at the end of the book.

**4.3.1** Explain briefly what is meant by the following terms:

(a) a eutectic reaction;

(b) a eutectoid reaction.

**4.3.2** The phase diagram for the copper-antimony system is shown below. The phase diagram contains the intermetallic compound marked "X" on the diagram. Determine the chemical formula of this compound. The atomic weights of copper and antimony are 63.54 and 121.75 respectively.

**Answer**

$Cu_2Sb$.

**4.3.3** The copper-antimony phase diagram contains two eutectic reactions and one eutectoid reaction. For each reaction:

(a) identify the phases involved;

(b) give the compositions of the phases;

(c) give the temperature of the reaction.

**Answers**

eutectic at 650°C: L(31% Sb) = $\alpha$(12% Sb) + $\beta$(32% Sb).

eutectic at 520°C: L(77% Sb) = $Cu_2Sb$ + δ(98% Sb).
eutectoid at 420°C: β(42% Sb) = ε(38% Sb) + $Cu_2Sb$.

**3.4** A copper-antimony alloy containing 95 weight% antimony is allowed to cool from 650°C to room temperature. Describe the different phase changes which take place as the alloy is cooled and make labelled sketches of the microstructure to illustrate your answer.

**3.5** Sketch a graph of temperature against time for a copper-antimony alloy containing 95 weight% antimony over the range 650°C to 500°C and account for the shape of your plot.

# Chapter 4.4

# Physical properties of metals

## 4.4.1 Introduction

The ways in which any material interacts and responds to various forms of energy are of prime interest to scientists and, in the context of engineering, provide the essential base for design and innovation. The energy acting on a material may derive from force fields (gravitational, electric, magnetic), electromagnetic radiation (heat, light, X-rays), high-energy particles, etc. The responses of a material, generally referred to as its physical properties, are governed by the structural arrangement of atoms/ions/molecules in the material. The theme of the structure–property relation which has run through previous chapters is developed further. Special attention will be given to the diffusion of atoms/ions within materials because of the importance of thermal behavior during manufacture and service. In this brief examination, which will range from density to superconductivity, the most important physical properties of materials are considered.

## 4.4.2 Density

This property, defined as the mass per unit volume of a material, increases regularly with increasing atomic numbers in each subgroup. The reciprocal of the density is the specific volume $v$, while the product of $v$ and the relative atomic mass $W$ is known as the atomic volume $\Omega$. The density may be determined by the usual 'immersion' method, but it is instructive to show how X-rays can be used. For example, a powder photograph may give the lattice parameter of an fcc metal, say copper, as 0.36 nm. Then $1/(3.6 \times 10^{-10})^3$ or $2.14 \times 10^{28}$ cells of this size (0.36 nm) are found in a cube of 1 m edge length. The total number of atoms in 1 m³ is then $4 \times 2.14 \times 10^{28} = 8.56 \times 10^{28}$ since an fcc cell contains four atoms. Furthermore, the mass of a copper atom is 63.57 times the mass of a hydrogen atom (which is $1.63 \times 10^{-24}$ g) so that the mass of 1 m³ of copper, i.e. the density, is $8.56 \times 10^{28} \times 63.57 \times 1.63 \times 10^{-24} = 8900$ kg m$^{-3}$.

On alloying, the density of a metal changes. This is because the mass of the solute atom differs from that of the solvent, and also because the lattice parameter usually changes on alloying. The parameter change may often be deduced from Vegard's law, which assumes that the lattice parameter of a solid solution varies linearly with atomic concentration, but numerous deviations from this ideal behavior do exist.

The density clearly depends on the mass of the atoms, their size and the way they are packed. Metals are dense because they have heavy atoms and close packing; ceramics have lower densities than metals because they contain light atoms, either C, N or O; polymers have low densities because they consist of light atoms in chains. Figure 4.4-1 shows the spread in density values for the different material classes. Such 'material property charts', as developed by Ashby, are useful when selecting materials during engineering design.

## 4.4.3 Thermal properties

### 4.4.3.1 Thermal expansion

If we consider a crystal at absolute zero temperature, the ions sit in a potential well of depth $E_{r0}$ below the energy of a free atom (Figure 4.4-3). The effect of raising the

# CHAPTER 4.4 Physical properties of metals

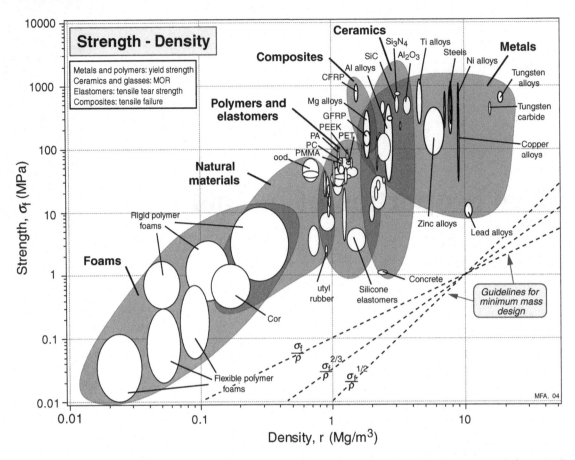

**Figure 4.4-1** Strength σ, plotted against density, ρ (yield strength for metals and polymers, compressive strength for ceramics, tear strength for elastomers and tensile strength for composites). The guide lines of constant $\sigma/\rho$, $\sigma^{2/3}/\rho$ and $\sigma^{1/2}/\rho$ are used in minimum weight, yield-limited, design (after Ashby, 1992).

temperature of the crystal is to cause the ions to oscillate in this asymmetrical potential well about their mean positions. As a consequence, this motion causes the energy of the system to rise, increasing with increasing amplitude of vibration. The increasing amplitude of vibration also causes an expansion of the crystal, since as a result of the sharp rise in energy below $r_0$ the ions as they vibrate to and fro do not approach much closer than the equilibrium separation, $r_0$, but separate more widely when moving apart. When the distance $r$ is such that the atoms are no longer interacting, the material is transformed to the gaseous phase, and the energy to bring this about is the energy of evaporation.

The change in dimensions with temperature is usually expressed in terms of the linear coefficient of expansion $\alpha$, given by $\alpha = (1/l)(dl/dT)$, where $l$ is the original length of the specimen and $T$ is the absolute temperature. Because of the anisotropic nature of crystals, the value of $\alpha$ usually varies with the direction of measurement and even in a particular crystallographic direction the dimensional change with temperature may not always be uniform.

Phase changes in the solid state are usually studied by *dilatometry*. The change in dimensions of a specimen can be transmitted to a sensitive dial gauge or electrical transducer by means of a fused silica rod. When a phase transformation takes place, because the new phase usually occupies a different volume to the old phase, discontinuities are observed in the coefficient of thermal expansion $\alpha$–$T$ curve. Some of the 'nuclear metals' which exist in many allotropic forms, such as uranium and plutonium, show a negative coefficient along one of the crystallographic axes in certain of their allotropic modifications.

The change in volume with temperature is important in many metallurgical operations such as casting, welding and heat treatment. Of particular importance is the volume change associated with the melting or, alternatively, the freezing phenomenon, since this is responsible for many of the defects, both of a macroscopic and microscopic size, which exist in crystals. Most metals increase their volume by about 3% on melting, although those metals which have crystal structures of lower coordination, such as bismuth, antimony or gallium, contract on melting. This volume change is quite small, and while the liquid structure is more open than the solid structure, it is clear that the liquid state resembles the solid state more closely than it does the gaseous phase. For the simple metals the latent heat of melting, which is merely the work done in separating the atoms from the close-packed structure of the solid to the more open liquid structure, is only about one-thirtieth of the latent heat of evaporation, while the electrical and thermal conductivities are reduced only to three-quarters to one-half of the solid-state values.

### Worked example

In copper, what percentage of the volume change which occurs as the specimen is heated from room temperature to its melting point is due to the increased vacancy concentration, assuming that the vacancy concentration at the melting point (1083°C) is $\sim 10^{-4}$? (Linear thermal expansion coefficient of copper $\alpha$ is $16.5 \times 10^{-6} \text{K}^{-1}$.)

### Solution

At the melting point (1083°C) vacancy concentration $= 10^{-4}$, i.e. one vacancy every $10^4$ atom sites.

On heating to melting point, $\Delta$ Volume/initial volume $= 3\alpha \Delta T$

$$= 3 \times 16.5 \times 10^{-6} \times (1083 - rT)$$
$$= 49.5 \times 10^{-6} \times 1060$$
$$= 5.25 \times 10^{-2}.$$

Fractional change due to vacancies $= \dfrac{10^{-4}}{5.25 \times 10^{-2}}$

$= 0.19 \times 10^{-2} \approx 0.2\%.$

## 4.4.3.2 Specific heat capacity

The *specific heat* is another thermal property important in the processing operations of casting or heat treatment, since it determines the amount of heat required in the process. Thus, the specific heat (denoted by $C_p$, when dealing with the specific heat at constant pressure) controls the increase in temperature, $dT$, produced by the addition of a given quantity of heat, $dQ$, to one gram of matter, so that $dQ = C_p dT$.

The specific heat of a metal is due almost entirely to the vibrational motion of the ions. However, a small part of the specific heat is due to the motion of the free electrons, which becomes important at high temperatures, especially in transition metals with electrons in incomplete shells.

The classical theory of specific heat assumes that an atom can oscillate in any one of three directions, and hence a crystal of $N$ atoms can vibrate in $3N$ independent normal modes, each with its characteristic frequency. Furthermore, the mean energy of each normal mode will be $kT$, so that the total vibrational thermal energy of the metal is $E = 3NkT$. In solid and liquid metals, the volume changes on heating are very small and, consequently, it is customary to consider the specific heat at constant volume. If $N$, the number of atoms in the crystal, is equal to the number of atoms in a gram-atom (i.e. Avogadro number), the heat capacity per gram-atom, i.e. the atomic heat, at constant volume is given by

$$C_v \left(\frac{dQ}{dT}\right)_v = \frac{dE}{dT} = 3Nk = 24.95 \text{J K}^{-1}.$$

In practice, of course, when the specific heat is experimentally determined, it is the specific heat at constant pressure, $C_p$, which is measured, not $C_v$, and this is given by

$$C_p \left(\frac{dE + PdV}{dT}\right)_p = \frac{dH}{dT},$$

where $H = E + PV$ is known as the heat content or enthalpy, $C_p$ is greater than $C_v$ by a few percent because some work is done against interatomic forces when the crystal expands, and it can be shown that

$$C_p - C_v = 9\alpha^2 VT/\beta,$$

where $\alpha$ is the coefficient of linear thermal expansion, $V$ is the volume per gram-atom and $\beta$ is the compressibility.

Dulong and Petit were the first to point out that the specific heat of most materials, when determined at sufficiently high temperatures and corrected to apply to constant volume, is approximately equal to $3R$, where **R** is the gas constant. However, deviations from the 'classical' value of the atomic heat occur at low temperatures, as shown in Figure 4.4-2a. This deviation is readily accounted for by the quantum theory, since the vibrational energy must then be quantized in multiples of $h\nu$, where **h** is Planck's constant and $\nu$ is the characteristic frequency of the normal mode of vibration.

According to the quantum theory, the mean energy of a normal mode of the crystal is

$$E(\nu) = \frac{1}{2}h\nu + \{h\nu/\exp(h\nu/kT) - 1\},$$

where $\frac{1}{2}h\nu$ represents the energy a vibrator will have at the absolute zero of temperature, i.e. the zero-point energy. Using the assumption made by Einstein (1907) that all vibrations have the same frequency (i.e. all atoms vibrate independently), the heat capacity is

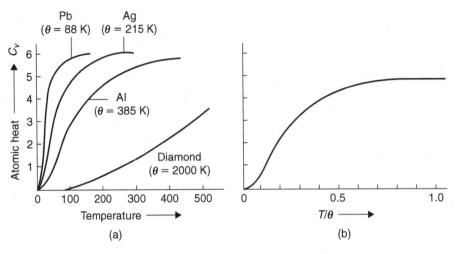

**Figure 4.4-2** The variation of atomic heat with temperature.

$$C_v = (dE/dT)_v$$
$$= 3Nk(h\nu/kT)^2[\exp(h\nu/kT)/\{\exp(h\nu/kT) - 1\}^2].$$

This equation is rarely written in such a form because most materials have different values of $\nu$. It is more usual to express $\nu$ as an equivalent temperature defined by $\Theta_E = h\nu/k$, where $\Theta_E$ is known as the Einstein characteristic temperature. Consequently, when $C_v$ is plotted against $T/\Theta_E$, the specific heat curves of all pure metals coincide and the value approaches zero at very low temperatures and rises to the classical value of $3Nk = 3R \cong 25.2 \text{ J g}^{-1}$ at high temperatures.

Einstein's formula for the specific heat is in good agreement with experiment for $T \gtrsim \Theta_E$, but is poor for low temperatures where the practical curve falls off less rapidly than that given by the Einstein relationship. However, the discrepancy can be accounted for, as shown by Debye, by taking account of the fact that the atomic vibrations are not independent of each other. This modification to the theory gives rise to a Debye characteristic temperature $\Theta_D$, which is defined by

$$k\Theta_D = h\nu_D,$$

where $\nu_D$ is Debye's maximum frequency. Figure 4.4-2b shows the atomic heat curves of Figure 4.4-2a plotted against $T/\Theta_D$; in most metals for low temperatures ($T/\Theta_D \ll 1$) a $T^3$ law is obeyed, but at high temperatures the free electrons make a contribution to the atomic heat which is proportional to $T$ and this causes a rise of C above the classical value.

### 4.4.3.3 The specific heat curve and transformations

The specific heat of a metal varies smoothly with temperature, as shown in Figure 4.4-2a, provided that no phase change occurs. On the other hand, if the metal undergoes a structural transformation the specific heat curve exhibits a discontinuity, as shown in Figure 4.4-3. If the phase change occurs at a fixed temperature, the metal undergoes what is known as a first-order transformation; for example, the α to γ, γ to δ and δ to liquid phase changes in iron shown in Figure 4.4-3a. At the transformation temperature the latent heat is absorbed without a rise in temperature, so that the specific heat ($dQ/dT$) at the transformation temperature is infinite. In some cases, known as transformations of the second order, the phase transition occurs over a range of temperature (e.g. the order–disorder transformation in alloys), and is associated with a specific heat peak of the form shown in Figure 4.4-3b. Obviously the narrower the temperature range $T_1$–$T_c$, the sharper is the specific heat peak, and in the limit when the total change occurs at a single temperature, i.e. $T_1 = T_c$, the specific heat becomes infinite and equal to the latent heat of transformation. A second-order transformation also occurs in iron (see Figure 4.4-3a), and in this case is due to a change in ferromagnetic properties with temperature.

### 4.4.3.4 Free energy of transformation

In Section 4.4.3.2 it was shown that any structural changes of a phase could be accounted for in terms of the variation of free energy with temperature. The relative

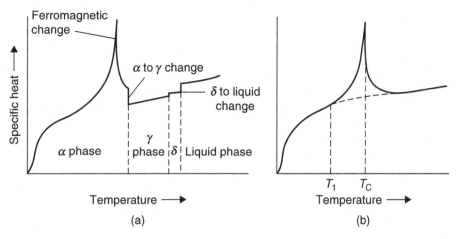

**Figure 4.4-3** The effect of solid-state transformations on the specific heat–temperature curve.

magnitude of the free energy value governs the stability of any phase, and from Figure 4.3-9a it can be seen that the free energy G at any temperature is in turn governed by two factors: (1) the value of G at 0 K, $G_0$, and (2) the slope of the G–T curve, i.e. the temperature dependence of free energy. Both of these terms are influenced by the vibrational frequency, and consequently the specific heat of the atoms, as can be shown mathematically. For example, if the temperature of the system is raised from $T$ to $T + dT$ the change in free energy of the system dG is

$$dG = dH - TdS - SdT$$
$$= C_p dT - T(C_p dT/T) - SdT$$
$$= -SdT,$$

so that the free energy of the system at a temperature $T$ is

$$G = G_0 - \int_0^T SdT.$$

At the absolute zero of temperature, the free energy $G_0$ is equal to $H_0$, and then

$$G = H_0 - \int_0^T SdT,$$

which if $S$ is replaced by $\int_0^T (C_p/T)dT$ becomes

$$G = H_0 - \int_0^T \left[ \int_0^T (C_p/T)dT \right] dT. \quad (4.4.1)$$

Equation (4.4.1) indicates that the free energy of a given phase decreases more rapidly with rise in temperature the larger its specific heat. The intersection of the free energy–temperature curves, shown in Figure 2.8a, therefore takes place because the low-temperature phase has a smaller specific heat than the higher-temperature phase.

At low temperatures the second term in equation (4.4.1) is relatively unimportant, and the phase that is stable is the one which has the lowest value of $H_0$, i.e. the most close-packed phase which is associated with a strong bonding of the atoms. However, the more strongly bound the phase, the higher is its elastic constant, the higher the vibrational frequency and consequently the smaller the specific heat (see Figure 4.4-2a). Thus, the more weakly bound structure, i.e. the phase with the higher $H_0$ at low temperature, is likely to appear as the stable phase at higher temperatures. This is because the second term in equation (4.4.1) now becomes important and G decreases more rapidly with increasing temperature, for the phase with the largest value of $\int(C_p/T)dT$. From Figure 4.4-2b it is clear that a large $\int(C_p/T)dT$ is associated with a low characteristic temperature and hence with a low vibrational frequency, such as is displayed by a metal with a more open structure and small elastic strength. In general, therefore, when phase changes occur the more close-packed structure usually exists at the low temperatures and the more open structures at the high temperatures. From this viewpoint a liquid, which possesses no long-range structure, has a higher entropy than any solid phase, so that ultimately all metals must melt at a sufficiently high temperature, i.e. when the TS term outweighs the H term in the free energy equation.

The sequence of phase changes in such metals as titanium, zirconium, etc. is in agreement with this prediction and, moreover, the alkali metals, lithium and sodium, which are normally bcc at ordinary temperatures, can be transformed to fcc at sub-zero temperatures. It is interesting to note that iron, being bcc ($\alpha$-iron) even at low temperatures and fcc ($\gamma$-iron) at high temperatures, is an exception to this rule. In this case, the stability of the bcc structure is thought to be associated with its ferromagnetic properties. By having a bcc structure the interatomic distances are of the correct value for the exchange interaction to allow the electrons to adopt parallel spins (this is a condition for magnetism). While this state is one of low entropy it is also one of minimum internal energy, and in the lower temperature ranges this is the factor which governs the phase stability, so that the bcc structure is preferred.

Iron is also of interest because the bcc structure, which is replaced by the fcc structure at temperatures above 910°C, reappears as the δ-phase above 1400°C. This behavior is attributed to the large electronic specific heat of iron, which is a characteristic feature of most transition metals. Thus, the Debye characteristic temperature of γ-iron is lower than that of α-iron and this is mainly responsible for the α to γ transformation. However, the electronic specific heat of the α-phase becomes greater than that of the γ-phase above about 300°C and eventually at higher temperatures becomes sufficient to bring about the return to the bcc structure at 1400°C.

## 4.4.4 Diffusion

### 4.4.4.1 Diffusion laws

Some knowledge of diffusion is essential in understanding the behavior of materials, particularly at elevated temperatures. A few examples include such commercially important processes as annealing, heat treatment, the age hardening of alloys, sintering, surface hardening, oxidation and creep. Apart from the specialized diffusion processes, such as grain boundary diffusion and diffusion down dislocation channels, a distinction is frequently drawn between diffusion in pure metals, homogeneous alloys and inhomogeneous alloys. In a pure material self-diffusion can be observed by using radioactive tracer atoms. In a homogeneous alloy diffusion of each component can also be measured by a tracer method, but in an inhomogeneous alloy diffusion can be determined by chemical analysis merely from the broadening of the interface between the two metals as a function of time. Inhomogeneous alloys are common in metallurgical practice (e.g. cored solid solutions) and in such cases diffusion always occurs in such a way as to produce a macroscopic flow of solute atoms down the concentration gradient. Thus, if a bar of an alloy, along which there is a concentration gradient (Figure 4.4-4), is heated for a few hours at a temperature where atomic migration is fast, i.e. near the melting point, the solute atoms are redistributed until the bar becomes uniform in

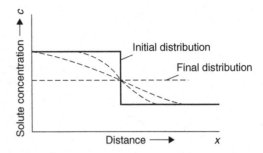

**Figure 4.4-4** Effect of diffusion on the distribution of solute in an alloy.

composition. This occurs even though the individual atomic movements are random, simply because there are more solute atoms to move down the concentration gradient than there are to move up. This fact forms the basis of Fick's law of diffusion, which is

$$dn/dt = -Ddc/dx. \qquad (4.4.2)$$

Here the number of atoms diffusing in unit time across unit area through a unit concentration gradient is known as the diffusivity or diffusion coefficient,[1] $D$. It is usually expressed as units of $cm^2\,s^{-1}$ or $m^2\,s^{-1}$ and depends on the concentration and temperature of the alloy.

To illustrate, we may consider the flow of atoms in one direction $x$, by taking two atomic planes A and B of unit area separated by a distance $b$, as shown in Figure 4.4-5. If $c_1$ and $c_2$ are the concentrations of diffusing atoms in these two planes ($c_1 > c_2$), the corresponding number of such atoms in the respective planes is $n_1 = c_1 b$ and $n_2 = c_2 b$. If the probability that any one jump in the $+x$ direction is $p_x$, then the number of jumps per unit time made by one atom is $p_x v$, where $v$ is the mean frequency with which an atom leaves a site irrespective of directions. The number of diffusing atoms leaving A and arriving at B in unit time is $(p_x v c_1 b)$ and the number making the reverse transition is $(p_x v c_2 b)$ so that the net gain of atoms at B is

$$p_x v b(c_1 - c_2) = J_x,$$

with $J_x$ the flux of diffusing atoms. Setting $c_1 - c_2 = -b(dc/dx)$, this flux becomes:

$$J_x = -p_x v_v b^2 (dc/dx) = -\tfrac{1}{2} v b^2 (dc/dx) \qquad (4.4.3)$$
$$= -D(dc/dx).$$

In cubic lattices, diffusion is isotropic and hence all six orthogonal directions are equally likely, so that $p_x = \tfrac{1}{6}$. For simple cubic structures $b = a$ and thus

---

[1] The conduction of heat in a still medium also follows the same laws as diffusion.

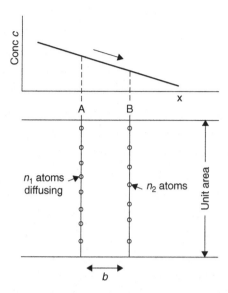

**Figure 4.4-5** Diffusion of atoms down a concentration gradient.

$$D_x = D_y = D_z = \frac{1}{6}va^2 = D, \quad (4.4.4)$$

whereas in fcc structures $b = a/\sqrt{2}$ and $D = \frac{1}{12}va^2$, and in bcc structures $D = \frac{1}{24}va^2$.

Fick's first law only applies if a steady state exists in which the concentration at every point is invariant, i.e. $(dc/dt) = 0$ for all $x$. To deal with non-stationary flow in which the concentration at a point changes with time, we take two planes A and B, as before, separated by unit distance and consider the rate of increase of the number of atoms $(dc/dt)$ in a unit volume of the specimen; this is equal to the difference between the flux into and that out of the volume element. The flux across one plane is $J_x$ and across the other $J_x + dJ_x$, the difference being $-dJ_x$. We thus obtain Fick's second law of diffusion:

$$\frac{dc}{dt} = \frac{dJ_x}{dx} = \frac{d}{dx}\left(D_x \frac{dc}{dx}\right). \quad (4.4.5)$$

When $D$ is independent of concentration this reduces to

$$\frac{dc}{dt} = D_x \frac{d^2c}{dx^2} \quad (4.4.6)$$

and in three dimensions becomes

$$\frac{dc}{dt} = \frac{d}{dx}\left(D_x \frac{dc}{dx}\right) + \frac{d}{dy}\left(D_y \frac{dc}{dy}\right) + \frac{d}{dz}\left(D_z \frac{dc}{dZ}\right).$$

An illustration of the use of the diffusion equations is the behavior of a diffusion couple, where there is a sharp interface between pure metal and an alloy. Figure 4.4-4 can be used for this example and as the solute moves from alloy to the pure metal, the way in which the concentration varies is shown by the dotted lines. The solution to Fick's second law is given by

$$c = \frac{c_0}{2}\left[1 - \frac{2}{\sqrt{\pi}}\int_0^{x/[2\sqrt{(Dt)}]} \exp(-y^2)dy\right], \quad (4.4.7)$$

where $c_0$ is the initial solute concentration in the alloy and $c$ is the concentration at a time $t$ at a distance $x$ from the interface. The integral term is known as the Gauss error function (erf $(y)$) and as $y \to \infty$, erf $(y) \to 1$. It will be noted that at the interface where $x = 0$, then $c = c_0/2$, and in those regions where the curvature $\partial^2c/\partial x^2$ is positive the concentration rises, in those regions where the curvature is negative the concentration falls, and where the curvature is zero the concentration remains constant.

This particular example is important because it can be used to model the depth of diffusion after time $t$, e.g. in the case hardening of steel, providing the concentration profile of the carbon after a carburizing time $t$, or dopant in silicon. Starting with a constant composition at the surface, the value of $x$ where the concentration falls to half the initial value, i.e. $1 - \text{erf}(y) = \frac{1}{2}$, is given by $x = \sqrt{(Dt)}$. Thus, knowing $D$ at a given temperature, the time to produce a given depth of diffusion can be estimated.

The diffusion equations developed above can also be transformed to apply to particular diffusion geometries. If the concentration gradient has spherical symmetry about a point, c varies with the radial distance $r$ and, for constant $D$,

$$\frac{dc}{dt} = D\left(\frac{d^2c}{dr^2} + \frac{2}{r}\frac{dc}{dr}\right). \quad (4.4.8)$$

When the diffusion field has radial symmetry about a cylindrical axis, the equation becomes

$$\frac{dc}{dt} = D\left(\frac{d^2c}{dr^2} + \frac{1}{r}\frac{dc}{dr}\right) \quad (4.4.9)$$

and the steady-state condition $(dc/dt) = 0$ is given by

$$\frac{d^2c}{dr^2} + \frac{1}{r}\frac{dc}{dr} = 0, \quad (4.4.10)$$

which has a solution $c = A \ln(r) + B$. The constants $A$ and $B$ may be found by introducing the appropriate boundary conditions and for $c = c_0$ at $r = r_0$ and $c = c_1$ at $r = r_1$ the solution becomes

$$c = \frac{c_0 \ln(r_1/r) + c_1 \ln(r/r_0)}{\ln(r_1/r_0)}.$$

The flux through any shell of radius $r$ is $-2\pi rD(dc/dr)$ or

$$J = -\frac{2\pi D}{\ln(r_1/r_0)}(c_1 - c_0). \qquad (4.4.11)$$

Diffusion equations are of importance in many diverse problems and in Chapter 3 are applied to the diffusion of vacancies from dislocation loops and the sintering of voids.

### 4.4.4.2 Mechanisms of diffusion

The transport of atoms through the lattice may conceivably occur in many ways. The term 'interstitial diffusion' describes the situation when the moving atom does not lie on the crystal lattice, but instead occupies an interstitial position. Such a process is likely in interstitial alloys where the migrating atom is very small (e.g. carbon, nitrogen or hydrogen in iron). In this case, the diffusion process for the atoms to move from one interstitial position to the next in a perfect lattice is not defect-controlled. A possible variant of this type of diffusion has been suggested for substitutional solutions in which the diffusing atoms are only temporarily interstitial and are in dynamic equilibrium with others in substitutional positions. However, the energy to form such an interstitial is many times that to produce a vacancy and, consequently, the most likely mechanism is that of the continual migration of vacancies. With vacancy diffusion, the probability that an atom may jump to the next site will depend on: (1) the probability that the site is vacant (which in turn is proportional to the fraction of vacancies in the crystal) and (2) the probability that it has the required activation energy to make the transition. For self-diffusion, where no complications exist, the diffusion coefficient is therefore given by

$$D = \frac{1}{6}a^2 f\nu \exp[(S_f + S_m)/k]$$
$$\times \exp[-E_f/kT]\exp[-E_m/kT]$$
$$= D_0 \exp[-(E_f + E_m)/kT].$$

The factor $f$ appearing in $D_0$ is known as a correlation factor and arises from the fact that any particular diffusion jump is influenced by the direction of the previous jump. Thus, when an atom and a vacancy exchange places in the lattice there is a greater probability of the atom returning to its original site than moving to another site, because of the presence there of a vacancy; $f$ is 0.80 and 0.78 for fcc and bcc lattices, respectively. Values for $E_f$ and $E_m$ are discussed in Chapter 3: $E_f$ is the energy of formation of a vacancy, $E_m$ the energy of migration, and the sum of the two energies, $Q = E_f + E_m$, is the activation energy for self-diffusion[2] $E_d$.

In alloys, the problem is not so simple and it is found that the self-diffusion energy is smaller than in pure metals. This observation has led to the suggestion that in alloys the vacancies associate preferentially with solute atoms in solution; the binding of vacancies to the impurity atoms increases the effective vacancy concentration near those atoms so that the mean jump rate of the solute atoms is much increased. This association helps the solute atom on its way through the lattice but, conversely, the speed of vacancy migration is reduced because it lingers in the neighborhood of the solute atoms, as shown in Figure 4.4-6. The phenomenon of association is of fundamental importance in all kinetic studies, since the mobility of a vacancy through the lattice to a vacancy sink will be governed by its ability to escape from the impurity atoms which trap it. This problem has been mentioned in Chapter 3.

When considering diffusion in alloys it is important to realize that in a binary solution of A and B the diffusion coefficients $D_A$ and $D_B$ are generally not equal. This inequality of diffusion was first demonstrated by Kirkendall using an α-brass/copper couple (Figure 4.4-7). He noted that if the position of the interfaces of the couple were marked (e.g. with fine W or Mo wires), during diffusion the markers move towards each other, showing that the zinc atoms diffuse out of the alloy more rapidly than copper atoms diffuse in. This being the case, it is not surprising that several workers have shown that porosity develops in such systems on that side of the interface from which there is a net loss of atoms.

The Kirkendall effect is of considerable theoretical importance, since it confirms the vacancy mechanism of diffusion. This is because the observations cannot easily be accounted for by any other postulated mechanisms of diffusion, such as direct place exchange, i.e. where neighboring atoms merely change place with each other. The Kirkendall effect is readily explained in terms of

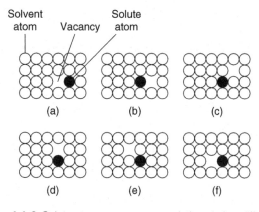

**Figure 4.4-6** Solute atom–vacancy association during diffusion.

---

[2] The entropy factor $\exp[(S_f + S_m)/k]$ is usually taken to be unity.

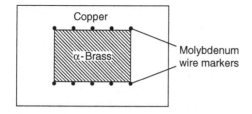

**Figure 4.4-7** α-Brass–copper couple for demonstrating the Kirkendall effect.

vacancies, since the lattice defect may interchange places more frequently with one atom than the other. The effect is also of some practical importance, especially in the fields of metal-to-metal bonding, sintering and creep.

### 4.4.4.3 Factors affecting diffusion

The two most important factors affecting the diffusion coefficient $D$ are temperature and composition. Because of the activation energy term the rate of diffusion increases with temperature according to equation (4.4.12), while each of the quantities $D$, $D_0$ and $Q$ varies with concentration; for a metal at high temperatures $Q \approx 20RT_m$, $D_0$ is $10^{-5}$ to $10^{-3}$ m$^2$ s$^{-1}$, and $D \cong 10^{-12}$ m$^2$ s$^{-1}$. Because of this variation of diffusion coefficient with concentration, the most reliable investigations into the effect of other variables necessarily concern self-diffusion in pure metals.

Diffusion is a structure-sensitive property and therefore $D$ is expected to increase with increasing lattice irregularity. In general, this is found experimentally. In metals quenched from a high temperature the excess vacancy concentration $\approx 10^9$ leads to enhanced diffusion at low temperatures, since $D = D_0 c_v \exp(-E_m/kT)$ where $c_v$ is the vacancy concentration. Grain boundaries and dislocations are particularly important in this respect and produce enhanced diffusion. Diffusion is faster in the cold-worked state than in the annealed state, although recrystallization may take place and tend to mask the effect. The enhanced transport of material along dislocation channels has been demonstrated in aluminum, where voids connected to a free surface by dislocations anneal out at appreciably higher rates than isolated voids. Measurements show that surface and grain boundary forms of diffusion also obey Arrhenius equations, with lower activation energies than for volume diffusion, i.e. $Q_{vol} \geq 2Q_{g \cdot b} \; 2Q_{surface}$ This behavior is understandable in view of the progressively more open atomic structure found at grain boundaries and external surfaces. It will be remembered, however, that the relative importance of the various forms of diffusion does not entirely depend on the relative activation energy or diffusion coefficient values. The amount of material transported by any diffusion process is given by Fick's law and for a given composition gradient also depends on the effective area through which the atoms diffuse. Consequently, since the surface area (or grain boundary area) to volume ratio of any polycrystalline solid is usually very small, it is only in particular phenomena (e.g. sintering, oxidation, etc.) that grain boundaries and surfaces become important. It is also apparent that grain boundary diffusion becomes more competitive the finer the grain and the lower the temperature. The lattice feature follows from the lower activation energy, which makes it less sensitive to temperature change. As the temperature is lowered, the diffusion rate along grain boundaries (and also surfaces) decreases less rapidly than the diffusion rate through the lattice. The importance of grain boundary diffusion and dislocation pipe diffusion is discussed again in Chapter 6 in relation to deformation at elevated temperatures, and is demonstrated convincingly in the deformation maps (see Figure 6.67), where the creep field is extended to lower temperatures when grain boundary (Coble creep) rather than lattice diffusion (Herring–Nabarro creep) operates.

Because of the strong binding between atoms, pressure has little or no effect, but it is observed that with extremely high pressure on soft metals (e.g. sodium) an increase in $Q$ may result. The rate of diffusion also increases with decreasing density of atomic packing. For example, self-diffusion is slower in fcc iron or thallium than in bcc iron or thallium when the results are compared by extrapolation to the transformation temperature. This is further emphasized by the anisotropic nature of $D$ in metals of open structure. Bismuth (rhombohedral) is an example of a metal in which $D$ varies by $10^6$ for different directions in the lattice; in cubic crystals $D$ is isotropic.

### 4.4.5 Anelasticity and internal friction

For an elastic solid it is generally assumed that stress and strain are directly proportional to one another, but in practice the elastic strain is usually dependent on time as well as stress, so that the strain lags behind the stress; this is an anelastic effect. On applying a stress at a level below the conventional elastic limit, a specimen will show an initial elastic strain $\varepsilon_e$ followed by a gradual increase in strain until it reaches an essentially constant value, $\varepsilon_e + \varepsilon_{an}$, as shown in Figure 4.4-8. When the stress is removed the strain will decrease, but a small amount remains, which decreases slowly with time. At any time $t$ the decreasing anelastic strain is given by the relation $\varepsilon = \varepsilon_{an} \exp(-t/\tau)$, where $\tau$ is known as the relaxation time, and is the time taken for the anelastic strain to decrease to $1/e \cong 36.79\%$ of its initial value. Clearly, if $\tau$ is large, the

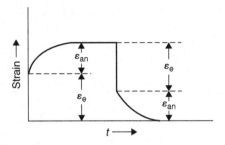

**Figure 4.4-8** Anelastic behavior.

strain relaxes very slowly, while if small the strain relaxes quickly.

In materials under cyclic loading this anelastic effect leads to a decay in amplitude of vibration and therefore a dissipation of energy by internal friction. Internal friction is defined in several different but related ways. Perhaps the most common uses the logarithmic decrement $\delta = \ln(A_n/A_{n+1})$, the natural logarithm of successive amplitudes of vibration. In a forced vibration experiment near a resonance, the factor $(\omega_2 - \omega_1)/\omega_0$ is often used, where $\omega_1$ and $\omega_2$ are the frequencies on the two sides of the resonant frequency $\omega_0$ at which the amplitude of oscillation is $1/\sqrt{2}$ of the resonant amplitude. Also used is the specific damping capacity $\Delta E/E$, where $\Delta E$ is the energy dissipated per cycle of vibrational energy $E$, i.e. the area contained in a stress–strain loop. Yet another method uses the phase angle $\alpha$ by which the strain lags behind the stress, and if the damping is small it can be shown that

$$\tan \alpha = \frac{\delta}{\pi} = \frac{1}{2\pi}\frac{\Delta E}{E} = \frac{\omega_2 - \omega_1}{\omega_0} = Q^{-1}. \quad (4.4.12)$$

By analogy with damping in electrical systems, $\tan \alpha$ is often written equal to $Q^{-1}$.

There are many causes of internal friction arising from the fact that the migration of atoms, lattice defects and thermal energy are all time-dependent processes. The latter gives rise to thermoelasticity and occurs when an elastic stress is applied to a specimen too fast for the specimen to exchange heat with its surroundings and so cools slightly. As the sample warms back to the surrounding temperature it expands thermally, and hence the dilatation strain continues to increase after the stress has become constant.

The diffusion of atoms can also give rise to anelastic effects in an analogous way to the diffusion of thermal energy giving thermoelastic effects. A particular example is the stress-induced diffusion of carbon or nitrogen in iron. A carbon atom occupies the interstitial site along one of the cell edges, slightly distorting the lattice tetragonally. Thus, when iron is stretched by a mechanical stress, the crystal axis oriented in the direction of the stress develops favored sites for the occupation of the interstitial atoms relative to the other two axes. Then if the stress is oscillated, such that first one axis and then another is stretched, the carbon atoms will want to jump from one favored site to the other. Mechanical work is therefore done repeatedly, dissipating the vibrational energy and damping out the mechanical oscillations. The maximum energy is dissipated when the time per cycle is of the same order as the time required for the diffusional jump of the carbon atom.

The simplest and most convenient way of studying this form of internal friction is by means of a Kê torsion pendulum, shown schematically in Figure 4.4-9. The specimen can be oscillated at a given frequency by adjusting the moment of inertia of the torsion bar. The energy loss per cycle $\Delta E/E$ varies smoothly with the frequency according to the relation:

$$\frac{\Delta E}{E} = 2\left(\frac{\Delta E}{E}\right)_{max}\left[\frac{\omega\tau}{1+(\omega\tau)^2}\right]$$

and has a maximum value when the angular frequency of the pendulum equals the relaxation time of the process; at low temperatures around room temperature this is interstitial diffusion. In practice, it is difficult to vary the angular frequency over a wide range and thus it is easier to keep $\omega$ constant and vary the relaxation time. Since the migration of atoms depends strongly on temperature according to an Arrhenius-type equation, the relaxation time $\tau_1 = 1/\omega_1$ and the peak occurs at a temperature $T_1$. For a different frequency value $\omega_2$ the peak occurs at a different temperature $T_2$, and so on (see Figure 4.4-10). It is thus possible to ascribe an activation energy $\Delta H$ for the internal process producing the damping by plotting ln $\tau$ versus $1/T$, or from the relation:

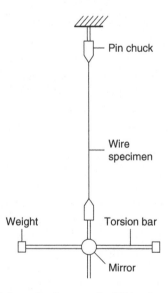

**Figure 4.4-9** Schematic diagram of a Kê torsion pendulum.

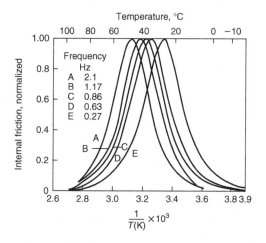

**Figure 4.4-10** Internal friction as a function of temperature for Fe with C in solid solution at five different pendulum frequencies (from Wert and Zener, 1949; by permission of the American Physical Society).

$$\Delta H = R \frac{\ln(\omega_2/\omega_1)}{1/T_1 - 1/T_2}.$$

In the case of iron the activation energy is found to coincide with that for the diffusion of carbon in iron. Similar studies have been made for other metals. In addition, if the relaxation time is $\tau$ the mean time an atom stays in an interstitial position is $(\frac{3}{2})\tau$, and from the relation $D = \frac{1}{24}a^2\nu$ for bcc lattices derived previously the diffusion coefficient may be calculated directly from

$$D = \frac{1}{36}\left(\frac{a^2}{\tau}\right).$$

Many other forms of internal friction exist in metals arising from different relaxation processes to those discussed above, and hence occurring in different frequency and temperature regions. One important source of internal friction is that due to stress relaxation across grain boundaries. The occurrence of a strong internal friction peak due to grain boundary relaxation was first demonstrated on polycrystalline aluminum at 300°C by Kê and has since been found in numerous other metals. It indicates that grain boundaries behave in a somewhat viscous manner at elevated temperatures and grain boundary sliding can be detected at very low stresses by internal friction studies. The grain boundary sliding velocity produced by a shear stress $\tau$ is given by $v = \tau d/\eta$ and its measurement gives values of the viscosity $\eta$ which extrapolate to that of the liquid at the melting point, assuming the boundary thickness to be $d \cong 0.5$ nm.

Movement of low-energy twin boundaries in crystals, domain boundaries in ferromagnetic materials, and dislocation bowing and unpinning all give rise to internal friction and damping.

## 4.4.6 Ordering in alloys

### 4.4.6.1 Long-range and short-range order

An ordered alloy may be regarded as being made up of two or more interpenetrating sublattices, each containing different arrangements of atoms. Moreover, the term 'superlattice' would imply that such a coherent atomic scheme extends over large distances, i.e. the crystal possesses long-range order. Such a perfect arrangement can exist only at low temperatures, since the entropy of an ordered structure is much lower than that of a disordered one, and with increasing temperature the degree of long-range order, $S$, decreases until at a critical temperature $T_c$ it becomes zero; the general form of the curve is shown in Figure 4.4-11. Partially ordered structures are achieved by the formation of small regions (domains) of order, each of which are separated from each other by domain or anti-phase domain boundaries, across which the order changes phase (Figure 4.4-12). However, even when long-range order is destroyed, the tendency for unlike atoms to be neighbors still exists, and short-range order results above $T_c$. The transition from complete disorder to complete order is a nucleation and growth process and may be likened to the annealing of a cold-worked structure. At high temperatures well above $T_c$, there are more than the random number of AB atom pairs, and with the lowering of temperature small nuclei of order continually form and disperse in an otherwise disordered matrix. As the temperature, and hence thermal agitation, is lowered these regions of order become more extensive, until at $T_c$ they begin to link together and the alloy consists of an interlocking mesh of small ordered regions. Below $T_c$ these domains absorb each other (cf. grain growth) as a result of anti-phase

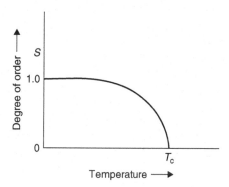

**Figure 4.4-11** Influence of temperature on the degree of order.

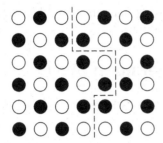

**Figure 4.4-12** An anti-phase domain boundary.

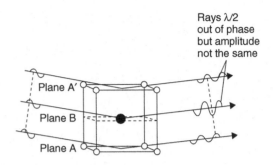

**Figure 4.4-13** Formation of a weak 1 0 0 reflection from an ordered lattice by the interference of diffracted rays of unequal amplitude.

domain boundary mobility until long-range order is established.

Some order–disorder alloys can be retained in a state of disorder by quenching to room temperature, while in others (e.g. β-brass) the ordering process occurs almost instantaneously. Clearly, changes in the degree of order will depend on atomic migration, so that the rate of approach to the equilibrium configuration will be governed by an exponential factor of the usual form, i.e. Rate$=Ae^{-Q/RT}$. However, Bragg has pointed out that the ease with which interlocking domains can absorb each other to develop a scheme of long-range order will also depend on the number of possible ordered schemes the alloy possesses. Thus, in β-brass only two different schemes of order are possible, while in fcc lattices such as $Cu_3Au$ four different schemes are possible and the approach to complete order is less rapid.

### 4.4.6.2 Detection of ordering

The determination of an ordered superlattice is usually done by means of the X-ray powder technique. In a disordered solution every plane of atoms is statistically identical and, as discussed in Chapter 4, there are reflections missing in the powder pattern of the material. In an ordered lattice, on the other hand, alternate planes become A-rich and B-rich, respectively, so that these 'absent' reflections are no longer missing but appear as extra superlattice lines. This can be seen from Figure 4.4-13: while the diffracted rays from the A planes are completely out of phase with those from the B planes their intensities are not identical, so that a weak reflection results.

Application of the structure factor equation indicates that the intensity of the superlattice lines is proportional to $|F^2| = S^2(f_A - f_B)^2$, from which it can be seen that in the fully disordered alloy, where $S = 0$, the superlattice lines must vanish. In some alloys, such as copper–gold, the scattering factor difference $(f_A - f_B)$ is appreciable and the superlattice lines are therefore quite intense and easily detectable. In other alloys, however, such as iron–cobalt, nickel–manganese and copper–zinc, the term $(f_A - f_{B,})$ is negligible for X-rays and the superlattice lines are very weak; in copper–zinc, for example, the ratio of the intensity of the superlattice lines to that of the main lines is only about 1:3500. In some cases special X-ray techniques can enhance this intensity ratio; one method is to use an X-ray wavelength near to the absorption edge when an anomalous depression of the $f$-factor occurs which is greater for one element than for the other. As a result, the difference between $f_A$ and $f_B$ is increased. A more general technique, however, is to use neutron diffraction, since the scattering factors for neighboring elements in the Periodic Table can be substantially different. Conversely, as Table 4.4-1 indicates, neutron diffraction is unable to show the existence of superlattice lines in $Cu_3Au$, because the scattering amplitudes of copper and gold for neutrons are approximately the same, although X-rays show them up quite clearly.

Sharp superlattice lines are observed as long as order persists over lattice regions of about $10^{-3}$ mm, large enough to give coherent X-ray reflections. When long-range order is not complete the superlattice lines become broadened, and an estimate of the domain size can be obtained from a measurement of the line breadth, as discussed in Chapter 4. Figure 4.4-14 shows variation of order S and domain size as determined from the intensity and breadth of powder diffraction lines. The domain sizes determined from the Scherrer line-broadening formula are in very good agreement with those observed by TEM. Short-range order is much more difficult to detect but nowadays direct measuring devices allow weak X-ray intensities to be measured more accurately, and as a result considerable information on the nature of short-range order has been obtained by studying the intensity of the diffuse background between the main lattice lines.

High-resolution transmission microscopy of thin metal foils allows the structure of domains to be examined directly. The alloy CuAu is of particular interest, since it has a face-centered tetragonal structure, often referred to as CuAu 1 below 380°C, but between 380°C and the disordering temperature of 410°C it has the CuAu 11 structures shown in Figure 4.4-15. The (0 0 2) planes are again alternately gold and copper, but halfway

Table 4.4-1 Radii (nm) of electronic orbits of atoms of transition metals of first long period (after Slater, Quantum Theory of Matter).

| Element | 3d | 4s | Atomic radius in metal (nm) |
|---|---|---|---|
| Sc | 0.061 | 0.180 | 0.160 |
| Ti | 0.055 | 0.166 | 0.147 |
| V  | 0.049 | 0.152 | 0.136 |
| Cr | 0.045 | 0.141 | 0.128 |
| Mn | 0.042 | 0.131 | 0.128 |
| Fe | 0.039 | 0.122 | 0.128 |
| Co | 0.036 | 0.114 | 0.125 |
| Ni | 0.034 | 0.107 | 0.125 |
| Cu | 0.032 | 0.103 | 0.128 |

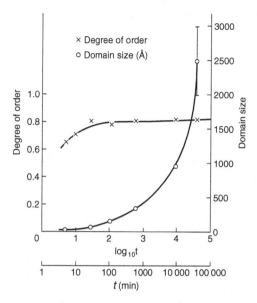

**Figure 4.4-14** Degree of order (x) and domain size (o) during isothermal annealing at 350° C after quenching from 465° C (after Morris, Besag and Smallman, 1974; courtesy of Taylor & Francis).

along the *a*-axis of the unit cell the copper atoms switch to gold planes and vice versa. The spacing between such periodic anti-phase domain boundaries is five unit cells or about 2nm, so that the domains are easily resolvable in TEM, as seen in Figure 4.4-16a. The isolated domain boundaries in the simpler superlattice structures such as CuAu 1, although not in this case periodic, can also be revealed by electron microscopy, and an example is shown in Figure 4.4-16b. Apart from static observations of these superlattice structures, annealing experiments inside the microscope also allow the effect of temperature on the structure to be examined directly. Such observations have shown that the transition from CuAu 1 to CuAu 11 takes place, as predicted, by the nucleation and growth of anti-phase domains.

## Worked example

The X-ray diffractometer data given below were obtained from a partially ordered 75 at.% Cu/25 at.% Au alloy (*a*-spacing = 0.3743 nm), using CuK$_\alpha$ radiation ($\lambda_{\text{average}}$ = 0.15418 nm). Using this and the other information provided, calculate the ordering parameter $S$ for this alloy.

| Diffraction Peak | Integrated intensity (counts × 10³) |
|---|---|
| {1 0 0} | 715 |
| {2 0 0} | 1660 |

| $\frac{\sin\theta}{\lambda}$ (nm$^{-1}$) | 0.0 | 1.0 | 2.0 | 3.0 | 4.0 |
|---|---|---|---|---|---|
| $f_{Cu}$ | 29 | 25.9 | 21.6 | 17.9 | 15.2 |
| $f_{Au}$ | 79 | 73.6 | 65.0 | 57.0 | 49.7 |

Lorentz polarization factor = $\dfrac{1+\cos^2 2\theta}{\sin^2\theta \cos\theta}$.

## Solution

Bragg's law: $\lambda = 2d \sin\theta$, $\lambda = 0.15418$ nm

$$\frac{\sin\theta}{\lambda} = \frac{1}{2d_{100}} = \frac{1}{2 \times 0.3743} = 1.335,$$
so $\theta_{100} = 11.886°$.

$$\frac{\sin\theta}{\lambda} = \frac{1}{2d_{200}} = \frac{1}{0.3743} = 2.672,$$
so $\theta_{200} = 24.325°$

(LPF)$_{100}$ = 44.28; (LPF)$_{200}$ = 9.29.
Plot $f_{Cu}$ and $f_{Au}$ vs $\frac{\sin\theta}{\lambda}$ and read off appropriate values. This gives:
$f_{Cu}$ = 24.6 for (1 0 0)
$f_{Cu}$ = 19.4 for (2 0 0)
$f_{Au}$ = 71.0 for (1 0 0)
$f_{Au}$ = 60.2 for (2 0 0)
$F_{100} = 60S(f_{Au} - f_{Cu})$ and $F_{200} = f_{Au} + 3f_{Cu}$.

$S$ (degree of order) = $\frac{P_A - C_A}{1 - C_A}$, where $P_A$ = probability of A sites filled by A atoms, and $C_A$ = atom fraction of A atoms.

$$\frac{I_{100}}{I_{200}} = \frac{S^2(f_{Au} - f_{Cu})^2 \times (\text{LPF})_{100}}{(f_{Au} + 3f_{Cu})^2 \times (\text{LPF})_{200}}.$$

Substitute in expression:

$$\frac{715}{1660} = \frac{S^2 \times (46.4)^2 \times 44.28}{(118.4)^2 \times 9.29},$$

so $S = 0.77$.

**Figure 4.4-15** One unit cell of the orthorhombic superlattice of CuAu, i.e. CuAu 11 (from Pashley and Presland, 1958–9; courtesy of the Institute of Materials, Minerals and Mining).

### 4.4.6.3 Influence of ordering on properties

#### 4.4.6.3.1 Specific heat

The order–disorder transformation has a marked effect on the specific heat, since energy is necessary to change atoms from one configuration to another. However, because the change in lattice arrangement takes place over a range of temperature, the specific heat–temperature curve will be of the form shown in Figure 4.4-3b. In practice the excess specific heat, above that given by Dulong and Petit's law, does not fall sharply to zero at $T_c$ owing to the existence of short-range order, which also requires extra energy to destroy it as the temperature is increased above $T_c$.

#### 4.4.6.3.2 Electrical resistivity

As discussed in Chapter 3, any form of disorder in a metallic structure (e.g. impurities, dislocations or point defects) will make a large contribution to the electrical resistance. Accordingly, superlattices below $T_c$ have a low electrical resistance, but on raising the temperature the resistivity increases, as shown in Figure 4.4.17a for ordered $Cu_3Au$. The influence of order on resistivity is further demonstrated by the measurement of resistivity as a function of composition in the copper–gold alloy system. As shown in Figure 4.4.17b, at composition near $Cu_3Au$ and CuAu, where ordering is most complete, the resistivity is extremely low, while away from these stoichiometric compositions the resistivity increases; the quenched (disordered) alloys given by the dotted curve also have high resistivity values.

#### 4.4.6.3.3 Mechanical properties

The mechanical properties are altered when ordering occurs. The change in yield stress is not directly related to the degree of ordering, however, and in fact $Cu_3Au$ crystals have a lower yield stress when well ordered than when only partially ordered. Experiments show that such effects can be accounted for if the maximum strength as a result of ordering is associated with critical domain size. In the alloy $Cu_3Au$, the maximum yield strength is exhibited by quenched samples after an annealing treatment of 5 min at 350°C, which gives a domain size of 6 nm (see Figure 4.4-14). However, if the alloy is well ordered and the domain size larger, the hardening is insignificant. In some alloys such as CuAu or CuPt, ordering produces a change of crystal structure and the resultant lattice strains can also lead to hardening. Thermal agitation is the most common means of destroying long-range order, but other methods (e.g. deformation) are equally effective. Figure 4.4.17c shows that cold work has a negligible effect upon the resistivity of the quenched (disordered) alloy but considerable influence on the well-annealed (ordered) alloy. Irradiation by neutrons or electrons also markedly affects the ordering (see Chapter 3).

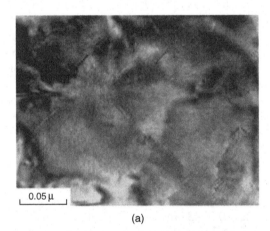

(a)

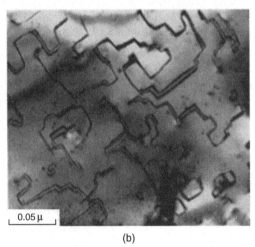

(b)

**Figure 4.4-16** Electron micrographs of (a) CuAu 11 and (b) CuAu 1 (from Pashley and Presland, 1958–9; courtesy of the Institute of Materials, Minerals and Mining).

# Physical properties of metals — CHAPTER 4.4

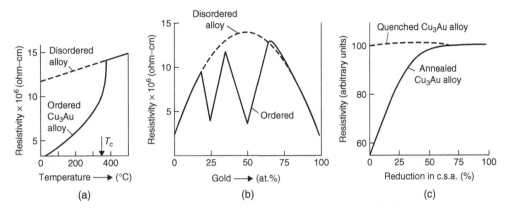

**Figure 4.4-17** Effect of temperature (a), composition (b), and deformation (c) on the resistivity of copper–gold alloys (after Barrett, 1952; courtesy of McGraw-Hill).

#### 4.4.6.3.4 Magnetic properties

The order–disorder phenomenon is of considerable importance in the application of magnetic materials. The kind and degree of order affects the magnetic hardness, since small ordered regions in an otherwise disordered lattice induce strains which affect the mobility of magnetic domain boundaries (see Section 4.4.8.4).

### 4.4.7 Electrical properties

#### 4.4.7.1 Electrical conductivity

One of the most important electronic properties of metals is the electrical conductivity, $\kappa$, and the reciprocal of the conductivity (known as the resistivity, $\rho$) is defined by the relation $R = \rho l / A$, where $R$ is the resistance of the specimen, $l$ is the length and $A$ is the cross-sectional area.

A characteristic feature of a metal is its high electrical conductivity, which arises from the ease with which the electrons can migrate through the lattice. The high thermal conduction of metals also has a similar explanation, and the Wiedmann–Franz law shows that the ratio of the electrical and thermal conductivities is nearly the same for all metals at the same temperature.

Since conductivity arises from the motion of conduction electrons through the lattice, resistance must be caused by the scattering of electron waves by any kind of irregularity in the lattice arrangement. Irregularities can arise from any one of several sources, such as temperature, alloying, deformation or nuclear irradiation, since all will disturb, to some extent, the periodicity of the lattice. The effect of temperature is particularly important and, as shown in Figure 4.4.18, the resistance increases linearly with temperature above about 100 K up to the melting point. On melting, the resistance increases markedly because of the exceptional disorder of the liquid state. However, for some metals such as bismuth, the resistance actually decreases, owing to the fact that the special zone structure which makes bismuth a poor conductor in the solid state is destroyed on melting.

In most metals the resistance approaches zero at absolute zero, but in some (e.g. lead, tin and mercury) the resistance suddenly drops to zero at some finite critical temperature above 0 K. Such metals are called superconductors. The critical temperature is different for each metal but is always close to absolute zero; the highest critical temperature known for an element is 8 K for niobium. Superconductivity is now observed at much higher temperatures in some intermetallic compounds and in some ceramic oxides (see Section 4.4.7.5).

An explanation of electrical and magnetic properties requires a more detailed consideration of electronic structure than that briefly outlined in Chapter 1. There the concept of band structure was introduced and the

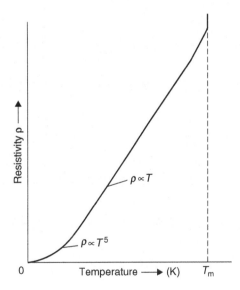

**Figure 4.4-18** Variation of resistivity with temperature.

electron can be thought of as moving continuously through the structure with an energy depending on the energy level of the band it occupies. The wave-like properties of the electron were also mentioned. For the electrons the regular array of atoms on the metallic lattice can behave as a three-dimensional diffraction grating, since the atoms are positively charged and interact with moving electrons. At certain wavelengths, governed by the spacing of the atoms on the metallic lattice, the electrons will experience strong diffraction effects, the results of which are that electrons having energies corresponding to such wavelengths will be unable to move freely through the structure. As a consequence, in the bands of electrons, certain energy levels cannot be occupied and therefore there will be energy gaps in the otherwise effectively continuous energy spectrum within a band.

The interaction of moving electrons with the metal ions distributed on a lattice depends on the wavelength of the electrons and the spacing of the ions in the direction of movement of the electrons. Since the ionic spacing will depend on the direction in the lattice, the wavelength of the electrons suffering diffraction by the ions will depend on their direction. The kinetic energy of a moving electron is a function of the wavelength according to the relationship:

$$E = \mathbf{h}^2/2m\lambda^2. \qquad (4.4.13)$$

Since we are concerned with electron energies, it is more convenient to discuss interaction effects in terms of the reciprocal of the wavelength. This quantity is called the wave number and is denoted by $k$.

In describing electron–lattice interactions it is usual to make use of a vector diagram in which the direction of the vector is the direction of motion of the moving electron and its magnitude is the wave number of the electron. The vectors representing electrons having energies which, because of diffraction effects, cannot penetrate the lattice, trace out a three-dimensional surface known as a Brillouin zone. Figure 4.4-19a shows such a zone for a face-centered cubic lattice. It is made up of plane faces which are, in fact, parallel to the most widely spaced planes in the lattice, i.e. in this case the {1 1 1} and {2 0 0} planes. This is a general feature of Brillouin zones in all lattices.

For a given direction in the lattice, it is possible to consider the form of the electron energies as a function of wave number. The relationship between the two quantities as given from equation (4.4.13) is

$$E = \mathbf{h}^2 k^2/2m, \qquad (4.4.14)$$

which leads to the parabolic relationship shown as a broken line in Figure 4.4-19b. Because of the existence of a Brillouin zone at a certain value of $k$, depending on the lattice direction, there exists a range of energy values which the electrons cannot assume. This produces a distortion in the form of the $E-k$ curve in the neighborhood of the critical value of $k$ and leads to the existence of a series of energy gaps, which cannot be occupied by electrons. The $E-k$ curve showing this effect is given as a continuous line in Figure 4.4-19b.

The existence of this distortion in the $E-k$ curve, due to a Brillouin zone, is reflected in the density of states–energy curve for the free electrons. As previously stated, the density of states–energy curve is parabolic in shape,

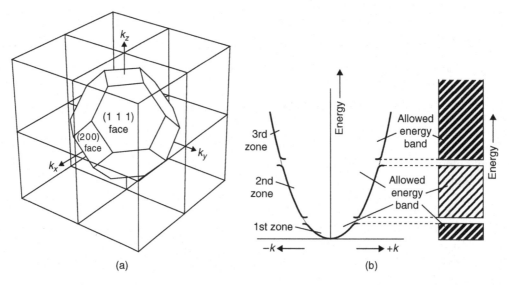

Figure 4.4-19 Schematic representation of a Brillouin zone in a metal.

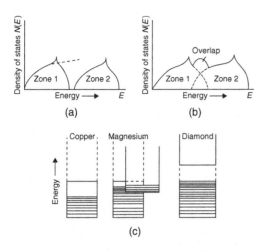

**Figure 4.4-20** Schematic representation of Brillouin zones.

but it departs from this form at energies for which Brillouin zone interactions occur. The result of such interactions is shown in Figure 4.4-20a, in which the broken line represents the $N(E)$–$E$ curve for free electrons in the absence of zone effects and the full line is the curve where a zone exists. The total number of electrons needed to fill the zone of electrons delineated by the full line in Figure 4.4-20a is $2N$, where $N$ is the total number of atoms in the metal. Thus, a Brillouin zone would be filled if the metal atoms each contributed two electrons to the band. If the metal atoms contribute more than two per atom, the excess electrons must be accommodated in the second or higher zones.

In Figure 4.4-20a the two zones are separated by an energy gap, but in real metals this is not necessarily the case, and two zones can overlap in energy in the $N(E)$–$E$ curves so that no such energy gaps appear. This overlap arises from the fact that the energy of the forbidden region varies with direction in the lattice, and often the energy level at the top of the first zone has a higher value in one direction than the lowest energy level at the bottom of the next zone in some other direction. The energy gap in the $N(E)$–$E$ curves, which represent the summation of electronic levels in all directions, is then closed (Figure 4.4-20b).

For electrical conduction to occur, it is necessary that the electrons at the top of a band should be able to increase their energy when an electric field is applied to materials so that a net flow of electrons in the direction of the applied potential, which manifests itself as an electric current, can take place. If an energy gap between two zones of the type shown in Figure 4.4-20a occurs, and if the lower zone is just filled with electrons, then it is impossible for any electrons to increase their energy by jumping into vacant levels under the influence of an applied electric field, unless the field strength is sufficiently great to supply the electrons at the top of the filled band with enough energy to jump the energy gap. Thus, metallic conduction is due to the fact that in metals the number of electrons per atom is insufficient to fill the band up to the point where an energy gap occurs. In copper, for example, the 4s valency electrons fill only one-half of the outer s-band. In other metals (e.g. Mg) the valency band overlaps a higher energy band and the electrons near the Fermi level are thus free to move into the empty states of a higher band. When the valency band is completely filled and the next higher band, separated by an energy gap, is completely empty, the material is either an insulator or a semiconductor. If the gap is several electron-volts wide, such as in diamond, where it is 7 eV, extremely high electric fields would be necessary to raise electrons to the higher band and the material is an insulator. If the gap is small enough, such as 1–2 eV as in silicon, then thermal energy may be sufficient to excite some electrons into the higher band and also create vacancies in the valency band; such a material is a semiconductor. In general, the lowest energy band which is not completely filled with electrons is called a conduction band and the band containing the valency electrons the valency band. For a conductor the valency band is also the conduction band. The electronic state of a selection of materials of different valencies is presented in Figure 4.4-20c. Although all metals are relatively good conductors of electricity, they exhibit among themselves a range of values for their resistivities. There are a number of reasons for this variability. The resistivity of a metal depends on the density of states of the most energetic electrons at the top of the band, and the shape of the $N(E)$–$E$ curve at this point.

In the transition metals, for example, apart from producing the strong magnetic properties, great strength and high melting point, the $d$-band is also responsible for the poor electrical conductivity and high electronic specific heat. When an electron is scattered by a lattice irregularity it jumps into a different quantum state, and it will be evident that the more vacant quantum states there are available in the same energy range, the more likely will be the electron to deflect at the irregularity. The high resistivities of the transition metals may therefore be explained by the ease with which electrons can be deflected into vacant $d$-states. Phonon-assisted s–d scattering gives rise to the non-linear variation of $\rho$ with temperature observed at high temperatures. The high electronic specific heat is also due to the high density of states in the unfilled $d$-band, since this gives rise to a considerable number of electrons at the top of the Fermi distribution which can be excited by thermal activation. In copper, of course, there are no unfilled levels at the top of the $d$-band into which electrons can go, and consequently both the electronic specific heat and electrical resistance are low. The conductivity also depends on the degree to which the electrons are scattered by the ions of the metal which are thermally vibrating, and by impurity atoms or other defects present in the metal.

Insulators can also be modified either by the application of high temperatures or by the addition of impurities. Clearly, insulators may become conductors at elevated temperatures if the thermal agitation is sufficient to enable electrons to jump the energy gap into the unfilled zone above.

### 4.4.7.2 Semiconductors

Some materials have an energy gap small enough to be surmounted by thermal excitation. In such intrinsic semiconductors, as they are called, the current carriers are electrons in the conduction band and holes in the valency band in equal numbers. The relative position of the two bands is as shown in Figure 4.4-21. The motion of a hole in the valency band is equivalent to the motion of an electron in the opposite direction. Alternatively, conduction may be produced by the presence of impurities, which either add a few electrons to an empty zone or remove a few from a full one. Materials which have their conductivity developed in this way are commonly known as semiconductors. Silicon and germanium containing small amounts of impurity have semiconducting properties at ambient temperatures and, as a consequence, they are frequently used in electronic transistor devices. Silicon normally has completely filled zones, but becomes conducting if some of the silicon atoms, which have four valency electrons, are replaced by phosphorus, arsenic or antimony atoms, which have five valency electrons. The extra electrons go into empty zones, and as a result silicon becomes an $n$-type semiconductor, since conduction occurs by negative carriers. On the other hand, the addition of elements of lower valency than silicon, such as aluminum, removes electrons from the filled zones, leaving behind 'holes' in the valency band structure. In this case silicon becomes $p$-type semiconductor, since the movement of electrons in one direction of the zone is accompanied by a movement of 'holes' in the other, and consequently they act as if they were positive carriers. The conductivity may be expressed as the product of (1) the number of charge carriers, $n$, (2) the charge carried by each (i.e. $e = 1.6 \times 10^{-19}$ C) and (3) the mobility of the carrier, $\mu$.

A pentavalent impurity which donates conduction electrons without producing holes in the valency band is called a donor. The spare electrons of the impurity atoms are bound in the vicinity of the impurity atoms in energy levels known as the donor levels, which are near the conduction band. If the impurity exists in an otherwise intrinsic semiconductor the number of electrons in the conduction band becomes greater than the number of holes in the valency band and, hence, the electrons are the majority carriers and the holes the minority carriers. Such a material is an $n$-type extrinsic semiconductor (see Figure 4.4-22a).

Trivalent impurities in Si or Ge show the opposite behavior, leaving an empty electron state, or hole, in the valency band. If the hole separates from the so-called acceptor atom an electron is excited from the valency band to an acceptor level $\Delta E \approx 0.01$ eV. Thus, with impurity elements such as Al, Ga or In creating holes in the valency band in addition to those created thermally, the majority carriers are holes and the semiconductor is of the $p$-type extrinsic form (see Figure 4.4-22b). For a semiconductor where both electrons and holes carry current the conductivity is given by

$$k = n_e e \mu_e + n_h e \mu_h, \qquad (4.4.15)$$

where $n_e$ and $n_h$ are, respectively, the volume concentration of electrons and holes, and $\mu_e$ and $\mu_h$ and the mobilities of the carriers, i.e. electrons and holes.

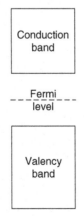

**Figure 4.4-21** Schematic diagram of an intrinsic semiconductor, showing the relative positions of the conduction and valency bands.

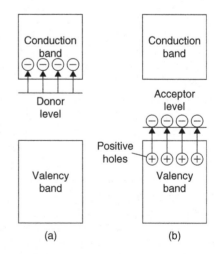

**Figure 4.4-22** Schematic energy band structure $n$-type (a) and $p$-type (b) semiconductors.

## Worked example

The conductivity of a semiconductor is given by $k = n_e e \mu_e + n_h e \mu_h$. What values of the carrier concentration $n_e$ and $n_h$ give minimum conductivity at a given temperature? Determine $n_e/n_h$ if $\mu_e/\mu_h = 3$.

## Solution

$\kappa = n_e e \mu_e + n_h e \mu_h$ and $n_e n_h = n_i^2$

$\therefore \kappa = n_e e \mu_e + n_i^2 e \mu_h / n_e$.

For minimum $\kappa, d\kappa/dn_e = 0 = e\mu_e - n_i^2 e \mu_h / n_e^2$

$\therefore n_e^2 = n_i^2 \mu_h / \mu_e$ or $n_e = n_i \sqrt{\mu_h/\mu_e}$.

$n_h = n_i^2 / n_e = n_i \sqrt{\mu_e/\mu_h}$; $n_e/n_h = \mu_h/\mu_e = 1/3$.

Semiconductor materials are extensively used in electronic devices such as the *p–n* rectifying junction, transistor (a double-junction device) and the tunnel diode. Semiconductor regions of either *p*- or *n*-type can be produced by carefully controlling the distribution and impurity content of Si or Ge single crystals, and the boundary between *p*- and *n*-type extrinsic semiconductor materials is called a *p–n* junction. Such a junction conducts a large current when the voltage is applied in one direction, but only a very small current when the voltage is reversed. The action of a *p–n* junction as a rectifier is shown schematically in Figure 4.4-23. The junction presents no barrier to the flow of minority carriers from either side, but since the concentration of minority carriers is low, it is the flow of majority carriers which must be considered. When the junction is biased in the forward direction, i.e. *n*-type made negative and the *p*-type positive, the energy barrier opposing the flow of majority carriers from both sides of the junction is reduced. Excess majority carriers enter the *p* and *n* regions, and these recombine continuously at or near the junction to allow large currents to flow. When the junction is reverse biased, the energy barrier opposing the flow of majority carriers is raised, few carriers move and little current flows.

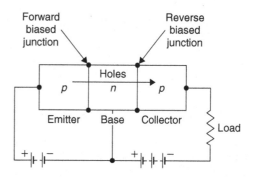

**Figure 4.4-24** Schematic diagram of a p–n–p transistor.

A transistor is essentially a single crystal with two *p–n* junctions arranged back to back to give either a *p–n–p* or *n–p–n* two-junction device. For a *p–n–p* device the main current flow is provided by the positive holes, while for an *n–p–n* device the electrons carry the current. Connections are made to the individual regions of the *p–n–p* device, designated emitter, base and collector respectively, as shown in Figure 4.4-24, and the base is made slightly negative and the collector more negative relative to the emitter. The emitter–base junction is therefore forward biased and a strong current of holes passes through the junction into the *n*-layer which, because it is thin ($10^{-2}$ mm), largely reach the collector base junction without recombining with electrons. The collector–base junction is reverse biased and the junction is no barrier to the passage of holes; the current through the second junction is thus controlled by the current through the first junction. A small increase in voltage across the emitter–base junction produces a large injection of holes into the base and a large increase in current in the collector, to give the amplifying action of the transistor.

Many varied semiconductor materials such as InSb and GaAs have been developed apart from Si and Ge. However, in all cases very high purity and crystal perfection is necessary for efficient semiconducting operations and, to produce the material, zone-refining techniques are used. Semiconductor integrated circuits are extensively used in microelectronic equipment and these are produced by vapor deposition through masks on to a single Si-slice, followed by diffusion of the deposits into the base crystal.

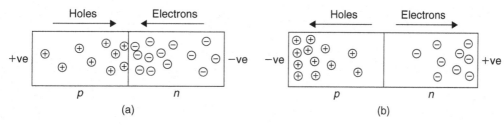

**Figure 4.4-23** Schematic illustration of *p–n* junction rectification with forward bias (a) and reverse bias (b).

Doped ceramic materials are used in the construction of *thermistors*, which are semiconductor devices with a marked dependence of electrical resistivity upon temperature. The change in resistance can be quite significant at the critical temperature. Positive temperature coefficient (PTC) thermistors are used as switching devices, operating when a control temperature is reached during a heating process. PTC thermistors are commonly based on barium titanate. Conversely, negative temperature coefficient (NTC) thermistors are based on oxide ceramics and can be used to signal a desired temperature change during cooling; the change in resistance is much more gradual and does not have the step characteristic of the PTC types.

Doped zinc oxide does not exhibit the linear voltage–current relation that one expects from Ohm's law. At low voltage, the resistivity is high and only a small current flows. When the voltage increases there is a sudden decrease in resistance, allowing a heavier current to flow. This principle is adopted in the *varistor*, a voltage-sensitive on/off switch. It is wired in parallel with high-voltage equipment and can protect it from transient voltage 'spikes' or overload.

### 4.4.7.3 Hall effect

When dealing with electrical properties it is often useful to know the type of charge carriers, their concentrations and mobility. Electrical conductivity measurements do not provide such information but they may be given by means of Hall effect measurements. The Hall effect is produced when a magnetic field is applied perpendicular to the direction of motion of a charge carrier whereby a force is exerted on the carrier particle perpendicular to both the magnetic field and the direction of the particle motion.

Electrons (or holes) moving in a particular direction, $x$, producing a current $I_x$ will be deflected in a direction $y$ by a magnetic field $B_z$. Thus, a Hall voltage $V_H$ will be established in the $y$-direction, the magnitude of which depends on $I_x$, $B_z$ and the specimen thickness $d$ according to

$$V_H = R_H I_x B_z / d,$$

where $R_H$ is the Hall coefficient for a given material. When conduction is by electrons, $R_H$ is negative and equal to $1/ne$. The mobility may be determined if the conductivity $k$ is known from the equation:

$$\mu_e = R_H k.$$

### Worked example

An $n$-type germanium sample is 2 mm wide and 0.2 mm thick. A current of 10 mA is passed longitudinally through the sample and a magnetic field of 0.1 Wb m$^{-2}$ is directed parallel to the thickness. The magnitude of the Hall voltage developed is 1.0 mV. Calculate the magnitude of the Hall constant and the concentration of electrons.

### Solution

Hall effect is to have:

(Hall electric field $E$) $\propto$ (applied magnetic field $B$) $\times$ (current density $J$) where $E$, $B$ and $J$ are mutually perpendicular.

Hall constant is defined as $R_H = \frac{E}{BJ}$.

Magnetic field $B = 0.1$ Wb m$^{-2}$

Electric field $E = \dfrac{\text{voltage}}{\text{gap}} = \dfrac{1.0 \times 10^{-3}}{0.2 \times 10^{-3}} = 5 \text{V m}^{-1}$.

Current density $J = \dfrac{\text{current}}{\text{area}}$

$$= \dfrac{10 \times 10^{-3}}{2 \times 10^{-3} \times 0.2 \times 10^{-3}}$$

$$= 2.5 \times 10^4 \text{A m}^{-2}$$

$$\therefore R_H = \dfrac{5}{0.1 \times 2.5 \times 10^4} = 2 \times 10^{-3} \text{m}^3 \text{C}^{-1}$$

Electron concentration $n = \dfrac{1}{R_H e}$

$$= \dfrac{1}{2 \times 10^{-3} \times 1.6 \times 10^{-19}}$$

$$= 3.1 \times 10^{21} \text{m}^{-3}.$$

### Worked example

A silicon crystal is doped with indium for which the electron acceptor level is 0.16 eV above the top of the valence band. The energy gap of silicon is 1.10 eV and the effective masses of electrons and holes are $0.26 m_0$ and $0.39 m_0$ respectively ($m_0 = 9.1 \times 10^{-31}$ kg is the rest mass of an electron). What impurity concentration would cause the Fermi level to coincide with the impurity level at 300 K and what fraction of the acceptor levels will be filled? What are the majority and minority carrier concentration in the crystal?

### Solution

Putting energy level at top of valence band $E_v = 0$, then bottom of conduction band $E_c = 1.10$ eV (band gap) and the required Fermi level $E_f = 0.16$ eV.

Concentration (number per m$^3$) of electrons (minority carriers) is given by:

$$n = 2\left(\frac{2\pi m_e^* kT}{h^2}\right)^{3/2} \exp\left(\frac{E_f - E_c}{kT}\right),$$

where $m_e^*$ = effective mass of electrons = $0.26 m_0$.

Concentration of holes (majority carriers) is given by:

$$p = 2\left(\frac{2\pi m_h^* kT}{h^2}\right)^{3/2} \exp\left(\frac{E_v - E_f}{kT}\right),$$

where $m_h^*$ = effective mass of holes = $0.39 m_0$.

$$\therefore n = 2 \times \left(\frac{2\pi \times 0.26 \times 9.1 \times 10^{-31} \times 1.38 \times 10^{-23} \times 300}{(6.62 \times 10^{-34})^2}\right)^{3/2}$$
$$\times \exp\left(\frac{-0.94 \times 1.6 \times 10^{-19}}{1.38 \times 10^{-23} \times 300}\right)$$
$$= 5.4 \times 10^8 \text{m}^{-3}$$

$$p = 2 \times \left(\frac{2\pi \times 0.39 \times 9.1 \times 10^{-31} \times 1.38 \times 10^{-23} \times 300}{(6.62 \times 10^{-34})^2}\right)^{3/2}$$
$$\times \exp\left(\frac{-0.16 \times 1.6 \times 10^{-19}}{1.38 \times 10^{-23} \times 300}\right)$$
$$= 1.25 \times 10^{22} \text{m}^{-3}.$$

Since $E_f$ coincides with the acceptor level, half the acceptors are ionized and therefore the total acceptor concentration = $2p = 2.5 \times 10^{22}$ m$^{-3}$.

### 4.4.7.4 Superconductivity

At low temperatures (<20 K) some metals have zero electrical resistivity and become superconductors. This superconductivity disappears if the temperature of the metal is raised above a critical temperature $T_c$, if a sufficiently strong magnetic field is applied or when a high current density flows. The critical field strength $H_c$, current density $J_c$ and temperature $T_c$ are interdependent. Figure 4.4-25 shows the dependence of $H_c$ on temperature for a number of metals; metals with high $T_c$ and $H_c$ values, which include the transition elements, are known as hard superconductors, those with low values such as Al, Zn, Cd, Hg and white-Sn are soft superconductors. The curves are roughly parabolic and approximate to the relation $H_c = H_0\left[1 - (T/T_c)^2\right]$, where $H_0$ is the critical field at 0 K; $H_0$ is about $1.6 \times 10^5 A_0$ m$^{-1}$ for Nb.

Superconductivity arises from conduction electron–electron attraction resulting from a distortion of the lattice through which the electrons are traveling; this is

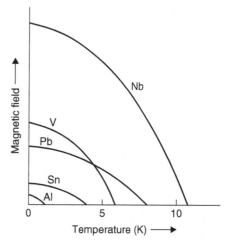

**Figure 4.4-25** Variation of critical field $H_c$ as a function of temperature for several pure metal superconductors.

clearly a weak interaction, since for most metals it is destroyed by thermal activation at very low temperatures. As the electron moves through the lattice it attracts nearby positive ions, thereby locally causing a slightly higher positive charge density. A nearby electron may in turn be attracted by the net positive charge, the magnitude of the attraction depending on the electron density, ionic charge and lattice vibrational frequencies, such that under favorable conditions the effect is slightly stronger than the electrostatic repulsion between electrons. The importance of the lattice ions in superconductivity is supported by the observation that different isotopes of the same metal (e.g. Sn and Hg) have different $T_c$ values proportional to $M^{-1/2}$, where $M$ is the atomic mass of the isotope. Since both the frequency of atomic vibrations and the velocity of elastic waves also varies as $M^{-1/2}$, the interaction between electrons and lattice vibrations (i.e. electron–phonon interaction) must be at least one cause of superconductivity.

The theory of superconductivity indicates that the electron–electron attraction is strongest between electrons in pairs, such that the resultant momentum of each pair is exactly the same and the individual electrons of each pair have opposite spin. With this particular form of ordering the total electron energy (i.e. kinetic and interaction) is lowered and effectively introduces a finite energy gap between this organized state and the usual, more excited state of motion. The gap corresponds to a thin shell at the Fermi surface, but does not produce an insulator or semiconductor, because the application of an electric field causes the whole Fermi distribution, together with gap, to drift to an unsymmetrical position, so causing a current to flow. This current remains even when the electric field is removed, since the scattering which is necessary to alter the displaced Fermi distribution is suppressed.

At 0 K all the electrons are in paired states but, as the temperature is raised, pairs are broken by thermal activation, giving rise to a number of normal electrons in equilibrium with the superconducting pairs. With increasing temperature the number of broken pairs increases until at $T_c$ they are finally eliminated together with the energy gap; the superconducting state then reverts to the normal conducting state. The superconductivity transition is a second-order transformation and a plot of $C/T$ as a function of $T^2$ deviates from the linear behavior exhibited by normal conducting metals, the electronic contribution being zero at 0 K. The main theory of superconductivity, due to the attempts of Bardeen, Cooper and Schrieffer (BCS) to relate $T_c$ to the strength of the interaction potential, the density of states at the Fermi surface and to the average frequency of lattice vibration involved in the scattering, provides some explanation for the variation of $T_c$ with the $e/a$ ratio for a wide range of alloys, as shown in Figure 4.4-26. The main effect is attributable to the change in density of states with $e/a$ ratio. Superconductivity is thus favored in compounds of polyvalent atoms with crystal structures having a high density of states at the Fermi surface. Compounds with high $T_c$ values, such as Nb$_3$Sn (18.1 K), Nb$_3$Al (17.5 K), V$_3$Si (17.0 K) and V$_3$Ga (16.8 K), all crystallize with the $\beta$-tungsten structure and have an $e/a$ ratio close to 4.7; $T_c$ is very sensitive to the degree of order and to deviation from the stoichiometric ratio, so values probably correspond to the non-stoichiometric condition.

The magnetic behavior of superconductivity is as remarkable as the corresponding electrical behavior, as shown in Figure 4.4-27 by the Meissner effect for an ideal (structurally perfect) superconductor. It is observed for a specimen placed in a magnetic field ($H < H_c$), which is then cooled down below $T_c$, that magnetic lines of force are pushed out. The specimen is a perfect diamagnetic material with zero inductance as well as zero resistance. Such a material is termed an ideal type I superconductor. An ideal type II superconductor behaves similarly at low field strengths, with $H < H_{c1} < H_c$, but then allows a gradual penetration of the field returning to the normal state when penetration is complete at $H > H_{c2} > H_c$. In detail, the field actually penetrates to a small extent in type I superconductors when it is below $H_c$ and in type II superconductors when $H$ is below $H_{c1}$, and decays away at a penetration depths $\approx$ 100–10 nm.

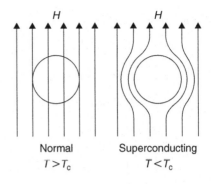

**Figure 4.4-27** The Meissner effect, shown by the expulsion of magnetic flux when the specimen becomes superconducting.

The observation of the Meissner effect in type I superconductors implies that the surface between the normal and superconducting phases has an effective positive energy. In the absence of this surface energy, the specimen would break up into separate fine regions of superconducting and normal material to reduce the work done in the expulsion of the magnetic flux. A negative surface energy exists between the normal and superconducting phases in a type II superconductor and hence the superconductor exists naturally in a state of finely separated superconducting and normal regions. By adopting a 'mixed state' of normal and superconducting regions the volume of interface is maximized, while at the same time keeping the volume of normal conduction as small as possible. The structure of the mixed state is believed to consist of lines of normal phases parallel to the applied field through which the field lines run, embedded in a superconducting matrix. The field falls off with distances from the center of each line over the characteristic distance $\lambda$, and vortices or whirlpools of supercurrents flow around each line; the flux line, together with its current vortex, is called a fluxoid. At $H_{c1}$, fluxoids appear in the specimen and increase in number as the magnetic field is raised. At $H_{c2}$, the fluxoids completely fill the cross-section of the sample

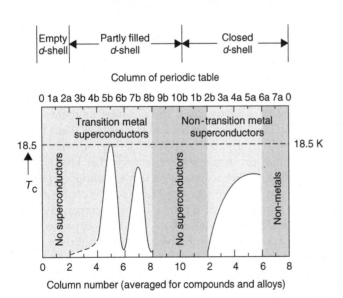

**Figure 4.4-26** The variation of $T_c$ with position in the periodic table (from Rose, Shepard and Wulff 1966; courtesy of John Wiley and Sons).

and type II superconductivity disappears. Type II superconductors are of particular interest because of their high critical fields, which makes them potentially useful for the construction of high-field electromagnetics and solenoids. To produce a magnetic field of ≈ 10 T with a conventional solenoid would cost more than 10 times that of a superconducting solenoid wound with Nb₃Sn wire. By embedding Nb wire in a bronze matrix it is possible to form channels of Nb₃Sn by interdiffusion. The conventional installation would require considerable power, cooling water and space, whereas the superconducting solenoid occupies little space, has no steady-state power consumption and uses relatively little liquid helium. It is necessary, however, for the material to carry useful currents without resistance in such high fields, which is not usually the case in annealed homogeneous type II superconductors. Fortunately, the critical current density is extremely sensitive to microstructure and is markedly increased by precipitation hardening, cold work, radiation damage, etc., because the lattice defects introduced pin the fluxoids and tend to immobilize them. Figure 4.4-28 shows the influence of metallurgical treatment on the critical current density.

### 4.4.7.5 Oxide superconductors

In 1986 a new class of 'warm' superconductors, based on mixed ceramic oxides, was discovered by J. G. Bednorz and K. A. Müller. These lanthanum–copper oxide superconductors had a $T_c$ around 35 K, well above liquid hydrogen temperature. Since then, three mixed oxide families have been developed with much higher $T_c$ values, all around 100 K. Such materials give rise to

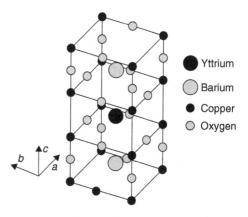

**Figure 4.4-29** Structure of 1–2–3 compound; the unit cell of the 90 K superconducting perovskite, $YBa_2Cu_3O_{7-x}$, where $x \sim 0$ (by courtesy of P. J. Hirst, Superconductivity Research Group, University of Birmingham, UK).

optimism for superconductor technology; first, in the use of liquid nitrogen rather than liquid hydrogen and, second, in the prospect of producing a room temperature superconductor.

The first oxide family was developed by mixing and heating the three oxides $Y_2O_3$, $BaO$ and $CuO$. This gives rise to the mixed oxide $YBa_2Cu_3O_{7-x}$, sometimes referred to as 1–2–3 compound or YBCO. The structure is shown in Figure 4.4-29 and is basically made by stacking three perovskite-type unit cells one above the other; the top and bottom cells have barium ions at the center and copper ions at the corners, the middle cell has yttrium at the center. Oxygen ions sit halfway along the cell edges but planes, other than those containing barium, have some missing oxygen ions (i.e. vacancies denoted by $x$ in the oxide formula). This structure therefore has planes of copper and oxygen ions containing vacancies, and copper–oxygen ion chains perpendicular to them. YBCO has a $T_c$ value of about 90 K, which is virtually unchanged when yttrium is replaced by other rare earth elements. The second family of oxides are Bi–Ca–Sr–Cu–$O_x$ materials with the metal ions in the ratio of 2:1:1:1, 2:1:2:2 or 2:2:2:3, respectively. The 2:1:1:1 oxide has only one copper–oxygen layer between the bismuth–oxygen layers, the 2:1:2:2 two and the 2:2:2:3 three, giving rise to an increasing $T_c$ up to about 105 K. The third family is based on Tl–Ca–Ba–Cu–O with a 2:2:2:3 structure having three copper–oxygen layers and a $T_c$ of about 125 K.

While these oxide superconductors have high $T_c$ values and high critical magnetic field ($H_c$) values, they unfortunately have very low values of $J_c$, the critical current density. A high $J_c$ is required if they are to be used for powerful superconducting magnets. Electrical applications are therefore unlikely until the $J_c$ value can be raised by several orders of magnitude comparable to

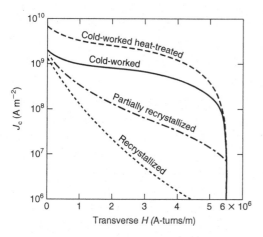

**Figure 4.4-28** The effect of processing on the $J_c$–$H$ curve of an Nb–25% Zr alloy wire which produces a fine precipitate and raises $J_c$ (from Rose, Shepard and Wulff 1966; courtesy of John Wiley and Sons).

those of conventional superconductors, i.e. $10^6$ A cm$^{-2}$. The reason for the low $J_c$ is thought to be largely due to the grain boundaries in polycrystalline materials, together with dislocations, voids and impurity particles. Single crystals show $J_c$ values around $10^5$ A cm$^{-2}$ and textured materials, produced by melt growth techniques, about $10^4$ A cm$^{-2}$, but both processes have limited commercial application. Electronic applications appear to be more promising, since it is in the area of thin (1 μm) films that high $J_c$ values have been obtained. By careful deposition control, epitaxial and single-crystal films having $J_c \geq 10^6$ A cm$^{-2}$ with low magnetic field dependence have been produced.

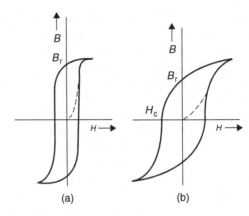

**Figure 4.4-30** B–H curves for soft (a) and hard (b) magnets.

## 4.4.8 Magnetic properties

### 4.4.8.1 Magnetic susceptibility

When a metal is placed in a magnetic field of strength $H$, the field induced in the metal is given by

$$B = H + 4\pi I, \quad (4.4.16)$$

where $I$ is the intensity of magnetization. The quantity $I$ is a characteristic property of the metal, and is related to the susceptibility per unit volume of the metal, which is defined as

$$\kappa = I/H. \quad (4.4.17)$$

The susceptibility is usually measured by a method which depends upon the fact that when a metal specimen is suspended in a non-uniform transverse magnetic field, a force proportional to $\kappa V \cdot H \cdot dH/dx$, where $V$ is the volume of the specimen and $dH/dx$ is the field gradient measured transversely to the lines of force, is exerted upon it. This force is easily measured by attaching the specimen to a sensitive balance, and one type commonly used is that designed by Sucksmith. In this balance the distortion of a copper–beryllium ring, caused by the force on the specimen, is measured by means of an optical or electromechanical system. Those metals for which $\kappa$ is negative, such as copper, silver, gold and bismuth, are repelled by the field and are termed diamagnetic materials. Most metals, however, have positive $\kappa$ values (i.e. they are attracted by the field) and are either paramagnetic (when $\kappa$ is small) or ferromagnetic (when $\kappa$ is very large). Only four pure metals–iron, cobalt and nickel from the transition series, and gadolinium from the rare earth series–are ferromagnetic ($\kappa \approx 1000$) at room temperature, but there are several ferromagnetic alloys and some contain no metals which are themselves ferromagnetic. The Heusler alloy, which contains manganese, copper and aluminum, is one example; ferromagnetism is due to the presence of one of the transition metals.

The ability of a ferromagnetic metal to concentrate the lines of force of the applied field is of great practical importance, and while all such materials can be both magnetized and demagnetized, the ease with which this can be achieved usually governs their application in the various branches of engineering. Materials may be generally classified either as magnetically soft (temporary magnets) or as magnetically hard (permanent magnets), and the difference between the two types of magnet may be inferred from Figure 4.4-30. Here, $H$ is the magnetic field necessary to induce a field of strength $B$ inside the material. Upon removal of the field $H$, a certain residual magnetism $B_r$, known as the remanence residual, is left in the specimen, and a field $H_c$, called the coercive force, must be applied in the opposite direction to remove it. A soft magnet is one which is easy both to magnetize and to demagnetize and, as shown in Figure 4.4-30a, a low value of $H$ is sufficient to induce a large field $B$ in the metal, while only a small field $H_c$ is required to remove it; a hard magnet is a material that is magnetized and demagnetized with difficulty (Figure 4.4-30b).

### 4.4.8.2 Diamagnetism and paramagnetism

Diamagnetism is a universal property of the atom, since it arises from the motion of electrons in their orbits around the nucleus. Electrons moving in this way represent electrical circuits and it follows from Lenz's law that this motion is altered by an applied field in such a manner as to set up a repulsive force. The diamagnetic contribution from the valency electrons is small, but from a closed shell it is proportional to the number of electrons in it and to the square of the radius of the 'orbit'. In many

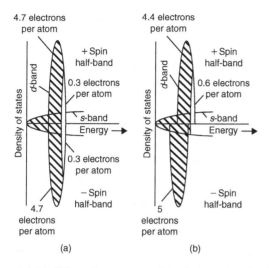

**Figure 4.4-31** Schematic representation of paramagnetic nickel (a) and ferromagnetic nickel (b) (after Raynor, 1988; by courtesy of Inst. of Materials).

metals this diamagnetic effect is outweighed by a paramagnetic contribution, the origin of which is to be found in the electron spin. Each electron behaves like a small magnet and in a magnetic field can take up one of two orientations, either along the field or in the other opposite direction, depending on the direction of the electron spin. Accordingly, the energy of the electron is either decreased or increased and may be represented conveniently by the band theory. Thus, if we regard the band of energy levels as split into two halves (Figure 4.4-31a), each half associated with electrons of opposite spin, it follows that, in the presence of the field, some of the electrons will transfer their allegiance from one band to the other until the Fermi energy level is the same in both. It is clear, therefore, that in this state there will be a larger number of electrons which have their energy lowered by the field than have their energy raised. This condition defines paramagnetism, since there will be an excess of unpaired spins which give rise to a resultant magnetic moment.

It is evident that an insulator will not be paramagnetic, since the bands are full and the lowered half-band cannot accommodate those electrons which wish to 'spill over' from the raised half-band. On the other hand, it is not true, as one might expect, that conductors are always paramagnetic. This follows because in some elements the natural diamagnetic contribution outweighs the paramagnetic contribution; in copper, for example, the newly filled $d$-shell gives rise to a larger diamagnetic contribution.

## 4.4.8.3 Ferromagnetism

The theory of ferromagnetism is difficult and at present not completely understood. Nevertheless, from the electron theory of metals it is possible to build up a band picture of ferromagnetic materials which goes a long way to explain not only their ferromagnetic properties, but also the associated high resistivity and electronic specific heat of these metals compared to copper. In recent years considerable experimental work has been done on the electronic behavior of the transition elements, and this suggests that the electronic structure of iron is somewhat different to that of cobalt and nickel.

Ferromagnetism, like paramagnetism, has its origin in the electron spin. In ferromagnetic materials, however, permanent magnetism is obtained and this indicates that there is a tendency for electron spins to remain aligned in one direction even when the field has been removed. In terms of the band structure this means that the half-band associated with one spin is automatically lowered when the vacant levels at its top are filled by electrons from the top of the other (Figure 4.4-31b); the change in potential energy associated with this transfer is known as the exchange energy. Thus, while it is energetically favorable for a condition in which all the spins are in the same direction, an opposing factor is the Pauli Exclusion Principle, because if the spins are aligned in a single direction many of the electrons will have to go into higher quantum states with a resultant increase in kinetic energy. In consequence, the conditions for ferromagnetism are stringent, and only electrons from partially filled $d$ or $f$ levels can take part. This condition arises because only these levels have (1) vacant levels available for occupation and (2) a high density of states, which is necessary if the increase in kinetic energy accompanying the alignment of spins is to be smaller than the decrease in exchange energy.

Both of these conditions are fulfilled in the transition and rare-earth metals, but of all the metals in the long periods only the elements iron, cobalt and nickel are ferromagnetic at room temperature, gadolinium just above RT ($T_c \approx 16°C$) and the majority are, in fact, strongly paramagnetic. This observation has led to the conclusion that the exchange interactions are most favorable, at least for the iron group of metals, when the ratio of the atomic radius to the radius of the unfilled shell, i.e. the $d$-shell, is somewhat greater than 3 (see Table 4.4.1). As a result of this condition it is hardly surprising that there are a relatively large number of ferromagnetic alloys and compounds, even though the base elements themselves are not ferromagnetic.

In ferromagnetic metals the strong interaction results in the electron spins being spontaneously aligned, even in the absence of an applied field. However, a specimen of iron can exist in an unmagnetized condition because such an alignment is limited to small regions, or domains, which statistically oppose each other. These domains are distinct from the grains of a polycrystalline metal and in general there are many domains in a single grain, as shown

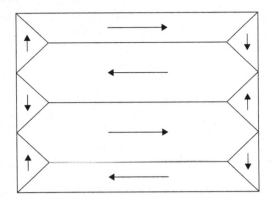

**Figure 4.4-32** Simple domain structure in a ferromagnetic material. The arrows indicate the direction of magnetization in the domains.

in Figure 4.4-32. Under the application of a magnetic field the favorably oriented domains grow at the expense of the others by the migration of the domain boundaries until the whole specimen appears fully magnetized. At high field strengths it is also possible for unfavorably oriented domains to 'snap-over' into more favorable orientations quite suddenly, and this process, which can often be heard using sensitive equipment, is known as the Barkhausen effect.

The state in which all the electron spins are in complete alignment is possible only at low temperatures. As the temperature is raised the saturation magnetization is reduced, falling slowly at first and then increasingly rapidly, until a critical temperature, known as the Curie temperature, is reached. Above this temperature, $T_c$, the specimen is no longer ferromagnetic, but becomes paramagnetic, and for the metals iron, cobalt and nickel this transition occurs at 780°C, 1075°C and 365°C respectively. Such a cooperative process may be readily understood from thermodynamic reasoning, since the additional entropy associated with the disorder of the electron spins makes the disordered (paramagnetic) state thermodynamically more stable at high temperatures. This behavior is similar to that shown by materials which undergo the order–disorder transformation and, as a consequence, ferromagnetic metals exhibit a specific heat peak of the form previously shown (see Figure 4.4-3b).

A ferromagnetic crystal in its natural state has a domain structure. From Figure 4.4-32 it is clear that by dividing itself into domains the crystal is able to eliminate those magnetic poles which would otherwise occur at the surface. The width of the domain boundary or Bloch wall is not necessarily small, however, and in most materials is of the order of 100 atoms in thickness. By having a wide boundary the electron spins in neighboring atoms are more nearly parallel, which is a condition required to minimize the exchange energy. On the other hand, within any one domain the direction of magnetization is parallel to a direction of easy magnetization (i.e. $\langle 1\,0\,0\rangle$ in iron, $\langle 1\,1\,1\rangle$ in nickel and $\langle 0\,0\,1\rangle$ in cobalt) and as one passes across a boundary the direction of magnetization rotates away from one direction of easy magnetization to another. To minimize this magnetically disturbed region, the crystal will try to adopt a boundary which is as thin as possible. Consequently, the boundary width adopted is one of compromise between the two opposing effects, and the material may be considered to possess a magnetic interfacial or surface energy.

### 4.4.8.4 Magnetic alloys

The work done in moving a domain boundary depends on the energy of the boundary, which in turn depends on the magnetic anisotropy. The ease of magnetization also depends on the state of internal strain in the material and the presence of impurities. Both these latter factors affect the magnetic 'hardness' through the phenomenon of magnetostriction, i.e. the lattice constants are slightly altered by the magnetization so that a directive influence is put upon the orientation of magnetization of the domains. Materials with internal stresses are hard to magnetize or demagnetize, while materials free from stresses are magnetically soft. Hence, since internal stresses are also responsible for mechanical hardness, the principle which governs the design of magnetic alloys is to make permanent magnetic materials as mechanically hard and soft magnets as mechanically soft as possible.

Magnetically soft materials are used for transformer laminations and armature stampings, where a high permeability and a low hysteresis are desirable: iron–silicon or iron–nickel alloys are commonly used for this purpose. In the development of magnetically soft materials it is found that those elements which form interstitial solid solutions with iron are those which broaden the hysteresis loop most markedly. For this reason, it is common to remove such impurities from transformer iron by vacuum melting or hydrogen annealing. However, such processes are expensive and, consequently, alloys are frequently used as 'soft' magnets, particularly iron-silicon and iron–nickel alloys (because silicon and nickel both reduce the amount of carbon in solution). The role of Si is to form a $\gamma$-loop and hence remove transformation strains and also improve orientation control. In the production of iron–silicon alloys the factors which are controlled include the grain size, the orientation difference from one grain to the next, and the presence of non-magnetic inclusions, since all are major sources of coercive force. The coercive force increases with decreasing grain size because the domain pattern in the neighborhood of a grain boundary is complicated owing to the orientation difference between two adjacent grains. Complex domain patterns can also arise at the free surface of the metal unless these are parallel to a direction of easy magnetization. Accordingly, to minimize the coercive force, rolling and annealing schedules are adopted to produce a preferred oriented material with a strong 'cube texture', i.e. one with two $\langle 1\ 0\ 0\rangle$ directions in the plane of the sheet (see Chapter 6). This procedure is extremely important, since transformer material is used in the form of thin sheets to minimize eddy-current losses. Fe–Si–B in the amorphous state is finding increasing application in transformers.

The iron–nickel series, *Permalloys*, present many interesting alloys and are used chiefly in communication engineering, where a high permeability is a necessary condition. The alloys in the range 40–55% nickel are characterized by a high permeability and at low field strengths this may be as high as 15 000 compared with 500 for annealed iron. The 50% alloy, *Hypernik*, may have a permeability which reaches a value of 70 000, but the highest initial and maximum permeability occurs in the composition range of the $FeNi_3$ superlattice, provided the ordering phenomenon is suppressed. An interesting development in this field is in the heat treatment of the alloys while in a strong magnetic field. By such a treatment, the permeability of *Permalloy* 65 has been increased to about 260 000. This effect is thought to be due to the fact that, during alignment of the domains, plastic deformation is possible and magnetostrictive strains may be relieved.

Magnetically hard materials are used for applications where a 'permanent' magnetic field is required, but where electromagnets cannot be used, such as in electric clocks, meters, etc. Materials commonly used for this purpose include *Alnico* (Al–Ni–Co) alloys, *Cunico* (Cu–Ni–Co) alloys, ferrites (barium and strontium), samarium–cobalt alloys ($SmCo_5$ and $Sm_2(Co, Fe, Cu, Zr)_{17}$) and *Neomax* ($Nd_2Fe_{14}B$). The *Alnico* alloys have high remanence but poor coercivities; the ferrites have rather low remanence but good coercivities, together with very cheap raw material costs. The rare-earth magnets have a high performance but are rather costly, although the Nd-based alloys are cheaper than the Sm-based ones.

In the development of magnetically hard materials, the principle is to obtain, by alloying and heat treatment, a matrix containing finely divided particles of a second phase. These fine precipitates, usually differing in lattice parameter from the matrix, set up coherency strains in the lattice which affect the domain boundary movement. Alloys of copper–nickel–iron, copper–nickel–cobalt and aluminum–nickel–cobalt are of this type. An important advance in this field is to make the particle size of the alloy so small, i.e. less than 100 nm diameter, that each grain contains only a single domain. Then magnetization can occur only by the rotation of the direction of magnetization *en bloc*. *Alnico* alloys containing 6–12% Al, 14–25% Ni, 0–35% Co, 0–8% Ti, 0–6% Cu in 40–70% Fe depend on this feature and are the most commercially important permanent magnet materials. They are precipitation-hardened alloys and are heat treated to produce rod-like precipitates (30 nm × 100 nm) lying along $\langle 1\ 0\ 0\rangle$ in the bcc matrix. During magnetic annealing the rods form along the $\langle 1\ 0\ 0\rangle$ axis nearest to the direction of the field, when the remanence and coercivity are markedly increased; $Sm_2(Co, Fe, Cu, Zr)_{17}$ alloys also rely on the pinning of magnetic domains by fine precipitates. Clear correlation exists between mechanical hardness and intrinsic coercivity. $SmCo_5$ magnets depend on the very high magnetocrystalline anisotropy of this compound and the individual grains are single-domain particles. The big advantage of these magnets over the *Alnico* alloys is their much higher coercivities.

The Heusler alloys, copper–manganese–aluminum, are of particular interest because they are made up from non-ferromagnetic metals and yet exhibit ferromagnetic properties. The magnetism in this group of alloys is associated with the compound $Cu_2MnAl$, evidently because of the presence of manganese atoms. The compound has the $Fe_3Al$-type superlattice when quenched from 800°C, and in this state is ferromagnetic, but when the alloy is slowly cooled it has a $\gamma$-brass structure and is non-magnetic, presumably because the correct exchange forces arise from the lattice rearrangement on ordering. A similar behavior is found in both the copper–manganese–gallium and the copper–manganese–indium systems.

The order–disorder phenomenon is also of magnetic importance in many other systems. As discussed previously, when ordering is accompanied by a structural change, i.e. cubic to tetragonal, coherency strains are set up which often lead to magnetic hardness. In FePt, for example, extremely high coercive forces are produced by rapid cooling. However, because the change in mechanical properties accompanying the transformation is found to be small, it has been suggested that the hard magnetic properties in this alloy are due to the small particle-size effect, which arises from the finely laminated state of the structure.

While the much cheaper but lower performance magnets such as ferrites have a significant market share of applications, the rare-earth (RE) magnets have revolutionized the properties and applications of permanent magnets. A parameter which illustrates the potential of these materials is the maximum energy product $(BH)_{max}$. The larger the value of $(BH)_{max}$ the smaller the volume of magnet required to produce a given magnetic flux. This is illustrated in Figure 4.4-33, where the neodymium–iron–boron materials have $(BH)_{max}$ in excess of 400 kJ m$^{-3}$, an order of magnitude stronger than the

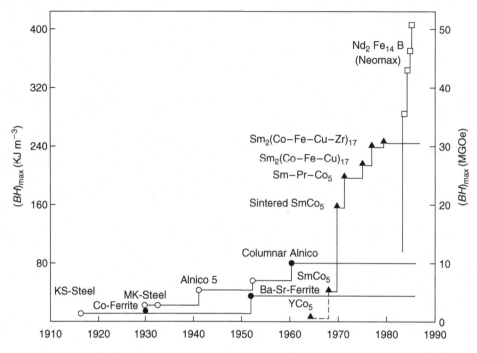

**Figure 4.4-33** The variation of $(BH)_{max}$ with time over this century (courtesy of I. R. Harris).

ferrites. In the drive for miniaturization the $Nd_2Fe_{14}B$ materials are unrivaled. They are also finding applications where a very strong permanent field is required, such as MRI scanners. The main process route for these magnets, shown in Figure 4.4-34, consists of powdering the coarse-grained cast ingot, aligning the fine powder along the easy magnetization axis, compacting and then sintering to produce a fully dense magnet. The alloy is very active with respect to hydrogen gas so that, on exposure at room temperature and around 1 bar pressure, the bulk alloy absorbs the hydrogen, particularly in the grain boundary region. The differential and overall volume expansion results in the bulk alloy 'decrepitating' (separating into parts with a crying sound) into very friable particulate matter which consists of fine, grain boundary debris and grains of $Nd_2Fe_{14}B$ which are of the order of ~ 100 $\mu$m in size. The phenomenon of 'hydrogen decrepitation' (HD) has been incorporated into the process route. Apart from economically producing powder, the HD powder is extremely friable, which substantially aids the subsequent jet-milling process. With this modified processing route substantial savings of between 15% and 25% can be achieved in the cost of magnet production. The majority of NdFeB magnets are now made by this process.

### 4.4.8.5 Anti-ferromagnetism and ferrimagnetism

Apart from the more usual dia-, para- and ferromagnetic materials, there are certain substances which are termed anti-ferromagnetic; in these, the net moments of neighboring atoms are aligned in opposite directions, i.e. anti-parallel. Many oxides and chlorides of the transition metals are examples, including both chromium and $\alpha$-manganese, and also manganese–copper alloys. Some of the relevant features of anti-ferromagnetism are similar in many respects to ferromagnetism, and are summarized as follows:

1. In general, the magnetization directions are aligned parallel or anti-parallel to crystallographic axes, e.g. in MnI and CoO the moment of the $Mn^{2+}$ and $Co^{2+}$ ions are aligned along a cube edge of the unit cell. The common directions are termed directions of anti-ferromagnetism.

2. The degree of long-range anti-ferromagnetic ordering progressively decreases with increasing temperature and becomes zero at a critical temperature, $T_n$, known as the Néel temperature; this is the anti-ferromagnetic equivalent of the Curie temperature.

3. An anti-ferromagnetic domain is a region in which there is only one common direction of anti-ferromagnetism; this is probably affected by lattice defects and strain.

The most characteristic property of an anti-ferromagnetic material is that its susceptibility $\chi$ shows a maximum as a function of temperature, as shown in Figure 4.4-35a. As the temperature is raised from 0 K the interaction which leads to anti-parallel spin alignment becomes less effective until at $T_n$ the spins are free.

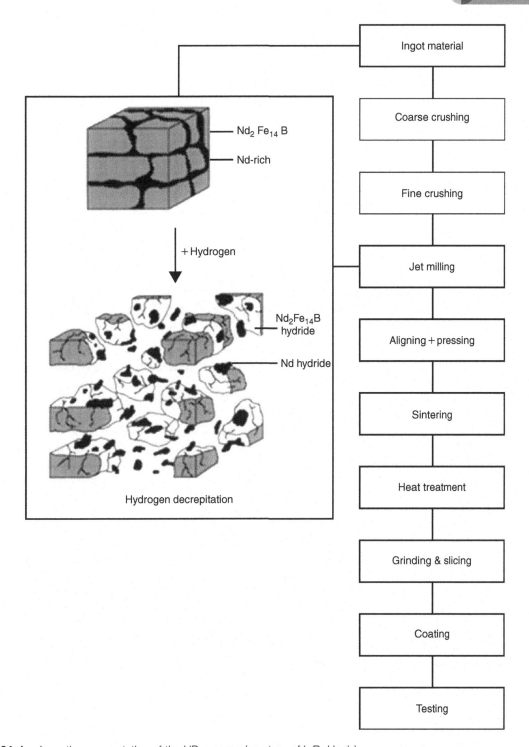

**Figure 4.4-34** A schematic representation of the HD process (courtesy of I. R. Harris).

Similar characteristic features are shown in the resistivity curves due to scattering as a result of spin disorder. However, the application of neutron diffraction techniques provides a more direct method of studying anti-ferromagnetic structures, as well as giving the magnetic moments associated with the ions of the metal. There is a magnetic scattering of neutrons in the case of certain magnetic atoms, and owing to the different scattering amplitudes of the parallel and anti-parallel atoms, the possibility arises of the existence of super-lattice lines in the anti-ferromagnetic state. In manganese oxide (MnO), for example, the parameter of the

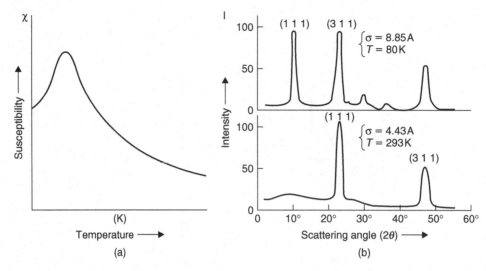

**Figure 4.4-35** (a) Variation of magnetic susceptibility with temperature for an anti-ferromagnetic material, (b) Neutron diffraction pattern from the anti-ferromagnetic powder MnO above and below the critical temperature for ordering (after Shull and Smart, 1949).

magnetic unit cell is 0.885 nm, whereas the chemical unit cell (NaCl structure) is half this value, 0.443 nm. This atomic arrangement is analogous to the structure of an ordered alloy and the existence of magnetic superlattice lines below the Néel point (122 K) has been observed, as shown in Figure 4.4-35b.

Some magnetic materials have properties which are intermediate between those of anti-ferromagnetic and ferromagnetic. This arises if the moments in one direction are unequal in magnitude to those in the other, as, for example, in magnetite ($Fe_3O_4$), where the ferrous and ferric ions of the $FeO \cdot Fe_2O_3$ compound occupy their own particular sites. Néel has called this state *ferrimagnetism* and the corresponding materials are termed ferrites. Such materials are of importance in the field of electrical engineering because they are ferromagnetic without being appreciably conducting; eddy-current troubles in transformers are therefore not so great. Strontium ferrite is extensively used in applications such as electric motors, because of these properties and low material costs.

## 4.4.9 Dielectric materials

### 4.4.9.1 Polarization

Dielectric materials, usually those which are covalent or ionic, possess a large energy gap between the valence band and the conduction band. These materials exhibit high electrical resistivity and have important applications as insulators, which prevent the transfer of electrical charge, and capacitors, which store electrical charge. Dielectric materials also exhibit piezoelectric and ferroelectric properties.

Application of an electric field to a material induces the formation of dipoles (i.e. atoms or groups of atoms with an unbalanced charge or moment), which become aligned with the direction of the applied field. The material is then polarized. This state can arise from several possible mechanisms – electronic, ionic or molecular, as shown in Figure 4.4-36a–c. With electronic polarization, the electron clouds of an atom are displaced with respect to the positively charged ion core setting up an

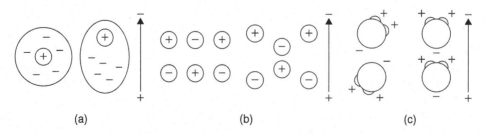

**Figure 4.4-36** Application of field to produce polarization by electronic (a), ionic (b) and molecular (c) mechanisms.

electric dipole with moment $\mu_e$. For ionic solids in an electric field, the bonds between the ions are elastically deformed and the anion–cation distances become closer or further apart, depending on the direction of the field. These induced dipoles produce polarization and may lead to dimensional changes. Molecular polarization occurs in molecular materials, some of which contain natural dipoles. Such materials are described as polar and for these the influence of an applied field will change the polarization by displacing the atoms and thus changing the dipole moment (i.e. atomic polarizability) or by causing the molecule as a whole to rotate to line up with the imposed field (i.e. orientation polarizability). When the field is removed these dipoles may remain aligned, leading to permanent polarization. Permanent dipoles exist in asymmetrical molecules such as $H_2O$, organic polymers with asymmetric structure and ceramic crystals without a center of symmetry.

### 4.4.9.2 Capacitors and insulators

In a capacitor the charge is stored in a dielectric material which is easily polarized and has a high electrical resistivity $\sim 10^{11}$ V A$^{-1}$ m to prevent the charge flowing between conductor plates. The ability of the material to polarize is expressed by the permittivity $\varepsilon$, and the relative permittivity or dielectric constant $\kappa$ is the ratio of the permittivity of the material and the permittivity of a vacuum $\varepsilon_0$ While a high $\kappa$ is important for a capacitor, a high dielectric strength or breakdown voltage is also essential. The dielectric constant $\kappa$ values for vacuum, water, polyethylene, *Pyrex* glass, alumina and barium titanate are 1, 78.3, 2.3, 4, 6.5 and 3000 respectively.

Structure is an important feature of dielectric behavior. Glassy polymers and crystalline materials have a lower $\kappa$ than their amorphous counterparts. Polymers with asymmetric chains have a high $\kappa$ because of the strength of the associated molecular dipole; thus, polyvinyl chloride (PVC) and polystyrene (PS) have larger $\kappa$ values than polyethylene (PE). $BaTiO_3$ has an extremely high $\kappa$ value because of its asymmetrical structure. Frequency response is also important in dielectric applications and depends on the mechanism of polarization. Materials which rely on electronic and ionic dipoles respond rapidly to frequencies of $10^{13}$–$10^{16}$Hz, but molecular polarization solids, which require groups of atoms to rearrange, respond less rapidly. Frequency is also important in governing dielectric loss due to heat and usually increases when one of the contributions to polarization is prevented. This behavior is common in microwave heating of polymer adhesives; preferential heating in the adhesive due to dielectric losses starts the thermosetting reaction. For moderate increases, raising the voltage and temperature increases the polarizability and leads to a higher dielectric constant. Nowadays, capacitor dielectrics combine materials with different temperature dependence to yield a final product with a small linear temperature variation. These materials are usually titanates of Ba, Ca, Mg, Sr and rare-earth metals.

For an insulator, the material must possess a high electrical resistivity, a high dielectric strength to prevent breakdown of the insulator at high voltages, a low dielectric loss to prevent heating, and small dielectric constant to hinder polarization and hence charge storage. Materials increasingly used are alumina, aluminum nitride, glass-ceramics, steatite porcelain and glasses.

### 4.4.9.3 Piezoelectric materials

When stress is applied to certain materials an electric polarization is produced proportional to the stress applied. This is the well-known piezoelectric effect. Conversely, dilatation occurs on application of an electric field. Common materials displaying this property are quartz, $BaTiO_3$, $Pb(Ti, Zr)O_3$ or PZT and Na or $LiNbO_3$. For quartz, the piezoelectric constant d relating strain $\varepsilon$ to field strength $F (\varepsilon = d \times F)$ is $2.3 \times 10^{-12}$ mV$^{-1}$, whereas for PZT it is $250 \times 10^{-12}$ mV$^{-1}$. The piezoelectric effect is used in transducers, which convert sound waves to electric fields or vice versa. Applications range from microphones, where a few millivolts are generated, to military devices creating several kilovolts and from small sub-nanometer displacements in piezoelectrically deformed mirrors to large deformations in power transducers.

### 4.4.9.4 Pyroelectric and ferroelectric materials

Some materials, associated with low crystal symmetry, are observed to acquire an electric charge when heated; this is known as pyroelectricity. Because of the low symmetry, the centers of gravity of the positive and negative charges in the unit cell are separated, producing a permanent dipole moment. Moreover, alignment of individual dipoles leads to an overall dipole moment which is non-zero for the crystal. Pyroelectric materials are used as detectors of electromagnetic radiation in a wide band from ultraviolet to microwave, in radiometers and in thermometers sensitive to changes of temperature as small as $6 \times 10^{-6}$°C. Pyroelectric TV camera tubes have also been developed for long-wavelength infrared imaging and are useful in providing

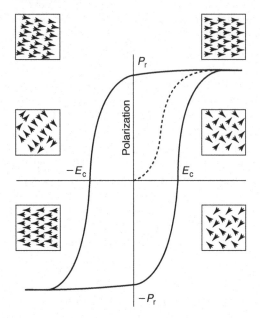

**Figure 4.4-37** Hysteresis loop for ferroelectric materials, showing the influence of electric field $E$ on polarization $P$.

## 4.4.10 Optical properties

### 4.4.10.1 Reflection, absorption and transmission effects

The optical properties of a material are related to the interaction of the material with electromagnetic radiation, particularly visible light. The electromagnetic spectrum is shown in Figure 4.4-1, from which it can be seen that the wavelength $\lambda$ varies from $10^4$ m for radio waves down to $10^{-14}$ m for $\gamma$-rays and the corresponding photon energies vary from $10^{-10}$ to $10^8$ eV.

Photons incident on a material may be reflected, absorbed or transmitted. Whether a photon is absorbed or transmitted by a material depends on the energy gap between the valency and conduction bands and the energy of the photon. The band structure for metals has no gap and so photons of almost any energy are absorbed by exciting electrons from the valency band into a higher energy level in the conduction band. Metals are thus opaque to all electromagnetic radiation from radio waves, through the infrared, the visible to the ultraviolet, but are transparent to high-energy X-rays and $\gamma$-rays. Much of the absorbed radiation is re-emitted as radiation of the same wavelength (i.e. reflected). Metals are both opaque and reflective, and it is the wavelength distribution of the reflected light, which we see, that determines the color of the metal. Thus, copper and gold reflect only a certain range of wavelengths and absorb the remaining photons, i.e. copper reflects the longer-wavelength red light and absorbs the shorter-wavelength blue. Aluminum and silver are highly reflective over the complete range of the visible spectrum and appear silvery.

Because of the gaps in their band structure, non-metals may be transparent. The wavelength for visible light varies from about 0.4 to 0.7 $\mu$m so that the maximum band-gap energy for which absorption of visible light is possible is given by $E = hc/\lambda = (4.13 \times 10^{-15}) \times (3 \times 10^8)/(4 \times 10^{-7}) = 3.1$ eV. The minimum band-gap energy is 1.8 eV. Thus, if the photons have insufficient energy to excite electrons in the material to a higher energy level, they may be transmitted rather than absorbed and the material is transparent. In high-purity ceramics and polymers, the energy gap is large and these materials are transparent to visible light. In semiconductors, electrons can be excited into acceptor levels or out of donor levels and phonons having sufficient energy to produce these transitions will be absorbed. Semiconductors are therefore opaque to short wavelengths and transparent to long. The band structure is influenced by crystallinity and hence materials such as glasses and polymers may be transparent in the amorphous state but opaque when crystalline.

visibility through smoke. Typical materials are strontium barium niobate and PZT with $Pb_2FeNbO_6$ additions to broaden the temperature range of operation.

Ferroelectric materials are those which retain a net polarization when the field is removed and is explained in terms of the residual alignment of permanent dipoles. Not all materials that have permanent dipoles exhibit ferroelectric behavior because these dipoles become randomly arranged as the field is removed so that no net polarization remains. Ferroelectrics are related to the pyroelectrics; for the former materials the direction of spontaneous polarization can be reversed by an electric field (Figure 4.4-37), whereas for the latter this is not possible. This effect can be demonstrated by a polarization versus field hysteresis loop similar in form and explanation to the $B$–$H$ magnetic hysteresis loop (see Figure 4.4-30). With increasing positive field all the dipoles align to produce a saturation polarization. As the field is removed a remanent polarization $P_r$ remains due to a coupled interaction between dipoles. The material is permanently polarized, and a coercive field $E_c$ has to be applied to randomize the dipoles and remove the polarization.

Like ferromagnetism, ferroelectricity depends on temperature and disappears above an equivalent Curie temperature. For $BaTiO_3$, ferroelectricity is lost at 120°C when the material changes crystal structure. By analogy with magnetism there is also a ferroelectric analog of anti-ferromagnetism and ferrimagnetism. $NaNbO_3$, for example, has a $T_c$ of 640°C and anti-parallel electric dipoles of unequal moments characteristic of a ferrielectric material.

High-purity non-metallics such as glasses, diamond or sapphire ($Al_2O_3$) are colorless but are changed by impurities. For example, small additions of $Cr^{3+}$ ions ($Cr_2O_3$) to $Al_2O_3$ produce a ruby color by introducing impurity levels within the band gap of sapphire which give rise to absorption of specific wavelengths in the visible spectrum. Coloring of glasses and ceramics is produced by addition of transition metal impurities which have unfilled $d$-shells. The photons easily interact with these ions and are absorbed; $Cr^{3+}$ gives green, $Mn^{2+}$ yellow and $Co^{2+}$ blue–violet coloring.

In photochromic sunglasses the energy of light quanta is used to produce changes in the ionic structure of the glass. The glass contains silver ($Ag^+$) ions as a dopant which are trapped in the disordered glass network of silicon and oxygen ions: these are excited by high-energy quanta (photons) and change to metallic silver, causing the glass to darken (i.e. light energy is absorbed). With a reduction in light intensity, the silver atoms re-ionize. These processes take a small period of time relying on absorption and non-absorption of light.

### 4.4.10.2 Optical fibers

Modern communication systems make use of the ability of optical fibers to transmit light signals over large distances. Optical guidance by a fiber is produced (see Figure 4.4-38) if a core fiber of refractive index $n_1$ is surrounded by a cladding of slightly lower index $n_2$ such that total internal reflection occurs confining the rays to the core: typically the core is about 100 $\mu$m and $n_1 - n_2 \approx 10^{-2}$. With such a simple optical fiber, interference occurs between different modes, leading to a smearing of the signals. Later designs use a core in which the refractive index is graded, parabolically, between the core axis and the interface with the cladding. This design enables modulated signals to maintain their coherency. In vitreous silica, the refractive index can be modified by additions of dopants such as $P_2O_5$ and $GeO_2$ which raise it and $B_2O_5$ and F which lower it. Cables are sheathed to give strength and environmental protection; PE and PVC are commonly used for limited fire-hazard conditions.

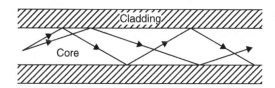

Figure 4.4-38 Optical guidance in a multimode fiber.

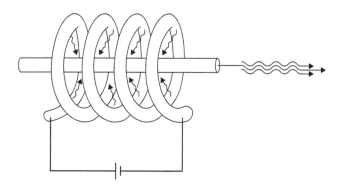

Figure 4.4-39 Schematic diagram of a laser.

### 4.4.10.3 Lasers

A laser (Light Amplification by Stimulated Emission of Radiation) is a powerful source of coherent light (i.e. monochromatic and all in phase). The original laser material, still used, is a single crystal rod of ruby, i.e. $Al_2O_3$ containing dopant $Cr^{3+}$ ions in solid solution. Nowadays, lasers can be solid, liquid or gaseous materials and ceramics, glasses and semiconductors. In all cases, electrons of the laser material are excited into a higher energy state by some suitable stimulus (see Figure 4.4-39). In a device this is produced by the photons from a flash tube, to give an intense source of light surrounding the rod to concentrate the energy into the laser rod. Alternatively, an electrical discharge in a gas is used. The ends of the laser rod are polished flat and parallel, and then silvered such that one end is totally reflecting and the other partially transmitting.

In the ruby laser, a xenon flash lamp excites the electrons of the $Cr^{3+}$ ions into higher energy states. Some of these excited electrons decay back to their ground state directly and are not involved in the laser process. Other electrons decay into a metastable intermediate state before further emission returns them to the ground state. Some of the electrons in the metastable state will be spontaneously emitted after a short (~ms) rest period. Some of the resultant photons are retained in the rod because of the reflectivity of the silvered ends and stimulate the release of other electrons from the metastable state. Thus, one photon releases another such that an avalanche of emissions is triggered, all in phase with the triggering photon. The intensity of the light increases as more emissions are stimulated until a very high intensity, coherent, collimated 'burst' of light is transmitted through the partially silvered end, lasting a few nanoseconds and with considerable intensity.

### 4.4.10.4 Ceramic 'windows'

Many ceramics, usually oxides, have been prepared in optically transparent or translucent forms (translucent means that incident light is partly reflected and partly transmitted). Examples include aluminum oxide, magnesium oxide, their double oxide or spinel, and chalcogenides (zinc sulfide, zinc selenide). Very pure raw materials of fine particle size are carefully processed to eliminate voids and control grain sizes. Thus, translucent alumina is used for the arc tube of high-pressure sodium lamps; a grain size of 25 $\mu$m gives the best balance of translucency and mechanical strength.

Ceramics are also available to transmit electromagnetic radiation with wavelengths which lie below or above the visible range of 400–700 nm (e.g. infrared, microwave, radar, etc.). Typical candidate materials for development include vitreous silica, cordierite glass-ceramics and alumina.

### 4.4.10.5 Electro-optic ceramics

Certain special ceramics combine electrical and optical properties in a unique manner. Lead lanthanum zirconium titanate, known as PLZT, is a highly transparent ceramic which becomes optically birefringent when electrically charged. This phenomenon is utilized as a switching mechanism in arc-welding goggles, giving protection against flash blindness. The PLZT plate is located between two 'crossed' sheets of polarizing material. A small impressed d.c. voltage on the PLZT plate causes it to split the incident light into two rays vibrating in different planes. One of these rays can pass through the inner polar sheet and enter the eye. A sudden flash of light will activate photodiodes in the goggles, reduce the impressed voltage and cause rapid darkening of the goggles.

## Problems

**4.4.1** Assuming that the vacancy concentration of a close packed metal is $10^{-4}$ at its melting point and that $D_0 = 10^{-4}$ m$^2$ s$^{-1}$, where $D = D_0 \exp(-E_D/kT)$ and $D$ is the self-diffusion coefficient, answer questions (i)–(vi), which relate to diffusion by a vacancy mechanism in a close-packed metal:

(i) What are the vacancy concentrations at $\frac{1}{4}$, $\frac{1}{2}$ and $\frac{3}{4}$ $T_m$ (in K)?

(ii) Estimate the diffusion coefficient of the vacancies at $\frac{1}{4}$, $\frac{1}{2}$ and $\frac{3}{4}$ $T_m$.

(iii) Estimate the self-diffusion coefficient for the metal at $\frac{1}{4}$, $\frac{1}{2}$ and $\frac{3}{4}$ $T_m$.

(iv) How far does a vacancy diffuse at $T_m/2$ in 1 hour?

(v) How far does an atom diffuse at $T_m/2$ in 1 hour?

(vi) If copper melts at 1065°C, estimate $E_f$. (Boltzmann's constant $k = 8.6 \times 10^{-5}$ eV K$^{-1}$.)

**4.4.2** The diffusivity of lithium in silicon is $10^{-9}$ m$^2$ s$^{-1}$ at 1376 K and $10^{-10}$ m$^2$ s$^{-1}$ at 968 K. What are the values of $E_D$ and $D_0$ for diffusion of lithium in silicon? ($E_D$ is the activation energy for diffusion in J mol$^{-1}$ and $R = 8.314$ J mol$^{-1}$ K$^{-1}$.)

**4.4.3** The melting endotherm of a sample of an impure material has been analyzed to determine the fraction, $f$, of sample melted at each temperature $T_s$:

| $f$ | 0.099 | 0.122 | 0.164 | 0.244 | 0.435 |
|---|---|---|---|---|---|
| $T_s$(K) | 426.0 | 426.5 | 427.0 | 427.5 | 428.0 |

The fraction, $f$, of the sample melted at temperature $T_s$ is given by

$$T_s = T_0 - \Delta T/f,$$

where $\Delta T = (T_0 - T_m)$, $T_0$ is the melting point of the pure and $T_m$ the impure sample. The van't Hoff equation,

$$\Delta T = \left(\frac{RT_0^2}{\Delta H_f}\right) x_2,$$

relates $\Delta T$ to mole fraction of impurity present, where $\mathbf{R}$ is the gas constant (8.31 JK$^{-1}$ mol$^{-1}$), $\Delta H_f$ is the enthalpy of fusion and $x_2$ the mole fraction of impurity.

If the enthalpy of fusion of the pure material is 25.5 kJ mol$^{-1}$, use the above data to determine graphically the lowering of the melting point and hence determine the mole fraction of impurity present in the sample.

**4.4.4** The resistivity of intrinsic germanium is 0.028 $\Omega$m at 385 K and $2.74 \times 10^{-4}$ $\Omega$m at 714K. Assume that the hole and electron mobilities both vary as $T^{-3/2}$.

(a) Determine the band-gap energy $E_g$.

(b) At what wavelength would you expect the onset of optical absorption?

**4.4.5** The magnetic susceptibility ($\chi$) of iron is temperature dependent according to $\chi \propto 1/(T - T_c)$, where $T_c$ is the Curie temperature. At 900°C, $\chi$ has a value of $2.5 \times 10^{-4}$. $T_c$ for iron is 770°C. Determine the susceptibility at 800°C.

**4.4.6** The current flowing around a superconducting loop of wire decays according to

$$i(t) = i(0)e^{-\frac{R}{L}t},$$

where $R$ = resistance and $L$ = self-inductance. What is the largest resistance a 1 m diameter loop of superconducting wire, 1 mm² cross-sectional area, can sustain if it is to maintain a current flow of 1 A for one year without appreciable loss (< 1%)? (Given: a loop with diameter $2a$ and wire thickness $2r$ has a self-inductance $L = \mu_0 a[\ln(\frac{8a}{r}) - \frac{7}{4}]$, where $\mu_0 = 4\pi \times 10^{-7}$.)

## Further reading

Anderson, J. C., Leaver, K. D., Rawlins, R. D. and Alexander, J. M. (1990). *Materials Science*. Chapman & Hall, London.

Braithwaite, N. and Weaver, G. (eds) (1990). *Open University Materials in Action Series*. Butterworths, London.

Cullity, B. D. (1972). *Introduction to Magnetic Materials*. Addison-Wesley, Wokingham.

Hume-Rothery, W. and Coles, B. R. (1946, 1969). *Atomic Theory for Students of Metallurgy*. Institute of Metals, London.

Porter, D. A. and Easterling, K. E. (1992). *Phase Transformations in Metals and Alloys*, 2nd edn. Van Nostrand Reinhold, Wokingham.

Raynor, G. V. (1947, 1988). *Introduction to Electron Theory of Metals*. Institute of Metals, London.

Shewmon, P. G. (1989). *Diffusion in Solids*. Minerals, Metals and Materials Soc., Warrendale, USA.

Swalin, R. A. (1972). *Thermodynamics of Solids*. Wiley, Chichester.

Warn, J. R. W. (1969). *Concise Chemical Thermodynamics*. Van Nostrand, New York.

# Chapter 4.5

# Mechanical properties of metals

## 4.5.1 Mechanical testing procedures

### 4.5.1.1 Introduction

Real crystals, however carefully prepared, contain lattice imperfections which profoundly affect those properties sensitive to structure. Careful examination of the mechanical behavior of materials can give information on the nature of these atomic defects. In some branches of industry the common mechanical tests, such as tensile, hardness, impact, creep and fatigue tests, may be used not to study the 'defect state', but to check the quality of the product produced against a standard specification. Whatever its purpose, the mechanical test is of importance in the development of both materials science and engineering properties. It is inevitable that a large number of different machines for performing the tests are in general use. This is because it is often necessary to know the effect of temperature and strain rate at different levels of stress depending on the material being tested. Consequently, no attempt is made here to describe the details of the various testing machines. The elements of the various tests are outlined below.

### 4.5.1.2 The tensile test

In a tensile test the ends of a test piece are fixed into grips, one of which is attached to the load-measuring device on the tensile machine and the other to the straining device. The strain is usually applied by means of a motor-driven crosshead and the elongation of the specimen is indicated by its relative movement. The load necessary to cause this elongation may be obtained from the elastic deflection of either a beam or proving ring, which may be measured by using hydraulic, optical or electromechanical methods. The last method (where there is a change in the resistance of strain gauges attached to the beam) is, of course, easily adapted into a system for autographically recording the load–elongation curve.

The load–elongation curves for both polycrystalline mild steel and copper are shown in Figure 4.5-1a and b. The corresponding stress (load per unit area, $P/A$) versus strain (change in length per unit length, $dl/l$) curves may be obtained knowing the dimensions of the test piece. At low stresses the deformation is elastic, reversible and obeys Hooke's law with stress linearly proportional to strain. The proportionality constant connecting stress and strain is known as the elastic modulus and may be either (a) the elastic or Young's modulus, $E$, (b) the rigidity or shear modulus $\mu$, or (c) the bulk modulus $K$, depending on whether the strain is tensile, shear or hydrostatic compressive, respectively. Young's modulus, bulk modulus, shear modulus and Poisson's ratio $v$, the ratio of lateral contractions to longitudinal extension in uniaxial tension, are related according to

$$K = \frac{E}{3(1-2v)}, \quad \mu = \frac{E}{2(1+v)}, \quad E = \frac{9K\mu}{3K+\mu}. \quad 4.5.1$$

In general, the elastic limit is an ill-defined stress, but for impure iron and low-carbon steels the onset of plastic deformation is denoted by a sudden drop in load,

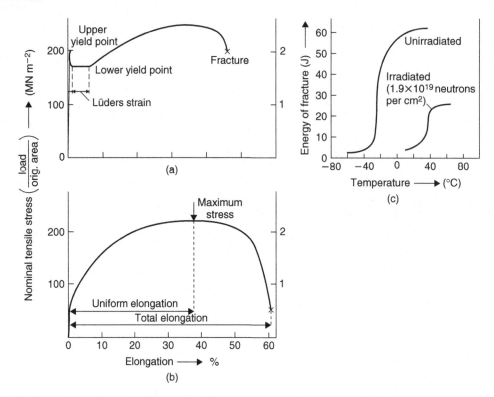

**Figure 4.5-1** Stress–elongation curves for: (a) impure iron, (b) copper, (c) ductile–brittle transition in mild steel (after Churchman, Mogford and Cottrell, 1957).

indicating both an upper and lower yield point.[1] This yielding behavior is characteristic of many metals, particularly those with bcc structure containing small amounts of solute element (see Section 4.5.4.6). For materials not showing a sharp yield point, a conventional definition of the beginning of plastic flow is the 0.1% proof stress, in which a line is drawn parallel to the elastic portion of the stress–strain curve from the point of 0.1% strain.

For control purposes the tensile test gives valuable information on the tensile strength (TS = maximum load/original area) and ductility (percentage reduction in area or percentage elongation) of the material. When it is used as a research technique, however, the exact shape and fine details of the curve, in addition to the way in which the yield stress and fracture stress vary with temperature, alloying additions and grain size, are probably of greater significance.

The increase in stress from the initial yield up to the TS indicates that the specimen hardens during deformation (i.e. work hardens). On straining beyond the TS the metal still continues to work-harden, but at a rate too small to compensate for the reduction in cross-sectional area of the test piece. The deformation then becomes unstable, such that as a localized region of the gauge length strains more than the rest, it cannot harden sufficiently to raise the stress for further deformation in this region above that to cause further strain elsewhere. A neck then forms in the gauge length, and further deformation is confined to this region until fracture. Under these conditions, the reduction in area $(A_0 - A_1)/A_0$, where $A_0$ and $A_1$ are the initial and final areas of the neck, gives a measure of the localized strain, and is a better indication than the strain to fracture measured along the gauge length.

True stress–true strain curves are often plotted to show the work hardening and strain behavior at large strains. The true stress $\sigma$ is the load $P$ divided by the area $A$ of the specimen at that particular stage of strain, and the total true strain in deforming from initial length $l_0$ to length $l_1$ is $\varepsilon = \int_{l_0}^{l_1} (dl/l) = \ln(l_1/l_0)$. The true stress–strain curves often fit the Ludwig relation $\sigma = k\varepsilon^n$, where $n$ is a work-hardening coefficient $\approx 0.1$–$0.5$ and $k$ the strength coefficient. Plastic instability, or necking, occurs when an increase in strain produces no increase in load supported by the specimen, i.e. $dP = 0$, and hence since $P = \sigma A$, then

$$dP = A d\sigma + \sigma dA = 0$$

---

[1] Load relaxations are obtained only on 'hard' beam Polanyi-type machines, where the beam deflection is small over the working load range. With 'soft' machines, those in which the load-measuring device is a soft spring, rapid load variations are not recorded because the extensions required are too large, while in dead-loading machines no load relaxations are possible. In these latter machines sudden yielding will show as merely an extension under constant load.

defines the instability condition. During deformation, the specimen volume is essentially constant (i.e. $dV = 0$) and from

$$dV = d(lA) = Adl + ldA = 0,$$

we obtain

$$\frac{d\sigma}{\sigma} = -\frac{dA}{A} = \frac{dl}{l} = d\varepsilon. \quad (4.5.2)$$

Thus, necking occurs at a strain at which the slope of the true stress–true strain curve equals the true stress at that strain, i.e. $d\sigma/d\varepsilon = \sigma$. Alternatively, since $k\varepsilon^n = \sigma = d\sigma/d\varepsilon = nk\varepsilon^{n-1}$, then $\varepsilon = n$ and necking occurs when the true strain equals the strain-hardening exponent. The instability condition may also be expressed in terms of the conventional (nominal strain),

$$\frac{d\sigma}{d\varepsilon} = \frac{d\sigma}{d\varepsilon_n}\frac{d\varepsilon_n}{d\varepsilon} = \frac{d\sigma}{d\varepsilon_n}\left(\frac{dl/l_0}{dl/l}\right) = \frac{d\sigma}{d\varepsilon_n}\frac{l}{l_0}$$

$$= \frac{d\sigma}{d\varepsilon_n}(1 + \varepsilon_n) = \sigma, \quad (4.5.3)$$

which allows the instability point to be located using Considère's construction (see Figure 4.5-2), by plotting the true stress against nominal strain and drawing the tangent to the curve from $\varepsilon_n = -1$ on the strain axis. The point of contact is the instability stress and the tensile strength is $\sigma/(1 + \varepsilon_n)$.

Tensile specimens can also give information on the type of fracture exhibited. Usually in poly-crystalline metals transgranular fractures occur (i.e. the fracture surface cuts through the grains) and the 'cup-and-cone' type of fracture is extremely common in really ductile metals such as copper. In this, the fracture starts at the center of the necked portion of the test piece and at first grows roughly perpendicular to the tensile axis, so forming the 'cup', but then, as it nears the outer surface, it turns into a 'cone' by fracturing along a surface at about 45° to the tensile axis. In detail the 'cup' itself consists of many irregular surfaces at about 45° to the tensile axis, which gives the fracture a fibrous appearance. Cleavage is also a fairly common type of transgranular fracture, particularly in materials of bcc structure when tested at low temperatures. The fracture surface follows certain crystal planes (e.g. {1 0 0}

planes), as is shown by the grains revealing large bright facets, but the surface also appears granular, with 'river lines' running across the facets where cleavage planes have been torn apart. Intercrystalline fractures sometimes occur, often without appreciable deformation. This type of fracture is usually caused by a brittle second phase precipitating out around the grain boundaries, as shown by copper containing bismuth or antimony.

### 4.5.1.3 Indentation hardness testing

The hardness of a metal, defined as the resistance to penetration, gives a conveniently rapid indication of its deformation behavior. The hardness tester forces a small sphere, pyramid or cone into the surface of the metals by means of a known applied load, and the hardness number (Brinell or Vickers diamond pyramid) is then obtained from the diameter of the impression. The hardness may be related to the yield or tensile strength of the metal, since during the indentation, the material around the impression is plastically deformed to a certain percentage strain. The Vickers hardness number (VPN) is defined as the load divided by the pyramidal area of the indentation, in kgf mm$^{-2}$, and is about three times the yield stress for materials which do not work harden appreciably. The Brinell hardness number (BHN) is defined as the stress $P/A$, in kgf mm$^{-2}$, where $P$ is the load and $A$ the surface area of the spherical cap forming the indentation. Thus,

$$\text{BHN} = P/\left(\frac{\pi}{2}D^2\right)\{1 - [1 - (d/D)^2]^{1/2}\},$$

where $d$ and $D$ are the indentation and indentor diameters respectively. For consistent results the ratio $d/D$ should be maintained constant and small. Under these conditions soft materials have similar values of BHN and VPN. Hardness testing is of importance in both control work and research, especially where information on brittle materials at elevated temperatures is required.

### 4.5.1.4 Impact testing

A material may have a high tensile strength and yet be unsuitable for shock loading conditions. To determine this the impact resistance is usually measured by means of the notched or unnotched Izod or Charpy impact test. In this test a load swings from a given height to strike the specimen, and the energy dissipated in the fracture is measured. The test is particularly useful in showing the decrease in ductility and impact strength of materials of bcc structure at moderately low temperatures. For example, carbon steels have a relatively high ductile–brittle transition temperature (Figure 4.5-1c) and, consequently, they may be used with safety at sub-zero temperatures only if the transition temperature is

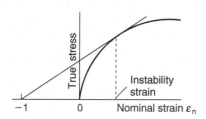

**Figure 4.5-2** Considère's construction.

lowered by suitable alloying additions or by refining the grain size. Nowadays, increasing importance is given to defining a fracture toughness parameter $K_c$ for an alloy, since many alloys contain small cracks which, when subjected to some critical stress, propagate; $K_c$ defines the critical combination of stress and crack length. Brittle fracture is discussed more fully in Chapter 7.

### 4.5.1.5 Creep testing

Creep is defined as plastic flow under constant stress, and although the majority of tests are carried out under constant load conditions, equipment is available for reducing the loading during the test to compensate for the small reduction in cross-section of the specimen. At relatively high temperatures creep appears to occur at all stress levels, but the creep rate increases with increasing stress at a given temperature. For the accurate assessment of creep properties, it is clear that special attention must be given to the maintenance of the specimen at a constant temperature, and to the measurement of the small dimensional changes involved. This latter precaution is necessary, since in many materials a rise in temperature by a few tens of degrees is sufficient to double the creep rate. Figure 4.5-3a shows the characteristics of a typical creep curve and, following the instantaneous strain caused by the sudden application of the load, the creep process may be divided into three stages, usually termed primary or transient creep, secondary or steady-state creep and tertiary or accelerating creep. The characteristics of the creep curve often vary, however, and the tertiary stage of creep may be advanced or retarded if the temperature and stress at which the test is carried out is high or low respectively (see Figure 4.5-3b and c). Creep is discussed more fully in Section 4.5.9.

### 4.5.1.6 Fatigue testing

The fatigue phenomenon is concerned with the premature fracture of metals under repeatedly applied low stresses, and is of importance in many branches of engineering (e.g. aircraft structures). Several different types of testing machines have been constructed in which the stress is applied by bending, torsion, tension or compression, but all involve the same principle of subjecting the material to constant cycles of stress. To express the characteristics of the stress system, three properties are usually quoted: these include (1) the maximum range of stress, (2) the mean stress and (3) the time period for the stress cycle. Four different arrangements of the stress cycle are shown in Figure 4.5-4 but the reverse and the repeated cycle tests (e.g. 'push–pull') are the most common, since they are the easiest to achieve in the laboratory.

The standard method of studying fatigue is to prepare a large number of specimens free from flaws, and to subject them to tests using a different range of stress, $S$, on each group of specimens. The number of stress cycles, $N$, endured by each specimen at a given stress level is recorded and plotted, as shown in Figure 4.5-5. This S–N diagram indicates that some metals can withstand indefinitely the application of a large number of stress reversals, provided the applied stress is below a limiting stress known as the endurance limit. For certain ferrous materials when they are used in the absence of corrosive conditions the assumption of a safe working range of stress seems justified, but for non-ferrous materials and for steels when they are used in corrosive conditions a definite endurance limit cannot be defined. Fatigue is discussed in more detail in Section 4.5.11.

## 4.5.2 Elastic deformation

It is well known that metals deform both elastically and plastically. Elastic deformation takes place at low stresses and has three main characteristics, namely (1) it is reversible, (2) stress and strain are linearly proportional to each other according to Hooke's Law, and (3) it is usually small (i.e. <1% elastic strain).

The stress at a point in a body is usually defined by considering an infinitesimal cube surrounding that point and the forces applied to the faces of the cube by the surrounding material. These forces may be resolved into components parallel to the cube edges and when divided by the area of a face give the nine stress components shown in Figure 4.5-6. A given component $\sigma_{ij}$ is the force acting in the $j$-direction per unit area of face normal to the $i$-direction. Clearly, when $i = j$ we have normal stress components (e.g. $\sigma_{xx}$) which may be either tensile (conventionally positive) or compressive (negative), and when $i \neq j$ (e.g. $\sigma_{xy}$ or $\tau_{xy}$) the stress components are shear. These shear stresses exert couples on the cube and to prevent rotation of the

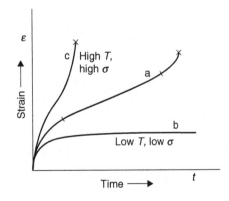

**Figure 4.5-3** Typical creep curves.

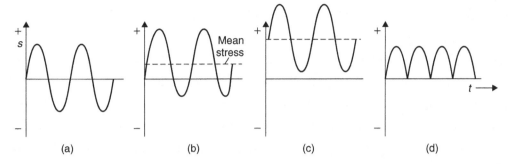

**Figure 4.5-4** Alternative forms of stress cycling: (a) reversed; (b) alternating (mean stress ≠ zero), (c) fluctuating and (d) repeated.

cube the couples on opposite faces must balance, and hence $\sigma_{ij} = \sigma_{ji}$.[2] Thus, stress has only six independent components.

When a body is strained, small elements in that body are displaced. If the initial position of an element is defined by its coordinates $(x,y,z)$ and its final position by $(x+u, y+v, z+w)$, then the displacement is $(u, v, w)$. If this displacement is constant for all elements in the body, no strain is involved, only a rigid translation. For a body to be under a condition of strain the displacements must vary from element to element. A uniform strain is produced when the displacements are linarly proportional to distance. In one dimension then $u = ex$, where $e = du/dx$ is the coefficient of proportionality or nominal tensile strain. For a three-dimensional uniform strain, each of the three components $u$, $v$, $w$ is made a linear function in terms of the initial elemental coordinates, i.e.

$$u = e_{xx}x + e_{xy}y + e_{xz}z$$
$$v = e_{yx}x + e_{yy}y + e_{yz}z$$
$$w = e_{zx}x + e_{zy}y + e_{zz}z.$$

The strains $e_{xx} = du/dx$, $e_{yy} = dv/dy$, $e_{zz} = dw/dz$ are the tensile strains along the $x$-, $y$- and $z$-axes, respectively. The strains $e_{xy}$, $e_{yz}$, etc. produce shear strains and in some cases a rigid-body rotation. The rotation produces no strain and can be allowed for by rotating the reference axes (see Figure 4.5-7). In general, therefore, $e_{ij} = \varepsilon_{ij} + \omega_{ij}$, with $\varepsilon_{ij}$ the strain components and $\omega_{ij}$ the rotation components. If, however, the shear strain is defined as the angle of shear, this is twice the corresponding shear strain component, i.e. $\gamma = 2\varepsilon_{ij}$. The strain tensor, like the stress tensor, has nine components, which are usually written as:

$$\begin{array}{ccc} \varepsilon_{xx} & \varepsilon_{xy} & \varepsilon_{xz} \\ \varepsilon_{yx} & \varepsilon_{yy} & \varepsilon_{yz} \\ \varepsilon_{zx} & \varepsilon_{zy} & \varepsilon_{zz} \end{array} \quad \text{or} \quad \begin{array}{ccc} \varepsilon_{xx} & \frac{1}{2}\gamma_{xy} & \frac{1}{2}\gamma_{xz} \\ \frac{1}{2}\gamma_{yx} & \varepsilon_{yy} & \frac{1}{2}\gamma_{yz} \\ \frac{1}{2}\gamma_{zx} & \frac{1}{2}\gamma_{zy} & \varepsilon_{zz}, \end{array}$$

where $\varepsilon_{xx}$, etc. are tensile strains and $\gamma_{xy}$, etc. are shear strains. All the simple types of strain can be produced from the strain tensor by setting some of the components equal to zero. For example, a pure dilatation (i.e. change of volume without change of shape) is obtained when $\varepsilon_{xx} = \varepsilon_{yy} = \varepsilon_{zz}$ and all other components are zero. Another example is a uniaxial tensile test when the tensile strain along the $x$-axis is simply $e = \varepsilon_{xx}$. However, because of the strains introduced by lateral contraction, $\varepsilon_{yy} = -\nu e$ and $\varepsilon_{zz} = -\nu e$, where $\nu$ is Poisson's ratio; all other components of the strain tensor are zero.

At small elastic deformations, the stress is linearly proportional to the strain. This is Hooke's law and in its simplest form relates the uniaxial stress to the uniaxial strain by means of the modulus of elasticity. For a general situation, it is necessary to write Hooke's law as a linear relationship between six stress components and the six strain components, i.e.

$$\sigma_{xx} = c_{11}\varepsilon_{xx} + c_{12}\varepsilon_{yy} + c_{13}\varepsilon_{zz} + c_{14}\gamma_{yz} + c_{15}\gamma_{zx} + c_{16}\gamma_{xy}$$
$$\sigma_{yy} = c_{21}\varepsilon_{xx} + c_{22}\varepsilon_{yy} + c_{23}\varepsilon_{zz} + c_{24}\gamma_{yz} + c_{25}\gamma_{zx} + c_{26}\gamma_{xy}$$
$$\sigma_{zz} = c_{31}\varepsilon_{xx} + c_{32}\varepsilon_{yy} + c_{33}\varepsilon_{zz} + c_{34}\gamma_{yz} + c_{35}\gamma_{zx} + c_{36}\gamma_{xy}$$
$$\tau_{yz} = c_{41}\varepsilon_{xx} + c_{42}\varepsilon_{yy} + c_{43}\varepsilon_{zz} + c_{44}\gamma_{yz} + c_{45}\gamma_{zx} + c_{46}\gamma_{xy}$$
$$\tau_{zx} = c_{51}\varepsilon_{xx} + c_{52}\varepsilon_{yy} + c_{53}\varepsilon_{zz} + c_{54}\gamma_{yz} + c_{55}\gamma_{zx} + c_{56}\gamma_{xy}$$
$$\tau_{xy} = c_{61}\varepsilon_{xx} + c_{62}\varepsilon_{yy} + c_{63}\varepsilon_{zz} + c_{64}\gamma_{yz} + c_{65}\gamma_{zx} + c_{66}\gamma_{xy}.$$

The constants $c_{11}$, $c_{12}$, ..., $c_{ij}$ are called the elastic stiffness constants.[3]

---

[2] The nine components of stress $\sigma_{ij}$ form a second-rank tensor usually written as: $\begin{array}{ccc} \sigma_{xx} & \sigma_{xy} & \sigma_{xz} \\ \sigma_{yx} & \sigma_{yy} & \sigma_{yz} \\ \sigma_{zx} & \sigma_{zy} & \sigma_{zz} \end{array}$ and is known as the stress tensor.

[3] Alternatively, the strain may be related to the stress, e.g. $\varepsilon_x = s_{11}\sigma_{xx} + s_{12}\sigma_{yy} + s_{13}\sigma_{zz} + \ldots$, in which case the constants $s_{11}, s_{12}, \ldots, s_{ij}$ are called elastic compliances.

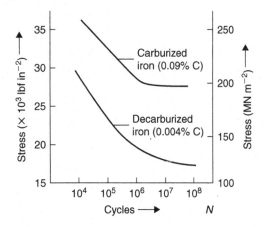

**Figure 4.5-5** S–N curve for carburized and decarburized iron.

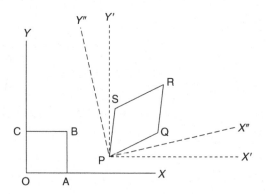

**Figure 4.5-7** Deformation of a square OABC to a parallelogram PQRS involving: (i) a rigid-body translation OP allowed for by redefining new axes $X'Y'$, (ii) a rigid-body rotation allowed for by rotating the axes to $X'''Y'''$, and (iii) a change of shape involving both tensile and shear strains.

Taking account of the symmetry of the crystal, many of these elastic constants are equal or become zero. Thus, in cubic crystals there are only three independent elastic constants $c_{11}$, $c_{12}$ and $c_{44}$ for the three independent modes of deformation. These include the application of (1) a hydrostatic stress $p$ to produce a dilatation $\Theta$ given by

$$p = -\frac{1}{3}(c_{11} + 2c_{12})\Theta = -\kappa\Theta,$$

where $\kappa$ is the bulk modulus, (2) a shear stress on a cube face in the direction of the cube axis defining the shear modulus $\mu = c_{44}$, and (3) a rotation about a cubic axis defining a shear modulus $\mu_1 = (c_{11} - c_{12})/2$. The ratio $\mu/\mu_1$ is the elastic anisotropy factor and in elastically isotropic crystals it is unity with $2c_{44} = c_{11} - c_{12}$; the constants are all interrelated, with $c_{11} = \kappa + 4\mu/3$, $c_{12} = \kappa - 2\mu/3$ and $c_{44} = \mu$.

Table 4.5-1 shows that most metals are far from isotropic and, in fact, only tungsten is isotrospic; the alkali metals and $\beta$-compounds are mostly anisotropic.

Generally, $2c_{44} > (c_{11} - c_{12})$ and, hence, for most elastically anisotropic metals $E$ is maximum in the $\langle 1\,1\,1 \rangle$ and minimum in the $\langle 1\,0\,0 \rangle$ directions. Molybdenum and niobium are unusual in having the reverse anisotropy when $E$ is greatest along $\langle 1\,0\,0 \rangle$ directions. Most commercial materials are polycrystalline, and consequently they have approximately isotropic properties. For such materials the modulus value is usually independent of the direction of measurement because the value observed is an average for all directions, in the various crystals of the specimen. However, if during manufacture a preferred orientation of the grains in the polycrystalline specimen occurs, the material will behave, to some extent, like a single crystal and some 'directionality' will take place.

**Table 4.5-1** Elastic constants of cubic crystals (GNm$^{-2}$)

| Metal | $C_{11}$ | $C_{12}$ | $C_{44}$ | $2c_{44}/(c_{11}-c_{12})$ |
|---|---|---|---|---|
| Na | 006.0 | 004.6 | 005.9 | 8.5 |
| K | 004.6 | 003.7 | 002.6 | 5.8 |
| Fe | 237.0 | 141.0 | 116.0 | 2.4 |
| W | 501.0 | 198.0 | 151.0 | 1.0 |
| Mo | 460.0 | 179.0 | 109.0 | 0.77 |
| Al | 108.0 | 62.0 | 28.0 | 1.2 |
| Cu | 170.0 | 121.0 | 75.0 | 3.3 |
| Ag | 120.0 | 90.0 | 43.0 | 2.9 |
| Au | 186.0 | 157.0 | 42.0 | 3.9 |
| Ni | 250.0 | 160.0 | 118.0 | 2.6 |
| $\beta$-Brass | 129.1 | 109.7 | 82.4 | 8.5 |

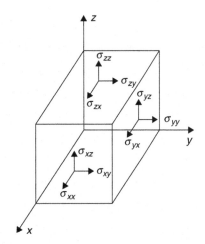

**Figure 4.5-6** Normal and shear stress components.

## 4.5.3 Plastic deformation

### 4.5.3.1 Slip and twinning

The limit of the elastic range cannot be defined exactly but may be considered to be that value of the stress below which the amount of plasticity (irreversible deformation) is negligible, and above which the amount of plastic deformation is far greater than the elastic deformation. If we consider the deformation of a metal in a tensile test, one or other of two types of curve may be obtained. Figure 4.5-1a shows the stress–strain curve characteristic of iron, from which it can be seen that plastic deformation begins abruptly and continues initially with no increase in stress. Figure 4.5-1b shows a stress–strain curve characteristic of copper, from which it will be noted that the transition to the plastic range is gradual. No abrupt yielding takes place and in this case the stress required to start macroscopic plastic flow is known as the flow stress.

Once the yield or flow stress has been exceeded plastic or permanent deformation occurs, and this is found to take place by one of two simple processes, slip (or glide) and twinning. During slip, shown in Figure 4.5-8a, the top half of the crystal moves over the bottom half along certain crystallographic planes, known as slip planes, in such a way that the atoms move forward by a whole number of lattice vectors; as a result the continuity of the lattice is maintained. During twinning (Figure 4.5-8b) the atomic movements are not whole lattice vectors, and the lattice generated in the deformed region, although the same as the parent lattice, is oriented in a twin relationship to it. It will also be observed that, in contrast to slip, the sheared region in twinning occurs over many atom planes, the atoms in each plane being moved forward by the same amount relative to those of the plane below them.

### 4.5.3.2 Resolved shear stress

All working processes such as rolling, extrusion, forging, etc. cause plastic deformation and, consequently, these operations will involve the processes of slip or twinning outlined above. The stress system applied during these working operations is often quite complex, but for plastic deformation to occur the presence of a shear stress is essential. The importance of shear stresses becomes clear when it is realized that these stresses arise in most processes and tests even when the applied stress itself is not a pure shear stress. This may be illustrated by examining a cylindrical crystal of area $A$ in a conventional tensile test under a uniaxial load $P$. In such a test, slip occurs on the slip plane, shown shaded in Figure 4.5-9 the area of which is $A/\cos\phi$, where $\phi$ is the angle between the normal to the plane OH and the axis of tension. The applied force $P$ is spread over this plane and may be resolved into a force normal to the plane along OH, $P\cos\phi$, and a force along OS, $P\sin\phi$. Here, OS is the line of greatest slope in the slip plane and the force $P\sin\phi$ is a shear force. It follows that the applied stress (force/area) is made up of two stresses, a normal stress $(P/A)\cos^2\phi$ tending to pull the atoms apart, and a shear stress $(P/A)\cos\phi\sin\phi$ trying to slide the atoms over each other.

In general, slip does not take place down the line of greatest slope unless this happens to coincide with the crystallographic slip of direction. It is necessary, therefore, to know the resolved shear stress on the slip plane and in the slip direction. Now, if OT is taken to represent the slip direction, the resolved shear stress will be given by

$$\sigma = P\cos\phi\,\sin\phi\,\cos\chi/A,$$

where $\chi$ is the angle between OS and OT. Usually this formula is written more simply as

$$\sigma = P\cos\phi\,\cos\lambda/A, \qquad (4.5.4)$$

where $\lambda$ is the angle between the slip direction OT and the axis of tension. It can be seen that the resolved shear stress has a maximum value when the slip plane is inclined at 45° to the tensile axis, and becomes smaller for

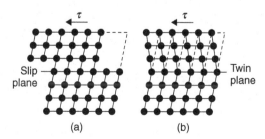

**Figure 4.5-8** Slip and twinning in a crystal.

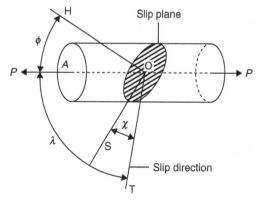

**Figure 4.5-9** Relation between the slip plane, slip direction and the axis of tension for a cylindrical crystal.

angles either greater than or less than 45°. When the slip plane becomes more nearly perpendicular to the tensile axis ($\phi > 45°$), it is easy to imagine that the applied stress has a greater tendency to pull the atoms apart than to slide them. When the slip plane becomes more nearly parallel to the tensile axis ($\phi < 45°$), the shear stress is again small but in this case it is because the area of the slip plane, $A/\cos\phi$, is correspondingly large.

A consideration of the tensile test in this way shows that it is shear stresses which lead to plastic deformation, and for this reason the mechanical behavior exhibited by a material will depend, to some extent, on the type of test applied. For example, a ductile material can be fractured without displaying its plastic properties if tested in a state of hydrostatic or triaxial tension, since under these conditions the resolved shear stress on any plane is zero. Conversely, materials which normally exhibit a tendency to brittle behavior in a tensile test will show ductility if tested under conditions of high shear stresses and low tension stresses. In commercial practice, extrusion approximates closely to a system of high shear stress, and it is common for normally brittle materials to exhibit some ductility when deformed in this way (e.g. when extruded).

## Worked example

A single crystal of iron is pulled along [1 2 3]. Which is the first slip system to operate?

### Solution

| Slip plane | Cos $\phi$ | Burgers vector | Cos $\lambda$ | Schmid factor $\times \sqrt{28} \times \sqrt{42}$ |
|---|---|---|---|---|
| (1 1 0) | $3/\sqrt{28}$ | $1/2[1\bar{1}1]$ | $2/\sqrt{42}$ | 6 |
|  |  | $1/2[1\bar{1}\bar{1}]$ | $-4/\sqrt{42}$ | $-12$ |
| (1 $\bar{1}$ 0) | $-1/\sqrt{28}$ | $1/2[1 1 1]$ | $6/\sqrt{42}$ | $-6$ |
|  |  | $1/2[1 1 \bar{1}]$ | 0 | 0 |
| (1 0 1) | $4/\sqrt{28}$ | $1/2[1 1 \bar{1}]$ | 0 | 0 |
|  |  | $1/2[1\bar{1}\bar{1}]$ | $-4/\sqrt{42}$ | $-16$ ← |
| (1 0 $\bar{1}$) | $-2/\sqrt{28}$ | $1/2[1 1 1]$ | $6/\sqrt{42}$ | $-12$ |
|  |  | $1/2[1\bar{1}\bar{1}]$ | $2/\sqrt{42}$ | $-4$ |
| (0 1 1) | $5/\sqrt{28}$ | $1/2[1 1 \bar{1}]$ | 0 | 0 |
|  |  | $1/2[\bar{1}1\bar{1}]$ | $-2/\sqrt{42}$ | $-10$ |
| (0 1 $\bar{1}$) | $-1/\sqrt{28}$ | $1/2[1 1 1]$ | $6/\sqrt{42}$ | $-6$ |
|  |  | $1/2[\bar{1}1 1]$ | $4/\sqrt{42}$ | $-4$ |

Slip will occur on (1 0 1)[1 $\bar{1}$ $\bar{1}$] first.

## 4.5.3.3 Relation of slip to crystal structure

An understanding of the fundamental nature of plastic deformation processes is provided by experiments on single crystals only, because if a polycrystalline sample is used the result obtained is the average behavior of all the differently oriented grains in the material. Such experiments with single crystals show that, although the resolved shear stress is a maximum along lines of greatest slope in planes at 45° to the tensile axis, slip occurs preferentially along certain crystal planes and directions. Three well-established laws governing the slip behavior exist, namely: (1) the direction of slip is almost always that along which the atoms are most closely packed, (2) slip usually occurs on the most closely packed plane, and (3) from a given set of slip planes and directions, the crystal operates on that system (plane and direction) for which the resolved shear stress is largest. The slip behavior observed in fcc metals shows the general applicability of these laws, since slip occurs along $\langle 1\,1\,0\rangle$ directions in $\{1\,1\,1\}$ planes. In cph metals slip occurs along $\langle 1\,1\,\bar{2}\,0\rangle$ directions, since these are invariably the closest packed, but the active slip plane depends on the value of the axial ratio. Thus, for the metals cadmium and zinc, $c/a$ is 1.886 and 1.856 respectively, the planes of greatest atomic density are the $\{0\,0\,0\,1\}$ basal planes and slip takes place on these planes. When the axial ratio is appreciably smaller than the ideal value of $c/a = 1.633$ the basal plane is not so closely packed, nor so widely spaced, as in cadmium and zinc, and other slip planes operate. In zirconium ($c/a = 1.589$) and titanium ($c/a = 1.587$), for example, slip takes place on the $\{1\,0\,\bar{1}\,0\}$ prism planes at room temperature and on the $\{1\,0\,\bar{1}\,1\}$ pyramidal planes at higher temperatures. In magnesium the axial ratio ($c/a = 1.624$) approximates to the ideal value, and although only basal slip occurs at room temperature, at temperatures above 225°C slip on the $\{1\,0\,2\,1\}$ planes has also been observed. Bcc metals have a single well-defined close-packed $\langle 111\rangle$ direction, but several planes of equally high density of packing, i.e. $\{1\,1\,2\}$, $\{1\,1\,0\}$ and $\{1\,2\,3\}$. The choice of slip plane in these metals is often influenced by temperature and a preference is shown for $\{1\,1\,2\}$ below $T_m/4$, $\{1\,1\,0\}$ from $T_m/4$ to $T_m/2$ and $\{1\,2\,3\}$ at high temperatures, where $T_m$ is the melting point. Iron often slips on all the slip planes at once in a common $\langle 1\,1\,1\rangle$ slip direction, so that a slip line (i.e. the line of intersection of a slip plane with the outer surface of a crystal) takes on a wavy appearance.

## 4.5.3.4 Law of critical resolved shear stress

This law states that slip takes place along a given slip plane and direction when the shear stress reaches a critical value. In most crystals the high symmetry of atomic arrangement provides several crystallographic equivalent planes and directions for slip (i.e. cph crystals have three systems made up of one plane containing three directions, fcc crystals have 12 systems made up of four planes each with three directions, while bcc crystals have many systems) and in such cases slip occurs first on that plane and along that direction for which the maximum stress acts (law 3 above). This is most easily demonstrated by testing in tension a series of zinc single crystals. Then, because zinc is cph in structure only one plane is available for the slip process and the resultant stress–strain curve will depend on the inclination of this plane to the tensile axis. The value of the angle $\phi$ is determined by chance during the process of single-crystal growth, and consequently all crystals will have different values of $\phi$, and the corresponding stress–strain curves will have different values of the flow stress, as shown in Figure 4.5-10a. However, because of the criterion of a critical resolved shear stress, a plot of resolved shear stress (i.e. the stress on the glide plane in the glide direction) versus strain should be a common curve, within experimental error, for all the specimens. This plot is shown in Figure 4.5-10b. The importance of a critical shear stress may be demonstrated further by taking the crystal which has its basal plane oriented perpendicular to the tensile axis, i.e. $\phi = 0°$, and subjecting it to a bend test. In contrast to its tensile behavior, where it is brittle it will now appear ductile, since the shear stress on the slip plane is only zero for a tensile test and not for a bend test. On the other hand, if we take the crystal with its basal plane oriented parallel to the tensile axis (i.e. $\phi = 90°$) this specimen will appear brittle whatever stress system is applied to it. For this crystal, although the shear force is large, owing to the large area of the slip plane, $A/\cos\phi$, the resolved shear stress is always very small and insufficient to cause deformation by slipping.

## 4.5.3.5 Multiple slip

The fact that slip bands, each consisting of many slip lines, are observed on the surface of deformed crystals shows that deformation is inhomogeneous, with extensive slip occurring on certain planes, while the crystal planes lying between them remain practically undeformed. Figure 4.5-11a and b shows such a crystal in which the set of planes shear over each other in the slip direction. In a tensile test, however, the ends of a crystal are not free to move 'sideways' relative to each other, since they are constrained by the grips of the tensile machine. In this case, the central portion of the crystal is altered in orientation, and rotation of both the slip plane and slip direction into the axis of tension occurs, as shown in Figure 4.5-11c. This behavior is more conveniently demonstrated on a stereographic projection of the crystal by considering the rotation of the tensile axis relative to the crystal rather than vice versa. This is illustrated in Figure 4.5-12a for the deformation of a crystal with fcc structure. The tensile axis, $P$, is shown in the unit triangle and the angles between $P$ and $[\bar{1}01]$, and $P$ and $(111)$ are equal to $\lambda$ and $\phi$ respectively. The active slip system is the $(111)$ plane and the $[\bar{1}01]$ direction, and as deformation proceeds the change in orientation is represented by the point, $P$, moving along the zone, shown broken in Figure 4.5-12a, towards $[\bar{1}01]$, i.e. $\lambda$ decreasing and $\phi$ increasing.

As slip occurs on the one system, the primary system, the slip plane rotates away from its position of maximum resolved shear stress until the orientation of the crystal reaches the $[001]$–$[\bar{1}11]$ symmetry line. Beyond this point, slip should occur equally on both the primary system and a second system (the conjugate system) $(\bar{1}\bar{1}1)$–$[011]$, since these two systems receive equal components of shear stress. Subsequently, during the process of multiple or duplex slip the lattice will rotate so as to keep

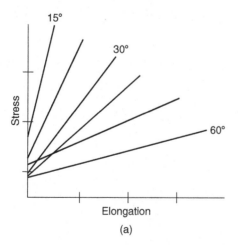 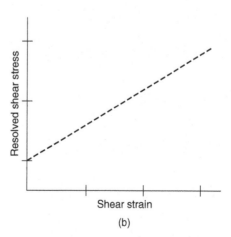

**Figure 4.5-10** Schematic representation of variation of stress versus elongation with orientation of basal plane (a) and constancy of revolved shear stress (b).

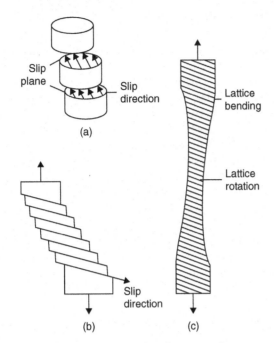

intersect that on the primary system, and to do this is presumably more difficult than to 'fit' a new slip plane in the relatively undeformed region between those planes on which slip has already taken place. This intersection process is more difficult in materials which have a low stacking-fault energy (e.g. α-brass).

### 4.5.3.6 Relation between work hardening and slip

The curves of Figure 4.5-1 show that following the yield phenomenon a continual rise in stress is required to continue deformation, i.e. the flow stress of a deformed metal increases with the amount of strain. This resistance of the metal to further plastic flow as the deformation proceeds is known as work hardening. The degree of work hardening varies for metals of different crystal structure, and is low in hexagonal metal crystals such as zinc or cadmium, which usually slip on one family of planes only. The cubic crystals harden rapidly on working but even in this case when slip is restricted to one slip system (see the curve for specimen A, Figure 4.5-13), the coefficient of hardening, defined as the slope of the plastic portion of the stress–strain curve, is small. Thus, this type of hardening, like overshoot, must be associated with the interaction which results from slip on intersecting families of planes. This interaction will be dealt with more fully in Section 4.5.6.2.

## 4.5.4 Dislocation behavior during plastic deformation

### 4.5.4.1 Dislocation mobility

The ease with which crystals can be plastically deformed at stresses many orders of magnitude less than the theoretical strength ($\tau_t = \mu b/2\pi a$) is quite remarkable, and

**Figure 4.5-11** (a) and (b) show the slip process in an unconstrained single crystal; (c) illustrates the plastic bending in a crystal gripped at its ends.

equal stresses on the two active systems, and the tensile axis moves along the symmetry line towards $[\bar{1}\,1\,2]$. This behavior agrees with early observations on virgin crystals of aluminum and copper, but not with those made on certain alloys, or pure metal crystals given special treatments (e.g. quenched from a high temperature or irradiated with neutrons). Results from the latter show that the crystal continues to slip on the primary system after the orientation has reached the symmetry line, causing the orientation to overshoot this line, i.e. to continue moving towards $[\bar{1}\,0\,1]$, in the direction of primary slip. After a certain amount of this additional primary slip the conjugate system suddenly operates, and further slip concentrates itself on this system, followed by overshooting in the opposite direction. This behavior, shown in Figure 4.5-12b, is understandable when it is remembered that slip on the conjugate system must

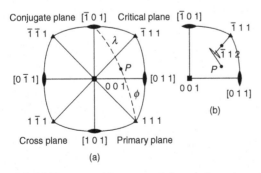

**Figure 4.5-12** Stereographic representation of slip systems in fcc crystals (a) and overshooting of the primary slip system (b).

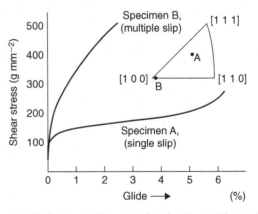

**Figure 4.5-13** Stress–strain curves for aluminum deformed by single and multiple slip (after Lücke and Lange, 1950).

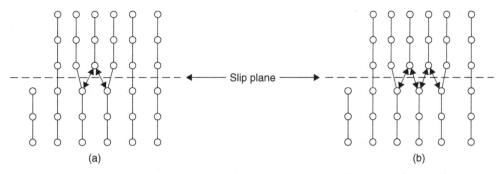

**Figure 4.5-14** Diagram showing structure of edge dislocation during gliding from equilibrium (a) to metastable position (b).

due to the mobility of dislocations. Figure 4.5-14a shows that as a dislocation glides through the lattice it moves from one symmetrical lattice position to another and at each position the dislocation is in neutral equilibrium, because the atomic forces acting on it from each side are balanced. As the dislocation moves from these symmetrical lattice positions some imbalance of atomic forces does exist, and an applied stress is required to overcome this lattice friction. As shown in Figure 4.5-14b, an intermediate displacement of the dislocation also leads to an approximately balanced force system.

The lattice friction depends rather sensitively on the dislocation width $w$ and has been shown by Peierls and Nabarro to be given by

$$\tau \approx \mu \exp[-2\pi w/b] \qquad (4.5.5)$$

for the shear of a rectangular lattice of interplanar spacing $a$ with $w = \mu b/2\pi(1 - v)\tau_t = a/(1 - v)$. The friction stress is therefore often referred to as the Peierls–Nabarro stress. The two opposing factors affecting $w$ are (1) the elastic energy of the crystal, which is reduced by spreading out the elastic strains, and (2) the misfit energy, which depends on the number of misaligned atoms across the slip plane. Metals with close-packed structures have extended dislocations and hence $w$ is large. Moreover, the close-packed planes are widely spaced, with weak alignment forces between them (i.e. have a small $b/a$ factor). These metals have highly mobile dislocations and are intrinsically soft. In contrast, directional bonding in crystals tends to produce narrow dislocations, which leads to intrinsic hardness and brittleness. Extreme examples are ionic and ceramic crystals and the covalent materials such as diamond and silicon. The bcc transition metals display intermediate behavior (i.e. intrinsically ductile above room temperatures but brittle below).

Direct measurements of dislocation velocity $v$ have now been made in some crystals by means of the etch-pitting technique; the results of such an experiment are shown in Figure 4.5-15. Edge dislocations move faster than screws, because of the frictional drag of jogs on screws, and the velocity of both varies rapidly with applied stress $\tau$ according to an empirical relation of the form $v = (\tau/\tau_0)^n$, where $\tau_0$ is the stress for unit speed and $n$ is an index which varies for different materials. At high stresses the velocity may approach the speed of elastic waves $\approx 10^3$ m s$^{-1}$. The index $n$ is usually low ($< 10$) for intrinsically hard, covalent crystals such as Ge, $\approx 40$ for bcc crystals and high ($\approx 200$) for intrinsically soft fcc crystals. It is observed that a critical applied stress is required to start the dislocations moving and denotes the onset of microplasticity. A macroscopic tensile test is

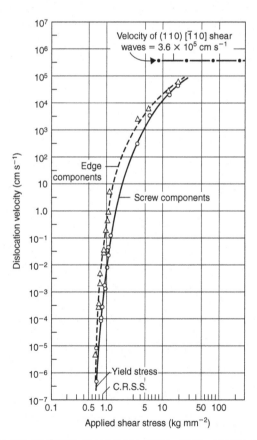

**Figure 4.5-15** Stress dependence of the velocity of edge and screw dislocations in lithium fluoride (from Johnston and Gilman, 1959; courtesy of the American Physical Society).

a relatively insensitive measure of the onset of plastic deformation and the yield or flow stress measured in such a test is related not to the initial motion of an individual dislocation, but to the motion of a number of dislocations at some finite velocity, e.g. ~10 nm s$^{-1}$, as shown in Figure 4.5-16a. Decreasing the temperature of the test or increasing the strain rate increases the stress level required to produce the same finite velocity (see Figure 4.5-16b), i.e. displacing the velocity–stress curve to the right. Indeed, hardening the material by any mechanism has the same effect on the dislocation dynamics. This observation is consistent with the increase in yield stress with decreasing temperature or increasing strain rate. Most metals and alloys are hardened by cold working or by placing obstacles (e.g. precipitates) in the path of moving dislocations to hinder their motion. Such strengthening mechanisms increase the stress necessary to produce a given finite dislocation velocity in a similar way to that found by lowering the temperature.

### 4.5.4.2 Variation of yield stress with temperature and strain rate

The high Peierls–Nabarro stress, which is associated with materials with narrow dislocations, gives rise to a short-range barrier to dislocation motion. Such barriers are effective only over an atomic spacing or so, hence thermal activation is able to aid the applied stress in overcoming them. Thermal activation helps a portion of the dislocation to cross the barrier after which glide then proceeds by the sideways movement of kinks. (This process is shown in Figure 4.5-28, Section 4.5.4.8) Materials with narrow dislocations therefore exhibit a significant temperature sensitivity; intrinsically hard materials rapidly lose their strength with increasing temperature, as shown schematically in Figure 4.5-17a. In this diagram the (yield stress/modulus) ratio is plotted against $T/T_m$ to remove the effect of modulus which decreases with temperature. Figure 4.5-17b shows that materials which exhibit a strong temperature-dependent yield stress also exhibit a high strain-rate sensitivity, i.e. the higher the imposed strain rate, the higher the yield stress. This arises because thermal activation is less effective at the faster rate of deformation.

In bcc metals a high lattice friction to the movement of a dislocation may arise from the dissociation of a dislocation on several planes. As discussed in Chapter 3, when a screw dislocation with Burgers vector $a/2[1\,1\,1]$ lies along a symmetry direction it can dissociate on three crystallographically equivalent planes. If such a dissociation occurs, it will be necessary to constrict the dislocation before it can glide in any one of the slip planes. This constriction will be more difficult to make as the temperature is lowered so that the large temperature dependence of the yield stress in bcc metals, shown in

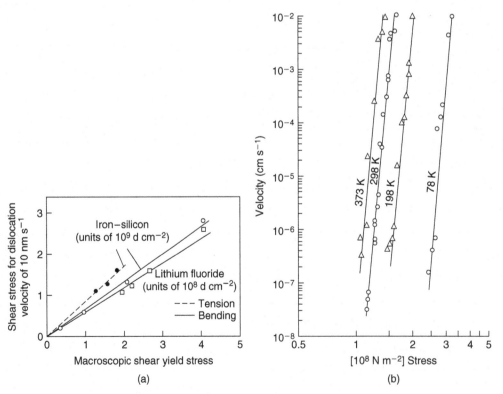

**Figure 4.5-16** (a) Correlation between stress to cause dislocation motion and the macro-yield stresses of crystals, (b) Edge dislocation motions in Fe–3% Si crystals (after Stein and Low, 1960; courtesy of the American Physical Society).

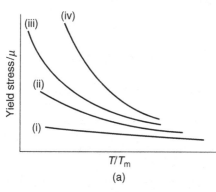 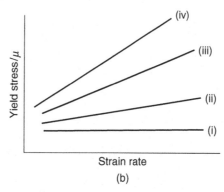

**Figure 4.5-17** Variation of yield stress with: (a) temperature and (b) strain rate for crystals with (i) fcc, (ii) bcc, (iii) ionic-bonded and (iv) covalent-bonded structure.

Figure 4.5-17a and also Figure 4.5-29, may be due partly to this effect. In fcc metals the dislocations lie on {1 1 1} planes, and although a dislocation will dissociate in any given (1 1 1) plane, there is no direction in the slip plane along which the dislocation could also dissociate on other planes; the temperature dependence of the yield stress is small, as shown in Figure 4.5-17a. In cph metals the dissociated dislocations moving in the basal plane will also have a small Peierls force and be glissile with low temperature dependence. However, screw dislocations moving on non-basal planes (i.e. prismatic and pyramidal planes) may have a high Peierls force because they are able to extend in the basal plane, as shown in Figure 4.5-18. Hence, constrictions will once again have to be made before the screw dislocations can advance on non-basal planes. This effect contributes to the high critical shear stress and strong temperature dependence of non-basal glide observed in this crystal system, as mentioned in Chapter 3.

### 4.5.4.3 Dislocation source operation

When a stress is applied to a material the specimen plastically deforms at a rate governed by the strain rate of the deformation process (e.g. tensile testing, rolling, etc.) and the strain rate imposes a particular velocity on the mobile dislocation population. In a crystal of dimensions $L_1 \times L_2 \times 1$ cm, shown in Figure 4.5-19, a dislocation with velocity $v$ moves through the crystal in time $t = L_1/v$ and produces a shear strain $b/L_2$, i.e. the strain rate is $bv/L_1L_2$. If the density of glissible dislocations is $\rho$, the total number of dislocations which become mobile in the crystal is $\rho L_1 L_2$ and the overall strain rate is thus given by

$$\gamma = \frac{b}{L_2}\frac{v}{L_1}\rho L_1 L_2 = \rho b v. \qquad (4.5.6)$$

At conventional strain rates (e.g. 1 s$^{-1}$) the dislocations would be moving at quite moderate speeds of a few cm s$^{-1}$ if the mobile density $\approx 10^7$ cm$^{-2}$. During high-speed deformation the velocity approaches the limiting velocity. The shear strain produced by these dislocations is given by

$$\gamma = \rho b \bar{x}, \qquad (4.5.7)$$

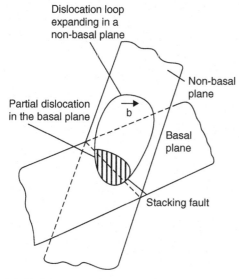

**Figure 4.5-18** Dissociation in the basal plane of a screw dislocation moving on a non-basal glide plane.

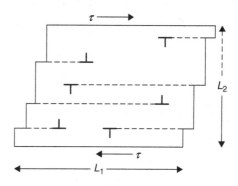

**Figure 4.5-19** Shear produced by gliding dislocations.

where $\bar{x}$ is the average distance a dislocation moves. If the distance $x \simeq 10^{-4}$ cm (the size of an average sub-grain) the maximum strain produced by $\rho \approx 10^7$ is about ($10^7 \times 3 \times 10^{-8} \times 10^{-4}$), which is only a fraction of 1%. In practice, shear strains > 100% can be achieved, and hence to produce these large strains many more dislocations than the original ingrown dislocations are required. To account for the increase in number of mobile dislocations during straining the concept of a dislocation source has been introduced. The simplest type of source is that due to Frank and Read, and accounts for the regenerative multiplication of dislocations. A modified form of the Frank–Read source is the multiple cross-glide source, first proposed by Koehler, which, as the name implies, depends on the cross-slip of screw dislocations and is therefore more common in metals of intermediate and high stacking-fault energy.

Figure 4.5-20 shows a Frank–Read source consisting of a dislocation line fixed at the nodes A and B (fixed, for example, because the other dislocations that join the nodes do not lie in slip planes). Because of its high elastic energy ($\approx 4$ eV per atom plane threaded by a dislocation) the dislocation possesses a line tension tending to make it shorten its length as much as possible (position 1, Figure 4.5-20). This line tension $T$ is roughly equal to $\alpha\mu b^2$, where $\mu$ is the shear modulus, $b$ the Burgers vector and $\alpha$ a constant usually taken to be about $\frac{1}{2}$. Under an applied stress the dislocation line will bow out, decreasing its radius of curvature until it reaches an equilibrium position in which the line tension balances the force due to the applied stress. Increasing the applied stress causes the line to decrease its radius of curvature further until it becomes semicircular (position 2). Beyond this point it has no equilibrium position so it will expand rapidly, rotating about the nodes and taking up the succession of forms indicated by 3, 4 and 5. Between stages 4 and 5 the two parts of the loop below AB meet and annihilate each other to form a complete dislocation loop, which expands into the slip plane and a new line source between A and B. The sequence is then repeated and one unit of slip is produced by each loop that is generated.

To operate the Frank–Read source the force applied must be sufficient to overcome the restoring force on the dislocation line due to its line tension. Referring to Figure 4.5-21, this would be $2Td\theta/2 > \tau bld\theta/2$, and if $T \sim \mu b^2/2$ the stress to do this is about $\mu b/l$, where $\mu$ and $b$ have their usual meaning and $l$ is the length of the Frank–Read source; the substitution of typical values ($\mu = 4 \times 10^{10}$N m$^{-2}$, $b = 2.5 \times 10^{-10}$m and $l = 10^{-6}$m) into this estimate shows that a critical shear stress of about 10 MPa is required. This value is somewhat less than, but of the same order as, that observed for the yield stress of virgin pure metal single crystals. Another source mechanism involves multiple cross-slip, as shown in Figure 4.5-22. It depends on the Frank–Read principle but does not require a dislocation segment to be anchored by nodes. Thus, if part of a moving screw dislocation undergoes double cross-slip the two pieces of edge dislocation on the cross-slip plane effectively act as anchoring points for a new source. The loop expanding on the slip plane parallel to the original plane may operate as a Frank–Read source, and any loops produced may in turn cross slip and become a source. This process therefore not only increases the number of dislocations on the original slip plane, but also causes the slip band to widen.

The concept of the dislocation source accounts for the observation of slip bands on the surface of deformed metals. The amount of slip produced by the passage of a single dislocation is too small to be observable as a slip line or band under the light microscope. To be resolved it must be at least 300 nm in height and hence $\approx$ 1000 dislocations must have operated in a given slip band. Moreover, in general, the slip band has considerable width, which tends to support the operation of the cross-glide source as the predominant mechanism of dislocation multiplication during straining.

### Worked example

A Frank–Read source is operated by an applied stress of magnitude $10^{-4}\mu$, where $\mu$ is the shear modulus. If the limiting speed of a dislocation is $10^3$ ms$^{-1}$, show that the source could nucleate a slip band which is observed in the light microscope to form in about $10^{-6}$ s.

### Solution

To observe a slip band in the light microscope, which has a resolution of approximately the wavelength of light, requires about 1000 dislocations to have emanated from the source.

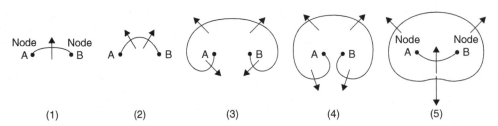

**Figure 4.5-20** Successive stages in the operation of a Frank–Read source. The plane of the paper is assumed to be the slip plane.

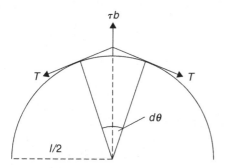

**Figure 4.5-21** Geometry of Frank–Read source used to calculate the stress to operate.

The stress to operate the source is $\tau = \mu b/\ell$, where $\mu$ is the shear modulus, $b$ the Burgers vector and $\ell$ the source length. Time to produce one dislocation loop takes $t \sim \ell/v$, where $v$ is the dislocation velocity. Thus, to nucleate a slip band of 1000 dislocations requires total time

$$t = 10^3 \ell/v = \frac{10^3 \mu b}{\tau v} = \frac{10^3 \mu \times 2.5 \times 10^{-10}}{10^{-4} \mu \times 10^3}$$
$$= 2.5 \times 10^{-6} \text{s}.$$

### 4.5.4.4 Discontinuous yielding

In some materials the onset of macroscopic plastic flow begins in an abrupt manner with a yield drop in which the applied stress falls, during yielding, from an upper to a lower yield point. Such yield behavior is commonly found in iron containing small amounts of carbon or nitrogen as impurity. The main characteristics of the yield phenomenon in iron may be summarized as follows.

#### 4.5.4.4.1 Yield point

A specimen of iron during tensile deformation (Figure 4.5-23a, curve 1) behaves elastically up to a certain high load A, known as the upper yield point, and then it suddenly yields plastically. The important feature to note from this curve is that the stress required to maintain plastic flow immediately after yielding has started is lower than that required to start it, as shown by the fall in load from A to B (the lower yield point). A yield point elongation to C then occurs, after which the specimen work hardens and the curve rises steadily and smoothly.

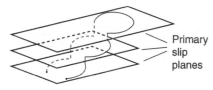

**Figure 4.5-22** Cross-slip multiplication source.

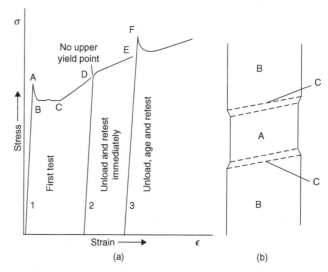

**Figure 4.5-23** Schematic representation of strain ageing (a) and Lüders band formation (b).

#### 4.5.4.4.2 Overstraining

The yield point can be removed temporarily by applying a small preliminary plastic strain to the specimen. Thus, if after reaching the point D, for example, the specimen is unloaded and a second test is made fairly soon afterwards, a stress–strain curve of type 2 will be obtained. The specimen deforms elastically up to the unloading point D, and the absence of a yield point at the beginning of plastic flow is characteristic of a specimen in an overstrained condition.

#### 4.5.4.4.3 Strain-age hardening

If a specimen which has been overstrained to remove the yield point is allowed to rest, or age, before retesting, the yield point returns, as shown in Figure 4.5-23a, curve 3. This process, which is accompanied by hardening (as shown by the increased stress, EF, to initiate yielding) is known as strain ageing or, more specifically, strain-age hardening. In iron, strain ageing is slow at room temperature but is greatly speeded up by annealing at a higher temperature. Thus, a strong yield point returns after an ageing treatment of only a few seconds at 200°C, but the same yield point will take many hours to develop if ageing is carried out at room temperature.

#### 4.5.4.4.4 Lüders band formation

Closely related to the yield point is the formation of Lüders bands. These bands are markings on the surface of the specimen which distinguish those parts of the specimen that have yielded, A, from those which have not, B. Arrival at the upper yield point is indicated by the formation of one or more of these bands and, as the specimen passes through the stage of the yield point elongation, these bands spread along the specimen and coalesce until the entire gauge length has been covered. At this stage the whole of the material within the gauge length has been overstrained, and the yield point elongation is complete. The growth of a Lüders band is shown diagrammatically in Figure 4.5-23b. It should be noted that the band is a macroscopic band crossing all the grains in the cross-section of a polycrystalline specimen, and thus the edges of the band are not necessarily the traces of individual slip planes. A second point to observe is that the rate of plastic flow in the edges of a band can be very high, even in an apparently slow test; this is because the zones, marked C in Figure 4.5-23b, are very narrow compared with the gauge length.

These Lüders bands frequently occur in drawing and stamping operations, when the surface markings in relief are called stretcher strains. These markings are unsightly in appearance and have to be avoided on many finished products. The remedy consists of overstraining the sheet prior to pressing operations, by means of a temper roll, or roller leveling, pass so that the yield phenomenon is eliminated. It is essential, once this operation has been performed, to carry out pressing before the sheet has time to strain-age; the use of a 'non-ageing' steel is an alternative remedy.

These yielding effects are influenced by the presence of small amounts of carbon or nitrogen atoms interacting with dislocations. The yield point can be removed by annealing at 700°C in a wet hydrogen atmosphere, and cannot subsequently be restored by any strain-ageing treatment. Conversely, exposing the decarburized specimen to an atmosphere of dry hydrogen containing a trace of hydrocarbon at 700°C for as little as one minute restores the yield point. The carbon and nitrogen atoms can also be removed from solution in other ways – for example, by adding to the iron such elements as molybdenum, manganese, chromium, vanadium, niobium or titanium, which have a strong affinity for forming carbides or nitrides in steels. For this reason, these elements are particularly effective in removing the yield point and producing a non-strain-ageing steel.

The carbon/nitrogen atoms are important in the yielding process because they interact with the dislocations and immobilize them. This locking of the dislocations is brought about because the strain energy due to the distortion of a solute atom can be relieved if it fits into a structural region where the local lattice parameter approximates to that of the natural lattice parameter of the solute. Such a condition will be brought about by the segregation of solute atoms to the dislocations, with large substitutional atoms taking up lattice positions in the expanded region and small ones in the compressed region; small interstitial atoms will tend to segregate to interstitial sites below the half-plane. Thus, where both dislocations and solute atoms are present in the lattice, interactions of the stress field can occur, resulting in a lowering of the strain energy of the system. This provides a driving force tending to attract solute atoms to dislocations and if the necessary time for diffusion is allowed, a solute atom 'atmosphere' will form around each dislocation.

When a stress is applied to a specimen in which the dislocations are locked by carbon atoms the dislocations are not able to move at the stress level at which free dislocations are normally mobile. With increasing stress, yielding occurs when dislocations suddenly become mobile, either by breaking away from the carbon atmosphere or by nucleating fresh dislocations at stress concentrations. At this high stress level the mobile dislocation density increases rapidly. The lower yield stress is then the stress at which free dislocations continue to move and produce plastic flow. The overstrained condition corresponds to the situation where the mobile dislocations, brought to rest by unloading the specimen, are set in motion again by reloading before the carbon atmospheres have time to develop by diffusion. If, however, time is allowed for diffusion to take place, new atmospheres can re-form and immobilize the dislocations again. This is the strain-aged condition when the original yield characteristics reappear.

The upper yield point in conventional experiments on polycrystalline materials is the stress at which initially yielded zones trigger yield in adjacent grains. As more and more grains are triggered, the yield zones spread across the specimen and form a Lüders band.

The propagation of yield is thought to occur when a dislocation source operates and releases an avalanche of dislocations into its slip plane, which eventually pile up at a grain boundary or other obstacle. The stress concentration at the head of the pile-up acts with the applied stress on the dislocations of the next grain and operates the nearest source, so that the process is repeated in the next grain. The applied shear stress $\sigma_y$ at which yielding propagates is given by

$$\sigma_y = \sigma_i + (\sigma_c r^{1/2})d^{-1/2}, \qquad (4.5.8)$$

where $r$ is the distance from the pile-up to the nearest source, $2d$ is the grain diameter and $\sigma_c$ is the stress required to operate a source which involves unpinning a dislocation $\tau_c$ at that temperature. Equation (4.5.8) reduces to the Hall–Petch equation $\sigma_y = \sigma_i + k_y d^{-1/2}$, where $\sigma_i$ is the 'friction' stress term and $k_y$ the grain size dependence parameter ($=m^2 \tau_c r^{1/2}$) discussed in Section 4.5.4.11.

### 4.5.4.5 Yield points and crystal structure

The characteristic feature of discontinuous yielding is that at the yield point the specimen goes from a condition where the availability of mobile dislocations is limited to one where they are in abundance, the increase in mobile density largely arising from dislocation multiplication at the high stress level. A further feature is that not all the dislocations have to be immobilized to observe a yield drop. Indeed, this is not usually possible because specimen handling, non-axial loading, scratches, etc. give rise to stress concentrations that provide a small local density of mobile dislocations (i.e. pre-yield microstrain).

For materials with a high Peierls–Nabarro (P–N) stress, yield drops may be observed even when they possess a significant mobile dislocation density. A common example is that observed in silicon; this is an extremely pure material with no impurities to lock dislocations, but usually the dislocation density is quite modest ($10^7$ m m$^{-3}$) and possesses a high P–N stress.

When these materials are pulled in a tensile test, the overall strain rate $\dot{\gamma}$ imposed on the specimen by the machine has to be matched by the motion of dislocations according to the relation $\dot{\gamma} = \rho b v$. However, because $\rho$ is small the individual dislocations are forced to move at a high speed $v$, which is only attained at a high stress level (the upper yield stress) because of the large P–N stress. As the dislocations glide at these high speeds, rapid multiplication occurs and the mobile dislocation density increases rapidly. Because of the increased value of the term $\rho$, a lower average velocity of dislocations is then required to maintain a constant strain rate, which means a lower glide stress. The stress that can be supported by the specimen thus drops during initial yielding to the lower yield point, and does not rise again until the dislocation–dislocation interactions caused by the increased $\rho$ produce a significant work hardening.

In the fcc metals, the P–N stress is quite small and the stress to move a dislocation is almost independent of velocity up to high speeds. If such metals are to show a yield point, the density of mobile dislocations must be reduced virtually to zero. This can be achieved as shown in Figure 4.5-24 by the tensile testing of whisker crystals which are very perfect. Yielding begins at the stress required to create dislocations in the perfect lattice, and the upper yield stress approaches the theoretical yield strength. Following multiplication, the stress for glide of these dislocations is several orders of magnitude lower.

Bcc transition metals such as iron are intermediate in their plastic behavior between the fcc metals and diamond cubic Si and Ge. Because of the significant P–N stress these bcc metals are capable of exhibiting a sharp yield point, even when the initial mobile dislocation density is not zero, as shown by the calculated curves of Figure 4.5-25. However, in practice, the dislocation density of well-annealed pure metals is about $10^{10}$ m m$^{-3}$ and too high for any significant yield drop without an element of dislocation locking by carbon atoms.

It is evident that discontinuous yielding can be produced in all the common metal structures provided the

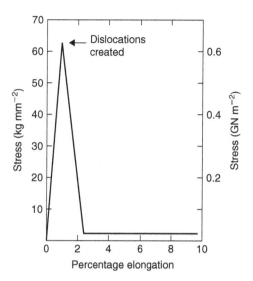

Figure 4.5-24 Yield point in a copper whisker.

appropriate solute elements are present and correct testing procedure adopted. The effect is particularly strong in the bcc metals and has been observed in α-iron, molybdenum, niobium, vanadium and β-brass, each containing a strongly interacting interstitial solute element. The hexagonal metals (e.g. cadmium and zinc) can also show the phenomenon provided interstitial nitrogen atoms are added. The copper- and aluminum-based fcc alloys also exhibit yielding behavior, but often to a lesser degree. In this case it is substitutional atoms (e.g. zinc in α-brass and copper in aluminum alloys) which are responsible for the phenomenon (see Section 4.5.4.7).

## Worked example

A low-carbon steel exhibits a yield point when tensile tested at a strain rate of 1 s$^{-1}$. If the density of mobile dislocations before and after the yield phenomenon is $10^{11}$ and $10^{14}$ m$^{-2}$ respectively, estimate:

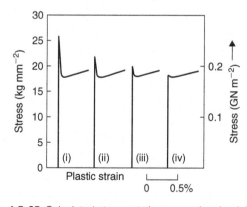

Figure 4.5-25 Calculated stress–strain curves showing influence of initial dislocation density on the yield drop in iron for $n = 35$ with: (i) $10^1$ cm$^{-2}$, (ii) $10^3$ cm$^{-2}$, (iii) $10^5$ cm$^{-2}$ and (iv) $10^7$ cm$^{-2}$ (after Hahn, 1962).

(i) The dislocation velocity at the upper yield point, and

(ii) The magnitude of the yield drop.

(Take the lattice parameter of the alloy to be 0.28 nm and the stress dependence of the dislocation velocity to have an exponent $n = 35$.)

### Solution

(i) Plastic strain rate is $\dot{\varepsilon} = \phi b \rho_m \bar{v}$. Thus, at the upper yield point (uyp),

$$1 = 0.5 \times \frac{\sqrt{3}}{2} \times 2.8 \times 10^{-10} \times 10^{11} \times \bar{v},$$

since $b = (\sqrt{3}/2)a$ and $\phi \sim 0.5$.

$$\therefore \bar{v} = 8.06 \times 10^{-2} \, \text{m s}^{-1}.$$

(ii) The average velocity $\bar{v} = (\sigma/\sigma_0)^n$. Since $\dot{\varepsilon}$ is the same at both upper and lower yield points, then

$$0.5 \times 10^{11} \times (\sigma_{\text{uyp}}/\sigma_0)^{35}$$
$$= 0.5 \times 10^{14} \times (\sigma_{\text{lyp}}/\sigma_0)^{35}$$

$$(\sigma_{\text{uyp}}/\sigma_{\text{lyp}})^{35} = 10^3.$$

The ratio of upper to lower yield points, $\sigma_{\text{uyp}}/\sigma_{\text{lyp}} = 10^{3/35} = 1.2$.

## 4.5.4.6 Discontinuous yielding in ordered alloys

Discontinuous yield points have been observed in a wide variety of $A_3B$-type alloys. Figure 4.5-26 shows the development of the yield point in $Ni_3Fe$ on ageing. The addition of Al speeds up the kinetics of ordering and therefore the onset of the yield point. Ordered materials deform by superdislocation motion and the link between yield points and superdislocations is confirmed by the observation that, in $Cu_3Au$, for example, a transition from groups of single dislocations to more randomly arranged superdislocation pairs takes place at $\sim S = 0.7$ (see Chapter 3), and this coincides with the onset of a large yield drop and rapid rise in work hardening.

Sharp yielding may be explained by at least two mechanisms, namely (1) cross-slip of the superdislocation onto the cube plane to lower the APB energy effectively pinning it and (2) dislocation locking by rearrangement of the APB on ageing. The shear APB between a pair of superdislocations is likely to be

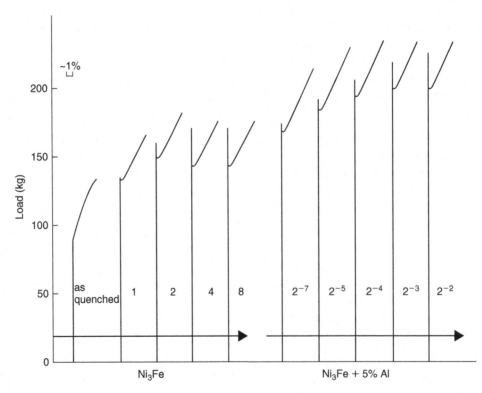

**Figure 4.5-26** Development of a yield point with ageing at 490°C for the times in days indicated. (a) $Ni_3Fe$. (b) $Ni_3Fe + 5\%$ Al. The tests are at room temperature.

energetically unstable, since there are many like bonds across the interface and thermal activation will modify this sharp interface by atomic rearrangement. This APB-locking model will give rise to sharp yielding because the energy required by the lead dislocation in creating sharp APB is greater than that released by the trailing dislocation initially moving across diffuse APB. Experimental evidence favors the APB model and weak-beam electron microscopy (see Figure 4.5-27) shows that the superdislocation separation for a shear APB corresponds to an energy of 48 ± 5 mJm$^{-2}$, whereas a larger dislocation separation corresponding to an APB energy of 25 ± 3 mJ m$^{-2}$ was observed for a strained and aged $Cu_3Au$.

### 4.5.4.7 Solute–dislocation interaction

Iron containing carbon or nitrogen shows very marked yield point effects and there is a strong elastic interaction between these solute atoms and the dislocations. The solute atoms occupy interstitial sites in the lattice and produce large tetragonal distortions, as well as large-volume expansions. Consequently, they can interact with both shear and hydrostatic stresses and can lock screw as well as edge dislocations. Strong yielding behavior is also expected in other bcc metals, provided they contain interstitial solute elements. On the other hand, in the case of fcc metals the arrangement of lattice positions around either interstitial or substitutional sites is too symmetrical to allow a solute atom to produce an asymmetrical distortion, and the atmosphere locking of screw dislocations, which requires a shear stress interaction, would appear to be impossible. Then by this argument, since the screw dislocations are not locked, a drop in stress at the yield point should not be observed. Nevertheless, yield points are observed in fcc materials and one reason for this is that unit dislocations in fcc metals dissociate into pairs of partial dislocations which are elastically coupled by a stacking fault. Moreover, since their Burgers vectors intersect at 120° there is no orientation of the line of the pair for which both can be pure screws. At least one of them must have a substantial edge component, and a locking of this edge component by hydrostatic interactions should cause a locking of the pair, although it will undoubtedly be weaker.

In its quantitative form the theory of solute atom locking has been applied to the formation of an atmosphere around an edge dislocation due to hydrostatic interaction. Since hydrostatic stresses are scalar quantities, no knowledge is required in this case of the orientation of the dislocation with respect to the interacting solute atom, but it is necessary in calculating shear stresses interactions.[4] Cottrell and Bilby have shown that if the introduction of a solute atom causes a volume change $\Delta v$ at some point in the lattice where the hydrostatic pressure of the stress field is $p$, the interaction energy is

$$V = p\Delta v = K\Theta\Delta v, \quad (4.5.9)$$

where $K$ is the bulk modulus and $\Theta$ is the local dilatation strain. The dilatation strain at a point $(R, \theta)$ from a positive edge dislocation is $b(1 - 2\nu) \times \sin \theta/2\pi R(1 - \nu)$, and substituting $K = 2\mu(1 + \nu)/3(1 - 2\nu)$, where $\mu$ is the shear modulus and $\nu$ Poisson's ratio, we get the expression:

$$V_{(R,\theta)} = b(1+\nu)\mu\Delta v \sin \theta/3\pi R(1-\nu)$$
$$= A \sin \theta/R. \quad (4.1.10)$$

This is the interaction energy at a point whose polar coordinates with respect to the center of the dislocation are $R$ and $\theta$. We note that $V$ is positive on the upper side ($0 < \theta < \pi$) of the dislocation for a large atom ($\Delta v > 0$) and negative on the lower side, which agrees with the qualitative picture of a large atom being repelled from the compressed region and attracted into the expanded one.

It is expected that the site for the strongest binding energy $V_{max}$ will be at a point $\theta = 3\pi/2$, $R = r_0 \approx b$; and using known values of $\mu$, $\nu$ and $\Delta v$ in equation (4.5.10), we obtain $A \approx 3 \times 10^{-29}$ N m$^2$ and $V_{max} \approx 1$ eV for carbon or nitrogen in $\alpha$-iron. This value is almost certainly too high because of the limitations of the interaction energy equation in describing conditions near the center of a dislocation, and a more realistic value obtained from experiment (e.g. internal friction experiments) is $V_{max} \approx \frac{1}{2}$ to $\frac{3}{4}$ eV. For a substitutional solute atom such as zinc in copper, $\Delta v$ is not only smaller but also easier to calculate from lattice parameter measurements. Thus, if $r$ and $r(1 + \varepsilon)$ are the atomic radii of the solvent and solute respectively, where $\varepsilon$ is the misfit value, the volume change $\Delta v$ is $4\pi r^3 \varepsilon$ and equation (4.5.10) becomes

$$V = 4(1+\nu)\mu b\varepsilon r^3 \sin \theta/3(1-\nu)R$$
$$= A \sin \theta/R. \quad (4.5.11)$$

Taking the known values $\mu = 40$ GN m$^{-2}$, $\nu = 0.36$, $b = 2.55 \times 10^{-10}$ m, $r_0$ and $\varepsilon = 0.06$, we find $A \approx 5 \times 10^{-30}$ N m$^2$, which gives a much lower binding energy, $V_{max} = \frac{1}{8}$ eV.

---

[4] To a first approximation a solute atom does not interact with a screw dislocation, since there is no dilatation around the screw; a second-order dilatation exists, however, which gives rise to a non-zero interaction falling off with distance from the dislocation according to $1/r^2$. In real crystals, anisotropic elasticity will lead to first-order size effects, even with screw dislocations, and hence a substantial interaction is to be expected.

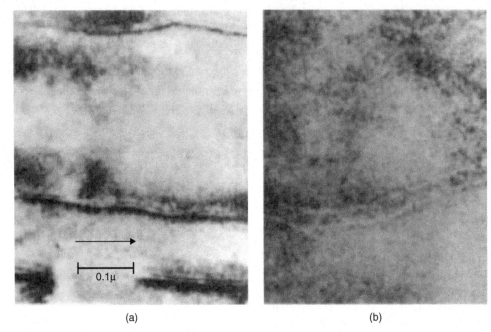

**Figure 4.5-27** Weak-beam micrographs showing separation of superdislocation partials in $Cu_3Au$. (a) As deformed. (b) After ageing at 225°C (after Morris and Smallman, 1975).

The yield phenomenon is particularly strong in iron because an additional effect is important; this concerns the type of atmosphere a dislocation gathers round itself, which can be either condensed or dilute. During the strain-ageing process, migration of the solute atoms to the dislocation occurs and two important cases arise. First, if all the sites at the center of the dislocation become occupied the atmosphere is then said to be condensed; each atom plane threaded by the dislocation contains one solute atom at the position of maximum binding, together with a diffuse cloud of other solute atoms further out. If, on the other hand, equilibrium is established before all the sites at the center are saturated, a steady state must be reached in which the probability of solute atoms leaving the center can equal the probability of their entering it. The steady-state distribution of solute atoms around the dislocations is then given by the relation

$$C_{(R,\theta)} = c_0 \exp[V_{(R,\theta)}/\mathbf{k}T],$$

where $c_0$ is the concentration far from a dislocation, $\mathbf{k}$ is Boltzmann's constant, $T$ is the absolute temperature and $C$ the local impurity concentration at a point near the dislocation where the binding energy is $V$. This is known as the dilute or Maxwellian atmosphere. Clearly, the form of an atmosphere will be governed by the concentration of solute atoms at the sites of maximum binding energy $V_{max}$, and for a given alloy (i.e. $c_0$ and $V_{max}$ fixed) this concentration will be

$$C_{V_{max}} = c_0 \exp(V_{max}/\mathbf{k}T) \quad (4.5.12)$$

as long as $C_{V_{max}}$ is less than unity. The value of $C_{V_{max}}$ depends only on the temperature, and as the temperature is lowered $C_{V_{max}}$ will eventually rise to unity. By definition the atmosphere will then have passed from a dilute to a condensed state. The temperature at which this occurs is known as the condensation temperature $T_c$, and can be obtained by substituting the value $C_{V_{max}} = 1$ in equation (4.5.12) when

$$T_c = V_{max}/\mathbf{k} \ln(1/c_0). \quad (4.5.13)$$

Substituting the value of $V_{max}$ for iron, i.e. $\frac{1}{2}$ eV, in this equation we find that only a very small concentration of carbon or nitrogen is necessary to give a condensed atmosphere at room temperature, and with the usual concentration strong yielding behavior is expected up to temperatures of about 400°C.

In the fcc structure, although the locking between a solute atom and a dislocation is likely to be weaker, condensed atmospheres are still possible if this weakness can be compensated for by sufficiently increasing the concentration of the solution. This may be why examples of yielding in fcc materials have been mainly obtained from alloys. Solid solution alloys of aluminum usually contain less than 0.1 at.% of solute element, and these show yielding in single crystals only at low temperature (e.g. liquid nitrogen temperature, $-196°C$), whereas supersaturated alloys show evidence of strong yielding even in polycrystals at room temperature; copper dissolved in aluminum has a misfit value $\varepsilon \approx 0.12$, which corresponds to $V_{max} = \frac{1}{4}$ eV, and from equation

(4.5.13) it can be shown that a 0.1 at.% alloy has a condensation temperature $T_c = 250$ K. Copper-based alloys, on the other hand, usually form extensive solid solutions and, consequently, concentrated alloys may exhibit strong yielding phenomena.

The best-known example is α-brass and, because $V_{max} \approx \frac{1}{8}$ eV, a dilute alloy containing 1 at.% zinc has a condensation temperature $T_c \approx 300$ K. At low zinc concentrations (1–10%) the yield point in brass is probably solely due to the segregation of zinc atoms to dislocations. At higher concentrations, however, it may also be due to short-range order.

### 4.5.4.8 Dislocation locking and temperature

The binding of a solute atom to a dislocation is short range in nature, and is effective only over an atomic distance or so (Figure 4.5-28). Moreover, the dislocation line is flexible and this enables yielding to begin by throwing forward a small length of dislocation line, only a few atomic spacings long, beyond the position marked $x_2$. The applied stress then separates the rest of the dislocation line from its anchorage by pulling the sides of this loop outward along the dislocation line, i.e. by double kink movement. Such a breakaway process would lead to a yield stress which depends sensitively on temperature, as shown in Figure 4.5-29a. It is observed, however, that $k_y$, the grain-size dependence parameter in the Hall–Petch equation, in most annealed bcc metals is almost independent of temperature down to the range (< 100 K) where twinning

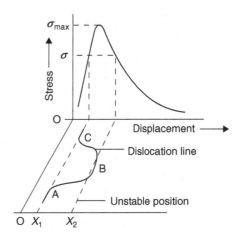

**Figure 4.5-28** Stress–displacement curve for the breakaway of a dislocation from its atmosphere (after Cottrell, 1957; courtesy of the Institution of Mechanical Engineers).

occurs, and that practically all the large temperature dependence is due to $\sigma_i$ (see Figure 4.5-29b). To explain this observation it is argued that when locked dislocations exist initially in the material, yielding starts by unpinning them if they are weakly locked (this corresponds to the condition envisaged by Cottrell–Bilby), but if they are strongly locked it starts instead by the creation of new dislocations at points of stress concentration. This is an athermal process and thus $k_y$ is almost independent of temperature. Because of the rapid diffusion of interstitial elements the conventional annealing and normalizing treatments should commonly produce strong locking. In support of this theory, it is observed that $k_y$ is dependent

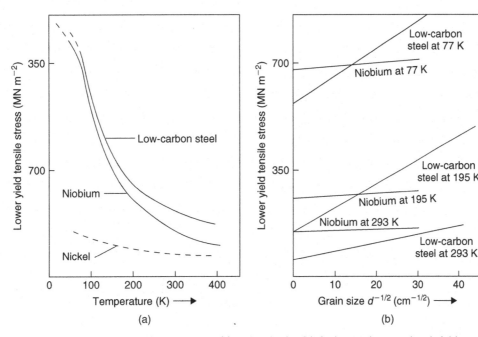

**Figure 4.5-29** Variation of lower yield stress with temperature (a) and grain size (b), for low-carbon steel and niobium; the curve for nickel is shown in (a) for comparison (after Adams, Roberts and Smallman, 1960; Hull and Mogford, 1958).

on temperature in the very early stages of ageing following either straining or quenching, but on subsequent ageing $k_y$ becomes temperature independent. The interpretation of $k_y$ therefore depends on the degree of ageing.

Direct observations of crystals that have yielded show that the majority of the strongly anchored dislocations remain locked and do not participate in the yielding phenomenon. Thus, large numbers of dislocations are generated during yielding by some other mechanism than breaking away from Cottrell atmospheres, and the rapid dislocation multiplication, which can take place at the high stress levels, is now considered the most likely possibility. Prolonged ageing tends to produce coarse precipitates along the dislocation line and unpinning by bowing out between them should easily occur before grain boundary creation. This unpinning process would also give $k_y$ independent of temperature.

### 4.5.4.9 Inhomogeneity interaction

A different type of elastic interaction can exist which arises from the different elastic properties of the solvent matrix and the region near a solute. Such an inhomogeneity interaction has been analyzed for both a rigid and a soft spherical region; the former corresponds to a relatively hard impurity atom and the latter to a vacant lattice site. The results indicate that the interaction energy is of the form $B/r^2$, where $B$ is a constant involving elastic constants and atomic size. It is generally believed that the inhomogeneity effect is small for solute–dislocation interactions but dominates the size effect for vacancy–dislocation interaction. The kinetics of ageing support this conclusion.

### 4.5.4.10 Kinetics of strain ageing

Under a force $F$ an atom migrating by thermal agitation acquires a steady drift velocity $v = DF/kT$ (in addition to its random diffusion movements) in the direction of $F$, where $D$ is the coefficient of diffusion. The force attracting a solute atom to a dislocation is the gradient of the interaction energy $dV/dr$ and hence $v = (D/kT)(A/r^2)$. Thus, atoms originally at a distance $r$ from the dislocation reach it in a time given approximately by

$$t = r/v = r^3 kT/AD.$$

After this time $t$ the number of atoms to reach unit length of dislocation is

$$n(t) = \pi r^2 c_0 = \pi c_0 [AD/kT] t]^{2/3},$$

where $c_0$ is the solute concentration in uniform solution in terms of the number of atoms per unit volume. If $\rho$ is the density of dislocations (cm cm$^{-3}$) and $f$ the fraction of the original solute which has segregated to the dislocation in time $t$, then

$$f = \pi \rho [(AD/kT)t]^{2/3}. \qquad (4.5.14)$$

This expression is valid for the early stages of ageing, and may be modified to fit the later stages by allowing for the reduction in the matrix concentration as ageing proceeds, such that the rate of flow is proportional to the amount left in the matrix,

$$df/dt = \pi \rho (AD/kT)^{2/3}(2/3)t^{-1/3}(1-f),$$

which when integrated gives

$$f = 1 - \exp\{-\pi \rho [(AD/kT)t]^{2/3}\}. \qquad (4.5.15)$$

This reduces to the simpler equation (4.5.14) when the exponent is small, and is found to be in good agreement with the process of segregation and precipitation on dislocations in several bcc metals. For carbon in α-Fe, Harper determined the fraction of solute atom still in solution using an internal friction technique and showed that log $(1-f)$ is proportional to $t^{2/3}$; the slope of the line is $\pi \rho (AD/kT)$ and evaluation of this slope at a series of temperatures allows the activation energy for the process to be determined from an Arrhenius plot. The value obtained for α-iron is 84 kJ mol$^{-1}$, which is close to that for the diffusion of carbon in ferrite.

The inhomogeneity interaction is considered to be the dominant effect in vacancy–dislocation interactions, with $V = -B/r^2$, where $B$ is a constant; this compares with the size effect for which $V = -A/r$ would be appropriate for the interstitial–dislocation interaction. It is convenient, however, to write the interaction energy in the general form $V = -A/r^n$ and, hence, following the treatment previously used for the kinetics of strain ageing, the radial velocity of a point defect towards the dislocation is

$$V = (D/kT)(nA/r^{n+1}). \qquad (4.5.16)$$

The number of a particular point defect species that reaches the dislocation in time $t$ is

$$n(t) = \pi r^2 c_0$$
$$= \pi c_0 [ADn(n+2)/kT]^{2/(n+2)} t^{2/(n+2)} \qquad (4.5.17)$$

and when $n = 2$ then $n(t) \propto t^{1/2}$, and when $n = 1$, $n(t) \propto t^{2/3}$. Since the kinetics of ageing in quenched copper follow $t^{1/2}$ initially, the observations confirm the importance of the inhomogeneity interaction for vacancies.

### 4.5.4.11 Influence of grain boundaries on plasticity

It might be thought that when a stress is applied to a polycrystalline metal, every grain in the sample deforms as if it were an unconstrained single crystal. This is not the case, however, and the fact that the aggregate does not deform in this manner is indicated by the high yield stress of polycrystals compared with that of single crystals. This increased strength of polycrystals immediately poses the question: is the hardness of a grain caused by the presence of the grain boundary or by the orientation difference of the neighboring grains? It is now believed that the latter is the case but that the structure of the grain boundary itself may be of importance in special circumstances, such as when brittle films, due to bismuth in copper or cementite in steel, form around the grains or when the grains slip past each other along their boundaries during high-temperature creep. The importance of the orientation change across a grain boundary to the process of slip has been demonstrated by experiments on 'bamboo'-type specimens, i.e. where the grain boundaries are parallel to each other and all perpendicular to the axis of tension. Initially, deformation occurs by slip only in those grains most favorably oriented, but later spreads to all the other grains as those grains which are deformed first work harden. It is then found that each grain contains wedge-shaped areas near the grain boundary, as shown in Figure 4.5-30a, where slip does not operate, which indicates that the continuance of slip from one grain to the next is difficult. From these observations it is natural to enquire what happens in a completely polycrystalline metal where the slip planes must in all cases make contact with a grain boundary. It will be clear that the polycrystalline aggregate must be stronger because, unlike the deformation of bamboo-type samples, where it is not necessary to raise the stress sufficiently high to operate those slip planes which made contact with a grain boundary, all the slip planes within any grain of a polycrystalline aggregate make contact with a grain boundary, but nevertheless have to be operated. The importance of the grain size on a strength is emphasized by Figure 4.5-29b, which shows the variation in lower yield stress, $\sigma_y$, with grain diameter, $2d$, for low-carbon steel. The smaller the grain size, the higher the yield strength according to a relation of the form

$$\sigma_y = \sigma_i + kd^{-1/2}, \quad (4.5.18)$$

where $\sigma_i$, is a lattice friction stress and $k$ a constant usually denoted $k_y$ to indicate yielding. Because of the difficulties experienced by a dislocation in moving from one grain to another, the process of slip in a polycrystalline aggregate does not spread to each grain by forcing a dislocation through the boundary. Instead, the slip band which is held up at the boundary gives rise to a stress concentration at the head of the pile-up group of dislocations which acts with the applied stress and is sufficient to trigger off sources in neighboring grains. If $\tau_i$

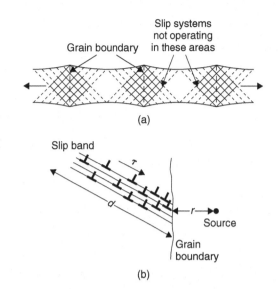

**Figure 4.5-30** (a) Grain-boundary blocking of slip. (b) Blocking of a slip band by a grain boundary.

is the stress a slip band could sustain if there were no resistance to slip across the grain boundary, i.e. the friction stress, and $\tau$ the higher stress sustained by a slip band in a polycrystal, then $(\tau - \tau_i)$ represents the resistance offered by the boundary, which reaches a limiting value when slip is induced in the next grain. The influence of grain size can be explained if the length of the slip band is proportional to $d$, as shown in Figure 4.5-30b. Thus, since the stress concentration a short distance $r$ from the end of the slip band is proportional to $(d/4r)^{1/2}$, the maximum shear stress at a distance $r$ ahead of a slip band carrying an applied stress $\tau$ in a polycrystal is given by $(\tau - \tau_i)[d/4r]^{1/2}$ and lies in the plane of the slip band. If this maximum stress has to reach a value $\tau_{max}$ to operate a new source at a distance $r$, then

$$(\tau - \tau_i)[d/4r]^{1/2} = \tau_{max}$$

or, rearranging,

$$\tau = \tau_i + (\tau_{max} 2r^{1/2})d^{-1/2},$$

which may be written as

$$\tau = \tau_i + k_s d^{-1/2}.$$

It then follows that the tensile flow curve of a polycrystal is given by

$$\sigma = m(\tau_i + k_s d^{-1/2}), \quad (4.5.19)$$

where $m$ is the orientation factor relating the applied tensile stress $\sigma$ to the shear stress, i.e. $\sigma = m\tau$. For a single crystal the $m$-factor has a minimum value of 2 as discussed, but in polycrystals deformation occurs in less favorably oriented grains and sometimes (e.g. hexagonal, intermetallics, etc.) on 'hard' systems, and so the $m$-factor is significantly higher. From equation (4.5.18) it can be seen that $\sigma_i = m\tau_i$ and $k = mk_s$.

While there is an orientation factor on a macroscopic scale in developing the critical shear stress within the various grains of a polycrystal, so there is a local orientation factor in operating a dislocation source ahead of a blocked slip band. The slip plane of the sources will not, in general, lie in the plane of maximum shear stress, and hence $\tau_{max}$ will need to be such that the shear stress, $\tau_c$, required to operate the new source must be generated in the slip plane of the source. In general, the local orientation factor dealing with the orientation relationship of adjacent grains will differ from the macroscopic factor of slip plane orientation relative to the axis of stress, so that $\tau_{max} = \frac{1}{2}m'\tau_c$. For simplicity, however, it will be assumed $m' = m$ and hence the parameter $k$ in the Petch equation is given by $k = m^2 \tau_c r^{1/2}$.

It is clear from the above treatment that the parameter $k$ depends essentially on two main factors. The first is the stress to operate a source dislocation, and this depends on the extent to which the dislocations are anchored or locked by impurity atoms. Strong locking implies a large $\tau_c$ and hence a large $k$; the converse is true for weak locking. The second factor is contained in the parameter $m$, which depends on the number of available slip systems. A multiplicity of slip systems enhances the possibility for plastic deformation and so implies a small $k$. A limited number of slip systems available would imply a large value of $k$. It then follows, as shown in Figure 4.5-31 that for (1) fcc metals, which have weakly locked dislocations and a multiplicity of slip systems, $k$ will generally be small, i.e. there is only a small grain size dependence of the flow stress, for (2) cph metals, $k$ will be large because of the limited slip systems, and for (3) bcc metals, because of the strong locking, $k$ will be large.

Each grain does not deform as a single crystal in simple slip, since, if this were so, different grains would then deform in different directions with the result that voids would be created at the grain boundaries. Except in high-temperature creep, where the grains slide past each other along their boundaries, this does not happen and each grain deforms in coherence with its neighboring grains. However, the fact that the continuity of the metal is maintained during plastic deformation must mean that each grain is deformed into a shape that is dictated by the deformation of its neighbors. Such behavior will, of course, require the operation of several slip systems, and von Mises has shown that to allow this unrestricted change of shape of a grain requires at least five independent shear modes. The deformation of metal crystals with cubic structure easily satisfies this condition so that the polycrystals of these metals usually exhibit

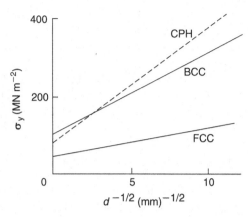

**Figure 4.5-31** Schematic diagram showing the grain-size dependence of the yield stress for crystals of different crystal structure.

considerable ductility, and the stress–strain curve generally lies close to that of single crystals of extreme orientations deforming under multiple slip conditions. The hexagonal metals do, however, show striking differences between their single-crystal and polycrystalline behavior. This is because single crystals of these metals deform by a process of basal plane slip, but the three shear systems (two independent) which operate do not provide enough independent shear mechanisms to allow unrestricted changes of shape in polycrystals. Consequently, to prevent gaps opening up at grain boundaries during the deformation of polycrystals, some additional shear mechanisms, such as non-basal slip and mechanical twinning, must operate. Hence, because the resolved stress for non-basal slip and twinning is greater than that for basal-plane slip, yielding in a polycrystal is prevented until the applied stress is high enough to deform by these mechanisms.

### 4.5.4.12 Superplasticity

A number of materials, particularly two-phase eutectic or eutectoid alloys, have been observed to exhibit large elongations ($\approx 1000\%$) without fracture, and such behavior has been termed superplasticity. Several metallurgical factors have been put forward to explain superplastic behavior and it is now generally recognized that the effect can be produced in materials either (1) with a particular structural condition or (2) tested under special test conditions. The particular structural condition is that the material has a very fine grain size and the presence of a two-phase structure is usually of importance in maintaining this fine grain size during testing. Materials which exhibit superplastic behavior under special test conditions are those for which a phase boundary moves through the strained material during the test (e.g. during temperature cycling).

In general, the superplastic material exhibits a high strain-rate sensitivity. Thus, the plastic flow of a solid may be represented by the relation

$$\sigma = K\dot{\varepsilon}^m, \quad (4.5.20)$$

where $\sigma$ is the stress, $\dot{\varepsilon}$ the strain rate and $m$ an exponent generally known as the strain-rate sensitivity. When $m = 1$ the flow stress is directly proportional to strain rate and the material behaves as a Newtonian viscous fluid, such as hot glass. Superplastic materials are therefore characterized by high $m$-values, since this leads to increased stability against necking in a tensile test. Thus, for a tensile specimen length $l$ with cross-sectional area $A$ under an applied load $P$, then $dl/l = -dA/A$ and, introducing the time factor, we obtain

$$\dot{\varepsilon} = -(1/A)dA/dt$$

and if, during deformation, the equation $\sigma = K\dot{\varepsilon}^m$ is obeyed, then

$$dA/dt = (P/K)^{1/m} A^{\{(1-(1/m)\}}. \quad (4.5.21)$$

For most metals and alloys $m \approx 0.1-0.2$ and the rate at which $A$ changes is sensitively dependent on $A$, and hence once necking starts the process rapidly leads to failure. When $m = 1$, the rate of change of area is independent of $A$ and, as a consequence, any irregularities in specimen geometry are not accentuated during deformation. The resistance to necking therefore depends sensitively on $m$, and increases markedly when $m \gtrsim 0.5$. Considering, in addition, the dependence of the flow stress on strain, then

$$\sigma = K^1 \varepsilon^n \dot{\varepsilon}^m \quad (4.5.22)$$

and, in this case, the stability against necking depends on a factor $(1 - n - m)/m$, but $n$-values are not normally very high. Superplastic materials such as Zn–Al eutectoid, Pb–Sn eutectic, Al–Cu eutectic, etc. have $m$ values approaching unity at elevated temperatures.

The total elongation increases as $m$ increases and, with increasing microstructural fineness of the material (grain size or lamella spacing), the tendency for superplastic behavior is increased. Two-phase structures are advantageous in maintaining a fine grain size during testing, but exceptionally high ductilities have been produced in several commercially pure metals (e.g. Ni, Zn and Mg), for which the fine grain size was maintained during testing at a particular strain rate and temperature.

It follows that there must be several possible conditions leading to superplasticity. Generally, it is observed metallographically that the grain structure remains remarkably equiaxed during extensive deformation and that grain boundary sliding is a common deformation mode in several superplastic alloys. While grain boundary sliding can contribute to the overall deformation by relaxing the five independent mechanisms of slip, it cannot give rise to large elongations without bulk flow of material (e.g. grain boundary migration). In polycrystals, triple junctions obstruct the sliding process and give rise to a low $m$-value. Thus, to increase the rate sensitivity of the boundary shear it is necessary to lower the resistance to sliding from barriers, relative to the viscous drag of the boundary; this can be achieved by grain boundary migration. Indeed, it is observed that superplasticity is controlled by grain boundary diffusion.

The complete explanation of superplasticity is still being developed, but it is already clear that, during deformation, individual grains or groups of grains with suitably aligned boundaries will tend to slide. Sliding continues until obstructed by a protrusion in a grain boundary, when the local stress generates dislocations which slip across the blocked grain and pile up at the opposite boundary until the back stress prevents further generation of dislocations and thus further sliding. At the temperature of the test, dislocations at the head of the pile-up can climb into and move along grain boundaries to annihilation sites. The continual replacement of these dislocations would permit grain boundary sliding at a rate governed by the rate of dislocation climb, which in turn is governed by grain boundary diffusion. It is important that any dislocations created by local stresses are able to traverse yielded grains and this is possible only if the 'dislocation cell size' is larger than, or at least of the order of, the grain size, i.e. a few microns. At high strain rates and low temperatures the dislocations begin to tangle and form cell structures, and superplasticity then ceases.

The above conditions imply that any metal in which the grain size remains fine during deformation could behave superplastically; this conclusion is borne out in practice. The stability of grain size can, however, be achieved more readily with a fine microduplex structure, as observed in some Fe–20Cr–6Ni alloys when hot-worked to produce a fine dispersion of austenite and ferrite. Such stainless steels have an attractive combination of properties (strength, toughness, fatigue strength, weldability and corrosion resistance) and, unlike the usual range of two-phase stainless steels, have good hot workability if 0.5Ti is added to produce a random distribution of TiC rather than $Cr_{23}C_6$ at ferrite–austenite boundaries.

Superplastic forming is now an established and growing industry, largely using vacuum forming to produce intricate shapes with high draw ratios. Two alloys which have achieved engineering importance are *Supral* (containing Al–6Cu–0.5Zr) and *IMI 318* (containing Ti–6Al–4V). *Supral* is deformed at 460°C and *IMI 318* at 900°C under argon. Although the process is slow, the loads required are also low and the process can be advantageous in the press-forming field to replace some of the present expensive and complex forming technology.

## 4.5.5 Mechanical twinning

### 4.5.5.1 Crystallography of twinning

Mechanical twinning plays only a minor part in the deformation of the common metals such as copper or aluminum, and its study has consequently been neglected. Nevertheless, twinning does occur in all the common crystal structures under some conditions of deformation. Table 4.5-2 shows the appropriate twinning elements for the common structures.

The geometrical aspects of twinning can be represented with the aid of a unit sphere, shown in Figure 4.5-32. The twinning plane $k_1$ intersects the plane of the drawing in the shear direction $\eta_1$. On twinning, the unit sphere is distorted to an ellipsoid of equal volume, and the shear plane $k_1$ remains unchanged during twinning, while all other planes become tilted. Distortion of planes occurs in all cases except $k_1$ and $k_2$. The shear strain, $s$, at unit distance from the twinning plane is related to the angle between $k_1$ and $k_2$. Thus, the amount of shear is fixed by the crystallographic nature of the two undistorted planes. In the bcc lattice, the two undistorted planes are the (1 1 2) and (1 1 $\bar{2}$) planes, displacement occurring in a [1 1 1] direction a distance of 0.707 lattice vectors. The twinning elements are thus:

| $K_1$ | $K_2$ | $\eta_1$ | $\eta_2$ | Shear |
|---|---|---|---|---|
| (1 1 2) | (1 1 $\bar{2}$) | [1 1 $\bar{1}$] | [1 1 1] | 0.707 |

where $k_1$ and $k_2$ denote the first and second undistorted planes respectively, and $\eta_1$ and $\eta_2$ denote directions lying in $k_1$ and $k_2$, respectively, perpendicular to the line of intersection of these planes. $k_1$ is also called the

| Table 4.5-2 Twinning elements for some common metals. | | | |
|---|---|---|---|
| **Structure** | **Plane** | **Direction** | **Metals** |
| Cph | {1 0 $\bar{1}$ 2} | ⟨1 0 $\bar{1}$ $\bar{1}$⟩ | Zn, Cd, Be, Mg |
| Bcc | {1 1 2} | ⟨1 1 1⟩ | Fe, β-brass, W, Ta, Nb, V, Cr, Mo |
| Fcc | {1 1 1} | ⟨1 1 2⟩ | Cu, Ag, Au, Ag–Au, Cu–Al |
| Tetragonal | {3 3 1} | – | Sn |
| Rhombohedral | {0 0 1} | – | Bi, As, Sb |

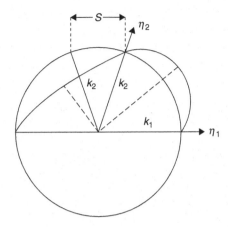

**Figure 4.5-32** Crystallography of twinning.

composition or twinning plane, while $\eta_1$ is called the shear direction. The twins consist of regions of crystal in which a particular set of {1 1 2} planes (the $k_1$ set of planes) is homogeneously sheared by 0.707 in a ⟨1 1 1⟩ direction (the $\eta_1$ direction). The same atomic arrangement may be visualized by a shear of 1.414 in the *reverse* ⟨111⟩ direction, but this larger shear has never been observed.

### 4.5.5.2 Nucleation and growth of twins

During the development of mechanical twins, thin lamellae appear very quickly ($\approx$ speed of sound) and these thicken with increasing stress by the steady movement of the twin interface. New twins are usually formed in bursts and are sometimes accompanied by a sharp audible click, which coincides with the appearance of irregularities in the stress–strain curve, as shown in Figure 4.5-33. The rapid production of clicks is responsible for the so-called twinning cry (e.g. in tin).

Although most metals show a general reluctance to twin, when tested under suitable conditions they can usually be made to do so. As mentioned in Section 4.5.3.1, the shear process involved in twinning must occur by the movement of partial dislocations and, consequently, the stress to cause twinning will depend not only on the line tension of the source dislocation, as in the case of slip, but also on the surface tension of the twin boundary. The stress to cause twinning is therefore usually greater than that required for slip, and at room temperature deformation will nearly always occur by slip in preference to twinning. As the deformation temperature is lowered the critical shear stress for slip increases and then, because the general stress level will be high, the process of deformation twinning is more likely.

Twinning is most easily achieved in metals of cph structure where, because of the limited number of slip systems, twinning is an essential and unavoidable mechanism of deformation in polycrystalline specimens (see Section 4.5.4.11), but in single crystals the

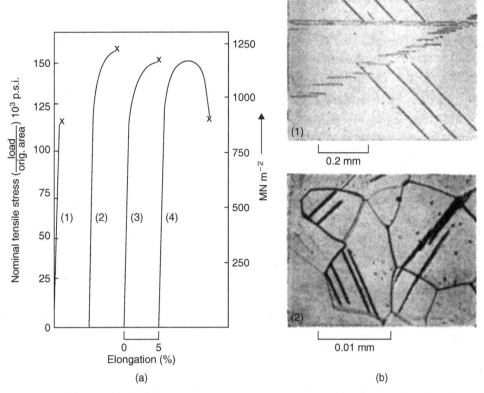

**Figure 4.5-33** (a) Effect of grain size on the stress–strain curves of specimens of niobium extended at a rate of $2.02 \times 10^{-4}$ s$^{-1}$ at 20 K: (1) grain size 2d = 1.414 mm, (2) grain size 2d = 0.312 mm, (3) grain size 2d = 0.0951 mm, (4) grain size 2d = 0.0476 mm. (b) Deformation twins in specimen 1 and specimen 3 extended to fracture. Etched in 95% HNO$_3$ + 5% HF (after Adams, Roberts and Smallman, 1960).

orientation of the specimen, the stress level and the temperature of deformation are all important factors in the twinning process. In metals of the bcc structure twinning may be induced by impact at room temperature or with more normal strain rates at low temperature, where the critical shear stress for slip is very high. In contrast, only a few fcc metals have been made to twin, even at low temperatures.

In zinc single crystals it is observed that there is no well-defined critical resolved shear stress for twinning such as exists for slip, and that a very high stress indeed is necessary to nucleate twins. In most crystals, slip usually occurs first and twin nuclei are then created by means of the very high stress concentration which exists at dislocation pile-ups. Once formed, the twins can propagate provided the resolved shear stress is higher than a critical value, because the stress to propagate a twin is much lower than that to nucleate it. This can be demonstrated by deforming a crystal oriented in such a way that basal slip is excluded, i.e. when the basal planes are nearly parallel to the specimen axis. Even in such an oriented crystal it is found that the stress to cause twinning is higher than that for slip on non-basal planes. In this case, non-basal slip occurs first, so that when a dislocation pile-up arises and a twin is formed, the applied stress is so high that an avalanche or burst of twins results.

It is also believed that in the bcc metals twin nucleation is more difficult than twin propagation. One possible mechanism is that nucleation is brought about by the stress concentration at the head of a piled-up array of dislocations produced by a burst of slip as a Frank–Read source operates. Such behavior is favored by impact loading, and it is well known that twin lamellae known as Neumann bands are produced this way in α-iron at room temperature. At normal strain rates, however, it should be easier to produce a slip burst suitable for twin nucleation in a material with strongly locked dislocations, i.e. one with a large $k$-value (as defined by equation (4.5.19)), than one in which the dislocation locking is relatively slight (small $k$-values). In this context it is interesting to note that both niobium and tantalum have a small $k$-value and, although they can be made to twin, do so with reluctance compared, for example, with α-iron.

In all the bcc metals the flow stress increases so rapidly with decreasing temperature (see Figure 4.5-29), that even with moderate strain rates ($10^{-4}$ s$^{-1}$) α-iron will twin at 77 K, while niobium with its smaller value of $k$ twins at 20 K. The type of Stress–strain behavior for niobium is shown in Figure 4.5-33a. The pattern of behavior is characterized by small amounts of slip interspersed between extensive bursts of twinning in the early stages of deformation. Twins, once formed, may themselves act as barriers, allowing further dislocation pile-up and further twin nucleation. The action of twins as barriers to slip dislocations could presumably account for the rapid work hardening observed at 20K.

Fcc metals do not readily deform by twinning, but it can occur at low temperatures, and even at 0°C, in favorably oriented crystals. The apparent restriction of twinning to certain orientations and low temperatures may be ascribed to the high shear stress attained in tests on crystals with these orientations, since the stress necessary to produce twinning is high. Twinning has been confirmed in heavily rolled copper. The exact mechanism for this twinning is not known, except that it must occur by the propagation of a half-dislocation and its associated stacking fault across each plane of a set of parallel (1 1 1) planes. For this process the half-dislocation must climb onto successive twin planes, as below for bcc iron.

### 4.5.5.3 Effect of impurities on twinning

It is well established that solid solution alloying favors twinning in fcc metals. For example, silver–gold alloys twin far more readily than the pure metals. Attempts have been made to correlate this effect with stacking-fault energy and it has been shown that the twinning stress of copper-based alloys increases with increasing stacking-fault energy. Twinning is also favored by solid solution alloying in bcc metals, and alloys of Mo–Re, W–Re and Nb–V readily twin at room temperature. In this case it has been suggested that the lattice frictional stress is increased and the ability to cross-slip reduced by alloying, thereby confining slip dislocations to bands where stress multiplication conducive to twin nucleation occurs.

### 4.5.5.4 Effect of prestrain on twinning

Twinning can be suppressed in most metals by a certain amount of prestrain; the ability to twin may be restored by an ageing treatment. It has been suggested that the effect may be due to the differing dislocation distribution produced under different conditions. For example, niobium will normally twin at −196°C, when a heterogeneous arrangement of elongated screw dislocations capable of creating the necessary stress concentrations is formed. Room temperature prestrain, however, inhibits twin formation as the regular network of dislocations produced provides more mobile dislocations and homogenizes the deformation.

### 4.5.5.5 Dislocation mechanism of twinning

In contrast to slip, the shear involved in the twinning process is homogeneous throughout the entire twinning region, and each atom plane parallel to the twinning plane moves over the one below it by only a fraction of a lattice spacing in the twinning direction. Nevertheless, mechanical twinning is thought to take place by a dislocation mechanism for the same reasons as slip, but the dislocations that cause twinning are partial and not unit dislocations. From the crystallography of the process it can be shown that twinning in the cph lattice, in addition to a simple shear on the twinning plane, must be accompanied by a localized rearrangement of the atoms, and furthermore, only in the bcc lattice does the process of twinning consist of a simple shear on the twinning plane (e.g. a twinned structure in this lattice can be produced by a shear of $1/\sqrt{2}$ in a $\langle 1\ 1\ 1\rangle$ direction on a $\{1\ 12\}$ plane).

An examination of Figure 4.5-8 shows that the main problem facing any theory of twinning is to explain how twinning develops homogeneously through successive planes of the lattice. This could be accomplished by the movement of a single twinning (partial) dislocation successively from plane to plane. One suggestion, similar in principle to the crystal growth mechanism, is the pole mechanism proposed by Cottrell and Bilby, illustrated in Figure 4.5-34a. Here, OA, OB and OC are dislocation lines. The twinning dislocation is OC, which produces the correct shear as it sweeps through the twin plane about its point of emergence O, and OA and OB form the pole dislocation, being partly or wholly of screw character with a pitch equal to the spacing of the twinning layers. The twinning dislocation rotates round the pole dislocation and, in doing so, not only produces a monolayer sheet of twinned crystal but also climbs up the 'pole' to the next layer. The process is repeated and a thick layer of twin is built up.

The dislocation reaction involved is as follows. The line AOB represents a unit dislocation with a Burgers vector $a/2[1\ 1\ 1]$ and that part OB of the line lies in the (1 1 2) plane. Then, under the action of stress, dissociation of this dislocation can occur according to the reaction:

$$a/2[1\ 1\ 1] \rightarrow a/3[1\ 1\ 2] + a/6[1\ 1\ \bar{1}].$$

The dislocation with vector $a/6[1\ 1\ \bar{1}]$ forms a line OC lying in one of the other {1 1 2} twin planes (e.g. the $(\bar{1}\ 2\ 1)$ plane) and produces the correct twinning shear. The line OB is left with a Burgers vector $a/3[1\ 1\ 2]$, which is of pure edge type and sessile in the (1 1 2) plane.

### 4.5.5.6 Twinning and fracture

It has been suggested that a twin, like a grain boundary, may present a strong barrier to slip and that a crack can be initiated by the pile-up of slip dislocations at the twin interface (see Figure 4.5-33). In addition, cracks may be initiated by the intersection of twins, and examples are

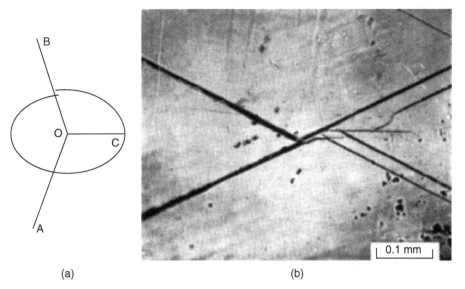

**Figure 4.5-34** (a) Diagram illustrating the pole mechanism of twinning, (b) The formation of a crack at a twin intersection in silicon–iron (after Hull, 1960).

common in molybdenum, silicon–iron (bcc) and zinc (cph). Figure 4.5-34b shows a very good example of crack nucleation in 3% silicon-iron; the crack has formed along an {0 0 1} cleavage plane at the intersection of two {1 1 2} twins, and part of the crack has developed along one of the twins in a zigzag manner while still retaining {0 0 1} cleavage facets.

In tests at low temperature on bcc and cph metals both twinning and fracture readily occur, and this has led to two conflicting views. First, that twins are nucleated by the high stress concentrations associated with fracture and, second, that the formation of twins actually initiates the fracture. It is probable that both effects occur.

## 4.5.6 Strengthening and hardening mechanisms

### 4.5.6.1 Point defect hardening

The introduction of point defects into materials to produce an excess concentration of either vacancies or interstitials often gives rise to a significant change in

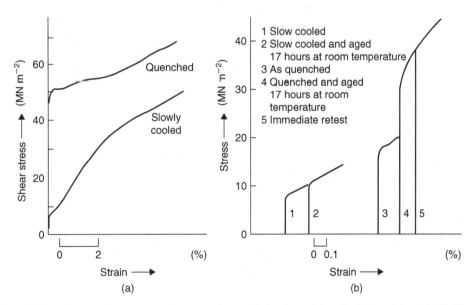

**Figure 4.5-35** Effect of quenching on the stress–strain curves from: (a) aluminum (after Maddin and Cottrell, 1955) and (b) gold (after Adams and Smallman, unpublished).

mechanical properties (Figures 4.5-35 and 4.5-36). For aluminum the shape of the Stress–strain curve is very dependent on the rate of cooling and a large increase in the yield stress may occur after quenching. We have already seen in Chapter 3 that quenched-in vacancies result in clustered vacancy defects and these may harden the material. Similarly, irradiation by high-energy particles may produce irradiation hardening (see Figure 4.5-36). Information on the mechanisms of hardening can be obtained from observation of the dependence of the lower yield stress on grain size. The results, reproduced in Figure 4.5-36b, show that the relation $\sigma_y = \sigma_i + k_y d^{-1/2}$, which is a general relation describing the propagation of yielding in materials, is obeyed.

This dependence of the yield stress, $\sigma_y$ on grain size indicates that the hardening produced by point defects introduced by quenching or irradiation is of two types: (1) an initial dislocation source hardening and (2) a general lattice hardening which persists after the initial yielding. The $k_y$ term would seem to indicate that the pinning of dislocations may be attributed to point defects in the form of coarsely spaced jogs, and the electron-microscope observations of jogged dislocations would seem to confirm this.

The lattice friction term $\sigma_i$ is clearly responsible for the general level of the Stress–strain curve after yielding and arises from the large density of dislocation defects. However, the exact mechanisms whereby loops and tetrahedra give rise to an increased flow stress is still controversial. Vacancy clusters are believed to be formed *in situ* by the disturbance introduced by the primary collision, and hence it is not surprising that neutron irradiation at 4 K hardens the material, and that thermal activation is not essential.

Unlike dispersion-hardened alloys, the deformation of irradiated or quenched metals is characterized by a low initial rate of work hardening (see Figure 4.5-35). This has been shown to be due to the sweeping out of loops and defect clusters by the glide dislocations, leading to the formation of cleared channels. Diffusion-controlled mechanisms are not thought to be important, since defect-free channels are produced by deformation at 4 K. The removal of prismatic loops, both unfaulted and faulted, and tetrahedra can occur as a result of the strong coalescence interactions with screws to form helical configurations and jogged dislocations when the gliding dislocations and defects make contact. Clearly, the sweeping-up process occurs only if the helical and jogged configurations can glide easily. Resistance to glide will arise from jogs not lying in slip planes and also from the formation of sessile jogs (e.g. Lomer–Cottrell dislocations in fcc crystals).

### 4.5.6.2 Work hardening

#### 4.5.6.2.1 Theoretical treatment

The properties of a material are altered by cold working, i.e. deformation at a low temperature relative to its melting point, but not all the properties are improved, for although the tensile strength, yield strength and hardness are increased, the plasticity and general ability to deform decreases. Moreover, the physical properties such as electrical conductivity, density and others are all lowered. Of these many changes in properties, perhaps the most outstanding are those that occur in the mechanical properties; the yield stress of mild steel, for example, may be raised by cold work from 170 up to 1050 MN m$^{-2}$.

Such changes in mechanical properties are, of course, of interest theoretically, but they are also of great importance in industrial practice. This is because the rate at which the material hardens during deformation influences both the power required and the method of working in the various shaping operations, while the

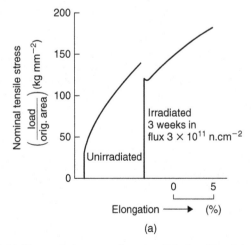
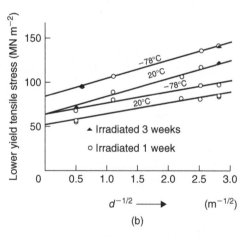

**Figure 4.5-36** (a) Stress–strain curves for unirradiated and irradiated fine-grained polycrystalline copper, tested at 20° C (b) Variation of yield stress with grain size and neutron dose (after Adams and Higgins, 1959).

magnitude of the hardness introduced governs the frequency with which the component must be annealed (always an expensive operation) to enable further working to be continued.

Since plastic flow occurs by a dislocation mechanism, the fact that work hardening occurs means that it becomes difficult for dislocations to move as the strain increases. All theories of work hardening depend on this assumption, and the basic idea of hardening, put forward by Taylor in 1934, is that some dislocations become 'stuck' inside the crystal and act as sources of internal stress, which oppose the motion of other gliding dislocations.

One simple way in which two dislocations could become stuck is by elastic interaction. Thus, two parallel-edge dislocations of opposite sign moving on parallel slip planes in any sub-grain may become stuck, as a result of the interaction discussed in Chapter 3. G. I. Taylor assumed that dislocations become stuck after traveling an average distance, $L$, while the density of dislocations reaches $\rho$, i.e. work hardening is due to the dislocations getting in each other's way. The flow stress is then the stress necessary to move a dislocation in the stress field of those dislocations surrounding it. This stress $\tau$ is quite generally given by

$$\tau = \alpha\mu b/l, \quad (4.5.23)$$

where $\mu$ is the shear modulus, $b$ the Burgers vector, $l$ the mean distance between dislocations, which is $\approx \rho^{-1/2}$, and $\alpha$ a constant; in the Taylor model $\alpha = 1/8\pi(1-\nu)$ where $\nu$ is Poisson's ratio. Figure 4.5-37 shows such a relationship for Cu–Al single crystals and polycrystalline Ag and Cu.

In his theory Taylor considered only a two-dimensional model of a cold-worked metal. However, because plastic deformation arises from the movement of dislocation loops from a source, it is more appropriate to assume that, when the plastic strain is $\gamma$, $N$ dislocation loops of side $L$ (if we assume for convenience that square loops are emitted) have been given off per unit volume. The resultant plastic strain is then given by

$$\gamma = NL^2 b \quad (4.5.24)$$

and $l$ by

$$l \approx [1/\rho^{1/2}] = [1/4LN]^{1/2}. \quad (4.5.25)$$

Combining these equations, the stress–strain relation

$$\tau = \text{const} \cdot (b/L)^{1/2}\gamma^{1/2} \quad (4.5.26)$$

is obtained. Taylor assumed $L$ to be a constant, i.e. the slip lines are of constant length, which results in a parabolic relationship between $\tau$ and $\gamma$.

Taylor's assumption that during cold work the density of dislocations increases has been amply verified, and indeed the parabolic relationship between stress and strain is obeyed, to a first approximation, in many polycrystalline aggregates where deformation in all grains takes place by multiple slip. Experimental work on single crystals shows, however, that the work or strain hardening curve may deviate considerably from parabolic behavior, and depends not only on crystal structure but also on other variables such as crystal orientation, purity and surface conditions (see Figures 4.5-38 and 4.5-39).

The crystal structure is important (see Figure 4.5-38) in that single crystals of some hexagonal metals slip only on one family of slip planes, those parallel to the basal plane, and these metals show a low rate of work hardening. The plastic part of the Stress–strain curve is also more nearly linear than parabolic, with a slope which is extremely small: this slope $(d\tau/d\gamma)$ becomes even smaller with increasing temperature of deformation. Cubic crystals, on the other hand, are capable of deforming in a complex manner on more than one slip system, and these metals normally show a strong work-hardening behavior. The influence of temperature depends on the stress level reached during deformation and on other factors which must be considered in greater

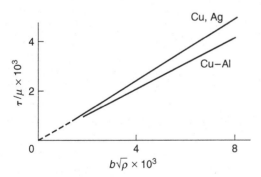

**Figure 4.5-37** Dependence of flow stress on (dislocation density)$^{1/2}$ for Cu, Ag and Cu–Al.

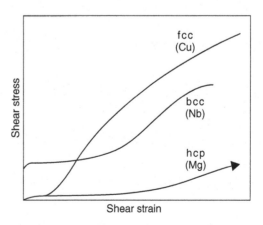

**Figure 4.5-38** Stress–strain curves of single crystals (after Hirsch and Mitchell, 1967; courtesy of the National Research Council of Canada).

detail. However, even in cubic crystals the rate of work hardening may be extremely small if the crystal is restricted to slip on a single slip system. Such behavior points to the conclusion that strong work hardening is caused by the mutual interference of dislocations gliding on intersecting slip planes.

Many theories of work hardening similar to that of Taylor exist but all are oversimplified, since work hardening depends not so much on individual dislocations as on the group behavior of large numbers of them. It is clear, therefore, that a theoretical treatment which would describe the complete stress–strain relationship is difficult, and consequently the present-day approach is to examine the various stages of hardening and then attempt to explain the mechanisms likely to give rise to the different stages. The work-hardening behavior in metals with a cubic structure is more complex than in most other structures because of the variety of slip systems available, and it is for this reason that much of the experimental evidence is related to these metals, particularly those with fcc structures.

### 4.5.6.2.2 Three-stage hardening

The stress–strain curve of a fcc single crystal is shown in Figure 4.5-39 and three regions of hardening are experimentally distinguishable. Stage I, or the easy glide region, immediately follows the yield point and is characterized by a low rate of work hardening $\theta_I$ up to several per cent glide; the length of this region depends on orientation, purity and size of the crystals. The hardening rate $(\theta_I/\mu) \sim 10^{-4}$ and is of the same order as for hexagonal metals. Stage II, or the linear hardening region, shows a rapid increase in work-hardening rate. The ratio $(\theta_{II}/\mu) = (d\tau/d\gamma)/\mu$ is of the same order of magnitude for all fcc metals, i.e. 1/300, although this is $\approx$ 1/150 for orientations at the corners of the stereographic triangle.

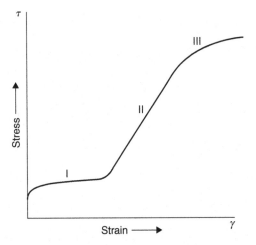

**Figure 4.5-39** Stress–strain curve showing the three stages of work hardening.

In this stage short slip lines are formed during straining quite suddenly, i.e. in a short increment of stress $\Delta\tau$, and thereafter do not grow either in length or intensity. The mean length of the slip lines, $L \approx 25$ μm, decreases with increasing strain. Stage III, or the parabolic hardening region, the onset of which is markedly dependent on temperature, exhibits a low rate of work hardening, $\theta_{III}$, and the appearance of coarse slip bands. This stage sets in at a strain which increases with decreasing temperature and is probably associated with the annihilation of dislocations as a consequence of cross-slip.

The low stacking-fault energy metals exhibit all three work-hardening stages at room temperature, but metals with a high stacking-fault energy often show only two stages of hardening. It is found, for example, that at 78 K aluminum behaves like copper at room temperature and exhibits all three stages, but at room temperature and above, Stage II is not clearly developed and Stage III starts before Stage II becomes at all predominant. This difference between aluminum and the noble metals is due not only to the difference in melting point, but also to the difference in stacking-fault energies, which affects the width of extended dislocations. The main effect of a change of temperature of deformation is, however, a change in the onset of Stage III; the lower the temperature of deformation, the higher is the stress $\tau_{III}$ corresponding to the onset of Stage III.

The Stage I easy glide region in cubic crystals, with its small linear hardening, corresponds closely to the hardening of cph crystals, where only one glide plane operates. It occurs in crystals oriented to allow only one glide system to operate, i.e. for orientations near the [0 1 1] pole of the unit triangle (Figure 4.5-12). In this case the slip distance is large, of the order of the specimen diameter, with the probability of dislocations slipping out of the crystal. Electron microscope observations have shown that the slip lines on the surface are very long ($\approx$ 1 mm) and closely spaced, and that the slip steps are small corresponding to the passage of only a few dislocations. This behavior obviously depends on such variables as sample size and oxide films, since these influence the probability of dislocations passing out of the crystal. It is also to be expected that the flow stress in easy glide will be governed by the ease with which sources begin to operate, since there is no slip on a secondary slip system to interfere with the movement of primary glide dislocations.

As soon as another glide system becomes activated there is a strong interaction between dislocations on the primary and secondary slip systems, which gives rise to a steep increase in work hardening. It is reasonable to expect that easy glide should end, and turbulent flow begin, when the crystal reaches an orientation for which two or more slip systems are equally stressed, i.e. for orientations on the symmetry line between [0 0 1] and [1 1 1]. However, easy glide generally ends before symmetrical orientations are reached and this is principally due to the formation of deformation bands to accommodate the rotation of the glide plane in fixed grips during tensile tests. This rotation leads to a high resolved stress on the secondary slip system, and its operation gives rise to those lattice irregularities which cause some dislocations to become 'stopped' in the crystal. The transformation to Stage II then occurs.

The characteristic feature of deformation in Stage II is that slip takes place on both the primary and secondary slip systems. As a result, several new lattice irregularities may be formed, which will include (1) forest dislocations, (2) Lomer–Cottrell barriers and (3) jogs produced either by moving dislocations cutting through forest dislocations or by forest dislocations cutting through source dislocations. Consequently, the flow stress $\tau$ may be identified, in general terms, with a stress which is sufficient to operate a source and then move the dislocations against (1) the internal elastic stresses from the forest dislocations, (2) the long-range stresses from groups of dislocations piled up behind barriers and (3) the factional resistance due to jogs. In a cold-worked metal all these factors may exist to some extent, but because a linear hardening law can be derived by using any one of the various contributory factors, there have been several theories of Stage II hardening, namely (1) the pile-up theory, (2) the forest theory and (3) the jog theory. All have been shown to have limitations in explaining various features of the deformation process, and have given way to a more phenomenological theory based on direct observations of the dislocation distribution during straining.

Observations on fcc and bcc crystals have revealed several common features of the microstructure, which include the formation of dipoles, tangles and cell structures with increasing strain. The most detailed observations have been made for copper crystals, and these are summarized below to illustrate the general pattern of behavior. In Stage I, bands of dipoles are formed (see Figure 4.5-40a), elongated normal to the primary Burgers vector direction. Their formation is associated with isolated forest dislocations and individual dipoles are about 1 μm in length and $\gtrsim$10 nm wide. Different patches are arranged at spacings of about 10 μm along the line of intersection of a secondary slip plane. With increasing strain in Stage I, the size of the gaps between the dipole clusters decreases and therefore the stress required to push dislocations through these gaps increases. Stage II begins (see Figure 4.5-40b) when the applied stress plus internal stress resolved on the secondary systems is sufficient to activate secondary sources near the dipole clusters. The resulting local secondary slip leads to local interactions between primary and secondary dislocations both in the gaps and in the clusters of dipoles, the gaps being filled with secondary dislocations and short lengths of other dislocations formed by interactions (e.g. Lomer–Cottrell dislocations in fcc crystals and $a\langle 1\,0\,0\rangle$-type dislocations in bcc crystals). Dislocation barriers are thus formed surrounding the original sources.

In Stage II (see Figure 4.5-40c) it is proposed that dislocations are stopped by elastic interaction when they pass too close to an existing tangled region with high dislocation density. The long-range internal stresses due to the dislocations piling up behind are partially relieved by secondary slip, which transforms the discrete pile-up into a region of high dislocation density containing secondary dislocation networks and dipoles. These regions of high dislocation density act as new obstacles to dislocation glide, and since every new obstacle is formed near one produced at a lower strain, two-dimensional dislocation structures are built up forming the walls of an irregular cell structure. With increasing strain the number of obstacles increases, the distance a dislocation glides decreases and therefore the slip line length decreases in Stage II. The structure remains similar throughout Stage II but is reduced in scale. The obstacles are in the form of ribbons of high densities of dislocations which, like pile-ups, tend to form sheets. The work-hardening rate depends mainly on the effective radius of the obstacles, and this has been considered in detail by Hirsch and co-workers and shown to be a constant fraction $k$ of the discrete pile-up length on the primary slip system. In general, the work-hardening rate is given by $\theta_{11} = k\mu/3\pi$ and, for an fcc crystal, the small variation in $k$ with orientation and alloying element is able to account for the variation of $\theta_{11}$ with those parameters.

The dislocation arrangement in metals with other structures is somewhat similar to that of copper, with differences arising from stacking-fault energy. In Cu–Al alloys the dislocations tend to be confined more to the active slip planes, the confinement increasing with decreasing $\gamma_{SF}$. In Stage I dislocation multipoles are formed as a result of dislocations of opposite sign on parallel nearby slip planes 'pairing up' with one another. Most of these dislocations are primaries. In Stage II the density of secondary dislocations is much less ($\approx \frac{1}{3}$) than that of the primary dislocations. The secondary slip occurs in bands and in each band slip on one particular secondary plane predominates. In niobium, a metal with high $\gamma_{SF}$, the dislocation distribution is rather similar to copper. In Mg, typical of cph metals, Stage I is extensive and the dislocations are mainly in the form of primary edge multipoles, but forest dislocations threading the primary slip plane do not appear to be generated.

From the curve shown in Figure 4.5-39 it is evident that the rate of work hardening decreases in the later stages of the test. This observation indicates that at a sufficiently high stress or temperature the dislocations held up in Stage II are able to move by a process which at lower stresses and temperature had been suppressed. The

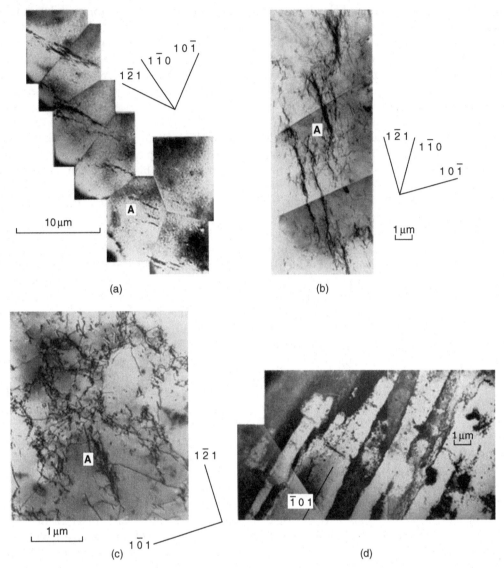

**Figure 4.5-40** Dislocation structure observed in copper single crystals deformed in tension to: (a) Stage I, (b) end of easy glide and beginning of Stage II, (c) top of Stage II and (d) Stage III (after Steeds, 1963; Crown copyright; reproduced by permission of the Controller, HM Stationery Office).

onset of Stage III is accompanied by cross-slip, and the slip lines are broad, deep and consist of segments joined by cross-slip traces. Electron metallographic observations on sections of deformed crystal inclined to the slip plane (see Figure 4.5-40d) show the formation of a cell structure in the form of boundaries, approximately parallel to the primary slip plane of spacing about 1–3μm, plus other boundaries extending normal to the slip plane as a result of cross-slip.

The simplest process which is in agreement with the experimental observations is that the screw dislocations held up in Stage II cross-slip and possibly return to the primary slip plane by double cross-slip. By this mechanism, dislocations can bypass the obstacles in their glide plane and do not have to interact strongly with them. Such behavior leads to an increase in slip distance and a decrease in the accompanying rate of work hardening. Furthermore, it is to be expected that screw dislocations leaving the glide plane by cross-slip may also meet dislocations on parallel planes and be attracted by those of opposite sign. Annihilation then takes place and the annihilated dislocation will be replaced, partly at least, from the original source. This process, if repeated, can lead to slip-band formation, which is also an important experimental feature of Stage III. Hardening in Stage III is then due to the edge parts of the loops which remain in the crystal and increase in density as the source continues to operate.

The importance of the value of the stacking-fault energy, $\gamma$, on the stress–strain curve is evident from its importance to the process of cross-slip. Low values of γ give rise to wide stacking-fault 'ribbons', and consequently cross-slip is difficult at reasonable stress levels. Thus, the screws cannot escape from their slip plane, the slip distance is small, the dislocation density is high and the transition from Stage II to Stage III is delayed. In aluminum the stacking-fault ribbon width is very small because γ has a high value, and cross-slip occurs at room temperature. Stage II is therefore poorly developed unless testing is carried out at low temperatures. These conclusions are in agreement with the observations of dislocation density and arrangement.

### 4.5.6.2.3 Work hardening in polycrystals

The dislocation structure developed during the deformation of fcc and bcc polycrystalline metals follows the same general pattern as that in single crystals: primary dislocations produce dipoles and loops by interaction with secondary dislocations, which give rise to local dislocation tangles gradually developing into three-dimensional networks of sub-boundaries. The cell size decreases with increasing strain, and the structural differences that are observed between various metals and alloys are mainly in the sharpness of the sub-boundaries. In bcc metals, and fcc metals with high stacking-fault energy, the tangles rearrange into sharp boundaries but in metals of low stacking-fault energy the dislocations are extended, cross-slip is restricted and sharp boundaries are not formed even at large strains. Altering the deformation temperature also has the effect of changing the dislocation distribution; lowering the deformation temperature reduces the tendency for cell formation, as shown in Figure 4.5-41. For a given dislocation distribution the dislocation density is simply related to the flow stress τ by an equation of the form:

$$\tau = \tau_0 + \alpha\mu b\rho^{1/2}, \tag{4.5.27}$$

where α is a constant at a given temperature $\approx 0.5$; $\tau_0$ for fcc metals is zero (see Figure 4.5-37). The work-hardening rate is determined by the ease with which tangled dislocations rearrange themselves and is high in materials with low γ, i.e. brasses, bronzes and austenitic steels, compared to Al and bcc metals. In some austenitic steels, work hardening may be increased and better sustained by a strain-induced phase transformation (see Chapter 7).

Grain boundaries affect work hardening by acting as barriers to slip from one grain to the next. In addition, the continuity criterion of polycrystals enforces complex slip in the neighborhood of the boundaries, which spreads across the grains with increasing deformation. This introduces a dependence of work-hardening rate on grain size which extends to several percent elongation. After this stage, however, the work-hardening rate is independent of grain size and for fcc polycrystals is about $\mu/40$, which, allowing for the orientation factors, is roughly comparable with that found in single crystals deforming in multiple slip. Thus, from the relations $\sigma = m\tau$ and $\varepsilon = \gamma/m$, the average resolved shear stress on a slip plane is rather less than half the applied tensile stress, and the average shear strain parallel to the slip plane is rather more than twice the tensile elongation. The polycrystal work-hardening rate is thus related to the single-crystal work-hardening rate by the relation

$$d\sigma/d\varepsilon = m^2 d\tau/d\gamma. \tag{4.5.28}$$

For bcc metals with a multiplicity of slip systems the ease of cross-slip $m$ is more nearly 2, so that the work-hardening rate is low. In polycrystalline cph metals the deformation is complicated by twinning, but in the absence of twinning $m \approx 6.5$, and hence the work-hardening rate is expected to be more than an order of magnitude greater than for single crystals, and also higher than the rate observed in fcc poly crystals, for which $m \approx 3$.

### 4.5.6.2.4 Dispersion-hardened alloys

On deforming an alloy containing incoherent, non-deformable particles, the rate of work hardening is much greater than that shown by the matrix alone. The dislocation density increases very rapidly with strain because the particles produce a turbulent and complex deformation pattern around them. The dislocations gliding in the matrix leave loops around particles either by bowing between the particles or by cross-slipping around them; both these mechanisms are discussed in Chapter 7. The stresses in and around particles may also be relieved by activating secondary slip systems in the matrix. All these dislocations spread out from the particle as strain proceeds and, by intersecting the primary glide plane, hinder primary dislocation motion and lead to

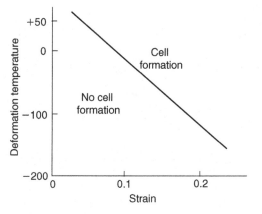

**Figure 4.5-41** Influence of deformation strain and temperature on the formation of a cell structure in α-iron.

intense work hardening. A dense tangle of dislocations is built up at the particle and a cell structure is formed with the particles predominantly in the cell walls.

At small strains ($\lesssim 1\%$), work hardening probably arises from the back-stress exerted by the few Orowan loops around the particles, as described by Fisher, Hart and Pry. The Stress–strain curve is reasonably linear with strain $\varepsilon$ according to

$$\sigma = \sigma_i + \alpha\mu f^{3/2}\varepsilon,$$

with the work hardening depending only on $f$, the volume fraction of particles. At larger strains the 'geometrically necessary' dislocations stored to accommodate the strain gradient which arises because one component deforms plastically more than the other determine the work hardening. A determination of the average density of dislocations around the particles with which the primary dislocations interact allows an estimate of the work-hardening rate, as initially considered by Ashby. Thus, for a given strain $\varepsilon$ and particle diameter $d$ the number of loops per particle is

$$n \sim \varepsilon d/b$$

and the number of particles per unit volume

$$N_v = 3f/4\pi r^3, \quad \text{or} \quad 6f/\pi d^3.$$

The total number of loops per unit volume is $nN_v$ and hence the dislocation density $\rho = nN_v\pi d = 6f\varepsilon/db$. The stress–strain relationship from equation (6.27) is then

$$\sigma = \sigma_i + \alpha\mu(fb/d)^{1/2}\varepsilon^{1/2} \tag{4.5.29}$$

and the work-hardening rate

$$d\sigma/d\varepsilon = \alpha'\mu(f/d)^{1/2}(b/\varepsilon)^{1/2}. \tag{4.5.30}$$

Alternative models taking account of the detailed structure of the dislocation arrays (e.g. Orowan, prismatic and secondary loops) have been produced to explain some of the finer details of dispersion-hardened materials. However, this simple approach provides a useful working basis for real materials. Some additional features of dispersion-strengthened alloys are discussed in Chapter 7.

### 4.5.6.2.5 Work hardening in ordered alloys

A characteristic feature of alloys with long-range order is that they work-harden more rapidly than in the disordered state. $\theta_{11}$ for Fe–Al with a B2 ordered structure is $\approx \mu/50$ at room temperature, several times greater than a typical fcc or bcc metal. However, the density of secondary dislocations in Stage II is relatively low and only about 1/100 of that of the primary dislocations. One mechanism for the increase in work-hardening rate is thought to arise from the generation of anti-phase domain boundary (APB) tubes. A possible geometry is shown in Figure 4.5-42a; the superdislocation partials shown each contain a jog produced, for example, by intersection with a forest dislocation, which are non-aligned along the direction of the Burgers vector. When the dislocation glides and the jogs move non-conservatively a tube of APBs is generated. Direct evidence for the existence of tubes from weak-beam electron microscope studies was first reported for Fe–30 at.% Al. The micrographs show faint lines along $\langle 1\,1\,1 \rangle$, the Burgers vector direction, and are about 3 nm in width. The images are expected to be weak, since the contrast arises from two closely spaced overlapping faults, the second effectively canceling the displacement caused by the first, and are visible only when superlattice reflections are excited. APB tubes have since been observed in other compounds.

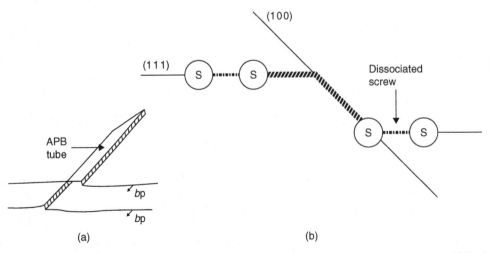

**Figure 4.5-42** Schematic diagram of superdislocation: (a) with non-aligned jogs which, after glide, produce an APB tube and (b) cross-slipped onto the cube plane to form a Kear–Wilsdorf (K–W) lock.

Theory suggests that jogs in superdislocations in screw orientations provide a potent hardening mechanism, estimated to be about eight times as strong as that resulting from pulling out of APB tubes on non-aligned jogs on edge dislocations. The major contributions to the stress to move a dislocation are (1) $\tau_s$, the stress to generate point defects or tubes, and (2) the interaction stress $\tau_i$ with dislocations on neighboring slip planes, and $\tau_s + \tau_i = \frac{3}{4}\alpha_s \mu (\rho_f/\rho_p)^\varepsilon$. Thus, with $\alpha_s = 1.3$ and provided $\rho_f/\rho_p$ is constant and small, linear hardening with the observed rate is obtained.

In crystals with $A_3B$ order only one rapid stage of hardening is observed compared with the normal three-stage hardening of fcc metals. Moreover, the temperature dependence of $\theta_{11}/\mu$ increases with temperature and peaks at $\sim 0.4T_m$. It has been argued that the APB tube model is unable to explain why anomalously high work-hardening rates are observed for those single-crystal orientations favorable for single slip on {1 1 1} planes alone. An alternative model to APB tubes has been proposed based on cross-slip of the leading unit dislocation of the superdislocation. If the second unit dislocation cannot follow exactly in the wake of the first, both will be pinned.

For alloys with $L1_2$ structure the cross-slip of a screw superpartial with $b = \frac{1}{2}[\bar{1}\,0\,1]$ from the primary (1 1 1) plane to the (0 1 0) plane was first proposed by Kear and Wilsdorf. The two $\frac{1}{2}[\bar{1}\,0\,1]$ superpartials, one on the (1 1 1) plane and the other on the (0 1 0) plane, are of course dissociated into $\langle 1\,1\,2 \rangle$-type partials and the whole configuration is sessile. This dislocation arrangement is known as a Kear–Wilsdorf (K–W) lock and is shown in Figure 4.5-42b. Since cross-slip is thermally activated, the number of locks and therefore the resistance to (1 1 1) glide will increase with increasing temperature. This could account for the increase in yield stress with temperature, while the onset of cube slip at elevated temperatures could account for the peak in the flow stress.

Cube cross-slip and cube slip has now been observed in a number of $L1_2$ compounds by TEM. There is some TEM evidence that the APB energy on the cube plane is lower than that on the (1 1 1) plane (see Chapter 8) to favor cross-slip, which would be aided by the torque, arising from elastic anisotropy, exerted between the components of the screw dislocation pair.

### 4.5.6.3 Development of preferred orientation

#### 4.5.6.3.1 Crystallographic aspects

When a polycrystalline metal is plastically deformed the individual grains tend to rotate into a common orientation. This preferred orientation is developed gradually with increasing deformation, and although it becomes extensive above about 90% reduction in area, it is still inferior to that of a good single crystal. The degree of texture produced by a given deformation is readily shown on a monochromatic X-ray transmission photograph, since the grains no longer reflect uniformly into the diffraction rings but only into certain segments of them. The results are usually described in terms of an ideal orientation, such as $[u, v, w]$ for the fibre texture developed by drawing or swaging, and $\{h\,k\,l\} \langle u\,v\,w \rangle$ for a rolling texture for which a plane of the form $(h\,k\,l)$ lies parallel to the rolling plane and a direction of the type $\langle u\,v\,w \rangle$ is parallel to the rolling direction. However, the scatter about the ideal orientation can only be represented by means of a pole figure, which describes the spread of orientation about the ideal orientation for a particular set of $(h\,k\,l)$ poles (see Figure 4.5-43).

In tension, the grains rotate in such a way that the movement of the applied stress axis is towards the operative slip direction as discussed in Section 4.5.3.5 and for compression the applied stress moves towards the

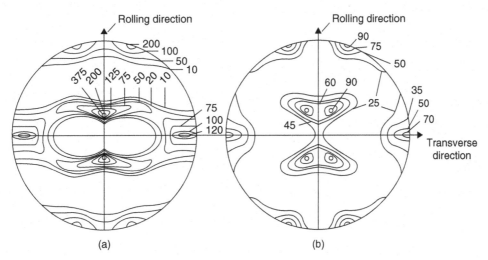

**Figure 4.5-43** (1 1 1) pole figures from copper (a) and $\alpha$-brass (b) after 95% deformation (intensities in arbitrary units).

slip plane normal. By considering the deformation process in terms of the particular stresses operating and applying the appropriate grain rotations it is possible to predict the stable end-grain orientation and hence the texture developed by extensive deformation. Table 4.5-3 shows the predominant textures found in different metal structures for both wires and sheet.

For fcc metals a marked transition in deformation texture can be effected either by lowering the deformation temperature or by adding solid solution alloying elements which lower the stacking-fault energy. The transition relates to the effect on deformation modes of reducing stacking-fault energy or thermal energy, deformation banding and twinning becoming more prevalent and cross-slip less important at lower temperatures and stacking-fault energies. This texture transition can be achieved in most fcc metals by alloying additions and by altering the rolling temperature. Al, however, has a high fault energy and because of the limited solid solubility it is difficult to lower by alloying. The extreme types of rolling texture, shown by copper and 70/30 brass, are given in Figure 4.5-43a and b.

In bcc metals there are no striking examples of solid solution alloying effects on deformation texture, the preferred orientation developed being remarkably insensitive to material variables. However, material variables can affect cph textures markedly. Variations in $c/a$ ratio alone cause alterations in the orientation developed, as may be appreciated by consideration of the twinning modes, and it is also possible that solid solution elements alter the relative values of critical resolved shear stress for different deformation modes. Processing variables are also capable of giving a degree of control in hexagonal metals. No texture, stable to further deformation, is found in hexagonal metals and the angle of inclination of the basal planes to the sheet plane varies continuously with deformation. In general, the basal plane lies at a small angle (<45°) to the rolling plane, tilted either towards the rolling direction (Zn, Mg) or towards the transverse direction (Ti, Zr, Be, Hf).

The deformation texture cannot, in general, be eliminated by an annealing operation even when such a treatment causes recrystallization. Instead, the formation of a new annealing texture usually results, which is related to the deformation texture by standard lattice rotations.

### 4.5.6.3.2 Texture hardening

The flow stress in single crystals varies with orientation according to Schmid's law and hence materials with a preferred orientation will also show similar plastic anisotropy, depending on the perfection of the texture. The significance of this relationship is well illustrated by a crystal of beryllium, which is cph and capable of slip only on the basal plane, a compressive stress approaching $\approx 2000$ MN m$^{-2}$ applied normal to the basal plane produces negligible plastic deformation. Polycrystalline beryllium sheet, with a texture such that the basal planes lie in the plane of the sheet, shows a correspondingly high strength in biaxial tension. When stretched uniaxially the flow stress is also quite high, when additional (prismatic) slip planes are forced into action even though the shear stress for their operation is five times greater than for basal slip. During deformation there is little thinning of the sheet, because the $\langle 1\,1\,\bar{2}\,0 \rangle$ directions are aligned in the plane of the sheet. Other hexagonal metals, such as titanium and zirconium, show less marked strengthening in uniaxial tension because prismatic slip occurs more readily, but resistance to biaxial tension can still be achieved. Applications of texture hardening lie in the use of suitably textured sheet for high biaxial strength, e.g. pressure vessels, dent resistance, etc. Because of the multiplicity of slip systems, cubic metals offer much less scope for texture hardening. Again, a consideration of single-crystal deformation gives the clue; for, whereas in a hexagonal crystal $m$ can vary from 2 (basal planes at 45° to the stress axis) to infinity (when the basal planes are normal), in an fcc crystal $m$ can vary only by a factor of 2 with orientation, and in bcc crystals the variation is rather less. In extending this approach to polycrystalline material certain assumptions have to be made about the mutual constraints between grains. One approach gives $m = 3.1$ for a random aggregate of fcc crystals and the calculated orientation dependence of $\sigma/\tau$ for fiber texture shows that a rod with $\langle 1\,1\,1 \rangle$ or $\langle 1\,1\,0 \rangle$ texture ($\sigma/\tau = 3.664$) is 20% stronger than a random structure; the cube texture ($\sigma/\tau = 2.449$) is 20% weaker.

If conventional mechanical properties were the sole criterion for texture-hardened materials, then it seems unlikely that they would challenge strong precipitation-hardened alloys. However, texture hardening has more subtle benefits in sheet metal forming in optimizing fabrication performance. The variation of strength in the plane of the sheet is readily assessed by tensile tests carried out in various directions relative to the rolling direction. In many sheet applications, however, the requirement is for through-thickness strength (e.g. to resist

**Table 4.5-3** Deformation textures in metals with common crystal structures.

| Structure | Wire (fiber texture) | Sheet (rolling texture) |
|---|---|---|
| Bcc | [1 1 0] | {1 1 2}$\langle 1\,\bar{1}\,0 \rangle$ to {1 0 0}$\langle 0\,1\,1 \rangle$ |
| Fcc | [1 1 1], [1 0 0] double fiber | {1 1 0}$\langle 1\,1\,2 \rangle$ to {3 5 1}$\langle 1\,1\,2 \rangle$ |
| Cph | [2 1 0] | {0 0 0 1}$\langle 1\,0\,0\,0 \rangle$ |

thinning during pressing operations). This is more difficult to measure and is often assessed from uniaxial tensile tests by measuring the ratio of the strain in the width direction to that in the thickness direction of a test piece. The strain ratio $R$ is given by

$$R = \varepsilon_w/\varepsilon_t = \ln(w_0/w)/\ln(t_0/t)$$
$$= \ln(w_0/w)/\ln(wL/w_0L_0), \quad (4.5.31)$$

where $w_0$, $L_0$ and $t_0$ are the original dimensions of width, length and thickness, and $w$, $L$ and $t$ are the corresponding dimensions after straining, which is derived assuming no change in volume occurs. The average strain ratio $\bar{R}$, for tests at various angles in the plane of the sheet, is a measure of the normal anisotropy, i.e. the difference between the average properties in the plane of the sheet and that property in the direction normal to the sheet surface. A large value of $R$ means that there is a lack of deformation modes oriented to provide strain in the through-thickness direction, indicating a high through-thickness strength.

In deep drawing, schematically illustrated in Figure 4.5-44, the dominant stress system is radial tension combined with circumferential compression in the drawing zone, while that in the base and lower cup wall (i.e. central stretch-forming zone) is biaxial tension. The latter stress is equivalent to a through-thickness compression, plus a hydrostatic tension which does not affect the state of yielding. Drawing failure occurs when the central stretch-forming zone is insufficiently strong to support the load needed to draw the outer part of the blank through the die. Clearly, differential strength levels in these two regions, leading to greater ease of deformation in the drawing zone compared with the stretching zone, would enable deeper draws to be made: this is the effect of increasing the $\bar{R}$ value, i.e. high through-thickness strength relative to strength in the plane of the sheet will favor drawability. This is confirmed in Figure 4.5-45, where deep drawability as determined by limiting drawing ratio (i.e. ratio of maximum drawable blank diameter to final cup diameter) is remarkably insensitive to ductility and, by inference from the wide range of materials represented in the figure, to absolute strength level. Here it is noted that for hexagonal metals slip occurs readily along $\langle 1\,1\,\bar{2}\,0\rangle$, thus contributing no strain in the $c$-direction, and twinning only occurs on $\{1\,0\,\bar{1}\,2\}$ when the applied stress nearly parallel to the $c$-axis is compressive for $c/a > \sqrt{3}$ and tensile for $c/a < \sqrt{3}$. Thus, titanium, $c/a < \sqrt{3}$, has a high strength in through-thickness compression, whereas Zn with $c/a < \sqrt{3}$ has low through-thickness strength when the basal plane is oriented parallel to the plane of the sheet. In contrast, hexagonal metals with $c/a > \sqrt{3}$ would have a high $R$ for $\{1\,0\,\bar{1}\,0\}$ parallel to the plane of the sheet.

Texture hardening is much less in the cubic metals, but fcc materials with $\{1\,1\,1\}\langle 1\,1\,0\rangle$ slip system and bcc with $\{1\,1\,0\}\langle 1\,1\,1\rangle$ are expected to increase $R$ when the texture has component with $\{1\,1\,1\}$ and $\{1\,1\,0\}$ parallel to the plane of the sheet. The range of values of $\bar{R}$ encountered in cubic metals is much less. Face-centered cubic metals have $\bar{R}$ ranging from about 0.3 for cube texture, $\{1\,0\,0\}\langle 0\,0\,1\rangle$, to a maximum, in textures so far attained, of just over 1.0. Higher values are sometimes obtained in body-centered cubic metals. Values of $\bar{R}$ in the range 1.4–1.8 obtained in aluminum-killed low-carbon steel are associated with significant improvements in deep-drawing performance compared with rimming steel, which has $\bar{R}$-values between 1.0 and 1.4. The highest values of $R$ in steels are associated with texture

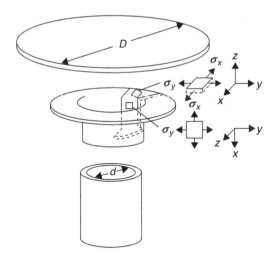

**Figure 4.5-44** Schematic diagram of the deep-drawing operations, indicating the stress systems operating in the flange and the cup wall. Limiting drawing ratio is defined as the ratio of the diameter of the largest blank which can satisfactorily complete the draw ($D_{max}$) to the punch diameter (d) (after Dillamore, Smallman and Wilson, 1969; courtesy of the Institute of Materials, Minerals and Mining).

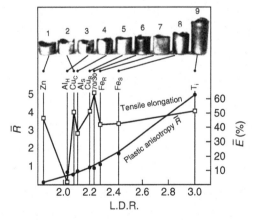

**Figure 4.5-45** Limiting draw ratios (LDR) as a function of average values of $R$ and of elongation to fracture measured in tensile tests at 0°, 45° and 90° to the rolling direction (after Wilson, 1966; courtesy of the Institute of Materials, Minerals and Mining).

components with {1 1 1} parallel to the surface, while crystals with {1 0 0} parallel to the surface have a strongly depressing effect on $\bar{R}$.

In most cases it is found that the $R$ values vary with testing direction and this has relevance in relation to the strain distribution in sheet metal forming. In particular, ear formation on pressings generally develops under a predominant uniaxial compressive stress at the edge of the pressing. The ear is a direct consequence of the variation in strain ratio for different directions of uniaxial stressing, and a large variation in $R$ value, where $\Delta R = (R_{max} - R_{min})$ generally correlates with a tendency to form pronounced ears. On this basis we could write a simple recipe for good deep-drawing properties in terms of strain ratio measurements made in a uniaxial tensile test as high $R$ and low $\Delta R$. Much research is aimed at improving forming properties through texture control.

## 4.5.7 Macroscopic plasticity

### 4.5.7.1 Tresca and von Mises criteria

In dislocation theory it is usual to consider the flow stress or yield stress of ductile metals under simple conditions of stressing. In practice, the engineer deals with metals under more complex conditions of stressing (e.g. during forming operations) and hence needs to correlate yielding under combined stresses with that in uniaxial testing. To achieve such a yield stress criterion it is usually assumed that the metal is mechanically isotropic and deforms plastically at constant volume, i.e. a hydrostatic state of stress does not affect yielding. In assuming plastic isotropy, macroscopic shear is allowed to take place along lines of maximum shear stress and crystallographic slip is ignored, and the yield stress in tension is equal to that in compression, i.e. there is no Bauschinger effect.

A given applied stress state in terms of the principal stresses $\sigma_1, \sigma_2, \sigma_3$ which act along three principal axes, $X_1, X_2$ and $X_3$, may be separated into the hydrostatic part (which produces changes in volume) and the deviatoric components (which produce changes in shape). It is assumed that the hydrostatic component has no effect on yielding and hence the more the stress state deviates from pure hydrostatic, the greater the tendency to produce yield. The stresses may be represented on a stress-space plot (see Figure 4.5-46a), in which a line equidistant from the three stress axes represents a pure hydrostatic stress state. Deviation from this line will cause yielding if the deviation is sufficiently large, and define a yield surface which has sixfold symmetry about the hydrostatic line. This arises because the conditions of isotropy imply equal yield stresses along all three axes, and the absence of the Bauschinger effect implies equal yield stresses along $\sigma_1$ and $-\sigma_1$. Taking a section through stress space, perpendicular to the hydrostatic line, gives the two simplest yield criteria satisfying the symmetry requirements corresponding to a regular hexagon and a circle.

The hexagonal form represents the Tresca criterion (see Figure 4.5-46c), which assumes that plastic shear takes place when the maximum shear stress attains a critical value $k$ equal to shear yield stress in uniaxial tension. This is expressed by

$$\tau_{max} = \frac{\sigma_1 - \sigma_3}{2} = k, \qquad (4.5.32)$$

where the principal stresses $\sigma_1 > \sigma_2 > \sigma_3$. This criterion is the isotropic equivalent of the law of resolved shear stress in single crystals. The tensile yield stress $Y = 2k$ is obtained by putting $\sigma_1 = Y, \sigma_2 = \sigma_3 = 0$.

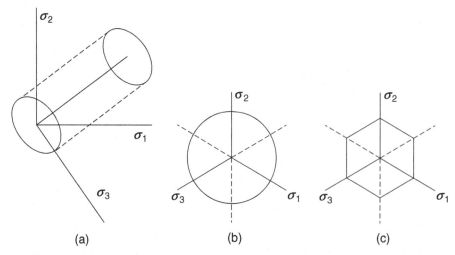

**Figure 4.5-46** Schematic representation of the yield surface with: (a) principal stresses $\sigma_1, \sigma_2$ and $\sigma_3$, (b) von Mises yield criterion and (c) Tresca yield criterion.

The circular cylinder is described by the equation:

$$(\sigma_1 - \sigma_2)^2 + (\sigma_2 - \sigma_3)^2 + (\sigma_3 - \sigma_1)^2 = \text{constant} \tag{4.5.33}$$

and is the basis of the von Mises yield criterion (see Figure 4.5-46b). This criterion implies that yielding will occur when the shear energy per unit volume reaches a critical value given by the constant. This constant is equal to $6k^2$ or $2Y^2$, where $k$ is the yield stress in simple shear, as shown by putting $\sigma_2 = 0$, $\sigma_1 = \sigma_3$, and $Y$ is the yield stress in uniaxial tension when $\sigma_2 = \sigma_3 = 0$. Clearly, $Y = \sqrt{3}k$ compared to $Y = 2k$ for the Tresca criterion and, in general, this is found to agree somewhat closer with experiment.

In many practical working processes (e.g. rolling), the deformation occurs under approximately plane strain conditions with displacements confined to the $X_1X_2$ plane. It does not follow that the stress in this direction is zero and, in fact, the deformation conditions are satisfied if $\sigma_3 = \frac{1}{2}(\sigma_1 + \sigma_2)$, so that the tendency for one pair of principal stresses to extend the metal along the $X_3$-axis is balanced by that of the other pair to contract it along this axis. Eliminating $\sigma_3$ from the von Mises criterion, the yield criterion becomes

$$(\sigma_1 - \sigma_2) = 2k$$

and the plane strain yield stress, i.e. when $\sigma_2 = 0$, given when

$$\sigma_1 = 2k = 2Y/\sqrt{3} = 1.15Y.$$

For plane strain conditions, the Tresca and von Mises criteria are equivalent and two-dimensional flow occurs when the shear stress reaches a critical value. The above condition is thus equally valid when written in terms of the deviatoric stresses $\sigma'_1$, $\sigma'_2$, $\sigma'_3$ defined by equations of the type:

$$\sigma'_1 = \sigma_1 - \frac{1}{3}(\sigma_1 + \sigma_2 + \sigma_3).$$

Under plane stress conditions, $\sigma_3 = 0$ and the yield surface becomes two-dimensional and the von Mises criterion becomes

$$\sigma_1^2 + \sigma_1\sigma_2 + \sigma_2^2 = 3k^2 = Y^2, \tag{4.5.34}$$

which describes an ellipse in the stress plane. For the Tresca criterion the yield surface reduces to a hexagon inscribed in the ellipse, as shown in Figure 4.5-47. Thus, when $\sigma_1$ and $\sigma_2$ have opposite signs, the Tresca criterion becomes $\sigma_1 - \sigma_2 = 2k = Y$ and is represented by the edges of the hexagon CD and FA. When they have the same sign, then $\sigma_1 = 2k = Y$ or $\sigma_2 = 2k = Y$ and defines the hexagon edges AB, BC, DE and EF.

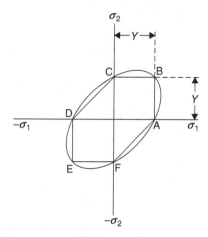

**Figure 4.5-47** The von Mises yield ellipse and Tresca yield hexagon.

### 4.5.7.2 Effective stress and strain

For an isotropic material, a knowledge of the uniaxial tensile test behavior, together with the yield function, should enable the stress–strain behavior to be predicted for any stress system. This is achieved by defining an effective stress–effective strain relationship such that if $\sigma = K\varepsilon^n$ is the uniaxial stress–strain relationship then we may write

$$\bar{\sigma} = K\bar{\varepsilon}^n \tag{4.5.35}$$

for any state of stress. The stress-strain behavior of a thin-walled tube with internal pressure is a typical example, and it is observed that the flow curves obtained in uniaxial tension and in biaxial torsion coincide when the curves are plotted in terms of effective stress and effective strain. These quantities are defined by:

$$\bar{\sigma} = \frac{\sqrt{2}}{2}[(\sigma_1 - \sigma_2)^2 + (\sigma_2 - \sigma_3)^2 + (\sigma_3 - \sigma_1)^2]^{1/2} \tag{4.5.36}$$

and

$$\bar{\varepsilon} = \frac{\sqrt{2}}{3}[(\varepsilon_1 - \varepsilon_2)^2 + (\varepsilon_2 - \varepsilon_3)^2 + (\varepsilon_3 - \varepsilon_1)^2]^{1/2}, \tag{4.5.37}$$

where $\varepsilon_1$, $\varepsilon_2$ and $\varepsilon_3$ are the principal strains, all of which reduce to the axial normal components of stress and strain for a tensile test. It should be emphasized, however, that this generalization holds only for isotropic media and for constant loading paths, i.e. $\sigma_1 = \alpha\sigma_2 = \beta\sigma_3$, where $\alpha$ and $\beta$ are constants independent of the value of $\sigma_1$.

## 4.5.8 Annealing

### 4.5.8.1 General effects of annealing

When a metal is cold worked, by any of the many industrial shaping operations, changes occur in both its physical and mechanical properties. While the increased hardness and strength which result from the working treatment may be of importance in certain applications, it is frequently necessary to return the metal to its original condition to allow further forming operations (e.g. deep drawing) to be carried out for applications where optimum physical properties, such as electrical conductivity, are essential. The treatment given to the metal to bring about a decrease of the hardness and an increase in the ductility is known as annealing. This usually means keeping the deformed metal for a certain time at a temperature higher than about one-third the absolute melting point.

Cold working produces an increase in dislocation density; for most metals $\rho$ increases from the value of $10^{10}$–$10^{12}$ lines m$^{-2}$ typical of the annealed state, to $10^{12}$–$10^{13}$ after a few percent deformation, and up to $10^{15}$–$10^{16}$ lines m$^{-2}$ in the heavily deformed state. Such an array of dislocations gives rise to a substantial strain energy stored in the lattice, so that the cold-worked condition is thermodynamically unstable relative to the undeformed one. Consequently, the deformed metal will try to return to a state of lower free energy, i.e. a more perfect state. In general, this return to a more equilibrium structure cannot occur spontaneously but only at elevated temperatures, where thermally activated processes such as diffusion, cross-slip and climb take place. Like all non-equilibrium processes the rate of approach to equilibrium will be governed by an Arrhenius equation of the form:

$$\text{Rate} = A \exp[-Q/kT],$$

where the activation energy Q depends on impurity content, strain, etc.

The formation of atmospheres by strain ageing is one method whereby the metal reduces its excess lattice energy, but this process is unique in that it usually leads to a further increase in the structure-sensitive properties rather than a reduction to the value characteristic of the annealed condition. It is necessary, therefore, to increase the temperature of the deformed metal above the strain-ageing temperature before it recovers its original softness and other properties.

The removal of the cold-worked condition occurs by a combination of three processes, namely: (1) recovery, (2) recrystallization and (3) grain growth. These stages have been successfully studied using light microscopy, transmission electron microscopy or X-ray diffraction; mechanical property measurements (e.g. hardness); and physical property measurements (e.g. density, electrical resistivity and stored energy). Figure 4.5-48 shows the change in some of these properties on annealing. During the recovery stage the decrease in stored energy and electrical resistivity is accompanied by only a slight lowering of hardness, and the greatest simultaneous change in properties occurs during the primary recrystallization stage. However, while these measurements are no doubt striking and extremely useful, it is necessary to understand them to correlate such studies with the structural changes by which they are accompanied.

### 4.5.8.2 Recovery

This process describes the changes in the distribution and density of defects with associated changes in physical and mechanical properties which take place in worked crystals before recrystallization or alteration of orientation occurs. It will be remembered that the structure of a cold-worked metal consists of dense dislocation networks, formed by the glide and interaction of dislocations, and, consequently, the recovery stage of annealing is chiefly concerned with the rearrangement of these dislocations to reduce the lattice energy and does not involve the migration of large-angle boundaries. This rearrangement of the dislocations is assisted by thermal activation. Mutual annihilation of dislocations is one process.

When the two dislocations are on the same slip plane, it is possible that as they run together and annihilate they will have to cut through intersecting dislocations on other planes, i.e. 'forest' dislocations. This recovery process will therefore be aided by thermal fluctuations, since the activation energy for such a cutting process is small. When the two dislocations of opposite sign are not on the same slip plane, climb or cross-slip must first occur, and both processes require thermal activation.

One of the most important recovery processes which leads to a resultant lowering of the lattice strain energy is rearrangement of the dislocations into cell walls. This

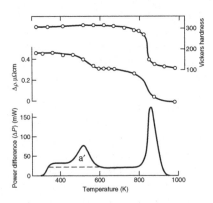

**Figure 4.5-48** Rate of release of stored energy ($\Delta P$), increment in electrical resistivity ($\Delta \rho$) and hardness (VPN) for specimens of nickel deformed in torsion and heated at 6 K min$^{-1}$ (Clareborough, Hargreaves and West, 1955).

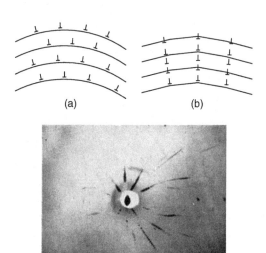

**Figure 4.5-49** (a) Random arrangement of excess parallel edge dislocations. (b) Alignment into dislocation walls. (c) Laue photograph of polygonized zinc (after Cahn, 1949; courtesy of Institute of Materials, Minerals and Mining).

process in its simplest form was originally termed polygonization and is illustrated schematically in Figure 4.5-49, whereby dislocations all of one sign align themselves into walls to form small-angle or sub-grain boundaries. During deformation a region of the lattice is curved, as shown in Figure 4.5-49a, and the observed curvature can be attributed to the formation of excess edge dislocations parallel to the axis of bending. On heating, the dislocations form a sub-boundary by a process of annihilation and rearrangement. This is shown in Figure 4.5-49b, from which it can be seen that it is the excess dislocations of one sign which remain after the annihilation process that align themselves into walls.

Polygonization is a simple form of sub-boundary formation and the basic movement is climb, whereby the edge dislocations change their arrangement from a horizontal to a vertical grouping. This process involves the migration of vacancies to or from the edge of the half-planes of the dislocations. The removal of vacancies from the lattice, together with the reduced strain energy of dislocations which results, can account for the large change in both electrical resistivity and stored energy observed during this stage, while the change in hardness can be attributed to the rearrangement of dislocations and to the reduction in the density of dislocations.

The process of polygonization can be demonstrated using the Laue method of X-ray diffraction. Diffraction from a bent single crystal of zinc takes the form of continuous radial streaks. On annealing, these asterisms break up into spots, as shown in Figure 4.5-49c, where each diffraction spot originates from a perfect polygonized sub-grain, and the distance between the spots represents the angular misorientation across the sub-grain boundary. Direct evidence for this process is observed in the electron microscope, where, in heavily deformed polycrystalline aggregates at least, recovery is associated with the formation of sub-grains out of complex dislocation networks by a process of dislocation annihilation and rearrangement. In some deformed metals and alloys the dislocations are already partially arranged in sub-boundaries, forming diffuse cell structures by dynamical recovery (see Figure 4.5-40). The conventional recovery process is then one in which these cells sharpen and grow. In other metals, dislocations are more uniformly distributed after deformation, with hardly any cell structure discernible, and the recovery process then involves formation, sharpening and growth of sub-boundaries. The sharpness of the cell structure formed by deformation depends on the stacking-fault energy of the metal, the deformation temperature and the extent of deformation (see Figure 4.5-41).

### 4.5.8.3 Recrystallization

The most significant changes in the structure-sensitive properties occur during the primary recrystallization stage. In this stage the deformed lattice is completely replaced by a new unstrained one by means of a nucleation and growth process, in which practically stress-free grains grow from nuclei formed in the deformed matrix. The orientation of the new grains differs considerably from that of the crystals they consume, so that the growth process must be regarded as incoherent, i.e. it takes place by the advance of large-angle boundaries separating the new crystals from the strained matrix.

During the growth of grains, atoms get transferred from one grain to another across the boundary. Such a process is thermally activated, as shown in Figure 4.5-50, and by the usual reaction-rate theory the frequency of atomic transfer one way is

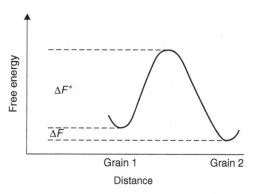

**Figure 4.5-50** Variation in free energy during grain growth.

$$\nu \exp\left(-\frac{\Delta F}{kT}\right) \; s^{-1} \quad (4.5.38)$$

and in the reverse direction

$$\nu \exp\left(-\frac{\Delta F^* + \Delta F}{kT}\right) \; s^{-1}, \quad (4.5.39)$$

where $\Delta F$ is the difference in free energy per atom between the two grains, i.e. supplying the driving force for migration, and $\Delta F^*$ is an activation energy. For each net transfer the boundary moves forward a distance $b$ and the velocity $v$ is given by

$$v = M \Delta F, \quad (4.5.40)$$

where $M$ is the mobility of the boundary, i.e. the velocity for unit driving force, and is thus

$$M = \frac{bv}{kT} \exp\left(\frac{\Delta S^*}{k}\right) \exp\left(-\frac{\Delta E^*}{kT}\right). \quad (4.5.41)$$

Generally, the open structure of high-angle boundaries should lead to a high mobility. However, they are susceptible to the segregation of impurities, low concentrations of which can reduce the boundary mobility by orders of magnitude. In contrast, special boundaries which are close to a CSL are much less affected by impurity segregation and hence can lead to higher relative mobility.

It is well known that the rate of recrystallization depends on several important factors, namely: (1) the amount of prior deformation (the greater the degree of cold work, the lower the recrystallization temperature and the smaller the grain size), (2) the temperature of the anneal (as the temperature is lowered, the time to attain a constant grain size increases exponentially[5]) and (3) the purity of the sample (e.g. zone-refined aluminum recrystallizes below room temperature, whereas aluminum of commercial purity must be heated several hundred degrees). The role these variables play in recrystallization will be evident once the mechanism of recrystallization is known. This mechanism will now be outlined.

Measurements, using the light microscope, of the increase in diameter of a new grain as a function of time at any given temperature can be expressed as shown in Figure 4.5-51. The diameter increases linearly with time until the growing grains begin to impinge on one another, after which the rate necessarily decreases. The classical interpretation of these observations is that nuclei form spontaneously in the matrix after a so-called nucleation time, $t_0$, and these nuclei then proceed to grow steadily

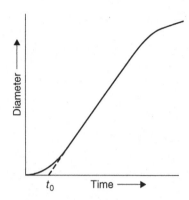

**Figure 4.5-51** Variation of grain diameter with time at a constant temperature.

as shown by the linear relationship. The driving force for the process is provided by the stored energy of cold work contained in the strained grain on one side of the boundary relative to that on the other side. Such an interpretation would suggest that the recrystallization process occurs in two distinct stages, i.e. first nucleation and then growth.

During the linear growth period the radius of a nucleus is $R = G(t - t_0)$, where $G$, the growth rate, is $dR/dt$ and, assuming the nucleus is spherical, the volume of the recrystallized nucleus is $4\pi/3 \; G^3(t-t_0)^3$. If the number of nuclei that form in a time increment $dt$ is $N \, dt$ per unit volume of unrecrystallized matrix, and if the nuclei do not impinge on one another, then for unit total volume

$$f = \frac{4}{3}\pi NG^3 \int_0^t (t - t_0)^3 dt$$

or

$$f = \frac{\pi}{3} G^3 t^4. \quad (4.5.42)$$

This equation is valid in the initial stages when $f \ll 1$. When the nuclei impinge on one another the rate of recrystallization decreases and is related to the amount untransformed $(1 - f)$ by

$$f = 1 - \exp\left(-\frac{\pi}{3} NG^3 t^4\right), \quad (4.5.43)$$

where, for short times, equation (4.5.42) reduces to equation (4.5.41). This Johnson-Mehl equation is expected to apply to any phase transformation where there is random nucleation, constant $N$ and $G$, and small $t_0$ In practice, nucleation is not random and the rate not constant, so that equation (4.5.43) will not strictly apply. For the case where the nucleation rate decreases exponentially, Avrami developed the equation:

---

[5] The velocity of linear growth of new crystals usually obeys an exponential relationship of the form $v = v_0 \exp[-Q/RT]$.

$$f = 1 - \exp(-kt^n), \quad (4.5.44)$$

where $k$ and $n$ are constants, with $n \approx 3$ for a fast and $n \approx 4$ for a slow decrease of nucleation rate. Provided there is no change in the nucleation mechanism, $n$ is independent of temperature but $k$ is very sensitive to temperature $T$; clearly, from equation (4.5.43), $k = \pi N G^3 / 3$, and both $N$ and $G$ depend on $T$.

An alternative interpretation is that the so-called incubation time $t_0$ represents a period during which small nuclei, of a size too small to be observed in the light microscope, are growing very slowly. This latter interpretation follows from the recovery stage of annealing. Thus, the structure of a recovered metal consists of sub-grain regions of practically perfect crystal and, thus, one might expect the 'active' recrystallization nuclei to be formed by the growth of certain sub-grains at the expense of others.

The process of recrystallization may be pictured as follows. After deformation, polygonization of the bent lattice regions on a fine scale occurs and this results in the formation of several regions in the lattice where the strain energy is lower than in the surrounding matrix; this is a necessary primary condition for nucleation. During this initial period when the angles between the sub-grains are small and less than one degree, the sub-grains form and grow quite rapidly. However, as the sub-grains grow to such a size that the angles between them become of the order of a few degrees, the growth of any given sub-grain at the expense of the others is very slow. Eventually one of the sub-grains will grow to such a size that the boundary mobility begins to increase with increasing angle. A large-angle boundary, $\theta \approx 30-40°$, has a high mobility because of the large lattice irregularities or 'gaps' which exist in the boundary transition layer. The atoms on such a boundary can easily transfer their allegiance from one crystal to the other. This sub-grain is then able to grow at a much faster rate than the other sub-grains which surround it and so acts as the nucleus of a recrystallized grain. The further it grows, the greater will be the difference in orientation between the nucleus and the matrix it meets and consumes, until it finally becomes recognizable as a new strain-free crystal separated from its surroundings by a large-angle boundary.

The recrystallization nucleus therefore has its origin as a sub-grain in the deformed microstructure. Whether it grows to become a strain-free grain depends on three factors: (1) the stored energy of cold work must be sufficiently high to provide the required driving force, (2) the potential nucleus should have a size advantage over its neighbors and (3) it must be capable of continued growth by existing in a region of high lattice curvature (e.g. transition band), so that the growing nucleus can quickly achieve a high-angle boundary. *In situ* experiments in the HVEM have confirmed these factors. Figure 4.5-52a shows the as-deformed substructure in the transverse section of rolled copper, together with the orientations of some selected areas. The sub-grains are observed to vary in width from 50 to 500 nm, and exist between regions 1 and 8 as a transition band across which the orientation changes sharply. On heating to 200°C, the sub-grain region 2 grows into the transition region and the orientation of the new grain well developed at 300°C is identical to the original sub-grain (Figure 4.5-52b).

With this knowledge of recrystallization the influence of several variables known to affect the recrystallization behavior of a metal can now be understood. Prior deformation, for example, will control the extent to which a region of the lattice is curved. The larger the deformation, the more severely will the lattice be curved and, consequently, the smaller will be the size of a growing sub-grain when it acquires a large-angle boundary. This must mean that a shorter time is necessary at any given temperature for the sub-grain to become an 'active' nucleus or, conversely, that the higher the annealing temperature, the quicker will this stage be reached. In some instances, heavily cold-worked metals recrystallize without any significant recovery owing to the formation of strain-free cells during deformation. The importance of impurity content on recrystallization temperature is also evident from the effect impurities have on obstructing sub-boundary dislocation and grain boundary mobility.

The intragranular nucleation of strain-free grains, as discussed above, is considered as abnormal sub-grain growth, in which it is necessary to specify that some sub-grains acquire a size advantage and are able to grow at the expense of the normal sub-grains. It has been suggested that nuclei may also be formed by a process involving the rotation of individual cells so that they coalesce with neighboring cells to produce larger cells by volume diffusion and dislocation rearrangement.

In some circumstances, intergranular nucleation is observed in which an existing grain boundary bows out under an initial driving force equal to the difference in free energy across the grain boundary. This strain-induced boundary migration is irregular and is from a grain with low strain (i.e. large cell size) to one of larger strain and smaller cell size. For a boundary to grow in this way the strain energy difference per unit volume across the boundary must be sufficient to supply the energy increase to bow out a length of boundary $\approx 1 \, \mu m$.

Segregation of solute atoms to, and precipitation on, the grain boundary tends to inhibit intergranular nucleation and gives an advantage to intragranular nucleation, provided the dispersion is not too fine. In general, the recrystallization behavior of two-phase alloys is extremely sensitive to the dispersion of the second phase. Small, finely dispersed particles retard recrystallization by reducing both the nucleation rate and the grain

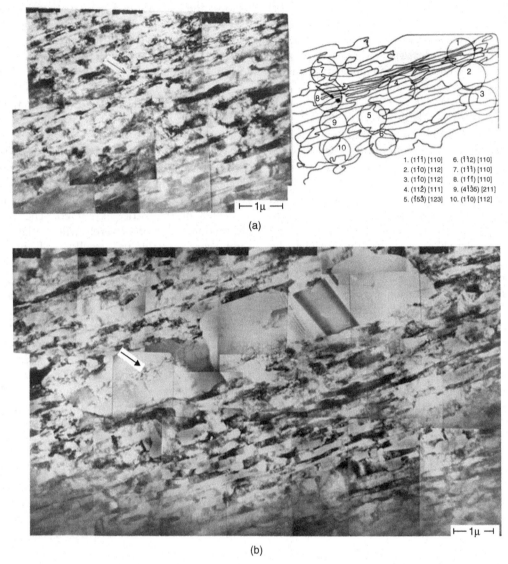

**Figure 4.5-52** Electron micrographs of copper. (a) Cold-rolled 95% at room temperature, transverse section. (b) Heated to 300°C in the HVEM.

boundary mobility, whereas large, coarsely dispersed particles enhance recrystallization by increasing the nucleation rate. During deformation, zones of high dislocation density and large misorientations are formed around non-deformable particles, and, on annealing, recrystallization nuclei are created within these zones by a process of polygonization by sub-boundary migration. Particle-stimulated nucleation occurs above a critical particle size, which decreases with increasing deformation. The finer dispersions tend to homogenize the microstructure (i.e. dislocation distribution), thereby minimizing local lattice curvature and reducing nucleation.

The formation of nuclei becomes very difficult when the spacing of second-phase particles is so small that each developing sub-grain interacts with a particle before it becomes a viable nucleus. The extreme case of this is SAP (sintered aluminum powder), which contains very stable, close-spaced oxide particles. These particles prevent the rearrangement of dislocations into cell walls and their movement to form high-angle boundaries, and hence SAP must be heated to a temperature very close to the melting point before it recrystallizes.

### 4.5.8.4 Grain growth

When primary recrystallization is complete (i.e. when the growing crystals have consumed all the strained material), the material can lower its energy further by reducing its total area of grain surface. With extensive annealing it is often found that grain boundaries straighten, small grains shrink and larger ones grow. The general phenomenon is known as grain growth, and the most important factor governing the process is the surface tension of the grain boundaries. A grain boundary has a surface tension, $T$ (= surface free energy per unit area), because its atoms have a higher free energy than those within the grains. Consequently, to reduce this energy a poly crystal will tend to minimize the area of its grain boundaries and, when this occurs, the configuration taken up by any set of grain boundaries (see Figure 4.5-53) will be governed by the condition that

$$T_A/\sin A = T_B \sin B = T_C/\sin C. \quad (4.5.45)$$

Most grain boundaries are of the large-angle type with their energies approximately independent of orientation, so that for a random aggregate of grains $T_A = T_B = T_C$ and the equilibrium grain boundary angles are each equal to 120°. Figure 4.5-53b shows an idealized grain in two dimensions surrounded by others of uniform size, and it can be seen that the equilibrium grain shape takes the form of a polygon of six sides with 120° inclusive angles. All polygons with either more or less than this number of sides cannot be in equilibrium. At high temperatures where the atoms are mobile, a grain with fewer sides will tend to become smaller, under the action of the grain boundary surface tension forces, while one with more sides will tend to grow.

Second-phase particles have a major inhibiting effect on boundary migration and are particularly effective in the control of grain size. The pinning process arises from surface tension forces exerted by the particle–matrix interface on the grain boundary as it migrates past the particle. Figure 4.5-54 shows that the drag exerted by the particle on the boundary, resolved in the forward direction, is

$$F = \pi r \gamma \sin 2\theta,$$

where $\gamma$ is the specific interfacial energy of the boundary; $F = F_{max} = \pi r \gamma$ when $\theta = 45°$. Now if there are $N$ particles per unit volume, the volume fraction is $4\pi r_s N/3$

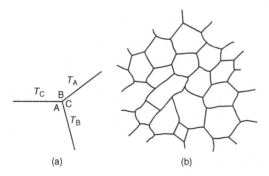

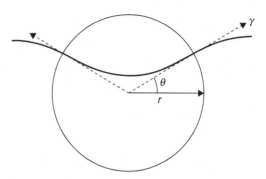

**Figure 4.5-54** Diagram showing the drag exerted on a boundary by a particle.

and the number $n$ intersecting unit area of boundary is given by

$$n = 3f/2\pi r^2. \quad (4.5.46)$$

For a grain boundary migrating under the influence of its own surface tension the driving force is $2\gamma/R$, where $R$ is the minimum radius of curvature and, as the grains grow, $R$ increases and the driving force decreases until it is balanced by the particle drag, when growth stops. If $R \sim d$, the mean grain diameter, then the critical grain diameter is given by the condition

$$nF \approx 2\gamma/d_{crit}$$

or

$$d_{crit} \approx 2\gamma(2\pi r^2/3f\pi r\gamma) = 4r/3f. \quad (4.5.47)$$

This Zener drag equation overestimates the driving force for grain growth by considering an isolated spherical grain. A heterogeneity in grain size is necessary for grain growth, and taking this into account gives a revised equation:

$$d_{crit} \approx \frac{\pi r}{3f}\left[\frac{3}{2} - \frac{2}{Z}\right], \quad (4.5.48)$$

where $Z$ is the ratio of the diameters of growing grains to the surrounding grains. This treatment explains the successful use of small particles in refining the grain size of commercial alloys.

During the above process growth is continuous and a uniform coarsening of the polycrystalline aggregate usually occurs. Nevertheless, even after growth has finished the grain size in a specimen which was previously severely cold-worked remains relatively small, because of the large number of nuclei produced by the working treatment. Exaggerated grain growth can often be induced, however, in one of two ways, namely: (1) by subjecting the specimen to a critical strain-anneal treatment or (2) by a process of secondary recrystallization. By applying a critical deformation (usually a few percent

**Figure 4.5-53** (a) Relation between angles and surface tensions at a grain boundary triple point. (b) Idealized polygonal grain structure.

strain) to the specimen the number of nuclei will be kept to a minimum, and if this strain is followed by a high-temperature anneal in a thermal gradient some of these nuclei will be made more favorable for rapid growth than others. With this technique, if the conditions are carefully controlled, the whole of the specimen may be turned into one crystal, i.e. a single crystal. The term secondary recrystallization describes the process whereby a specimen which has been given a primary recrystallization treatment at a low temperature is taken to a higher temperature to enable the abnormally rapid growth of a few grains to occur. The only driving force for secondary recrystallization is the reduction of grain boundary free energy, as in normal grain growth, and, consequently, certain special conditions are necessary for its occurrence. One condition for this 'abnormal' growth is that normal continuous growth is impeded by the presence of inclusions, as is indicated by the exaggerated grain growth of tungsten wire containing thoria, or the sudden coarsening of deoxidized steel at about 1000°C. A possible explanation for the phenomenon is that in some regions the grain boundaries become free (e.g. if the inclusions slowly dissolve or the boundary tears away) and as a result the grain size in such regions becomes appreciably larger than the average (Figure 4.5-55a). It then follows that the grain boundary junction angles between the large grain and the small ones that surround it will not satisfy the condition of equilibrium discussed above. As a consequence, further grain boundary movement to achieve 120° angles will occur, and the accompanying movement of a triple junction point will be as shown in Figure 4.5-55b. However, when the dihedral angles at each junction are approximately 120° a severe curvature in the grain boundary segments between the junctions will arise, and this leads to an increase in grain boundary area. Movement of these curved boundary segments towards their centers of curvature must then take place and this will give rise to the configuration shown in Figure 4.5-55c. Clearly, this sequence of events can be repeated and continued growth of the large grains will result.

The behavior of the dispersed phase is extremely important in secondary recrystallization and there are many examples in metallurgical practice where the control of secondary recrystallization with dispersed particles has been used to advantage. One example is in the use of Fe–3% Si in the production of strip for transformer laminations. This material is required with (1 1 0) [0 0 1] 'Goss' texture because of the [0 0 1] easy direction of magnetization, and it is found that the presence of MnS particles favors the growth of secondary grains with the appropriate Goss texture. Another example is in the removal of the pores during the sintering of metal and ceramic powders, such as alumina and metallic carbides. The sintering process is essentially one of vacancy creep involving the diffusion of vacancies from the pore of radius $r$ to a neighboring grain boundary, under a driving force $2\gamma_s/r$, where $\gamma_s$ is the surface energy. In practice, sintering occurs fairly rapidly up to about 95% full density because there is a plentiful association of boundaries and pores. When the pores become very small, however, they are no longer able to anchor the grain boundaries against the grain growth forces, and hence the pores sinter very slowly, since they are stranded within the grains some distance from any boundary. To promote total sintering, an effective dispersion is added. The dispersion is critical, however, since it must produce sufficient drag to slow down grain growth, during which a particular pore is crossed by several migrating boundaries, but not sufficiently large to give rise to secondary recrystallization when a given pore would be stranded far from any boundary.

The relation between grain size, temperature and strain is shown in Figure 4.5-56 for commercially pure aluminum. From this diagram it is clear that either a critical strain-anneal treatment or a secondary recrystallization process may be used for the preparation of perfect strain-free single crystals.

### 4.5.8.5 Annealing twins

A prominent feature of the microstructures of most annealed fcc metals and alloys is the presence of many straight-sided bands that run across grains. These

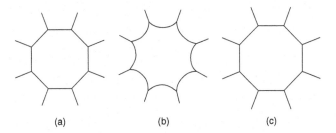

**Figure 4.5-55** Grain growth during secondary recrystallization.

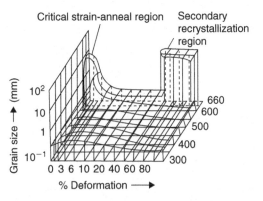

**Figure 4.5-56** Relation between grain size, deformation and temperature for aluminum (after Burgers, courtesy of Akademie-Verlags-Gesellschaft).

bands have a twinned orientation relative to their neighboring grain and are referred to as annealing twins (see Chapter 3). The parallel boundaries usually coincide with a (1 1 1) twinning plane with the structure coherent across it, i.e. both parts of the twin hold a single (1 1 1) plane in common.

As with formation of deformation twins, it is believed that a change in stacking sequence is all that is necessary to form an annealing twin. Such a change in stacking sequence may occur whenever a properly oriented grain boundary migrates. For example, if the boundary interface corresponds to a (1 1 1) plane, growth will proceed by the deposition of additional (1 1 1) planes in the usual stacking sequence **ABCABC**... If, however, the next newly deposited layer falls into the wrong position, the sequence **ABCABCB** is produced, which constitutes the first layer of a twin. Once a twin interface is formed, further growth may continue with the sequence in reverse order, **ABCABC|BACB**... until a second accident in the stacking sequence completes the twin band, **ABCABCBACBACBABC**. When a stacking error, such as that described above, occurs the number of nearest neighbors is unchanged, so that the ease of formation of a twin interface depends on the relative value of the interface energy. If this interface energy is low, as in copper, where $\gamma_{twin} < 20$ mJ m$^{-2}$ twinning occurs frequently while, if it is high, as in aluminum, the process is rare.

Annealing twins are rarely (if ever) found in cast metals because grain boundary migration is negligible during casting. Worked and annealed metals show considerable twin band formation; after extensive grain growth a coarse-grained metal often contains twins which are many times wider than any grain that was present shortly after recrystallization. This indicates that twin bands grow in width, during grain growth, by migration in a direction perpendicular to the (1 1 1) composition plane, and one mechanism whereby this can occur is illustrated schematically in Figure 4.5-57. This shows that a twin may form at the corner of a grain, since the grain boundary configuration will then have a lower interfacial energy. If this happens the twin will then be able to grow in width because one of its sides forms part of the boundary of the growing grain. Such a twin will continue to grow in width until a second mistake in the positioning of the atomic layers terminates it; a complete twin band is then formed. In copper and its alloys, $\gamma_{twin}/\gamma_{gb}$ is low and hence twins occur frequently, whereas in aluminum the corresponding ratio is very much higher and so twins are rare.

Twins may develop according to the model shown in Figure 4.5-58 where, during grain growth, a grain contact is established between grains C and D. Then if the orientation of grain D is close to the twin orientation of grain C, the nucleation of an annealing twin at the grain boundary, as shown in Figure 4.5-58d, will lower the total boundary energy. This follows because the twin/D interfaces will be reduced to about 5% of the normal grain boundary energy, the energies of the C/A and twin/A interfaces will be approximately the same, and the extra area of interface C/twin has only a very low energy. This model indicates that the number of twins per unit grain boundary area depends only on the number of new grain contacts made during grain growth, irrespective of grain size and annealing temperature.

### 4.5.8.6 Recrystallization textures

The preferred orientation developed by cold work often changes on recrystallization to a totally different preferred orientation. To explain this observation, Barrett and (later) Beck have put forward the 'oriented growth' theory of recrystallization textures, in which it is proposed that nuclei of many orientations initially form but, because the rate of growth of any given nucleus depends on the orientation difference between the matrix and growing crystal, the recrystallized texture will arise from those nuclei which have the fastest growth rate in the cold-worked matrix, i.e. those bounded by large-angle boundaries. It then follows that, because the matrix has a texture, all the nuclei which grow will have orientations that differ by 30–40° from the cold-worked texture. This explains why the new texture in fcc metals is often related to the old texture, by a rotation of approximately 30–40° around ⟨1 1 1⟩ axes, in bcc metals by 30° about ⟨1 1 0⟩ and in hcp by 30° about ⟨0 0 0 1⟩. However, while it is undoubtedly true that oriented growth provides a selection between favorably and unfavorably oriented nuclei, there are many observations to indicate that the initial nucleation is not entirely random. For instance, because of the crystallographic symmetry one would expect grains appearing in an fcc texture to be related to

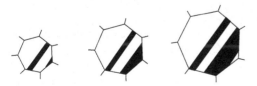

**Figure 4.5-57** Formation and growth of annealing twins (from Burke and Turnbull, 1952).

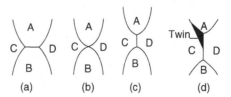

**Figure 4.5-58** Nucleation of an annealing twin during grain growth.

rotations about all four $\langle 1\,1\,1\rangle$ axes, i.e. eight orientations arising from two possible rotations about each of the four $\langle 1\,1\,1\rangle$ axes. All these possible orientations are rarely (if ever) observed.

To account for such observations, and for those cases where the deformation texture and the annealing texture show strong similarities, oriented nucleation is considered to be important. The oriented nucleation theory assumes that the selection of orientations is determined in the nucleation stage. It is generally accepted that all recrystallization nuclei pre-exist in the deformed matrix, as sub-grains, which become more perfect through recovery processes prior to recrystallization. It is thus most probable that there is some selection of nuclei determined by the representation of the orientations in the deformation texture, and that the oriented nucleation theory should apply in some cases. In many cases the orientations which are strongly represented in the annealing texture are very weakly represented in the deformed material. The most striking example is the 'cube' texture, $(1\,0\,0)\,[0\,0\,1]$, found in most fcc pure metals which have been annealed following heavy rolling reductions. In this texture, the cube axes are extremely well aligned along the sheet axes, and its behavior resembles that of a single crystal. It is thus clear that cube-oriented grains or sub-grains must have a very high initial growth rate in order to form the remarkably strong quasi-single-crystal cube texture. The percentage of cubically aligned grains increases with increased deformation, but the sharpness of the textures is profoundly affected by alloying additions. The amount of alloying addition required to suppress the texture depends on those factors which affect the stacking-fault energy, such as the lattice misfit of the solute atom in the solvent lattice, valency, etc., in much the same way as that described for the transition of a pure metal deformation texture.

In general, however, if the texture is to be altered a distribution of second phase must either be present before cold rolling or be precipitated during annealing. In aluminum, for example, the amount of cube texture can be limited in favor of retained rolling texture by limiting the amount of grain growth with a precipitate dispersion of Si and Fe. By balancing the components, earing can be minimized in drawn aluminum cups. In aluminum-killed steels AlN precipitation prior to recrystallization produces a higher proportion of grains with $\{1\,1\,1\}$ planes parallel to the rolling plane and a high $\overline{R}$-value suitable for deep drawing. The AlN dispersion affects sub-grain growth, limiting the available nuclei and increasing the orientation selectivity, thereby favoring the high-energy $\{1\,1\,1\}$ grains. Improved $\overline{R}$-values in steels in general are probably due to the combined effect of particles in homogenizing the deformed microstructure and in controlling the subsequent sub-grain growth. The overall effect is to limit the availability of nuclei with orientations other than $\{1\,1\,1\}$.

## 4.5.9 Metallic creep

### 4.5.9.1 Transient and steady-state creep

Creep is the process by which plastic flow occurs when a constant stress is applied to a metal for a prolonged period of time. After the initial strain $\varepsilon_0$ which follows the application of the load, creep usually exhibits a rapid transient period of flow (stage 1) before it settles down to the linear steady-stage stage 2, which eventually gives way to tertiary creep and fracture. Transient creep, sometimes referred to as $\beta$-creep, obeys a $t^{1/3}$ law. The linear stage of creep is often termed steady-state creep and obeys the relation

$$\varepsilon = \kappa t. \tag{4.5.49}$$

Consequently, because both transient and steady-state creep usually occur together during creep at high temperatures, the complete curve (Figure 4.5-59) during the primary and secondary stages of creep fits the equation

$$\varepsilon = \beta t^{1/3} + \kappa t \tag{4.5.50}$$

extremely well. In contrast to transient creep, steady-state creep increases markedly with both temperature and stress. At constant stress the dependence on temperature is given by

$$\dot{\varepsilon}_{ss} = d\varepsilon/dt = \text{const.}\exp[-Q/\mathbf{k}T], \tag{4.5.51}$$

where $Q$ is the activation energy for steady-state creep, while at constant temperature the dependence on stress $\sigma$ (compensated for modulus $E$) is

$$\dot{\varepsilon}_{ss} = \text{const.}(\sigma/E)^n. \tag{4.5.52}$$

Steady-state creep is therefore described by the equation:

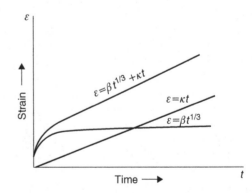

**Figure 4.5-59** Combination of transient and steady-state creep.

$$\dot{\varepsilon}_{ss} = A(\sigma/E)^n \exp(-Q/kT). \quad (4.5.53)$$

The basic assumption of the mechanism of steady-state creep is that during the creep process the rate of recovery $r$ (i.e. decrease in strength, $d\sigma/dt$) is sufficiently fast to balance the rate of work hardening $h = (d\sigma/d\varepsilon)$. The creep rate $(d\varepsilon/dt)$ is then given by

$$d\varepsilon/dt = (d\sigma/dt)/(d\sigma/d\varepsilon) = r/h. \quad (4.5.54)$$

To prevent work hardening, both the screw and edge parts of a glissile dislocation loop must be able to escape from tangled or piled-up regions. The edge dislocations will, of course, escape by climb, and since this process requires a higher activation energy than cross-slip, it will be the rate-controlling process in steady-state creep. The rate of recovery is governed by the rate of climb, which depends on diffusion and stress such that

$$r = A(\sigma/E)^p D = A(\sigma/E)^p D_0 \exp[-Q/kT],$$

where $D$ is a diffusion coefficient and the stress term arises because recovery is faster the higher the stress level and the closer dislocations are together. The work-hardening rate decreases from the initial rate $h_0$ with increasing stress, i.e. $h = h_0(E/\sigma)^q$, thus

$$\dot{\varepsilon}_{ss} = r/h = B(\sigma/E)^n D, \quad (4.5.55)$$

where $B$ $(=A/h_0)$ is a constant and $n$ $(= p + q)$ is the stress exponent.

The structure developed in creep arises from the simultaneous work hardening and recovery. The dislocation density $\rho$ increases with $\varepsilon$ and the dislocation network gets finer, since dislocation spacing is proportional to $\rho^{-1/2}$. At the same time, the dislocations tend to reduce their strain energy by mutual annihilation and rearrange to form low-angle boundaries, and this increases the network spacing. Straining then proceeds at a rate at which the refining action just balances the growth of the network by recovery; the equilibrium network size being determined by the stress. Although dynamical recovery can occur by cross-slip, the rate-controlling process in steady-state creep is climb, whereby edge dislocations climb out of their glide planes by absorbing or emitting vacancies; the activation energy is therefore that of self-diffusion. Structural observations confirm the importance of the recovery process to steady-state creep. These show that sub-grains form within the original grains and, with increasing deformation, the sub-grain angle increases while the dislocation density within them remains constant.[6] The climb process may, of course, be important in several different ways. Thus, climb may help a glissile dislocation to circumvent different barriers in the structure, such as a sessile dislocation, or it may lead to the annihilation of dislocations of opposite sign on different glide planes. Moreover, because creep-resistant materials are rarely pure metals, the climb process may also be important in allowing a glissile dislocation to get round a precipitate or move along a grain boundary. A comprehensive analysis of steady-state creep, based on the climb of dislocations, has been given by Weertman.

The activation energy for creep $Q$ may be obtained experimentally by plotting $\ln \dot{\varepsilon}_{ss}$ versus $1/T$, as shown in Figure 4.5-60. Usually above $0.5T_m$, $Q$ corresponds to the activation energy for self-diffusion $E_{SD}$, in agreement with the climb theory, but below $0.5T_m$, $Q < E_{SD}$, possibly corresponding to pipe diffusion. Figure 4.5-61 shows that three creep regimes may be identified and the temperature range where $Q=E_{SD}$ can be moved to higher temperatures by increasing the strain rate. Equation (4.55) shows that the stress exponent $n$ can be obtained experimentally by plotting $\ln \dot{\varepsilon}_{ss}$ versus $\ln \sigma$, as shown in Figure 4.5-62, where $n \approx 4$. While $n$ is generally about 4 for dislocation creep, Figure 4.5-63 shows that $n$ may vary considerably from this value depending on the stress regime; at low stresses (i.e. regime I) creep occurs not by dislocation glide and climb, but by stress-directed flow of vacancies.

## Worked example

Creep data for a light alloy are given in the table

| Stress (N mm$^{-2}$) | Temperature (K) | Minimum creep rate (s$^{-1}$) |
|---|---|---|
| 8.9 | 600 | $1 \times 10^{-5}$ |
| 5.0 | 600 | $1 \times 10^{-6}$ |
| 5.0 | 640 | $5 \times 10^{-6}$ |

Calculate the expected steady-state creep rate at a constant stress of 2.8 N mm$^{-2}$ at (i) 600K and (ii) 640 K.

## Solution

The creep equation: $\dot{\varepsilon}_{ss} = A\sigma^n \exp(-Q/\mathbf{R}T)$.

At constant $T = 600$ K,

$$n = \frac{\log\dot{\varepsilon}_1 - \log\dot{\varepsilon}_2}{\log\sigma_1 - \log\sigma_2}$$

$$= \frac{-5 - (-6)}{\log 8.9 - \log 5} = 4$$

Therefore, at 2.8 MPa, 600 K,

---

[6] Sub-grains do not always form during creep and in some metallic solid solutions where the glide of dislocations is restrained due to the dragging of solute atoms, the steady-state substructure is essentially a uniform distribution of dislocations.

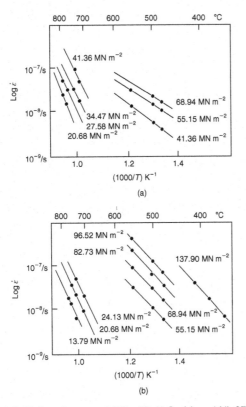

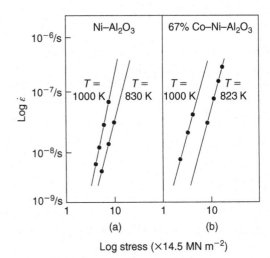

**Figure 4.5-62** Log $\dot{\varepsilon}$ versus log $\sigma$ for Ni–Al$_2$O$_3$ (a) and Ni–67Co–Al$_2$O$_3$ (b) (after Hancock, Dillamore and Smallman, 1972; courtesy of Institute of Materials, Minerals and Mining).

**Figure 4.5-60** Log $\dot{\varepsilon}$ versus $1/T$ for Ni–Al$_2$O$_3$ (a) and Ni–67Co–Al$_2$O$_3$ (b), showing the variation in activation energy above and below $0.5T_m$ (after Hancock, Dillamore and Smallman, 1972; courtesy of Institute of Materials, Minerals and Mining).

$$\ln 5 \times 10^{-6} = \ln(A\sigma^n) - \frac{Q}{RT_2}$$
$$\therefore Q = (\ln 5 \times 10^{-6} - \ln 10^{-6})$$
$$\times \frac{8.314}{1/600 - 1/640}$$
$$= 128.4 \times 10^3 \text{ J mol}^{-1}$$
$$= 128.4 \text{ kJ mol}^{-1}.$$

At constant $\sigma = 2.8$ MPa,

$$-5 - \log \dot{\varepsilon} = 4 \times (\log 8.9 - \log 2.8)$$
$$\therefore \dot{\varepsilon} = 0.98 \times 10^{-7} \text{s}^{-1}.$$

At constant $\sigma = 5.0$ MPa,

$$\ln 10^{-6} = \ln(A\sigma^n) - \frac{Q}{RT_1}$$

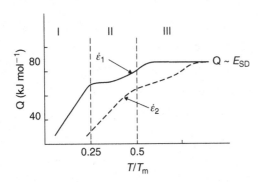

**Figure 4.5-61** Variation in activation energy $Q$ with temperature for aluminum.

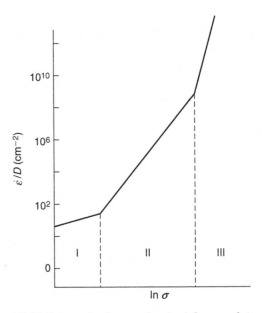

**Figure 4.5-63** Schematic diagram showing influence of stress on diffusion-compensated steady-state creep.

$$\ln \dot{\varepsilon} - \ln(0.98 \times 10^{-7}) = \frac{Q}{R}\left(\frac{1}{600} - \frac{1}{640}\right)$$
$$= \frac{128.4 \times 10^3}{8.314}$$
$$\times \left(\frac{1}{600} - \frac{1}{640}\right)$$
$$\therefore \dot{\varepsilon} = 4.9 \times 10^{-7}\,\text{s}^{-1}.$$

## 4.5.9.2 Grain boundary contribution to creep

In the creep of polycrystals at high temperatures, the grain boundaries themselves are able to play an important part in the deformation process due to the fact that they may (1) slide past each other or (2) create vacancies. Both processes involve an activation energy for diffusion and therefore may contribute to steady-state creep.

Grain boundary sliding during creep was inferred initially from the observation of steps at the boundaries, but the mechanism of sliding can be demonstrated on bi-crystals. Figure 4.5-64 shows a good example of grain boundary movement in a bi-crystal of tin, where the displacement of the straight grain boundary across its middle is indicated by marker scratches. Grain boundaries, even when specially produced for bi-crystal experiments, are not perfectly straight, and after a small amount of sliding at the boundary interface, movement will be arrested by protuberances. The grains are then locked, and the rate of slip will be determined by the rate of plastic flow in the protuberances. As a result, the rate of slip along a grain boundary is not constant with time, because the dislocations first form into piled-up groups, and later these become relaxed. Local relaxation may be envisaged as a process in which the dislocations in the pile-up climb towards the boundary. In consequence, the activation energy for grain boundary slip may be identified with that for steady-state creep. After climb, the dislocations are spread more evenly along the boundary, and are thus able to give rise to grain boundary migration, when sliding has temporarily ceased, which is proportional to the overall deformation.

A second creep process which also involves the grain boundaries is one in which the boundary acts as a source and sink for vacancies. The mechanism depends on the migration of vacancies from one side of a grain to another, as shown in Figure 4.5-65, and is often termed Herring–Nabarro creep, after the two workers who originally considered this process. If, in a grain of sides $d$ under a stress $\sigma$, the atoms are transported from faces BC and AD to the faces AB and DC the grain creeps in the direction of the stress. To transport atoms in this way involves creating vacancies on the tensile faces AB and DC and destroying them on the other compressive faces by diffusion along the paths shown.

On a tensile face AB the stress exerts a force $\sigma b^2$ (or $\sigma \Omega^{2/3}$) on each surface atom and so does work $\sigma b^2 \times b$ each time an atom moves forward one atomic spacing $b$ (or $\Omega^{1/3}$) to create a vacancy. The energy of vacancy formation at such a face is thus reduced to $(E_f - \sigma b^3)$ and the concentration of vacancies in equilibrium correspondingly increased to $c_\tau = \exp[(-E_f + \sigma b^3)/\text{k}.T] = c_0 \exp(\sigma b^3/\text{k}T)$. The vacancy concentration on the compressive faces will be reduced to $c_c = c_0 \exp(-\sigma b^3/\text{k}T)$. Vacancies will therefore flow down the concentration gradient, and the number crossing a face under tension to one under compression will be given by Fick's law as

$$\phi = -D_v d^2 (c_T - c_c)/\alpha d,$$

where $D_v$ is the vacancy diffusivity and $\alpha$ relates to the diffusion length. Substituting for $c_T$, $c_c$ and $D = (D_v c_0 b^3)$ leads to

$$\phi = 2dD \sinh(\sigma b^3/\text{k}T)/\alpha b^3.$$

Each vacancy created on one face and annihilated on the other produces a strain $\varepsilon = b^3/d^3$, so that the creep strain rate $\dot{\varepsilon} = \phi(b^3/d^3)$. At high temperatures and low stresses this reduces to

$$\dot{\varepsilon}_{\text{H--N}} = 2D\sigma b^3/\alpha d^2 b\text{k}T = B_{\text{H--N}} D\sigma\omega/d^2\text{k}T, \quad (4.5.56)$$

where the constant $B_{H-N} \sim 10$.

In contrast to dislocation creep, Herring–Nabarro creep varies linearly with stress and occurs at $T \approx 0.8 T_m$ with $\sigma \approx 10^6\,\text{N m}^{-2}$. The temperature range over which vacancy–diffusion creep is significant can be extended to

**Figure 4.5-64** Grain boundary sliding on a bi-crystal tin (after Puttick and King, 1952; courtesy of Institute of Materials, Minerals and Mining).

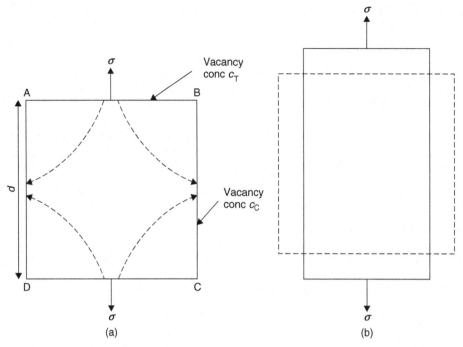

**Figure 4.5-65** Schematic representation of Herring–Nabarro creep; with $c_T > c_C$ vacancies flow from the tensile faces to the longitudinal faces (a) to produce creep, as shown in (b).

much lower temperatures (i.e. $T \approx 0.5T_m$) if the vacancies flow down the grain boundaries rather than through the grains. Equation (4.5.56) is then modified for Coble or grain boundary diffusion creep, and is given by

$$\dot{\varepsilon}_{Coble} = B_c D_{gb} \sigma \Omega \omega / kTd^3, \qquad (4.5.57)$$

where $\omega$ is the width of the grain boundary. Under such conditions (i.e. $T \approx 0.5$–$0.6T_m$ and low stresses), diffusion creep becomes an important creep mechanism in a number of high-technology situations, and has been clearly identified in magnesium-based canning materials used in gas-cooled reactors.

### 4.5.9.3 Tertiary creep and fracture

Tertiary creep and fracture are logically considered together, since the accelerating stage represents the initiation of conditions which lead to fracture. In many cases the onset of accelerating creep is an indication that voids or cracks are slowly but continuously forming in the material, and this has been confirmed by metallography and density measurements. The type of fracture resulting from tertiary creep is not transcrystalline but grain boundary fracture. Two types of grain boundary fracture have been observed. The first occurs principally at the triple point formed where three grain boundaries meet, and sliding along boundaries on which there is a shear stress produces stress concentrations at the point of conjunction sufficiently high to start cracks. However, under conditions of slow strain rate for long times, which would be expected to favor recovery, small holes form on grain boundaries, especially those perpendicular to the tensile axis, and these gradually grow and coalesce.

Second-phase particles play an important part in the nucleation of cracks and cavities by concentrating stress in sliding boundaries and at the intersection of slip bands with particles, but these stress concentrations are greatly reduced by plastic deformation by power-law creep and by diffusional processes. Cavity formation and early growth is therefore intimately linked to the creep process itself and the time-to-fracture correlates well with the minimum creep rate for many structural materials. Fracture occurs when the larger, more closely spaced cavities coalesce. Creep fracture is discussed further in Chapter 7.

### 4.5.9.4 Creep-resistant alloy design

The problem of the design of engineering creep-resistant alloys is complex, and the optimum alloy for a given service usually contains several constituents in various states of solution and precipitation. Nevertheless, it is worth considering some of the principles underlying creep-resistant behavior in the light of the preceding theories.

First, let us consider the strengthening of the solid solution by those mechanisms which cause dislocation locking and those which contribute to lattice friction hardening. The former include solute atoms interacting with (1) the dislocation or (2) the stacking fault. Friction hardening can arise from (1) the stress fields around individual atoms (i.e. the Mott–Nabarro effect), (2) clusters of solute atoms in solid solutions, (3) by increasing the separation of partial dislocations and so making climb, cross-slip and intersection more difficult, (4) by the solute atoms becoming attached to jogs and thereby impeding climb, and (5) by influencing the energies of formation and migration of vacancies. The alloy can also be hardened by precipitation, and it is significant that many of the successful industrial creep-resistant alloys are of this type (e.g. the nickel alloys, and both ferritic and austenitic steels).

The effectiveness of these various methods of conferring strength on the alloy will depend on the conditions of temperature and stress during creep. All the effects should play some part during fast primary creep, but during the slow secondary creep stage the impeding of dislocation movement by solute locking effects will probably be small. This is because modern creep-resistant alloys are in service up to temperatures of about two-thirds the absolute melting point $(T/T_m \approx \frac{2}{3})$ of the parent metal, whereas above about $T/T_m \approx \frac{1}{2}$ solute atoms will migrate as fast as dislocations. Hardening which relies on clusters will be more difficult to remove than that which relies upon single atoms and should be effective up to higher temperatures. However, for any hardening mechanism to be really effective, whether it is due to solute atom clusters or actual precipitation, the rate of climb and cross-slip past the barriers must be slow. Accordingly, the most probable role of solute alloying elements in modern creep-resistant alloys is in reducing the rate of climb and cross-slip processes. The three hardening mechanisms listed as 3,4 and 5 above are all effective in this way. From this point of view, it is clear that the best parent metals on which to base creep-resistant alloys will be those in which climb and cross-slip are difficult; these include the fcc and cph metals of low stacking-fault energy, for which the slip dislocations readily dissociate. Generally, the creep rate is described by the empirical relation

$$\dot{\varepsilon} = A(\sigma/E)^n (\gamma)^m D, \qquad (4.5.58)$$

where $A$ is a constant, $n$ and $m$ stress and fault energy exponents respectively, and $D$ the diffusivity; for fcc materials $m \approx 3$ and $n \approx 4$. The reason for the good creep strength of austenitic and Ni-based materials containing Co, Cr, etc. arises from their low fault energy and also because of their relatively high melting point when $D$ is small.

From the above discussion it appears that a successful creep-resistant material would be an alloy, the composition of which gives a structure with a hardened solid-solution matrix containing a sufficient number of precipitated particles to force glissile partial dislocations either to climb or to cross-slip to circumvent them. The constitution of the Nimonic alloys, which consist of a nickel matrix containing dissolved chromium, titanium, aluminum and cobalt, is in accordance with these principles, and since no large atomic size factors are involved it appears that one of the functions of these additions is to lower the stacking-fault energy and thus widen the separation of the partial dislocations. A second object of the titanium and aluminum alloy additions[7] is to produce precipitation, and in the Nimonic alloys much of the precipitate is $Ni_3Al$. This precipitate is isomorphous with the matrix, and while it has a parameter difference ($\approx \frac{1}{2}\%$) small enough to give a low interfacial energy, it is nevertheless sufficiently large to give a source of hardening. Thus, since the energy of the interface provides the driving force for particle growth, this low-energy interface between particle and matrix ensures a low rate of particle growth and hence a high service temperature.

Grain boundary precipitation is advantageous in reducing grain boundary sliding. Alternatively, the weakness of the grain boundaries may be eliminated altogether by using single-crystal material. Nimonic alloys used for turbine blades have been manufactured in single-crystal form by directional solidification (see Chapters 2 and 8).

Dispersions are effective in conferring creep strength by two mechanisms. First, the particle will hinder a dislocation and force it to climb and cross-slip. Second, and more important, is the retarding effect on recovery as shown by some dispersions, $CU-Al_2O_3$ (extruded), SAP (sintered alumina powder) and $Ni-ThO_2$, which retain their hardness almost to the melting point. A comparison of SAP with a 'conventional' complex aluminum alloy shows that at 250°C there is little to choose between them but at 400°C SAP is several times stronger. Generally, the dislocation network formed by strain hardening interconnects the particles and is thereby anchored by them. To do this effectively, the particle must be stable at the service temperature and remain finely dispersed. This depends on the solubility C, diffusion coefficient $D$ and interfacial energy $\gamma_1$, since the time to dissolve the particle is $t = r^4 kT/DC\gamma_1 R^2$. In precipitation-hardening alloys, C is appreciable and $D$ offers little scope for adjustment; great importance is therefore placed on $\gamma_1$ as for the $Ni_3$ (TiAl) phase in Nimonics, where it is very low.

Figure 4.5-62 shows that $n \approx 4$ both above and below $0.5T_m$ for the $Ni-Al_2O_3$ and $Ni-CO-Al_2O_3$ alloys that

---

[7] The chromium forms a spinel with NiO and hence improves the oxidation resistance.

Table 4.5-4 Experimentally determined parameters from creep of Ni–Al2O3 and Ni–Co–Al2O3 alloys.

| Alloy | Test temperature | | | | |
|---|---|---|---|---|---|
| | 773 K | | 1000 K | | |
| | $Q$ (kJmol$^{-1}$) | $A$ (s$^{-1}$) | $Q$ (kJmol$^{-1}$) | $A$ (s$^{-1}$) | $A/D_0$ |
| Ni | 85 | $1.67 \times 10^{16}$ | 276 | $1.1 \times 10^{28}$ | $5.5 \times 10^{28}$ |
| Ni–67% Co | 121 | $9.95 \times 10^{19}$ | 276 | $2.2 \times 10^{28}$ | $5.8 \times 10^{28}$ |

were completely recrystallized, which contrasts with values very much greater than 4 for extruded TD nickel and other dispersion-strengthened alloys[8] containing a dislocation substructure. This demonstrates the importance of substructure and probably indicates that in completely recrystallized alloys containing a dispersoid, the particles control the creep behavior, whereas in alloys containing a substructure the dislocation content is more important. Since $n \approx 4$ for the Ni– and Ni–Co–Al$_2$O$_3$ alloys in both temperature regimes, the operative deformation mechanism is likely to be the same, but it is clear from the activation energies, listed in Table 4.5-4 that the rate-controlling thermally activated process changes with temperature. The activation energy is greater at the higher temperature when it is also, surprisingly, composition (or stacking-fault energy) independent.

Such behavior may be explained, if it is assumed that the particles are bypassed by cross-slip (see Chapter 7) and this process is easy at all temperatures, but it is the climb of the edge segments of the cross-slipped dislocations that is rate controlling. At low temperatures, climb would proceed by pipe diffusion so that the composition dependence relates to the variation in the ease of pipe diffusion along dislocations of different widths. At high temperatures, climb occurs by bulk diffusion and the absence of any composition dependence is due to the fact that in these alloys the jog distribution is determined mainly by dislocation/particle interactions and not, as in single-phase alloys and in dispersion-strengthened alloys containing a substructure, by the matrix stacking-fault energy. The optimum creep resistance of dispersion-strengthened alloys is produced when a uniform dislocation network in a fibrous grain structure is anchored by the particles and recovery is minimized. Such a structure can reduce the creep rate by several orders of magnitude from that given in Figure 4.5-62, but it depends critically upon the working and heat treatment used in fabricating the alloy.

Second-phase particles can also inhibit diffusion creep. Figure 4.5-66 shows the distribution of particles before and after diffusion creep, and indicates that the longitudinal boundaries tend to collect precipitates as vacancies are absorbed and the boundaries migrate inwards, while the tensile boundaries acquire a PFZ. Such a structural change has been observed in Mg–0.5% Zr (*Magnox ZR55*) at 400°C and is accompanied by a reduced creep rate. It is not anticipated that diffusion is significantly affected by the presence of particles and hence the effect is thought to be due to the particles affecting the vacancy-absorbing capabilities of the grain boundaries. Whatever mechanism is envisaged for the annihilation of vacancies at a grain boundary, the climb-glide of grain boundary dislocations is likely to be involved and such a process will be hindered by the presence of particles.

## 4.5.10 Deformation mechanism maps

The discussion in this chapter has emphasized that, over a range of stress and temperature, an alloy is capable of deforming by several alternative and independent mechanisms, e.g. dislocation creep with either pipe diffusion at low temperatures and lattice diffusion at high temperatures being the rate-controlling mechanism, and diffusional creep with either grain-boundary diffusion or lattice diffusion being important. In a particular range of temperature, one of these mechanisms is dominant and it is therefore useful in engineering applications to identify the operative mechanism for a given stress–temperature condition, since it is ineffective to change the metallurgical factors to influence, for example, a component deforming by power-law creep controlled by pipe diffusion if the operative mechanism is one of Herring–Nabarro creep.

The various alternative mechanisms are displayed conveniently on a deformation mechanism map in which the appropriate stress, i.e. shear stress or equivalent stress, compensated by modulus on a log scale, is plotted against homologous temperature $T/T_m$ as shown in Figure 4.5-67 for nickel and a nickel-based superalloy with a grain size of 100 μm. By comparing the diagrams it

---

[8] To analyze these it is generally necessary to introduce a threshold (or friction) stress $\sigma_0$ that the effective stress is $(\sigma - \sigma_0)$.

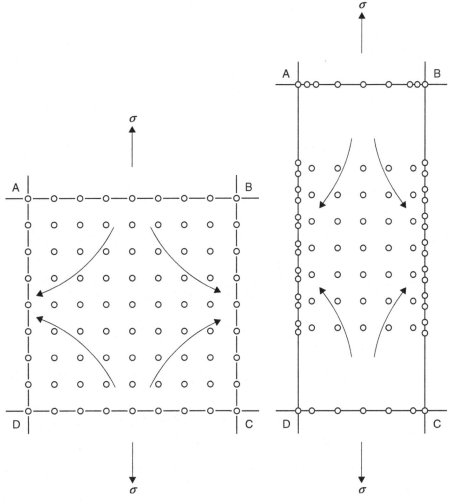

**Figure 4.5-66** Schematic diagram showing the distribution of second-phase particles before and after diffusion creep.

is evident that solid solution strengthening and precipitation hardening have raised the yield stress and reduced the dislocation creep field. The shaded boxes shown in Figure 4.5-67 indicate the typical stresses and temperatures to which a turbine blade would be subjected; it is evident that the mechanism of creep during operation has changed and, indeed, the creep rate is reduced by several orders of magnitude.

## 4.5.11 Metallic fatigue

### 4.5.11.1 Nature of fatigue failure

The term fatigue applies to the behavior of a metal which, when subjected to a cyclically variable stress of sufficient magnitude (often below the yield stress), produces a detectable change in mechanical properties. In practice, a large number of service failures are due to fatigue, and so engineers are concerned mainly with fatigue failure, where the specimen is actually separated into two parts. Some of these failures can be attributed to poor design of the component, but in some cases can be ascribed to the condition of the material. Consequently, the treatment of fatigue may be conveniently divided into three aspects: (1) engineering considerations, (2) gross metallurgical aspects, and (3) fine-scale structural and atomic changes.

The fatigue conditions which occur in service are usually extremely complex. Common failures are found in axles where the eccentric load at a wheel or pulley produces a varying stress which is a maximum in the skin of the axle. Other examples, such as the flexure stresses produced in aircraft wings and in undercarriages during ground taxiing, do, however, emphasize that the stress system does not necessarily vary in a regular sinusoidal manner. The series of aircraft disasters attributed to pressurized cabin failures is perhaps the most spectacular example of this type of fatigue failure.

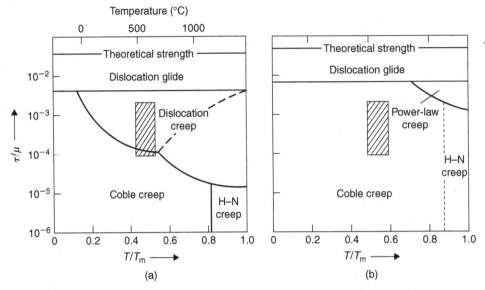

Figure 4.5-67 Deformation mechanism maps for nickel (a) and nickel-based superalloy (b) (after M. F. Ashby).

## 4.5.11.2 Engineering aspects of fatigue

In laboratory testing of materials the stress system is usually simplified, and both the Woehler and push–pull types of test are in common use. The results are usually plotted on the familiar $S-N$ curve (i.e. stress versus the number of cycles to failure, usually plotted on a logarithmic scale). Ferritic steels may be considered to exhibit a genuine fatigue limit with a fatigue ratio $S/TS \approx 0.5$. However, other materials, such as aluminum- or copper-based alloys, certainly those of the age-hardening variety, definitely do not show a sharp discontinuity in the $S-N$ curve. For these materials no fatigue limit exists and all that can be specified is the endurance limit at $N$ cycles. The importance of the effect is illustrated by the behavior of commercial aluminum-based alloys containing zinc, magnesium and copper. Such an alloy may have a TS of 617 MN m$^{-2}$ but the fatigue stress for a life of $10^8$ cycles is only 154 MN m$^{-2}$ (i.e. a fatigue ratio at $10^8$ cycles of 0.25).

The amplitude of the stress cycle to which the specimen is subjected is the most important single variable in determining its life under fatigue conditions, but the performance of a material is also greatly affected by various other conditions, which may be summarized as follows:

1. *Surface preparation.* Since fatigue cracks frequently start at or near the surface of the component, the surface condition is an important consideration in fatigue life. The removal of machining marks and other surface irregularities invariably improves the fatigue properties. Putting the surface layers under compression by shot peening or surface treatment improves the fatigue life.

2. *Effect of temperature.* Temperature affects the fatigue properties in much the same way as it does the tensile strength (TS); the fatigue strength is highest at low temperatures and decreases gradually with rising temperature. For mild steel the ratio of fatigue limit to TS remains fairly constant at about 0.5, while the ratio of fatigue limit to yield stress varies over much wider limits. However, if the temperature is increased above about 100°C, both the tensile strength and the fatigue strength of mild steel show an increase, reaching a maximum value between 200 and 400°C. This increase, which is not commonly found in other materials, has been attributed to strain ageing.

3. *Frequency of stress cycle.* In most metals the frequency of the stress cycle has little effect on the fatigue life, although lowering the frequency usually results in a slightly reduced fatigue life. The effect becomes greater if the temperature of the fatigue test is raised, when the fatigue life tends to depend on the total time of testing rather than on the number of cycles. With mild steel, however, experiments show that the normal speed effect is reversed in a certain temperature range and the number of cycles to failure increases with decrease in the frequency of the stress cycle. This effect may be correlated with the influence of temperature and strain rate on the TS. The temperature at which the tensile strength reaches a maximum depends on the rate of strain, and it is therefore not surprising that the temperature at which the fatigue strength reaches a maximum depends on the cyclic frequency.

4. *Mean stress.* For conditions of fatigue where the mean stress, i.e.

$$\Delta\sigma N_f^a = (\sigma_{max} + \sigma_{min})/2,$$

does not exceed the yield stress $\sigma_y$, then the relationship

$$\Delta\sigma N_f^a = \text{const.}, \qquad (4.5.59)$$

known as Basquin's law, holds over the range $10^2$ to $\approx 10^5$ cycles, i.e. $N$ less than the knee of the $S-N$ curve, where $a \approx \frac{1}{10}$ and $N_f$ is the number of cycles to failure. For low cycle fatigue with $\Delta\sigma > \sigma_y$, then Basquin's law no longer holds, but a reasonable relationship

$$\Delta\varepsilon_p N_j^b = D^b = \text{const.}, \qquad (4.5.60)$$

known as the Coffin–Manson law, is found, where $\Delta\varepsilon_p$ is the plastic strain range, $b \approx 0.6$ and $D$ is the ductility of the material. If the mean stress becomes tensile a lowering of the fatigue limit results. Several relationships between fatigue limit and mean stress have been suggested, as illustrated in Figure 4.5-68a. However, there is no theoretical reason why a material should follow any given relationship and the only safe rule on which to base design is to carry out prior tests on the material concerned to determine its behavior under conditions similar to those it will meet in service. Another common engineering relationship frequently used, known as Miner's concept of cumulative damage, is illustrated in Figure 4.5-68b. This hypothesis states that damage can be expressed in terms of the number of cycles applied divided by the number to produce failure at a given stress level. Thus, if a maximum stress of value $S_1$ is applied to a specimen for $n_1$ cycles which is less than the fatigue life $N_1$, and then the maximum stress is reduced to a value equal to $S_2$, the specimen is expected to fail after $n_2$ cycles, since according to Miner the following relationship will hold:

$$n_1/N_1 + n_2/N_2 + \ldots = \Sigma n/N = 1. \qquad (4.5.61)$$

5. *Environment.* Fatigue occurring in a corrosive environment is usually referred to as corrosion fatigue. It is well known that corrosive attack by a liquid medium can produce etch pits which may act as notches, but when the corrosive attack is simultaneous with fatigue stressing, the detrimental effect is far greater than just a notch effect. Moreover, from microscopic observations the environment appears to have a greater effect on crack propagation than on crack initiation. For most materials even atmospheric oxygen decreases the fatigue life by influencing the speed of crack propagation, and it is possible to obtain a relationship between fatigue life and the degree of vacuum in which the specimen has been held.

It is now well established that fatigue starts at the surface of the specimen. This is easy to understand in the Woehler test because, in this test, it is there that the stress is highest. However, even in push–pull fatigue, the surface is important for several reasons: (1) slip is easier at the surface than in the interior of the grains, (2) the environment is in contact with the surface and (3) any specimen misalignment will always give higher stresses at the surface. Accordingly, any alteration in surface properties must bring about a change in the fatigue properties. The best fatigue resistance occurs in materials with a worked surface layer produced by polishing with emery, shot-peening or skin-rolling the surface. This beneficial effect of a worked surface layer is principally due to the fact that the surface is put into compression, but the increased TS as a result of work hardening also plays a part. Electropolishing the specimen by removing the surface layers usually has a detrimental effect on the fatigue properties, but other common surface preparations such as nitriding and carburizing, both of which produce a surface layer which is in compression, may be beneficial. Conversely, such surface treatments as the decarburizing of steels and the

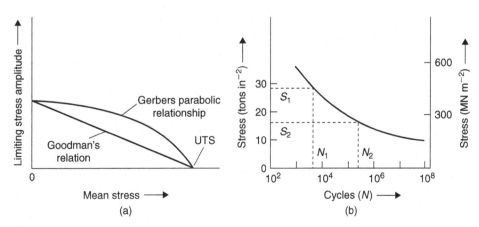

**Figure 4.5-68** Fatigue relationships.

cladding of aluminum alloys with pure aluminum increase their susceptibility to fatigue.

The alloy composition and thermal and mechanical history of the specimen are also of importance in the fatigue process. Any treatment which increases the hardness or yield strength of the material will increase the level of the stress needed to produce slip and, as we shall see later, since the fundamental processes of fatigue are largely associated with slip, this leads directly to an increase in fatigue strength. It is also clear that grain size is a relevant factor: the smaller the grain size, the higher is the fatigue strength at a given temperature.

The fatigue processes in stable alloys are essentially the same as those of pure metals but there is, of course, an increase in fatigue strength. However, the processes in unstable alloys and in materials exhibiting a yield point are somewhat different. In fatigue, as in creep, structural instability frequently leads to enhancement of the fundamental processes. In all cases the approach to equilibrium is more complete, so that in age-hardening materials, solution-treated specimens become harder and fully aged specimens become softer. The changes which occur are local rather than general, and are associated with the enhanced diffusion brought about by the production of vacancies during the fatigue test. Clearly, since vacancy mobility is a thermally activated process, such effects can be suppressed at sufficiently low temperatures.

In general, non-ferrous alloys do not exhibit the type of fatigue limit shown by mild steel. One exception to this generalization is the alloy aluminum/2–7% magnesium/0.5% manganese, and it is interesting to note that this alloy also has a sharp yield point and shows Lüders markings in an ordinary tensile test. Accordingly, it has been suggested that the fatigue limit occupies a similar place in the field of alternating stresses to that filled by the yield point in unidirectional stressing. Stresses above the fatigue limit readily unlock the dislocations from their solute atom atmospheres, while below the fatigue limit most dislocations remain locked. In support of this view, it is found that when the carbon and nitrogen content of mild steel is reduced, by annealing in wet hydrogen, striking changes take place in the fatigue limit (Figure 4.5-5) as well as in the sharp yield point.

## 4.5.11.3 Structural changes accompanying fatigue

Observations of the structural details underlying fatigue hardening show that in polycrystals large variations in slip-band distributions and the amount of lattice misorientation exist from one grain to another. Because of such variations it is difficult to typify structural changes, so that in recent years this structural work has been carried out more and more on single crystals; in particular, copper has received considerable attention as being representative of a typical metal. Such studies have now established that fatigue occurs as a result of slip, the direction of which changes with the stress cycle, and that the process continues throughout the whole of the test (shown, for example, by interrupting a test and removing the slip bands by polishing; the bands reappear on subsequent testing).

Moreover, four stages in the fatigue life of a specimen are distinguishable; these may be summarized as follows. In the early stages of the test, the whole of the specimen hardens. After about 5% of the life, slip becomes localized and persistent slip bands appear; they are termed persistent because they reappear and are not permanently removed by electropolishing. Thus, reverse slip does not continue throughout the whole test in the bulk of the metal (the matrix). Electron microscope observations show that metal is extruded from the slip bands and that fine crevices called intrusions are formed within the band. During the third stage of the fatigue life the slip bands grow laterally and become wider, and at the same time cracks develop in them. These cracks spread initially along slip bands, but in the later stages of fracture the propagation of the crack is often not confined to certain crystallographic directions and catastrophic rupture occurs. These two important crack growth stages, i.e. stage I in the slip band and stage II roughly perpendicular to the principal stress, are shown in Figure 4.5-69 and are influenced by the formation of localized (persistent) slip bands (i.e. PSBs). However, PSBs are not clearly defined in low stacking-fault energy, solid solution alloys.

Cyclic stressing therefore produces plastic deformation which is not fully reversible and the build-up of dislocation density within grains gives rise to fatigue hardening with an associated structure which is characteristic of the strain amplitude and the ability of the dislocations to cross-slip, i.e. temperature and SFE. The non-reversible flow at the surface leads to intrusions, extrusions and crack formation in PSBs. These two aspects will now be considered separately and in greater detail.

### 4.5.11.3.1 Fatigue hardening

If a single or polycrystalline specimen is subjected to many cycles of alternating stress, it becomes harder than a similar specimen extended unidirectionally by the same stress applied only once. This may be demonstrated by stopping the fatigue test and performing a static tensile test on the specimen when, as shown in Figure 4.5-70, the yield stress is increased. During the process, persistent slip bands appear on the surface of the specimen and it is in such bands that cracks eventually form. The behavior of a fatigue-hardened specimen has two unusual features when compared with an ordinary work-

# Mechanical properties of metals  CHAPTER 4.5

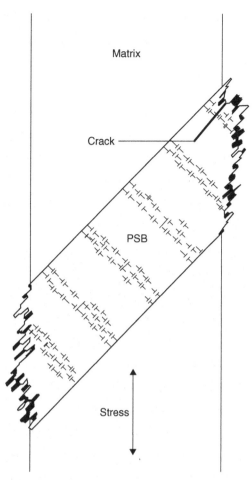

**Figure 4.5-69** Persistent slip band (PSB) formation in fatigue, and stage I and stage II crack growth.

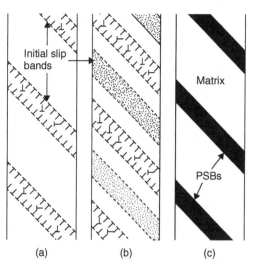

**Figure 4.5-71** Formation of persistent slip bands (PSBs) during fatigue.

hardened material. The fatigue-hardened material, having been stressed symmetrically, has the same yield stress in compression as in tension, whereas the work-hardened specimen (e.g. prestrained in tension) exhibits a Bauschinger effect, i.e. weaker in compression than tension. It arises from the fact that the obstacles behind the dislocation are weaker than those resisting further dislocation motion, and the pile-up stress causes it to slip back under a reduced load in the reverse direction. The other important feature is that the temperature

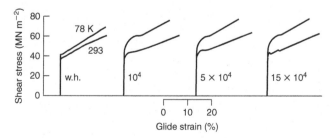

**Figure 4.5-70** Stress–strain curves for copper after increasing amounts of fatigue testing (after Broom and Ham, 1959).

dependence of the hardening produced by fatigue is significantly greater than that of work hardening and, because of the similarity with the behavior of metals hardened by quenching and by irradiation, it has been attributed to the effect of vacancies and dislocation loops created during fatigue.

At the start of cyclic deformation the initial slip bands (Figure 4.5-71a) consist largely of primary dislocations in the form of dipole and multipole arrays; the number of loops is relatively small because the frequency of cross-slip is low. As the specimen work-hardens slip takes place between the initial slip bands, and the new slip bands contain successively more secondary dislocations because of the internal stress arising from nearby slip bands (Figure 4.5-71b). When the specimen is completely filled with slip bands, the specimen has work hardened and the softest regions are now those where slip occurred originally, since these bands contain the lowest density of secondary dislocations. Further slip and the development of PSBs takes place within these original slip bands, as shown schematically in Figure 4.5-71c.

As illustrated schematically in Figure 4.5-72, TEM of copper crystals shows that the main difference between the matrix and the PSBs is that in the matrix the dense arrays of edge dislocation (di- and multipoles) are in the form of large veins occupying about 50% of the volume, whereas they form a 'ladder'-type structure within walls occupying about 10% of the volume in PSBs. The PSBs are the active regions in the fatigue process while the matrix is associated with the inactive parts of the specimen between the PSBs. Steady-state deformation then takes place by the to-and-fro glide of the same dislocations in the matrix, whereas an equilibrium between dislocation multiplication and annihilation exists in the PSBs. Multiplication occurs by bowing-out of the walls and annihilation takes place by interaction with edge

217

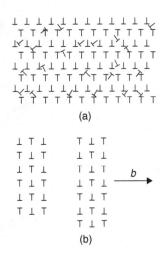

**Figure 4.5-72** Schematic diagram showing vein structure of matrix (a) and ladder structure of PSBs (b).

dislocations of opposite sign ($\approx 75b$ apart) on glide planes in the walls and of screw dislocations ($\approx 200b$ apart) on glide planes in the low-dislocation channels, the exact distance depending on the ease of cross-slip.

### 4.5.11.4 Crack formation and fatigue failure

Extrusions, intrusions and fatigue cracks can be formed at temperatures as low as 4 K where thermally activated movement of vacancies does not take place. Such observations indicate that the formation of intrusions and cracks cannot depend on either chemical or thermal action and the mechanism must be a purely geometrical process which depends on cyclic stressing.

Two general mechanisms have been suggested. The first, the Cottrell 'ratchet' mechanism, involves the use of two different slip systems with different directions and planes of slip, as is shown schematically in Figure 4.5-73. The most favored source (e.g. $S_1$ in Figure 4.5-73a) produces a slip step on the surface at P during a tensile half-cycle. At a slightly greater stress in the same half-cycle, the second source $S_2$ produces a second step at Q (Figure 4.5-73b). During the compression half-cycle, the source $S_1$ produces a surface step of opposite sign at P' (Figure 4.5-73c), but, owing to the displacing action of $S_2$, this is not in the same plane as the first and thus an intrusion is formed. The subsequent operation of $S_2$ produces an extrusion at QQ' (Figure 4.5-73d) in a similar manner. Such a mechanism requires the operation of two slip systems and, in general, predicts the occurrence of intrusions and extrusions with comparable frequency, but not in the same slip band.

The second mechanism, proposed by Mott, involves cross-slip resulting in a column of metal extruded from the surface and a cavity is left behind in the interior of the crystal. One way in which this could happen is by the cyclic movement of a screw dislocation along a closed circuit of crystallographic planes, as shown in Figure 4.5-74. During the first half-cycle the screw dislocation glides along two faces ABCD and BB' C' C of the band, and during the second half-cycle returns along the faces B' C' A' D and A' D' DA. Unlike the Cottrell mechanism this process can be operated with a single slip direction, provided cross-slip can occur.

Neither mechanism can fully explain all the experimental observations. The interacting slip mechanism predicts the occurrence of intrusions and extrusions with comparable frequency but not, as is often found, in the same slip band. With the cross-slip mechanism, there is no experimental evidence to show that cavities exist beneath the material being extruded. It may well be that different mechanisms operate under different conditions.

In a polycrystalline aggregate the operation of several slip modes is necessary and intersecting slip unavoidable. Accordingly, the widely differing fatigue behavior of metals may be accounted for by the relative ease with which cross-slip occurs. Thus, those factors which affect the onset of stage III in the work-hardening curve will also be important in fatigue, and conditions suppressing cross-slip would, in general, increase the resistance to fatigue failure, i.e. low stacking-fault energy and low temperatures. Aluminum would be expected to have poor fatigue properties on this basis, but the unfavorable fatigue characteristics of the high-strength aluminum

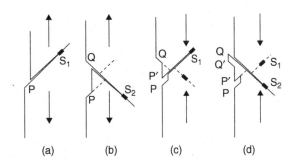

**Figure 4.5-73** Formation of intrusions and extrusions (after Cottrell, 1959; courtesy of John Wiley and Sons).

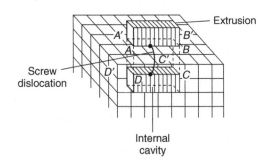

**Figure 4.5-74** Formation of an extrusion and associated cavity by the Mott mechanism.

alloys is probably also due to the unstable nature of the alloy and to the influence of vacancies.

In pure metals and alloys, transgranular cracks initiate at intrusions in PSBs or at sites of surface roughness associated with emerging planar slip bands in low SFE alloys. Often the microcrack forms at the PSB/matrix interface where the stress concentration is high. In commercial alloys containing inclusions or second-phase particles, the fatigue behavior depends on the particle size. Small particles $\approx 0.1$ μm can have beneficial effects by homogenizing the slip pattern and delaying fatigue-crack nucleation. Larger particles reduce the fatigue life by both facilitating crack nucleation by slip band/particle interaction and increasing crack growth rates by interface decohesion and voiding within the plastic zone at the crack tip. The formation of voids at particles on grain boundaries can lead to intergranular separation and crack growth. The preferential deformation of 'soft' precipitate-free zones (PFZs) associated with grain boundaries in age-hardened alloys also provides a mechanism of intergranular fatigue-crack initiation and growth. To improve the fatigue behavior it is therefore necessary to avoid PFZs and obtain a homogeneous deformation structure and uniform precipitate distribution by heat treatment; localized deformation in PFZs can be restricted by a reduction in grain size.

From the general appearance of a typical fatigue fracture, shown in Figure 4.5-75, one can distinguish two distinct regions. The first is a relatively smooth area, through which the fatigue crack has spread slowly. This area usually has concentric marks about the point of origin of the crack which correspond to the positions at which the crack was stationary for some period. The remainder of the fracture surface shows a typically rough

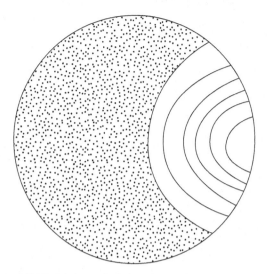

**Figure 4.5-75** A schematic fatigue fracture.

transcrystalline fracture where the failure has been catastrophic. Electron micrographs of the relatively smooth area show that this surface is covered with more or less regular contours perpendicular to the direction of the propagation front. These fatigue striations represent the successive positions of the propagation front and are spaced further apart the higher the velocity of propagation. They are rather uninfluenced by grain boundaries and in metals where cross-slip is easy (e.g. mild steel or aluminum) may be wavy in appearance. Generally, the lower the ductility of the material, the less well defined are the striations.

Stage II growth is rate controlling in the fatigue failure of most engineering components, and is governed by the stress intensity at the tip of the advancing crack. The

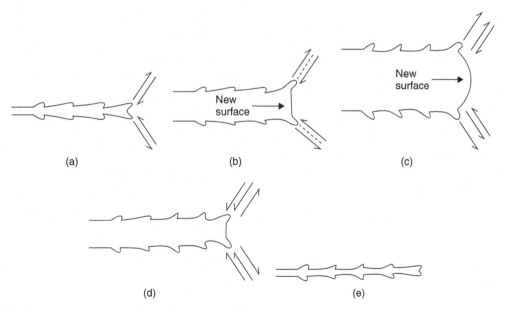

**Figure 4.5-76** Schematic illustration of the formation of fatigue striations.

striations seen on the fracture surface may form by a process of plastic blunting at the tip of the crack, as shown in Figure 4.5-76. In (a) the crack under the tensile loading part of the cycle generates shear stresses at the tip. With increasing tensile load the crack opens up and a new surface is created (b), separation occurs in the slip band and 'ears' are formed at the end of the crack. The plastic deformation causes the crack to be both extended and blunted (c). On the compressive part of the cycle the crack begins to close (d). The shear stresses are reversed and with increasing load the crack almost closes (e). In this part of the cycle the new surface folds and the ears correspond to the new striations on the final fracture surface. A one-to-one correlation therefore exists between the striations and the opening and closing with ear formation. Crack growth continues in this manner until it is long enough to cause the final instability when either brittle or ductile (due to the reduced cross-section not being able to carry the load) failure occurs. In engineering alloys, rather than pure metals, which contain inclusions or second-phase particles, cracking or voiding occurs ahead of the original crack tip rather than in the ears when the tensile stress or strain reaches a critical value. This macroscopic stage of fracture is clearly of importance to engineers in predicting the working life of a component and has been successfully treated by the application of fracture mechanics, as discussed in Chapter 7.

### 4.5.11.5 Fatigue at elevated temperatures

At ambient temperature the fatigue process involves intracrystalline slip and surface initiation of cracks, followed by transcrystalline propagation. When fatigued at elevated temperatures $>0.5\ T_m$, pure metals and solid solutions show the formation of discrete cavities on grain boundaries, which grow, link up and finally produce failure. It is probable that vacancies produced by intracrystalline slip give rise to a supersaturation which causes the vacancies to condense on those grain boundaries that are under a high shear stress where the cavities can be nucleated by a sliding or ratchet mechanism. It is considered unlikely that grain boundary sliding contributes to cavity growth, increasing the grain size decreases the cavity growth because of the change in boundary area. *Magnox* (Mg) and alloys used in nuclear reactors up to $0.75T_m$ readily form cavities, but the high-temperature nickel-base alloys do not show intergranular cavity formation during fatigue at temperatures within their normal service range, because intracrystalline slip is inhibited by $\gamma'$ precipitates. Above about $0.7\ T_m$, however, the $\gamma'$ precipitates coarsen or dissolve and fatigue then produce cavities and eventually cavity failure.

## Problems

**4.5.1** (a) In the diagram the dislocation line is taken to point into the paper. Mark on the diagram the Burgers vector. Under the action of the shear stress shown, which way would the dislocation move? (b) If $\underline{u} = [1\ 1\ \bar{2}]$, $\underline{b} = 1/2[\bar{1}\ 1\ 0]$ and $n = (1\ 1\ 1)$, and the shear stress shown is replaced by a uniaxial compressive stress $\sigma$ along $[2\ 0\ 1]$, deduce in which direction the dislocation would move.
What would be the magnitude of the resolved shear stress on the dislocation?

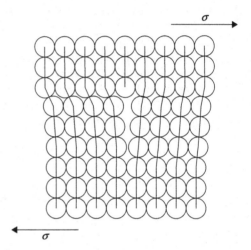

**4.5.2** A single crystal of aluminum is pulled along $[1\ 1\ 0]$. Which slip system or slip systems operate first?

**4.5.3** Estimate the shear stress at the upper yield point and the yield drop (shear stress) that occurs when the mobile dislocation density increases by two orders of magnitude from the initial density of $10^5$ cm$^{-2}$. (Take the strain rate to be $10^{-3}$ s$^{-1}$, $\tau_o$ the stress for unit dislocation velocity of 1 cm s$^{-1}$ to be $2.8 \times 10^4$ N cm$^{-2}$, $n$ to be 20 and $b$, the Burgers vector, to be $2 \times 10^{-8}$ cm.)

**4.5.4** The strengthening of a polycrystalline metal is provided by grain refinement and dispersion of particles. The tensile yield stress of the metal is 400 MPa when the grain size is 0.32 mm and 300 MPa when $d = 1$ mm. Calculate the average distance between the particles. Assume the shear modulus of the metal $\mu = 80$ GPa and $b = 0.25$ nm.

**4.5.5** A steel with a grain size of 25 μm has a yield stress of 200 MPa and with a grain size of 9 μm a yield stress of 300 MPa. A dispersion of non-deformable particles is required to raise the strength to 500 MPa in a steel with grain size 100 μm. What would be the required dispersion spacing? (Assume the shear modulus $\mu = 80$ GPa and the Burgers vector $b = 0.2$ nm.)

**4.5.6** The deformation mechanism map given in the figure below shows three fields of creep for each of which the creep rate $\dot{\varepsilon}\,(s^{-1})$ is represented by an expression of the form $\dot{\varepsilon} = A\sigma^n \exp(-Q/\mathbf{R}T)$. The constant $A$ is $1.5 \times 10^5$, $5.8 \times 10^5/d^2$ and $10^{-9}/d^3$ for dislocation creep, Herring–Nabarro creep and Coble creep respectively ($d =$ grain size in m), while the stress exponent $n$ is 5, 1 and 1 and the activation energy $Q$ (kJ mol$^{-1}$) 550, 550 and 400. The stress $\sigma$ is in MPa. Assuming that the grain size of the material is 1 mm and given the gas constant $\mathbf{R} = 8.3$ Jmol$^{-1}$K$^{-1}$:

(i) Label the three creep fields
(ii) Calculate the stress level $\sigma$ in MPa of the boundary AB
(iii) Calculate the temperature (K) of the boundary AC.

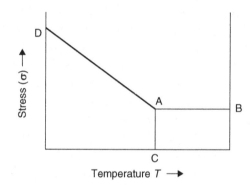

**4.5.7** During the strain ageing of a mild steel specimen, the yield point returned after 1302, 420, 90 and 27 seconds when aged at 50, 65, 85 and 100°C respectively. Determine the activation energy for the diffusion of carbon in α-iron.

**4.5.8** In a high-temperature application an alloy is observed to creep at an acceptable steady-state rate under a stress of 70 MPa at a temperature of 1250 K. If metallurgical improvements would allow the alloy to operate at the same creep rate but at a higher stress level of 77 MPa, estimate the new temperature at which the alloy would operate under the original stress conditions. (Take stress exponent $n$ to be 5, and activation energy for creep $Q$ to be 200 kJ mol$^{-1}$.)

**4.5.9** Cyclic fatigue of an aluminum alloy showed it failed under a stress range $\Delta\sigma = 280$ MPa after $10^5$ cycles, and for a range 200 MPa after $10^7$ cycles. Using Basquin's law, estimate the life of the component subjected to a stress range of 150 MPa.

# Further reading

Argon, A. (1969). *The Physics of Strength and Plasticity*. MIT Press, Cambridge, MA.

Cottrell, (1964). *Mechanical Properties of Matter*. John Wiley, Chichester.

Cottrell, A. H. (1964). *The Theory of Crystal Dislocations*. Blackie, Glasgow.

*Dislocations and Properties of Real Metals* (1984). Conf. Metals Society.

Evans, R.W. and Wilshire, B. (1993). *Introduction to Creep*. Institute of Materials, London.

Freidel, J. (1964). *Dislocations*. Pergamon Press, London.

Hirsch, P. B. (ed.) (1975). *The Physics of Metals. 2. Defects*. Cambridge University Press, Cambridge.

Hirth, J. P. and Lothe, J. (1984). *Theory of Dislocations*. McGraw-Hill, New York.

# Section Five

**Production, forming and joining of metals**

# Chapter 5.1

# Production, forming and joining of metals

## Introduction

Figure 5.1-1 shows the main routes that are used for processing raw metals into finished articles. Conventional forming methods start by *melting* the basic metal and then *casting* the liquid into a mould. The casting may be a large prism-shaped ingot, or a continuously cast "strand", in which case it is *worked* to standard sections (e.g. sheet, tube) or *forged* to shaped components. Shaped components are also made from standard sections by *machining* or *sheet metal-working*. Components are then assembled into finished articles by *joining* operations (e.g. welding) which are usually carried out in conjunction with *finishing* operations (e.g. grinding or painting). Alternatively, the casting can be made to the final shape of the component, although some light machining will usually have to be done on it.

Increasing use is now being made of alternative processing routes. In *powder metallurgy* the liquid metal is atomised into small droplets which solidify to a fine powder. The powder is then *hot pressed* to shape (as we shall see in Chapter 19, hot-pressing is the method used for shaping high-technology ceramics). *Melt spinning* (Chapter 9) gives high cooling rates and is used to make amorphous alloys. Finally, there are a number of specialised processes in which components are formed directly from metallic compounds (e.g. *electro forming* or *chemical vapour deposition*).

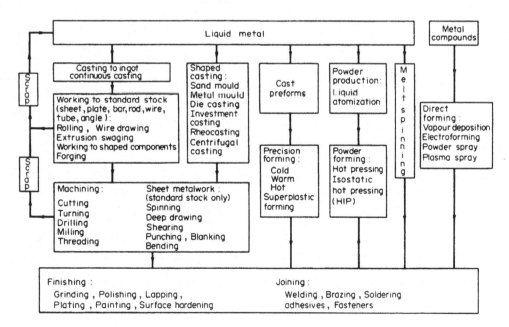

**Figure 5.1-1** Processing routes for metals.

*Engineering Materials and Processes Desk Reference*; ISBN: 9781856175869
Copyright © 2009 Elsevier Ltd; All rights of reproduction, in any form, reserved.

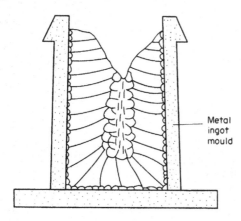

**Figure 5.1-2** Typical ingot structure.

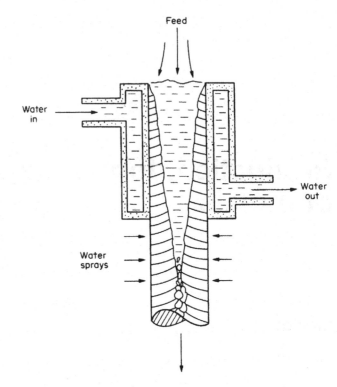

**Figure 5.1-3** Continuous casting.

It is not our intention here to give a comprehensive survey of the forming processes listed in Fig. 5.1-1. This would itself take up a whole book, and details can be found in the many books on production technology. Instead, we look at the underlying principles, and relate them to the characteristics of the materials that we are dealing with.

## Casting

We have already looked at casting structures in Chapter 9. Ingots tend to have the structure shown in Fig. 5.1-2. When the molten metal is poured into the mould, *chill crystals* nucleate on the cold walls of the mould and grow inwards. But the chill crystals are soon overtaken by the much larger columnar grains. Finally, nuclei are swept into the remaining liquid and these grow to produce *equiaxed* grains at the centre of the ingot. As the crystals grow they reject dissolved impurities into the remaining liquid, causing *segregation*. This can lead to bands of solid impurities (e.g. iron sulphide in steel) or to gas bubbles (e.g. from dissolved nitrogen). And because most metals contract when they solidify, there will be a substantial *contraction cavity* at the top of the ingot as well (Fig. 5.1-2).

These *casting defects* are not disastrous in an ingot. The top, containing the cavity, can be cut off. And the gas pores will be squashed flat and welded solid when the white-hot ingot is sent through the rolling mill. But there are still a number of disadvantages in starting with ingots. Heavy segregation may persist through the rolling operations and can weaken the final product.* And a great deal of work is required to roll the ingot down to the required section.

Many of these problems can be solved by using *continuous casting* (Fig. 5.1-3). Contraction cavities do not form because the mould is continuously topped up with liquid metal. Segregation is reduced because the columnar grains grow over smaller distances. And, because the product has a small cross-section, little work is needed to roll it to a finished section.

Shaped castings must be poured with much more care than ingots. Whereas the structure of an ingot will be greatly altered by subsequent working operations, the structure of a shaped casting will directly determine the strength of the finished article. Gas pores should be avoided, so the liquid metal must be *degassed* to remove dissolved gases (either by adding reactive chemicals or – for high-technology applications - casting in a vacuum). *Feeders* must be added (Fig. 5.1-4) to make up the contraction. And inoculants should be added to *refine* the grain size (Chapter 9). This is where powder metallurgy is useful. When atomised droplets solidify, contraction is immaterial. Segregation is limited to the size of the powder particles (2 to 150 $\mu$m); and the small powder size will give a small grain size in the hot-pressed product.

Shaped castings are usually poured into moulds of sand or metal (Fig. 5.1-4). The first operation in sand casting is to make a *pattern* (from wood, metal or plastic) shaped like the required article. Sand is rammed around the pattern and the mould is then split to remove the pattern. Passages are cut through the sand for ingates and

---

*Welded joints are usually in a state of high *residual stress*, and this can tear a steel plate apart if it happens to contain layers of segregated impurity.

# Production, forming and joining of metals
## CHAPTER 5.1

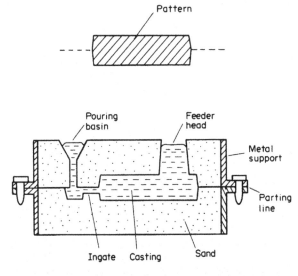

**Figure 5.1-4** Sand casting. When the casting has solidified it is removed by destroying the sand mould. The casting is then "fettled" by cutting off the ingate and the feeder head.

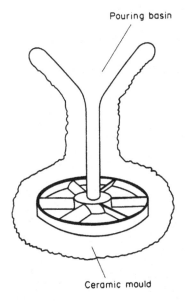

**Figure 5.1-6** Investment casting.

risers. The mould is then re-assembled and poured. When the casting has gone solid it is removed by destroying the mould. Metal moulds are machined from the solid. They must come apart in enough places to allow the casting to be removed. They are costly, but can be used repeatedly; and they are ideal for pressure die casting (Fig. 5.1-5), which gives high production rates and improved accuracy. Especially intricate castings cannot be made by these methods: it is impossible to remove a complex pattern from a sand mould, and impossible to remove a complex casting from a metal one! This difficulty can be overcome by using *investment casting* (Fig. 5.1-6). A wax pattern is coated with a ceramic slurry. The slurry is dried to give it strength, and is then fired (as Chapter 19 explains, this is just how we make ceramic cups and plates). During firing the wax burns out of the ceramic mould to leave a perfectly shaped mould cavity.

## Working processes

The working of metals and alloys to shape relies on their great *plasticity*: they can be deformed by large percentages, especially in compression, without breaking. But the *forming pressures* needed to do this can be large – as high as $3\sigma_y$, or even more, depending on the geometry of the process.

We can see where these large pressures come from by modelling a typical forging operation (Fig. 5.1-7). In order to calculate the forming pressure at a given position $x$ we apply a force $f$ to a movable section of the forging die. If we break the forging up into four separate pieces we can arrange for it to deform when the movable die sections are pushed in. The sliding of one piece over another requires a shear stress $k$ (the shear yield stress). Now the work needed to push the die sections in must equal the work needed to shear the pieces of the forging over one another. The work done on each die section is $f \times u$, giving a total work input of $2fu$. Each sliding interface has area $\sqrt{2}(d/2)L$.

The sliding force at each interface is thus $\sqrt{2}(d/2)L \times k$. Each piece slides a distance $(\sqrt{2})u$ relative to its neighbour. The work absorbed at each interface is thus $\sqrt{2}(d/2)Lk(\sqrt{2})u$; and there are four interfaces. The work balance thus gives

$$2fu = 4\sqrt{2}(d/2)Lk(\sqrt{2})u = 4dLku, \qquad (5.1.1)$$

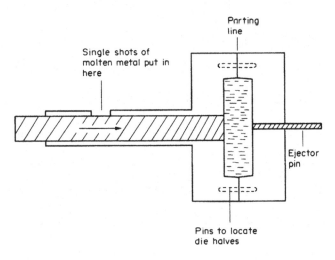

**Figure 5.1-5** Pressure die casting.

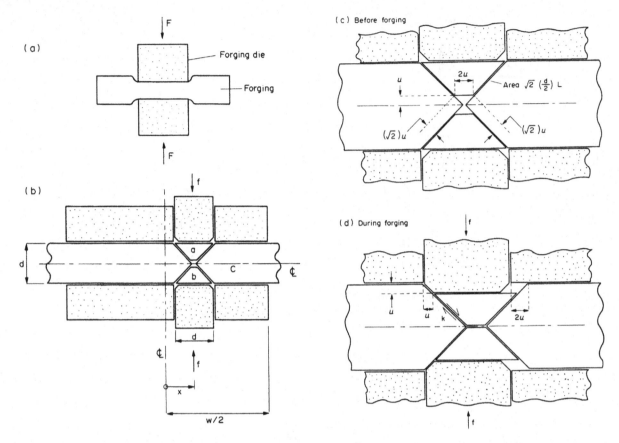

**Figure 5.1-7** A typical forging operation. **(a)** Overall view. **(b)** to **(d)** Modelling the plastic flow. We assume that flow only takes place in the plane of the drawing. The third dimension, measured perpendicular to the drawing, is $L$.

or

$$f = 2dLk. \quad (5.1.2)$$

The forming pressure, $p_f$, is then given by

$$p_f = \frac{f}{dL} = 2k = \sigma_y \quad (5.1.3)$$

which is just what we would expect.

We get a quite different answer if we include the friction between the die and the forging. The extreme case is one of sticking friction: the coefficient of friction is so high that a shear stress $k$ is needed to cause sliding between die and forging. The total area between the dies and piece $c$ is given by

$$2\left\{\left(\frac{W}{2}\right) - \left(x + \frac{d}{2}\right)\right\}L = (w - 2x - d)L. \quad (5.1.4)$$

Piece $c$ slides a distance $2u$ relative to the die surfaces, absorbing work of amount

$$(w - 2x - d)Lk2u. \quad (5.1.5)$$

Pieces $a$ and $b$ have a total contact area with the dies of $2dL$. They slide a distance $u$ over the dies, absorbing work of amount

$$2dLku. \quad (5.1.6)$$

The overall work balance is now

$$2fu = 4dLku + 2(w - 2x - d)Lku + 2dLku \quad (5.1.7)$$

or

$$f = 2Lk\left(d + \frac{w}{2} - x\right). \quad (5.1.8)$$

The forming pressure is then

$$p_f = \frac{f}{dL} = \sigma_y\left\{1 + \frac{(w/2) - x}{d}\right\}. \quad (5.1.9)$$

This equation is plotted in Fig. 5.1-8: $p_f$ increases linearly from a value of $\sigma_y$ at the edge of the die to a maximum of

$$p_{max} = \sigma_y\left(1 + \frac{w}{2d}\right) \quad (5.1.10)$$

at the centre.

# Production, forming and joining of metals   CHAPTER 5.1

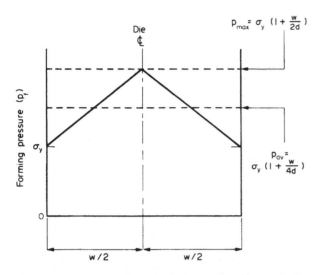

**Figure 5.1-8** How the forming pressure varies with position in the forging.

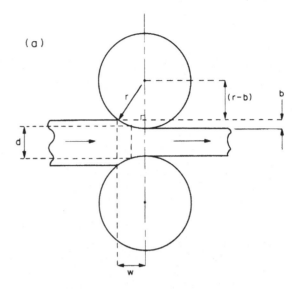

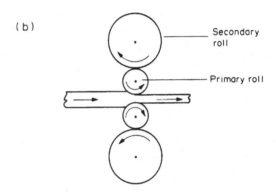

**Figure 5.1-9 (a)** In order to minimise the effects of friction, rolling operations should be carried out with minimum values of $w/d$. **(b)** Small rolls give small $w/d$ values, but they may need to be supported by additional secondary rolls.

It is a salutory exercise to put some numbers into eqn (5.1.10): if $w/d = 10$, then $p_{max} = 6\sigma_y$. Pressures of this magnitude are likely to deform the metal-forming tools themselves — clearly an undesirable state of affairs. The problem can usually be solved by heating the workpiece to $\approx 0.7\,T_m$ before forming, which greatly lowers $\sigma_y$. Or it may be possible to change the geometry of the process to reduce $w/d$. *Rolling* is a good example of this. From Fig. 5.1-9 we can write

$$(r-b)^2 + w^2 = r^2. \qquad (5.1.11)$$

Provided $b \ll 2r$ this can be expanded to give

$$w = \sqrt{2rb}. \qquad (5.1.12)$$

Thus

$$\frac{w}{d} = \frac{\sqrt{2rb}}{d} = \left(\frac{2r}{d}\right)^{1/2}\left(\frac{b}{d}\right)^{1/2}. \qquad (5.1.13)$$

Well-designed rolling mills therefore have rolls of small diameter. However, as Fig. 5.1-9 shows, these may need to be supported by additional secondary rolls which do not touch the workpiece. In fact, aluminium cooking foil is rolled by primary rolls the diameter of a pencil, backed up by a total of 18 secondary rolls.

## Recovery and recrystallisation

When metals are forged, or rolled, or drawn to wire, they *work-harden*. After a deformation of perhaps 80% a limit is reached, beyond which the metal cracks or fractures. Further rolling or drawing is possible if the metal is *annealed* (heated to about $0.6\,T_m$). During annealing, old, deformed grains are replaced by new, undeformed grains, and the working can be continued for a further 80% or so.

Figure 5.1-10 shows how the microstructure of a metal changes during plastic working and annealing. If the metal has been annealed to begin with (Fig. 5.1-10a) it will have a comparatively low dislocation density (about $10^{12}\,\text{m}^{-2}$) and will be relatively soft and ductile. Plastic working (Fig. 5.1-10b) will greatly increase the dislocation density (to about $10^{15}\,\text{m}^{-2}$). The metal will work-harden and will lose ductility. Because each dislocation strains the lattice the deformed metal will have a large strain energy (about $2\,\text{MJ m}^{-3}$). Annealing gives the atoms enough thermal energy that they can move under the driving force of this strain energy. The first process to occur is *recovery* (Fig. 5.1-10c). Because the strain fields of the closely spaced dislocations interact, the total strain energy can be reduced by rearranging the dislocations into low-angle grain boundaries. These boundaries form the surfaces of irregular *cells* — small

229

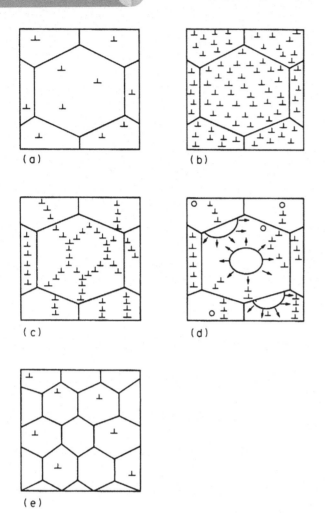

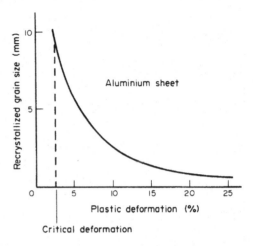

**Figure 5.1-11** Typical data for recrystallised grain size as a function of prior plastic deformation. Note that, below a critical deformation, there is not enough strain energy to nucleate the new strain-free grains. This is just like the critical undercooling needed to nucleate a solid from its liquid (see Fig. 7.1-4).

**Figure 5.1-10** How the microstructure of a metal is changed by plastic working and annealing. (a) If the starting metal has already been annealed it will have a comparatively low dislocation density. (b) Plastic working greatly increases the dislocation density. (c) Annealing leads initially to recovery – dislocations move to low-energy positions. (d) During further annealing new grains nucleate and grow. (e) The fully recrystallised metal consists entirely of new undeformed grains.

volumes which are relatively free of dislocations. During recovery the dislocation density goes down only slightly: the hardness and ductility are almost unchanged. The major changes come from *recrystallisation*. New grains nucleate and grow (Fig. 5.1-10d) until the whole of the metal consists of undeformed grains (Fig. 5.1-10e). The dislocation density returns to its original value, as do the values of the hardness and ductility.

Recrystallisation is not limited just to getting rid of work-hardening. It is also a powerful way of controlling the grain size of worked metals. Although single crystals are desirable for a few specialised applications (see Chapter 9) the metallurgist almost always seeks a fine grain size. To begin with, fine-grained metals are stronger and tougher than coarse-grained ones. And large grains can be undesirable for other reasons. For example, if the grain size of a metal sheet is comparable to the sheet thickness, the surface will rumple when the sheet is pressed to shape; and this makes it almost impossible to get a good surface finish on articles such as car-body panels or spun aluminium saucepans.

The ability to control grain size by recrystallisation is due to the general rule (e.g. Chapter 11) that the harder you drive a transformation, the finer the structure you get. In the case of recrystallisation this means that the greater the prior plastic deformation (and hence the stored strain energy) the finer the recrystallised grain size (Fig. 5.1-11). To produce a fine-grained sheet, for example, we simply reduce the thickness by about 50% in a cold rolling operation (to give the large stored strain energy) and then anneal the sheet in a furnace (to give the fine recrystallised structure).

## Machining

Most engineering components require at least some machining: turning, drilling, milling, shaping, or grinding. The cutting tool (or the abrasive particles of the grinding wheel) parts the chip from the workpiece by a process of plastic shear (Fig. 5.1-12). Thermodynamically, all that is required is the energy of the two new surfaces created when the chip peels off the surface; in reality, the work done in the plastic shear (a strain of order 1) greatly exceeds this minimum necessary energy. In addition, the friction is very high ($\mu \approx 0.5$) because the chip surface which bears against the tool is freshly formed, and free from adsorbed films which could reduce adhesion. This friction can be reduced by generous lubrication with water-soluble *cutting fluids*, which also cool the tool. Free

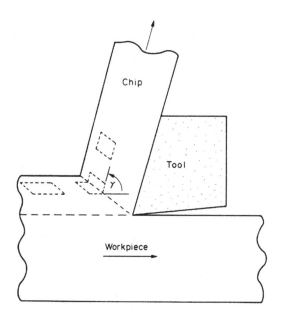

**Figure 5.1-12** Machining.

*cutting* alloys have a built-in lubricant which smears across the tool face as the chip forms: lead in brass, manganese sulphide in steel.

Machining is expensive – in energy, wasted material and time. Forming routes which minimise or avoid machining result in considerable economies.

## Joining

Many of the processes used to join one metal to another are based on casting. We have already looked at *fusion welding* (Fig. 13.6). The most widely used welding process is arc welding: an electric arc is struck between an electrode of filler metal and the workpieces, providing the heat needed to melt the filler and fuse it to the parent plates. The electrode is coated with a *flux* which melts and forms a protective cover on the molten metal. In *submerged arc* welding, used for welding thick sections automatically, the arc is formed beneath a pool of molten flux. In *gas welding* the heat source is an oxyacetylene flame. In *spot welding* the metal sheets to be joined are pressed together between thick copper electrodes and fused together locally by a heavy current. Small, precise welds can be made using either an electron beam or a laser beam as the heat source.

Brazing and soldering are also fine-scale casting processes. But they use filler metals which melt more easily than the parent metal. The filler does not join to the parent metal by fusion (melting together). Instead, the filler spreads over, or wets, the solid parent metal and, when it solidifies, becomes firmly stuck to it. True metal-to-metal contact is essential for good wetting. Before brazing, the parent surfaces are either mechanically abraded or acid pickled to remove as much of the surface oxide film as possible. Then a flux is added which chemically reduces any oxide that forms during the heating cycle. Specialised brazing operations are done in a vacuum furnace which virtually eliminates oxide formation.

Adhesives, increasingly used in engineering applications, do not necessarily require the application of heat. A thin film of epoxy, or other polymer, is spread on the surfaces to be joined, which are then brought together under pressure for long enough for the adhesive to polymerise or set. Special methods are required with adhesives, but they offer great potential for design.

Metal parts are also joined by a range of *fasteners*: rivets, bolts, or tabs. In using them, the stress concentration at the fastener or its hole must be allowed for: fracture frequently starts at a fastening point.

## Surface engineering

Often it is the properties of a surface which are critical in an engineering application. Examples are components which must withstand wear; or exhibit low friction; or resist oxidation or corrosion. Then the desired properties can often be achieved by creating a thin surface layer with good (but expensive) properties on a section of poorer (but cheaper) metal, offering great economies of production.

Surface treatments such as *carburising* or *nitriding* give hard surface layers, which give good wear and fatigue resistance. In carburising, a steel component is heated into the austenite region. Carbon is then diffused into the surface until its concentration rises to 0.8% or more. Finally the component is quenched into oil, transforming the surface into hard martensite. Steels for nitriding contain aluminium: when nitrogen is diffused into the surface it reacts to form aluminium nitride, which hardens the surface by precipitation hardening. More recently *ion implantation* has been used: foreign ions are accelerated in a strong electric field and are implanted into the surface. Finally, laser heat treatment has been developed as a powerful method for producing hard surfaces. Here the surface of the steel is scanned with a laser beam. As the beam passes over a region of the surface it heats it into the austenite region. When the beam passes on, the surface it leaves behind is rapidly quenched by the cold metal beneath to produce martensite.

## Energy-efficient forming

Many of the processes used for working metals are energy-intensive. Large amounts of energy are needed to melt metals, to roll them to sections, to machine them or to weld them together. Broadly speaking, the more steps there are between raw metal and finished article (see Fig. 5.1-1) then the greater is the cost of production. There is thus a big incentive to minimise the number of processing stages and to maximise the efficiency of the remaining operations. This is not new. For centuries, lead sheet for organ pipes has been made in a single-stage casting operation. The Victorians were the pioneers of pouring intricate iron castings which needed the minimum of machining. Modern processes which are achieving substantial energy savings include the single-stage casting of thin wires or ribbons (melt spinning, see Chapter 9) or the spray deposition of "atomised" liquid metal to give semi-finished seamless tubes. But modifications of conventional processes can give useful economies too. In examining a production line it is always worth questioning whether a change in processing method could be introduced with economic benefits.

# Examples

**5.1.1** Estimate the percentage volume contraction due to solidification in pure copper. Use the following data: $T_m = 1083°C$; density of solid copper at $20°C = 8.96$ Mg m$^{-3}$; average coefficient of thermal expansion in the range 20 to $1083°C = 20.6$M K$^{-1}$; density of liquid copper at $T_m = 8.00$ Mgm$^{-3}$.

## Answer

5%.

**5.1.2** A silver replica of a holly leaf is to be made by investment casting. (A natural leaf is coated with ceramic slurry which is then dried and fired. During firing the leaf burns away, leaving a mould cavity.) The thickness of the leaf is 0.4 mm. Calculate the liquid head needed to force the molten silver into the mould cavity. It can be assumed that molten silver does not wet the mould walls.

[Hint: the pressure needed to force a non-wetting liquid into a parallel-sided cavity of thickness $t$ is given by

$$p = \frac{T}{(t/2)}$$

where $T$ is the surface tension of the liquid.] The density and surface tension of molten silver are 9.4 Mg m$^{-3}$ and 0.90 Nm$^{-1}$.

## Answer

49 mm.

**5.1.3** Aluminium sheet is to be rolled according to the following parameters: starting thickness 1 mm, reduced thickness 0.8 mm, yield strength 100 MPa. What roll radius should be chosen to keep the forming pressure below 200 MPa?

## Answer

16.2 mm, or less.

**5.1.4** Aluminium sheet is to be rolled according to the following parameters: sheet width 300 mm, starting thickness 1mm, reduced thickness 0.8 mm, yield strength 100 MPa, maximum forming pressure 200 MPa, roll radius 16.2 mm, roll length 300 mm. Calculate the force $F$ that the rolling pressure will exert on each roll.

[Hint: use the average forming pressure, $p_{av}$, shown in Fig. 5.1-8.]

The design states that the roll must not deflect by more than 0.01 mm at its centre. To achieve this bending stiffness, each roll is to be backed up by one secondary roll as shown in Fig. 5.1-9b. Calculate the secondary roll radius needed to meet the specification. The central deflection of the secondary roll is given by

$$\delta = \frac{5FL^3}{384\ EI}$$

where $L$ is the roll length and $E$ is the Young's modulus of the roll material. $I$, the second moment of area of the roll section, is given by

$$I = \pi r_s^4/4$$

where $r_s$ is the secondary roll radius. The secondary roll is made from steel, with $E = 210$ GPa. You may neglect the bending stiffness of the primary roll.

## Answers

$F = 81$ kN; $r_s = 64.5$ mm.

**5.1.5** Copper capillary fittings are to be used to solder copper water pipes together as shown below:

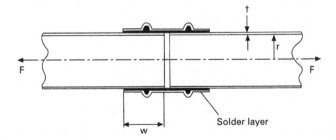

The joint is designed so that the solder layer will yield in shear at the same axial load $F$ that causes the main tube to fail by tensile yield. Estimate the required value of $W$, given the following data: $t = 1$ mm; $\sigma_y$ (copper) $= 120$ MPa; $\sigma_y$(solder) $= 10$ MPa.

### Answer

24 mm.

**5.1.6** A piece of plain carbon steel containing 0.2 wt% carbon was case-carburised to give a case depth of 0.3 mm. The carburising was done at a temperature of 1000°C. The Fe–C phase diagram shows that, at this temperature, the iron can dissolve carbon to a maximum concentration of 1.4 wt%. Diffusion of carbon into the steel will almost immediately raise the level of carbon in the steel to a constant value of 1.4 wt% just beneath the surface of the steel. However, the concentration of carbon well below the surface will increase more slowly toward the maximum value of 1.4wt% because of the time needed for the carbon to diffuse into the interior of the steel.

The diffusion of carbon into the steel is described by the time-dependent diffusion equation

$$C(x,t) = (C_s - C_0)\left\{1 - \mathrm{erf}\left(\frac{x}{2\sqrt{Dt}}\right)\right\} + C_0.$$

The symbols have the meanings: $C$, concentration of carbon at a distance $x$ below the surface after time $t$; $C_s$, 1.4 wt% C; $C_0$, 0.2 wt% C; $D$, diffusion coefficient for carbon in steel. The "error function", $\mathrm{erf}(y)$, is given by

$$\mathrm{erf}(y) = \frac{2}{\sqrt{\pi}}\int_0^y e^{-Z^2}\,dZ.$$

The following table gives values for this integral.

| $y$ | 0 | 0.1 | 0.2 | 0.3 | |
|---|---|---|---|---|---|
| erf($y$) | 0 | 0.11 | 0.22 | 0.33 | |
| $y$ | 0.4 | 0.5 | 0.6 | 0.7 | |
| erf($y$) | 0.43 | 0.52 | 0.60 | 0.68 | |
| $y$ | 0.8 | 0.9 | 1.0 | 1.1 | 1.2 |
| erf($y$) | 0.74 | 0.80 | 0.84 | 0.88 | 0.91 |
| $y$ | 1.3 | 1.4 | 1.5 | ∞ | |
| erf($y$) | 0.93 | 0.95 | 0.97 | 1.00 | |

The diffusion coefficient may be taken as

$$D = 9 \times 10^{-6}\,\mathrm{m^2 s^{-1}}\exp\left\{\frac{-125\,\mathrm{kJ\ mol^{-1}}}{RT}\right\}$$

where $R$ is the gas constant and $T$ is the absolute temperature.

Calculate the time required for carburisation, if the depth of the case is taken be the value of $x$ for which $C = 0.5$ wt% carbon.

### Answer

8.8 minutes.

**5.1.7** Using the equations and tabulated error function data from Example 5.1.6, show that the expression $x = \sqrt{Dt}$ gives the distance over which the concentration of the diffusion profile halves.

**5.1.8** Describe the processes of recovery and recrystallisation that occur during the high-temperature annealing of a work-hardened metal. How does the grain size of the fully recrystallised metal depend on the initial amount of work hardening? Mention some practical situations in which recrystallisation is important.

**5.1.9** A bar of cold-drawn copper had a yield strength of 250 MPa. The bar was later annealed at 600°C for 5 minutes. The yield strength after annealing was 50 MPa. Explain this change. (Both yield strengths were measured at 20°C.)

# Section Six

## Light alloys

# Chapter 6.1

# Light alloys

## Introduction

No fewer than 14 pure metals have densities $\leq 4.5$ Mg m$^{-3}$ (see Table 6.1-1). Of these, titanium, aluminium and magnesium are in common use as structural materials. Beryllium is difficult to work and is toxic, but it is used in moderate quantities for heat shields and structural members in rockets. Lithium is used as an alloying element in aluminium to lower its density and save weight on airframes. Yttrium has an excellent set of properties and, although scarce, may eventually find applications in the nuclear-powered aircraft project. But the majority are unsuitable for structural use because they are chemically reactive or have low melting points.*

Table 6.1-2 shows that alloys based on aluminium, magnesium and titanium may have better stiffness/weight and strength/weight ratios than steel. Not only that; they are also corrosion resistant (with titanium exceptionally so); they are non-toxic; and titanium has good creep properties. So although the light alloys were originally developed for use in the aerospace industry, they are now much more widely used. The dominant use of aluminium alloys is in building and construction: panels, roofs, and frames. The second-largest consumer is the container and packaging industry; after that come transportation systems (the fastest-growing sector, with aluminium replacing steel and cast iron in cars and mass-transit systems); and the use of aluminium as an electrical conductor. Magnesium is lighter but more expensive. Titanium alloys are mostly used in aerospace applications where the temperatures are too high for aluminium or magnesium; but its extreme corrosion resistance makes it attractive in chemical engineering, food processing and bio-engineering. The growth in the use of these alloys is rapid: nearly 7% per year, higher than any other metals, and surpassed only by polymers.

The light alloys derive their strength from *solid solution hardening*, *age* (or *precipitation*) *hardening*, and *work hardening*. We now examine the principles behind each hardening mechanism, and illustrate them by drawing examples from our range of generic alloys.

## Solid solution hardening

When other elements dissolve in a metal to form a solid solution they make the metal harder. The solute atoms differ in size, stiffness and charge from the solvent atoms. Because of this the randomly distributed solute atoms interact with dislocations and make it harder for them to move. The theory of solution hardening is rather complicated, but it predicts the following result for the yield strength

$$\sigma_y \propto \varepsilon_s^{3/2} C^{1/2}, \qquad (6.1.1)$$

where C is the solute concentration. $\varepsilon_s$ is a term which represents the "mismatch" between solute and solvent atoms. The form of this result is just what we would expect: badly matched atoms will make it harder for

---

* There are, however, many *non-structural* applications for the light metals. Liquid sodium is used in large quantities for cooling nuclear reactors and in small amounts for cooling the valves of high-performance i.c. engines (it conducts heat 143 times better than water but is less dense, boils at 883°C, and is safe as long as it is kept in a sealed system.) Beryllium is used in windows for X-ray tubes. Magnesium is a catalyst for organic reactions. And the reactivity of calcium, caesium and lithium makes them useful as residual gas scavengers in vacuum systems.

*Engineering Materials and Processes Desk Reference*; ISBN: 9781856175869
Copyright © 2009 Elsevier Ltd; All rights of reproduction, in any form, reserved.

# CHAPTER 6.1  Light alloys

Table 6.1-1 The light metals

| Metal | Density (Mg m$^{-3}$) | $T_m$(°C) | Comments |
|---|---|---|---|
| Titanium | 4.50 | 1667 | High $T_m$—excellent creep resistance. |
| Yttrium | 4.47 | 1510 | Good strength and ductility; scarce. |
| Barium | 3.50 | 729 | |
| Scandium | 2.99 | 1538 | Scarce. |
| Aluminium | 2.70 | 660 | |
| Strontium | 2.60 | 770 | Reactive in air/water. |
| Caesium | 1.87 | 28.5 | Creeps/melts; very reactive in air/water. |
| Beryllium | 1.85 | 1287 | Difficult to process; very toxic. |
| Magnesium | 1.74 | 649 | |
| Calcium | 1.54 | 839 | Reactive in air/water. |
| Rubidium | 1.53 | 39 | Creep/melt; very reactive in air/water. |
| Sodium | 0.97 | 98 | |
| Potassium | 0.86 | 63 | |
| Lithium | 0.53 | 181 | |

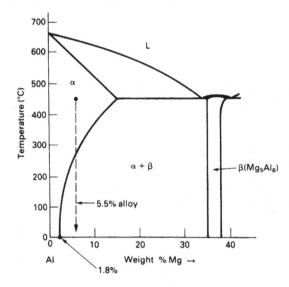

Figure 6.1-1 The aluminium end of the Al–Mg phase diagram.

## (a) Hold at 450°C ("solution heat treat")

This puts the 5.5% alloy into the single phase ($\alpha$) field and all the Mg will dissolve in the Al to give a random substitutional solid solution.

## (b) Cool moderately quickly to room temperature

The phase diagram tells us that, below 275°C, the 5.5% alloy has an *equilibrium* structure that is two-phase, $\alpha$ + Mg$_5$Al$_8$. If, then, we cool the alloy *slowly* below 275°C, Al and Mg atoms will diffuse together to form precipitates of the intermetallic compound Mg$_5$Al$_8$. However, below 275°C, diffusion is slow and the C-curve for the precipitation reaction is well over to the right (Fig. 6.1-2). So if we cool the 5.5% alloy moderately quickly we will miss the nose of the C-curve. None of the Mg will be taken out of solution as Mg$_5$Al$_8$, and we will end up with a supersaturated solid solution at room

dislocations to move than well-matched atoms; and a large population of solute atoms will obstruct dislocations more than a sparse population.

Of the generic aluminium alloys (see Fig. 6.1-4), the 5000 series derives most of its strength from solution hardening. The Al–Mg phase diagram (Fig. 6.1-1) shows why: at room temperature aluminium can dissolve up to 1.8 wt% magnesium at equilibrium. In practice, Al–Mg alloys can contain as much as 5.5 wt% Mg in solid solution at room temperature – a supersaturation of 5.5 − 1.8 = 3.7 wt%. In order to get this supersaturation the alloy is given the following schedule of heat treatments.

Table 6.1-2 Mechanical properties of structural light alloys

| Alloy | Density $\rho$ (Mg m$^{-3}$) | Young's modulus $E$ (GPa) | Yield strength $\sigma_y$ (MPa) | $E/\rho$ | $E^{1/2}/\rho$ | $E^{1/3}/\rho$ | $\sigma_y/\rho$ | Creep temperature (°C) |
|---|---|---|---|---|---|---|---|---|
| Al alloys | 2.7 | 71 | 25–600 | 26 | 3.1 | 1.5 | 9–220 | 150–250 |
| Mg alloys | 1.7 | 45 | 70–270 | 25 | 4.0 | 2.1 | 41–160 | 150–250 |
| Ti alloys | 4.5 | 120 | 170–1280 | 27 | 2.4 | 1.1 | 38–280 | 400–600 |
| (Steels) | (7.9) | (210) | (220–1600) | 27 | 1.8 | 0.75 | 28–200 | (400–600) |

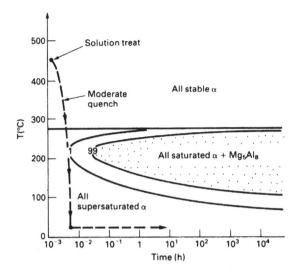

**Figure 6.1-2** Semi-schematic TTT diagram for the precipitation of $Mg_5Al_8$ from the Al–5.5 wt% Mg solid solution.

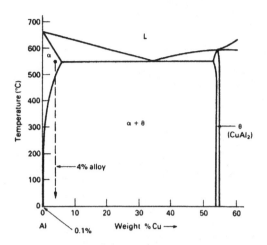

**Figure 6.1-3** The aluminium end of the Al–Cu phase diagram.

temperature. As Table 6.1-3 shows, this supersaturated Mg gives a substantial increase in yield strength.

Solution hardening is not confined to 5000 series aluminium alloys. The other alloy series all have elements dissolved in solid solution; and they are all solution strengthened to some degree. But most aluminium alloys owe their strength to fine precipitates of intermetallic compounds, and solution strengthening is not dominant as it is in the 5000 series. Turning to the other light alloys, the most widely used titanium alloy (Ti–6 Al 4V) is dominated by solution hardening (Ti effectively dissolves about 7 wt% Al, and has complete solubility for V). Finally, magnesium alloys can be solution strengthened with Li, Al, Ag and Zn, which dissolve in Mg by between 2 and 5 wt%.

## Age (precipitation) hardening

When the phase diagram for an alloy has the shape shown in Fig. 6.1-3 (a solid solubility that decreases markedly as the temperature falls), then the potential for *age* (or *precipitation*) *hardening* exists. The classic example is the Duralumins, or 2000 series aluminium alloys, which contain about 4% copper.

The Al–Cu phase diagram tells us that, between 500°C and 580°C, the 4% Cu alloy is single phase: the Cu dissolves in the Al to give the random substitutional solid solution $\alpha$. Below 500°C the alloy enters the two-phase field of $\alpha + CuAl_2$. As the temperature decreases the amount of $CuAl_2$ increases, and at room temperature the equilibrium mixture is 93 wt% $\alpha$ + 7 wt% $CuAl_2$. Figure 6.1-4a shows the microstructure that we would get by cooling an Al–4 wt% Cu alloy *slowly* from 550°C to room temperature. In slow cooling the driving force for the precipitation of $CuAl_2$ is small and the nucleation rate is low. In order to accommodate the equilibrium amount of $CuAl_2$ the few nuclei that do form grow into large precipitates of $CuAl_2$ spaced well apart. Moving dislocations find it easy to avoid the precipitates and the alloy is rather soft. If, on the other hand, we cool the alloy rather *quickly*, we produce a much finer structure (Fig. 6.1-4b). Because the driving force is large the nucleation rate is high. The precipitates, although small, are closely spaced: they get in the way of moving dislocations and make the alloy harder.

There are limits to the precipitation hardening that can be produced by direct cooling: if the cooling rate is too high we will miss the nose of the C-curve for the precipitation reaction and will not get any precipitates at all! But large increases in yield strength *are* possible if we *age harden* the alloy.

To age harden our Al–4 wt% Cu alloy we use the following schedule of heat treatments.

(a) Solution heat treat at 550°C. This gets all the Cu into solid solution.

| Table 6.1-3 Yield strengths of 5000 series (Al–Mg) alloys | | | |
|---|---|---|---|
| **Alloy** | **(wt% Mg)** | $\sigma_y$ **(MPa) (annealed condition)** | |
| 5005 | 0.8 | 40 | |
| 5050 | 1.5 | 55 | |
| 5052 | 2.5 | 90 | |
| 5454 | 2.7 | 120 | supersaturated |
| 5083 | 4.5 | 145 | |
| 5456 | 5.1 | 160 | |

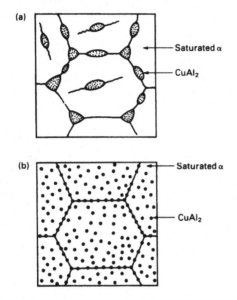

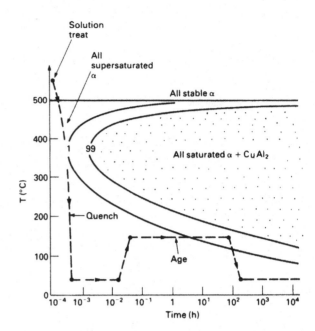

**Figure 6.1-4** Room temperature microstructures in the Al + 4 wt% Cu alloy. (**a**) Produced by slow cooling from 550°C. (**b**) Produced by moderately fast cooling from 550°C. The precipitates in (**a**) are large and far apart. The precipitates in (**b**) are small and close together.

(**b**) Cool rapidly to room temperature by quenching into water or oil ("quench").* We will miss the nose of the C-curve and will end up with a highly supersaturated solid solution at room temperature (Fig. 6.1-5).

(**c**) Hold at 150°C for 100 hours ("age"). As Fig. 6.1-5 shows, the supersaturated $\alpha$ will transform to the equilibrium mixture of saturated $\alpha$ + $CuAl_2$. But it will do so under a very high driving force and will give a very fine (and very strong) structure.

Figure 6.1-5, as we have drawn it, is oversimplified. Because the transformation is taking place at a low temperature, where the atoms are not very mobile, it is not easy for the $CuAl_2$ to separate out in one go. Instead, the transformation takes place in four distinct stages. These are shown in Figs 6.1-6(a)–(e). The progression may appear rather involved but it is a good illustration of much of the material in the earlier chapters. More importantly, each stage of the transformation has a direct effect on the yield strength.

Four separate hardening mechanisms are at work during the ageing process:

## (a) Solid solution hardening

At the start of ageing the alloy is mostly strengthened by the 4 wt% of copper that is trapped in the supersaturated

**Figure 6.1-5** TTT diagram for the precipitation of $CuAl_2$ from the Al + 4 wt% Cu solid solution. Note that the *equilibrium* solubility of Cu in Al at room temperature is only 0.1 wt% (see Fig. 6.1-3). The quenched solution is therefore carrying 4/0.1 = 40 times as much Cu as it wants to.

$\alpha$. But when the GP zones form, almost all of the Cu is removed from solution and the solution strengthening virtually disappears (Fig. 6.1-7).

## (b) Coherency stress hardening

The coherency strains around the GP zones and $\theta''$ precipitates generate stresses that help prevent dislocation movement. The GP zones give the larger hardening effect (Fig. 6.1-7)

## (c) Precipitation hardening

The precipitates can obstruct the dislocations directly. But their effectiveness is limited by two things: dislocations can either *cut through* the precipitates, or they can *bow around* them (Fig. 6.1-8).

Resistance to cutting depends on a number of factors, of which the shearing resistance of the precipitate lattice is only one. In fact the cutting stress *increases* with ageing time (Fig. 6.1-7).

Bowing is easier when the precipitates are far apart. During ageing the precipitate spacing increases from 10 nm to 1 $\mu$m and beyond (Fig. 6.1-9). The

---

* The C-curve nose is ≈150°C higher for Al-4 Cu than for Al-5.5 Mg (compare Figs 6.1-5 and 6.1-2). Diffusion is faster, and a more rapid quench is needed to miss the nose.

Light alloys CHAPTER 6.1

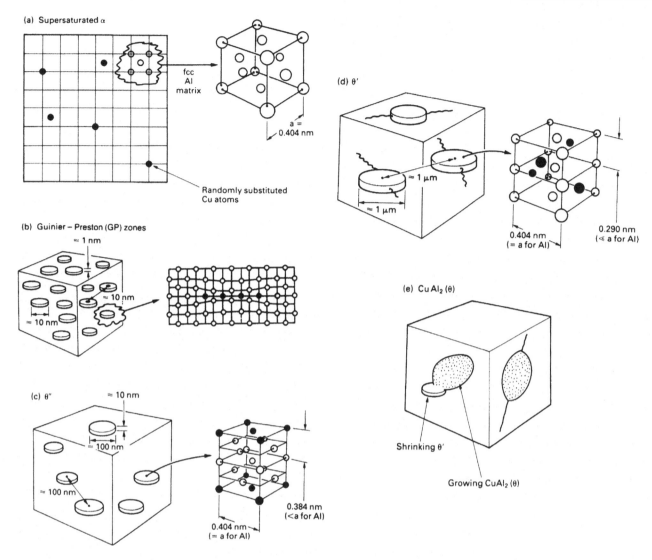

**Figure 6.1-6** Stages in the precipitation of CuAl$_2$. Disc-shaped GP zones (**b**) nucleate homogeneously from supersaturated solid solution (**a**). The disc faces are perfectly coherent with the matrix. The disc edges are also coherent, but with a large *coherency strain*. (**c**) Some of the GP zones grow to form precipitates called $\theta''$. (The remaining GP zones dissolve and transfer Cu to the growing $\theta''$ by diffusion through the matrix.) Disc faces are perfectly coherent. Disc edges are coherent, but the mismatch of lattice parameters between the $\theta''$ and the Al matrix generates coherency strain. (**d**) Precipitates called $\theta'$ nucleate at matrix dislocations. The $\theta''$ precipitates all dissolve and transfer Cu to the growing $\theta'$. Disc faces are still perfectly coherent with the matrix. But disc edges are now *incoherent*. Neither faces nor edges show coherency strain, but for different reasons. (**e**) Equilibrium CuAl$_2$ ($\theta$) nucleates at grain boundaries and at $\theta'$–matrix interfaces. The $\theta'$ precipitates all dissolve and transfer Cu to the growing $\theta$. The CuAl$_2$ is completely *incoherent* with the matrix. Because of this it grows as *rounded* rather than disc-shaped particles.

bowing stress therefore decreases with ageing time (Fig. 6.1-7).

The four hardening mechanisms add up to give the overall variation of yield strength shown in Fig. 6.1-7. *Peak strength is reached if the transformation is stopped at $\theta''$.* If the alloy is aged some more the strength will *decrease*; and the only way of recovering the strength of an overaged alloy is to solution-treat it at 550°C, quench, and start again! If the alloy is not aged for long enough, then it will not reach peak strength; but this can be put right by more ageing.

Although we have chosen to age our alloy at 150°C, we could, in fact, have aged it at any temperature below 180°C (see Figure 6.1-10). The lower the ageing temperature, the longer the time required to get peak hardness. In practice, the ageing time should be long enough to give good control of the heat treatment operation without being too long (and expensive).

Finally, Table 6.1-4 shows that copper is not the only alloying element that can age-harden aluminium. Magnesium and titanium can be age hardened too, but not as much as aluminium.

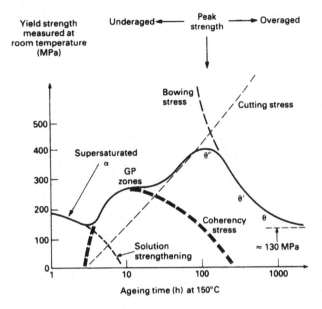

**Figure 6.1-7** The yield strength of quenched Al–4 wt% Cu changes dramatically during ageing at 150°C.

## Work hardening

Commercially pure aluminium (1000 series) and the non-heat-treatable aluminium alloys (3000 and 5000 series) are usually work hardened. The work hardening superimposes on any solution hardening, to give considerable extra strength (Table 6.1-5).

Work hardening is achieved by cold rolling. The yield strength increases with strain (reduction in thickness) according to

$$\sigma_y = A\varepsilon^n, \qquad (6.1.2)$$

where $A$ and $n$ are constants. For aluminium alloys, $n$ lies between 1/6 and 1/3.

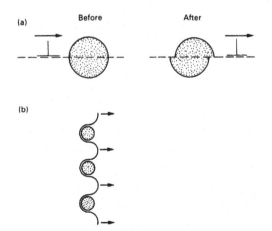

**Figure 6.1-8** Dislocations can get past precipitates by (**a**) cutting or (**b**) bowing.

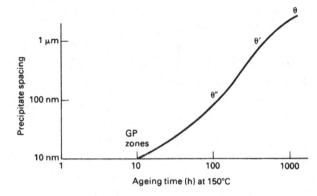

**Figure 6.1-9** The gradual increase of particle spacing with ageing time.

## Thermal stability

Aluminium and magnesium melt at just over 900 K. Room temperature is 0.3 $T_m$, and 100°C is 0.4 $T_m$. Substantial diffusion can take place in these alloys if they are used for long periods at temperatures approaching 80–100°C. Several processes can occur to reduce the yield strength: loss of solutes from supersaturated solid solution, overageing of precipitates and recrystallisation of cold-worked microstructures.

This lack of *thermal stability* has some interesting consequences. During supersonic flight frictional heating can warm the skin of an aircraft to 150°C. Because of this, Rolls-Royce had to develop a special age-hardened aluminium alloy (RR58) which would not over-age during the lifetime of the Concorde supersonic airliner. When aluminium cables are fastened to copper busbars in power circuits contact resistance heating at the

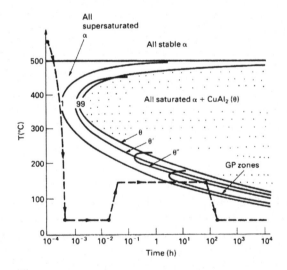

**Figure 6.1-10** Detailed TTT diagram for the Al–4 wt% Cu alloy. We get peak strength by ageing to give $\theta''$. The lower the ageing temperature, the longer the ageing time. Note that GP zones do not form above 180°C: if we age above this temperature we will fail to get the peak value of yield strength.

| Table 6.1-4 Yield strengths of heat-treatable alloys ||||
|---|---|---|---|
| Alloy series | Typical composition (wt%) | $\sigma_y$ (MPa) Slowly cooled | $\sigma_y$ (MPa) Quenched and aged |
| 2000 | Al + 4 Cu + Mg, Si, Mn | 130 | 465 |
| 6000 | Al + 0.5 Mg 0.5 Si | 85 | 210 |
| 7000 | Al + 6 Zn + Mg, Cu, Mn | 300 | 570 |

| Table 6.1-5 Yield strengths of work-hardened aluminium alloys ||||
|---|---|---|---|
| Alloy number | $\sigma_y$ (MPa) Annealed | "Half hard" | "Hard" |
| 1100 | 35 | 115 | 145 |
| 3005 | 65 | 140 | 185 |
| 5456 | 140 | 300 | 370 |

junction leads to interdiffusion of Cu and Al. Massive, brittle plates of $CuAl_2$ form, which can lead to joint failures; and when light alloys are welded, the properties of the heat-affected zone are usually well below those of the parent metal.

## Examples

**6.1.1** An alloy of Al–4 weight% Cu was heated to 550°C for a few minutes and was then quenched into water. Samples of the quenched alloy were aged at 150°C for various times before being quenched again. Hardness measurements taken from the re-quenched samples gave the following data:

| Ageing time (h) | 0 | 10 | 100 | 200 | 1000 |
|---|---|---|---|---|---|
| Hardness (MPa) | 650 | 950 | 1200 | 1150 | 1000 |

Account briefly for this behaviour.
Peak hardness is obtained after 100 h at 150°C. Estimate how long it would take to get peak hardness at (a) 130°C, (b) 170°C.
[Hint: use Fig. 6.1-10.]

### Answers
(a) $10^3$ h; (b) 10 h.

**6.1.2** A batch of 7000 series aluminium alloy rivets for an aircraft wing was inadvertently over-aged. What steps can be taken to reclaim this batch of rivets?

**6.1.3** Two pieces of work-hardened 5000 series aluminium alloy plate were butt welded together by arc welding. After the weld had cooled to room temperature, a series of hardness measurements was made on the surface of the fabrication. Sketch the variation in hardness as the position of the hardness indenter passes across the weld from one plate to the other. Account for the form of the hardness profile, and indicate its practical consequences.

**6.1.4** One of the major uses of aluminium is for making beverage cans. The body is cold-drawn from a single slug of 3000 series non-heat treatable alloy because this has the large ductility required for the drawing operation. However, the top of the can must have a much lower ductility in order to allow the ring-pull to work (the top must tear easily). Which alloy would you select for the top from Table 6.1-5? Explain the reasoning behind your choice. Why are non-heat treatable alloys used for can manufacture?

**6.1.5** A sample of Al–4 wt% Cu was cooled slowly from 550°C to room temperature. The yield strength of the slowly cooled sample was 130 MPa. A second sample of the alloy was quenched into cold water from 550°C and was then aged at 150°C for 100 hours. The yield strength of the quenched-and-aged sample was 450 MPa. Explain the difference in yield strength. [Both yield strengths were measured at 20°C.]

# Section Seven

## Plastics

# Chapter 7.1

# Introduction to plastics

## 7.1.1 Introduction

This chapter encourages the reader to familiarize themself with plastics. It aims to open the reader's eyes to design features in familiar products, and to relate these features to polymer processes. The dismantling exercises can be adapted to suit different courses; for students on a biomaterials course, blood sugar monitors, asthma inhalers, or blood apheresis units can be dismantled. For those on a sports/materials course, the components of a running shoe could be considered. Product examination can be tackled at different levels. The level described here is suitable at the start of a degree course. Later, when most of the topics in the book have been studied, more complex tasks can be tackled – improving the design of an existing product, with reselection of materials and processing route.

There are some polymer identification exercises, using simple equipment. This would make the reader familiar with the appearance of the main plastics. Professional methods of polymer identification, such as differential scanning calorimetry, Fourier transform infrared (FTIR) spectroscopy and optical microscopy, may be dealt with later in degree courses.

This book explores the characteristic properties of polymers and attempts to explain them in terms of microstructure.

## 7.1.2 Dismantling consumer products

Using familiar products, the aim is to note component shapes, to see how they are assembled, and measure the variation in thickness. Recycling can also be considered; the ease of dismantling depends on whether the product was intended to be repaired, or to be scrapped if faulty. Screws may be hidden under adhesive labels, and the location of snap-fit parts may be difficult to find.

### 7.1.2.1 Plastic kettle

A new plastic kettle can be bought for less than £30, or a discarded one used. Preferably use a cordless kettle, which can be lifted from the powered base. The following four activities can be extended if necessary, by consideration of aesthetics, weight, and ease of filling and pouring.

*Briefly touch the kettle's outer surface when the water is boiling*

Although the initial temperature of the kettle's outer surface may be 90°C, the low thermal conductivity of the plastic body compared with that of your finger, means that the skin surface temperature takes more than a minute to reach an equilibrium value, and this value is c. 50°C. *With a dry finger* touch the kettle's outer surface for less than 5 s. If you have access to a digital thermometer with a fine thermocouple probe, tape the thermocouple to the outer surface of the kettle and check the temperature. What can you deduce about the thermal conductivity of the plastic?

*Measure the thickness of the body at a range of locations*

Dismantle the kettle and make a vertical section through the body with a hacksaw. Use callipers to measure the body thickness at a range of locations, and mark the values on the plastic.

Over what range does the thickness vary? Figure 7.1-1 shows a typical section. Check how the colouring is achieved. If there is no paint layer on the outside,

*Engineering Materials and Processes Desk Reference*; ISBN: 9781856175869
Copyright © 2009 Elsevier Ltd; All rights of reproduction, in any form, reserved.

# CHAPTER 7.1 Introduction to plastics

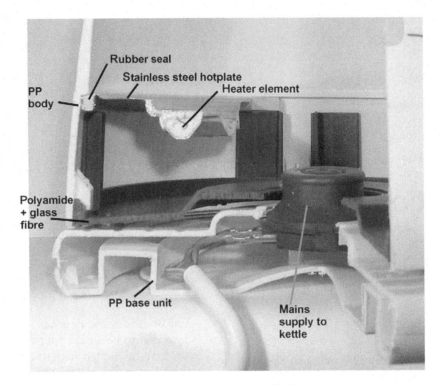

**Figure 7.1-1** Section of a plastic kettle and powered base unit (most of electrical heater was removed from the kettle).

the colour must be integral (for pigments, www.specialchem4polymer.com).

*Examine the electrical insulation in the base unit*

Note how the electrical conductors are insulated from the parts that are handled. A metal-bodied kettle must have separate insulation wherever mains-connected parts are attached; however plastic is an electrical insulator. Figure 7.1-1 shows the mains power connections in the base. The coloured insulation (live, neutral and earth) of the braided copper wires is plasticised PVC.

*Examine the linking mechanism for the heater switch*

Identify the mechanism that connects the on/off switch to the internal contact switch that applies mains power to the heater unit. Identify the thermostat that detects the boiling of the kettle, and note how it switches off the power. Figure 7.1-2 shows a typical arrangement.

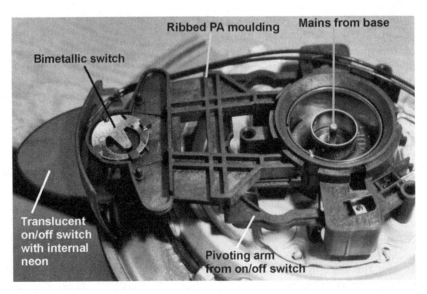

**Figure 7.1-2** Underside of the heater unit inside the kettle, showing the power switch and switch mechanism.

Introduction to plastics          CHAPTER 7.1

**Figure 7.1-3** Mechanism that locks the spools when the tape is not being played, as seen inside a video cassette.

## 7.1.2.2 VHS video cassette

Video cassettes are becoming obsolete with the increasing use of DVDs, so one such cassette could be sacrificed.

*Check how it has been assembled*

Dismantle the cassette by unscrewing the five screws that fasten the two halves together, using a small Phillips-type screwdriver. Lift off the top of the cassette. If there is a clear plastic window, that allows the tape levels to be seen, check how it is attached to the main body. Count the number of parts. After dismantling, see how easy it is to reassemble!

*Identify the plastic springs that lock the spools*

When a cassette is removed from a recorder, the tape spools are locked to prevent the unwinding of the tape. When it is inside the recorder, a pin presses through a flap at the base of the cassette, causing a lever to operate on two plastic mouldings (Fig. 7.1-3). They engage with slots in the rim of the tape spools. Check, by

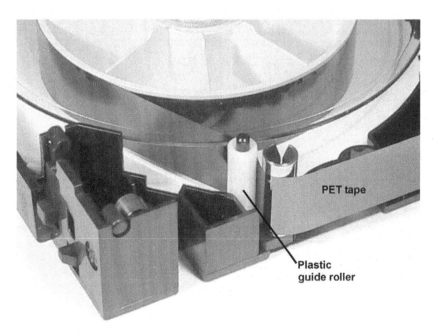

**Figure 7.1-4** PET tape in a VHS cassette passes round a plastic guide roller and a fixed metal cylinder.

249

pressing with a finger, that the springs can be easily bent. They are made of the engineering thermoplastic polyoxymethylene.

*Measure the tensile strength of the tape*

Unwind some of the 13 mm wide coated PET tape and measure the thickness (0.02 mm) with callipers. A loop of tape can withstand a tensile force of about 60 N before it yields and about 80 N before it fails in tension. Check this by using a spring balance on a loop of tape, and calculate its tensile strength (approximately 150 MPa). It must bend around cylinders of diameter 5 and 6 mm (Fig. 7.1-4), so it must have a very low bending stiffness. It must resist wear as it is dragged over the stationary metal cylinders. It must be dimensionally stable, so that the coating is not damaged.

### 7.1.2.3 Stackable plastic chair

This has a polypropylene (PP) seat, with welded tubular-steel legs. Use a Phillips screwdriver to remove the four screws that attach seat to the legs. These self-tapping screws (Fig. 7.1-5) with sharp, widely spaced threads, are much longer than the typical 4 mm thick seat. When screwed into a moulded cylindrical boss on the hidden side of the seat, the threads cut grooves in the initially smooth plastic. It is supported by four or more buttresses, to prevent bending loads causing failure, where the boss joins the seat. Measure the thickness of the buttresses.

Note the texture on the upper surface of the chair (Fig. 7.1-6), whereas the lower surface is smooth. How has this texture been achieved? Is it a reproduction of the mould surface texture, or has it been produced by a post-moulding operation?

### 7.1.2.4 Telephone handset

An old handset from an office may be available for dismantling. The numbers for dialling are printed on separate thermoplastic mouldings, each mounted on a domed rubber spring (Fig. 7.1-7). The domes depress with a click as the side walls buckle, and act as electrical switches. A layer of carbon-black filled rubber on the base comes into contact with copper tracks on the printed circuit board (PCB). The PCB consists of a polyester resin plus woven fibreglass (GRP) composite, which is also an insulator. The copper tracks on the PCB lead to holes where components are mounted; the PCB must tolerate the temperature of molten solder without distortion.

### 7.1.2.5 Summary

Having completed the dismantling exercises, try to add to the following list. *Plastics have advantages over metals of being*

1. self-coloured, by adding about 0.1% of dispersed pigment. There are no painting costs, and the product maintains its colour if scratched.

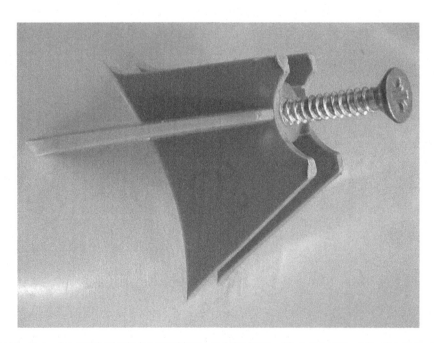

**Figure 7.1-5** Self-tapping screw for attached tubular metal legs, and the boss with buttresses under the seat of a PP stacking chair.

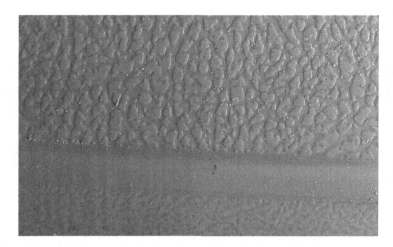

**Figure 7.1-6** Texture on the upper surface of a PP chair.

2. electrical insulators. There is no need for insulating layers between live parts and the body of product, and assembly is simplified.
3. thermal insulators. This conserves energy, and touching a kettle body will not cause scalds.
4. of low density, so lightweight products can be made.
5. impact resistant, with a high yield strain, so thin panels do not dent if locally loaded.

*Plastics have advantages over ceramics or glass of being*

1. tough, so that the impacts are unlikely to cause brittle fractures.

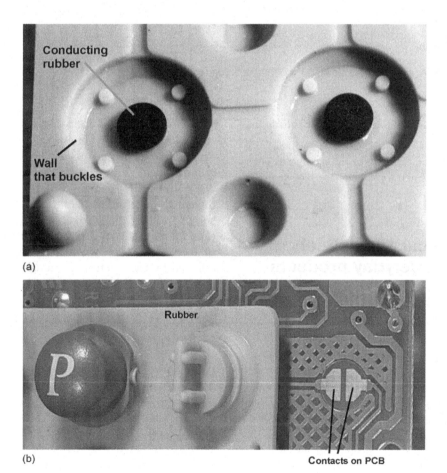

**Figure 7.1-7** Views from both sides of an injection-moulded rubber switch from a telephone.

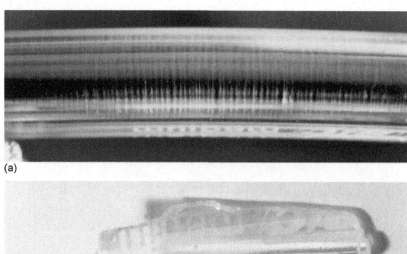

**Figure 7.1-8** (a) Crazes in, (b) broken pieces of, a PS Biro after a bending experiment.

2. low melting point, so the energy costs for processing are low.
3. capable of being moulded into complex shapes with the required final dimensions (they are 'net-shape', with no final machining stage).

## 7.1.3 Mechanical and optical properties of everyday products

Several disposable plastic products are considered, to illustrate mechanical and optical properties.

### 7.1.3.1 Crazing and fracture of a biro

Find a Bic biro (or a similar ballpoint pen) with a transparent polystyrene body. Hold it up towards a light source and bend it, using the thumbs as the inner and the forefingers as the outer loading points. Make sure that the curved portion is away from you and not aimed at anyone else. Deform the biro by about 10 mm and hold this for about 30 s, then release the load. The biro should return to its original shape, showing that large elastic strains can occur. Tilt the biro against the light and look for parallel reflective planes (Fig. 7.1-8a). These are called *crazes*.

Continue the loading until the body fractures. Although the ink tube will trap the broken pieces of the body, it is likely that a small piece(s) of PS might detach (Fig. 7.1-8b). Do not do the experiment without the ink tube, as pieces can fly off at speed. The strain energy released by the fracture is enough to create more than one fracture surface.

### 7.1.3.2 Ductile yielding of low-density polyethylene strapping

Low-density polyethylene (LDPE) strapping, cut from 0.42 mm thick film, is used to hold four packs of drink cans together. If pulled slowly with the hands, parts of the strapping undergo tensile necking followed by cold drawing of the thin region (Fig. 7.1-9). Mark parallel lines at 5 mm intervals across the LDPE before the experiment. Note the extension ratio in the neck, and how the shoulder of the neck moves into the un-necked region.

**Figure 7.1-9** Necking and cold drawing of LDPE strapping from a four-pack of drink tins

## 7.1.3.3 Optical properties of a CD and polyethylene film

This requires a laser pointer and a CD. Observe safety precautions: do not aim the laser beam at anyone's eyes. Aim it, at approximately normal incidence, at the side of the CD that appears silvered. When the beam hits the tracks near the centre of the disc, a diffraction pattern is created (Fig. 7.1-10). This pattern is a two-dimensional analogue of X-ray diffraction from a three-dimensional crystal.

If the laser beam hits the main part of the disc, there are just two diffraction peaks, in addition to the directly reflected beam. These are caused by the regular track spacing in the radial direction. As the circumferential pits are irregularly spaced along each track, this part of the disc acts as a one-dimensional diffraction grating. The diffraction pattern is used to keep the reading head on the track. If you scratch off part of the label and the underlying metallized layer, the CD will be transparent in this region. Hence the material, polycarbonate, is transparent.

Macro-bubbles, used inside cardboard boxes for the shock-resistant packaging of goods (Fig. 7.1-11), are manufactured from 200 mm wide tubular polyethylene film approximately 0.05 mm. The tube is inflated with air then welded at approximately 100 mm intervals. Place a macro-bubble on top of a printed page with a range of font sizes, and note the smallest font size that you can read. High-density polyethylene (HDPE) bubbles scatter light more than LDPE, so it is more difficult to read the text. If an HDPE bubble is lifted by about 20 mm, it is impossible to read the text.

## 7.1.3.4 Degradation of polymers in sunlight

Visit a beach and collect plastic articles that have been there for a couple of years. Apart from foam and hollow air-filled products, there will be polyethylene (PE) or polypropylene (PP) products, which are less dense than water. Note how colours have faded, the surface has become opaque, and the product has started to crack.

## 7.1.3.5 Viscoelasticity of a foam bed

Acquire some 'slow recovery' foam such as *Confor* (samples are often given away by bedding showrooms). Compress the surface with one hand for a minute, and then observe how long it takes for the indentations in the foam to disappear. Repeat the exercise after the foam has been placed in a refrigerator (5°C) when it will be much stiffer, or after it has been placed in an oven at 60°C

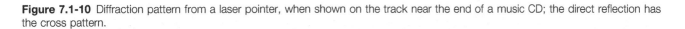

**Figure 7.1-10** Diffraction pattern from a laser pointer, when shown on the track near the end of a music CD; the direct reflection has the cross pattern.

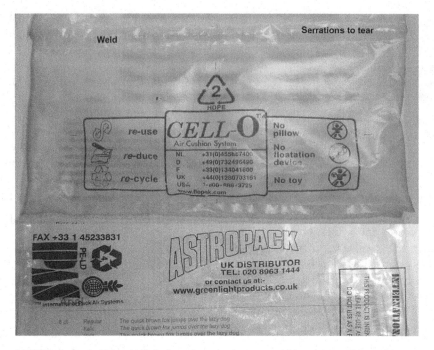

**Figure 7.1-11** LDPE and HDPE macro-bubbles (deflated) on top of a test page with a range of font sizes.

(when it will be much less stiff and will recover quickly). This shows that the strong viscoelastic response only occurs in a temperature range where the polymer is leathery; close to its glass transition temperature.

## 7.1.4 Identifying plastics

Make a collection of food packaging: a milk bottle, a carbonated drink bottle, a supermarket carrier bag, a near-transparent lidded container for food, a margarine container.

Note how plastic bottles have replaced glass for soft drinks, milk, ketchup, etc. One-trip plastic bottles are essential for the sales of bottled water, while they have replaced metal cans for many products. Even containers apparently made from paper (such as Tetrapak) rely on an inner polyethylene layer to protect the paper from the liquid contents.

**Table 7.1-1** Recycling marks for polymers

| No. | Legend | Polymer |
| --- | --- | --- |
| 1 | PET | Polyethylene terephthalate |
| 2 | HDPE | High-density polyethylene |
| 3 | PVC | Polyvinyl chloride |
| 4 | LDPE | Low-density polyethylene |
| 5 | PP | Polypropylene |
| 6 | PS | Polystyrene |

Use the methods below to identify which plastics are used in one or more products.

### 7.1.4.1 Recycling marks

Recycling marks on products (Fig. 7.1-11) allow the common plastics to be identified (Table 7.1-1). Sometimes numbers are used in place of the abbreviation for the polymer name.

### 7.1.4.2 Product appearance, if unpigmented

*Translucent* products are semi-crystalline, e.g. PE. Some thin (<1mm) or highly oriented products appear transparent, in spite of being semi-crystalline (e.g. PET bottles), since the crystals are too small to scatter light. Some thicker PP products appear translucent, but thin mouldings, especially if the PP is nucleated, will appear nearly transparent.

*Transparent* mouldings thicker than 1 mm will be one of the glassy polymers (PVC, PS, PC, etc.). If a thin film of molten, unpigmented plastic is opaque to light it is likely to be filled.

### 7.1.4.3 Density

An electronic densitometer can measure the density of small (<10g) pieces, using Archimedes principle. The pieces are first weighed in air, then again, while suspended

Introduction to plastics  CHAPTER 7.1

**Table 7.1-2** Polymer densities and transition temperatures

| Abbreviations | Polymer | Density (kgm$^{-3}$) | $T_g$ (°C) | $T_m$ (°C) | Event if bent through 90° |
|---|---|---|---|---|---|
| **Semi-crystalline plastics** | | | | | |
| P4MP | Poly (4-methyl-pentene-1) | 830 | 25 | 238 | Semi-brittle |
| PP | Polypropylene | 900–910 | −10 | 170 | Whitens |
| LDPE | Low-density polyethylene | 920–925 | −120 | 120 | Ductile |
| MDPE | Medium density polyethylene | 935–945 | −120 | 130 | Ductile |
| HDPE | High-density polyethylene | 955–965 | −120 | 140 | Ductile |
| PA 6 | Polyamide 6 | 1120–1150 | 50 | 228 | Ductile |
| PA 66 | Polyamide 6,6 | 1130–1160 | 57 | 265 | Ductile |
| PET | Polyethylene terephthalate | 1336–1340 | 80 | 260 | Ductile |
| POM | Polyoxymethylene (Acetal) | 1410 | −85 | 170 | Semi-brittle |
| PVDC | Polyvinylidene chloride | 1750 | −18 | 205 | |
| PTFE | Polytetrafluoro ethylene | 2200 | −73 | 332 | Ductile |
| **Glassy plastics** | | | | | |
| PS | Polystyrene | 1050 | 100 | | Brittle |
| SAN | Styrene acrylonitrile copolymer | 1080 | 100 | | |
| ABS | Acrylonitrile butadiene styrene copolymer | 990–1100 | 100 | | Whitens |
| PC | Polycarbonate | 1200 | 145 | | Ductile |
| PVCu | Polyvinyl chloride unplasticised | 1410 | 80 | | Ductile |
| PMMA | Polymethyl methacrylate | 1190 | 105 | | Brittle |

$T_m$, crystal melting temperature; $T_g$, glass transition temperature.

in water. Table 7.1-2 gives the densities and melting points of the main polymers. They are arranged in classes, in order of increasing density. The density of semi-crystalline plastics increases with crystallinity, so a range is given. If a significant amount of a reinforcing or toughening material is added, the density changes, making it more difficult to identify the polymer.

### 7.1.4.4 Melting temperatures

Table 7.1-2 shows the temperature $T_m$ at which the crystalline phase melts, or, for non-crystalline polymers, the glass transition temperature $T_g$ at which the glass changes into a melt. Samples can be dragged across the surface of metal hotplates, set to a range of temperatures. However, when the polymer is just above $T_m$, some polymers leave a streak of melt, while others of higher viscosity just deform. Therefore, transition temperatures can be overestimated.

### 7.1.4.5 Young's modulus

Estimate the order of magnitude of the Young's modulus of a flat part of the product by flexing it. This works best if a standard sized (say 100 mm long, 20 mm wide, 2 mm thick) beam is cut from the product and loaded in three-point bending, since the bending stiffness varies with the cube of the thickness. LDPE is of a much lower Young's modulus (c. 100 MPa) than most other plastics (1-3 GPa), and the surface can be marked with a finger nail.

## 7.1.5 Product features related to processing

The aim is to recognise design features associated with processes. Both the product shape and surface marks provide clues for process identification.

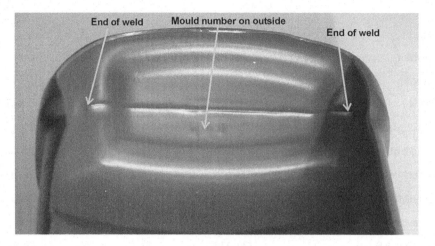

**Figure 7.1-12** Sectioned blow-moulded bottle, showing the weld line at the base.

## 7.1.5.1 Blow mouldings

These are hollow containers, usually with an opening of smaller diameter than the body. Both ends of the moulding may be cut off to produce a tubular product, or one end cut off for a bucket-shaped container. The wall thickness varies with position, and there is a weld line across the closed end of the container (Fig. 7.1-12). Sometimes near-parallel lines are visible on the inner surface. These indicate the extrusion direction when the parison (tubular preform) emerged from a die. Look for the weld location on the base of an HDPE milk bottle; this aligns with the external surface line from the mould split. Section the milk bottle vertically, in a plane perpendicular to the weld line, and measure the thickness at the weld line compared with elsewhere. Note the threads in the neck are corrugated.

Stiff HDPE tool boxes can be created by allowing the two sides of the blow moulding to come into contact at some locations and form welds (Fig. 7.1-13). The 0.4 mm thick hinge region is created by pressing the HDPE with metal bars, and the 'click shut' catches are also part of the blow moulding.

## 7.1.5.2 Extruded products

These have a constant cross section. Examples are domestic gutters or down-pipe, or (replacement) window frames, made from PVC. Look through a length of an extrusion, towards a window, for markings parallel to the extrusion direction, which have come from the die. The outer surface is in contact with a sizing die, whereas the inner surface cools in air and can change shape slightly. The pipe wall provides the bending stiffness and resistance against weathering. Pipes for cable TV in the UK are green with a corrugated exterior, but a smooth inner wall. Figure 7.1-14 shows

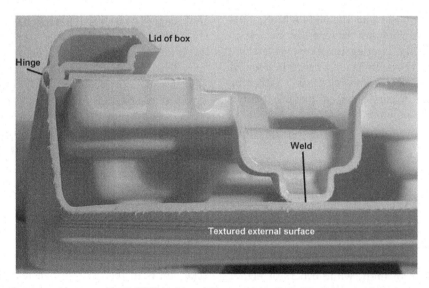

**Figure 7.1-13** Section through a blow-moulded HDPE tool box (45 mm thick). Both lid and base of the box are hollow, with reinforcing welds at intervals interior.

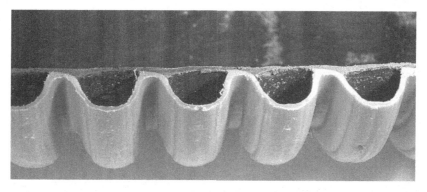

**Figure 7.1-14** Extruded HDPE pipe, with corrugations at 18 mm intervals in the outer layer, for buried electric cables.

a pipe for electrical cable, with an outer red corrugated layer bonded to a smooth inner black layer. Such pipes offer maximum resistance to crushing by soil loads for a given weight of polymer.

### 7.1.5.3 Injection mouldings

Injection mouldings can contain T-junctions (where *ribs* meet a surface) and holes. Figures 7.1-1 - 7.1-7 show injection moulded parts. The point where the *sprue*, which feeds the melt into the mould cavity, has been removed should be visible as a slightly rough, often circular region. On the concave side of the product, circular surface marks indicate the location of *ejector pins*, which push the cold moulding from the mould. Figure 7.1-15 of a moulded box, shows the ejector pin marks on the inside of the box, and a moulded-in hinge between the two halves.

Consider the polypropylene seat of a stackable chair. The seat sides are 'bent over', providing a place to grip the seat. These sides provide bending stiffness to the seat. You can prove this if you can cut off the side parts; if you lean back in the chair, it flexes excessively at the back/seat junction. The seat surface has a moulded-in texture (Fig. 7.1-5), to increase the coefficient of friction with your clothes, and to disguise scratches. Note how dust build-up and scratches spoil the appearance of the hidden side, which is smooth.

### 7.1.5.4 Thermoformed products

These tend to be curved panels, or shallow containers. They have a variable thickness, since only the convex side contacts a metal die. They can be as thin as 0.1 mm, since a sheet of melt is stretched before contact with the cold mould, or as thick as 10 mm. There will be no signs of any injection point or ejector pins. Typical examples are disposable coffee cups (Fig. 7.1-16a), margarine containers, baths and shower trays.

Use a sharp pair of scissors to cut a section through the cup, and note the corrugations which increase the bending stiffness of the wall. The corrugations also provide grip and reduce heat transfer to the fingers. Note the variation in wall thickness.

Either use a hot air blower (for paint stripping), or put the cup on a layer of aluminium foil in an oven at 120°C and note the gradual shape reversion to a nearly-flat sheet (Fig. 7.1-16b). The thermoforming process involved the elastic stretching of a sheet of polymer melt, and this orientation was frozen into the cup when it cooled. On

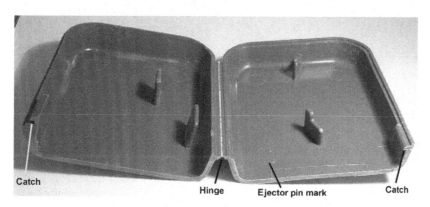

**Figure 7.1-15** Section through an injection-moulded PP box for a micrometer. The moulded-in hinge has whitened in use.

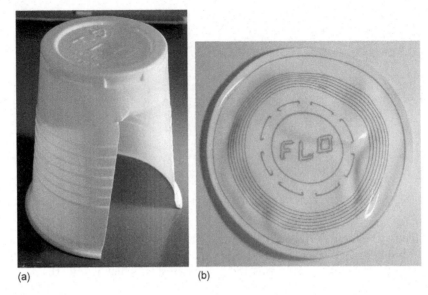

(a) (b)

**Figure 7.1-16** Section through a thermoformed PS disposable coffee cup, with shallow corrugations in the 0.2 mm thick sidewall. The corrugations were outlined in felt-tip then the cup heat reverted to 0.8 mm thick sheet.

reheating, the plastic attempts to return to its original shape.

### 7.1.5.5 Blown film

The blown film process creates a continuous tube of film, usually less than 0.25 mm thick, which is flattened and rolled up. It can be cut into lengths and welded to produce products such as supermarket carrier bags or protective bubbles (Fig. 7.1-11). For a carrier bag, determine where the film has been welded and where it has been cut, or folded. Support the handles on a spring balance, then use sand (or tins) as the loading medium and determine the tensile strength of the polyethylene in the handle region.

### 7.1.5.6 Injection-blow moulded bottles

Compare a PET carbonated drink bottle with an HDPE milk bottle. The moulded neck threads of the PET bottle (Fig. 7.1-17) have T-sections. The internal pressure of 4 bar in the carbonated drink bottle can only be resisted by a lightweight bottle if the polymer is oriented to increase its strength. However, the milk bottle is under no internal pressure, so a lower cost material (HDPE) and process can be used. Note the location of the injection

**Figure 7.1-17** Details of the neck region of an injection-blow moulded PET bottle. The bottle on the left has been heat treated at 120 $^\circ$C, while the neck of a preform is shown on the right.

sprue in the centre of the base of the PET bottle. Try placing an empty PET bottle in an oven at 120°C for 1 h. Note how it shrinks both in length and diameter, showing that the PET had been biaxially oriented. The base and the neck become milky in appearance, due to the crystallisation of these initially amorphous regions. The shrinkage in these regions is relatively low. The main part of the bottle remains clear, since it was already semi-crystalline.

## 7.1.6 Summary

Hopefully you are now familiar with the appearance and some typical properties of the commodity thermoplastics, and can recognise how some products have been made. You are now ready to study the microstructure and processing of polymers in more detail, and to find out how the properties can be related to the microstructure.

# Chapter 7.2

# General properties of plastics

## 7.2.1 Introduction

It would be difficult to imagine our modern world without plastics. Today they are an integral part of everyone's lifestyle with applications varying from commonplace domestic articles to sophisticated scientific and medical instruments. Nowadays designers and engineers readily turn to plastics because they offer combinations of properties not available in any other materials. Plastics offer advantages such as lightness, resilience, resistance to corrosion, colour fastness, transparency, ease of processing, etc., and although they have their limitations, their exploitation is limited only by the ingenuity of the designer.

The term *plastic* refers to a family of materials which includes nylon, polyethylene and PTFE just as zinc, aluminium and steel fall within the family of *metals*. This is an important point because just as it is accepted that zinc has quite different properties from steel, similarly nylon has quite different properties from PTFE. Few designers would simply specify *metal* as the material for a particular component so it would be equally unsatisfactory just to recommend *plastic*. This analogy can be taken still further because in the same way that there are different grades of steel there are also different grades of, say, polypropylene. In both cases the good designer will recognise this and select the most appropriate material and grade on the basis of processability, toughness, chemical resistance, etc.

It is usual to think that plastics are a relatively recent development but in fact, as part of the larger family called *polymers*, they are a basic ingredient of animal and plant life. Polymers are different from metals in the sense that their structure consists of very long chain-like molecules. Natural materials such as silk, shellac, bitumen, rubber and cellulose have this type of structure. However, it was not until the 19th century that attempts were made to develop a synthetic polymeric material and the first success was based on cellulose. This was a material called *Parkesine*, after its inventor Alexander Parkes, and although it was not a commercial success it was a start and it led to the development of *Celluloid*. This material was an important breakthrough because it became established as a good replacement for natural materials which were in short supply – for example, ivory for billiard balls.

During the early 20th century there was considerable interest in these new synthetic materials. Phenol-formaldehyde (*Bakelite*) was introduced in 1909, and at about the time of the Second World War materials such as nylon, polyethylene and acrylic (*Perspex*) appeared on the scene. Unfortunately many of the early applications for plastics earned them a reputation as being cheap substitutes. It has taken them a long time to overcome this image but nowadays the special properties of plastics are being appreciated, which is establishing them as important materials in their own right. The ever increasing use of plastics in all kinds of applications means that it is essential for designers and engineers to become familiar with the range of plastics available and the types of performance characteristics to be expected so that these can be used to the best advantage.

This chapter is written as a general introduction to design with plastics. It outlines the range of plastics available, describes the type of behaviour which they exhibit and illustrates the design process involved in selecting the best plastic for a particular application.

*Engineering Materials and Processes Desk Reference*; ISBN: 9781856175869
Copyright © 2009 Elsevier Ltd; All rights of reproduction, in any form, reserved.

## 7.2.2 Polymeric materials

Synthetic large molecules are made by joining together thousands of small molecular units known as **monomers**. The process of joining the molecules is called **polymerisation** and the number of these units in the long molecule is known as the **degree of polymerisation**. The names of many polymers consist of the name of the monomer with the suffix **poly-**. For example, the polymers polypropylene and polystyrene are produced from propylene and styrene respectively.

It is an unfortunate fact that many students and indeed design engineers are reluctant to get involved with plastics because they have an image of complicated materials with structures described by complex chemical formulae. In fact it is not necessary to have a detailed knowledge of the structure of plastics in order to make good use of them. Perfectly acceptable designs are achieved provided one is familiar with their performance characteristics in relation to the proposed service conditions. An awareness of the structure of plastics can assist in understanding why they exhibit a time-dependent response to an applied force, why acrylic is transparent and stiff whereas polyethylene is opaque and flexible, etc., but it is not necessary for one to be an expert in polymer chemistry in order to use plastics.

The words **polymers** and **plastics** are often taken as synonymous but in fact there is a distinction. The polymer is the pure material which results from the process of polymerisation and is usually taken as the family name for materials which have long chain-like molecules (and this includes rubbers). Pure polymers are seldom used on their own and it is when additives are present that the term plastic is applied. Polymers contain additives for a number of reasons. The following list outlines the purpose of the main additives used in plastics.

**Antistatic agents.** Most polymers, because they are poor conductors of current, build up a charge of static electricity. Antistatic agents attract moisture from the air to the plastic surface, improving its surface conductivity and reducing the likelihood of a spark or a discharge.

**Coupling agents.** Coupling agents are added to improve the bonding of the plastic to inorganic filler materials, such as glass fibres. A variety of silanes and titanates are used for this purpose.

**Fillers.** Some fillers, such as short fibres or flakes of inorganic materials, improve the mechanical properties of a plastic. Others, called *extenders*, permit a large volume of a plastic to be produced with relatively little actual resin. Calcium carbonate, silica and clay are frequently used extenders.

**Flame retardants.** Most polymers, because they are organic materials, are flammable. Additives that contain chlorine, bromine, phosphorous or metallic salts reduce the likelihood that combustion will occur or spread.

**Lubricants.** Lubricants such as wax or calcium stearate reduce the viscosity of the molten plastic and improve forming characteristics.

**Pigments.** Pigments are used to produce colours in plastics.

**Plasticisers.** Plasticisers are low molecular weight materials which alter the properties and forming characteristics of the plastic. An important example is the production of flexible grades of polyvinyl chloride by the use of plasticisers.

**Reinforcement.** The strength and stiffness of polymers are improved by adding fibres of glass, carbon, etc.

**Stabilisers.** Stabilisers prevent deterioration of the polymer due to environmental factors. Antioxidants are added to ABS, polyethylene and polystyrene. Heat stabilisers are required in processing polyvinyl chloride. Stabilisers also prevent deterioration due to ultra-violet radiation.

There are two important classes of plastics.

### (a) Thermoplastic materials

In a thermoplastic material the very long chain-like molecules are held together by relatively weak Van der Waals forces. A useful image of the structure is a mass of randomly distributed long strands of sticky wool. When the material is heated the intermolecular forces are weakened so that it becomes soft and flexible and eventually, at high temperatures, it is a viscous melt. When the material is allowed to cool it solidifies again. This cycle of softening by heat and solidifying on cooling can be repeated more or less indefinitely and is a major advantage in that it is the basis of most processing methods for these materials. It does have its drawbacks, however, because it means that the properties of thermoplastics are heat sensitive. A useful analogy which is often used to describe these materials is that, like candle wax, they can be repeatedly softened by heat and will solidify when cooled.

Examples of thermoplastics are polyethylene, polyvinyl chloride, polystyrene, nylon, cellulose acetate, acetal, polycarbonate, polymethyl methacrylate and polypropylene.

An important subdivision within the thermoplastic group of materials is related to whether they have a **crystalline** (ordered) or an **amorphous** (random) structure. In practice, of course, it is not possible for a moulded plastic to have a completely crystalline structure due to the complex physical nature of the molecular chains. Some plastics, such as polyethylene and nylon, can achieve a high degree of crystallinity but they are probably more accurately described as *partially crystalline* or *semi-crystalline*. Other plastics such as acrylic and polystyrene are always amorphous. The presence of crystallinity in those plastics capable of crystallising is very dependent on their thermal history and hence on the processing conditions used to produce the moulded article. In turn, the mechanical properties of the moulding are very sensitive to whether or not the plastic possesses crystallinity.

In general, plastics have a higher density when they crystallise due to the closer packing of the molecules. Typical characteristics of crystalline and amorphous plastics are shown below.

| Amorphous | Crystalline |
|---|---|
| • *Broad softening range* – thermal agitation of the molecules breaks down the weak secondary bonds. The rate at which this occurs throughout the formless structure varies producing broad temperature range for softening. | • *Sharp melting point* – the regular close-picked structure results in most of the secondary bonds being broken down at the same time. |
| • *Usually transparent* – the looser structure transmits light so the material appears transparent. | • *Usually opaque* – the difference in refractive indices between the two phases (amorphous and crystalline) causes interference so the material appears translucent or opaque. |
| • *Low shrinkage* – all thermoplastics are processed in the amorphous state. On solidification, the random arrangement of molecules produces little volume change and hence low shrinkage. | • *High shrinkage* – as the material solidifies from the amorphous state the polymers take up a closely packed, highly aligned structure. This produces a significant volume change manifested as high shrinkage. |
| • *Low chemical resistance* – the more open random structure enables chemicals to penetrate deep into the material and to destroy many of the secondary bonds. | • *High chemical resistance* – the tightly packed structure prevents chemical attack deep within the material. |
| • *Poor fatigue and wear resistance* – the random structure contributes little to fatigue or wear properties. | • *Good fatigue and wear resistance* – the uniform structure is responsible for good fatigue and wear properties. |

| Examples of amorphous and crystalline thermoplastics ||
|---|---|
| **Amorphous** | **Crystalline** |
| Polyvinyl Chloride (PVC) | Polyethylene (PE) |
| Polystyrene (PS) | Polypropylene (PP) |
| Polycarbonate (PC) | Polyamide (PA) |
| Acrylic (PMMA) | Acetal (POM) |
| Acrylonitrile-butadiene-styrene (ABS) | Polyester (PEW, PBTF') |
| Polyphenylene (PPO) | Fluorocarbons (PTFE, PFA, FEP and ETFE) |

## (b) Thermosetting Plastics

A thermosetting plastic is produced by a chemical reaction which has two stages. The first stage results in the formation of long chain-like molecules similar to those present in thermoplastics, but still capable of further reaction. The second stage of the reaction (**cross-linking** of chains) takes place during moulding, usually under the application of heat and pressure. The resultant moulding will be rigid when cooled but a close network structure has been set up within the material. During the second stage the long molecular chains have been interlinked by strong bonds so that the material cannot be softened again by the application of heat. If excess heat is applied to these materials they will char and degrade. This type of behaviour is analogous to boiling an egg. Once the egg has cooled and is hard, it cannot be softened again by the application of heat.

Since the cross-linking of molecules is by strong chemical bonds, thermosetting materials are characteristically quite rigid materials and their mechanical properties are not heat sensitive. Examples of thermosets are phenol formaldehyde, melamine formaldehyde, urea formaldehyde, epoxies and some polyesters.

## 7.2.3 Plastics available to the designer

Plastics, more than any other design material, offer such a wide spectrum of properties that they must be given serious consideration in most component designs. However, this does not mean that there is sure to be a plastic with the correct combination of properties for every application. It simply means that the designer must have an awareness of the properties of the range of plastics available and keep an open mind. One of the most common faults in design is to be guided by pre-conceived notions. For example, an initial commitment to plastics based on an irrational approach is itself a serious design fault. A good design always involves a judicious selection of a material from the whole range available, including non-plastics. Generally, in fact, it is only against a background of what other materials have to offer that the full advantages of plastics can be realised.

In the following sections most of the common plastics will be described briefly to give an idea of their range of properties and applications. However, before going on to this it is worthwhile considering briefly several of the special categories into which plastics are divided.

### 7.2.3.1 Engineering plastics

Many thermoplastics are now accepted as engineering materials and some are distinguished by the loose description **engineering plastics.** The term probably originated as a classification distinguishing those that could be substituted satisfactorily for metals such as aluminium in small devices and structures from those with inadequate mechanical properties. This demarcation is clearly artificial because the properties on which it is based are very sensitive to the ambient temperature, so that a thermoplastic might be a satisfactory substitute for a metal at a particular temperature and an unsatisfactory substitute at a different one.

A useful definition of an engineering material is that it is able to support loads more or less indefinitely. By such a criterion thermoplastics are at a disadvantage compared with metals because they have low time-dependent moduli and inferior strengths except in rather special circumstances. However, these rather important disadvantages are off-set by advantages such as low density, resistance to many of the liquids that corrode metals and above all, easy processability. Thus, where plastics compete successfully with other materials in engineering applications it is usually because of a favourable balance of properties rather than because of an outstanding superiority in some particular respect, although the relative ease with which they can be formed into complex shapes tends to be a particularly dominant factor. In addition to conferring the possibility of low production costs, this ease of processing permits imaginative designs that often enable plastics to be used as a superior alternative to metals rather than merely as a tolerated substitute.

Currently the materials generally regarded as making up the **engineering plastics** group are Nylon, acetal, polycarbonate, modified polyphenylene oxide (PPO), thermoplastic polyesters, polysulphone and polyphenylene sulphide. The newer grades of polypropylene also possess good basic *engineering* performance and this would add a further 0.5 m tonnes. And then there is unplasticised polyvinyl chloride (uPVC) which is widely used in industrial pipework and even polyethylene, when used as an artificial hip joint for example, can come into the reckoning. Hence it is probably unwise to exclude any plastic from consideration as an engineering material even though there is a sub-group specifically entitled for this area of application.

In recent years a whole new generation of high performance engineering plastics have become commercially available. These offer properties far superior to anything available so far, particularly in regard to high temperature performance, and they open the door to completely new types of application for plastics.

The main classes of these new materials are

**(i) Polyarylethers and polyarylthioethers**

polyarylethersulphones (PES)
polyphenylene sulphide (PPS)
polyethernitrile (PEN)
polyetherketones (PEK and PEEK)

**(ii) Polyimides and polybenzimidazole**

polyetherimide (PEI)
thermoplastic polyimide (PI)
polyamideimide (PAI)

**(iii) Fluropolymers**

fluorinated ethylene propylene (FEP)
perfluoroalkoxy (PFA)

A number of these materials offer service temperatures in excess of 200°C and fibre-filled grades can be used above 300°C.

### 7.2.3.2 Thermosets

In recent years there has been some concern in the thermosetting material industry that usage of these materials is on the decline. Certainly the total market for thermoset compounds has decreased in Western Europe. This has happened for a number of reasons. One is the image that thermosets tend to have as old-fashioned materials with outdated, slow production methods. Other reasons include the arrival of high temperature engineering plastics and miniaturisation in the electronics industry. However, thermosets are now fighting back and have a very much improved image as colourful, easy-flow moulding materials with a superb range of properties.

Phenolic moulding materials, together with the subsequently developed easy-flowing, granular thermosetting materials based on urea, melamine, unsaturated polyester (UP) and epoxide resins, today provide the backbone of numerous technical applications on account of their non-melting, high thermal and chemical resistance, stiffness, surface hardness, dimensional stability and low flammability. In many cases, the combination of properties offered by thermosets cannot be matched by competing engineering thermoplastics such as polyamides, polycarbonates, PPO, PET, PBT or acetal, nor by the considerably more expensive products such as polysulphone, polyethersulphone and PEEK.

### 7.2.3.3 Composites

One of the key factors which make plastics attractive for engineering applications is the possibility of property enhancement through fibre reinforcement. Composites produced in this way have enabled plastics to become acceptable in, for example, the demanding aerospace and automobile industries. Currently in the USA these industries utilise over 100,000 tonnes of reinforced plastics out of a total consumption of over one million tonnes.

Both thermoplastics and thermosets can reap the benefit of fibre reinforcement although they have developed in separate market sectors. This situation has arisen due to fundamental differences in the nature of the two classes of materials, both in terms of properties and processing characteristics.

Thermosetting systems, hampered on the one hand by brittleness of the crosslinked matrix, have turned to the use of long, indeed often continuous, fibre reinforcement but have on the other hand been able to use the low viscosity state at impregnation to promote maximum utilization of fibre properties. Such materials have found wide application in large area, relatively low productivity, moulding. On the other hand, the thermoplastic approach with the advantage of toughness, but unable to grasp the benefit of increased fibre length, has concentrated on the short fibre, high productivity moulding industry. It is now apparent that these two approaches are seeking routes to move into each other's territory. On the one hand the traditionally long-fibre based thermoset products are accepting a reduction in properties through reduced fibre length, in order to move into high productivity injection moulding, while thermoplastics, seeking even further advances in properties, by increasing fibre length, have moved into long-fibre injection moulding compounds and finally into truly structural plastics with continuous, aligned fibre thermoplastic composites such as the advanced polymer composite **(APC)** developed by IC1 and the stampable glass mat reinforced thermoplastics **(GMT)** developed in the USA.

Glass fibres are the principal form of reinforcement used for plastics because they offer a good combination of strength, stiffness and price. Improved strengths and stiffnesses can be achieved with other fibres such as aramid **(Kevlar)** or carbon fibres but these are expensive. The latest developments also include the use of hybrid systems to get a good balance of properties at an acceptable price. For example, the impact properties of carbon-fibre composites can be improved by the addition of glass fibres and the stiffness of gfrp can be increased by the addition of carbon fibres.

Another recent development is the availability of reinforced plastics in a form very convenient for moulding. One example is polyester dough and sheet moulding compounds (DMC and SMC respectively). DMC, as the name suggests, has a dough-like consistency and consists of short glass fibres (15–20%) and fillers (up to 40%) in a polyester resin. The specific gravity is in the range 1.7–2.1. SMC consists of a polyester resin impregnated with glass fibres (20%–30%). It is supplied as a sheet wound into a roll with a protective polythene film on each side of the sheet. The specific gravity is similar to that of DMC and both materials are usually formed using heat and pressure in a closed mould.

### 7.2.3.4 Structural foam

The concept of structural foams offers an unusual but exciting opportunity for designers. Many plastics can be foamed by the introduction of a blowing agent so that when moulded the material structure consists of a cellular rigid foam core with a solid tough skin. This type of structure is of course very efficient in material terms and offers an excellent strength-to-weight ratio.

The foam effect is achieved by the dispersion of inert gas throughout the molten resin directly before moulding. Introduction of the gas is usually carried out by pre-blending the resin with a chemical blowing agent which releases gas when heated, or by direct injection of the gas (usually nitrogen). When the compressed gas/resin mixture is rapidly injected into the mould cavity, the gas expands explosively and forces the material into all parts of the mould. An internal cellular structure is thus formed within a solid skin.

Polycarbonate, polypropylene and modified PPO are popular materials for structural foam moulding. One of the main application areas is housings for business equipment and domestic appliances because the number of component parts can be kept to the absolute minimum due to integral moulding of wall panels, support brackets, etc. Other components include vehicle body panels and furniture.

Structural foam mouldings may also include fibres to enhance further the mechanical properties of the material. Typical performance data for foamed polypropylene relative to other materials is given in Table 7.2-1.

### 7.2.3.5 Elastomers

Conventional rubbers are members of the polymer family in that they consist of long chain-like molecules. These chains are coiled and twisted in a random manner and have sufficient flexibility to allow the material to undergo very large deformations. In the *green* state the rubber would not be able to recover fully from large deformations because the molecules would have undergone irreversible sliding past one another. In order to prevent this sliding, the molecules are anchored together by a curing **(vulcanisation)** process. Thus the molecules are cross-linked in a way similar to that which occurs in thermosets. This linking does not detract from the random disposition of the molecules nor their coiled and twisted nature so that when the rubber is deformed the molecules stretch and unwind but do not slide. Thus when the applied force is removed the rubber will snap back to its original shape.

Vulcanised rubbers possess a range of very desirable properties such as resilience, resistance to oils, greases and ozone, flexibility at low temperatures and resistance to many acids and bases. However, they require careful (slow) processing and they consume considerable amounts of energy to facilitate moulding and vulcanisation. These disadvantages led to the development of **thermoplastic rubbers (elastomers)**. These are materials which exhibit the desirable physical characteristics of rubber but with the ease of processing of thermoplastics.

**Table 7.2-1** Comparison of structural foams based on various grades of polypropylene with some traditional materials

| | Unfilled copolymer | | 40% talc-filled homopolymer | | 30% coupled glass-reinforced | | Chipboard | Pine | Aluminium | Mild steel |
|---|---|---|---|---|---|---|---|---|---|---|
| | Solid | Foam | Solid | Foam | Solid | Foam | | | | |
| Flexural modulus MN/m$^2$ | 1.4 | 1.2 | 4.4 | 2.5 | 6.7 | 3.5 | 2.3 | 7.9 | 70 | 207 |
| Specific gravity | 0.905 | 0.72 | 1.24 | 1.00 | 1.12 | 0.90 | 0.650 | 0.641 | 2.7 | 7.83 |
| Relative thickness at equivalent rigidity | 1 | 1.05 | 0.68 | 0.81 | 0.59 | 0.74 | 0.85 | 0.56 | 0.27 | 0.19 |
| Relative weight at equivalent rigidity | 1 | 0.84 | 0.94 | 0.90 | 0.74 | 0.73 | 0.61 | 0.40 | 0.81 | 1.65 |

At present there are five types of thermoplastic rubber (TPR). Three of these, the polyurethane, the styrenic and the polyester are termed segmented block copolymers in that they consist of thermoplastic molecules grafted to the rubbery molecules. At room temperature it is the thermoplastic molecules which clump together to anchor the rubbery molecules. When heat is applied the thermoplastic molecules are capable of movement so that the material may be shaped using conventional thermoplastic moulding equipment.

The olefinic type of TPR is the latest development and is different in that it consists of fine rubber particles in a thermoplastic matrix as shown in Fig. 7.2-1. The matrix is usually polypropylene and it is this which melts during processing to permit shaping of the material. The rubber filler particles then contribute the flexibility and resilience to the material. The other type of TPR is the polyamide and the properties of all five types are summarised in Table 7.2-4.

### 7.2.3.6 Polymer alloys

The development of new polymer alloys has caused a lot of excitement in recent years but in fact the concept has been around for a long time. Indeed one of the major commercial successes of today, ABS, is in fact an alloy of acrylonitrile, butadiene and styrene. The principle of alloying plastics is similar to that of alloying metals – to achieve in one material the advantages possessed by several others. The recent increased interest and activity in the field of polymer alloys has occurred as a result of several new factors. One is the development of more sophisticated techniques for combining plastics which were previously considered to be incompatible. Another is the keen competition for a share of new market areas such as automobile bumpers, body panels etc. These applications call for combinations of properties not previously available in a single plastic and it has been found that it is less expensive to combine existing plastics than to develop a new monomer on which to base the new plastic.

In designing an alloy, polymer chemists choose candidate resins according to the properties, cost, and/or processing characteristics required in the end product. Next, compatibility of the constituents is studied, tested, and either optimised or accommodated.

Certain polymers have come to be considered standard building blocks of the polyblends. For example, impact strength may be improved by using polycarbonate, ABS and polyurethanes. Heat resistance is improved by using polyphenylene oxide, polysulphone, PVC, polyester (PET and PBT) and acrylic. Barrier properties are improved by using plastics such as ethylene vinyl alchol (EVA). Some modern plastic alloys and their main characteristics are given in Table 7.2-2.

### 7.2.3.7 Liquid crystal polymers

Liquid crystal polymers (LCP) are a recent arrival on the plastics materials scene. They have outstanding dimensional stability, high strength, stiffness, toughness and chemical resistance all combined with ease of processing. LCPs are based on thermoplastic aromatic

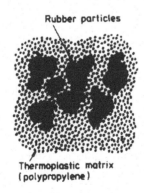

**Fig. 7.2-1** Typical structure of olefinic TPR.

## General properties of plastics — CHAPTER 7.2

| Table 7.2-2 Typical plastic alloys | |
|---|---|
| **Alloy** | **Features** |
| PVC/acrylic | Tough with good flame and chemical resistance |
| PVC/ABS | Easily processed with good impact and flame resistance |
| Polycarbonate/ABS | Hard with high heat distortion temperature and good notch impact strength |
| ABS/Polysulphone | Less expensive than unmodified polysulphone |
| Polyphenylene oxide/HIPS | Improved processability, reduced cost |
| SAN/olefin | Good weatherability |
| Nylon/elastomer | Improved notched impact strength |
| Modified amorphous nylon | Easily processed with excellent surface finish and toughness |
| Polycarbonate/PBT | Tough engineering plastic |

polyesters and they have a highly ordered structure even in the molten state. When these materials are subjected to stress the molecular chains slide over one another but the ordered structure is retained. It is the retention of the highly crystalline structure which imparts the exceptional properties to LCPs.

## Typical characteristics of some important plastics

### (a) Semi-crystalline plastics

#### Low density polyethylene (LDPE)
This is one of the most widely used plastics. It is characterised by a density in the range 918–935 kg/m$^3$ and is very tough and flexible. Its major application is in packaging film although its outstanding dielectric properties means it is also widely used as an electrical insulator. Other applications include domestic ware, tubing, squeeze bottles and cold water tanks.

#### Linear low density polyethylene (LLDPE)
This new type of polyethylene was introduced in 1977. LLDPE is produced by a low pressure process and it has a regular structure with short chain branches. Depending on the cooling rate from the melt, the material forms a structure in which the molecules are linked together. Hence for any given density, LLDPE is stiffer than LDPE and exhibits a higher yield strength and greater ductility. Although the different melt processing characteristics of LLDPE take a little getting used to, this new material has taken over traditional LDPE markets.

#### High density polyethylene (HDPE)
This material has a density in the range 935–965 kg/m$^3$ and is more crystalline than LDPE. It is also slightly more expensive but as it is much stronger and stiffer it finds numerous applications in such things as dustbins, bottle crates, general purpose fluid containers and pipes.

One of the most exciting recent developments in this sector has been the introduction to the marketplace of metallocene-based polyethylenes. Metallocenes have been recognised as suitable catalysts for the manufacture of polyethylenes since the 1950s. However, it is only recently that their use has been perfected. Their big advantage is that they are single site catalysts so that the polymer molecules which are produced tend to be all the same – a fact which offers an array of superior properties. Traditional catalysts for polyethylene (Ziegler Natta catalysts) are multi-sited so that they produce polymers with short, medium and long molecules. In the new metallocene grades of polyethylene, the absence of low molecular weight species results in low extractables, a narrow melting range and free-flowing material even at low densities. The absence of high molecular weight species contributes excellent melting point control, clarity and improved flexibility/toughness at low temperatures.

Metallocene-based polyethylene does not offer the lower production costs associated with LLDPE. Hence there will be a price premium for the new materials but this is felt to be justified in view of their improved property profile.

#### Cross-linked polyethylene (XLPE)
Some thermoplastic materials such as polyethylene can have their structure altered so that the molecular chains become cross-linked and the material then behaves like a thermoset. In the case of polyethylene, a range of cross-linking methods are available. These include the use of radiation, peroxides and silanes. In some cases the cross-linking can occur during moulding whereas in other cases the end-product shape is created before the cross-linking is initiated. The action of cross-linking has a number of beneficial effects including improved stress crack resistance, improved creep resistance, better chemical resistance, improved toughness and better general thermomechanical stability.

#### Polypropylene (PP)
Polypropylene is an extremely versatile plastic and is available in many grades and also as a copolymer (ethylene/propylene). It has the lowest density of all thermoplastics (in the order of 900 kg/m$^3$ and this combined with strength, stiffness and excellent fatigue and chemical resistance make it attractive in many situations. These include crates, small machine parts, car components (fans, fascia panels etc), chair shells, cabinets for

TV, tool handles, etc. Its excellent fatigue resistance is utilised in the moulding of integral hinges (e.g. accelerator pedals and forceps/tweezers). Polypropylene is also available in fibre form (for ropes, carpet backing) and as a film (for packaging).

### Polyamides (nylon)

There are several different types of nylon (e.g. nylon 6, nylon 66, nylon 11) but as a family their characteristics of strength, stiffness and toughness have earned them a reputation as *engineering plastics*. Table 7.2-3 compares the relative merits of light metal alloys and nylon.

Typical applications for nylon include small gears, bearings, bushes, sprockets, housings for power tools, terminal blocks and slide rollers. An important design consideration is that nylon absorbs moisture which can affect its properties and dimensional stability. Glass reinforcement reduces this problem and produces an extremely strong, impact resistant material. Another major application of nylon is in fibres which are notoriously strong. The density of nylon is about 1100 kg/m$^3$.

### Acetals

The superior properties of acetal in terms of its strength, stiffness and toughness have also earned it a place as an engineering plastic. It is more dense than nylon but in many respects their properties are similar and they can be used for the same types of light engineering application. A factor which may favour acetal in some cases is its relatively low water absorption. The material is available as both a homopolymer and a copolymer. The former is slightly stronger and stiffer whereas the copolymer has improved high temperature performance. This latter feature makes this material very attractive for hot water plumbing applications and as the body for electric kettles.

### Polytetrafluoroethylene (PTFE)

The major advantages of this material are its excellent chemical resistance and its extremely low coefficient of friction. Not surprisingly its major area of application is in bearings particularly if the environment is aggressive. It is also widely used in areas such as insulating tapes, gaskets,

**Table 7.2-3** Comparison between die casting alloys and nylons

| Points for comparison | Die casting alloys | Nylon |
| --- | --- | --- |
| Cost of raw material tonne | Low | High |
| Cost of mould | High | Can be lower – no higher |
| Speed of component production | Slower than injection moulding of nylon | Lower component production costs |
| Accuracy of component | Good | Good |
| Post moulding operations | Finishing – painting. Paint chips off easily | Finishing – not required – painting not required. Compounded colour retention permanent. |
| Surface hardness | Low – scratches easily | Much higher. Scratch resistant. |
| Rigidity | Good to brittleness | Glass reinforced grades as good or better |
| Elongation | Low | GR grades comparable unfilled grades excellent |
| Toughness (flexibility) | Low | GR grades comparable unfilled grades excellent |
| Impact | Low | All grades good |
| Notch sensitivity | Low | Low |
| Youngs modulus (E) | Consistent | Varies with load |
| General mechanical properties | Similar to GR grades of 66 nylon | Higher compressive strength |
| Heat conductivity | High | Low |
| Electrical insulation | Low | High |
| Weight | High | Low |
| Component assembly | Snap fits difficult | Very good |

pumps, diaphragms and of course non-stick coatings on cooking utensils.

### Thermoplastic polyesters

These linear polyesters are highly crystalline and exhibit toughness, strength, abrasion resistance, low friction, chemical resistance and low moisture absorption. Polyethylene terephthalate (PET) has been available for many years but mainly as a fibre (e.g. Terylene). As a moulding material it was less attractive due to processing difficulties but these were overcome with the introduction of polybutylene terephthalate (PBT). Applications include gears, bearings, housings, impellers, pulleys, switch parts, bumper extensions, etc. and of course PET is now renowned for its success as a replacement for glass in beverage bottles. PBT does not have such a high performance specification as PET but it is more readily moulded.

### Polyetheretherketone

This material, which is more commonly known as PEEK, is one of the new generation plastics which offer the possibility of high service temperatures. It is crystalline in nature which accounts in part for its high resistance to attack from acids, alkalis and organic solvents. It is easily processed and may be used continuously at 200°C where it offers good abrasion resistance, low flammability, toughness, strength and good fatigue resistance. Its density is 1300 kg/m$^3$. Applications include wire coatings, electrical connections, fans, impellers, fibres, etc.

## (b) Amorphous plastics

### Polyvinyl chloride (PVC)

This material is the most widely used of the amorphous plastics. It is available in two forms – plasticised or unplasticised. Both types are characterised by good weathering resistance, excellent electrical insulation properties, good surface properties and they are self extinguishing. Plasticised PVC is flexible and finds applications in wire covering, floor tiles, toy balls, gloves and rainwear. Unplasticised PVC (uPVC) is hard, tough, strong material which is widely used in the building industry. For example, pipes, gutters, window frames and wall claddings are all made in this material. The familiar credit cards are also made from uPVC.

### Polymethyl methacrylate (PMMA)

This material has exceptional optical clarity and resistance to outdoor exposure. It is resistant to alkalis, detergents, oils and dilute acids but is attacked by most solvents. Its peculiar property of total internal reflection is useful in advertising signs and some medical applications.

Typical uses include illuminated notices, control panels, dome-lights, lighting diffusers, baths, face guards, nameplates, lenses and display models.

### Polystyrene (PS)

Polystyrene is available in a range of grades which generally vary in impact strength from brittle to very tough. The non-pigmented grades have crystal clarity and overall their low cost coupled with ease of processing makes them used for such things as model aircraft kits, vending cups, yoghurt containers, light fittings, coils, relays, disposable syringes and casings for ballpoint pens. Polystyrene is also available in an expanded form which is used for such things as ceiling tiles and is excellent as a packaging material and thermal insulator.

### Acrylonitrile-butadiene-styrene (ABS)

ABS materials have superior strength, stiffness and toughness properties to many plastics and so they are often considered in the category of engineering plastics. They compare favourably with nylon and acetal in many applications and are generally less expensive. However, they are susceptible to chemical attack by chlorinated solvents, esters, ketones, acids and alkalis.

Typical applications are housings for TV sets, telephones, fascia panels, hair brush handles, luggage, helmets and linings for refrigerators.

### Polycarbonates

These materials also come within the category of engineering plastics and their outstanding feature is extreme toughness. They are transparent and have good temperature resistance but are attacked by alkaline solutions and hydrocarbon solvents. Typical applications include vandal-proof street lamp covers, baby feeding bottles, machine housings and guards, camera parts, electrical components, safety equipment and compact discs.

### Polyethersulphone

This material is one of the new high temperature plastics. It is recommended for load bearing applications up to 180°C. Even without flame retardants it offers low flammability and there is little change in dimensions of electrical properties in the temperature range 0–200°C. It is easily processed on conventional moulding equipment. Applications include aircraft heating ducts, terminal blocks, engine manifolds, bearings, grilles, tool handles, non-stick coatings.

### Modified polyphenylene oxide (PPO)

The word *modified* in this material refers to the inclusion of high impact polystyrene to improve processability and reduce the cost of the basic PPO. This material offers a range of properties which make it attractive for a whole range of applications. For example, it may be used at 100–150°C where it is rigid, tough and strong with good creep resistance and hydrolytic stability. Water absorption is very small and there is excellent dimensional stability. Applications include business machine parts, flow valves, headlight parts, engine manifolds, fascia panels, grilles, pump casings, hair dryer housings, etc.

### (c) Thermoplastic rubbers

There are five types of thermoplastic rubbers currently available. These are based on (i) olefinics (e.g. *Alcryn, Santoprene*) (ii) polyurethanes (e.g. *Elastollan, Caprolan, Pellethane*) (iii) polyesters (e.g. *Hytrel, Arnitel*) (iv) styrenics (e.g. *Solprene, Cariflex*) and (v) polyamides (e.g. *Pebax, Dinyl*) Some typical properties are given in Table 7.2-4.

### (d) Thermosetting plastics

#### Aminos

There are two basic types of amino plastics–urea formaldehyde and melamine formaldehyde. They are hard, rigid materials with good abrasion resistance and their mechanical characteristics are sufficiently good for continuous use at moderate temperatures (up to 100°C). Urea formaldehyde is relatively inexpensive but moisture absorption can result in poor dimensional stability. It is generally used for bottle caps, electrical switches, plugs, utensil handles and trays. Melamine formaldehyde has lower water absorption and improved temperature and chemical resistance. It is typically used for tableware, laminated worktops and electrical fittings.

#### Phenolics

Phenol-formaldehyde (*Bakelite*) is one of the oldest synthetic materials available. It is a strong, hard, brittle material with good creep resistance and excellent electrical properties. Unfortunately the material is only available in dark colours and it is susceptible to attack by alkalis and oxidising agents. Typical applications are domestic electrical fittings, saucepan handles, fan blades, smoothing iron handles and pump parts.

#### Polyurethanes

This material is available in three forms - rigid foam, flexible foam and elastomer. They are characterised by high strength and good chemical and abrasion resistance. The rigid foam is widely used as an insulation material, the flexible foam is an excellent cushion material for furniture and the elastomeric material is used in solid tyres and shock absorbers.

#### Polyesters

The main application of this material is as a matrix for glass fibre reinforcement. This can take many forms and is probably most commonly known as a DIY type material used for the manufacture of small boats, chemical containers, tanks and repair kits for cars, etc.

#### Epoxides

Epoxy resins are more expensive than other equivalent thermosets (e.g. polyesters) but they can generally outperform these materials due to better toughness, less shrinkage during curing, better weatherability and lower moisture absorption. A major area of application is in the aircraft industry because of the combination of properties offered when they are reinforced with fibres. They have an operating temperature range of −25 to 150°C.

## 7.2.4 Selection of plastics

The previous section has given an indication of the range of plastics available to the design engineer. The important

**Table 7.2-4** Physical characteristics of thermoplastic rubbers

| Type | Olefinic | Polyurethane | Polyester | Styrenic | Polyamide |
|---|---|---|---|---|---|
| Hardness (Shore A–D) | 60A to 60D | 60A to 60D | 40D to 72D | 30A to 45D | 40D to 63D |
| Resilience (%) | 30 to 40 | 40 to 50 | 43 to 62 | 60 to 70 | – |
| Tensile strength (MN/m$^2$) | 8 to 20 | 30 to 55 | 21 to 45 | 25 to 45 | – |
| Resistances | | | | | |
| Chemicals | F | P/G | E | E | P/E |
| Oils | F | E | E | F | – |
| Solvents | P/F | F | G | P | P/E |
| Weathering | E | G | E | P/E | E |
| Specific gravity | 0.97–1.34 | 1.11–1.21 | 1.17–1.25 | 0.93–1.0 | 1.0–1.12 |
| Service temperature (°C) | −50–130 | −40–130 | −65–130 | −30–120 | −65–130 |

KEY: P = poor, F = fair, G = good. E = excellent

question then arises *How do we decide which plastic, if any, is best for a particular application?* Material selection is not as difficult as it might appear but it does require an awareness of the general behaviour of plastics as a group, as well as a familiarity with the special characteristics of individual plastics.

The first and most important steps in the design process are to define clearly the purpose and function of the proposed product and to identify the service environment. Then one has to assess the suitability of a range of candidate materials. The following are generally regarded as the most important characteristics requiring consideration for most engineering components.

(1) mechanical properties – strength, stiffness, specific strength and stiffness, fatigue and toughness, and the influence of high or low temperatures on these properties;

(2) corrosion susceptibility and degradation;

(3) wear resistance and frictional properties;

(4) special properties, for example, thermal, electrical, optical and magnetic properties, damping capacity, etc;

(5) moulding and/or other methods of fabrication;

(6) total costs attributable to the selected material and manufacturing route.

In the following sections these factors will be considered briefly in relation to plastics.

## 7.2.4.1 Mechanical properties

### Strength and stiffness

Thermoplastic materials are viscoelastic which means that their mechanical properties reflect the characteristics of both viscous liquids and elastic solids. Thus when a thermoplastic is stressed it responds by exhibiting viscous flow (which dissipates energy) and by elastic displacement (which stores energy). The properties of viscoelastic materials are time, temperature and strain rate dependent. Nevertheless the conventional stress–strain test is frequently used to describe the (short-term) mechanical properties of plastics. It must be remembered, however, that the information obtained from such tests may only be used for an initial sorting of materials. It is not suitable, or intended, to provide design data which must usually be obtained from long term tests.

In many respects the stress–strain graph for a plastic is similar to that for a metal (see Fig. 7.2-2).

At low strains there is an elastic region whereas at high strains there is a nonlinear relationship between stress and strain and there is a permanent element to the strain.

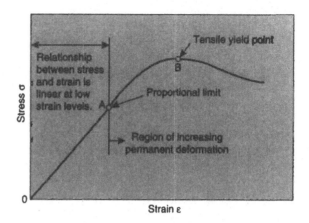

**Fig. 7.2-2** Typical stress–strain graph for plastics.

In the absence of any specific information for a particular plastic, design strains should normally be limited to 1%. Lower values ($\simeq 0.5\%$) are recommended for the more brittle thermoplastics such as acrylic, polystyrene and values of 0.2–0.3% should be used for thermosets.

The effect of material temperature is illustrated in Fig. 7.2-3. As temperature is increased the material becomes more flexible and so for a given stress the material deforms more. Another important aspect to the behaviour of plastics is the effect of strain rate. If a thermoplastic is subjected to a rapid change in strain it appears stiffer than if the same maximum strain were applied but at a slower rate. This is illustrated in Fig. 7.2-4.

It is important to realise also that within the range of grades that exist for a particular plastic, there can be significant differences in mechanical properties. For example, with polypropylene for each 1 kg/m$^3$ change in density there is a corresponding 4% change in modulus. Fig. 7.2-5 illustrates the typical variation which occurs for the different grades of ABS. It may be seen that very

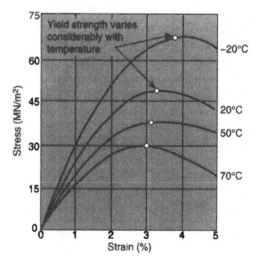

**Fig. 7.2-3** Effect of material temperature on stress–strain behaviour of plastics.

# CHAPTER 7.2  General properties of plastics

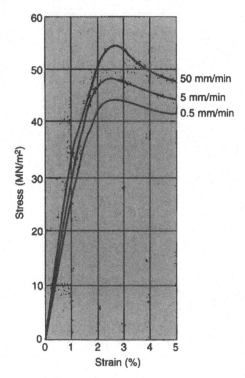

**Fig. 7.2-4** Effect of strain rate on stress–strain behaviour of plastics.

often a grade of material selected for some specific desirable feature (e.g. high impact strength) results in a decrease in some other property of the material (e.g. tensile strength).

The stiffness of a plastic is expressed in terms of a modulus of elasticity. Most values of elastic modulus quoted in technical literature represent the slope of

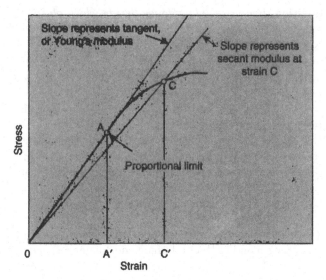

**Fig. 7.2-6** Tangent and secant modulus.

a tangent to the stress–strain curve at the origin (see Fig. 7.2-6). This is often referred to as Youngs modulus, $E$, but it should be remembered that for a plastic this will not be a constant and, as mentioned earlier, is only useful for quality control purposes, not for design. Since the tangent modulus at the origin is sometimes difficult to determine precisely, a secant modulus is often quoted to remove any ambiguity. A selected strain value of, say 2% (point $C'$, Fig. 7.2-6) enables a precise point, C, on the stress–strain curve to be identified. The slope of a line through C and 0 is the secant modulus. Typical short-term mechanical properties of plastics are given in Table 7.2-5. These are given for illustration purposes. For each type of plastic there are many different grades and a wide variety of properties are possible. The literature supplied by the manufacturers should be consulted in specific instances.

## Material selection for strength

If, in service, a material is required to have a certain strength in order to perform its function satisfactorily then a useful way to compare the structural efficiency of a range of materials is to calculate their strength desirability factor.

Consider a structural member which is essentially a beam subjected to bending (Fig. 7.2-7). Irrespective of the precise nature of the beam loading the maximum stress, $\sigma$, in the beam will be given by

$$\sigma = \frac{M_{max}(d/2)}{I} = \frac{M_{max}(d/2)}{bd^2/12} \qquad (7.2.1)$$

Assuming that we are comparing different materials on the basis that the mean length, width and loading is

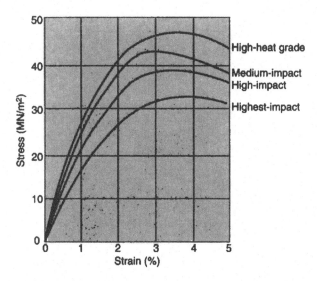

**Fig. 7.2-5** Effect of grade on mechanical properties of ABS.

### General properties of plastics — CHAPTER 7.2

**Table 7.2-5** Short-term properties of some important plastics

| Material | Density (kg/m$^3$) | Tensile strength (MN/m$^2$) | Flexural modulus (GN/m$^2$) | % elongation at break | Price* |
|---|---|---|---|---|---|
| ABS (high impact) | 1040 | 38 | 2.2 | 8 | 2.1 |
| Acetal (homopolymer) | 1420 | 68 | 2.8 | 40 | 3.5 |
| Acetal (copolymer) | 1410 | 70 | 2.6 | 65 | 3.3 |
| Acrylic | 1180 | 70 | 2.9 | 2 | 2.5 |
| Cellulose acetate | 1280 | 30 | 1.7 | 30 | 3.2 |
| CAB | 1190 | 25 | 1.3 | 60 | – |
| Epoxy | 1200 | 70 | 3.0 | 3 | 8.3 |
| Modified PPO | 1060 | 45 | 2.3 | 70 | – |
| Nylon 66 | 1140 | 70 | 2.8 | 60 | 3.9 |
| Nylon 66 (33% glass) | 1380 | 115 | 5.1 | 4 | 4.0 |
| PEEK | 1300 | 62 | 3.8 | 4 | 42 |
| PEEK (30% carbon) | 1400 | 240 | 14 | 1.6 | 44 |
| PET | 1360 | 75 | 3 | 70 | 3.0 |
| PET (36% glass) | 1630 | 180 | 12 | 3 | 3.5 |
| Phenolic (mineral filled) | 1690 | 55 | 8.0 | 0.8 | 1.25 |
| Polyamideimide | 1400 | 185 | 4.5 | 12 | 67 |
| Polycarbonate | 1150 | 65 | 2.8 | 100 | 4.2 |
| Polyetherimide | 1270 | 105 | 3.3 | 60 | – |
| Polyethersulphone | 1370 | 84 | 2.6 | 60 | 13.3 |
| Polyimide | 1420 | 72 | 2.5 | 8 | 150 |
| Polypropylene | 905 | 33 | 1.5 | 150 | 1 |
| Polysulphone | 1240 | 70 | 2.6 | 80 | 11 |
| Polystyrene | 1050 | 40 | 3.0 | 1.5 | 1.1 |
| Polythene (LD) | 920 | 10 | 0.2 | 400 | 0.83 |
| Polythene (HD) | 950 | 32 | 1.2 | 150 | 1.1 |
| PTFE | 2100 | 25 | 0.5 | 200 | 13.3 |
| PVC (rigid) | 1400 | 50 | 3.0 | 80 | 0.88 |
| PVC (flexible) | 1300 | 14 | 0.007 | 300 | 0.92 |
| SAN | 1080 | 72 | 3.6 | 2 | 1.8 |
| DMC (polyester) | 1800 | 40 | 9.0 | 2 | 1.5 |
| SMC (polyester) | 1800 | 70 | 11.0 | 3 | 1.3 |

\* On a weight basis, relative to polypropylene.

**Fig. 7.2-7** Beam subjected to bending.

fixed but the beam depth is variable then equation (7.2.1) may be written as

$$\sigma = \beta_1/d^2 \qquad (7.2.2)$$

where $\beta_1$ is a constant.

But the weight, $w$, of the beam is given by

$$w = \rho b d L \qquad (7.2.3)$$

So substituting for $d$ from (7.2.2) into (7.2.3)

$$w = \beta_2 \rho/\sigma^{1/2} \qquad (7.2.4)$$

where $\beta_2$ is the same constant for all materials.

Hence, if we adopt loading/weight as a desirability factor, $D_f$, then this will be given by

$$D_f = \frac{\sigma_y^{1/2}}{\rho} \qquad (7.2.5)$$

where $\sigma_y$ and $\rho$ are the strength and density values for the materials being compared.

Similar desirability factors may be derived for other geometries such as struts, columns etc. This concept is taken further later where material costs are taken into account and Tables 7.2-11 and 7.2-12 give desirability factors for a range of loading configurations and materials.

## Material selection for stiffness

If in the service of a component it is the deflection, or stiffness which is the limiting factor rather than strength, then it is necessary to look for a different desirability factor in the candidate materials. Consider the beam situation described above. This time, irrespective of the loading, the deflection, $\delta$, will be given by

$$\delta = \alpha_1 \left(\frac{WL^3}{EI}\right) \qquad (7.2.6)$$

where $\alpha_1$ is a constant and $W$ represents the loading.

The stiffness may then be expressed as

$$\frac{W}{\delta} = \left(\frac{1}{\alpha_1}\right)\frac{EI}{L^3}$$

$$\frac{W}{\delta} = \alpha_2(Ed^3) \qquad (7.2.7)$$

where $\alpha_2$ is a constant and again it is assumed that the beam width and length are the same in all cases.

Once again the beam weight will be given by equation (7.2.3) so substituting for $d$ from equation (7.2.7)

$$w = \alpha_3 \rho/E^{1/3} \qquad (7.2.8)$$

Hence, the desirability factor, $D_f$, expressed as maximum stiffness for minimum weight will be given by

$$D_f = \frac{E^{1/3}}{\rho} \qquad (7.2.9)$$

where $E$ is the elastic modulus of the material in question and $\rho$ is the density. As before a range of similar factors can be derived for other structural elements and these are illustrated in Section 7.2.4.6. (Tables 7.2-11 and 7.2-12) where the effect of material cost is also taken into account. Note also that since for plastics the modulus, $E$, is not a constant it is often necessary to use a long-term (creep) modulus value in equation (7.2.9) rather than the short-term quality control value usually quoted in trade literature.

### Ductility

A load-bearing device or component must not distort so much under the action of the service stresses that its function is impaired, nor must it fail by rupture, though local yielding may be tolerable. Therefore, high modulus and high strength, with ductility, is the desired combination of attributes. However, the inherent nature of plastics is such that high modulus tends to be associated with low ductility and steps that are taken to improve the one cause the other to deteriorate. The major effects are summarised in Table 7.2-6. Thus it may be seen that there is an almost inescapable rule by which increased modulus is accompanied by decreased ductility and vice versa.

### Creep and recovery behaviour

Plastics exhibit a time-dependent strain response to a constant applied stress. This behaviour is called creep. In a similar fashion if the stress on a plastic is removed it exhibits a time dependent recovery of strain back towards its original dimensions. This is illustrated in Fig. 7.2-8.

### Stress relaxation

Another important consequence of the viscoelastic nature of plastics is that if they are subjected to a particular strain and this strain is held constant it is found

## General properties of plastics — CHAPTER 7.2

**Table 7.2-6** Balance between stiffness and ductility in thermoplastics

| | Effect on | |
|---|---|---|
| | **Modulus** | **Ductility** |
| Reduced temperature | increase | decrease |
| Increased straining rate | increase | decrease |
| Multiaxial stress field | increase | decrease |
| Incorporation of plasticizer | decrease | increase |
| Incorporation of rubbery phase | decrease | increase |
| Incorporation of glass fibres | increase | decrease |
| Incorporation of particulate filler | increase | decrease |

that as time progresses, the stress necessary to maintain this strain decreases. This is termed stress relaxation and is of vital importance in the design of gaskets, seals, springs and snap-fit assemblies.

### Creep rupture

When a plastic is subjected to a constant tensile stress its strain increases until a point is reached where the material fractures. This is called creep rupture or, occasionally, static fatigue. It is important for designers to be aware of this failure mode because it is a common error, amongst those accustomed to dealing with metals, to assume that if the material is capable of withstanding the applied (static) load in the short term then there need be no further worries about it. This is not the case with plastics where it is necessary to use long-term design data, particularly because some plastics which are tough at short times tend to become embrittled at long times.

### Fatigue

Plastics are susceptible to brittle crack growth fractures as a result of cyclic stresses, in much the same way as metals are. In addition, because of their high damping and low thermal conductivity, plastics are also prone to thermal softening if the cyclic stress or cyclic rate is high. The plastics with the best fatigue resistance are polypropylene, ethylene-propylene copolymer and PVDF.

### Toughness

By toughness we mean the resistance to fracture. Some plastics are inherently very tough whereas others are inherently brittle. However, the picture is not that simple because those which are nominally tough may become embrittled due to processing conditions, chemical attack, prolonged exposure to constant stress, etc. Where toughness is required in a particular application it is very important therefore to check carefully the service conditions in relation to the above type of factors. At room temperature the toughest unreinforced plastics include nylon 66, LDPE, LLDPE, EVA and polyurethane structural foam. At sub-zero temperatures it is necessary to consider plastics such as ABS, polycarbonate and EVA.

## 7.2.4.2 Degradation

### Physical or chemical attack

Although one of the major features which might prompt a designer to consider using plastics is corrosion

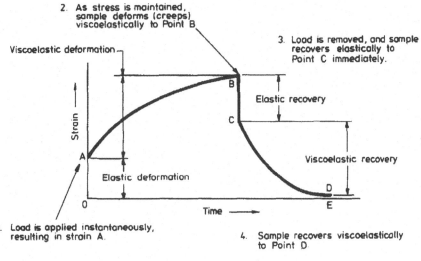

**Fig. 7.2-8** Typical creep and recovery behaviour of a plastic.

resistance, nevertheless plastics are susceptible to chemical attack and degradation. As with metals, it is often difficult to predict the performance of a plastic in an unusual environment so it is essential to check material specifications and where possible carry out proving trials. Clearly, in the space available here it is not possible to give precise details on the suitability of every plastic in every possible environment. Therefore the following sections give an indication of the general causes of polymer degradation to alert the designer to a possible problem.

The degradation of a plastic occurs due to a breakdown of its chemical structure. It should be recognised that this breakdown is not necessarily caused by concentrated acids or solvents. It can occur due to apparently innocuous mediums such as water **(hydrolysis)**, or oxygen **(oxidation)**. Degradation of plastics is also caused by heat, stress and radiation. During moulding the material is subjected to the first two of these and so it is necessary to incorporate stabilisers and antioxidants into the plastic to maintain the properties of the material. These additives also help to delay subsequent degradation for an acceptably long time.

As regards the general behaviour of polymers, it is widely recognised that crystalline plastics offer better environmental resistance than amorphous plastics. This is as a direct result of the different structural morphology of these two classes of material. Therefore engineering plastics which are also crystalline e.g. nylon 66 are at an immediate advantage because they can offer an attractive combination of load-bearing capability and an inherent chemical resistance. In this respect the arrival of crystalline plastics such as PEEK and polyphenylene sulfide (PPS) has set new standards in environmental resistance, albeit at a price. At room temperature there is no known solvent for PPS, and PEEK is only attacked by 98% sulphuric acid.

### Weathering

This generally occurs as a result of the combined effect of water absorption and exposure to ultra-violet radiation (u-v). Absorption of water can have a plasticizing action on plastics which increases flexibility but ultimately (on elimination of the water) results in embrittlement, while u-v causes breakdown of the bonds in the polymer chain. The result is general deterioration of physical properties. A loss of colour or clarity (or both) may also occur. Absorption of water reduces dimensional stability of moulded articles. Most thermoplastics, in particular cellulose derivatives, are affected, and also polyethylene, PVC, and nylons.

### Oxidation

This is caused by contact with oxidising acids, exposure to u-v, prolonged application of excessive heat, or exposure to weathering. It results in a deterioration of mechanical properties (embrittlement and possibly stress cracking), increase in power factor, and loss of clarity. It affects most thermoplastics to varying degrees, in particular polyolefins, PVC, nylons, and cellulose derivatives.

### Environmental stress cracking (ESC)

In some plastics, brittle cracking occurs when the material is in contact with certain substances whilst under stress. The stress may be externally applied in which case one would be prompted to take precautions. However, internal or residual stresses introduced during processing are probably the more common cause of ESC. Most organic liquids promote ESC in plastics but in some cases the problem can be caused by a liquid which one would not regard as an aggressive chemical. The classic example of ESC is the brittle cracking of polyethylene washing-up bowls due to the residual stresses at the moulding gate coupled with contact with the aqueous solution of washing-up liquid. Although direct attack on the chemical structure of the plastic is not involved in ESC the problem can be alleviated by controlling structural factors. For example, the resistance of polyethylene is very dependent on density, crystallinity, melt flow index (MFI) and molecular weight. As well as polyethylene, other plastics which are prone to ESC are ABS and polystyrene.

The mechanism of ESC is considered to be related to penetration of the promoting substance at surface defects which modifies the surface energy and promotes fracture.

### 7.2.4.3 Wear resistance and frictional properties

There is a steady rate of increase in the use of plastics in bearing applications and in situations where there is sliding contact e.g. gears, piston rings, seals, cams, etc. The advantages of plastics are low rates of wear in the absence of conventional lubricants, low coefficients of friction, the ability to absorb shock and vibration and the ability to operate with low noise and power consumption. Also when plastics have reinforcing fibres they offer high strength and load carrying ability. Typical reinforcements include glass and carbon fibres and fillers include PTFE and molybdenum disulphide in plastics such as nylon, polyethersulphone (PES), polyphenylene sulfide (PPS), polyvinylidene fluoride (PVDF) and polyetheretherketone (PEEK).

The friction and wear of plastics are extremely complex subjects which depend markedly on the nature of the application and the properties of the material. The frictional properties of plastics differ considerably

General properties of plastics    CHAPTER 7.2

Table 7.2-7 Coefficients of friction and relative wear rates for plastics

| Material | Coefficient of friction | | Relative wear rate |
|---|---|---|---|
| | Static | Dynamic | |
| Nylon | 0.2 | 0.28 | 33 |
| Nylon/glass | 0.24 | 0.31 | 13 |
| Nylon/carbon | 0.1 | 0.11 | 1 |
| Polycarbonate | 0.31 | 0.38 | 420 |
| Polycarbonate/glass | 0.18 | 0.20 | 5 |
| Polybutylene terephthalate (PBT) | 0.19 | 0.25 | 35 |
| PBT/glass | 0.11 | 0.12 | 2 |
| Polyphenylene sulfide (PPS) | 0.3 | 0.24 | 90 |
| PPS/glass | 0.15 | 0.17 | 19 |
| PPS/carbon | 0.16 | 0.15 | 13 |
| Acetal | 0.2 | 0.21 | – |
| PTFE | 0.04 | 0.05 | – |

from those of metals. Even reinforced plastics have modulus values which are much lower than metals. Hence metal/thermoplastic friction is characterised by adhesion and deformation which results in frictional forces that are not proportional to load but rather to speed. Table 7.2-7 gives some typical coefficients of friction for plastics.

The wear rate of plastics is governed by several mechanisms. The primary one is adhesive wear which is characterised by fine particles of polymer being removed from the surface. This is a small-scale effect and is a common occurrence in bearings which are performing satisfactorily. However, the other mechanism is more serious and occurs when the plastic becomes overheated to the extent where large troughs of melted plastic are removed. Table 7.2-7 shows typical primary wear rates for different plastics, the mechanism of wear is complex and the relative wear rates may change depending on specific circumstances.

In linear bearing applications the suitability of a plastic is usually determined from its PV rating. This is the product of P (the bearing load divided by the projected bearing area) and V (the linear shaft velocity). Fig. 7.2-9 shows the limiting PV lines for a range of plastics – combinations of P and V above the lines are not permitted. The PV ratings may be increased if the bearing is lubricated or the mode of operation is intermittent. The PV rating will be decreased if the operating temperature is increased. Correction factors for these variations may be obtained from material/bearing manufacturers. The plastics with the best resistance to wear are ultra high molecular weight polyethylene (used in hip joint replacements) and PTFE lubricated versions of nylon, acetal and PBT. It is not recommended to use the same plastic for both mating surfaces in applications such as gear wheels.

### 7.2.4.4 Special properties

#### Thermal properties

Before considering conventional thermal properties such as conductivity it is appropriate to consider briefly the effect of temperature on the mechanical properties of plastics. It was stated earlier that the properties of plastics are markedly temperature dependent. This is as a result of their molecular structure. Consider first an amorphous plastic in which the molecular chains have a random configuration. Inside the material, even though it is not possible to view them, we know that the molecules are in a state of continual motion. As the material is heated up the molecules receive more energy and there is an increase in their relative movement. This makes the material more flexible. Conversely if the material is cooled down then molecular mobility decreases and the material becomes stiffer.

With plastics there is a certain temperature, called the **glass transition temperature,** $T_g$, below which the material behaves like glass, i.e. it is hard and rigid. As can be seen from Table 7.2-8 the value for $T_g$ for a particular plastic is not necessarily a low temperature. This immediately helps to explain some of the differences which we observe in plastics. For example, at room temperature polystyrene and acrylic are below their respective $T_g$ values and hence we observe these materials in their glassy state. Note, however, that in contrast, at room temperature, polyethylene is above its glass transition temperature and so we observe a very flexible material. When cooled below its $T_g$ it then becomes a hard, brittle solid, Plastics can have several transitions.

The main $T_g$ is called the glass–rubber transition and signifies a change from a flexible, tough material to a glassy state in which the material exhibits stiffness, low creep and toughness although with a sensitivity to notches. At lower temperatures there is then a secondary transition characterised by a change to a hard, rigid, brittle state.

It should be noted that although Table 7.2-8 gives specific values of $T_g$ for different polymers, in reality the glass transition temperature is not a material constant. As

Table 7.2-8 Typical thermal properties of materials

| Material | Densty (kg/m$^3$) | Specific heat (kJ/kg K) | Thermal conductivity (W/m/K) | Coeff. of therm exp ($\mu$m/m/°C) | Thermal diffusivity (m$^2$/s) × 10$^{-7}$ | Glass transition temp, $T_g$(°C) | Max. operating, temp (°C) |
|---|---|---|---|---|---|---|---|
| ABS | 1040 | 1.3 | 0.25 | 90 | 1.7 | 115 | 70 |
| Acetal (homopolymer) | 1420 | 1.5 | 0.2 | 80 | 0.7 | −85 | 85 |
| Acetal (copolymer) | 1410 | 1.5 | 0.2 | 95 | 0.72 | −85 | 90 |
| Acrylic | 1180 | 1.5 | 0.2 | 70 | 1.09 | 105 | 50 |
| Cellulose acetate | 1280 | 1.6 | 0.15 | 100 | 1.04 | – | 60 |
| CAB | 1190 | 1.6 | 0.14 | 100 | 1.27 | – | 60 |
| Epoxy | 1200 | 0.8 | 0.23 | 70 | – | – | 130 |
| Modified PPO | 1060 | – | 0.22 | 60 | – | – | 120 |
| Nylon 66 | 1140 | 1.7 | 0.24 | 90 | 1.01 | 56 | 90 |
| Nylon 66 (33% glass) | 1380 | 1.6 | 0.52 | 30 | 1.33 | – | 100 |
| PEEK | 1300 | – | – | 48 | – | 143 | 204 |
| PEEK (30% carbon) | 1400 | – | – | 14 | – | – | 255 |
| PET | 1360 | 1.0 | 0.2 | 90 | – | 75 | 110 |
| PET (36% glass) | 1630 | – | – | 40 | – | – | 150 |
| Phenolic (glass filled) | 1700 | – | 0.5 | 18 | – | – | 185 |
| Polyamide-imide | 1400 | – | 0.25 | 36 | – | 260 | 210 |
| Polycarbonate | 1150 | 1.2 | 0.2 | 65 | 1.47 | 149 | 125 |
| Polyester | 1200 | 1.2 | 0.2 | 100 | – | – | – |
| Polyetherimide | 1270 | – | 0.22 | 56 | – | 200 | 170 |
| Polyethersulphone | 1370 | – | 1.18 | 55 | – | 230 | 180 |
| Polyimide | 1420 | – | – | 45 | – | 400 | 260 |
| Polyphenylene sulfide | 1340 | – | – | 49 | – | 85 | 150 |
| Polypropylene | 905 | 2.0 | 0.20 | 100 | 0.65 | −10 | 100 |
| Polysulphone | 1240 | 1.3 | – | 56 | – | 180 | 170 |
| Polystyrene | 1050 | 1.3 | 0.15 | 80 | 0.6 | 100 | 50 |
| Polythene (LD) | 920 | 2.2 | 0.24 | 200 | 1.17 | −120 | 50 |
| Polythene (HP) | 950 | 2.2 | 0.25 | 120 | 1.57 | −120 | 55 |
| PTFE | 2100 | 1.0 | 0.25 | 140 | 0.7 | −113 | 250 |
| PVC (rigid) | 1400 | 0.9 | 0.16 | 70 | 1.16 | 80 | 50 |
| PVC (flexible) | 1300 | 1.5 | 0.14 | 140 | 0.7 | 80 | 50 |
| SAN | 1080 | 1.3 | 0.17 | 70 | 0.81 | 115 | 60 |

### Table 7.2-8 Typical thermal properties of materials—Cont'd

| Material | Densty (kg/m³) | Specific heat (kJ/kg K) | Thermal conductivity (W/m/K) | Coeff. of therm exp (μm/m/°C) | Thermal diffusivity (m²/s) × 10⁻⁷ | Glass transition temp, $T_g$(°C) | Max. operating, temp (°C) |
|---|---|---|---|---|---|---|---|
| DMC (polyester) | 1800 | – | 0.2 | 20 | – | – | 130 |
| SMC (polyester) | 1800 | – | 0.2 | 20 | – | – | 130 |
| Polystyrene foam | 32 | – | 0.032 | – | – | – | – |
| PU foam | 32 | – | 0.032 | – | – | – | – |
| Stainless steel | 7855 | 0.49 | 90 | 10 | – | – | 800 |
| Nickel chrome alloy | 7950 | – | 12 | 14 | – | – | 900 |
| Zinc | 7135 | 0.39 | 111 | 39 | – | – | – |
| Copper | 8940 | 0.39 | 400 | 16 | – | – | – |

with many other properties of polymers it will depend on the testing conditions used to obtain it.

In the so-called crystalline plastics the structure consists of both crystalline (ordered) regions and amorphous (random) regions. When these materials are heated there is again increased molecular mobility but the materials remain relatively stiff due to the higher forces between the closely packed molecules. When the crystalline plastics have their temperature reduced they exhibit a glass transition temperature associated with the amorphous regions. At room temperature polypropylene, for example, is quite rigid and tough, not because it is below its $T_g$ but because of the strong forces between the molecules in the crystalline regions. When it is cooled below −10°C it becomes brittle because the amorphous regions go below their $T_g$.

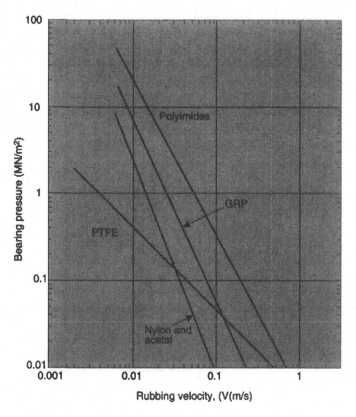

**Fig. 7.2-9** Typical P–V ratings for plastics rubbing on steel.

# CHAPTER 7.2  General properties of plastics

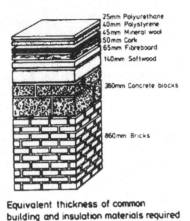

**Fig. 7.2-10** Comparative thermal conductivities for a range of materials.

In the past a major limitation to the use of plastics materials in the engineering sector has been temperature. This limitation arises not only due to the reduction in mechanical properties at high temperatures, including increased propensity to creep, but also due to limitations on the continuous working temperature causing permanent damage to the material as a result of thermal and oxidative degradation. Significant gains in property retention at high temperatures with crystalline polymers have been derived from the incorporation of fibrous reinforcement, but the development of new polymer matrices is the key to further escalation of the useful temperature range.

Table 7.2-8 indicates the service temperatures which can be used with a range of plastics. It may be seen that there are now commercial grades of unreinforced plastics rated for continuous use at temperatures in excess of 200°C. When glass or carbon fibres are used the service temperatures can approach 300°C.

The other principal thermal properties of plastics which are relevant to design are thermal conductivity and coefficient of thermal expansion. Compared with most materials, plastics offer very low values of thermal conductivity, particularly if they are foamed. Figure 7.2-10 shows comparisons between the thermal conductivity of a selection of metals, plastics and building materials. In contrast to their low conductivity, plastics have high coefficients of expansion when compared with metals. This is illustrated in Fig. 7.2-11 and Table 7.2-8 gives fuller information on the thermal properties of plastics and metals.

## Electrical properties

Traditionally plastics have established themselves in applications which require electrical insulation. PTFE and polyethylene are among the best insulating materials available. The material properties which are particularly relevant to electrical insulation are *dielectric strength, resistance and tracking*.

The insulating property of any insulator will break down in a sufficiently strong electric field. The dielectric strength is defined as the electric strength (V/m) which an insulating material can withstand. For plastics the dielectric strength can vary from 1 to 1000 MV/m. Materials may be compared on the basis of their relative permittivity (or dielectric constant). This is the ratio of the permittivity of the material to the permittivity of a vacuum. The ability of a material to resist the flow of electricity is determined by its volume resistivity, measured in ohm m. Insulators are defined as having volume resistivities greater than about $10^4$ ohm m. Plastics are well above this, with values ranging from about $10^8$ to $10^{16}$ ohm m. These compare with a value of about $10^{-8}$

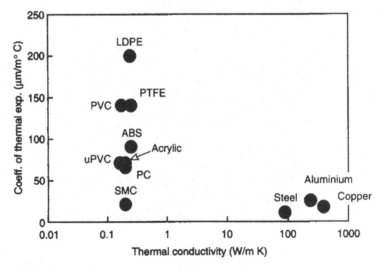

**Fig. 7.2-11** Typical thermal properties of plastics.

ohm m for copper. Although plastics are good insulators, local breakdown may occur due to tracking. This is the name given to the formation of a conducting path (arc) across the surface of the polymer. It can be caused by surface contamination (for example dust and moisture) and is characterised by the development of carbonised destruction of the surface carrying the arc. Plastics differ greatly in their propensity to tracking – PTFE, acetal, acrylic and PP/PE copolymers offer very good resistance.

It is interesting to note that although the electrical insulation properties of plastics have generally been regarded as one of their major advantages, in recent years there has been a lot of research into the possibility of conducting plastics. This has been recognised as an exciting development area for plastics because electrical conduction if it could be achieved would offer advantages in designing against the build up of static electricity and in shielding of computers, etc from electro-magnetic interference (EMI). There have been two approaches – coating or compounding. In the former the surface of the plastic is treated with a conductive coating (e.g. carbon or metal) whereas in the second, fillers such as brass, aluminium or steel are incorporated into the plastic. It is important that the filler has a high aspect ratio (length:diameter) and so fibres or flakes of metal are used. There has also been some work done using glass fibres which are coated with a metal before being incorporated into the plastic. Since the fibre aspect ratio is critical in the performance of conductive plastics there can be problems due to breaking up of fibres during processing. In this regard thermosetting plastics have an advantage because their simpler processing methods cause less damage to the fibres. Conductive grades of DMC are now available with resistivities as low as $7 \times 10^{-3}$ ohm m.

## Optical properties

The optical properties of a plastic which are important are refraction, transparency, gloss and light transfer. The reader is referred to BS 4618:1972 for precise details on these terms. Table 7.2-9 gives data on the optical properties of a selection of plastics. Some plastics may be optically clear (e.g. acrylic, cellulosics and ionomers) whereas others may be made transparent. These include epoxy, polycarbonate, polyethylene, polypropylene, polystyrene, polysulphone and PVC.

**Table 7.2-9** Typical properties of plastics

| Material | Refractive index | Light transmission | Dispersive power |
|---|---|---|---|
| Acrylic | 1.49 | 92 | 58 |
| Polycarbonate | 1.59 | 89 | 30–35 |
| Polystyrene | 1.59 | 88 | 31 |
| CAB | 1.49 | 85 | – |
| SAN | 1.57 | – | 36 |
| Nylon 66 | 1.54 | 0 | – |

## Flammability

The fire hazard associated with plastics has always been difficult to assess and numerous tests have been devised which attempt to grade materials as regards flammability by standard small scale methods under controlled but necessarily artificial conditions. Descriptions of plastics as *self-extinguishing*, *slow burning*, *fire retardant* etc. have been employed to describe their behaviour under such standard test conditions, but could never be regarded as predictions of the performance of the material in real fire situations, the nature and scale of which can vary so much.

Currently there is a move away from descriptions such as *fire-retardant* or *self-extinguishing* because these could imply to uninformed users that the material would not burn. The most common terminology for describing the flammability characteristics of plastics is currently the **Critical Oxygen Index (COI)**. This is defined as the minimum concentration of oxygen, expressed as volume per cent, in a mixture of oxygen and nitrogen that will just support combustion under the conditions of test. Since air contains 21% oxygen, plastics having a COI of greater than 0.21 are regarded as self-extinguishing. In practice a higher threshold (say 0.27) is advisable to allow for unforeseen factors in a particular fire hazard

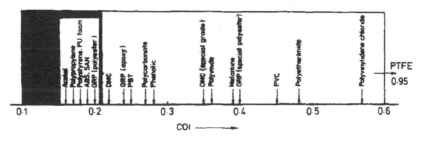

**Fig. 7.2-12** Oxygen index values for plastics.

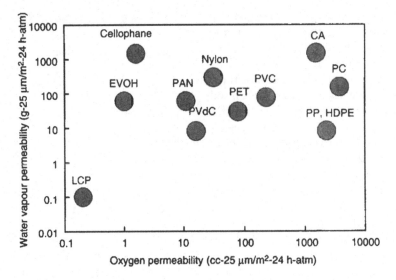

**Fig. 7.2-13** Permeability data for a range of plastics.

situation. Fig. 7.2-12 shows the typical COI values for a range of plastics.

### Permeability

The low density of plastics is an advantage in many situations but the relatively loose packing of the molecules means that gases and liquids can permeate through the plastic. This can be important in many applications such as packaging or fuel tanks. It is not possible to generalise about the performance of plastics relative to each other or in respect to the performance of a specific plastic in contact with different liquids and gases.

Some plastics are poor at offering resistance to the passage of fluids through them whereas others are excellent. Their relative performance may be quantified in terms of a permeation constant, $k$, given by

$$k = \frac{Qd}{At\,p} \qquad (7.2.10)$$

where

$Q$ = volume of fluid passing through the plastic
$d$ = thickness of plastic
$A$ = exposed area
$t$ = time
$p$ = pressure difference across surfaces of plastic.

The main fluids of interest with plastics are oxygen and water vapour (for packaging applications) and $CO_2$ (for carbonated drinks applications). Figure 7.2-13 and Figure. 7.2-14 illustrate the type of behaviour exhibited by a range of plastics. In some cases it is necessary to use

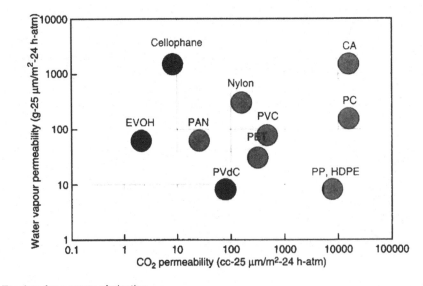

**Fig. 7.2-14** Permeability data for a range of plastics.

## 7.2.4.5 Processing

A key decision in designing with plastics is the processing method employed. The designer must have a thorough knowledge of processing methods because plastics are unique in design terms in that they offer a wider choice of conversion techniques than are available for any other material. A simple container for example could be made by injection moulding, blow moulding or rotational moulding; light fittings could be thermoformed or injection moulded. In this brief introduction to designing with plastics it is not possible to do justice to the range of processing methods available for plastics.

## 7.2.4.6 Costs

It is a popular misconception that plastics are cheap materials. They are not. On a weight basis most plastics are more expensive than steel and only slightly less expensive than aluminium. Prices for plastics can range from about £600 per tonne for polypropylene to about £25,000 per tonne for carbon fibre reinforced PEEK. Table 7.2-5 compares the costs of a range of plastics.

However, it should always be remembered that it is bad design practice to select materials on the basis of cost per unit weight. In the mass production industries, in particular, the raw material cost is of relatively little importance. It is the *in-position* cost which is all important. The in-position cost of a component is the sum of several independent factors i.e. raw material costs, fabrication costs and performance costs.

It is in the second two of these cost components that, in relation to other materials, plastics can offer particular advantages. Fabrication costs include power, labour, consumables, etc and Table 7.2-10 shows that, in terms of the overall energy consumption, plastics come out much better than metals. Performance costs relate to servicing, warranty claims, etc. On this basis plastics can be very attractive to industries manufacturing consumer products because they can offer advantages such as colour fastness, resilience, toughness, corrosion resistance and uniform quality – all features which help to ensure a reliable product.

However, in general these fabrication and performance advantages are common to all plastics and so a decision has to be made in regard to which plastic would be best for a particular application. Rather than compare the basic raw material costs it is better to use a cost index on the basis of the cost to achieve a certain performance. Consider again the material selection procedures illustrated in Section 7.2.4.1 in relation to strength and stiffness.

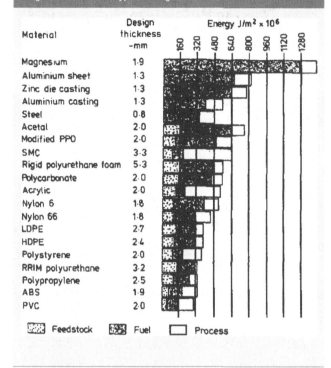

Table 7.2-10 The energy required to manufacture and process a range of materials at typical design thickness

multiple layers of plastics because no single plastic offers the combination of price, permeation resistance, printability, etc. required for the application. When multilayers are used, an overall permeation constant for the composite wall may be obtained from

$$\frac{1}{k} = \frac{1}{d}\sum_{i=1}^{i=N}\frac{d_i}{k_i} \qquad (7.2.11)$$

Table 7.2-11 Desirability factors for some common loading configurations

| Component | Desirability factor, $D_f$ | |
|---|---|---|
| | Strength basis | Stiffness basis |
| Rectangular beam with fixed width | $\sigma_y^{1/2}/\rho C$ | $E^{1/3}/\rho C$ |
| Struts or ties | $\sigma_y/\rho C$ | $E/\rho C$ |
| Thin wall cylinders under pressure | $\sigma_y/\rho C$ | – |
| Thin wall shafts in tension | $\tau_m/\rho C$ | $G/\rho C$ |
| Long rods in compression (buckling) | – | $E^{1/2}/\rho C$ |

## CHAPTER 7.2　General properties of plastics

**Table 7.2-12** Desirability factors for a range of materials

| Material | Density, $\rho$ kg/m$^3$ | Proof of fracture stress $\sigma_y$ (MN/m$^2$) | Modulus $E$ (GN/m$^2$) | $\dfrac{\sigma_y}{\rho}$ | $\dfrac{E}{\rho}$ | $\dfrac{\sigma_y^{1/2}}{\rho}$ ($\times 10^{-3}$) | $\dfrac{E^{1/2}}{\rho}$ ($\times 10^{-3}$) | $\dfrac{E^{1/3}}{\rho}$ ($\times 10^{-3}$) |
|---|---|---|---|---|---|---|---|---|
| Aluminium (pure) | 2700 | 90 | 70 | 0.033 | 0.026 | 3.51 | 3.12 | 1.53 |
| Aluminium alloy | 2810 | 500 | 71 | 0.178 | 0.025 | 7.95 | 3.0 | 1.47 |
| Stainless steel | 7855 | 980 | 185 | 0.125 | 0.024 | 4.0 | 1.73 | 0.73 |
| Titanium alloy | 4420 | 900 | 107 | 0.204 | 0.024 | 6.78 | 2.34 | 1.07 |
| Spruce | 450 | 35 | 9 | 0.078 | 0.020 | 13.15 | 6.67 | 4.62 |
| GRP (80% unidirectional glass in polyester) | 2000 | 1240 | 48 | 0.62 | 0.024 | 17.6 | 3.46 | 1.82 |
| CFRP (60% unidirectional fibres in epoxy) | 1500 | 1050 | 189 | 0.7 | 0.126 | 21.6 | 9.16 | 3.82 |
| Nylon 66 | 1140 | 70 | 0.78* | 0.061 | $6.8 \times 10^{-4}$ | 8.34 | 0.77 | 0.81 |
| ABS | 1040 | 35 | 1.2* | 0.034 | $11.5 \times 10^{-4}$ | 5.68 | 1.05 | 1.02 |
| Polycarbonate | 1150 | 60 | 2.0* | 0.052 | $17.4 \times 10^{-4}$ | 6.73 | 1.23 | 1.09 |
| PEEK (+ 30% C) | 1450 | 215 | 15.5 | 0.19 | 0.011 | 10.1 | 2.7 | 1.72 |

\* 1500 h creep modules

### Selection for strength at minimum cost

If the cost of a material is C per unit weight then from equation (7.2.3) the cost of the beam considered in the analysis would be

$$C_b = (\rho b d L) C \quad (7.2.12)$$

Substituting for $d$ from (7.2.2) then the cost of the beam on a strength basis would be

$$C_b = \beta_3 \left( \dfrac{\rho C}{\sigma_y^{1/2}} \right) \quad (7.2.13)$$

where $\beta_3$ is a constant, which will be the same for all materials. Therefore we can define a cost factor, $C_f$, where

$$C_f = \left( \dfrac{\rho C}{\sigma_y^{1/2}} \right) \quad (7.2.14)$$

which should be minimised in order to achieve the best combination of price and performance. Alternatively we may take the reciprocal of $C_f$ to get a desirability factor, $D_f$,

$$D_f = \left( \dfrac{\sigma_y^{1/2}}{\rho C} \right) \quad (7.2.15)$$

and this may be compared to $D_f$ given by equation (7.2.5).

### Selection for stiffness at minimum cost

Using equation (7.2.7) and an analysis similar to above it may be shown that on the basis of stiffness and cost, the desirability factor, $D_f$, is given by

$$D_f = \left( \dfrac{E^{1/3}}{\rho C} \right) \quad (7.2.16)$$

Tables 7.2-11 and 7.2-12 give desirability factors for configurations other than the beam analysed above and typical numerical values of these factors for a range of materials.

# Bibliography

Waterman, N.A. *The Selection and Use of Engineering Materials*, Design Council, London (1979)

Crane, F.A.C. and Charles, J.A. *Selection and Use of Engineering Materials* Butterworths, London (1984)

Crawford, R.J. *Plastics and Rubber–Engineering Design and Application*, MEP, London (1985)

Powell, P.C. *Engineering with Polymers*, Chapman and Hall, London (1984)

Hall, C. *Polymer Materials*, Macmillan, London (1981)

Birley, A.W. and Scott, M.J. *Plastic Materials: Properties and Applications*, Leonard Hall, Glasgow (1982)

Benham, P.P., Crawford, R.J. and Armstrong, C.G. *Mechanics of Engineering Materials*, Longmans (1996)

Lancaster, J.K. *Friction and Wear of Plastics*, Chapter 14 in Polymer Science edited by A.D. Jenkins (vol 2), North-Holland Publ. Co., London (192)

Bartenev, G.M. and Lavrentev, V.V. *Friction and Wear in Polymers*, Elsevier Science Publ. Co., Amsterdam (1981)

Schwartz, S.S. and Goodman, S.H. *Plastics Materials and Processes*, Van Nostrand Reinhold, New York (1982)

Blythe, A.R. *Electrical Properties of Polymers*, Cambridge Univ. Press (1980)

Van Krevelen, D.W. *Properties of Polymers* 2nd Edition, Elsevier, Amsterdam (1976)

Mills, N. *Plastics*, Edward Arnold, London (1986)

Kemmish, D.J. High performance engineering plastics. *RAPRA Review Reports* 8, 2 (1995)

Oswald, T.A. and Menges, G. *Materials Science of Polymers for Engineers*, Hanser, Munich (1995)

Birley, A.W., Haworth, B. and Batchelor, J. *Physics of Plastics*, Hanser, Munich (1992)

Belofsky, H. *Plastics: Product Design and Process Engineering*, Hanser, Munich (1995)

Gruenwald, G. *Plastics: How Structure Determines Properties*, Hanser, Munich (1992)

Dominghaus, H. *Plastics for Engineers*, Hanser, New York (1993)

Chanier, J.M. *Polymeric Materials and Processing*, Hanser, New York (1990)

Progelhof, R.C. and Throne, J.L. *Polymer Engineering Principles*, Hanser, New York (1993)

# Chapter 7.3

# Processing of plastics

## 7.3.1 Introduction

One of the most outstanding features of plastics is the ease with which they can be processed. In some cases semi-finished articles such as sheets or rods are produced and subsequently fabricated into shape using conventional methods such as welding or machining. In the majority of cases, however, the finished article, which may be quite complex in shape, is produced in a single operation. The processing stages of heating, shaping and cooling may be continuous (e.g. production of pipe by extrusion) or a repeated cycle of events (eg production of a telephone housing by injection moulding) but in most cases the processes may be automated and so are particularly suitable for mass production. There is a wide range of processing methods which may be used for plastics. In most cases the choice of method is based on the shape of the component and whether it is thermoplastic or thermosetting. It is important therefore that throughout the design process, the designer must have a basic understanding of the range of processing methods for plastics since an ill-conceived shape or design detail may limit the choice of moulding methods.

In this chapter each of the principal processing methods for plastics is described and where appropriate a Newtonian analysis of the process is developed. Although most polymer melt flows are in fact Non-Newtonian, the simplified analysis is useful at this stage because it illustrates the approach to the problem without concealing it by mathematical complexity. In practice the simplified analysis may provide sufficient accuracy for the engineer to make initial design decisions and at least it provides a quantitative aspect which assists in the understanding of the process.

## 7.3.2 Extrusion

### 7.3.2.1 General features of single screw extrusion

One of the most common methods of processing plastics is **extrusion** using a screw inside a barrel as illustrated in Fig. 7.3-1. The plastic, usually in the form of granules or powder, is fed from a hopper on to the screw. It is then conveyed along the barrel where it is heated by conduction from the barrel heaters and shear due to its movement along the screw flights. The depth of the screw channel is reduced along the length of the screw so as to compact the material. At the end of the extruder the melt passes through a die to produce an extrudate of the desired shape. As will be seen later, the use of different dies means that the extruder screw/barrel can be used as the basic unit of several processing techniques.

Basically an extruder screw has three different zones.

### (a) Feed zone

The function of this zone is to preheat the plastic and convey it to the subsequent zones. The design of this section is important since the constant screw depth must supply sufficient material to the metering zone so as not to starve it, but on the other hand not supply so much material that the metering zone is overrun. The optimum design is related to the nature and shape of the feed stock, the geometry of the screw and the frictional properties of the screw and barrel in relation to the plastic. The frictional behaviour of the feed-stock material has a considerable influence on the rate of melting which can be achieved.

*Engineering Materials and Processes Desk Reference*; ISBN: 9781856175869
Copyright © 2009 Elsevier Ltd.; All rights of reproduction, in any form, reserved.

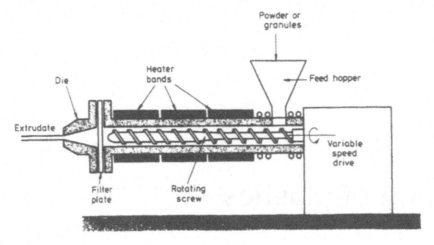

Fig. 7.3-1 Schematic view of single screw extruder.

### (b) Compression zone

In this zone the screw depth gradually decreases so as to compact the plastic. This compaction has the dual role of squeezing any trapped air pockets back into the feed zone and improving the heat transfer through the reduced thickness of material.

### (c) Metering zone

In this section the screw depth is again constant but much less than the feed zone. In the metering zone the melt is homogenised so as to supply, at a constant rate, material of uniform temperature and pressure to the die. This zone is the most straightforward to analyse since it involves a viscous melt flowing along a uniform channel.

The pressure build-up which occurs along a screw is illustrated in Fig. 7.3-2 The lengths of the zones on a particular screw depend on the material to be extruded. With nylon, for example, melting takes place quickly so that the compression of the melt can be performed in one pitch of the screw. PVC on the other hand is very heat sensitive and so a compression zone which covers the whole length of the screw is preferred.

As plastics can have quite different viscosities, they will tend to behave differently during extrusion. Fig. 7.3-3 shows some typical outputs possible with different plastics in extruders with a variety of barrel diameters.

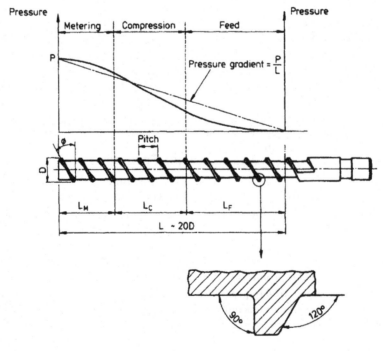

Fig. 7.3-2 Typical zones on a extruder screw.

# Processing of plastics — CHAPTER 7.3

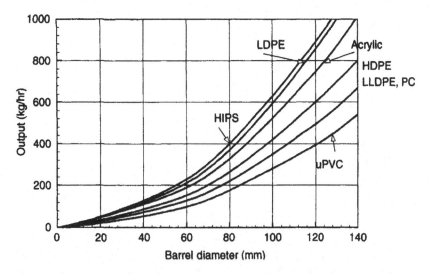

**Fig. 7.3-3** Typical extruder outputs for different plastics.

This diagram is to provide a general idea of the ranking of materials – actual outputs may vary ±25% from those shown, depending on temperatures, screw speeds, etc.

In commercial extruders, additional zones may be included to improve the quality of the output. For example there may be a mixing zone consisting of screw flights of reduced or reversed pitch. The purpose of this zone is to ensure uniformity of the melt and it is sited in the metering section. Fig. 7.3-4 shows some designs of mixing sections in extruder screws.

Some extruders also have a venting zone. This is principally because a number of plastics are hygroscopic – they absorb moisture from the atmosphere. If these materials are extruded wet in conventional equipment the quality of the output is not good due to trapped water vapour in the melt. One possibility is to pre-dry the feedstock to the extruder but this is expensive and can lead to contamination. Vented barrels were developed to overcome these problems. As shown in Fig. 7.3-5, in the first part of the screw the granules are taken in and melted, compressed and homogenised in the usual way. The melt pressure is then reduced to atmospheric pressure in the decompression zone. This allows the volatiles to escape from the melt through a special port in the barrel. The melt is then conveyed along the barrel to a second compression zone which prevents air pockets from being trapped.

The venting works because at a typical extrusion temperature of 250°C the water in the plastic exists as a vapour at a pressure of about 4 MN/m$^2$. At this pressure it will easily pass out of the melt and through the exit orifice. Note that since atmospheric pressure is

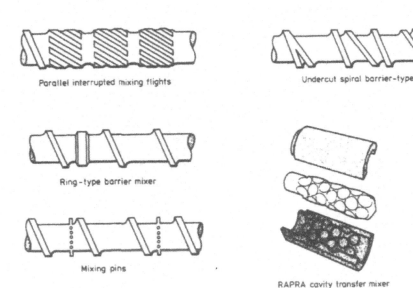

**Fig. 7.3-4** Typical designs of mixing zones.

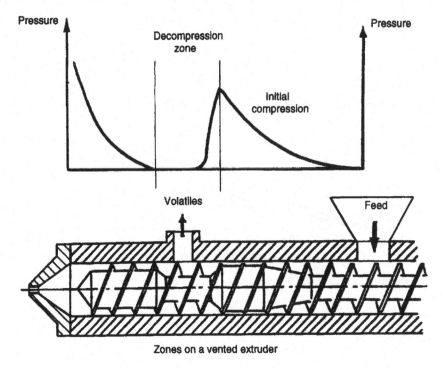

**Fig. 7.3-5** Zones on a vented extruder.

about 0.1 MN/m² the application of a vacuum to the exit orifice will have little effect on the removal of volatiles.

Another feature of an extruder is the presence of a gauze filter after the screw and before the die. This effectively filters out any inhomogeneous material which might otherwise clog the die. These *screen packs* as they are called, will normally filter the melt to 120–150/μm. However, there is conclusive evidence to show that even smaller particles than this can initiate cracks in plastic extrudates e.g. polyethylene pressure pipes. In such cases it has been found that fine melt filtration ($\approx 450/$μm) can significantly improve the performance of the extrudate.

Since the filters by their nature tend to be flimsy they are usually supported by a breaker plate. As shown in Fig. 7.3-6 this consists of a large number of countersunk holes to allow passage of the melt whilst preventing dead spots where particles of melt could gather. The breaker

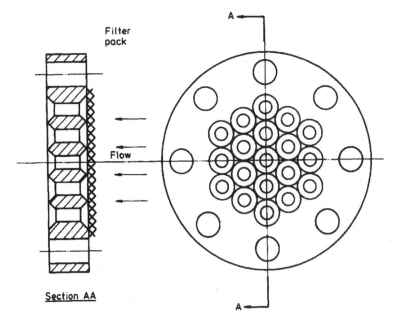

**Fig. 7.3-6** Breaker plate with filter pack.

plate also conveniently straightens out the spiralling melt flow which emerges from the screw. Since the fine mesh on the filter will gradually become blocked it is periodically removed and replaced. In many modern extruders, and particularly with the fine filter systems referred to above, the filter is changed automatically so as not to interrupt continuous extrusion.

It should also be noted that although it is not their primary function, the breaker plate and filter also assist the build-up of back pressure which improves mixing along the screw. Since the pressure at the die is important, extruders also have a valve after the breaker plate to provide the necessary control.

## 7.3.2.2 Mechanism of flow

As the plastic moves along the screw, it melts by the following mechanism. Initially a thin film of molten material is formed at the barrel wall. As the screw rotates, it scrapes this film off and the molten plastic moves down the front face of the screw flight. When it reaches the core of the screw it sweeps up again, setting up a rotary movement in front of the leading edge of the screw flight. Initially the screw flight contains solid granules but these tend to be swept into the molten pool by the rotary movement. As the screw rotates, the material passes further along the barrel and more and more solid material is swept into the molten pool until eventually only melted material exists between the screw flights.

As the screw rotates inside the barrel, the movement of the plastic along the screw is dependent on whether or not it adheres to the screw and barrel. In theory there are two extremes. In one case the material sticks to the screw only and therefore the screw and material rotate as a solid cylinder inside the barrel. This would result in zero output and is clearly undesirable. In the second case the material slips on the screw and has a high resistance to rotation inside the barrel. This results in a purely axial movement of the melt and is the ideal situation. In practice the behaviour is somewhere between these limits as the material adheres to both the screw and the barrel. The useful output from the extruder is the result of a drag flow due to the interaction of the rotating screw and stationary barrel. This is equivalent to the flow of a viscous liquid between two parallel plates when one plate is stationary and the other is moving. Superimposed on this is a flow due to the pressure gradient which is built up along the screw. Since the high pressure is at the end of the extruder the pressure flow will reduce the output. In addition, the clearance between the screw flights and the barrel allows material to leak back along the screw and effectively reduces the output. This leakage will be worse when the screw becomes worn.

The external heating and cooling on the extruder also plays an important part in the melting process. In high output extruders the material passes along the barrel so quickly that sufficient heat for melting is generated by the shearing action and the barrel heaters are not required. In these circumstances it is the barrel cooling which is critical if excess heat is generated in the melt. In some cases the screw may also be cooled. This is not intended to influence the melt temperature but rather to reduce the frictional effect between the plastic and the screw. In all extruders, barrel cooling is essential at the feed pocket to ensure an unrestricted supply of feedstock.

The thermal state of the melt in the extruder is frequently compared with two ideal thermodynamic states. One is where the process may be regarded as **adiabatic.** This means that the system is fully insulated to prevent heat gain or loss from or to the surroundings. If this ideal state was to be reached in the extruder it would be necessary for the work done on the melt to produce just the right amount of heat without the need for heating or cooling. The second ideal case is referred to as **isothermal.** In the extruder this would mean that the temperature at all points is the same and would require immediate heating or cooling from the barrel to compensate for any loss or gain of heat in the melt. In practice the thermal processes in the extruder fall somewhere between these ideals. Extruders may be run without external heating or cooling but they are not truly adiabatic since heat losses will occur. Isothermal operation along the whole length of the extruder cannot be envisaged if it is to be supplied with relatively cold granules. However, particular sections may be near isothermal and the metering zone is often considered as such for analysis.

## 7.3.2.3 Analysis of flow in extruder

As discussed in the previous section, it is convenient to consider the output from the extruder as consisting of three components – drag flow, pressure flow and leakage. The derivation of the equation for output assumes that in the metering zone the melt has a constant viscosity and its flow is isothermal in a wide shallow channel. These conditions are most likely to be approached in the metering zone.

### (a) Drag flow

Consider the flow of the melt between parallel plates as shown in Fig. 7.3-7a.

For the small element of fluid ABCD the volume flow rate $dQ$ is given by

$$dQ = V \cdot dy \cdot dx \qquad (7.3.1)$$

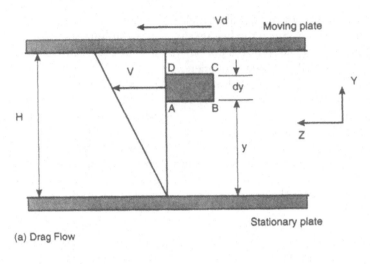

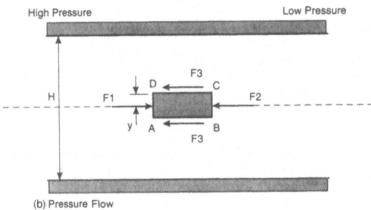

**Fig. 7.3-7** Melt flow between parallel plates.

Assuming the velocity gradient is linear, then

$$V = V_d \left[ \frac{y}{H} \right]$$

Substituting in (7.3.1) and integrating over the channel depth, $H$, then the total drag flow, $Q_d$, is given by

$$Q_d = \int_0^H \int_0^T \frac{V_d y}{H} dy \cdot dx$$

$$Q_d = \frac{1}{2} THV_d \quad (7.3.2)$$

This may be compared to the situation in the extruder where the fluid is being dragged along by the relative movement of the screw and barrel. Fig. 7.3-8 shows the position of the element of fluid and (7.3.2) may be modified to include terms relevant to the extruder dimensions. For example

$$V_d = \pi DN \cos\phi$$

where $N$ is the screw speed (in revolutions per unit time).

$$T = (\pi D \tan\phi - e) \cos\phi$$

So

$$Q_d = \frac{1}{2} (\pi D \tan\phi - e)(\pi DN \cos^2\phi) H$$

In most cases the term, $e$, is small in comparison with $(\pi D \tan\phi)$ so this expression is reduced to

$$Q_d = \frac{1}{2} \pi^2 D^2 NH \sin\phi \cos\phi \quad (7.3.3)$$

Note that the shear rate in the metering zone will be given by $V_d/H$.

### (b) Pressure flow

Consider the element of fluid shown in Fig. 7.3-7b. The forces are

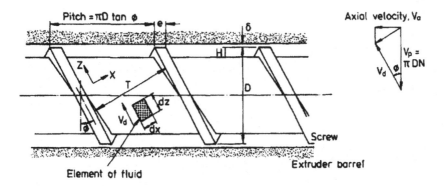

**Fig. 7.3-8** Details of extruder screw. In both cases, AB = dz, element width = dx and channel width = T.

$$F_1 = \left(P + \frac{\partial P}{\partial z} \cdot dz\right) dy\, dx$$

$$F_2 = P \cdot dy\, dx$$

$$F_3 = \tau_y\, dz\, dx$$

where $P$ is pressure and $d\tau$ is the shear stress acting on the element. For steady flow these forces are in equilibrium so they may be equated as follows:

$$F_1 = F_2 + 2F_3$$

which reduces to

$$y\frac{dP}{dz} = \tau_y \tag{7.3.4}$$

Now for a Newtonian fluid, the shear stress, $\tau_y$, is related to the viscosity, $\eta$, and the shear rate, $\gamma$, by the equation

$$\tau_y = \eta\dot{\gamma} = \eta\frac{dV}{dy}$$

Using this in equation (7.3-4)

$$y\frac{dP}{dz} = \eta\frac{dV}{dy}$$

Integrating

$$\int_0^v dV = \frac{1}{\eta}\frac{dP}{dz}\int_{H/2}^y y\, dy$$

So

$$V = \frac{1}{\eta}\frac{dP}{dz}\left(\frac{y^2}{2} - \frac{H^2}{8}\right) \tag{7.3.5}$$

Also, for the element of fluid of depth, $dy$, at distance, $y$, from the centre line (and whose velocity is $V$) the elemental flow rate, $dQ$, is given by

$$dQ = VT\, dy$$

This may be integrated to give the pressure flow, $Q_p$

$$Q_p = 2\int_0^{H/2}\frac{1}{\eta}\frac{dP}{dz}\cdot T\left(\frac{y^2}{2} - \frac{H^2}{8}\right)dy$$

$$Q_p = -\frac{1}{12\eta}\frac{dP}{dz}\cdot TH^3 \tag{7.3.6}$$

Referring to the element of fluid between the screw flights as shown in Fig. 7.3-8, this equation may be rearranged using the following substitutions. Assuming $e$ is small, $T = \pi D \tan\phi \cdot \cos\phi$

Also

$$\sin\phi = \frac{dL}{dz} \text{ so } \frac{dP}{dz} = \frac{dP}{dL}\sin\phi$$

Thus the expression for $Q_p$ becomes

$$Q_p = -\frac{\pi D H^3 \sin^2\phi}{12\eta}\cdot\frac{dP}{dL} \tag{7.3.7}$$

### (c) Leakage

The leakage flow may be considered as flow through a wide slit which has a depth, $\delta$, a length ($e \cos\varphi$) and a width of ($\pi D/\cos\varphi$). Since this is a pressure flow, the derivation is similar to that described in (b). For convenience therefore the following substitutions may be made in 7.3.6.

$$h = \delta$$
$$T = \pi D/\cos\phi$$

Pressure gradient $= \dfrac{\Delta P}{e \cos\phi}$ (see Fig. 7.3-9)

So the leakage flow, $Q_L$, is given by

$$Q_L = \frac{\pi^2 D^2 \delta^3}{12\eta e}\tan\phi\frac{dP}{dL} \tag{7.3.8}$$

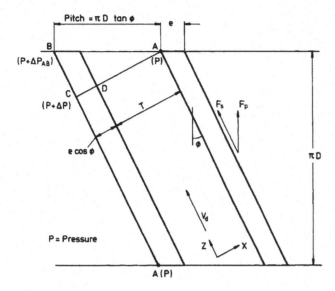

**Fig. 7.3-9** Development of screw.

A factor is often required in this equation to allow for eccentricity of the screw in the barrel. Typically this increases the leakage flow by about 20%.

The total output is the combination of drag flow, back pressure flow and leakage. So from (7.3.3), (7.3.7) and (7.3.8)

$$Q = \frac{1}{2}\pi^2 D^2 \ NH \ \sin\phi \cos\phi - \frac{\pi D H^3 \sin^2\phi}{12\eta}\frac{dP}{dL}$$
$$- \frac{\pi^2 D^2 \delta^3}{12\eta e} \tan\phi \frac{dP}{dL} \quad (3.1.9)$$

For many practical purposes sufficient accuracy is obtained by neglecting the leakage flow term. In addition the pressure gradient is often considered as linear so

$$\frac{dP}{dL} = \frac{P}{L}$$

where '$L$' is the length of the extruder. In practice the length of an extruder screw can vary between 17 and 30 times the diameter of the barrel. The shorter the screw the cooler the melt and the faster the moulding cycle. In the above analysis, it is the melt flow which is being considered and so the relevant pressure gradient will be that in the metering zone. However, as shown in Fig. 7.3-2 this is often approximated by $P/L$. If all other physical dimensions and conditions are constant then the variation of output with screw flight angle, $\phi$, can be studied. As shown in Fig. 7.3-10 the maximum output would be obtained if the screw flight angle was about 35°C. In practice a screw flight angle of 17.7° is frequently used because

(i) this is the angle which occurs if the pitch of the screw is equal to the diameter and so it is convenient to manufacture,

(ii) for a considerable portion of the extruder length, the screw is acting as a solids conveying device and it is known that the optimum angle in such cases is 17° to 20°.

It should also be noted that in some cases correction factors, $F_d$, and $F_p$ are applied to the drag and pressure flow terms. They are to allow for edge effects and are solely dependent on the channel width, $T$, and channel depth, $h$, in the metering zone. Typical values are illustrated in Fig. 7.3-11.

### 7.3.2.4 Extruder/die characteristics

From equation (7.3.9) it may be seen that there are two interesting situations to consider. One is the case of free

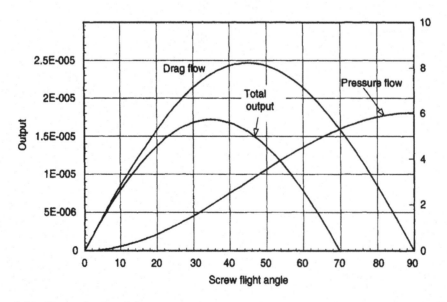

**Fig. 7.3-10** Variation of drag flow and pressure flow.

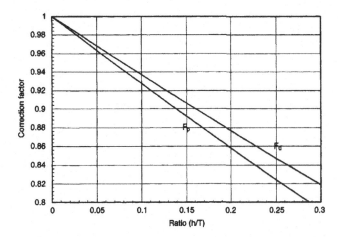

**Fig. 7.3-11** Flow correction factors as a function of screw geometry.

discharge where there is no pressure build up at the end of the extruder so

$$Q = Q_{max} = \frac{1}{2}\pi^2 D^2 NH \sin\phi \cos\phi \qquad (7.3.10)$$

The other case is where the pressure at the end of the extruder is large enough to stop the output. From (7.3.9) with $Q = 0$ and ignoring the leakage flow

$$P = P_{max} = \frac{6\pi DLN\eta}{H^2 \tan\phi} \qquad (7.3.11)$$

In Fig. 7.3-12 these points are shown as the limits of the screw characteristic. It is interesting to note that when a die is coupled to the extruder their requirements are conflicting. The extruder has a high output if the pressure at its outlet is low. However, the outlet from the extruder is the inlet to the die and the output of the latter increases with inlet pressure. As will be seen later the output, $Q$, of a Newtonian fluid from a die is given by a relation of the form

$$Q = KP \qquad (7.3.12)$$

where $K = \frac{\pi R^4}{8\eta L_d}$ for a capillary die of radius $R$ and length $L_d$.

Equation 7.3.12 enables the die characteristics to be plotted on Fig. 7.3-12 and the intersection of the two characteristics is the operating point of the extruder. This plot is useful in that it shows the effect which changes in various parameters will have on output. For example, increasing screw speed, $N$, will move the extruder characteristic upward. Similarly an increase in the die radius, $R$, would increase the slope of the die characteristic and in both cases the extruder output would increase.

The operating point for an extruder/die combination may also be determined from equations (7.3.9) and (7.3.12) – ignoring leakage flow

$$Q = \frac{1}{2}\pi^2 D^2 NH \sin\phi \cos\phi - \frac{\pi DH^3 \sin^2\phi}{12\eta}\frac{P}{L} = \frac{\pi R^4}{8\eta L_d}\cdot P$$

So for a capillary die, the pressure at the operating point is given by

$$P_{OP} = \left\{\frac{2\pi\eta D^2 NH \sin\phi \cos\phi}{(R^4/2L_d) + (DH^3 \sin^2\phi)/3L}\right\} \qquad (7.3.13)$$

### 7.3.2.5 Other die geometries

For other die geometries it is necessary to use the appropriate form of equation (7.3.12). For other geometries it is possible to use the empirical equation which was developed by Boussinesq. This has the form

$$Q = \frac{Fbd^3}{12\eta L_d}\cdot P \qquad (7.3.14)$$

where

$b$ is the greatest dimension of the cross-section
$d$ is the least dimension of the cross-section
$F$ is a non-dimensional factor as given in Fig. 7.3-13.

Using equation (7.3.14) it is possible to modify the expression for the operating pressure to the more general form

$$P_{OP} = \left\{\frac{2\pi\eta D^2 NH \sin\phi \cos\phi}{\frac{Fbd^3}{3\pi L_d} + (DH^3 \sin^2\phi/3L)}\right\} \qquad (7.3.15)$$

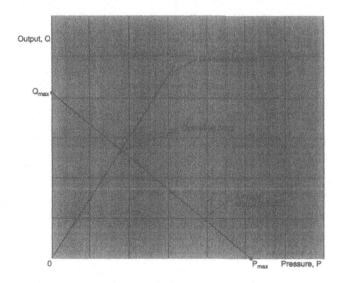

**Fig. 7.3-12** Extruder and die characteristics.

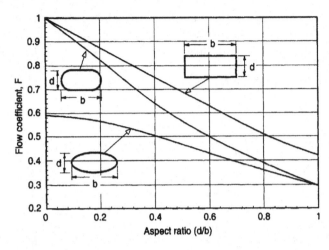

**Fig. 7.3-13** Flow coefficient as a function of channel geometry.

For a capillary die, one may obtain a value of $F$ from Fig. 7.3-13 as 0.295 and substituting $b = d = 2R$, this equation reduces to the same form as equation (7.3.13).

## Example 7.3.1

A single screw extruder is to be designed with the following characteristics.

L/D ratio = 24, screw flight angle = 17.7°
Max. screw speed = 100 rev/min, screw diameter = 40 mm
flight depth (metering zone) = 3 mm.

If the extruder is to be used to process polymer melts with a maximum melt viscosity of 500 Ns/m², calculate a suitable wall thickness for the extruder barrel based on the von Mises yield criterion. The tensile yield stress for the barrel metal is 925 MN/m² and a factor of safety of 2.5 should be used.

### Solution

The maximum pressure which occurs in the extruder barrel is when there is no output. Therefore the design needs to consider this worst case blockage situation. As given by equation (7.3.11)

$$P_{max} = \frac{6\pi DL\eta}{H^2 \tan\phi}$$

$$= \frac{6\pi \times 40 \times (24 \times 10) \times (100/60) \times 500}{(3)^2 \tan 17.7°}$$

$$= 210 \text{ MN/m}^2$$

The von Mises criterion relates the tensile yield stress of a material to a state of multi-axial stress in a component made from the material. In a cylinder (the barrel of the extruder in this case), the principal stresses which exist as a result of an internal pressure are

$$\text{hoop stress}, \sigma_1 = \frac{P_{max}D}{2h}$$

$$\text{axial stress}, \sigma_2 = \frac{P_{max}D}{4h}$$

where $h$ = wall thickness of the barrel.

The von Mises criterion simply states that yielding (failure) will occur if

$$\left(\frac{\sigma_Y}{FS}\right)^2 \leq \sigma_1^2 + \sigma_2^2 - \sigma_1\sigma_2$$

where

$\sigma_Y$ = tensile yield stress of material
$FS$ = factor of safety.

In this case, therefore

$$\left(\frac{925}{2.5}\right)^2 = \left(\frac{(210)40}{2h}\right)^2 + \left(\frac{(210)40}{4h}\right)^2 - \frac{(210)^2(40)^2}{8h^2}$$

$$h = 9.8 \text{ mm}$$

Hence a barrel wall thickness of 10 mm would be appropriate.

## Example 7.3.2

A single screw extruder is to be used to manufacture a nylon rod 5 mm in diameter at a production rate of 1.5 m/min. Using the following information, calculate the required screw speed.

| Nylon | Extruder | Die |
|---|---|---|
| Viscosity = 42.0 Ns/m² | Diameter = 30 mm | Length = 4 mm |
| Density (solid) = 1140 kg/m³ | Length = 750 mm | Diameter = 5 mm |
| Density (melt) = 790 kg/m³ | Screw flight angle = 17.7° | |
| | Metering channel depth = 2.5 mm | |

Die swelling effects may be ignored and the melt viscosity can be assumed to be constant.

### Solution

The output rate of solid rod

$= \text{speed} \times \text{cross-sectional area}$

$= 1.5 \times \pi \left(2.5 \times 10^{-3}\right)^2 / 60 = 49.1 \times 10^{-6} m^3/s$

As the solid material is more dense than the melt, the melt flow rate must be greater in the ratio of the solid/melt densities. Therefore

$$\text{Melt flow rate through die} = 49.1 \times 10^{-6} \left(\frac{1140}{790}\right)$$
$$= 70.8 \times 10^{-6} \, \text{m}^3/\text{s}$$

The pressure necessary to achieve this flow rate through the die is obtained from

$$Q = \frac{\pi P R^4}{8 \eta L_d}$$
$$P = \frac{8 \times 420 \times 4 \times 10^{-3} \times 70.8 \times 10^{-6}}{\pi (2.5 \times 10^{-3})^4} = 7.8 \, \text{MN/m}^2$$

At the operating point, the die output and the extruder output will be the same. Hence

$$Q = 70.8 \times 10^{-6}$$
$$= \frac{1}{2}\pi^2(30 \times 10^{-3})^2 N(2.5 \times 10^{-3}) \sin 17.7 \cos 17.7$$
$$- \frac{\pi(30 \times 10^{-3})(2.6 \times 10^{-3})^3 \sin 17}{12 \times 420} \left(\frac{7.8 \times 10^6}{0.75}\right)$$
$$N = 22 \, \text{rev/min}$$

## 7.3.2.6 General features of twin screw extruders

In recent years there has been a steady increase in the use of extruders which have two screws rotating in a heated barrel. These machines permit a wider range of possibilities in terms of output rates, mixing efficiency, heat generation, etc compared with a single screw extruder. The output of a twin screw extruder can be typically three times that of a single screw extruder of the same diameter and speed. Although the term 'twin-screw' is used almost universally for extruders having two screws,

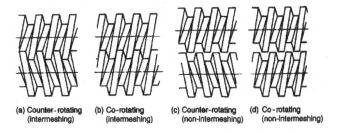

**Fig. 7.3-14** Different types of twin screw extruder

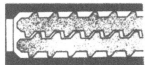
(a) Non-conjugated screws showing some passages around each screw

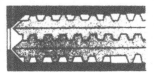
(b) Conjugated screws showing closed passages around each screw

**Fig. 7.3-15** Two types of twin screw extruder.

the screws need not be identical. There are in fact a large variety of machine types. Fig. 7.3-14 illustrates some of the possibilities with counter-rotating and co-rotating screws. In addition the screws may be conjugated or non-conjugated. A non-conjugated screw configuration is one in which the screw flights are a loose fit into one another so that there is ample space for material between the screw flights (see Fig. 7.3-15).

In a counter-rotating twin screw extruder the material is sheared and pressurised in a mechanism similar to calendering (see Section 7.3.5), i.e. the material is effectively squeezed between counter-rotating rolls. In a co-rotating system the material is transferred from one screw to the other in a figure-of-eight pattern as shown in Fig. 7.3-16. This type of arrangement is particularly suitable for heat sensitive materials because the material is conveyed through the extruder quickly with little possibility of entrapment. The movement around the screws is slower if the screws are conjugated but the propulsive action is greater.

Table 7.3-1 compares the single screw extruder with the main types of twin screw extruders.

## 7.3.2.7 Processing methods based on the extruder

Extrusion is an extremely versatile process in that it can be adapted, by the use of appropriate dies, to produce a wide range of products. Some of the more common of these production techniques will now be described.

### (a) Granule production/compounding

In the simplest case an extruder may be used to convert polymer formulations and additives into a form (usually granules) which is more convenient for use in other processing methods, such as injection moulding. In the extruder the feedstock is melted, homogenised and forced through a capillary shaped die. It emerges as a continuous lace which is cooled in a long water bath so that it may be chopped into short granules and packed into sacks. The haul-off apparatus shown in Fig. 7.3-17 is used to draw down the extrudate to the required dimensions. The granules are typically 3 mm diameter and about 4 mm long. In most cases a multi-hole die is used to increase the production rate.

**Table 7.3-1** Comparison of single-screw, co-rotating and counter-rotating twin-screw extruders

| Type | Single screw | Co-rotating screw | | Counter-rotating twin screw |
|---|---|---|---|---|
| | | Low speed type | High speed type | |
| Principle | Friction between cylinder and materials and the same between material and screw | Mainly depend on the frictional action as in the case of single screw extruder | | Forced mechanical conveyance based on gear pump principle |
| Conveying efficiency | Low | Medium | | High |
| Mixing efficiency | Low | Medium/High | | High |
| Shearing action | High | Medium | High | Low |
| Self-cleaning effect | Slight | Medium/High | High | Low |
| Energy efficiency | Low | Medium/High | | High |
| Heat generation | High | Medium | High | Low |
| Temp distribution | Wide | Medium | Narrow | Narrow |
| Max. revolving speed (rpm) | 100–300 | 25–35 | 250–300 | 35–45 |
| Max. effective length of screw L/D | 30–32 | 7–18 | 30–40 | 10–21 |

### (b) Profile production

Extrusion, by its nature, is ideally suited to the production of continuous lengths of plastic mouldings with a uniform cross-section. Therefore as well as producing the laces as described in the previous section, the simple operation of a die change can provide a wide range of profiled shapes such as pipes, sheets, rods, curtain track, edging strips, window frames, etc (see Fig. 7.3-18).

The successful manufacture of profiled sections depends to a very large extent on good die design. Generally this is not straightforward, even for a simple cross-section such as a square, due to the interacting effects of post-extrusion swelling and the flow characteristics of complex viscoelastic fluids. Most dies are designed from experience to give approximately the correct shape and then *sizing* units are used to control precisely the desired shape. The extrudate is then cooled as quickly as possible. This is usually done in a water bath the length of which depends on the section and the material being cooled. For example, longer baths are needed for crystalline plastics since the recrystallisation is exothermic.

The storage facilities at the end of the profile production line depend on the type of product (see Fig. 7.3-19). If it is rigid then the cooled extrudate may be cut to size on a guillotine for stacking. If the extrudate is flexible then it can be stored on drums.

### (c) Film blowing

Although plastic sheet and film may be produced using a slit die, by far the most common method nowadays is the film blowing process illustrated in Fig. 7.3-20. The molten plastic from the extruder passes through an annular die and emerges as a thin tube. A supply of air

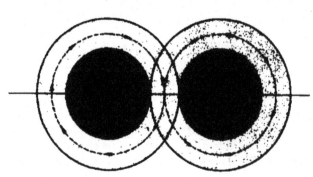

**Fig. 7.3-16** Material flow path with co-rotating screws.

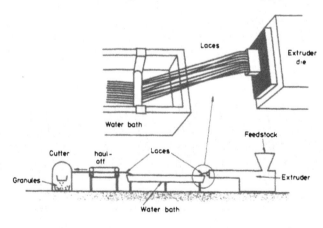

**Fig. 7.3-17** Use of extruder to produce granules.

Processing of plastics    CHAPTER 7.3

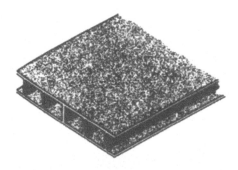

**Fig. 7.3-18** (a) Extruded panel sections (b) Extruded window profile.

to the inside of the tube prevents it from collapsing and indeed may be used to inflate it to a larger diameter. Initially the bubble consists of molten plastic but a jet of air around the outside of the tube promotes cooling and at a certain distance from the die exit, a freeze line can be identified. Eventually the cooled film passes through collapsing guides and nip rolls before being taken off to storage drums or, for example, gussetted and cut to length for plastic bags. Most commercial systems are provided with twin storage facilities so that a full drum may be removed without stopping the process.

The major advantage of film blowing is the ease with which biaxial orientation can be introduced into the film. The pressure of the air in the bubble determines the *blow-up* and this controls the circumferential orientation. In addition, axial orientation may be introduced by increasing the nip roll speed relative to the linear velocity of the bubble. This is referred to as *draw-down*.

It is possible to make a simple estimate of the orientation in blown film by considering only the effects due to the inflation of the bubble. Since the volume flow rate is the same for the plastic in the die and in the bubble, then for unit time

$$\pi D_d h_d L_d = \pi D_b h_b L_b$$

where $D$, $h$ and $L$ refer to diameter, thickness and length respectively and the subscript '$d$' is for the die and '$b$' is for the bubble.

So the orientation in the machine direction, $O_{MD}$, is given by

$$O_{MD} = \frac{L_b}{L_d} = \frac{D_d h_d}{h_b D_b} = \frac{h_d}{h_b B_R}$$

where $B_R$ = blow-up ratio ($D_b/D_d$)

Also the orientation in the transverse direction, $O_{TD}$, is given by

$$O_{TD} = \frac{D_b}{D_d} = B_R$$

Therefore the ratio of the orientations may be expressed as

$$\frac{O_{MD}}{O_{TD}} = \frac{h_d}{h_b (B_R)^2} \qquad (7.3.16)$$

### Example 7.3.3

A plastic shrink wrapping with a thickness of 0.05 mm is to be produced using an annular die with a die gap of 0.8 mm. Assuming that the inflation of the bubble dominates the orientation in the film, determine the blow-up ratio required to give uniform biaxial orientation.

### Solution

Since $O_{MD} = O_{TD}$
then the blow-up ratio,

$$B_R = \sqrt{\frac{h_d}{h_b}} = \sqrt{\frac{0.8}{0.05}} = 4$$

Common blow-up ratios are in the range 1.5 to 4.5

This example illustrates the simplified approach to film blowing. Unfortunately in practice the situation is more complex in that the film thickness is influenced by

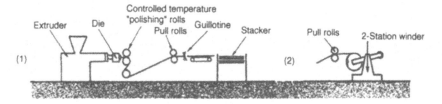

**Fig. 7.3-19 (a)** Sheet extrusion (1) thick sheet (2) thin sheet.

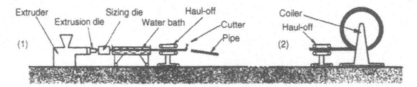

**Fig. 7.3-19(b)** Pipe extrusion (1) rigid pipe (2) flexible pipe.

draw-down, relaxation of induced stresses/strains and melt flow phenomena such as die swell. In fact the situation is similar to that described for blow moulding (see below) and the type of analysis outlined in that section could be used to allow for the effects of die swell. However, since the most practical problems in film blowing require iterative type solutions involving melt flow characteristics, volume flow rates, swell ratios, etc the study of these is delayed until Chapter 5 where a more rigorous approach to polymer flow has been adopted.

### (d) Blow moulding

This process evolved originally from glass blowing technology. It was developed as a method for producing hollow plastic articles (such as bottles and barrels) and although this is still the largest application area for the process, nowadays a wide range of technical mouldings can also be made by this method e.g. rear spoilers on cars and videotape cassettes. There is also a number of variations on the original process but we will start by considering the conventional extrusion blow moulding process.

### Extrusion Blow Moulding

Initially a molten tube of plastic called the *Parison* is extruded through an annular die. A mould then closes round the parison and a jet of gas inflates it to take up the shape of the mould. This is illustrated in Fig. 7.3-21a. Although this process is principally used for the production of bottles (for washing-up liquid, disinfectant, soft drinks, etc.) it is not restricted to small hollow articles. Domestic cold water storage tanks, large storage drums and 200 gallon containers have been blow-moulded. The main materials used are PVC, polyethylene, polypropylene and PET.

The conventional extrusion blow moulding process may be continuous or intermittent. In the former method the extruder continuously supplies molten polymer through the annular die. In most cases the mould assembly moves relative to the die. When the mould has closed around the parison, a hot knife separates the latter from the extruder and the mould moves away for inflation, cooling and ejection of the moulding. Meanwhile the next parison will have been produced and this mould may move back to collect it or, in multi-mould systems, this would have been picked up by another mould. Alternatively in

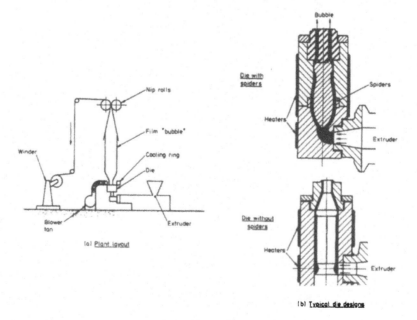

**Fig. 7.3-20** Film blowing process.

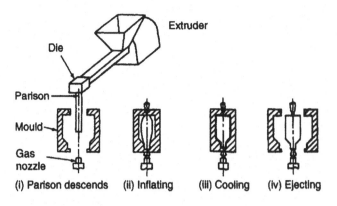

**Fig. 7.3-21** Stages in blow moulding.

some machines the mould assembly is fixed and the required length of parison is cut off and transported to the mould by a robot arm.

In the intermittent processes, single or multiple parisons are extruded using a reciprocating screw or ram accumulator. In the former system the screw moves forward to extrude the parisons and then screws back to prepare the charge of molten plastic for the next shot. In the other system the screw extruder supplies a constant output to an accumulator. A ram then pushes melt from the accumulator to produce a parison as required.

Although it may appear straightforward, in fact the geometry of the parison is complex. In the first place its dimensions will be greater than those of the die due to the phenomenon of post extrusion swelling. Secondly there may be deformities (e.g. curtaining) due to flow defects. Thirdly, since most machines extrude the parison vertically downwards, during the delay between extrusion and inflation, the weight of the parison causes sagging or draw-down. This sagging limits the length of articles which can be produced from a free hanging parison. The complex combination of swelling and thinning makes it difficult to produce articles with a uniform wall thickness. This is particularly true when the cylindrical parison is inflated into an irregularly shaped mould because the uneven drawing causes additional thinning. In most cases therefore to blow mould successfully it is necessary to program the output rate or die gap to produce a controlled non-uniform distribution of thickness in the parison which will give a uniform thickness in the inflated article.

During moulding, the inflation rate and pressure must be carefully selected so that the parison does not burst. Inflation of the parison is generally fast but the overall cycle time is dictated by the cooling of the melt when it touches the mould. Various methods have been tried in order to improve the cooling rate e.g. injection of liquid carbon dioxide, cold air or high pressure moist air. These usually provide a significant reduction in cycle times but since the cooling rate affects the mechanical properties and dimensional stability of the moulding it is necessary to try to optimise the cooling in terms of production rate and quality.

Extrusion blow moulding is continually developing to be capable of producing even more complex shapes. These include unsymmetrical geometries and double wall mouldings. In recent years there have also been considerable developments in the use of in-the-mould transfers. This technology enables labels to be attached to bottles and containers as they are being moulded. Fig. 7.3-22 illustrates three stages in the blow moulding of a complex container.

### Analysis of blow moulding

As mentioned previously, when the molten plastic emerges from the die it swells due to the recovery of elastic deformations in the melt. It will be shown later that the following relationship applies:

$$B_{SH} = B_{ST}^2$$

where

$B_{SH}$ = swelling of the thickness ($= h_1/h_d$)

$B_{ST}$ = swelling of the diameter ($= D_1/D_d$)

therefore

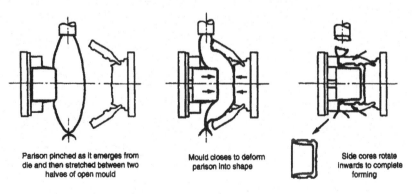

**Fig. 7.3-22** Stages in blow moulding of complex hollow container.

$$\frac{h_1}{h_d} = \left(\frac{D_1}{D_d}\right)^2$$

$$h_1 = h_d(B_{ST})^2$$

Now consider the situation where the parison is inflated to fill a cylindrical die of diameter, $D_m$. Assuming constancy of volume and neglecting draw-down effects, then from Fig. 7.3-23

$$\pi D_1 h_1 = \pi D_m h$$

$$h = \frac{D_1}{D_m} h_1$$

$$= \frac{D_1}{D_m}(h_d \cdot B_{ST}^2)$$

$$= \frac{B_{ST} \cdot D_d}{D_m}(h_d \cdot B_{ST}^2)$$

$$h = B_{ST}^3 h_d \left(\frac{D_d}{D_m}\right) \quad (7.3.18)$$

This expression therefore enables the thickness of the moulded article to be calculated from a knowledge of the die dimensions, the swelling ratio and the mould diameter. The following example illustrates the use of this analysis.

### Example 7.3.2

A blow moulding die has an outside diameter of 30 mm and an inside diameter of 27 mm. The parison is inflated with a pressure of 0.4 MN/m² to produce a plastic bottle of diameter 50 mm. If the extrusion rate used causes a thickness swelling ratio of 2, estimate the wall thickness of the bottle. Comment on the suitability of the production conditions if melt fracture occurs at a stress of 6 MN/m².

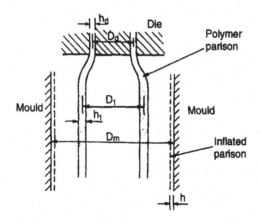

**Fig.7.3-23** Analysis of blow moulding.

### Solution

From equation (7.3.18)

$$\text{wall thickness, } h = B_{ST}^3 h_d \left(\frac{D_d}{D_m}\right)$$

Now

$$h_d = \tfrac{1}{2}(30 - 27) = 1.5 \text{ mm}$$

$$B_{ST} = \sqrt{B_{SH}} = \sqrt{2} = 1.414$$

$$D_d = \tfrac{1}{2}(30 + 27) = 28.5 \text{ mm}$$

So

$$h = (1.414)^3 (1.5)\left(\frac{28.5}{50}\right) = 2.42 \text{ mm}$$

The maximum stress in the inflated parison will be the hoop stress, $\sigma_\theta$, which is given by

$$\sigma_\theta = \frac{PD_m}{2h} = \frac{0.4 \times 50}{2 \times 2.42}$$
$$= 4.13 \text{ MN/m}^2$$

Since this is less than the melt fracture stress (6 MN/m²) these production conditions would be suitable.

### Extrusion stretch blow moulding

Molecular orientation has a very large effect on the properties of a moulded article. During conventional blow moulding the inflation of the parison causes molecular orientation in the hoop direction. However, bi-axial stretching of the plastic before it starts to cool in the mould has been found to provide even more significant improvements in the quality of blow-moulded bottles. Advantages claimed include improved mechanical properties, greater clarity and superior permeation characteristics. Cost savings can also be achieved through the use of lower material grades or thinner wall sections.

Biaxial orientation may be achieved in blow moulding by

(a) stretching the extruded parison longitudinally before it is clamped by the mould and inflated. This is based on the Neck Ring process developed as early as the 1950s. In this case, molten plastic is extruded into a ring mould which forms the neck of the bottle and the parison is then stretched. After the mould closes around the parison, inflation of the

## Processing of plastics CHAPTER 7.3

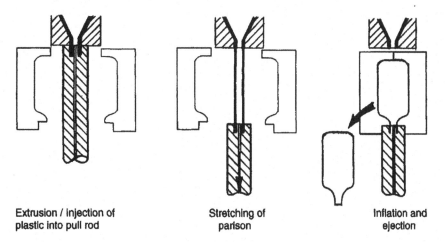

**Fig. 7.3-24** Neck ring stretch blow moulding.

bottle occurs in the normal way. The principle is illustrated in Fig. 7.3-24.

**(b)** producing a preform 'bottle' in one mould and then stretching this longitudinally prior to inflation in the full size bottle mould. This is illustrated in Fig. 7.3-25.

### Injection stretch blow moulding

This is another method which is used to produce biaxially oriented blow moulded containers. However, as it involves injection moulding, the description of this process will be considered in more detail later (Section 7.3.3.9).

### (e) Extrusion coating processes

There are many applications in which it is necessary to put a plastic coating on to paper or metal sheets and the extruder provides an ideal way of doing this. Normally a thin film of plastic is extruded from a slit die and is immediately brought into contact with the medium to be coated. The composite is then passed between rollers to ensure proper adhesion at the interface and to control the thickness of the coating (see Fig. 7.3-26).

Another major type of coating process is wire covering. The tremendous demand for insulated cables in the electrical industry means that large tonnages of plastic are used in this application. Basically a bare wire, which may be heated or have its surface primed, is drawn through a special die attached to an extruder (see Fig. 7.3-27). The drawing speed may be anywhere between 1 m/min and 1000 m/min depending on the diameter of the wire. When the wire emerges from the die it has a coating of plastic, the thickness of which depends on the speed of the wire and the extrusion conditions. It then passes into a cooling trough which may extend for a linear distance of several hundred metres. The coated wire is then wound on to storage drums.

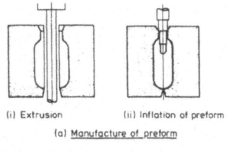
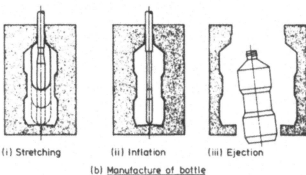

**Fig. 7.3-25** Extrusion stretch blow moulding.

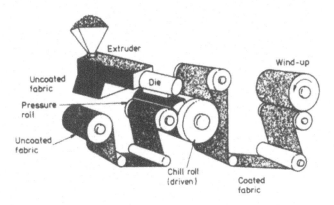

**Fig. 7.3-26** Extrusion coating process.

303

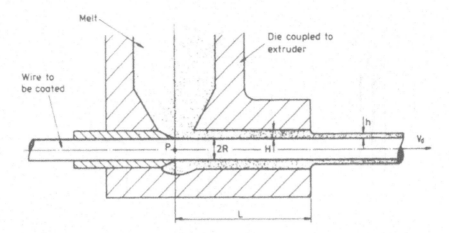

**Fig. 7.3-27** Wire Covering Die.

Wire covering can be analysed in a very similar manner to that described for extrusion. The coating on the wire arises from two effects:

**(a)** *Drag Flow* due to the movement of the wire

**(b)** *Pressure Flow* due to the pressure difference between the extruder exit and the die exit.

From (7.3.2) the drag flow, $Q_d$, is given by

$$Q_d = \frac{1}{2}THV_d \text{ where } T = 2\pi\left(R + \frac{h}{2}\right)$$

From (7.3.6) the pressure flow, $Q_p$, is given by

$$Q_p = \frac{1}{12\eta}\frac{dP}{dz} \cdot TH^3$$

So combining these two equations, the total output, Q, is given by

$$Q = \frac{1}{2}THV_d + \frac{TH^3}{12\eta} \cdot \frac{P}{L} \quad (7.3.19)$$

This must be equal to the volume of coating on the wire so

$$Q = \pi V_d((R+h)^2 - R^2)$$
$$Q = \pi V_d h(2R+h)$$

Combining equations (7.3.19) and (7.3.20)

$$\pi V_d h(2R+h) = \frac{1}{2}THV_d + \frac{TH^3}{12\eta} \cdot \frac{P}{L}$$

from which

$$P = \frac{6\eta L V_d}{H^3}(2H - H) \quad (7.3.21)$$

This is an expression for the pressure necessary at the extruder exit and therefore enables the appropriate extrusion conditions to be set.

### (f) Recent developments in extrusion technology

#### (i) Co-Extrusion

As a result of the wide range of requirements which occur in practice it is not surprising that in many cases there is no individual plastic which has the correct combination of properties to satisfy a particular need. Therefore it is becoming very common in the manufacture of articles such as packaging film, yoghurt containers, refrigerator liners, gaskets and window frames that a multi-layer plastic composite will be used. This is particularly true for extruded film and thermoforming sheets (see Section 7.3.4). In co-extrusion two or more polymers are combined in a single process to produce a multilayer film. These co-extruded films can either be produced by a blown film or a cast film process as illustrated in Fig 7.3-28(a) and (b). The cast process using a slot die and chill roll to cool the film, produces a film with good clarity and high gloss. The film blowing process, however, produces a stronger film due to the transverse orientation which can be introduced and this process offers more flexibility in terms of film thickness.

In most cases there is insufficient adhesion between the basic polymers and so it is necessary to have an adhesive film between each of the layers. Recent investigations of co-extrusion have been centred on methods of avoiding the need for the adhesive layer. The

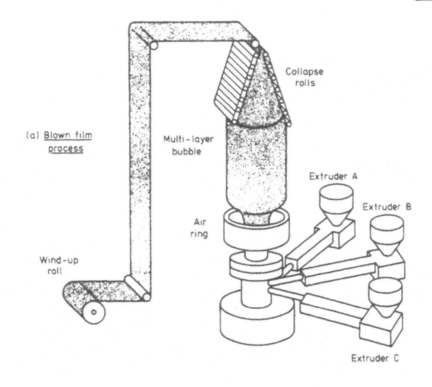

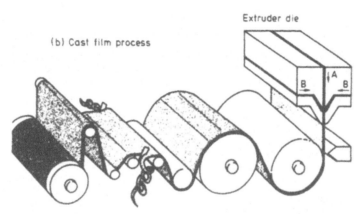

**Fig. 7.3-28** Co-extrusion of plastic film.

most successful seems to be the development of reactive bonding processes in which the co-extruded layers are chemically cross-linked together.

The main reason for producing multi-layer co-extruded films is to get matedais with better barrier properties - particularly in regard to gas permeation. The following Table shows the effects which can be achieved. Data on permeability of plastics are also given in Figs 1.13 and 1.14.

*(ii) Highly oriented grids:*
Net-like polymer grids have become an extremely important development - particularly to civil engineers. The attraction in civil engineering applications is that the open grid structure permits soil particles to interlock through the apertures thus providing an extremely strong reinforcement to the soil. These geogrids under the trade name 'Tensar' are now widely used for road and runway construction, embankment supports, landslide repairs, etc.

The polymer grid achieves its very high strength due to the orientation of the polymer molecules during its manufacture. The process of manufacture is illustrated in Fig. 7.3-29. An extruded sheet, produced to a very fine tolerance and with a controlled structure, has a pattern of holes stamped into it. The hole shapes and pattern can be altered depending on the performance required of the finished product. The perforated sheet is then stretched in one direction to give thin sections of highly orientated polymer with the tensile strength of mild steel. This type

Table 7.3.2 Transmission rates for a range of plastics

| Polymer | Layer distribution ($\mu$m) | Density (kg/m³) | Transmission rates Oxygen (cm³/m² 24 hr atm) | Water vapour (g/m² 24 hr) |
|---|---|---|---|---|
| ABS | 1000 | 1050 | 30 | 2 |
| uPVC | 1000 | 1390 | 5 | 0.75 |
| Polypropylene | 1000 | 910 | 60 | 0.25 |
| PET | 1000 | 1360 | 1 | 2 |
| LDPE | 1000 | 920 | 140 | 0.5 |
| HDPE | 1000 | 960 | 60 | 0.3 |
| PS/EVOH*/PE | 825/25/150 | 1050 | 5† | 1.6 |
| PS/PVdC/PE | 825/50/125 | 1070 | 1 | 0.4 |
| PP/EVOH/PP | 300/40/660 | 930 | 1† | 0.25 |

\* EVOH ethyl vinyl alcohol.
† Depends on humidity.

of grid can be used in applications where uniaxial strength is required. In other cases, where biaxial strength is necessary, the sheet is subjected to a second stretching operation in the transverse direction. The advantages of highly oriented grids are that they are light and very easy to handle. The advantage of obtaining a highly oriented molecular structure is also readily apparent when one compares the stiffness of a HDPE grid ($\simeq 10$ GN/m²) with the stiffness of unoriented HDPE ($\simeq 1$ GN/m²).

*(iii) Reactive Extrusion*

The most recent development in extrusion is the use of the extruder as a 'mini-reactor'. Reactive extrusion is the name given to the process whereby the plastic is manufactured in the extruder from base chemicals and once produced it passes through a die of the desired shape. Currently this process is being used the manufacture of low tonnage materials (<5000 tonnes p.a.) where the cost of a full size reactor run could not be justified. In the future it may be simply part of the production line.

## 7.3.3 Injection moulding

### 7.3.3.1 Introduction

One of the most common processing methods for plastics is injection moulding. Nowadays every home, every vehicle, every office, every factory contains a multitude of different types of articles which have been injection moulded. These include such things as electric drill casings, yoghurt cartons, television housings, combs, syringes, paint brush handles, crash helmets, gearwheels, typewriters, fascia panels, reflectors, telephones, briefcases – the list is endless.

The original injection moulding machines were based on the pressure die casting technique for metals. The first machine is reported to have been patented in the United States in 1872, specifically for use with Celluloid. This was an important invention but probably before its time because in the following years very few developments in injection moulding processes were reported and it was not until the 1920s, in Germany, that a renewed interest was taken in the process. The first German machines were very simple pieces of equipment and relied totally on manual operation. Levers were used to clamp the mould and inject the melted plastic with the result that the pressures which could be attained were not very high. Subsequent improvements led to the use of pneumatic cylinders for clamping the injection which not only lifted some of the burden off the operator but also meant that higher pressures could be used.

The next major development in injection moulding, i.e. the introduction of hydraulically operated machines, did not occur until the late 1930s when a wide range of thermoplastics started to become available. However, these machines still tended to be hybrids based on die casting technology and the design of injection moulding machines for plastics was not taken really seriously until the 1950s when a new generation of equipment was developed.

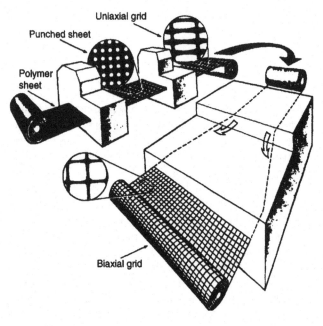

**Fig. 7.3-29** Tensar manufacturing process.

These machines catered more closely for the particular properties of polymer melts and modem machines are of the same basic design although of course the control systems are very much more sophisticated nowadays.

In principle, injection moulding is a simple process. A thermoplastic, in the form of granules or powder, passes from a feed hopper into the barrel where it is heated so that it becomes soft. It is then forced through a nozzle into a relatively cold mould which is clamped tightly closed. When the plastic has had sufficient time to become solid the mould opens, the article is ejected and the cycle is repeated. The major advantages of the process include its versatility in moulding a wide range of products, the ease with which automation can be introduced, the possibility of high production rates and the manufacture of articles with close tolerances. The basic injection moulding concept can also be adapted for use with thermosetting materials.

## 7.3.3.2 Details of the process

The earliest injection moulding machines were of the plunger type as illustrated in Fig. 7.3-30 and there are still many of these machines in use today. A pre-determined quantity of moulding material drops from the feed hopper into the barrel. The plunger then conveys the material along the barrel where it is heated by conduction from the external heaters. The material is thus plasticised under pressure so that it may be forced through the nozzle into the mould cavity. In order to split up the mass of material in the barrel and improve the heat transfer, a torpedo is fitted in the barrel as shown.

Unfortunately there are a number of inherent disadvantages with this type of machine which can make it difficult to produce consistent moulding. The main problems are:

(a) There is little mixing or homogenisation of the molten plastic.

(b) It is difficult to meter accurately the shot size. Since metering is on a volume basis, any variation in the density of the material will alter the shot weight.

(c) Since the plunger is compressing material which is in a variety of forms (varying from a solid granule to a viscous melt) the pressure at the nozzle can vary quite considerably from cycle to cycle.

(d) The presence of the torpedo causes a significant pressure loss.

(e) The flow properties of the melt are pressure sensitive and since the pressure is erratic, this amplifies the variability in mould filling.

Some of the disadvantages of the plunger machine may be overcome by using a pre-plasticising system. This type of machine has two barrels. Raw material is fed into the first barrel where an extruder screw or plunger plasticises the material and feeds it through a non-return valve into the other barrel. A plunger in the second barrel then forces the melt through a nozzle and into the mould. In this system there is much better homogenisation because the melt has to pass through the small opening connecting the two barrels. The shot size can also be metered more accurately since the volume of material fed to the second barrel can be controlled by a limit switch on its plunger. Another advantage is that there is no longer a need for the torpedo on the main injection cylinder.

However, nowadays this type of machine is seldom used because it is considerably more complicated and

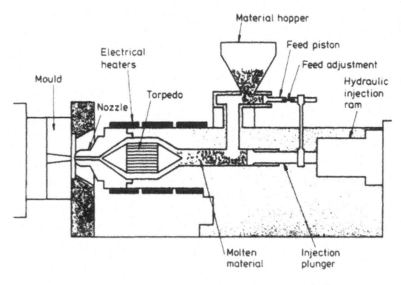

**Fig. 7.3-30** Plunger type injection moulding machine.

more expensive than necessary. One area of application where it is still in use is for large mouldings because a large volume of plastic can be plasticised prior to injection using the primary cylinder plunger.

For normal injection moulding, however, the market is now dominated by the reciprocating screw type of injection moulding machine. This was a major breakthrough in machine design and yet the principle is simple. An extruder type screw in a heated barrel performs a dual role. On the one hand it rotates in the normal way to transport, melt and pressurize the material in the barrel but it is also capable, whilst not rotating, of moving forward like a plunger to inject melt into the mould. A typical injection moulding machine cycle is illustrated in Fig. 7.3-31. It involves the following stages:

**(a)** After the mould closes, the screw (not rotating) pushes forward to inject melt into the cooled mould. The air inside the mould will be pushed out through small vents at the furthest extremities of the melt flow path,

**(b)** When the cavity is filled, the screw continues to push forward to apply a holding pressure (see Fig. 7.3-31). This has the effect of squeezing extra melt into the cavity to compensate for the shrinkage of the plastic as it cools. This holding pressure is only effective as long as the gate(s) remain open.

**(c)** Once the gate(s) freeze, no more melt can enter the mould and so the screw-back commences. At this stage the screw starts to rotate and draw in new plastic from the hopper. This is conveyed to the front of the screw but as the mould cavity is filled with plastic, the effect is to push the screw backwards. This prepares the next shot by accumulating the desired amount of plastic in front of the screw. At a pre-set point in time, the screw stops rotating and the machine sits waiting for the solidification of the moulding and runner system to be completed.

**(d)** When the moulding has cooled to a temperature where it is solid enough to retain its shape, the mould opens and the moulding is ejected. The

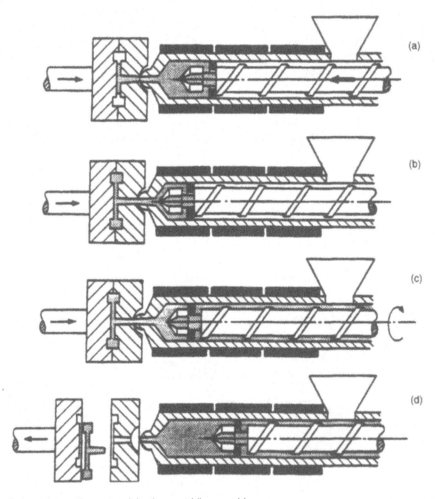

**Fig. 7.3-31** Typical cycle in reciprocating screw injection moulding machine.

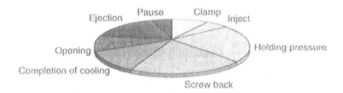

**Fig. 7.3-32** Stages during injection moulding.

mould then closes and the cycle is repeated (see Fig. 7.3-32).

There are a number of important features in reciprocating screw injection moulding machines and these will now be considered in turn.

### Screws

The screws used in these machines are basically the same as those described earlier for extrusion. The compression ratios are usually in the range 2.5:1 to 4:1 and the most common L/D ratios are in the range 15 to 20. Some screws are capable of injecting the plastic at pressures up to 200 MN/m$^2$. One important difference from an extruder screw is the presence of a back-flow check valve at the end of the screw as illustrated in Fig. 7.3-33. The purpose of this valve is to stop any back flow across the flights of the screw when it is acting as a plunger. When material is being conveyed forward by the rotation of the screw, the valve opens as shown. One exception is when injection moulding heat-sensitive materials such as PVC. In such cases there is no check valve because this would provide sites where material could get clogged and would degrade.

### Barrels and heaters

These are also similar to those in extruder machines. In recent years, vented barrels have become available to facilitate the moulding of water sensitive plastics without the need for pre-drying. Water sensitivity in plastics can take several forms. If the plastic absorbs water then dimensional changes will occur, just as with wood or paper. The plastic will also be plasticised by the water so that there will be property changes such as a reduction in modulus and an increase in toughness. All these effects produced by water absorption are reversible.

Another event which may occur is *hydrolysis*. This is a chemical reaction between the plastic and water. It occurs extremely slowly at room temperature but can be significant at moulding temperatures. Hydrolysis causes degradation, reduction in properties (such as impact strength) and it is irreversible. Table 7.3.3 indicates the sensitivity of plastics to moisture. Note that generally extrusion requires a lower moisture content than injection moulding to produce good quality products.

### Nozzles

The nozzle is screwed into the end of the barrel and provides the means by which the melt can leave the barrel and enter the mould. It is also a region where the melt can be heated both by friction and conduction from a heater band before entering the relatively cold channels in the mould. Contact with the mould causes heat transfer from the nozzle and in cases where this is excessive it is advisable to withdraw the nozzle from the mould during the screw-back part of the moulding cycle. Otherwise the plastic may freeze off in the nozzle.

There are several types of nozzle. The simplest is an open nozzle as shown in Fig. 7.3-34a. This is used whenever possible because pressure drops can be minimised and there are no hold up points where the melt can stagnate and decompose. However, if the melt viscosity is low then leakage will occur from this type of nozzle particularly if the barrel/nozzle assembly retracts from the mould each cycle. The solution is to use a shut-off nozzle of which there are many types. Fig. 7.3-34b shows a nozzle which is shut off by external means. Fig. 7.3-34c shows a nozzle with a spring loaded needle valve which opens when the melt pressure exceeds a certain value or alternatively when the nozzle is pressed up against the mould. Most of the shut-off nozzles have the disadvantage that they restrict the flow of the material and provide undersirable stagnation sites. For this reason they should not be used with heat sensitive materials such as PVC.

### Clamping Systems

In order to keep the mould halves tightly closed when the melt is being injected under high pressures it is necessary to have a clamping system. This may be either (a) hydraulic or (b) mechanical (toggle) – or some combination of the two.

In the hydraulic system, oil under pressure is introduced behind a piston connected to the moving platen of the machine. This causes the mould to close and the clamp force can be adjusted so that there is no leakage of molten plastic from the mould.

The toggle is a mechanical device used to amplify force. Toggle mechanisms tend to be preferred for high

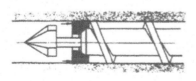
(a) Valve closed

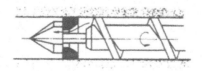
(b) Valve open

**Fig. 7.3-33** Typical check valve.

Table 7.3-3 Water sensitivity of some common plastics

| Drying not required (Materials do not hydrolyse) | Drying required | |
| --- | --- | --- |
| | Absorbs only | Hydrolyses |
| Polyethylene | Acrylic (0.02/0.08)* | PET (0.002/0.002) |
| Polypropylene | ABS (0.02/0.08) | Polycarbonate (0.01/0.02) |
| Polystyrene | SAN (0.02/0.08) | Nylon 66 (0.08/0.15) |
| PVC | | |

\* Required maximum moisture content for extrusion/injection moulding (%)

speed machines and where the clamping force is relatively small. The two main advantages of the toggle system are that it is more economical to run the small hydraulic cylinder and since the toggle is self locking it is not necessary to maintain the hydraulic pressure throughout the moulding cycle. On the other hand the toggle system has the disadvantages that there is no indication of the clamping force and the additional moving parts increase maintenance costs.

### 7.3.3.3 Moulds

In the simplest case an injection mould (or 'tool') consists of two halves into which the impression of the part to be moulded is cut. The mating surfaces of the mould halves are accurately machined so that no leakage of plastic can occur at the split line. If leakage does occur the flash on the moulding is unsightly and expensive to remove. A typical injection mould is illustrated in Fig. 7.3-35. It may by seen that in order to facilitate mounting the mould in the machine and cooling and ejection of the moulding, several additions are made to the basic mould halves. Firstly, backing plates permit the mould to be bolted on to the machine platens. Secondly, channels are machined into the mould to allow the mould temperature to be controlled. Thirdly, ejector pins are included to that the moulded part can be freed from the mould. In most cases the ejector pins are operated by the shoulder screw hitting a stop when the mould opens. The mould cavity is joined to the machine nozzle by means of the **sprue.** The sprue anchor pin then has the function of pulling the sprue away from the nozzle and ensuring that the moulded part remains on the moving half of the mould, when the mould opens. For multi-cavity moulds the impressions are joined to the sprue by **runners** – channels cut in one or both halves of the mould through which the plastic will flow without restriction. A narrow constriction between the runner and the cavity allows the moulding to be easily separated from the runner and sprue. This constriction is called the **gate.**

A production injection mould is a piece of high precision engineering manufactured to very close tolerances by skilled craftsmen. A typical mould can be considered to consist of (i) the cavity and core and (ii) the remainder of the mould (often referred to as the bolster). Of these two, the latter is the more straightforward because although it needs to be accurately made, in general, conventional machine tools can be used. The cavity and core, however, may be quite complex in shape and so they often need special techniques. These can include casting, electro-deposition, hobbing, pressure casting, spark erosion and NC machining.

Finishing and polishing the mould surfaces is also extremely important because the melt will tend to

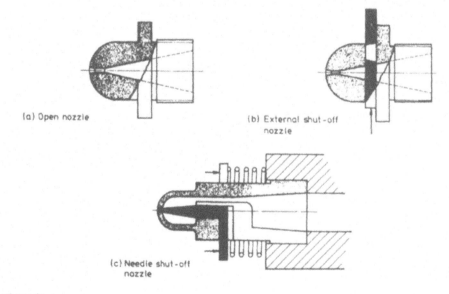

**Fig. 7.3-34** Types of nozzle.

Processing of plastics    CHAPTER 7.3

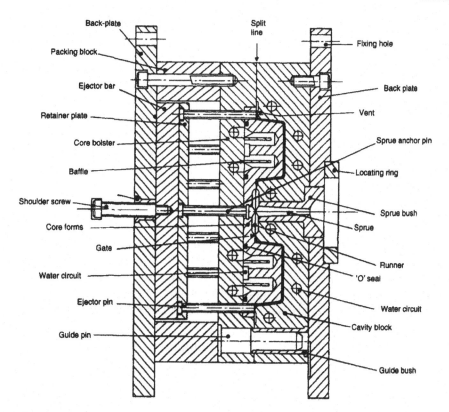

**Fig. 7.3-35** Details of injection mould.

reproduce every detail on the surface of the mould. Finally the mould will have to be hardened to make it stand up to the treatment it receives in service. As a result of all the time and effort which goes into mould manufacture, it is sometimes found that a very complex mould costs more than the moulding machine on which it is used. Several features of the mould are worthy of special mention.

### (a) Gates

As mentioned earlier the gate is the small orifice which connects the runner to the cavity. It has a number of functions. Firstly, it provides a convenient weak link by which the moulding can be broken off from the runner system. In some moulds the degating may be automatic when the mould opens. The gate also acts like a valve in that it allows molten plastic to fill the mould but being small it usually freezes off first. The cavity is thus sealed off from the runner system which prevents material being sucked out of the cavity during screw-back. As a general rule, small gates are preferable because no finishing is required if the moulding is separated cleanly from the runner. So for the initial trials on a mould the gates are made as small as possible and are only opened up if there are mould filling problems.

In a multi-cavity mould it is not always possible to arrange for the runner length to each cavity to be the same. This means that cavities close to the sprue would be filled quickly whereas cavities remote from the sprue receive the melt later and at a reduced pressure. To alleviate this problem it is common to use small gates close to the sprue and progressively increase the dimensions of the gates further along the runners. This has the effect of balancing the fill of the cavities. If a single cavity mould is multi-gated then here again it may be beneficial to balance the flow by using various gate sizes.

Examples of gates which are in common use are shown in Fig. 7.3-36. Sprue gates are used when the sprue bush can feed directly into the mould cavity as, for example, with single symmetrical moulding such as buckets. Pin gates are particularly successful because they cause high shear rates which reduce the viscosity of the plastic and so the mould fills more easily. The side gate is the most common type of gate and is a simple rectangular section feeding into the side of the cavity. A particular attraction of this type of gate is that mould filling can be improved by increasing the width of the gate but the freeze time is unaffected because the depth is unchanged.

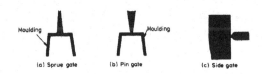

**Fig. 7.3-36** Types of gate.

311

### (b) Runners

The runner is the flow path by which the molten plastic travels from the sprue (i.e. the moulding machine) to the gates (i.e. the cavity). To prevent the runner freezing off prematurely, its surface area should be small so as to minimise heat transfer to the mould. However, the cross sectional area of the runner should be large so that it presents little resistance to the flow of the plastic but not so large that the cycle time needs to be extended to allow the runner to solidify for ejection. A good indication of the efficiency of a runner is, therefore, the ratio of its cross-sectional area to its surface area. For example, a semi-circular channel cut into one half of the mould is convenient to machine but it only has an area ratio of 0.153 D where D is the diameter of the semi-circle. A full round runner, on the other hand, has a ratio of 0.25 D. A square section also has this ratio but is seldom used because it is difficult to eject. A compromise is a trapezoidal section (cut into one half of the mould) or a hexagonal section.

### (c) Sprues

The sprue is the channel along which the molten plastic first enters the mould. It delivers the melt from the nozzle to the runner system. The sprue is incorporated in a hardened steel bush which has a seat designed to provide a good seal with the nozzle. Since it is important that the sprue is pulled out when the mould opens it is tapered as shown in Fig. 7.3-35 and there is a sprue pulling device mounted directly opposite the sprue entry. This can take many forms but typically it would be an undercut or reversed taper to provide a key for the plastic on the moving half of the mould. Since the sprue, like the runner system, is effectively waste it should not be made excessively long.

### (d) Venting

Before the plastic melt is injected, the cavity in the closed mould contains air. When the melt enters the mould, if the air cannot escape it become compressed. At worst this may affect the mould filling, but in any case the sudden compression of the air causes considerable heating. This may be sufficient to burn the plastic and the mould surface at local hot spots. To alleviate this problem, vents are machined into the mating surfaces of the mould to allow the air to escape. The vent channel must be small so that molten plastic will not flow along it and cause unsightly flash on the moulded article. Typically a vent is about 0.025 mm deep and several millimeters wide. Away from the cavity the depth of the vent can be increased so that there is minimum resistance to the flow of the gases out of the mould.

### (e) Mould temperature control

For efficient moulding, the temperature of the mould should be controlled and this is normally done by passing a fluid through a suitably arranged channel in the mould. The rate at which the moulding cools affects the total cycle time as well as the surface finish, tolerances, distortion and internal stresses of the moulded article. High mould temperatures improve surface gloss and tend to eliminate voids. However, the possibility of flashing is increased and sink marks are likely to occur. If the mould temperature is too low then the material may freeze in the cavity before it is filled. In most cases the mould temperatures used are a compromise based on experience.

### Example 7.3.5

The runner lay-out for an eight cavity mould is illustrated in Fig. 7.3-37. If the mould is to be designed so that the pressure at the gate is the same in all cases, determine the radius of the runner in section A. The flow may be assumed to be isothermal.

### Solution

Although this runner system is symmetrical, it is not balanced. If the runner had the same diameter throughout all sections, then the mouldings close to the sprue would fill first and would be over-packed before the outermost cavities were filled. In a good mould design, all the cavities fill simultaneously at the same pressure. In this case it is necessary to ensure that the pressure drop in Sections 1 and 3 is the same as the pressure drop in Section 2.

The pressure drop, $\Delta P$, for isothermal flow in a circular section channel is given by

$$\Delta P = \frac{8\eta L Q}{\pi R^4} \tag{7.3.22}$$

where

$\eta$ = viscosity of the plastic

$L$ = length of channel

$Q$ = volume flow rate

$R$ = radius of channel

If the volume flow rate towards point J is $q$ (ie the input at the sprue is $2q$) then at J the flow will split as follows:

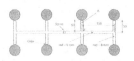

**Fig. 7.3-37** Lay-out for eight cavity mould.

Flow along runner 1 $= xq$

Flow along runner 2 $= 1/2(1-x)q$

Flow along runner 3 $= 1/2xq$

where

$$x = \frac{A_1}{A_1 + 2A_2} = \frac{R_1^2}{R_1^2 + 2R_2^2}$$

(A refers to the area of the relevant runner.)
Using equation (7.3.22) we can write

Pressure loss in runner 1 $= \dfrac{8\eta L_1 xq}{\pi R_1^4}$

Pressure loss in runner 2 $= \dfrac{8\eta L_2(1-x)q}{2\pi R_2^4}$

Pressure loss in runner 3 $= \dfrac{8\eta L_3 xq}{2\pi R_3^4}$

Thus, equating pressure losses after point J

$$\frac{8\eta L_2(1-x)q}{2\pi R_2^4} = \frac{8\eta L_1 xq}{\pi R_1^4} + \frac{8\eta L_3 xq}{2\pi R_3^4}$$

Substituting for $x$ and rearranging to get $R_2$

$$R_2 = \frac{R_1 R_3^2 \sqrt{2L_2}}{\sqrt{2L_1 R_3^4 + L_3 R_1^4}}$$

For the dimensions given:

$R_2 = 3.8$ mm

In practice there are a number of other factors to be taken into account. For example, the above analysis assumes that this plastic is Newtonian, ie that it has a constant viscosity, $\eta$. In reality the plastic melt is non-Newtonian so that the viscosity will change with the different shear rates in each of the three runner sections analysed. In addition, the melt flow into the mould will not be isothermal – the plastic melt immediately in contact with the mould will solidify. This will continuously reduce the effective runner cross-section for the melt coming along behind.

## Multi-daylight moulds

This type of mould, also often referred to as a three plate mould, is used when it is desired to have the runner system in a different plane from the parting line of the moulding. This would be the case in a multi-cavity mould where it was desirable to have a central feed to each

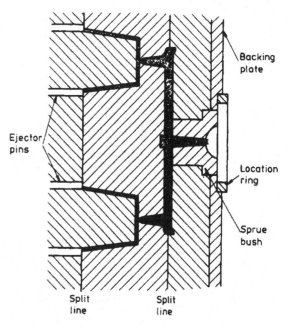

**Fig. 7.3-38** Typical 3-plate mould.

cavity (see Fig. 7.3-38). In this type of mould there is automatic degating and the runner system and sprue are ejected separately from the moulding.

## Hot runner moulds

The runners and sprues are necessary in a mould but they are not part of the end-product. Unfortunately, it is not economically viable to discard them so they must be reground for subsequent reprocessing. Regrinding is expensive and can introduce contamination into the material so that any system which avoids the accumulation of runners and sprues is attractive. A system has been developed to do this and it is really a logical extension of three plate moulding. In this system, strategically placed heaters and insulation in the mould keep the plastic in the runner at the injection temperature. During each cycle therefore the component is ejected but the melt in the runner channel is retained and injected into the cavity during the next shot. A typical mould layout is shown in Fig. 7.3-39.

Additional advantages of hot runner moulds are (i) elimination of trimming and (ii) possibility of faster cycle times because the runner system does not have to freeze off. However, these have to be weighed against the disadvantages of the system. Since the hot runner mould is more complex than a conventional mould it will be more expensive. Also there are many areas in the hot runner manifold where material can get trapped. This means that problems can be experienced during colour or grade changes because it is difficult to remove all of the previous material. As a practical point

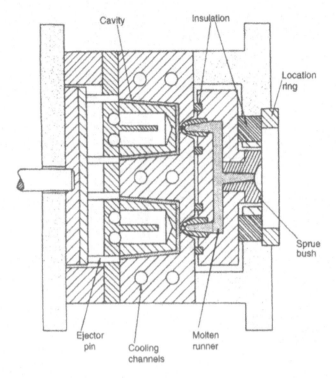

**Fig. 7.3-39** Layout of hot runner mould.

it should also be realised that the system only works as long as the runner remains molten. If the runner system freezes off then the hot runner manifold needs to be dismantled to remove the runners. Note also that hot runner system are not suitable for heat sensitive materials such as PVC.

### Insulated runner moulds

This is similar in concept to the hot runner mould system. In this case, instead of having a specially heated manifold in the mould, large runners (13–25 mm diameter) are used. The relatively cold mould causes a frozen skin to form in the runner which then insulates its core so that this remains molten. As in the previous case the runner remains in the mould when the moulding is ejected and the molten part of the runner is then injected into the cavity for the next shot. If an undue delay causes the whole runner to freeze off then it may be ejected and when moulding is restarted the insulation layer soon forms again. This type of system is widely used for moulding of fast cycling products such as flower pots and disposable goods. The main disadvantage of the system is that it is not suitable for polymers or pigments which have a low thermal stability or high viscosity, as some of the material may remain in a semi-molten form in the runner system for long periods of time.

A recent development of the insulated runner principle is the *distribution tube system*. This overcomes the possibility of freezing-off by insertion of heated tubes into the runners. However, this system still relies on a thick layer of polymer forming an insulation layer on the wall of the runner and so this system is not suitable for heat sensitive materials.

Note that both the insulated runner and the distribution tube systems rely on a cartridge heater in the gate area to prevent premature freezing off at the gate (see Fig. 7.3-40).

### Mould clamping force

In order to prevent 'flashing', i.e. a thin film of plastic escaping out of the mould cavity at the parting line, it is necessary to keep the mould tightly closed during injection of the molten plastic. Before setting up a mould on a machine it is always worthwhile to check that there is sufficient clamping force available on the machine. To do this it is necessary to be able to estimate what clamping force will be needed. The relationship between mould area and clamp requirements has occupied the minds of moulders for many years. Practical

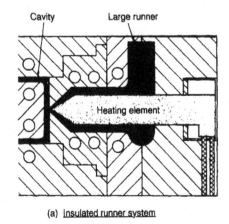
(a) Insulated runner system

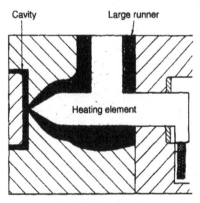
(b) Distributed runner system

**Fig. 7.3-40** Insulated and distributed runner systems.

experience suggests that the clamping pressure over the projected area of the moulding should be between 10 and 50 MN/m² depending on factors such as shape, thickness, and type of material. The mould clamping force may also be estimated in the following way. Consider the moulding of a disc which is centre gated as shown in Fig. 7.3-41a. The force on the shaded element is given by

$$\text{Force, } F = \int_0^R P_r 2\pi r\, dr \quad (7.3.23)$$

The cavity pressure will vary across the disc and it is necessary to make some assumption about this variation. Experimental studies have suggested that an empirical relation of the form

$$P_r = P_0\left(1 - \left(\frac{r}{R}\right)^m\right) \quad (7.3.24)$$

is most satisfactory. $P_0$ is the pressure at the gate and $m$ is a constant which is usually between 0.3 and 0.75. It will be shown later that '$m$' is in fact equal to $(1-n)$ where '$n$' is the index in the Power Law expression for polymer melt flow.

Substituting (7.3.24) in (7.3.23) then

$$F = \int_0^R P_0(1 - (\tfrac{r}{R})^m) 2\pi r\, dr$$

$$F = \pi R^2 P_0 \left(\frac{m}{m+2}\right) \quad (7.3.25)$$

This is a simple convenient expression for estimating the clamping force required for the disc. The same expression may also be used for more complex shapes where the projected area may be approximated as a circle. It will also give sufficiently accurate estimates for a square plate when the radius, $R$, in Fig. 7.3-41a is taken as half of the diagonal.

An alternative way of looking at this equation is that the clamping pressure, based on the projected area of the moulding, is given by

$$\text{Clamping pressure} = \left(\frac{m}{m+2}\right) \times \text{Injection pressure}$$

For any particular material the ratio $(m/(m+2))$ may be determined from the flow curves and it will be temperature and (to some extent) pressure dependent.

In practice the clamping pressure will also depend on the geometry of the cavity. In particular the flow ratio (flow length/channel lateral dimension) is important. Fig. 7.3-42 illustrates typical variations in the Mean Effective Pressure in the cavity for different thicknesses and flow ratios. The data used here is typical for easy flow materials such as polyethylene, polypropylene and polystyrene. To calculate the clamp force, simply multiply the appropriate Mean Effective Pressure by the projected area of the moulding. In practice it is prudent to increase this value by 10–20% due to the uncertainties associated with specific moulds.

For plastics other than the easy flow materials referred to above, it would be normal to apply a factor to allow for the higher viscosity. Typical viscosity factors are given below.

| Material | Viscosity Factor |
| --- | --- |
| Polyethylene, polypropylene, polystyrene | 1 |
| Nylon 66 | 1.2 → 1.4 |
| ABS | 1.3 → 1.4 |
| Acrylic | 1.5 → 1.7 |
| PVC | 1.6 → 1.8 |
| Polycarbonate | 1.7 → 2.0 |

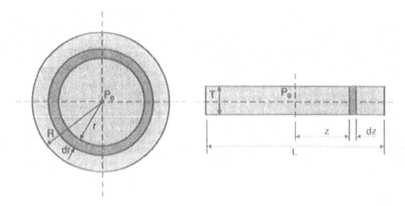

**Fig. 7.3-41** Clamp force analysis.

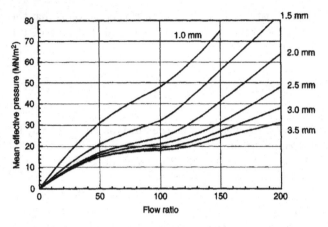

**Fig. 7.3-42** Clamping pressures for different cavity geometries (typical values for easy flow materials).

### Example 7.3.6

The mould shown in Fig. 7.3-35 produces four cup shaped ABS mouldings. The depth of the cups is 60 mm, the diameter at the is 90 mm and the wall thickness is 1.0 mm. The distance from the sprue to the cavity is 40 mm and the runner diameter is 6 mm. Calculate the clamp force necessary on the moulding machine and estimate how the clamp force would change if the mould was designed so as to feed the cups through a pin gate in the centre of the base (as illustrated in Fig. 7.3-38). The clamp pressure data in Fig. 7.3-42 should be used and the taper on the side of the cups may be ignored.

### Solution

(a) Within the cavity, the maximum flow length for the plastic melt will be from the gate, along the side of the cup and across the base of the cup, ie

Flow length $= 60 + 90 = 150$ mm

The thickness of the moulding is 1 mm, hence the flow ratio $= 150/1 = 150$. From Fig. 7.3-42 at this thickness and flow ratio, the mean effective pressure is 75 MN/m².

Allowing an extra 15% for uncertainties and applying the viscosity factor of 1.4 for ABS, then the appropriate mean effective pressure is $75 \times 1.15 \times 1.4 = 120$ MN/m². For each cavity, the projected area is $(\pi/4)(90)^2 = 6360$ mm² $= 6.36 \times 10^{-3}$ m².

Hence, clamp force per cavity $= 120 \times 6.36 \times 10^{-3}$
$= 763$ kN.

The projected area of the runners is $4 \times 40 \times 6 = 960$ mm².

Assuming that the mean effective pressure also applies to the runner system, then

clamp force for runners $= 120 \times 0.96 \times 10^{-3}$
$= 115$ kN

Hence total clamp force for 4 cavities and 1 runner system is given by

Total clamp force $= (4 \times 763) + 115 = 3167$ kN

The required clamp force is therefore 317 tonnes.

(b) If a pin gate in the middle of the base is used instead of an edge gate, then the flow ratio will be different. In this case

flow length $= \frac{1}{2}(90) + 60 = 105$

This is also the flow ratio, so from Fig. 7.3-42 the mean effective pressure is 50 MN/m². Applying the viscosity factor, etc as above, then

Clamp force per cavity $= 50 \times 1.15 \times 1.4$
$\times \left(\frac{\pi}{4}\right)(90)^2 \times 10^{-6}$
$= 512$ kN

In this case the runner system will be almost totally in the 'shadow' of the projected area of the cavities and so they can be ignored.

Hence, total clamp force $= 4 \times 512$
$= 2048$ kN $= 205$ tonnes.

Another common shape which is moulded is a thin rectangular strip. Consider the centre gated strip as shown in Fig. 7.3-41b. In the same way as before the clamping force, $F$, is given by

$$F = 2 \int_0^{L/2} P_z T \, dz \quad (7.3.26)$$

$$F = 2 \int_0^{L/2} P_0 \left(1 - \left(\frac{z}{L/2}\right)^m\right) T \, dz$$

i.e. $F = T P_0 L \left(\dfrac{m}{m+1}\right) \quad (7.3.27)$

## 7.3.3.4 Structural foam injection moulding

Foamed thermoplastic articles have a cellular core with a relatively dense (solid) skin. The foam effect is achieved by the dispersion of inert gas throughout the molten resin directly before moulding. Introduction of the gas is usually carried out either by pre-blending the resin with a chemical blowing agent which releases gas when heated or by direct injection of the gas (usually nitrogen).

When the compressed gas/resin mixture is rapidly injected into the mould cavity, the gas expands explosively and forces the material into all parts of the mould. The advantages of these types of foam moulding are

(a) for a given weight they are many times more rigid than a solid moulding

(b) they are almost completely free from orientation effects and the shrinkage is uniform

(c) very thick sections can be moulded without sink marks.

Foamed plastic articles may be produced with good results using normal screw-type injection moulding machines (see Fig.7.3-43a). However, the limitations on shot size, injection speed and platen area imposed by conventional injection equipment prevent the full large-part capabilities of structural foam from being realised. Specialised foam moulding machines currently in use can produce parts weighing in excess of 50 kg (see Fig. 7.3-43b).

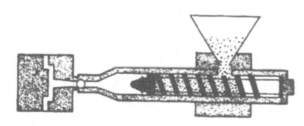

(a) Standard injection moulding press

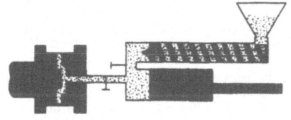

(b) Specialised foam moulding press

**Fig. 7.3-43** Structural foam moulding.

Wall sections in foam moulding are thicker than in solid material. Longer cycle times can therefore be expected due to both the wall thickness and the low thermal conductivity of the cellular material. In contrast, however, the injection pressures in foam moulding are low when compared with conventional injection moulding. This means that less clamping force is needed per unit area of moulding and mould costs are less because lower strength mould materials may be used.

## 7.3.3.5 Sandwich moulding

This is an injection moulding method which permits material costs to be reduced in large mouldings. In most mouldings it is the outer surface of an article which is important in terms of performance in service. If an article has to be thick in order that it will have adequate flexural stiffness then the material within the core of the article is wasted because its only function is to keep the outer surfaces apart. The philosophy of sandwich moulding is that two different materials (or two forms of the same material) should be used for the core and skin. That is, an expensive high performance material is used for the skin and a low-cost commodity or recycled plastic is used for the core. The way that this can be achieved is illustrated in Fig. 7.3-44.

Initially the skin material is injected but not sufficient to fill the mould. The core material is then injected and it flows laminarly into the interior of the core. This continues until the cavity is filled as shown in Fig.7.3-44c. Finally the nozzle valve rotates so that the skin material is injected into the sprue thereby clearing the valve of core material in preparation for the next shot. In a number of cases the core material is foamed to produce a sandwich section with a thin solid skin and a cellular core.

It is interesting that in the latest applications of sandwich moulding it is the core material which is being regarded as the critical component. This is to meet design requirements for computers, electronic equipment and some automotive parts. In these applications there is a growing demand for covers and housings with electromagnetic interference (EMI) shielding. The necessity of using a plastic with a high loading of conductive filler (usually carbon black) means that surface finish is poor and unattractive. To overcome this the sandwich moulding technique can be used in that a good quality surface can be moulded using a different plastic.

## 7.3.3.6 Gas Injection moulding

In recent years major developments have been made in the use of an inert gas to act as the core in an injection moulded plastic product. This offers many advantages

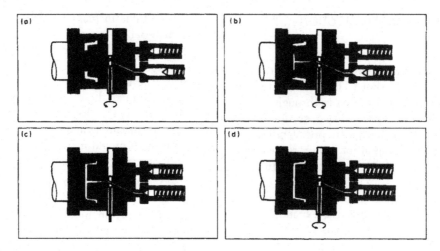

**Fig. 7.3-44** Stages in sandwich moulding process.

including greater stiffness/weight ratios and reduced moulded-in stresses and distortion.

The first stage of the cycle is the flow of molten polymer into the mould cavity through a standard feed system. Before this flow of polymer is complete, the injection of a predetermined quantity of gas into the melt begins through a special nozzle located within the cavity or feed system as shown in Fig. 7.3-45. The timing, pressure and speed of the gas injection is critical.

The pressure at the polymer gate remains high and, therefore, the gas chooses a natural path through the hotter and less viscous parts of the polymer melt towards the lower pressure areas. The flow of gas cores out a hollow centre extending from its point of entry towards the last point of fill. By controlling the amount of gas injected into the hollow core, the pressure on the cooling polymer is controlled and maintained until the moulding is packed. The final stage is the withdrawal of the gas nozzle, prior to mould opening, which allows the gas held in the hollow core to vent.

The gas injection process overcomes many of the limitations of injection mouldings such as moulded-in stress and distortion. These limitations are caused by laminar flow and variation in pressure throughout the moulding. With the gas injection process, laminar flow is considerably reduced and a uniform pressure is maintained. The difficulty of transmitting a very high pressure uniformly throughout a moulding can also cause inconsistent volumetric shrinkage of the polymer, and this leads to isolated surface sink marks. Whilst cycle times are comparable with those of conventional injection moulding, clamping forces are much lower. Also, by using gas to core out the polymer instead of mixing with it, gas-injection overcomes a number of shortcomings of the structural foam process. In particular there are no surface imperfections (caused by escaping gas bubbles in structural foam moulding) and cycle times are lower because thinner sections are being cooled.

### 7.3.3.7 Shear controlled orientation in injection moulding (SCORIM)

One of the major innovations in recent years is the use of pulsed pressure through the gates to introduce and control the orientation of the structure (or fillers) in injection moulded products. A special manifold is attached to the machine nozzle as illustrated in Fig. 7.3-46 . This diagram relates to the *double live feed* of melt although up to four pistons, capable of applying oscillating pressure may be used.

Shear controlled orientation in injection moulding (SCORIM) is based on the progressive application of macroscopic shears at the melt-solid interface during solidification in the moulding of a polymer matrix.

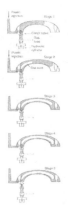

**Fig. 7.3-45** Stages in the gas injection moulding of an automotive handle (courtesy of Cinpress Ltd).

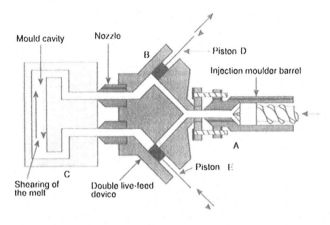
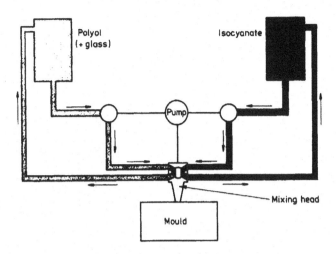

**Fig. 7.3-46** One embodiment of SCORIM where the device (B) for producing shear during solidification, by the action of pistons (D) and (E), is placed between the injection moulding machine barrel (A) and the mould (C) (Courtesy of Brunel University).

**Fig. 7.3-47** Schematic view of reaction injection moulding.

Macroscopic shears of specified magnitude and direction, applied at the melt-solid interface provide several advantages:

(i) Enhanced polymer matrix or fibre alignment by design in moulded polymers or fibre reinforced polymers.

(ii) Elimination of mechanical discontinuities that result from the initial mould filling process, including internal weld lines.

(iii) Reduction in the detrimental effects of a change in moulded section thickness.

(iv) Elimination or reduction in defects resulting from the moulding of thick sectioned components.

## 7.3.3.8 Reaction injection moulding

Although there have been for many years a number of moulding methods (such as hand lay-up of glass fibres in polyester and compression moulding of thermosets or rubber) in which the plastic material is manufactured at the same time as it is being shaped into the final article, it is only recently that this concept has been applied in an injection moulding type process. In Reaction Injection Moulding (RIM), liquid reactants are brought together just prior to being injected into the mould. In-mould polymerisation then takes place which forms the plastic at the same time as the moulding is being produced. In some cases reinforcing fillers are incorporated in one of the reactants and this is referred to as Reinforced Reaction Injection Moulding (RRIM)

The basic RIM process is illustrated in Fig. 7.3-47. A range of plastics lend themselves to the type of fast polymerisation reaction which is required in this process – polyesters, epoxies, nylons and vinyl monomers.

However, by far the most commonly used material is polyurethane. The components A and B are an isocyanate and a polyol and these are kept circulating in their separate systems until an injection shot is required. At this point the two reactants are brought together in the mixing head and injected into the mould.

Since the reactants have a low viscosity, the injection pressures are relatively low in the RIM process. Thus, comparing a conventional injection moulding machine with a RIM machine having the same clamp force, the RIM machine could produce a moulding with a much greater projected area (typically about 10 times greater). Therefore the RIM process is particularly suitable for large area mouldings such as car bumpers and body panels. Another consequence of the low injection pressures is that mould materials other than steel may be considered. Aluminium has been used successfully and this permits weight savings in large moulds. Moulds are also less expensive than injection moulds but they must not be regarded as cheap. RIM moulds require careful design and, in particular, a good surface finish because the expansion of the material in the mould during polymerisation causes every detail on the surface of the mould to be reproduced on the moulding.

## 7.3.3.9 Injection blow moulding

In Section 7.3.2.7 we considered the process of extrusion blow moulding which is used to produce hollow articles such as bottles. At that time it was mentioned that if molecular orientation can be introduced to the moulding then the properties are significantly improved. In recent years the process of injection blow moulding has been developed to achieve this objective. It is now very widely used for the manufacture of bottles for soft drinks.

The steps in the process are illustrated in Fig. 7.3-48. Initially a preform is injection moulded. This is

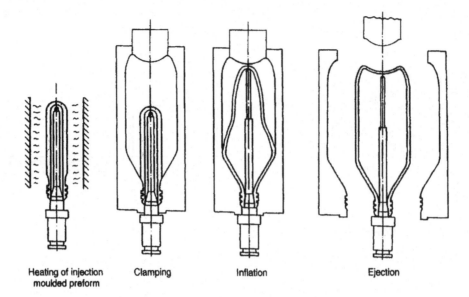

**Fig. 7.3-48** Injection blow moulding process.

subsequently inflated in a blow mould in order to produce the bottle shape. In most cases the second stage inflation step occurs immediately after the injection moulding step but in some cases the preforms are removed from the injection moulding machine and subsequently re-heated for inflation.

The advantages of injection blow moulding are that

(i) the injection moulded parison may have a carefully controlled wall thickness profile to ensure a uniform wall thickness in the inflated bottle.

(ii) it is possible to have intricate detail in the bottle neck.

(iii) there is no trimming or flash (compare with extrusion blow moulding).

A variation of this basic concept is the *Injection Orientation Blow Moulding* technique developed in the 1960s in the USA but upgraded for commercial use in the 1980s by AOKI in Japan. The principle is very similar to that described above and is illustrated in Fig. 7.3-49. It may be seen that the method essentially combines injection moulding, blow moulding and thermoforming to manufacture high quality containers.

### 7.3.3.10 Injection moulding of thermosetting materials

In the past the thought of injection moulding thermosets was not very attractive. This was because early trials had shown that the feed-stock was not of a consistent quality which meant that continual alterations to the machine settings were necessary. Also, any undue delays could cause premature curing of the resin and consequent blockages in the system could be difficult to remove. However, in recent years the processing characteristics of thermosets have been improved considerably so that injection moulding is likely to become one of the major production methods for these materials. The injection moulding of fibre reinforced thermosets, such as DMC (Section 7.3.10.2), is also becoming very common.

Nowadays, the injection moulder can be supplied with uniform quality granules which consist of partially polymerised resin, fillers and additives. The formulation of the material is such that it will flow easily in the barrel with a slow rate of polymerisation. The curing is then completed rapidly in the mould. In most respects the process is similar to the injection moulding of thermoplastics and the sequence of operations in a single cycle is as described earlier. For thermosets a special barrel and screw are used. The screw is of approximately constant depth over its whole length and there is no check valve which might cause material blockages (see Fig. 7.3-50). The barrel is only kept warm (80–110°C) rather than very hot as with thermoplastics because the material must not cure in this section of the machine. Also, the increased viscosity of the thermosetting materials means that higher screw torques and injection pressures (up to 200 MN/m$^2$ are needed).

On the mould side of the machine the major difference is that the mould is maintained very hot (150–200°C) rather than being cooled as is the case with thermoplastics. This is to accelerate the curing of the material once it has taken up the shape of the cavity. Another difference is that, as thermosetting materials are abrasive and require higher injection pressures, harder

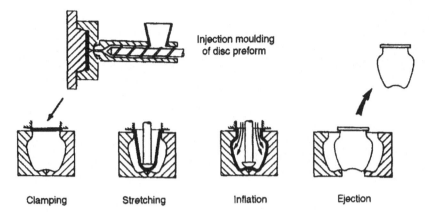

**Fig. 7.3-49** Injection orientation stretch blow moulding.

steels with extra wear resistance should be used for mould manufacture. As a result of the abrasive nature of the thermosets, hydraulic mould clamping is preferred to a toggle system because the inevitable dust from the moulding powder increases the wear in the linkages of the latter.

When moulding thermosetting articles, the problem of material wastage in sprues and runners is much more severe because these cannot be reused. It is desirable therefore to keep the sprue and runner sections of the mould cool so that these do not cure with the moulding. They can then be retained in the mould during the ejection stage and then injected into the cavity to form the next moulding. This is analogous to the hot runner system described earlier for thermoplastics.

The advantages of injection moulding thermosets are as follows:

**(a)** fast cyclic times (see Table 7.3-4)
**(b)** efficient metering of material
**(c)** efficient pre-heating of material
**(d)** thinner flash – easier finishing
**(e)** lower mould costs (fewer impressions).

## 7.3.4 Thermoforming

When a thermoplastic sheet is heated it becomes soft and pliable and the techniques for shaping this sheet are known as thermoforming. This method of manufacturing plastic articles developed in the 1950s but limitations such as poor wall thickness distribution and large peripheral waste restricted its use to simple packaging applications. In recent years, however, there have been major advances in machine design and material availability with the result that although packaging is still the major market sector for the process, a wide range of other products are made by thermoforming. These include aircraft window reveals, refrigerator liners, baths, switch panels, car bumpers, motorbike fairings etc.

The term 'thermoforming' incoroporates a wide range of possibilities for sheet forming but basically there are two sub-divisions – vacuum forming and pressure forming.

### (a) Vacuum forming

In this processing method a sheet of thermoplastic material is heated and then shaped by reducing the air pressure between it and a mould. The simplest type of

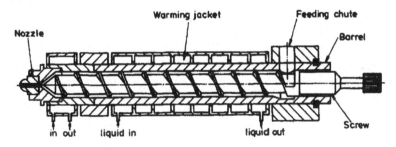

**Fig. 7.3-50** Injection moulding of thermosets and rubbers.

Table 7.3-4 For the same part, injection moulding of thermosets can offer up to 25% production increase and lower part-costs than compression.

| Compression moulding | Minutes |
|---|---|
| Open mould, unload piece | 0.105 |
| Mould cleaning | 0.140 |
| Close machine, start pressure | 0.100 |
| Moulding cycle time | 2.230 |
| Total compression cycle | 2.575 |
| *Injection moulding* | |
| Unload piece, open/close machine | 0.100 |
| Moulding cycle time | 1.900 |
| Total injection cycle | 2.000 |

vacuum forming is illustrated in Fig. 7.3-51a. This is referred to as *Negative Forming* and is capable of providing a depth of draw which is 1/3–1/2 of the maximum width. The principle is very simple. A sheet of plastic, which may range in thickness from 0.025 mm to 6.5 mm, is clamped over the open mould. A heater panel is then placed above the sheet and when sufficient softening has occurred the heater is removed and the vacuum is applied. For the thicker sheets it is essential to have heating from both sides.

In some cases Negative Forming would not be suitable because, for example, the shape formed in Fig. 7.3-51 would have a wall thickness in the corners which is considerably less than that close to the clamp. If this was not acceptable then the same basic shape could be produced by *Positive Forming*. In this case a male (positive) mould is pushed into the heated sheet before the vacuum is applied. This gives a better distribution of material and deeper shapes can be formed – depth to width ratios of 1:1 are possible. This thermoforming method is also

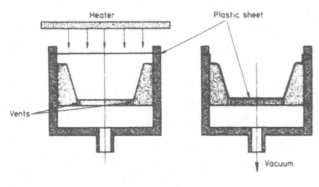

**Fig. 7.3-51** Vacuum forming process.

referred to as *Drape Forming*. Another alternative would be to have a female mould as in Fig. 7.3-51 but after the heating stage and before the vacuum is applied, a plug comes down and guides the sheet into the cavity. When the vacuum is applied the base of the moulding is subjected to less draw and the result is a more uniform wall thickness distribution. This is called *Plug Assisted Forming*. Note that both Positive Forming and Plug Assisted Forming effectively apply a pre-stretch to the plastic sheet which improves the performance of the material quite apart from the improved wall thickness distribution.

In the packaging industry *skin* and *blister* vacuum machines are used. Skin packaging involves the encapsulation of articles between a tight, flexible transparent skin and a rigid backing which is usually cardboard. Blister packs are preformed foils which are sealed to a rigid backing card when the goods have been inserted.

The heaters used in thermoforming are usually of the infra red type with typical loadings of between 10 and 30 kW/m$^2$. Normally extra heat is concentrated at the clamped edges of the sheet to compensate for the additional heat losses in this region. The key to successful vacuum forming is achieving uniform heating over the sheet. One of the major attractions of vacuum forming is that since only atmospheric pressure is used to do the shaping, the moulds do not have to be very strong. Materials such as plaster, wood and thermosetting resins have all been used successfully. However, in long production runs mould cooling becomes essential in which case a metal mould is necessary. Experience has shown that the most satisfactory metal is undoubtedly aluminium. It is easily shaped, has good thermal conductivity, can be highly polished and has an almost unlimited life.

Materials which can be vacuum formed satisfactorily include polystyrene, ABS, PVC, acrylic, polycarbonate, polypropylene and high and low density polyethylene. Co-extruded sheets of different plastics and multi-colour laminates are also widely used nowadays. One of the most recent developments is the thermoforming of crystallisable PET for high temperature applications such as oven trays. The PET sheet is manufactured in the amorphous form and then during thermoforming it is permitted to crystallise. The resulting moulding is thus capable of remaining stiff at elevated temperatures.

### (b) Pressure forming

This is generally similar to vacuum forming except that pressure is applied above the sheet rather than vacuum below it. The advantage of this is that higher pressures can be used to form the sheet. A typical system is illustrated in Fig. 7.3-52 and in recent times this has become attractive as an alternative to injection

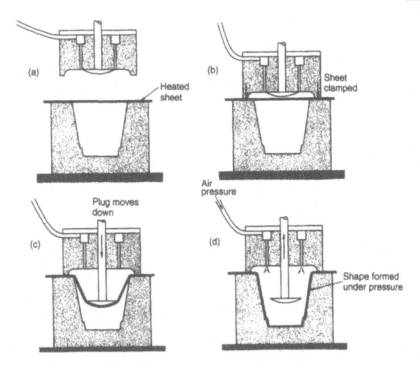

**Fig. 7.3-52** Pressure forming process.

moulding for moulding large area articles such as machine housings.

### (c) Matched die forming

A variation of thermoforming which does not involve gas pressure or vacuum is matched die forming. The concept is very simple and is illustrated in Fig. 7.3-53. The plastic sheet is heated as described previously and is then sandwiched between two halves of a mould. Very precise detail can be reproduced using this thermoforming method but the moulds need to be more robust than for the more conventional process involving gas pressure or vacuum.

### (d) Dual-sheet thermoforming

This technique, also known as Twin-Sheet Forming, is a recent development. It is essentially a hybrid of blow moulding and thermoforming. Two heated sheets are placed between two mould halves and clamped as shown in Fig. 7.3-54. An inflation tube at the parting line then injects gas under pressure so that the sheets are forced out against the mould. Alternatively, a vacuum can be drawn between the plastic sheet and the mould in each half of the system. This technique has interesting possibilities for further development and will compete with blow moulding, injection moulding and rotational moulding in a number of market sectors. It can be noted that the two mould halves can be of different shapes and the two plastic sheets could be of different materials, provided a good weld can be obtained at the parting line.

### 7.3.4.1 Analysis of thermoforming

If a thermoplastic sheet is softened by heat and then pressure is applied to one of the sides so as to generate a freely blown surface, it will be found that the shape so formed has a uniform thickness. If this was the case during thermoforming, then a simple volume balance between the original sheet and the final shape could provide the wall thickness of the end product.

$$A_i h_i = A_f h_f \qquad (7.3.28)$$

where $A$ = surface area, and $h$ = wall thickness ('i' and 'f' refer to initial and final conditions).

### Example 7.3.4

A rectangular box 150 mm long, 100 mm wide and 60 mm deep is to be thermoformed from a flat sheet 150 mm x 100 mm x 2 mm. Estimate the average thickness of the walls of the final product if (a) conventional vacuum forming is used and (b) plug assisted moulding is used (the plug being 140 mm x 90 mm).

### Solution

**(a)** The initial volume of the sheet is given by

$$A_i h_i = 150 \times 100 \times 2 = 3 \times 10^4 \text{ mm}^3$$

The surface area of the final product is

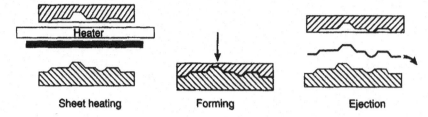

Fig. 7.3-53 Thermoforming between matched dies.

$$A_f = (150 \times 100) + 2(100 \times 60) + 2(150 \times 60)$$
$$= 4.5 \times 10^4 \text{ mm}^2$$

Therefore, from equation (7.3.28)

$$h_f = \frac{3 \times 10^4}{4.5 \times 10^4} = 0.67 \text{ mm}$$

**(b)** If plug assist is used then it could be assumed that over the area 140 mm × 90 mm, the wall thickness will remain at 2 mm. The volume of this part of the moulding will be

$$\text{Vol} = 140 \times 90 \times 2 = 2.52 \times 10^4 \text{ mm}^3$$

This would leave a volume of $(3 \times 10^4 - 2.52 \times 10^4)$ to form the walls. The area of the walls is

$$A_w = (2 \times 100 \times 60) + (2 \times 150 \times 60)$$
$$= 3 \times 10^4 \text{ mm}^2$$

This ignores a small area in the base of the box, outside the edges of the plug. Hence, the thickness of the walls in this case would be

$$h_w = \frac{(3 \times 10^4) - (2.52 \times 10^4)}{3 \times 10^4} = 0.16 \text{ mm}$$

These calculations can give a useful first approximation of the dimensions of a thermoformed part. However, they will not be strictly accurate because in a real situation, when the plastic sheet is being stretched down into the cold mould it will freeze off at whatever thickness it has reached when it touches the mould.

Consider the thermoforming of a plastic sheet of thickness, $h_0$, into a conical mould as shown in Fig. 7.3-55a. At this moment in time, $t$, the plastic is in contact with the mould for a distance, $S$, and the remainder of the sheet is in the form of a spherical dome of radius, $R$, and thickness, $h$. From the geometry of the mould the radius is given by

$$R = \frac{H - S \sin \alpha}{\sin \alpha \tan \alpha} \qquad (7.3.29)$$

Also the surface area, $A$, of the spherical bubble is given by

$$A = 2\pi R^2 (1 - \cos \alpha) \qquad (7.3.30)$$

At a subsequent time, $(t + dt)$, the sheet will be formed to the shape shown in Fig. 7.3-55b. The change in thickness of the sheet in this period of time may be estimated by assuming that the volume remains constant.

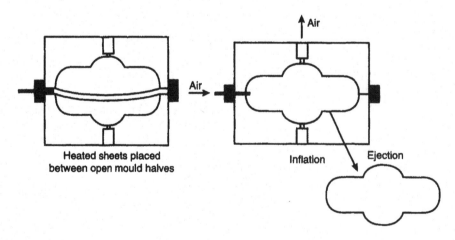

Fig. 7.3-54 Dual sheet forming.

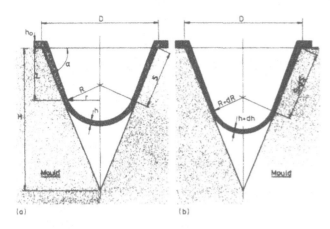

**Fig. 7.3-55** Analysis of thermo forming.

$$2\pi R^2(1-\cos\alpha)h = 2\pi(R+dR)^2(1-\cos\alpha)(h+dh) + 2\pi rh\, dS \sin\alpha$$

Substituting for $r(=R\sin\alpha)$ and for $R$ from (7.3.29) this equation may be reduced to the form

$$\frac{dh}{h} = \left[2-\left(\frac{\sin^2\alpha \tan\alpha}{1-\cos\alpha}\right)\right] \cdot \frac{\sin\alpha\, dS}{(H-S\sin\alpha)} \quad (7.3.31)$$

This equation may be integrated with the boundary condition that $h=h_1$ at $S=0$. As a result the thickness, $h$, at a distance, $S$, along the side of the conical mould is given by

$$h = h_1\left(\frac{H-S\sin\alpha}{H}\right)^{\sec\alpha-1} \quad (7.3.32)$$

Now consider again the boundary condition referred to above. At the point when the softened sheet first enters the mould it forms part of a spherical bubble which does not touch the sides of the cone. The volume balance is therefore

$$\left(\frac{D^2}{4}\right)h_0 = \frac{2(D/2)^2(1-\cos\alpha)h_1}{\sin^2\alpha}$$

So,

$$h_1 = \frac{\sin^2\alpha}{2(1-\cos\alpha)} \cdot h_0$$

Making the substitution for $h_1$ in (7.3.32)

$$h = \frac{\sin^2\alpha}{2(1-\cos\alpha)}\left[\frac{H-S\sin\alpha}{H}\right]^{\sec\alpha-1} \cdot h_0 \quad (7.3.33)$$

or

$$h/h_0 = \left(\frac{1+\cos\alpha}{2}\right)\left[\frac{H-L}{H}\right]^{\sec\alpha-1}$$

This equation may also be used to calculate the wall thickness distribution in deep truncated cone shapes but note that its derivation is only valid up to the point when the spherical bubble touches the centre of the base. Thereafter the analysis involves a volume balance with freezing-off on the base and sides of the cone.

### Example 7.3.8

A small flower pot as shown in Fig. 7.3-56 is to be thermoformed using negative forming from a flat plastic sheet 2.5 mm thick. If the diameter of the top of the pot is 70 mm, the diameter of the base is 45 mm and the depth is 67 mm estimate the wall thickness of the pot at a point 40 mm from the top. Calculate also the draw ratio for this moulding.

### Solution

(a) $\alpha = \tan^{-1}\left(\dfrac{67}{12.45}\right) = 79.4°$

Using the terminology from Fig. 7.3-39b

$H = 35\tan\alpha = 187.6$ mm

From equation (7.3.33)

$$h/h_0 = \left(\frac{1+\cos 79.4°}{2}\right)\left(\frac{187.6-40}{187.6}\right)^{(\sec 79.4)-1}$$

$$= 0.203$$

$$h = 0.203 \times 2.5 = 0.51 \text{mm}$$

(b) The *draw-ratio* for a thermoformed moulding is the ratio of the area of the product to the initial area of the sheet. In this case therefore

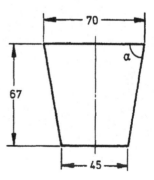

**Fig. 7.3-56** Thermoformed flower pot.

$$\text{Draw ratio} = \frac{\pi\sqrt{[(R-r)^2 + h^2](R+r)} + \pi r^2}{\pi R^2}$$

$$= \frac{\pi\sqrt{[(35-22.5)^2 + 67^2](35+22.5)} + \pi(22.5)^2}{\pi(35)^2}$$

$$= 3.6$$

## 7.3.5 Calendering

Calendering is a method of producing plastic film and sheet by squeezing the plastic through the gap (or 'nip') between two counter-rotating cylinders. The art of forming a sheet in this way can be traced to the paper, textile and metal industries. The first development of the technique for polymeric materials was in the middle 19th century when it was used for mixing additives into rubber. The subsequent application to plastics was not a complete success because the early machines did not have sufficient accuracy or control over such things as cylinder temperature and the gap between the rolls. Therefore acceptance of the technique as a viable production method was slow until the 1930s when special equipment was developed specifically for the new plastic materials. As well as being able to maintain accurately roll temperature in the region of 200°C these new machines had power assisted nip adjustment and the facility to adjust the rotational speed of each roll independently. These developments are still the main features of modern calendering equipment.

Calenders vary in respect of the number of rolls and of the arrangement of the rolls relative to one another. One typical arrangement is shown in Fig. 7.3-57 – the inverted L-type. Although the calendering operation as illustrated here looks very straightforward it is not quite as simple as that. In the production plant a lot of ancillary equipment is needed in order to prepare the plastic material for the calender rolls and to handle the sheet after the calendering operation. A typical sheet production unit would start with premixing of the polymer, plasticiser, pigment, etc in a ribbon mixer followed by gelation of the premix in a Banbury Mixer and/or a short screw extruder. At various stages, strainers and metal detectors are used to remove any foreign matter. These preliminary operations result in a material with a dough-like consistency which is then supplied to the calender rolls for shaping into sheets.

However, even then the process is not complete. Since the hot plastic tends to cling to the calender rolls it is necessary to peel it off using a high speed roll of smaller diameter located as shown in Fig. 7.3-57. When the sheet leaves the calender it passes between embossing rolls and then on to cooling drums before being trimmed and stored on drums. For thin sheets the speed of the winding drum can be adjusted to control the drawdown. Outputs vary in the range 0.1–2 m/s depending on the sheet thickness.

Calendering can achieve surprising accuracy on the thickness of a sheet. Typically the tolerance is ±0.005 mm but to achieve this it is essential to have very close control over roll temperatures, speeds and proximity. In addition, the dimensions of the rolls must be very precise. The production of the rolls is akin to the manufacture of an injection moulding tool in the sense that very high machining skills are required. The particular features of a calender roll are a uniform specified surface finish, minimal eccentricity and a special barrel profile ('crown') to compensate for roll deflection under the very high presurres developed between the rolls.

Since calendering is a method of producing sheet/film it must be considered to be in direct competition with extrusion based processes. In general, film blowing and die extrusion methods are preferred for materials such as polyethylene, polypropylene and polystyrene but calendering has the major advantage of causing very little thermal degradation and so it is widely used for heat sensitive materials such as PVC.

### 7.3.5.1 Analysis of calendering

A detailed analysis of the flow of molten plastic between two rotating rolls is very complex but fortunately sufficient accuracy for many purposes can be achieved by using a simple Newtonian model. The assumptions made are that

**(a)** the flow is steady and laminar

**(b)** the flow is isothermal

**(c)** the fluid is incompressible

**(d)** there is no slip between the fluid and the rolls.

If the clearance between the rolls is small in relation to their radius then at any section $x$ the problem may be analysed as the flow between parallel plates at a distance $h$ apart. The velocity profile at any section is thus made

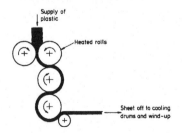

**Fig. 7.3-57** Typical arrangement of calender rolls.

up of a drag flow component and a pressure flow component.

For a fluid between two parallel plates, each moving at a velocity $V_d$, the drag flow velocity is equal to $V_d$. In the case of a calender with rolls of radius, $R$, rotating at a speed, $N$, the drag velocity will thus be given by $2\pi RN$.

The velocity component due to pressure flow between two parallel plates has already been determined in Section 7.3.2.3b.

$$V_p = \frac{1}{2\eta}\frac{dP}{dx}(y^2 - (h/2)^2)$$

Therefore the total velocity at any section is given by

$$V = V_d + \frac{1}{2\eta}\frac{dP}{dx}[y^2 - (h/2)^2]$$

Considering unit width of the calender rolls the total throughput, $Q$, is given by

$$Q = 2\int_0^{h/2} V dy$$
$$= 2\int_0^{h/2}\left[V_d + \frac{1}{2\eta}\frac{dP}{dx}(y^2 - (h/2)^2)\right]dy$$
$$= h\left(V_d - \frac{h^2}{12\eta}\frac{dP}{dx}\right) \tag{7.3.34}$$

Since the output is given by $V_d H$

then

$$V_d H = h\left(V_d - \frac{h^2}{12\eta}\frac{dP}{dx}\right) \tag{7.3.35}$$

From this it may be seen that $\dfrac{dP}{dx} = 0$ at $h = H$.

To determine the shape of the pressure profile it is necessary to express $h$ as a function of $x$. From the equation of a circle it may be seen that

$$h = H_0 + 2(R - (R^2 - x^2)^{1/2}) \tag{7.3.36}$$

However, in the analysis of calendering this equation is found to be difficult to work with and a useful approximation is obtained by expanding $(R^2 - x^2)^{1/2}$ using the binomial series and retaining only the first two terms. This gives

$$h = H_0\left(1 + \frac{x^2}{H_0 R}\right) \tag{7.3.37}$$

Therefore as shown earlier $dP/dx$ will be zero at

$$H = H_0\left(1 + \frac{x^2}{H_0 R}\right)$$
$$x = \pm\sqrt{(H - H_0)R'} \tag{7.3.38}$$

This gives a pressure profile of the general shape shown in Fig. 7.3-58. The value of the maximum pressure may be obtained by rearranging (7.3.35) and substituting for $h$ from (7.3.37)

$$\frac{dP}{dx} = \frac{12\eta V_d\left(H_0 - H + \dfrac{x^2}{R}\right)}{\left(H_0 + \dfrac{x^2}{R}\right)^3} \tag{7.3.39}$$

If this equation is integrated and the value of $x$ from (7.3.38) substituted then the maximum pressure may be obtained as

$$P_{max} = \frac{3\eta V_d}{H_0}\left(2\omega - \frac{(4H_0 - 3H)}{H_0}\right.$$
$$\left.\left(\omega + \sqrt{\frac{R}{H_0}}\tan^{-1}\sqrt{\left(\frac{H - H_0}{H}\right)}\right)\right) \tag{7.3.40}$$

where

$$\omega = \frac{\sqrt{(H - H_0)R}}{H} \tag{7.3.41}$$

### Example 7.3.9

A calender having rolls of diameter 0.4 m produces plastic sheet 2 m wide at the rate of 1300 kg/hour. If the nip between rolls is 10 mm and the exit velocity of the sheet is

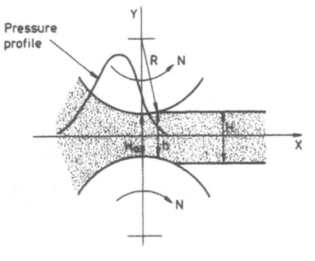

Fig 7.3.58 Melt flow between calender rolls.

0.01 m/s estimate the position and magnitude of the maximum pressure. The density of the material is 1400 kg/m³ and its viscosity is $10^4$ Ns/m².

**Solution**

Flow rate, Q = 1300 kg/hour = 0.258 x $10^{-3}$ m³/s
but
$Q = HWV_d$ where $W$ = width of sheet
So

$$H = \frac{0.258 \times 10^{-3}}{2 \times 0.01} = 12.9 \, \text{mm}$$

The distance upstream of the nip at which the pressure is a maximum is given by equation (7.3.38)

$$x = \sqrt{(12.9 - 10)200} = 24.08 \, \text{mm}$$

Also from (7.3.37)

$$P_{\max} = \frac{3 \times 10^4 \times 0.01}{10 \times 10^{-3}}\{(2 \times 1.865) \\ - 0.13[1.865 + (4.45)(0.494)]\} \\ = 96 \, \text{kN/m}^2$$

## 7.3.6 Rotational moulding

Rotational moulding, like blow moulding, is used to produce hollow plastic articles. However, the principles in each method are quite different. In rotational moulding a carefully weighed charge of plastic powder is placed in one half of a metal mould. The mould halves are then clamped together and heated in an oven. During the heating stage the mould is rotated about two axes at right angles to each other. After a time the plastic will be sufficiently softened to form a homogeneous layer on the surface of the mould. The latter is then cooled while still being rotated. The final stage is to take the moulded article from the mould.

The process was originally developed in the 1940s for use with vinyl plastisols in liquid form. It was not until the 1950s that polyethylene powders were successfully moulded in this way. Nowadays a range of materials such as nylon, polycarbonate, ABS, high impact polystyrene and polypropylene can be moulded but by far the most common material is polyethylene.

The process is attractive for a number of reasons. Firstly, since it is a low pressure process the moulds are generally simple and relatively inexpensive. Also the moulded articles can have a very uniform thickness, can contain reinforcement, are virtually strain free and their surface can be textured if desired. The use of this moulding method is growing steadily because although the cycle times are slow compared with injection or blow moulding, it can produce very large, thick walled articles which could not be produced economically by any other technique. Wall thicknesses of 10 mm are not a problem for rotationally moulded articles.

There is a variety of ways in which the cycle of events described above may be carried out. For example, in some cases (particularly for very large articles) the whole process takes place in one oven. However, a more common set-up is illustrated in Fig. 7.3-59. The mould is on the end of an arm which first carries the cold mould containing the powder into a heated oven. During heating the mould rotates about the arm (major) axis and also about its own (minor) axis (see Fig. 7.3-60). After a pre-set time in the oven the arm brings the mould into a cooling chamber. The

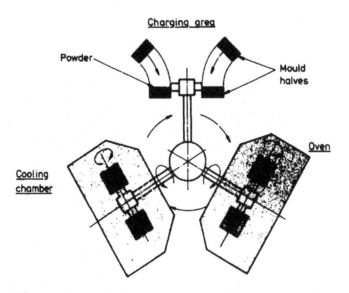

**Fig. 7.3-59** Typical rotational moulding process.

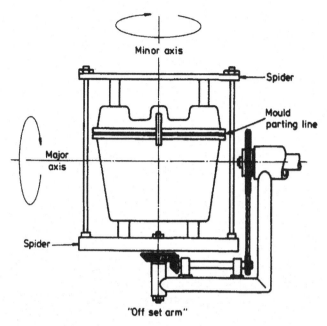

**Fig. 7.3-60** Typical 'off-set arm' rotation.

rate of cooling is very important. Clearly, fast cooling is desirable for economic reasons but this may cause problems such as warping. Normally therefore the mould is initially cooled using blown air and this is followed by a water spray. The rate of cooling has such a major effect on product quality that even the direction of the air jets on the mould during the initial gradual cooling stage can decide the success or otherwise of the process. As shown in Fig. 7.3-59 there are normally three arms (mould holders) in a complete system so that as one is being heated another is being cooled and so on. In many machines the arms are fixed rigidly together and so the slowest event (heating, cooling or charging/discharging) dictates when the moulds progress to the next station. In some modern machines, the arms are independent so that if cooling is completed then that arm can leave the cooling bay whilst the other arms remain in position.

It is important to realise that rotational moulding is not a centrifugal casting technique. The rotational speeds are generally below 20 rev/min with the ratio of speeds about the major and minor axes being typically 4 to 1. Also since all mould surfaces are not equidistant from the centre of rotation any centrifugal forces generated would tend to cause large variations in wall thickness. In fact in order to ensure uniformity of all thickness it is normal design practice to arrange that the point of intersection of the major and minor axis does not coincide with the centroid of the mould.

The heating of rotational moulds may be achieved using infra-red, hot liquid, open gas flame or hot-air convection. However, the latter method is the most common. The oven temperature is usually in the range 250–450°C and since the mould is cool when it enters the oven it takes a certain time to get up to a temperature which will melt the plastic. This time may be estimated as follows.

When the mould is placed in the heated oven, the heat input (or loss) per unit time must be equal to the change in internal energy of the material (in this case the mould).

$$hA(T_0 - T) = \rho C_p V \left(\frac{dT}{dt}\right) \quad (7.3.42)$$

where $h$ is the convective heat transfer coefficient
$A$ is the surface area of mould
$T_0$ is the temperature of the oven
$T_t$ is the temperature of the mould at time $t$
$\rho$ is the density of the mould material
$C_p$ is the specific heat of the mould material
$V$ is the volume of the walls of the mould
and $t$ is time

Rearranging this equation and integrating then

$$hA \int_0^t dt = \rho C_p V \int_{T_i}^{T_t} \frac{dT}{(T_0 - T)}$$

$$hAt = -\rho C_p V \log_e \left(\frac{T_0 - T_t}{T_0 - T_i}\right)$$

$$\left(\frac{T_0 - T_t}{T_0 - T_i}\right) = e^{-h\beta t/\rho C_p} \quad (7.3.43)$$

where $T_i$ is the initial temperature of the mould and $\beta$ is the surface area to volume ratio ($A/V$).

This equation suggests that there is an exponential rise in mould temperature when it enters the oven, and in practice this is often found to be the case.

Fig. 7.3-61 illustrates typical temperature profiles during the rotational moulding of polyethylene. With typical values of oven temperatures and data for an aluminium mould

$T_0 = 300°C, \quad T_i = 30°C, \quad T_t = 20°C$
$h = 22 \text{ W/m}^2\text{K} \quad C_p = 917 \text{ J/kg K}, \, p = 2700 \text{ kg/m}^3$

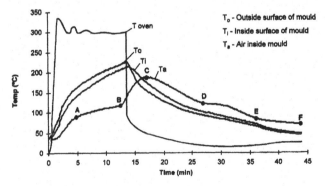

**Fig. 7.3-61** Temperature profiles during rotational moulding.

then for an aluminium cube mould 330 mm side and 6 mm thick, as was used to produce Fig. 7.3-61 then

$$t = \frac{-\rho C_p}{\beta h} \log_e\left\{\frac{T_o - T_t}{T_o - T_i}\right\} = \frac{-2700 \times 917}{1000 \times 22}$$
$$\log_e\left\{\frac{330 - 220}{330 - 30}\right\}$$
$$t = 1.9 \text{ minutes}$$

For a steel mould of the same dimensions and thickness, a quick calculation ($h = 11$ W/m$^2$K, $C_p = 480$ J/kg K and $\rho = 7850$ kg/m$^3$) shows that the steel mould would take three times longer to heat up. However, in practice, steel moulds are less than a third of the thickness of aluminium. Therefore, although aluminium has a better thermal conductivity, steel moulds tend to heat up more quickly because they are thinner.

It is important to note that the above calculation is an approximation for the time taken to heat the mould to any desired temperature. Fig. 7.3-61 shows that in practice it takes considerably longer for the mould temperature to get to 220°C. This is because although initially the mould temperature is rising at the rate predicted in the above calculation, once the plastic starts to melt, it absorbs a significant amount of the thermal energy input.

Fig. 7.3-61 illustrates that the mould temperature is quite different from the set oven temperature (330°C) or indeed the actual oven temperature, throughout the moulding cycle. An even more important observation is that in order to control the rotational moulding process it is desirable to monitor the temperature of the air inside the mould. This is possible because there is normally a vent tube through the mould wall in order to ensure equal pressures inside and outside the mould. This vent tube provides an easy access for a thermocouple to measure the internal air temperature.

The internal air temperature characteristic has a unique shape which shows clearly what is happening at all stages throughout the process. Up to point A in Fig. 7.3-61 there is simply powder tumbling about inside the mould. At point A the mould has become sufficiently hot that plastic starts to melt and stick to the mould. The melting process absorbs energy and so over the region AB, the internal air temperature rises less quickly. It may also be seen that the temperature of the mould now starts to rise less quickly. At B all the plastic has melted and so a larger proportion of the thermal energy input goes to heating the inner air. This temperature rises more rapidly again, at a rate similar to that in the initial phase of the process.

Over the region BC the melt is effectively sintering because at B it is a powdery mass loosely bonded together whereas at C it has become a uniform melt. The value of temperature at C is very important because if the oven period is too short, then the material will not have sintered properly and there will be an excess of pin-holes. These are caused where the powder particles have fused together and trapped a pocket of air. If the oven period is too long then the pin-holes will all have disappeared but thermal/oxidative degradation will have started at the inner surface of the moulding. Extensive tests have shown that this is a source of brittleness in the mouldings and so the correct choice of temperature at C is a very important quality control parameter. For most grades of polyethylene the optimum temperature is in the region of 200°C ±5°C.

Once the mould is removed from the oven the mould starts to cool at a rate determined by the type of cooling – blown air (slow) or water spray (fast). There may be a overshoot in the internal air temperature due to the thermal momentum of the melt. This overshoot will depend on the wall thickness of the plastic product. In Fig. 7.3-61 it may be seen that the inner air temperature continues to rise for several minutes after the mould has been taken out of the oven (at about 13.5 minutes).

During cooling, a point D is reached where the internal air temperature decreases less quickly for a period. This represents the solidification of the plastic and because this process is exothermic, the inner air cannot cool so quickly. Once solidification is complete, the inner air cools more rapidly again. Another kink (point E) may appear in this cooling curve and, if so, it represents the point where the moulding has separated from the mould wall. In practice this is an important point to keep consistent because it affects shrinkage, warpage, etc in the final product. Once the moulding separates from the mould, it will cool more slowly and will tend to be more crystalline, have greater shrinkage and lower impact strength.

Developments in rotational moulding are continuing, with the ever increasing use of features such as

**(i)** mould pressurisation (to consolidate the melt, remove pin-holes, reduce cycle times and provide more consistent mould release),

**(ii)** internal heating/cooling (to increase cycle times and reduce warpage effects).

In overall terms the disadvantages of rotational moulding are its relative slowness and the limited choice of plastics which are commercially available in powder form with the correct additive package. However, the advantages of rotational moulding in terms of stress-free moulding, low mould costs, fast lead times and easy control over wall thickness distribution (relative to blow moulding) means that currently rotational moulding is the fastest growing sector of the plastics processing industry. Typical annual growth rates are between 10 and 12% p.a.

### 7.3.6.1 Slush moulding

This is a method for making hollow articles using liquid plastics, particularly PVC plastisols. A shell-like mould is heated to a pre-determined temperature (typically 130°C for plastisols) and the liquid is then poured into the mould to completely fill it. A period of time is allowed to elapse until the required thickness of plastic gels. The excess liquid is then poured out and the plastic skin remaining in the mould is cured in an oven. The moulding is then taken from the mould.

It should be noted that when the plastisol liquid gels it has sufficient strength to remain in position on the inside surface of the mould. However, it has insufficient tear strength to be useful and so it has to go through the higher temperature curing stage to provide the necessary toughness and strength in the end-product. The mould is not rotated during slush moulding.

## 7.3.7 Compression moulding

Compression moulding is one of the most common methods used to produce articles from thermosetting plastics. The process can also be used for thermoplastics but this is less common – the most familiar example is the production of LP records. The moulding operation as used for thermosets is illustrated in Fig. 7.3-62. A pre-weighed charge of partially polymerised thermoset is placed in the lower half of a heated mould and the upper half is then forced down. This causes the material to be squeezed out to take the shape of the mould. The application of the heat and pressure accelerates the polymerisation of the thermoset and once the cross-linking ('curing') is completed the article is solid and may be ejected while still very hot. Mould temperatures are usually in the range of 130–200°C Cycle times may be long (possibly several minutes) so it is desirable to have multi-cavity moulds to increase production rates. As a result, moulds usually have a large projected area so the closing force needed could be in the region of 100–500 tonnes to give the 7–25 $MN/m^2$ cavity pressure needed. It should also be noted that compression moulding is also used for Dough Moulding Compounds (DMC) – these will be considered in Section 7.3.10.2

During compression moulding, the charge of material may be put into the mould either as a powder or a preformed 'cake'. In both cases the material is preheated to reduce the temperature difference between it and the mould. If the material is at a uniform temperature in the mould then the process may be analysed as follows.

Consider a 'cake' of moulding resin between the compression platens as shown in Fig. 7.3-63. When a constant force, $F$, is applied to the upper platen the resin flows as a result of a pressure gradient. If the flow is assumed Newtonian then the pressure flow equation derived in Section 7.3.2.3 may be used

$$\text{flow rate, } Q_p = \frac{1}{12\eta}\left(\frac{dP}{dz}\right) TH^3 \quad (7.3.6)$$

For the annular element of radius, $r$, in Fig. 7.3-63 it is more convenient to use cylindrical co-ordinates so this equation may be rewritten as

$$Q_p = \frac{1}{12\eta}\left(\frac{dP}{dr}\right) \cdot (2\pi r) H^3$$

Now if the top platen moves down by a distance, $dH$, the volume displaced is $(\pi r^2 dH)$ and the volume flow rate is $\pi r^2 (dH/dt)$. Therefore

$$\pi r^2 \left(\frac{dH}{dt}\right) = \frac{1}{12\eta}\left(\frac{dP}{dr}\right) \cdot (2\pi r) H^3$$

$$\frac{12\eta}{H^3} \cdot \frac{dH}{dt} = \frac{2}{r}\frac{dP}{dr} \quad (7.3.44)$$

This simple differential equation is separable and so each side may be solved in turn.
Let

$$\frac{2}{r}\frac{dP}{dr} = A \text{ where } A = f(H)$$

so

$$\int_r^P dP = \frac{A}{2}\int_R^r r\, dr$$

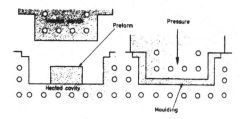

**Fig. 7.3-62** Principle of compression moulding.

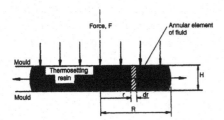

**Fig. 7.3-63** Analysis of compression moulding.

or

$$P = \frac{A}{4}(r^2 - R^2)$$

Now the force on the element is $2\pi r dr(P)$ so the total force, $F$, is given by integrating across the platen surface.

$$F = \int_0^R 2\pi r \left(\frac{A}{4}\right)(r^2 - R^2)dr = -\frac{\pi A R^4}{8}$$

This may be rearranged to give

$$A = -\frac{8F}{\pi R^4} = -\frac{8\pi F H^2}{V^2}$$

where $V = \pi R^2 H$
Substituting for $A$ in (7.3.44)

$$-\frac{8\pi F H^2}{V^2} = \frac{12\eta}{H^3}\frac{dH}{dt}$$

So

$$-\int_0^t \frac{2\pi F}{3\eta V^2}dt = \int_{H_0}^H \frac{dH\, 2\pi Ft}{H^5\, 3\eta V^2} = \frac{1}{4}\left(\frac{1}{H^4} - \frac{1}{H_0^4}\right)$$

Since $H_0 \gg H$ then $(1/H_0^4)$ may be neglected. As a result the compaction force $F$, is given by

$$F = \frac{3\eta V^2}{8\pi t H^4} \tag{7.3.45}$$

where $H$ is the platen separation at time, $t$.

### Example 7.3.10
A circular plate with a diameter of 0.3 m is to be compression moulded from phenol formaldehyde. If the preform is cylindrical with a diameter of 50 mm and a depth of 36 mm estimate the platen force needed to produce the plate in 10 seconds. The viscosity of the phenol may be taken as $10^3$ Ns/m$^2$.

**Solution**

Volume

$$V = \pi\left(\frac{50}{2}\right)^2 \times 36 = \pi\left(\frac{300}{2}\right)^2 H$$

So $\quad H = 1$ mm

From (7.3.45)

$$F = \frac{3\eta V^2}{8\pi t H^4} = \frac{3 \times 10^3 \times (\pi \times 625 \times 36)^2}{10^6 \times 8\pi \times 10 \times (1)^4}$$

$$= 59.6 \text{ kN}$$

## 7.3.8 Transfer moulding

Transfer moulding is similar to compression moulding except that instead of the moulding material being pressurized in the cavity, it is pressurized in a separate chamber and then forced through an opening and into a closed mould. Transfer moulds usually have multi-cavities as shown in Fig. 7.3-64. The advantages of transfer moulding are that the preheating of the material and injection through a narrow orifice improves the temperature distribution in the material and accelerates

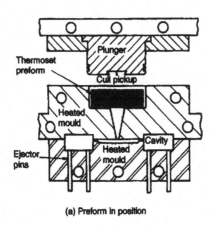
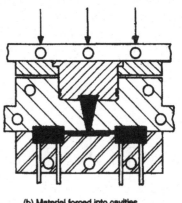

(a) Preform in position  (b) Material forced into cavities

**Fig. 7.3-64** Transfer moulding of thermosetting materials.

the crosslinking reaction. As a result the cycle times are reduced and there is less distortion of the mouldings. The improved flow of the material also means that more intricate shapes can be produced.

The success of transfer moulding prompted further developments in this area and clearly it was only a relatively small step to an injection moulding process for thermosets as described in Section 7.3.3.10.

## 7.3.9 Processing reinforced thermoplastics

Fibre reinforced thermoplastics can be processed using most of the conventional thermoplastic processing methods described earlier. Extrusion, rotational moulding, blow moulding and thermoforming of short fibre reinforced thermoplastics are all possible, but the most important commercial technique is injection moulding. In most respects this process is similar to the moulding of un-reinforced thermoplastics but there are a number of important differences. For example the melt viscosity of a reinforced plastic is generally higher than the un-reinforced material. As a result the injection pressures need to be higher, by up to 80% in some cases. In addition the cycle times are generally lower because the greater stiffness of the material allows it to be ejected from the mould at a higher temperature than normal. However, the increased stiffness can also hamper ejection from the mould so it is important to have adequate taper on side walls of the cavity and a sufficient number of strategically placed ejector pins. Where possible a reciprocating screw machine is preferred to a plunger machine because of the better mixing, homogenisation, metering and temperature control of the melt. However, particular attention needs to be paid to such things as screw speed and back pressure because these will tend to break up the fibres and thus affect the mechanical properties of the mouldings.

A practical difficulty which arises during injection moulding of reinforced plastics is the increased wear of the moulding machine and mould due to the abrasive nature of the fibres. However, if hardened tool steels are used in the manufacture of screws, barrels and mould cavities then the problem may be negligible.

An inherent problem with all of the above moulding methods is that they must, by their nature, use short fibres (typically 0.2–0.4 mm long). As a result the full potential of the reinforcing fibres is not realised. In recent years therefore, there have been a number of developments in reinforced thermoplastics to try to overcome these problems. One approach has been to produce continuous fibre tapes or mats which can be embedded in a thermoplastic matrix. The best known materials of this type are the Aromatic Polymer Composites (APC) and the glass mat reinforced thermoplastics (GMT). One of the most interesting of these consists of unidirectional carbon fibres in a matrix of polyetheretherketone (PEEK). The material comes in the form of a wide tape which may be arranged in layers in one half of a mould to align the unidirectional fibres in the desired directions. The assembly is then pressurised between the two matched halves of the heated mould. The result is a laminated thermoplastic composite containing continuous fibres aligned to give maximum strength and stiffness in the desired directions.

Another recent development has been the arrival of special injection moulding grades of thermoplastics containing long fibres. At the granule production stage the thermoplastic lace contains continuous fibres and to achieve this it is produced by pultrusion (see Section 7.3.10.3) rather than the conventional compounding extruder. The result is that the granules contain fibres of the same length as the granule ($\simeq 10$ mm).

These long fibres give better product performance although injection moulding machine modifications may be necessary to prevent fibre damage and reduce undesirable fibre orientation effects in the mould.

## 7.3.10 Processing reinforced thermosets

There is a variety of ways in which fibre reinforcement may be introduced into thermosetting materials and as a result there is a range of different methods used to process these materials. In many cases the reinforcement is introduced during the fabrication process so that its extent can be controlled by the moulder. Before looking at the possible manufacturing methods for fibre reinforced thermosetting articles it is worth considering the semantics of fibre technology. Because of their fibre form, reinforcing materials have borrowed some of their terminology from the textile industry.

**Filament** This is a single fibre which is continuous or at least very long compared with its diameter.

**Yarn or Roving** Continuous bundle of filaments generally fewer than 10,000 in number.

**Tow** A large bundle of fibres generally 10,000 or more, not twisted.

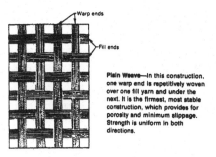

**Fig. 7.3-65** Plain weave fibre fabric.

**Fabric, Cloth or Mat** Woven strands of filament. The weave pattern used depends on the flexibility and balance of strength properties required in the warp and fill directions. Fig. 7.3-65 shows a plain weave in which the strength is uniform in both directions. The warp direction refers to the direction parallel to the length of the fabric. Fabrics are usually designated in terms of the number of yarns of filament per unit length of warp and fill direction.

**Chopped Fibres:** These may be subdivided as follows

Milled Fibres: These are finely ground or milled fibres. Lengths range from 30 to 3000 microns and the fibre (L/D) ratio is typically about 30. Fibres in this form are popular for closed mould manufacturing methods such as injection moulding.

Short Chopped Fibres: These are fibres with lengths up to about 6 mm. The fibre (L/D) ratio is typically about 800. They are more expensive than milled fibres but provide better strength and stiffness enhancement.

Long Chopped Fibres: These are chopped fibres with lengths up to 50 mm. They are used mainly in the manufacture of SMC and DMC (see Section 7.3.10.2).

**Chopped Strand Mat** This consists of strands of long chopped fibres deposited randomly in the form of a mat. The strands are held together by a resinous binder.

## Manufacturing methods

The methods used for manufacturing articles using fibre reinforced thermosets are almost as varied as the number of material variations that exist. They can, however, be divided into three main categories. These are manual, semiautomatic and automatic.

The *Manual* processes cover methods such as hand lay-up, spray-up, pressure bag and autoclave moulding.

The *Semi-Automatic* processes include processes such as cold pressing, hot pressing, compression moulding of SMC and DMC, resin injection.

The *Automatic* processes are those such as pultrusion, filament winding, centrifugal casting and injection moulding.

Typical market shares for the different methods are shown in Fig. 7.3-66.

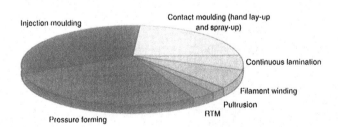

**Fig. 7.3-66** Typical market shares for composite moulding methods in Europe.

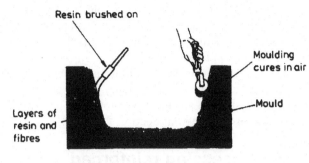

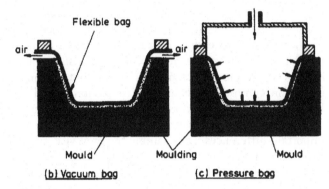

**Fig. 7.3-67** Hand lay-up techniques.

## 7.3.10.1 Manual processing methods

### (a) Hand Lay-Up:

This method is by far the most widely used processing method for fibre reinforced materials. In the UK it takes up about 40% of the FRP market. Its major advantage is

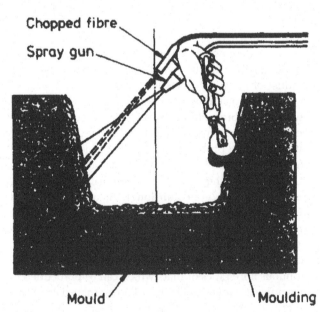

**Fig. 7.3-68** Spray-up technique.

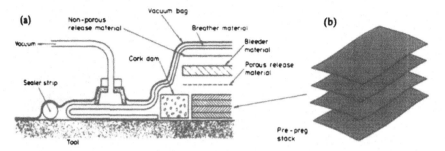

**Fig. 7.3-69** Diagrammatic cross section of a bagged lay-up.

that it is a very simple process so that very little special equipment is needed and the moulds may be made from plaster, wood, sheet metal or even FRP. The first step is to coat the mould with a release agent to prevent the moulding sticking to it. This is followed by a thin layer (approximately 0.3–0.4 mm) of pure resin (called a gelcoat) which has a number of functions. Firstly it conceals the irregular mesh pattern of the fibres and this improves the appearance of the product when it is taken from the mould. Secondly, and probably most important, it protects the reinforcement from attack by moisture which would tend to break down the fibre/resin interface. A tissue mat may be used on occasions to back up the gelcoat. This improves the impact resistance of the surface and also conceals the coarse texture of the reinforcement. However, it is relatively expensive and is only used if considered absolutely necessary.

When the gelcoat has been given time to partially cure the main reinforcement is applied. Initially a coat of resin (unsaturated polyester is the most common) is brushed on and this is followed by layers of tailored glass mat positioned by hand. As shown in Fig, 7.3-67 a roller is then used to consolidate the mat and remove any trapped air. The advantage of this technique is that the strength and stiffness of the composite can be controlled by building up the thickness with further layers of mat and resin as desired. Curing takes place at room temperature but heat is sometimes applied to accelerate this. Ideally any trimming should be carried out before the curing is complete because the material will still be sufficiently soft for knives or shears to be used. After curing, special cutting wheels may be needed.

Variations on this basic process are (i) *vacuum bag moulding* and (ii) *pressure bag moulding*. In the former process a flexible bag (frequently rubber) is clamped over the lay-up in the mould and a vacuum is applied between the moulding and the bag. This sucks the bag on to the moulding to consolidate the layers of reinforcement and resin. It also squeezes out trapped air and excess resin. The latter process is similar in principle except that pressure is applied above the bag instead of a vacuum below it. The techniques are illustrated in Fig. 7.3-67b and c.

**(b) Spray-Up:**

In this process, the preparatory stages are similar to the previous method but instead of using glass mats the reinforcement is applied using a spray gun. Roving is fed to a chopper unit and the chopped strands are sprayed on to the mould simultaneously with the resin (see Fig. 7.3-68). The thickness of the moulding (and hence the strength) can easily be built up in sections likely to be highly stressed. However, the success of the method depends to a large extent on the skill of the operator since he controls the overall thickness of the composite and also the glass/resin ratio.

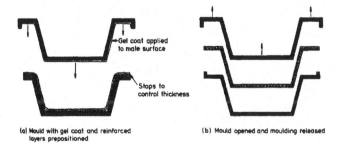

**Fig. 7.3-70** Basic cold press moulding process.

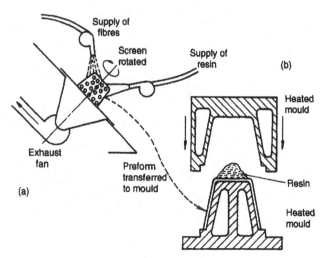

**Fig. 7.3-71** Pre-form moulding of GFRP.

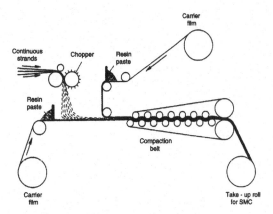

**Fig. 7.3-72** Manufacture of SMC material.

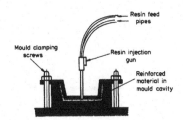

**Fig. 7.3-74** Resin injection process.

the flexible bag and the pre-preg stack, the latter will be squeezed tightly on to the mould. The whole assembly is then transferred to a very large oven (autoclave) for curing.

### (c) Autoclave Moulding:

In order to produce high quality, high precision mouldings for the aerospace industries, for example, it is necessary to have strict control over fibre alignment and consolidation of the fibres in the matrix. To achieve this, fabric 'pre-pregs' (i.e. a fabric consisting of woven fibre yarns pre-impregnated with the matrix material) are carefully arranged in layers in an open mould. The arrangement of the layers will determine the degree of anisotropy in the moulded article. A typical layer arrangement is shown in Fig. 7.3-69a. The pre-preg stack is then covered with a series of bleeder and breather sheets, as shown in Fig. 7.3-69b and finally with a flexible vacuum bag. When the air is extracted from between

## 7.3.10.2 Semi-automatic processing methods

### (a) Cold Press Moulding:

The basis of this process is to utilise pressure applied to two unheated halves of a mould to disperse resin throughout a prepared fabric stack placed in the mould. The typical procedure is as follows. Release agent and gelcoat are applied to the mould surfaces and the fibre fabric is laid into the lower part of the open mould. The activated resin is then poured on top of the mat and when the mould is closed the resin spreads throughout the reinforcement. High pressures are not necessary as the process relies on squeezing the resin throughout the reinforcement rather than forcing the composite into shape. A typical value of cycle time is about 10–15 minutes compared with several hours for hand lay-up methods. The process is illustrated in Fig. 7.3-70.

### (b) Hot Press Mouldings:

In this type of moulding the curing of the reinforced plastic is accelerated by the use of heat ($\simeq 180°C$) and

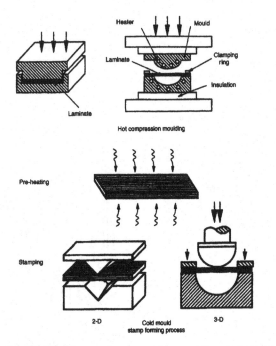

**Fig. 7.3-73** Other types of compression moulding and stamp forming.

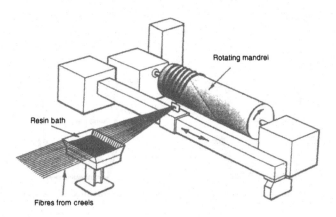

**Fig. 7.3-75** Filament winding of fibre composites.

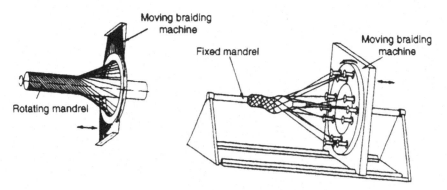

**Fig. 7.3-76** Two types of filament winding.

pressure ($\simeq 15$ MN/m$^2$). The general heading of Hot Press Moulding includes both preform moulding and compression moulding.

(i) *Pre-form Moulding:* This technique is particularly suitable for mass production and/or more complex shapes. There are two distinct stages. In the first a preform is made by, for example, spraying chopped fibres on to a perforated metal screen which has the general shape of the article to be moulded. The fibres are held on the screen by suction applied behind it (see Fig. 7.3-71). A resin binder is then sprayed on the mat and the resulting preform is taken from the screen and cured in an oven at about 150°C for several minutes. Other methods by which the preform can be made include tailoring a continuous fibre fabric to shape and using tape to hold it together. The preform is then transferred to the lower half of the heated mould and the activated resin poured on top. The upper half of the mould is then brought into position to press the composite into shape. The cure time in the mould depends on the temperature, varying typically from 1 minute at 150°C to 10 minutes at 80°C. If the mould was suitably prepared with release agent the moulding can then be ejected easily. This method would not normally be considered for short production runs because the mould costs are high.

(ii) *Compression Moulding* (see also Section 7.3.7): Sheet Moulding Compounds: SMC is supplied as a pliable sheet which consists of a mixture of chopped strand mat or chopped fibres (25% by weight) pre-impregnated with resin, fillers, catalyst and pigment. It is ready for moulding and so is simply placed between the halves of the heated mould. The application of pressure then forces the sheet to take up the contours of the mould. The beauty of the method is that the moulding is done 'dry' i.e. it is not necessary to pour on resins. Fig. 7.3-72 illustrates a typical method used to manufacture SMC material.

Dough Moulding Compounds: DMC (also known as BMC - Bulk Moulding Compound) is supplied as a dough or rope and is a mixture of chopped strands (20% by weight) with resin, catalyst and pigment. It flows readily and so may be formed into shape by compression or transfer moulding techniques. In compression moulding the charge of dough may be placed in the lower half of the heated mould, in a similar fashion to that illustrated in Fig. 7.3-50b although it is generally wise to preform it to the approximate shape of the cavity. When the mould is closed, pressure is applied causing the DMC to flow in all sections of the cavity. Curing generally takes a couple of minutes for mould temperatures in the region of 120°–160°C although clearly this also depends on the section thickness.

In general, SMC moulds less well than DMC on intricate shapes but it is particularly suitable for large shell-like mouldings – automotive parts such as body panels and fascia panels are ideal application areas. An engine inlet manifold manufactured from SMC has recently been developed in the UK. DMC finds its applications in the more complicated shapes such as business machine housings, electric drill bodies, etc. In France, a special moulding method, called ZMC, but based on DMC moulding concepts has been developed. Its most famous application to date is the rear door of the Citroen BX saloon and the process is currently under active consideration for the rear door of a VW saloon car. Injection moulding of DMC is also becoming common for intricately shaped articles (see Section 7.3.3.10).

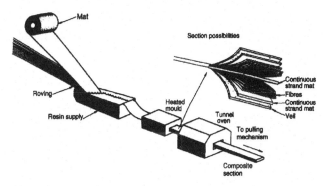

**Fig. 7.3-77** Pultrusion process.

Other types of compression moulding and stamp forming used for continuous fibre reinforced composites are illustrated in Fig. 7.3-73.

### (c) Resin Injection:

This is a cold mould process using relatively low pressures (approximately 450 kN/m$^2$). The mould surfaces are prepared with release agent and gelcoat before the reinforcing mat is arranged in the lower half of the mould. The upper half is then clamped in position and the activated resin is injected under pressure into the mould cavity. The advantage of this type of production method is that it reduces the level of skill needed by the operator because the quality of the mould will determine the thickness distribution in the moulded article (see Fig. 7.3-74). In recent times there has been a growing use of pre-formed fabric shells in the resin injection process. The pre-form is produced using one of the methods described above and this is placed in the mould. This improves the quality and consistency of the product and reinforcements varying from chopped strand mat to close weave fabric in glass, aramid, carbon or hybrids of these may be used. It is possible, with care, to achieve reinforcement loadings in the order of 65%.

### (d) Vacuum Injection:

This is a development of resin injection in which a vacuum is used to draw resin throughout the reinforcement. It overcomes the problem of voids in the resin/fibre laminate and offers faster cycle times with greater uniformity of product.

## 7.3.10.3 Automatic processes

### (a) Filament Winding:

In this method, continuous strands of reinforcementare used to gain maximum benefit from the fibre strength. In a typical process rovings or single strands are passed through a resin bath and then wound onto a rotating mandrel. By arranging for the fibres to traverse the mandrel at a controlled and/or programmed manner, as illustrated in Fig. 7.3-75, it is possible to lay down the reinforcement in any desired fashion. This enables very high strengths to be achieved and is particularly suited to pressure vessels where reinforcement in the highly stressed hoop direction is important.

In the past a limitation on this process was that it tended to be restricted to shapes which were symmetrical about an axis of rotation and from which the mandrel could be easily extracted. However, in recent years there have been major advances through the use of collapsible or expendable cores and in particular through the development of computer-controlled winding equipment. The latter has opened the door to a whole new range of products which can be filament wound – for example, space-frame structures. Braiding machines for complex shapes are shown in Fig. 7.3-76.

### (b) Centrifugal Casting:

This method is used for cylindrical products which can be rotated about their longitudinal axis. Resin and fibres are introduced into the rotating mould/mandrel and are thrown out against the mould surface. The method is particularly suited to long tubular structures which can have a slight taper e.g. street light columns, telegraph poles, pylons, etc.

### (c) Pultrusion:

This is a continuous production method similar in concept to extrusion. Woven fibre mats and/or rovings are drawn through a resin bath and then through a die to form some desired shape (for example a 'plank' as illustrated in Fig. 7.3-77). The profiled shape emerges from the die and then passes through a tunnel oven to accelerate the curing of the resin. The pultruded composite is eventually cut to length for storage. A wide range of pultruded shapes may be produced – U channels, I beams, aerofoil shapes, etc.

### (d) Injection Moulding:

The injection moulding process can also be used for fibre reinforced thermoplastics and thermosets, for example DMC materials. This offers considerable advantages over compression moulding due to the higher production speeds, more accurate metering and lower product costs which can be achieved. The injection moulding process for thermosets has already been dealt with in Section 7.3.3.8. See also the section on Reaction Injection Moulding (RIM) since this offers the opportunity to incorporate fibres.

## Bibliography

Fisher, E.G. *Extrusion of Plastics*, Newnes-Butterworth, 1976.

Schenkel, G. *Plastics Extrusion Technology and Practice*, Iliffe, 1966.

Fisher, E.G. *Blow Moulding of Plastics*, Iliffe, 1971.

Rubin, I. *Injection Moulding-Theory and Practice*, Wiley, 1972.

Holmes-Walker, W.A. *Polymer Conversion*, Applied Science Publishers, 1975.

Bown, J. *Injection Moulding of Plastic Components*, McGraw-Hill, 1979.

Dym, J.B. *Injection Moulds and Moulding*, Van Nostrand Rheinhold, 1979.

Pye, R.G.W. *Injection Mould Design*, George Godwin, 1978.

Elden, R.A. and Swann, A.D. *Calendering of Plastics*, Plastics Institute Monograph, Iliffe, 1971.

Rosenzweig, N., Markis, M., and Tadmar, Z. *Wall Thickness Distribution in Thermoforming*. Polym. Eng. Sci. October 1979 Vol 19 No. 13 pp 946–950.

Parker, F.J. *The Status of Thermoset Injection Moulding Today*, Progress in Rubber and Plastic Technology, October 1985, vol 1, No 4, pp 22–59.

Whelan, A. and Brydson, J.A. *Developments with Thermosetting Plastics* App. Sci. Pub. London, 1975.

Monk, J.F. *Thermosetting Plastics– Practical Moulding Technology*, George Godwin, London, 1981.

Rosato, D.V. ed. *Blow Moulding Handbook*, Hanser, Munich, 1989.

Hepburn, C. *Polyurethane Elastomers* (ch 6-RIM) Applied Science Publishers, London, 1982.

Martin, J., *Pultrusion* Ch 3 in 'Plastics Product Design Handbook- B' ed by E. Miller, Dekker Inc, New York, 1983.

Titow, W.V. and Lanham, B.J. *Reinforced Thermoplastics*, Applied Science Publisher, 1975.

Penn, W.S. *GRP Technology*, Maclaren, 1966.

Beck, R.D. *Plastic Product Design*, Van Nostrand Reinhold Co. 1980.

Schwartz, S.S. and Goodman, S.H. *Plastic Materials and Processes*, Van Nostrand, New York, 1982.

Crawford, R.J. *Rotational Moulding of Plastics*, Research Studies Press 2nd edition, 1996. 340 Processing of Plastics

Rosato, D.V. and Rosato, D.V. *Injection Molding Handbook*, 2nd edn, Chapman and Hall, New York, 1995.

Stevens, M.J. and Covas, J.A. *Extruder Principles and Operation*, 2nd edn, Chapman and Hall, London, 1995.

Michaeli, W. *Extrusion Dies*, Hanser, Munich, 1984.

Baird, D.G. and Collias, D.I. *Polymer Processing*, Butterworth-Heinemann, Newton, USA, 1995.

Michaeli, W. *Polymer Processing*, Hanser, Munich, 1992.

Lee, N. (ed.), *Plastic Blow Moulding Handbook*, Van Nostrand Reinhold, New York, 1990.

Throne, J.L. *Technology of Thermoforming*, Hanser, Munich, 1996.

Florian, J. *Practical Thermoforming*, Marcel Dekker, New York, 1987.

Boussinesq, M.J., J. *Math Pures et Appl.*, **2**, 13 1868 pp. 377–424.

Meng Hou, Lin Ye and Yiu-Wing Mai. Advances in processing of continuous fibre reinforced, composites. *Plastics, Rubber and Composites Proc. and Appl.*, **23**, 5 (1995) pp. 279–292.

Mitchell, P. (ed.) *Tool and Manufacturing Engineers Handbook*, Vol 8, 4th edition, Soc. Man. Eng., Michigan 1996.

# Questions

**7.3.1** In a particular extruder screw the channel depth is 2.4 mm, the screw diameter is 50 mm, the screw speed is 100 rev/min, the flight angle is 17° 42' and the pressure varies linearly over the screw length of 1000 mm from zero at entry to 20 MN/m$^2$ at the die entry. Estimate

(a) the drag flow

(b) the pressure flow

(c) the total flow.

The plastic has a viscosity of 200 Ns/m$^2$. Calculate also the shear rate in the metering zone.

**7.3.2** Find the operating point for the above extruder when it is combined with a die of length 40 mm and diameter 3 mm. What would be the effect on pressure and output if a plastic with viscosity 400 Ns/m$^2$ was used.

**7.3.3** A single screw extruder has the following dimensions:

screw length = 500 mm
screw diameter = 25 mm
flight angle = 17°42'
channel depth = 2 mm
channel width = 22 mm

If the extruder is coupled to a die which is used to produce two laces for subsequent granulation, calculate the output from the extruder/die combination when the screw speed is 100 rev/min. Each of the holes in the lace die is 1.5 mm diameter and 10 mm long and the viscosity of the melt may be taken as 400 Ns/m$^2$.

**7.3.4** An extruder is coupled to a die, the output of which is given by $(KP/\eta)$ where $P$ is the pressure drop across the die, $\eta$ is the viscosity of the plastic and $K$ is a constant. What are the optimum values of screw helix angle and channel depth to give maximum output from the extruder.

**7.3.5** 7A circular plate of diameter 0.5 m is to be moulded using a sprue gate in its centre. If the melt pressure is 50 MN/m$^2$ and the pressure loss coefficient is 0.6 estimate the clamping force required.

**7.3.6** The container shown at the top of p. 340 is injection moulded using a gate at point A. If the injection pressure at the nozzle is 140 MN/m$^2$ and the pressure loss coefficient, m, is 0.5, estimate (i) the flow ratio and (ii) the clamping force needed.

**7.3.7** Compare the efficiencies of the runners shown on p. 340.

**7.3.8** A calender having rolls of diameter 0.3 m produces plastic sheet 1 m wide at the rate of 2000 kg/hour. If the roll speed is 5 rev/minute and the nip between the rolls is 4.5 mm, estimate the position and magnitude of the maximum pressure. The density of the material is 1400 kg/m$^3$ and its viscosity is $1.5 \times 10^4$ Ns/m$^2$.

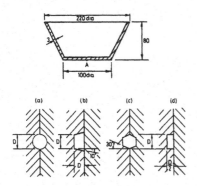

**7.3.9** A calender having rolls of 0.2 m diameter produces 2 mm thick plastic sheet at a linear velocity of 0.1 m/s. Investigate the effect of nips in the range 0.8 to 1.9 mm on the pressure profile. The viscosity is $10^3$ Ns/m$^2$.

**7.3.10** A hemispherical dome of 200 mm diameter has been vacuum formed from a flat sheet 4 mm thick. What is the thickness of the dome at the point furthest away from its diameter.

**7.3.11** A disposable tumbler which has the shape of a frustrum of a cone is to be vacuum formed from a flat plastic sheet 3 mm thick. If the diameter of the mouth of the tumbler is 60 mm, the diameter of the base is 40 mm and the depth is 60 mm estimate the wall thickness at (a) a point 35 mm from the top and (b) in the centre of the base.

**7.3.12** A blow moulding die which has an outside diameter of 40 mm and a die gap of 2 mm is used to produce a plastic bottle with a diameter of 70 mm. If the swelling ratio of the melt in the thickness direction is 1.8 estimate

(a) the parison dimensions

(b) the thickness of the bottle and

(c) a suitable inflation pressure if melt fracture occurs at a stress of 10 MN/m$^2$.

**7.3.13** A plastic film, 0.1 mm thick, is required to have its orientation in the transverse direction twice that in the machine direction. If the film blowing die has an outer diameter of 100 mm and an inner diameter of 98 mm estimate the blow-up ratio which will be required and the lay flat film width. Neglect extrusion induced effects and assume there is no draw-down.

**7.3.14** A molten polymer is to be coated on a cable at a speed of 0.5 m/s. The cable diameter is 15 mm and the coating thickness required is 0.3 mm. The die used has a length of 60 mm and an internal diameter of 16 mm. What pressure must be developed at the die entry if the viscosity of the polymer under these operating conditions is 100 Ns/m$^2$.

**7.3.15** During a rotational moulding operation an aluminium mould with a uniform thickness of 3 mm is put into an oven at 300°C If the initial temperature of the mould is 23°C, estimate the time taken for it to reach 250°C The natural convection heat transfer coefficient is 28.4 J/m$^2$s.K and the thermal diffusivity and conductivity of aluminium may be taken as 8.6 × 10$^{-5}$ m$^2$/s and 230.1 J/m.s.K respectively.

**7.3.16** A billet of PVC weighing 150 g is to be compression moulded into a long playing record of diameter 300 mm. If the maximum force which the press can apply is 100 kN estimate the time needed to fill the mould. The density and viscosity of the PVC may be taken as 1200 kg/m$^3$ and 10 Ns/m$^2$ respectively.

# Section Eight

## Ceramics and glasses

# Chapter 8.1

# Ceramics and glasses

## Introduction

If you have ever dropped a plate on the kitchen floor and seen it disintegrate, you might question whether ceramics have a role as load-bearing materials in engineering. But any friend with a historical perspective will enlighten you. Ceramic structures are larger and have survived longer than any other works. The great pyramid of Giza is solid ceramic (nearly 1,000,000 tonnes of it); so is the Parthenon, the Forum, the Great Wall of China. The first cutting tools and weapons were made of flint – a glass; and pottery from 5000 BC survives to the present day. Ceramics may not be as tough as metals, but for resistance to corrosion, wear, decay and corruption, they are unsurpassed.

Today, cement and concrete replace stone in most large structures. But cement, too, is a ceramic: a complicated but fascinating one. The understanding of its structure, and how it forms, is better now than it used to be, and has led to the development of special high-strength cement pastes which can compete with polymers and metals in certain applications.

But the most exciting of all is the development, in the past 20 years, of a range of high-performance engineering ceramics. They can replace, and greatly improve on, metals in many very demanding applications. Cutting tools made of sialons or of dense alumina can cut faster and last longer than the best metal tools. Engineering ceramics are highly wear-resistant: they are used to clad the leading edges of agricultural machinery like harrows, increasing the life by 10 times. They are inert and biocompatible, so they are good for making artificial joints (where wear is a big problem) and other implants. And, because they have high melting points, they can stand much higher temperatures than metals can: vast development programs in Japan, the US and Europe aim to put increasing quantities of ceramics into reciprocating engines, turbines and turbochargers. In the next decade the potential market is estimated at $1 billion per year. Even the toughness of ceramics has been improved: modern body-armour is made of plates of boron carbide or of alumina, sewn into a fabric vest.

The next six chapters of this book focus on ceramics and glasses: non-metallic, inorganic solids. Five classes of materials are of interest to us here:

**(a)** *Glasses*, all of them based on silica ($SiO_2$), with additions to reduce the melting point, or give other special properties.

**(b)** The traditional *vitreous ceramics*, or clay products, used in vast quantities for plates and cups, sanitary ware, tiles, bricks, and so forth.

**(c)** The new *high-performance ceramics*, now finding application for cutting tools, dies, engine parts and wear-resistant parts.

**(d)** *Cement and concrete:* a complex ceramic with many phases, and one of three essential bulk materials of civil engineering.

**(e)** *Rocks and minerals*, including ice.

As with metals, the number of different ceramics is vast. But there is no need to remember them all: the generic ceramics listed below (and which you *should* remember) embody the important features; others can be understood in terms of these. Although their properties differ widely, they all have one feature in common: they are intrinsically brittle, and it is this that dictates the way in which they can be used.

*Engineering Materials and Processes Desk Reference*; ISBN: 9781856175869
Copyright © 2009 Elsevier Ltd; All rights of reproduction, in any form, reserved.

They are, potentially or actually, cheap. Most ceramics are compounds of oxygen, carbon or nitrogen with metals like aluminium or silicon; all five are among the most plentiful and widespread elements in the Earth's crust. The processing costs may be high, but the ingredients are almost as cheap as dirt: dirt, after all, is a ceramic.

# The generic ceramics and glasses

## Glasses

*Glasses* are used in enormous quantities: the annual tonnage is not far below that of aluminium. As much as 80% of the surface area of a modern office block can be glass; and glass is used in a load-bearing capacity in car windows, containers, diving bells and vacuum equipment. All important glasses are based on silica ($SiO_2$). Two are of primary interest: common window glass, and the temperature-resisting borosilicate glasses. Table 8.1-1 gives details.

## Vitreous ceramics

Potters have been respected members of society since ancient times. Their products have survived the ravages of time better than any other; the pottery of an era or civilisation often gives the clearest picture of its state of development and its customs. Modern pottery, porcelain, tiles, and structural and refractory bricks are made by processes which, though automated, differ very little from those of 2000 years ago. All are made from clays, which are formed in the wet, plastic state and then dried and fired. After firing, they consist of crystalline phases (mostly silicates) held together by a glassy phase based, as always, on silica ($SiO_2$). The glassy phase forms and melts when the clay is fired, and spreads around the surface of the inert, but strong, crystalline phases, bonding them together. The important information is summarised in Table 8.1-2

## High-performance engineering ceramics

Diamond, of course, is the ultimate engineering ceramic; it has for many years been used for cutting tools, dies, rock drills, and as an abrasive. But it is expensive. The strength of a ceramic is largely determined by two characteristics: its *toughness* ($K_c$), and the size distribution of *microcracks* it contains. A new class of fully dense, high-strength ceramics is now emerging which combine a higher $K_c$ with a much narrower distribution of smaller microcracks, giving properties which make them competitive with metals, cermets, even with diamond, for cutting tools, dies, implants and engine parts. And (at least potentially) they are cheap. The most important are listed in Table 8.1-3.

## Cement and concrete

Cement and concrete are used in construction on an enormous scale, equalled only by structural steel, brick and wood. *Cement* is a mixture of a combination of lime (CaO), silica ($SiO_2$) and alumina ($Al_2O_3$), which sets when mixed with water. Concrete is sand and stones (aggregate) held together by a cement. Table 8.1-4 summarises the most important facts.

**Table 8.1-1** Generic glasses

| Glass | Typical composition (wt%) | Typical uses |
|---|---|---|
| Soda-lime glass | 70 $SiO_2$, 10 CaO, 15 $Na_2O$ | Windows, bottles, etc.; easily formed and shaped. |
| Borosilicate glass | 80 $SiO_2$, 15 $B_2O_3$, 5 $Na_2O$ | Pyrex; cooking and chemical glassware; high-temperature strength, low coefficient of expansion, good thermal shock resistance. |

**Table 8.1-2** Generic vitreous ceramics

| Ceramic | Typical composition | Typical uses |
|---|---|---|
| Porcelain China Pottery Brick | Made from clays: hydrous alumino-silicate such as $Al_2(Si_2O_5)(OH)_4$ mixed with other inert minerals. | Electrical insulators. Artware and tableware tiles. Construction; refractory uses. |

**Table 8.1-3** Generic high-performance ceramics

| Ceramic | Typical composition | Typical uses |
|---|---|---|
| Dense alumina | $Al_2O_3$ | Cutting tools, dies; wear-resistant surfaces, bearings; medical implants; engine and turbine parts; armour. |
| Silicon carbide, nitride | SiC, $Si_3N_4$ | |
| Sialons | e.g. $Si_2AlON_3$ | |
| Cubic zirconia | $ZrO_2$ + 5wt% MgO | |

| Table 8.1-4 Generic cements and concretes | | |
|---|---|---|
| Cement | Typical composition | Uses |
| Portland cement | CaO + $SiO_2$ + $Al_2O_3$ | Cast facings, walkways, etc. and as component of concrete. General construction. |

## Natural ceramics

Stone is the oldest of all construction materials and the most durable. The pyramids are 5000 years old; the Parthenon 2200. Stone used in a load-bearing capacity behaves like any other ceramic; and the criteria used in design with stone are the same. One natural ceramic, however, is unique. Ice forms on the Earth's surface in enormous volumes: the Antarctic ice cap, for instance, is up to 3 km thick and almost 3000 km across; something like $10^{13} m^3$ of pure ceramic. The mechanical properties are of primary importance in some major engineering problems, notably ice breaking, and the construction of offshore oil rigs in the Arctic. Table 8.1-5 lists the important natural ceramics.

## Ceramic composites

The great stiffness and hardness of ceramics can sometimes be combined with the toughness of polymers or metals by making composites. Glass- and carbon-fibre reinforced plastics are examples: the glass or carbon fibres stiffen the rather floppy polymer; but if a fibre fails, the crack runs out of the fibre and blunts in the ductile polymer without propagating across the whole section. Cermets are another example: particles of hard tungsten carbide bonded by metallic cobalt, much as gravel is bonded with tar to give a hard-wearing road surface (another ceramic-composite). Bone is a natural ceramic-composite: particles of hydroxyapatite (the ceramic) bonded together by collagen (a polymer). Synthetic ceramic–ceramic composites (like glass fibres in cement, or silicon carbide fibres in silicon carbide) are now under development and may have important high-temperature

| Table 8.1-5 Genetic natural ceramics | | |
|---|---|---|
| Ceramic | Composition | Typical uses |
| Limestone (marble) Sandstone Granite | Largely $CaCO_3$ Largely $SiO_2$ Aluminium silicates | Building foundations, construction. |
| Ice | $H_2O$ | Arctic engineering. |

| Table 8.1-6 Ceramics composites | | |
|---|---|---|
| Ceramic composite | Components | Typical uses |
| Fibre glass CFRP | Glass–polymer Carbon–polymer | High-performance structures. |
| Cermet | Tungsten carbide–cobalt | Cutting tools, dies. |
| Bone | Hydroxyapatite–collagen | Main structural material of animals. |
| New ceramic composites | Alumina–silicon carbide | High temperature and high toughness applications. |

application in the next decade. The examples are summarised in Table 8.1-6.

## Data for ceramics

Ceramics, without exception, are hard, brittle solids. When designing with metals, failure by plastic collapse and by fatigue are the primary considerations. For ceramics, plastic collapse and fatigue are seldom problems; it is brittle failure, caused by direct loading or by thermal stresses, that is the overriding consideration.

Because of this, the data listed in Table 8.1-7 for ceramic materials differ in emphasis from those listed for metals. In particular, the Table shows the *modulus of rupture* (the maximum surface stress when a beam breaks in bending) and the *thermal shock resistance* (the ability of the solid to withstand sudden changes in temperature). These, rather than the yield strength, tend to be the critical properties in any design exercise.

As before, the data presented here are approximate, intended for the first phase of design. When the choice has narrowed sufficiently, it is important to consult more exhaustive data compilations; and then to obtain detailed specifications from the supplier of the material you intend to use. Finally, if the component is a critical one, you should conduct your own tests. The properties of ceramics are more variable than those of metals: the same material, from two different suppliers, could differ in toughness and strength by a factor of two.

There are, of course, many more ceramics available than those listed here: alumina is available in many densities, silicon carbide in many qualities. As before, the structure-insensitive properties (density, modulus and melting point) depend little on quality – they do not vary by more than 10%. But the structure-sensitive properties (fracture toughness, modulus of rupture and some thermal properties including expansion) are much more variable. For these, it is essential to consult manufacturers' data sheets or conduct your own tests.

Table 8.1-7 Properties of ceramics

| Ceramic | Cost (UK£ (US$) tonne$^{-1}$) | Density (Mg m$^{-3}$) | Young's modulus (GPa) | Compressive strength (MPa) | Modulus of rupture (MPa) | Weibull exponent m | Time exponent n | Fracture toughness (MPa m$^{1/2}$) | Melting (softening) temperature (K) | Specific heat (J kg$^{-1}$ K$^{-1}$) | Thermal conductivity (W m$^{-1}$ K$^{-1}$) | Thermal expansion coefficient (MK$^{-1}$) | Thermal shock resistance (K) |
|---|---|---|---|---|---|---|---|---|---|---|---|---|---|
| **Glasses** | | | | | | | | | | | | | |
| Soda glass | 700 (1000) | 2.48 | 74 | 1000 | 50 | Assume 10 in design | 10 | 0.7 | (1000) | 990 | 1 | 8.5 | 84 |
| Borosilicate glass | 1000 (1400) | 2.23 | 65 | 1200 | 55 | | 10 | 0.8 | (1100) | 800 | 1 | 4.0 | 280 |
| *Pottery, etc.* | | | | | | | | | | | | | |
| Porcelain | 260–1000 (360–1400) | 2.3–2.5 | 70 | 350 | 45 | | – | 1.0 | (1400) | 800 | 1 | 3 | 220 |
| **High-performance engineering ceramics** | | | | | | | | | | | | | |
| Diamond | 4 × 10$^8$ (6 × 10$^8$) | 3.52 | 1050 | 5000 | – | – | – | – | – | 510 | 70 | 1.2 | 1000 |
| Dense alumina | Expensive at present. | 3.9 | 380 | 3000 | 300–400 | 10 | 10 | 3–5 | 2323 (1470) | 795 | 25.6 | 8.5 | 150 |
| Silicon carbide | | 3.2 | 410 | 2000 | 200–500 | 10 | 40 | – | 3110 – | 1422 | 84 | 4.3 | 300 |
| Silicon nitride | Potentially 350–1000 | 3.2 | 310 | 1200 | 300–850 | – | 40 | 4 | 2173 – | 627 | 17 | 3.2 | 500 |
| Zirconia | | 5.6 | 200 | 2000 | 200–500 | 10–21 | 10 | 4–12 | 2843 – | 670 | 1.5 | 8 | 500 |
| Sialons | (490–1400) | 3.2 | 300 | 2000 | 500–830 | 15 | 10 | 5 | – – | 710 | 20–25 | 3.2 | 510 |
| **Cement, etc.** | | | | | | | | | | | | | |
| Cement | 52 (73) | 2.4–2.5 | 20–30 | 50 | 7 | 12 | 40 | 0.2 | – | – | 1.8 | 10–14 | <50 |
| Concrete | 26 (36) | 2.4 | 30–50 | 50 | 7 | 12 | 40 | 0.2 | – | – | 2 | 10–14 | |
| **Rocks and ice** | | | | | | | | | | | | | |
| Limestone | Cost of mining and transport | 2.7 | 63 | 30–80 | 20 | – | – | 0.9 | – | – | – | 8 | ≈100 |
| Granite | | 2.6 | 60–80 | 65–150 | 23 | – | – | – | – | – | – | 8 | |
| Ice | | 0.92 | 9.1 | 6 | 1.7 | – | – | 0.12 | 273 (250) | – | – | – | |

# Examples

**8.1.1** What are the five main generic classes of ceramics and glasses? For each generic class:
(a) give one example of a specific component made from that class;
(b) indicate why that class was selected for the component.

**8.1.2** How do the unique characteristics of ceramics and glasses influence the way in which these materials are used?

**8.1.3** The glass walls of the Sydney Opera House are constructed from glass panels butted together with a 10 mm expansion gap between adjacent panels. The gap is injected with a flexible polymer which is cured in-situ and also sticks to the edge of the glass (see photograph). Taking a notional panel width of 3 m, and a maximum temperature variation of 40° C, calculate the minimum strain capacity required from the polymer. See Table 8.1-7 for the thermal expansion coefficient of soda glass. Apart from tensile fracture of the polymer, what additional mechanical failure mechanism is possible in the jointing system?

Glass walls of Sydney Opera House.

## Answer

10%

# Section Nine

**Composite materials**

# Chapter 9.1

# Composite materials

## Definition and classification

Composite materials are material systems that consist of a discrete constituent (the reinforcement) distributed in a continuous phase (the matrix) and that derive their distinguishing characteristics from the properties and behavior of their constituents, from the geometry and arrangement of the constituents, and from the properties of the boundaries (interfaces) between the constituents. Composites are classified either on the basis of the nature of the continuous (matrix) phase (polymer-matrix, metal-matrix, ceramic-matrix, and intermetallic-matrix composites), or on the basis of the nature of the reinforcing phase (particle reinforced, fiber reinforced, dispersion strengthened, laminated, etc.). The properties of the composite can be tailored, and new combinations of properties can be achieved. For example, inherently brittle ceramics can be toughened by combining different types of ceramics in a ceramic-matrix composite, and inherently ductile metals can be made strong and stiff by incorporating a ceramic reinforcement.

It is usually sufficient, and often desirable, to achieve a certain minimum level of reinforcement content in a composite. Thus, in creep-resistant dispersion-strengthened composites, the reinforcement volume fraction is maintained below 15% in order to preserve many of the useful properties of the matrix. Other factors, such as the shape, size, distribution of the reinforcement, and properties of the interface, are also important. The shape, size, amount, and type of the reinforcing phase to be used are dictated by the combination of properties desired in the composite. For example, applications requiring anisotropic mechanical properties (high strength and high stiffness along one particular direction) employ directionally aligned, high-strength continuous fibers, whereas for applications where strength anisotropy is not critical and strength requirements are moderate, relatively inexpensive particulates can be used as the reinforcing phase. Figure 9.1-1 shows some examples of continuous and discontinuous reinforcements developed for use in modern engineered composites.

## Fibers

Long, continuous fibers with a large aspect ratio (i.e., length-to-diameter ratio) of metals, ceramics, glasses and polymers are used to reinforce various types of matrices. A hard and strong material such as a ceramic in a fibrous form will have fewer strength-limiting flaws than the same material in a bulk form. As preexisting cracks lower the fracture strength of brittle ceramics, reducing the size and/or probability of occurrence of cracks will diminish the extent of strength loss in the ceramic, thus allowing the actual strength to approach the theoretical fracture strength in the absence of cracks, which is $\sim 0.1\,E$, where E is the elastic modulus. If the fiber diameter scales with the grain size of the material, then the fracture strength will be high. In other words, smaller the fiber diameter, greater is its fracture strength. In the case of continuous fibers, a critical minimum aspect ratio of the fiber is needed to transfer the applied load from the weaker matrix to the stronger fiber. Furthermore, a small diameter allows a stiff fiber to be bent for shaping a preform that is used as a precursor in composite fabrication. Many commercial fibers are flexible, and permit filament winding and weaving techniques to be used for making a preform. Very stiff fibers are, however, shaped into preforms by using a fugitive binder material. For example, an organic compound that cements the fibers in the desired

*Engineering Materials and Processes Desk Reference*; ISBN: 9781856175869
Copyright © 2009 Elsevier Ltd; All rights of reproduction, in any form, reserved.

# CHAPTER 9.1 Composite materials

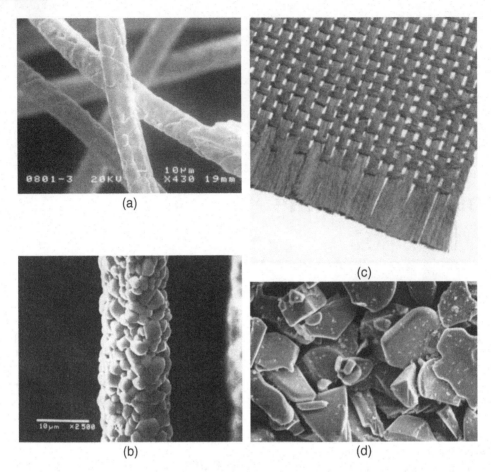

**Figure 9.1-1** (a) Scanning electron photomicrograph of sintered $TiO_2$ fiber (J. D. French and R. B. Cass, *Developing Innovative Ceramic Fibers*, www.ceramicbulletin.org, May 1998, pp. 61–65). (b) SEM photomicrograph of an individual PZT filament of 25 μm diameter (J. D. French and R. B. Cass, *Developing Innovative Ceramic Fibers*, www.ceramicbulletin.org, May 1998, pp. 61–65). (c) PZT fiber weave for a smart structure composite (J. D. French and R. B. Cass, *Developing Innovative Ceramic Fibers*,

preform shape may be used. The binder decomposes and is eliminated when the matrix material is combined with the preform to provide it support and rigidity. Selected examples of fibers used in composite matrices are briefly described below. For more details, the reader is referred to the book by Chawla referenced at the end of the chapter.

## Glass

*Glass* is a generic name for a family of ceramic fibers containing 50–60% silica (a glass former) in a solid solution that contains several other oxides such as $Al_2O_3$, CaO, MgO, $K_2O$. $Na_2O$, and $B_2O_3$, etc. Commercial glass fibers are classified as E-glass (for high electrical resistivity), S-glass (for high silica that imparts excellent high-temperature stability), and C-glass (for corrosion resistance).

Glass fiber is manufactured by melting the oxide ingredients in a furnace and then transferring the molten glass into a hot platinum crucible with a few hundred fine holes at its base. Molten glass flows through these holes and on cooling forms fine continuous filaments. The final fiber diameter is a function of the hole diameter in the platinum crucible, the viscosity of molten glass, and the liquid head in the crucible. The filaments are gathered into a strand, and a sizing is applied before the strand is wound on a drum. Glass is a brittle solid, and its strength is lowered by minute surface defects. The sizing protects the surface of glass filaments and also binds them into a strand. A common type of sizing contains polyvinyl acetate and a coupling agent that makes the strand compatible with various polymer matrices. The final fiber diameter is a function of the hole diameter in the platinum crucible, the viscosity of molten glass, and the liquid head in the reservoir. Another method to grow glass fibers makes use of a sol-gel-type chemical precipitation process. A sol containing fine colloidal particles is used as the precursor; due to their fine size, the particles remain suspended in the liquid vehicle and are stabilized against flocculation through ionic charge adsorption on the surface. The sol is gelled via pH adjustments, i.e., the liquid vehicle in the gel behaves as a highly viscous liquid, thus acquiring the physical characteristics of a solid. The gelling action occurs at room temperature. The gel is then drawn into fibers at high temperatures, that are lower than the temperatures used in conventional manufacture of glass fiber by melting. The Nextel fiber manufactured by the 3M company is a sol-gel-derived silica-based fiber. Moisture decreases the strength of glass fibers. They are also prone to static fatigue; that is, they cannot withstand loads for long periods of time. Glass fiber-reinforced plastics (GRPs) are widely used in the construction industry.

### Boron

Boron fibers are produced by vapor depositing boron on a fine filament, usually made from tungsten, carbon, or carbon-coated glass fiber. In one type of vapor deposition process, a boron hydride compound is thermally decomposed, and the boron vapor heterogeneously nucleates on the filament, thus forming a film. Such fibers are, however, not very strong or dense, owing to trapped vapor or gas that causes porosity and weakens the fiber. In an improved chemical vapor deposition (CVD) process, a halogen compound of boron is reduced by hydrogen gas at high temperatures, via the reaction $2BX_3 + 3H_2 \rightarrow 2B + 6HX$ ($X$ = Cl, Br, or I). Because of the high deposition temperatures involved, the precursor filament is usually tungsten. Fibers of consistently high quality are produced by this process, although the relatively high density of W filament slightly increases the fiber density. In the halide reduction process using $BCl_3$, a 10-to 12-$\mu$m-diameter W wire is pulled in a reaction chamber at one end through a mercury seal and out at the other end through another mercury seal. The mercury seals act as electrical contacts for resistance heating of the substrate wire when gases ($BCl_3 + H_2$) pass through the reaction chamber and react on the incandescent wire substrate to deposit boron coatings. The conversion efficiency of $BCl_3$ to B coating is only about 10%, and reuse of unreacted gas is important. Boron is also deposited on carbon monofilaments. A pyrolytic carbon coating is first applied to the carbon filament to accommodate the growth strains that result during boron deposition.

There is a critical temperature for obtaining a boron fiber with optimum properties and structure. The desirable amorphous (actually, microcrystalline with grain size of just a few nm) form of boron occurs below this critical temperature, whereas above this temperature there also occur crystalline forms of boron, which are undesirable from a mechanical properties viewpoint. Larger crystallites lower the mechanical strength of the fiber. Because of high deposition temperatures in CVD, diffusional processes are rapid, and this partially transforms the core region from pure W to a variety of boride phases such as $W_2B$, WB, $WB_4$, and others. As boron diffuses into the tungsten substrate to form borides, the core expands as much as 40% by volume, which results in an increase in the fiber diameter. This expansion generates residual stresses that can cause radial cracks and stress concentration in the fiber, thus lowering the fracture strength of the fiber. The average tensile strength of commercial boron fibers is about 3–4 GPa, and the modulus is 380–400 GPa. Usually a SiC coating is vapor-deposited onto the fiber to prevent any adverse reactions between B and the matrix such as Al at high temperatures.

### Carbon fiber

Carbon, which can exist in a variety of crystalline forms, is a light material (density: 2.268 g/cc). The graphitic form of carbon is of primary interest in making fibers. The other form of carbon is diamond, a covalent solid, with little flexibility and little scope to grow diamond fibers, although microcrystalline diamond coatings can be vapor-deposited on a fiberous substrate to grow coated diamond fibers. Carbon atoms in graphite are arranged in the form of hexagonal layers, which are attached to similar layers via van der Waals forces. The graphitic form is highly anisotropic, with widely different elastic modulus in the layer plane and along the c-axis of the unit cell (i.e., very high in-plane modulus and very low transverse modulus). The high-strength covalent bonds between carbon atoms in the hexagonal layer plane result in an extremely high modulus ($\sim$1000 GPa in single crystal) whereas the weak van der Waals bond between the neighboring layers results in a lower modulus (about one-half the modulus of pure Al) in that direction. In order to grow high-strength and high-modulus carbon fiber, a very high degree of preferred orientation of hexagonal planes along the fiber axis is needed.

The name *carbon fiber* is a generic one and represents a family of fibers all derived from carbonaceous precursors, and differing from one another in the size of the hexagonal sheets of carbon atoms, their stacking height, and the resulting crystalline orientations. These structural variations result in a wide range of physical and mechanical properties. For example, the axial tensile modulus can vary from 25 to 820 GPa, axial tensile strength from 500 to 5,000 MPa, and thermal conductivity from 4 to 1100 W/m.K, respectively. Carbon fibers of extremely high modulus are made by carbonization of organic precursor fibers followed by graphitization at high temperatures. The organic precursor fiber is generally a special long-chain polymer-based textile fiber (polyacrylonitrile or PAN and rayon, a thermosetting polymer) that can be carbonized without melting. Such fibers generally have poor mechanical properties because of a high degree of molecular disorder in polymer chains. Most processes of carbon fiber fabrication involve the following steps: (1) a stabilizing treatment (essentially an oxidation process) that enhances the thermal stability of the fibers and prevents the fiber from melting in the subsequent high-temperature treatment, and (2) a thermal treatment at 1000–1500°C called carbonization that removes noncarbon elements (e.g., $N_2$ and $H_2$). An optional thermal treatment called graphitization may be done at ~3000°C to further improve the mechanical properties of the carbon fiber by enabling the hexagonal crystalline sheets of graphite to increase their ordering. To produce high-modulus fiber, the orientation of the graphitic crystals or lamellae is improved by graphitization which consists of thermal and stretching treatments under rigorously controlled conditions. Besides the PAN and cellulosic (e.g., rayon) precursors, pitch is also used as a raw material to grow carbon fibers. Commercial pitches are mixtures of various organic compounds with an average molecular weight between 400 and 600. There are various sources of pitch; the three most commonly used are polyvinyl chloride (PVC), petroleum asphalt, and coal tar. The same processing steps (stabilization, carbonization, and optional graphitization) are involved in converting the pitch-based precursor into carbon fiber. Pitch-based raw materials are generally cheap, and the carbon fiber yield from pitch-based precursors is relatively high.

A recent innovation in carbon-based materials has been carbon nanotubes (CNT). Carbon nanotubes are relatively new materials—discovered in 1991—as a minor by-product of the carbon-arc process that is used to synthesize carbon's fullerene molecules. They present exciting possibilities for research and use. CNTs are a variant of their predecessor, fullerene carbon (with a geodesic dome arrangement of 60, 70, or even a few hundred C atoms in a molecule). Figure 9.1-2 shows a photograph of CNT. Single-walled CNT have been grown to an aspect ratio of $\sim 10^5$, with a length of about 100 $\mu$m, and therefore, from a composite mechanics standpoint, they can be considered as long, continuous fibers. The multiwalled CNT has an "onion-like" layered structure and is under extremely high internal stress, as evident from very small lattice spacing near the inner regions of the CNT. Carbon nanotubes (CNTs) have some remarkable properties, such as better electrical conductivity than copper, exceptional mechanical strength, and very high flexibility (with futuristic potential for use in even earthquake-resistant buildings and crash-resistant cars). There is already considerable interest in industry in using CNTs in chemical sensors, field emission elements, electronic interconnects in integrated nanotube circuits, hydrogen storage devices, temperature sensors and thermometers, and others. Because of the exceptional properties of CNTs (e.g., Young's modulus of CNT is 1–4 TeraPascals, TPa), there has been some interest in incorporating CNTs in polymers, ceramics, and metals. Owing to CNT's metallic or semiconducting character, incorporating CNT in polymer matrices permits attainment of an electrical conductivity sufficient to provide an electrostatic discharge at very low CNT concentrations. Similarly, extremely hard/and wear-resistant metal-matrix composites and tough ceramic-matrix composites are being developed. Since the discovery of

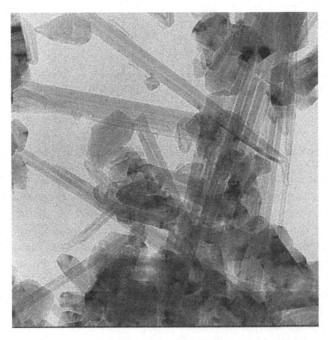

**Figure 9.1-2** Photograph of carbon nanotubes and polyhedral nanoparticles during fullerene production (R. Malhotra, R. S. Ruoff and D. C. Lorents, "Fullerene Materials," Advanced Materials & Processes, April 1995 p. 30). Reprinted with permission from ASM International, Materials Park, OH (www.asminternational.org).

CNTs in 1991, similar nanostructures were formed in other layered compounds such as BN, BCN, and $WS_2$, etc. For example, whereas CNTs are either metallic or semiconducting (depending on the shell helicity and diameter), BN nanotubes are insulating and could possibly serve as nanoshields for nanoconductors. Also, BN nanotubes are thermally more stable in oxidizing atmospheres than are CNTs and have comparable modulus. The strength of nanotubular materials can be increased by assembling them in the form of ropes, as has been done with CNT and BN nanotubes, with ropes made from single-walled CNTs being the strongest known material. The spacing between the individual nanotube strands in such a rope will be in the subnanometer range; for example, this spacing is ~0.34 nm in a rope made from multiwalled BN nanotubes, which is on the order of the (0001) lattice spacing in the hexagonal BN cell.

### Organic fibers

Because the covalent C-C bond is very strong, linear-chain polymers such as polyethylene can be made very strong and stiff by fully extending their molecular chains. A wide range of physical and mechanical properties can be attained by controlling the orientation of these polymer chains along the fiber axis and their order or crystallinity. Allied Corporation's Spectra 900 and Du Pont's aramid fiber Kevlar are two successful organic fibers widely used for composite strengthening. Aramid is an abbreviated name of a class of synthetic organic fibers that are aromatic polyamide compounds. Nylon is a generic name for any long-chain polyamide. Many highly sophisticated manufacturing techniques have been developed to fabricate the organic fibers for use in composites. These techniques include: tensile drawing, die drawing, hydrostatic extrusion, and gel spinning. A wide range of useful engineering properties is achieved in organic fibers depending on the chemical nature of the polymeric material, processing technique, and the control of process parameters. For example, high modulus polyethylene fibers with a modulus of 200 GPa, and Kevlar fibers with a modulus of 65–125 GPa and tensile strength of 2.8 GPa have been developed. Kevlar fibers have poor compression strength and should be used under compressive loading only as a hybrid fiber mixture, that is, as a combination of carbon fiber and Kevlar. One limitation of most organic fibers is that they degrade (lose color and strength) when exposed to visible or ultraviolet radiation, and a coating of a light-absorbing material is used to overcome this problem.

### Metallic fibers

Metals such as beryllium, tungsten, titanium, tantalum, and molybdenum, and alloys such as steels in the form of wires or fibers have high and very consistent tensile strength values as well as other attractive properties. Beryllium has a high modulus (300 GPa) and low density (1.8 g/cc) but also low strength (1300 MPa). Fine (0.1-mm) diameter steel wires with a high carbon (0.9%) content have very high strength (~5 GPa). Tungsten fibers have a very high melting point (3400°C) and are suited for heat-resistant applications. These various metallic fibers have been used as reinforcements in composite matrices based on metals (e.g., copper), concrete and polymers. For example, tungsten (density 19.3 g/cc) has been used as a reinforcement in advanced Ni- and Co-base superalloys for heat-resistant applications, and in Cu alloys for electrical contact applications. Similarly, steel wire is used to reinforce concrete and polymers (e.g., in steel belted tires). Other metallic reinforcements used in composite applications include ribbons and wires of rapidly quenched amorphous metallic alloys such as $Fe_{80}B_{20}$ and $Fe_{60}Cr_6Mo_6B_{28}$ having improved physical and mechanical properties.

### Ceramic fibers

Ceramic fibers such as single crystal sapphire, polycrystalline $Al_2O_3$, SiC, $Si_3N_4$, $B_4C$ and others have high strength at room- and elevated temperature, high modulus, excellent heat-resistance, and superior chemical stability against environmental attack. Both polymer pyrolysis and sol-gel techniques make use of organometallic compounds to grow ceramic fibers. Pyrolysis of polymers containing silicon, carbon, nitrogen, and boron under controlled conditions has been used to produce heat-resistant ceramic fibers such as SiC, $Al_2O_3$, $Si_3N_4$, BN, $B_4C$ and several others.

The commercial alumina fibers have a Young's modulus of 152–300 GPa and a tensile strength of 1.7 to 2.6 GPa. Alumina fibers are manufactured by companies such as Du Pont (fiber FP), Sumitomo Chemical (alumina-silica), and ICI (Saffil, $\delta$-alumina phase). Fiber FP is made by dry-spinning an aqueous slurry of fine alumina particles containing additives. The dry-spun yarn is subjected to two-step firing: low firing to control the shrinkage and flame-firing to improve the density of $\alpha$-alumina. A thin silica coating is generally applied to heal the surface flaws, giving higher tensile strength than uncoated fiber. The polymer pyrolysis route to make $Al_2O_3$ fibers makes use of dry-spinning of an organoaluminum compound to produce the ceramic precursor, followed by calcining of this precursor to obtain the final fiber. 3M Company uses a sol-gel route to synthesize an alumina fiber (containing silica and boria), called Nextel 312. The technique uses hydrolysis of a metal alkoxide, that is, a compound of the type $M(OR)_n$ where $M$ is the metal, $R$ is an organic compound, and $n$ is the metal valence. The process breaks the $M$-$OR$ bond and establishes the $MO$-$R$ to give the desired oxide. Hydrolysis of metal alkoxides creates sols that are spun and gelled. The gelled fiber is then densified at intermediate temperatures. The high surface energy of the fine pores of the gelled fiber permits low-temperature densification.

Silicon carbide fibers, whiskers and particulates are among the most widely used reinforcements in composites. SiC fiber is made using the CVD process. A dense coating of SiC is vapor-deposited on a tungsten or carbon filament heated to about 1300°C.

The deposition process involves high-temperature gaseous reduction of alkyl silanes (e.g., $CH_3SiCl_3$) by hydrogen. Typically, a gaseous mixture consisting of 70% $H_2$ and 30% silanes is introduced in the CVD reactor along with a 10–13 μm diameter tungsten or carbon filament.

The SiC-coated filament is wound on a spool, and the exhaust gases are passed through a condenser system to recover unused silanes. The CVD-coated SiC monofilament (~100–150 μm diameter) is mainly β-SiC with some α-SiC on the tungsten core. The SCS-6 fiber of AVCO Specialty Materials Company is a CVD SiC fiber with a gradient structure that is produced from the reaction of silicon- and carbon-containing compounds over a heated pyrolytic graphite-coated carbon core. The SCS-6 fiber is designed to have a carbon-rich outer surface that acts as a buffer layer between the fiber and the matrix metal in a composite, and the subsurface structure is graded to have stoichiometric SiC a few micrometers from the surface.

The SiC fiber obtained via the CVD process is thick (140 μm) and inflexible which presents difficulty in shaping the preform using mass production methods such as filament winding. A method, developed in Japan, to make fine and flexible continuous SiC fibers (Nicalon fibers) uses melt-spinning under $N_2$ gas of a silicon-based polymer such as polycarbosilane into a precursor fiber. This is followed by curing of the precursor fiber at 1000°C under $N_2$ to cross-link the molecular chains, making the precursor infusible during the subsequent pyrolysis at 1300°C in $N_2$ under mechanical stretch. This treatment converts the precursor into the inorganic SiC fiber. Nicalon fibers, produced using the above process, have high modulus (180–420 GPa) and high strength (~2 GPa).

Besides the SiC and $Al_2O_3$ fibers described in the preceding paragraphs, silicon nitride, boron carbide, and boron nitride are other useful ceramic fiber materials. $Si_3N_4$ fibers are produced by CVD using $SiCl_4$ and $NH_3$ as reactant gases, and forming the fiber as a coating onto a carbon or tungsten filament. In polymer-based synthesis of silicon nitride fibers, an organosilazane compound (i.e., a compound that has Si-NH-Si bonds) is pyrolyzed to give both SiC and $Si_3N_4$. Fibers of the oxidation-resistant material boron nitride are produced by melt-spinning a boric oxide precursor, followed by a nitriding treatment with ammonia that yields the BN fiber. A final thermal treatment eliminates residual oxides and stabilizes the high-purity BN phase. Boron carbide ($B_4C$) fibers are produced by the CVD process via the reaction of carbon yarn with $BCl_3$ and $H_2$ at high temperatures in a CVD reactor. In addition to the use of long and continuous fibers of different ceramic materials in composite matrices, vapor-phase grown ceramic whiskers have also been extensively used in composite materials. Whiskers are monocrystalline short ceramic fibers (aspect ratio ~50–10,000) having extremely high fracture strength values that approach the theoretical fracture strength of the material.

Figure 9.1-3 compares the room-temperature stress versus strain behavior of boron, Kevlar, and glass fibers; high-modulus graphite (HMG) fiber; and ceramic whiskers. The figure shows that whiskers are by far the strongest reinforcement, because of the absence of structural flaws, which results in their strength approaching the material's theoretical strength. Usually, however, there is considerable scatter in the strength properties of whiskers, and this becomes problematic in synthesizing composites with a narrow spread in their properties. Selected thermal and mechanical properties of some commercially available fibers are summarized in Table 9.1-1. The mechanical and physical properties such as elastic modulus (E) and coefficient of thermal expansion (CTE) of fibers are strongly orientation dependent, and usually exhibit significant disfferences in magnitude along the fiber axis and transverse to it. The high-temperature strength of some commercial silicon carbide fibers is compared in Figure 9.1-4. It can be noted that the fiber retains high strength to fairly high temperatures; for example, NLP 101 fiber retains a strength of 500 MPa at 1300°C, which is comparable to the room-temperature tensile strength of some high-strength, low-alloy steels.

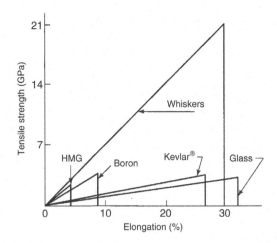

**Figure 9.1-3** Schematic comparison of stress–strain diagrams for common reinforcing fibers and whiskers (HMG, high-modulus graphite fiber) (A. Kelly, ed., *Concise Encyclopedia of Composite Materials*, Elsevier, 1994, p. 312.) Reprinted with permission from Elsevier.

**Table 9.1-1** Properties of selected reinforcement materials

| Reinforcement | $\rho$, kg·m$^{-3}$ | Diameter, $\mu$m | E, GPa | UTS, MPa | $C_p$, KJ·kg$^{-1}$·°K$^{-1}$ | k, W·m$^{-1}$·K$^{-1}$ | CTE, $10^{-6}$ °K$^{-1}$ |
|---|---|---|---|---|---|---|---|
| SiC whisker | 3200 | | 700 | 21,000 | 0.69 | | 2.5 |
| Si$_3$N$_4$ | 3100 | | 385 | 5,000–7,000 | | | |
| Sic (Nicalon) | 2550 | 10–20 | 180$^\parallel$ | 8,300 | 0.69 | 25 | 2.5–4.3 |
| Al$_2$O$_3$ (fiber FP) | 3950 | 20 | 385$^\parallel$ | 3,800 | | 37.7 | 8.3$^\parallel$ |
| Al$_2$O$_3$ (Saffil) | 3300 | | 285 | 1,500 | | | |
| Al$_2$O$_3$ (whisker) | 4000 | 10–25 | 700–1500 | 10,000–20,000 | 0.60 | 24 | 7.7 |
| E glass fiber | 2580 | 8–14 | 70$^\perp$ 70$^\parallel$ | 3,450 | 0.70 | 13 | 4.7$^{\parallel,\perp}$ |
| Borsic (SiC-coated B) | 2710 | 102–203 | 400 | 3,100 | 1.3 | 38 | 5.0 |
| PAN HM carbon fiber | 1950 | 7–10 | 390$^\parallel$ 12$^\perp$ | 2,200 | 0.71 | | $-0.5$ to $1.0^\parallel$ 7 to 12$^\perp$ |
| PAN HS carbon fiber | 1750 | 7–10 | 250$^\parallel$ 20$^\perp$ | 2,700 | 0.71 | 8 | $-0.5$ to $1.0^\parallel$ 7 to 12$^\perp$ |
| Aramid (Kevlar 49) | 1440 | 12 | 125$^\parallel$ | 2,800–3,500 | | | $-2$ to $-5^\parallel$ 59$^\perp$ |
| Sapphire | 4000 | | 470 | 2,000 | | | 6.2–6.8 |
| PRD 166 (Al$_2$O$_3$-ZrO$_2$) | 4200 | | 385 | 2,500 | | | |
| Nextel 480 | 3050 | | 224 | 2,275 | | | |

Note: $\rho$, density; E, modulus; UTS, ultimate tensile strength; k, thermal conductivity; CTE, coefficient of thermal expansion.
PAN HM carbon fiber: polyacrylonitrile-derived, high-modules graphite fiber.
PAN HS carbon fiber: polyacrylonitrile-derived, high-strength graphite fiber.
$^\parallel$ parallel to the fiber axis.
$^\perp$ perpendicular to the fiber axis.

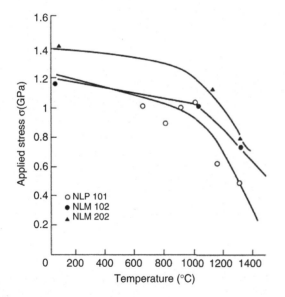

**Figure 9.1-4** High-temperature strength of some SiC fibers plotted as applied stress versus test temperature. (B. S. Mitchell, *An Introduction to Materials Engineering and Science for Chemical and Materials Engineers*, Wiley Interscience, Hoboken, NJ, 2004).

In addition to the synthetic fibers and whiskers, numerous low-cost, discontinuous fillers have been used in composites to conserve precious matrix materials at little expense to their engineering properties. These fillers include mica, sand, clay, talc, rice husk ash, fly ash, natural fibers (e.g., lingo-cellulosic fibers), recycled glass, and many others, including environmentally conscious biomorphic ceramics based on silicon carbide and silicon dioxide obtained from pyrolysis of natural wood. These various fillers and reinforcements permit a range of composite microstructures to be created that have a wide range of strength, stiffness, wear resistance, and other characteristics. Figure 9.1-5 shows the porous structure of pyrolyzed wood that has been used as a preform for impregnation with molten metals to create ceramic- or metal-matrix composites.

# Interface

Interfaces in composites are regions of finite dimensions at the boundary between the fiber and the matrix where compositional and structural discontinuities can occur

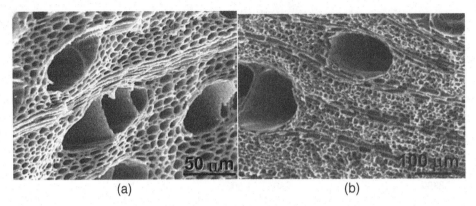

**Figure 9.1-5** (a) Scanning electron micrograph of a maple-derived carbonaceous preform showing porous structure. (M. Singh and J. A. Salem, Mechanical properties and microstructure of biomorphic silicon carbide ceramics fabricated from wood precursors, *Journal of the European Ceramic Society*, **22**, 2002, 2709–2717, Elsevier). (b) SEM micrograph of porous carbon preform made from mahogany wood perpendicular to the growth direction. (M. Singh and J. A. Salem, Mechanical properties and microstructure of biomorphic silicon carbide ceramics fabricated from wood precursors, *Journal of the European Ceramic Society*, **22**, 2002, 2709–2717, Elsevier). Reprinted with permission from Elsevier. Photo Courtesy of M. Singh, QSS Group, Inc., NASA Glenn Research Center, Cleveland, OH.

over distances varying from an atomic monolayer to over five orders of magnitude in thickness. Composite fabrication processes create interfaces between inherently dissimilar materials (e.g., ceramic fibers and metal matrices). The material incompatibility and the severe processing conditions generally needed for composite fabrication create interfaces that exist in a non-equilibrium state. Thus, interface evolution is thermodynamically ordained; however, interface design for properties by processing is essentially an outcome of a variety of kinetic phenomena (reaction kinetics, mass transport etc.). The nature of the interface that develops in composites during fabrication and subsequent service strongly influences the response of the composite to mechanical stresses and to thermal and corrosive environments. As the inherent properties of the fiber and matrix constituents in a composite are fixed, greatest latitude in designing bulk composite properties is realized by tailoring the interface (this is not strictly true, however, because processing conditions that lead to interface development also usually modify both the fiber properties as well as the metallurgy of the matrix). The development of an optimum interfacial bond between the fiber and the matrix is, therefore, a primary requirement for optimum performance of a composite. The nature and the properties of the interface (thickness, continuity, chemistry, strength, and adhesion) are determined by factors both intrinsic to the fiber and matrix materials (chemistry, crystallography, defect content), as well as extrinsic to them (time, temperature, pressure, atmosphere, and other process variables). One of the major goals in the study of interfaces in composites is to develop an understanding of and exercise control on the structure, chemistry, and properties of interfaces by judicious manipulation of processing conditions. To realize this goal, it is also necessary to develop and use techniques for mechanical, chemical, and structural characterization of interfaces in composites. Considerable progress has been made in characterizing and understanding interfaces at the microstructural, crystallographic, and atomic levels.

Bonding at the fiber-matrix interface develops from physical or chemical interactions, from frictional stresses due to irregular surface topography, and from residual stresses arising from the mismatch of coefficients of thermal expansion (CTE) of the fiber and the matrix materials. Both the fiber and matrix characteristics as well as the interface characteristics control the physical, mechanical, thermal, and chemical behavior of the composite. As an example, consider the fracture behavior of a composite in which fibers are the strengthening phase. The fracture in the composite can proceed at the interface, through the matrix, or through the reinforcement depending on their respective inherent mechanical properties and the defect population. If the matrix is weak relative to the fiber and the interface, it will fail by the usual crack nucleation and growth mechanism. If the matrix and the interface are strong, the load is transferred across the interface to the reinforcement, which will provide strengthening until a threshold stress is reached at which the composite will fail. Similarly, electrical, thermal, and other properties of a composite are determined by the properties of the fiber, matrix, and the interface.

Purely physical interactions (e.g., dispersion forces and electrostatic interactions) seldom dominate the interface behavior in composites. Usually, some chemical interaction between fiber and matrix aids interface growth and determines the interface behavior. Chemical interactions may involve adsorption, impurity segregation, diffusion, dissolution, precipitation and reaction layer formation. These interactions are abruptly terminated at the conclusion of composite fabrication, which renders the interface inherently unstable. Driven by a need for these interactions to proceed to completion and the interface to approach thermodynamic equilibrium, compositional and structural transformations at the interface continue after fabrication, usually with sluggish kinetics, via reaction paths that may involve intermediate non-equilibrium phases. Chemical interactions between the fiber and the matrix not only determine the interface properties and behavior, but may also modify the properties of the fiber and the metallurgy of the matrix. For example, the extent of fiber strength degradation and loss of age-hardening response because of chemical reactions in metal-matrix composites are directly related to the extent of interfacial reactions as reflected in the size of the reaction zone at the interface. The loss of age-hardening response is because of loss of chemically active solutes in the fiber-matrix reactions. It is, therefore, very important to control the processing conditions to design the interface for properties with minimum fiber degradation and little alterations in the metallurgy of the matrix. Because the rate of chemical reaction can be characterized in terms of temperature-dependent rate constants and activation energies, fundamental insights into the mechanisms of strength-limiting interfacial reactions can be derived.

Besides chemical interactions between the fiber and the matrix, the thermoelastic compatibility between the two is important, particularly if the fabrication and/or service involves significant temperature excursions. A large mismatch between the coefficients of thermal expansion (CTE) of the fiber and the matrix can give rise to appreciable thermoelastic stresses which may affect the adhesion at the interface. For example, these stresses can give rise to interfacial cracking if the matrix cannot accommodate these stresses by plastic flow. In such a case, stress-absorbing intermediate compliant layers are deposited at the interface to promote compatibility and reduce the tendency for cracking by reducing the CTE mismatch–induced stresses. Such layers may also provide protection to the reinforcement against excessive chemical attack in reactive matrices during fabrication and service.

A careful control of the fabrication conditions can enhance the interface strength without excessive fiber degradation. Usually, a moderate chemical interaction between fiber and matrix improves the wetting, assists liquid-state fabrication of composite, and enhances the strength of the interface, which in turn facilitates transfer of external stresses to the strengthening agent, i.e., the fiber. But an excessive chemical reaction would degrade the fiber strength (even though the interface strength may be high) and defeat the very purpose for which the fibers were incorporated in the monolith. In contrast, if toughening rather than strengthening is the objective, as in brittle ceramic-matrix composites, then creation of a weak rather than strong interface is desired so that crack deflection and frictional stresses during sliding of debonded fibers will permit realization of toughness. In such a case, recipes designed to improve the wetting by strong chemical reactions can induce too high a bond strength, which will, in turn, confer poor toughness on the composite. Thus, a delicate balance between several conflicting requirements is usually necessary to tailor the interface for a specific application with the aid of surface-engineering and processing science.

## Fiber strengthening

With a fiber residing in a matrix, the length of the fiber limits the distance over which bonding and load transfer are possible. For effective composite strengthening, the fiber must have a minimum critical length, $l_c$; fibers shorter than this length do not serve as load-bearing constituents. This critical length is

$$l_c = \frac{\sigma_f d}{2\tau_c} \tag{9.1.1}$$

where $d$ is the fiber diameter, $\sigma_f$ is the fracture strength of the fiber, and $\tau_c$ is the fiber–matrix bond strength. The critical length, $l_c$, is on the order of a few millimeters for most composites. Thus, for effective strengthening, the actual fiber length must exceed a few millimeters. In continuous fiber-reinforced composites, the tensile strength and the elastic modulus are strongly anisotropic. Usually, the strength of the composite along the fiber length (longitudinal strength) follows a simple rule of mixture (ROM), i.e., $\sigma_c = V_f \sigma_f + V_m \sigma_m$, where $\sigma_m$ and $\sigma_f$ are the strength of the matrix and reinforcement, respectively, and $V_f$ and $V_m$ are the fiber and matrix volume fractions. For composites reinforced with discontinuous, aligned fibers with a uniform distribution in the matrix and with the fiber length, $l$, greater than the critical length $l_c$, the composite strength, $\sigma_{cd}$, along the longitudinal direction is

$$\sigma_{cd} = \sigma_f V_f \left(1 - \frac{l_c}{2l}\right) + \sigma_m(1 - V_f) \tag{9.1.2}$$

If the fiber length is smaller than the critical length, then $\sigma_{cd}$ is given from

$$\sigma_{cd} = \frac{l\tau_c}{d} V_f + \sigma_m(1 - V_f) \tag{9.1.3}$$

where $d$ is the fiber diameter, and $\tau_c$ is the shear stress at the fiber surface, which for a plastically deformable matrix such as metals is the yield strength of the matrix, and for a brittle matrix (ceramics or polymers), is the frictional stress at the interface. Calculations based on Equation (9.1.2) show that for aligned, discontinuous fibers with $l > lc$, the loss in composite's strength relative to the case of continuous fibers will not be appreciable provided the stress concentration at the ends of the short fibers is negligible.

The elastic modulus of a continuous fiber-reinforced composite along the longitudinal (fiber) direction is given from the following ROM relationship:

$$E_c = V_f E_f + V_m E_m \quad (9.1.4)$$

where $E_m$ and $E_f$ are the Young's moduli of the matrix and reinforcement, respectively, and $V_f$ and $V_m$ are the volume fractions of the fiber and the matrix, respectively. The composite modulus, $E_{ct}$, transverse to the fiber direction is given from a relationship reminiscent of the electrical resistance of parallel circuit:

$$\frac{1}{E_{ct}} = \frac{V_f}{E_f} + \frac{V_m}{E_m} \quad (9.1.5)$$

For discontinuously reinforced composites,

$$E_{cd} = K E_f V_f + E_m V_m \quad (9.1.6)$$

where $K$ is called the fiber efficiency factor, and its value depends on the modulus ratio, $(E_f/E_m)$; for most cases $K$ is in the range 0.1–0.6. These relationships apply to situations where the fiber-matrix interface is devoid of reaction layers, such as in polymer-matrix composites.

## Polymer-matrix composites

### Matrix

Fiber-reinforced polymers are widely used as structural materials for relatively low-temperature use. Generally, polymers have lower strength and modulus than metals or ceramics but they are more resistant to chemical attack than metals. Figure 9.1-6 displays a schematic comparison of the strength characteristics of ceramics, metals, polymers, and elastomers. Prolonged exposure to UV light and some solvents can, however, cause polymer degradation. Polymers are giant, chainlike molecules or macromolecules, with covalently bonded carbon atoms as the backbone of the chain. Small-chain, low-molecular-weight organic molecules (monomers) are joined together via the process of polymerization, which converts monomers to polymers. Polymerization occurs either through condensation or through addition of a catalyst. In condensation polymerization, there occurs a stepwise reaction of molecules, and in each step a molecule of a simple compound, generally water, forms as a by-product. In addition to polymerization, monomers can be joined to form a polymer with the help of a catalyst without producing any by-products. For example, the linear addition of ethylene molecules ($CH_2$) results in polyethylene, with the final mass of polymer being the sum of monomer masses.

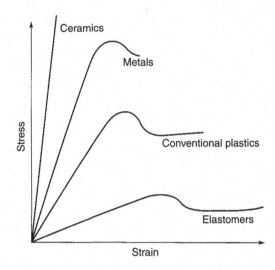

**Figure 9.1-6** Comparison of idealized stress–strain diagrams for metals, amorphous polymers, and elastomers. (B. S. Mitchell, *An Introduction to Materials Engineering and Science for Chemical and Materials Engineers*, Wiley-Interscience, Hoboken, NJ, 2004, p. 469).

Linear polymers consist of a long chain, often coiled or bent, of atoms with attached side groups (e.g., polyethylene, polyvinyl chloride, polymethyl metacrylate or PMMA). Branched polymers consist of side-branching of atomic chains. In cross-linked polymers, molecules of one chain are bonded (cross-linked) with those of another, thus forming a three-dimensional network. Cross-linking hinders sliding of molecules past one another, thus making the polymer strong and rigid. Ladder polymers form by linking linear polymers in a regular manner; ladder polymers are more rigid than linear polymers. Figure 9.1-7 illustrates these different types of polymers.

Unlike pure metals that melt at a fixed temperature, polymers show a range of temperatures over which crystallinity vanishes on heating. On cooling, polymer liquids contract just as metals do. In the case of amorphous polymers, this contraction continues below the melting point, $T_m$, of crystalline polymer to a temperature, $T_g$, called the glass transition temperature, at which the supercooled liquid polymer becomes extremely rigid owing to extremely high viscosity. The structure of the polymer below $T_g$ is essentially disordered, like that of a liquid. Figure 9.1-8 shows the changes in the specific volume as a function of temperature in a polymer. Many physical properties such as viscosity, heat capacity, modulus, and thermal expansion change abruptly at $T_g$. For example, Figure 9.1-9 displays the variation of the natural logarithm of elastic modulus as a function of temperature; the transitions from the glassy to rubbery, and rubbery to fluid states are accompanied by discontinuities in the modulus. The glass transition temperature, $T_g$, is a function of the structure of the polymer; for example, if a polymer has a rigid backbone structure and/

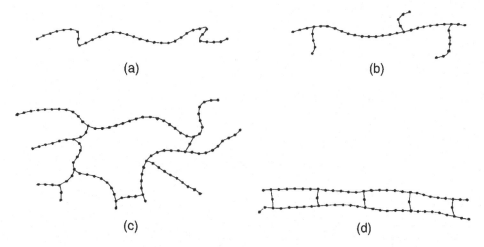

**Figure 9.1-7** Molecular chain configurations in polymers: (a) linear, (b) branched, (c) cross-linked, and (d) ladder. (K. K. Chawla, *Composite Material – Science & Engineering*, Springer-Verlag, New York, NY, 1987, p. 59).

or bulky branch groups, then $T_g$ will be quite high. The glass transition phenomenon is also observed in amorphous ceramics such as glasses. Glasses have a mixed ionic and covalent bonding, and are highly cross-linked. In fact, they may have a higher $T_g$ than polymers that have only covalent bonding and have less cross-linking. In amorphous polymers, there is no apparent order among the molecules and the molecular chains are arranged in a random manner. When polymers precipitate from dilute solutions, small, platelike single-crystalline regions called lamellae or crystallites form. In the lamellae, long molecular chains are folded in a regular manner, and many lamellae group together to form spherulites much like grains in metals.

Most linear polymers soften and melt upon heating. These are called thermoplastics and are readily shaped using liquid forming techniques. Examples include low- and high-density polyethylene, polystyrene, and PMMA.

When the molecules in a polymer are cross-linked in the form of a network, they do not soften on heating, and are called thermosetting polymers. Common thermosetting polymers include phenolic, polyester, polyurethane, and silicone. Thermosetting polymers decompose on heating. As noted, cross-linking makes sliding of molecules past one another difficult, thus making the polymer strong and rigid. A typical example is that of rubber cross-linked with sulfur, that is, vulcanized rubber. Vulcanized rubber has 10 times the strength of natural rubber.

There is another type of classification of polymers based on the type of repeating unit. If one type of repeating unit forms a polymer chain, it is called a homopolymer. In contrast, polymer chains having two different monomers form co-polymers. If the two different monomers are distributed randomly along the chain, then the polymer is called a regular or random co-

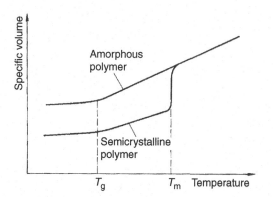

**Figure 9.1-8** Specific volume versus temperature relationship for amorphous and semicrystalline polymers ($T_g$, glass transition temperature; $T_m$, melting temperature). (K. K. Chawla, *Composite Material – Science & Engineering*, Springer-Verlag, New York, NY, 1987, p. 60).

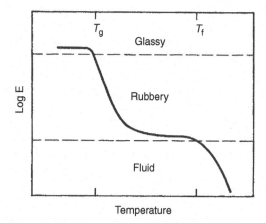

**Figure 9.1-9** Variation of modulus of an amorphous polymer with temperature. (K. K. Chawla, *Composite Material – Science & Engineering*, Springer-Verlag, New York, NY, 1987, p. 63).

# CHAPTER 9.1 Composite materials

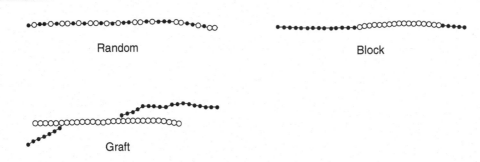

**Figure 9.1-10** Schematic representation of random, block, and graft copolymers. (K. K. Chawla, *Composite Material – Science & Engineering*, Springer-Verlag, New York, NY, 1987, p. 61)

polymer. If, however, a long sequence of one monomer is followed by a long sequence of another monomer, the polymer is called a block co-polymer. If a chain of one type of monomer has branches of another type, then a graft co-polymer is said to form. Figure 9.1-10 schematically illustrates these various polymer structures.

Molecular weight is a very important parameter that determines the properties of polymers. Many mechanical properties increase with increasing molecular weight (although polymer processing also becomes more difficult with increasing molecular weight). Another important parameter is degree of polymerization (DP), which is a measure of the number of basic units (mers) in a polymer. The molecular weight (MW) and the DP are related by $MW = DP \times (MW)u$, where $(MW)u$ is the molecular weight of the repeating unit. Because polymers contain various types of molecules, each of which has a different MW and DP, the molecular weight of the polymer is characterized by a distribution function. A narrower distribution indicates a homogeneous polymer with many repeating identical units whereas a broad distribution indicates that the polymer is composed of a large number of different species. In general, therefore, it is convenient to specify an average molecular weight or degree of polymerization. Unlike many common substances of low molecular weight, polymers can have very large molecular weights. For example, the molecular weights of natural rubber and polyethylene (a synthetic polymer) can be greater than $10^6$ and $10^5$, respectively. In addition, the molecular diameters of high-molecular-weight polymers can be over two orders of magnitude larger than ordinary substances such as water that have low molecular weights.

Polymers can be amorphous or partially crystalline, and the amount of crystallinity in a polymer can vary from 30 to 90%. The elastic modulus and strength of polymers increase with increasing crystallinity, although fully crystalline polymers are unrealistic. The long and complex molecular chains in a polymer are easily tangled, resulting in large segments of molecular chains that may remain trapped between crystalline regions and never reorganize themselves into an ordered molecular assembly characteristic of a fully crystalline state. The molecular arrangement in a polymer also influences the extent of crystallization; for example, linear molecules with small side groups can readily crystallize whereas branched-chain molecules with bulky side groups do not crystallize easily. Thus, linear high-density polyethylene can attain up to 90% crystallization whereas branched polyethylene can attain only about 65% crystallization.

## Properties of polymeric matrices

Glassy polymers follow Hooke's law and exhibit a linear elastic response to applied stress. The elastic strain in glassy polymers is less than 1%. In contrast, elastomers (rubbery polymers) show a nonlinear elastic behavior with a large elastic range (a few percent strain) as shown in Figure 9.1-6. The large elastic strain in elastomers is caused by a redistribution of the tangled molecular chains under an applied stress.

Highly cross-linked thermosetting resins such as polyesters, epoxies, and polyimides have high modulus and strength, but are also extremely brittle. The fracture energy (100–200 $J.m^{-2}$) of thermosetting resins is only slightly better than that of inorganic glasses (10–30 $J.m^{-2}$). In contrast, thermoplastic resins such as polymethylmetacrylate (PMMA) have fracture energies on the order of 1000 $J.m^{-2}$ because their large free volume facilitates absorption of the energy associated with crack propagation. The fracture toughness of hard and brittle thermosets such as epoxies and polyester resins is improved by distributing small (a few micrometers in size) and soft rubbery inclusions in the brittle matrix. The most common method is simple mechanical blending of the soft, elastomer particles and the thermoset resin, or copolymerization of a mixture of the two.

Polymers show a significant temperature dependence of their elastic modulus as shown in Figure 9.1-8. Below the glass transition temperature $T_g$, the polymer behaves as a hard and rigid solid, with an elastic modulus of about 5–7 GPa. Above $T_g$, the modulus drops significantly and the polymer exhibits a rubbery behavior. Upon heating the polymer to above its melting temperature $T_f$ at which the polymer becomes fluid, the modulus drops

# Composite materials CHAPTER 9.1

**Table 9.1-2** Mechanical properties of selected polymeric materials.

| Material | E, GPa | T.S., MPa | MOR, MPa | Y.S., MPa | % Elongation | Sp. Gr. | CTE, $10^{-5} \times °C^{-1}$ | $T_g$, C | $T_m$, C |
|---|---|---|---|---|---|---|---|---|---|
| LDPE | 0.17–0.28 | 8.3–31.4 | – | 9.0–14.5 | 100–650 | 0.92–0.93 | – | –100 | 120 |
| HDPE | 1.06–1.09 | 22.1–31.0 | – | 26.2–33.1 | 10–1200 | 0.95–0.96 | – | –115 | 130 |
| PVC | 2.40–4.10 | 40.7–51.7 | – | 0.47–44.8 | 40–80 | 1.30–1.58 | – | 80 | 212 |
| Teflon | 0.40–0.55 | 20.7–34.5 | – | – | 200–400 | 2.14–2.20 | – | 125 | 327 |
| Polystyrene | 2.28–3.28 | 35.9–51.7 | – | – | 1.2–2.5 | 1.04–1.05 | – | 100 | 240 |
| Polyester (PET) | 2.8–4.1 | 48.3–72.4 | 59.3 | – | 30–300 | 1.29–1.40 | – | 70 | 270 |
| Polycarbonate | 2.38 | 62.8–72.4 | 62.1 | – | 110–150 | 1.20 | – | 150 | 230 |
| Natural rubber | | 18–25 | | | 1000 | | | | |
| Butadiene styrene | | 1.5–2.3 | | | 3000 | | | | |
| Epoxy | | 35–85 | 15–35 | | | 1.38 | 8.0–11.0 | | |
| Polyimide | | 120 | 35 | | | 1.46 | 9.0 | | |
| PEEK | | 92 | 40 | | | 1.30 | – | | |
| Phenolics | | 50–55 | – | | | 1.30 | 4.5–11.0 | | |

Note: LDPE; low-density polyethylene; HDPE, high-density polyethylene; PVC, polyvinyl chloride; Teflon, polytetrafluoroethylene; E, tensile modulus; T.S., tensile strength; MOR, modulus of rupture (flexural strength); Y.S., yield strength; CTE, coefficient of thermal expansion; $T_g$, glass transition temperature; $T_m$, melting temperature.

abruptly. The glass transition temperature, melting temperature, and selected mechanical and thermal properties of common polymeric materials are listed in Table 9.1-2.

The thermal expansion of polymers, in particular thermoplastics, is strongly temperature-dependent, exhibits a non-linear increase with temperature, and is generally an order of magnitude or two greater than that of metals and ceramics. Therefore, in polymer composites the thermal expansion mismatch between inorganic fibers and polymer matrix is large, which may cause thermal stresses and cracking.

Both thermosets and thermoplastics are used as matrix materials for polymer composites. For example, polyesters, epoxies and polyimides are commonly used matrices in fiber-reinforced composites. Polyesters have fair resistance to water and various chemicals, as well as to aging, but they shrink between 4 and 8% on curing. In contrast, thermosetting epoxy resins have better moisture resistance, lower shrinkage on curing (about 3%), a higher maximum-use temperature, and good adhesion to glass fibers. Polyimides have a relatively high service-temperature range, 250–300°C but, they are brittle and have low fracture energies of 15–70 J·m$^{-2}$. Polymers degrade at high temperature and by moisture absorption, which causes swelling and a reduction in $T_g$. If the polymer is reinforced with fibers bonded to the matrix, then moisture absorption may cause severe internal stresses in the composite.

## Polymer composites

Polymer composites are fabricated using pultrusion and filament winding. In pultrusion (Figure 9.1-11), continuous fiber tows (i.e., fiber bundle with parallel strands of fibers) are impregnated with a thermoset resin and drawn through a die that forms the final component shape (e.g., tubes, rods, etc.). Hollow parts are made by pultruding the composite feedstock around a core or mandrel. The drawing rate and fiber volume fraction are controlled. The component is then cured in a preheated precision die to obtain the final shape and size. Glass-, carbon-, and aramid fiber-reinforced polyesters, vinyl esters, and epoxy resin matrix composites containing relatively large (40% or higher) volume fraction of aligned continuous fibers are produced via this method.

In filament winding, fiber strands or tows are first coated with a resin by passing them through a resin bath, and resin-coated fibers are automatically and continuously wound onto a mandrel for subsequent curing. The fiber volume fraction is controlled by the spacing between fiber strands and by the number of fiber layers

# CHAPTER 9.1    Composite materials

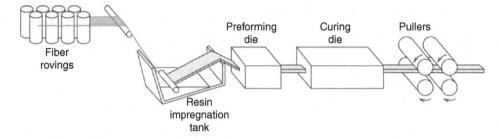

**Figure 9.1-11** Schematic diagram of the pultrusion process of making polymer composites. (W. D. Callister, Jr., *Materials Science and Engineering: An Introduction*, 6th ed., Wiley, New York, 2003, p. 555).

wound on the mandrel. The mechanical properties of the composite are influenced not only by the fiber volume fraction but also by the winding pattern (helical, circumferential, etc.). After the required number of fiber layers have been wound on the mandrel, the composite is cured in an oven and the mandrel removed to obtain a hollow composite object. Rocket motor castings, pipes, and pressure vessels are made using filament winding.

Another widely used method to make polymer-matrix composite consists of stacking layers of partially-cured thin (<1 mm) sheets of resin-impregnate and directionally aligned fibers called prepregs either manually (hand layup) or automatically in a three-dimensional sandwich structure. Prepregs are covered with a backing paper that is removed prior to lamination. Prepregs are made by sandwiching fiber tows between sheets of carrier paper that is coated with the resin matrix. On pressing the paper over fiber tows using heated rollers (a process called "calendering"), the resin melts and impregnates the fibers, thus forming a prepreg. A prepreg may be cut to various angles relative to the fiber axis to give prepregs of different orientations. For example, a prepreg with fibers parallel to the long dimension is a zero-degree lamina or ply, and a prepreg that is cut with fibers perpendicular to the long dimension is a 90-degree ply (intermediate angles are also used). Figure 9.1-12 shows stacking of prepregs in a 0°/±45°/90° orientation. The composite properties can be predicted from the theory of composite mechanics for a given stacking sequence and fiber orientation, which permits premeditated design of the composite for properties. For example, the coefficient of thermal expansion (CTE) of boron-epoxy and carbon-epoxy composites can be systematically varied between $-5 \times 10^{-6}$ and $30 \times 10^{-6}$ $°K^{-1}$, by controlling the ply angle as shown in Figure 9.1-13. After stacking the plies in the desired orientation, the final curing of the component is done under heat and pressure. For polymer composite with thermoplastic matrices, liquid-phase fabrication methods, such as injection molding, extrusion, and thermoforming, are used, and short, randomly oriented fibers can be dispersed for isotropic properties.

Figure 9.1-14 and Table 9.1-3 present selected mechanical property data on some polymer composites. The mechanical properties of polymer composites often degrade because of temperature and humidity effects (hygrothermal degradation). As stated earlier, moisture diffusion into a polymer matrix causes swelling, a decrease in the glass transition temperature of the matrix, and weakening of the interface. The resulting softening of the matrix and the interface can lead to composite failure at low stresses. Besides humidity and temperature,

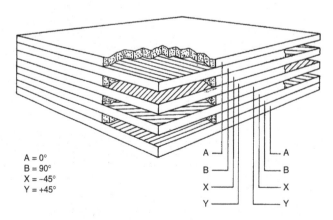

**Figure 9.1-12** Schematic of a laminated composite formed by stacking prepregs of different orientations (0°, +45°, −45°, 90°). (K. K. Chawla, *Composite Materials: Science and Engineering*, Springer, New York, NY, 1987, p. 91).

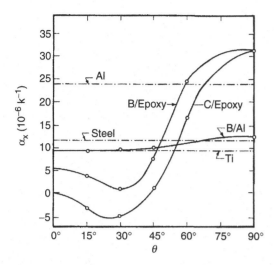

**Figure 9.1-13** Coefficient of thermal expansion as a function of fiber orientation for some polymer and metal composites, and metals and alloys. (Composite Materials in Engineering Design, American Society for Materials, 1973, p. 335). Reprinted with permission from ASM International, Materials Park, OH (www.asminternational.org).

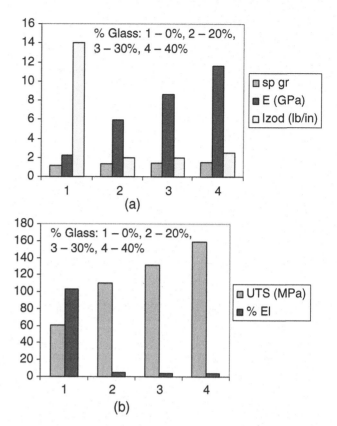

**Figure 9.1-14** (a) Specific gravity, Young's modulus (E), and impact strength of polycarbonate reinforced with randomly oriented glass fibers containing different glass fiber volume fractions. (Data are taken from *Materials Engineering's Materials Selector*, copyright Penton, IPC). (b) Percent elongation and ultimate tensile strength (UTS) of a glass fiber-reinforced epoxy composite with different glass fiber volume fractions. (Data are taken from *Materials Engineering's Materials Selector*, copyright Penton, IPC).

electromagnetic radiation primarily ultraviolet (UV) radiation, can also degrade the polymer composite because the UV radiation breaks the C-C bond in the polymer. Resistance to UV radiation is enhanced by adding carbon black to the polymer matrix.

## Ceramic-matrix composites

Ceramic-matrix composites surpass metal-matrix and polymer-matrix composites in elevated-temperature performance. As a result, there has been a great deal of interest in the development and testing of advanced ceramic composites based on carbides (SiC, TiC), oxides ($Al_2O_3$, $ZrO_2$, $SiO_2$), nitrides ($Si_3N_4$, BN, AlN), borides ($ZrB_2$, $HfB_2$), and glass and glass-ceramics for a number of applications. Some examples of ceramic-matrix composites are SiC/$Al_2O_3$, SiC/glass ceramic, SiC/$Si_3N_4$, TiC/$Si_3N_4$, $Al_2O_3$/$ZrO_2$, TiC/$Al_2O_3$, C/glass, C/SiC, and SiC/SiC. Thus, C/SiC composites produced by chemical vapor infiltration (CVI) have been developed for applications in hypersonic aircraft thermal structure, advanced rocket propulsion thrust chambers, cooled panels for nozzle ramps, and turbo pump blisks (blade+disc)/shaft attachments. Similarly, light-weight, and wear- and heat-resistant SiC/SiC composites have been proposed for applications in combustor liners, exhaust nozzles, re-entry thermal protection systems, radiant burners, hot gas filters, high-pressure heat exchangers, as well as in fusion reactors owing to their resistance to neutron flux. Fibers, whiskers, particulates, platelets, and laminates are the common reinforcement morphologies in ceramic-matrix composites.

Ceramic-matrix composites can also be designed for improved toughness. For example, in a laminated C/SiC composite, introduction of interlayers of graphite between SiC helps in deflecting cracks and improving the fracture toughness of SiC. The reinforcement may temporarily halt a crack, the interface may partially debond, thus allowing fiber pullout with some frictional resistance, or a weak interface may deflect the crack. Figure 9.1-15 shows a conceptual scheme of possible mechanisms that lead to toughening in ceramic-matrix composites. Figure 9.1-16 shows microstructural evidence of the mechanism of crack deflection and bridging by the reinforcing phase in a ceramic composite of 3Y-TZP matrix containing 15 vol% $Al_2O_3$ platelets of two nominal sizes. Another mechanism of toughening in ceramic composites is blunting of cracks due to volumetric changes accompanying solid-state phase transformations. Yttria-stabilized zirconia (YSZ) is an example of toughening by transformation in which dispersion of yttria particles toughens the zirconia through a stress-induced

**Table 9.1-3** Mechanical properties of polymer- and metal-matrix composites

| Composite | E, GCPa | Strength, MPa |
|---|---|---|
| B-Al (50% fiber) | 210$^{\parallel}$ | 1500$^{\parallel}$ |
|  | 150$^{\perp}$ | 140$^{\perp}$ |
| SiC-Al (50% fiber) | 310$^{\parallel}$ | 250$^{\parallel}$ |
|  | — | 105$^{\perp}$ |
| Fiber FP-Al-Li (60% fiber) | 262$^{\parallel}$ | 690$^{\parallel}$ |
|  | 152$^{\perp}$ | 172-207$^{\perp}$ |
| C-Al (30% fiber) | 160 | 690 |
| E-Glass-Epoxy (60% fiber) | 45$^{\parallel}$ | 1020$^{\parallel}$ |
|  | 12$^{\perp}$ | 40$^{\perp}$ |
| C-Epoxy (60% HS fiber) | 145$^{\parallel}$ | 1240$^{\parallel}$ |
|  | 10$^{\perp}$ | 41$^{\perp}$ |
| Aramid (Kevlar 49) - Epoxy (60% fiber) | 76$^{\parallel}$ | 1380$^{\parallel}$ |
|  | 5.5$^{\perp}$ | 30$^{\perp}$ |

Note: $^{\parallel}$, value parallel to the fiber axis. $^{\perp}$, value perpendicular to the fiber axis.

passing them through a slurry containing the matrix powders, a carrier liquid and binders, and wetting and dispersing agents. The coated fibers are wound on a drum, dried, cut, stacked, and consolidated at high temperatures. Fiber tows or preforms of unidirectionally aligned fibers can be impregnated with the slurry, and stacked in various orientation (e.g., cross-ply, angular ply) to develop the desired properties. Temperature, pressure, fiber arrangement and powder size distribution are some of the important process parameters in hot consolidation. Concurrent application of high pressures and high temperatures permits rapid densification and creation of a pore-free, fine-grained product by hot consolidation. However, fiber damage during hot consolidation, porosity and contamination from binder residues, and difficulty in producing complex shapes are some of the limitations of hot consolidation techniques. Ceramic-matrix composites such as SiC whisker-reinforced $Al_2O_3$, and ceramic fiber reinforced glass-matrix composites are produced by hot consolidation techniques. Ceramic composite laminates such as C/SiC are also produced by hot pressing. For example, SiC powders are mixed with a polymeric binder to a doughlike consistency, and pressed into thin sheets. These sheets are coated with graphite, stacked, and hot-pressed. In place of hot consolidation, cold compaction followed by sintering has also been used to fabricate various ceramic-matrix composites, although considerable shrinkage of the matrix during sintering often results in cracks and weakening of the matrix.

phase transformation. In addition, ceramic-matrix composites containing dispersions of fine metal particles (e.g., Cu or Cr in $Al_2O_3$) have better fracture toughness than the monolithic ceramic matrix.

Most ceramic-matrix composites are fabricated using the hot consolidation processes, although melt infiltration, chemical vapor infiltration, reaction bonding sol-gel processing, polymer pyrolysis, combustion synthesis, and electrophoretic deposition techniques are also used. In the hot consolidation technique, the fibers are first impregnated with the unconsolidated powder matrix by

In the melt infiltration method, a polymer precursor is pyrolyzed to yield a porous feedstock that is then infiltrated with a ceramic melt to produce high-density ceramic-matrix composite in a single step. The very high processing temperatures needed to melt ceramics and the very high viscosity of molten ceramics make the infiltration techniques energy intensive and rather difficult to use. In addition, undesirable chemical reactions at high processing temperatures, and very large thermal stresses due to solidification shrinkage and differential

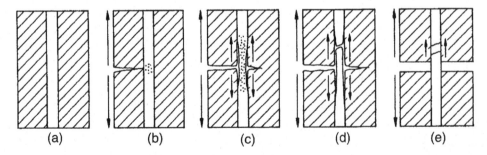

**Figure 9.1-15** Schematic diagram showing mechanism of toughness gain in a brittle-matrix composite. (a) original state with frictional gripping, (b) matrix crack being temporarily halted by the fiber, (c) debonding and crack deflection along the fiber-matrix interface as a result of interfacial shear and lateral contraction of the fiber and matrix, (d) continued debonding, fiber failure at a weak point, and further crack extension, (e) broken fiber ends are pulled out against frictional resistance of the interface. (B. Harris, *Met. Sci.*, **14**, 1980, p. 351).

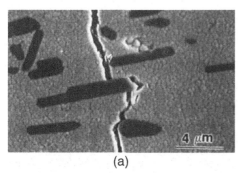 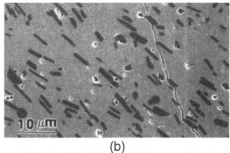

(a)                          (b)

**Figure 9.1-16** (a) SEM micrograph showing crack deflection and bridging in 3Y-TZP-$Al_2O_3$ composite containing 15 vol% of 10–15 μm platelets. (I. K. Cherian and W. M. Kriven, *American Ceramics Society Bulletin*, 80(12), December 2001, p. 57–67). (b) SEM micrograph showing crack deflection in 3Y-TZP-$Al_2O_3$ composite containing 15 vol % of 3–5 μm platelets. (I. K. Cherian and W. M. Kriven, *American Ceramics Society Bulletin*, 80(12), December 2001, 57–67).

contraction between the matrix and the reinforcement can cause matrix cracking and strength loss in the composite. Ceramic composites are also produced by reactive infiltration, which retains the overall simplicity of conventional infiltration but forms the composite at molten metal (rather than molten ceramic) temperatures. This reduces the energy consumption while taking advantage of the high fluidity of molten alloys to facilitate the infiltration. The infiltration of porous carbon preforms by molten Si or Si-Mo alloys has been used to form SiC/SiC/C composite. Fiber-reinforced organosilicon-based SiC composites are produced by first preparing a porous fibrous preform with some binder phase in it, followed by infiltration with polycarbosilanes at high temperatures and pressures, and polymerization. The infiltrated organosilicon polymer matrix is then thermally decomposed in an inert atmosphere between 800 and 1300 °K. The infiltration and thermal decomposition steps can be repeated to obtain composites of high density. Figure 9.1-17 shows the microstructure of some hot-pressed and melt-infiltrated ceramic-matrix composites, including an ultra-high-temperature ceramic composite consisting of a zirconium diboride ($ZrB_2$) matrix containing either SiC particles or both SiC particles and SCS-9a SiC fibers. Composites based on borides of refractory metals Zr, Ti, Hf, and Ta have high melting temperatures, high hardness, low volatility, and good thermal shock resistance and thermal conductivity, and can withstand ultra-high temperatures (1900–2500°C). These composites generally have good oxidation resistance, and further improvements are achieved through the use of additives such as SiC, which improves the oxidation resistance of diborides such as $ZrB_2$ and $HfB_2$. With SiC in $ZrB_2$, a thin, glassy layer of $SiO_2$ forms when the boride is exposed to an oxidizing atmosphere; the silica layer covers an inner layer of $ZrO_2$ on the surface of $ZrB_2$ base material. The outer glass layer provides good wettability and surface coverage, as well as oxidation resistance. Many such composites are still under development but hold promise for future deployment at ultra-high temperatures.

Figure 9.1-18 and Figure 9.1-19 show representative mechanical properties of ceramic-matrix composites. In Figure 9.1-18, the flexural strength and elastic modulus of a borosilicate glass matrix composite containing different volume fractions of aligned carbon fibers are plotted; both the flexural strength and stiffness of the composite increase with increasing volume fraction of C fiber. The effect of temperature on the flexural strength of a lithium aluminosilicate glass-ceramic (i.e., partially crystallized matrix) reinforced with SiC fibers is shown in Figure 9.1-19. Both unidirectional and two-dimensional reinforcement yields superior bend strength at all temperatures compared to the unreinforced matrix (although the transverse rupture strength with unidirectional fibers is very low in SiC/glass-ceramic composite). The useful temperature range over which acceptable strength levels are maintained in this composite extends to nearly 1100 °C (a slight increase in the bend strength near 1000 °C in Figure 9.1-19 is due to the severe oxidation of the composite, which increases the interface strength).

## Carbon-carbon composites

Carbon-carbon (C-C) composites consist of a carbon matrix reinforced with continuous carbon fibers. Carbon-carbon composites are extensively used for the nose cone and leading edges of the space shuttle, as solid-propellant rocket nozzles and exit cones, and as ablative nose tips and heat shield for ballistic missiles. The material has also been used in aircraft braking systems (e.g., in the Anglo-French *Concorde*) where the extremely high frictional torque generates intense heat, raising the brake disk temperature to 500 °C, and the interface temperature to 2000 °C. Likewise, the first wall tiles of thermonuclear fusion reactors also use C-C composites to cope with intense thermal loads. For the space exploration systems, C-C composites have been proposed for applications in radiator and heat management systems, turbine and turbopump housing, thrust cell jackets, and flanges.

# CHAPTER 9.1 Composite materials

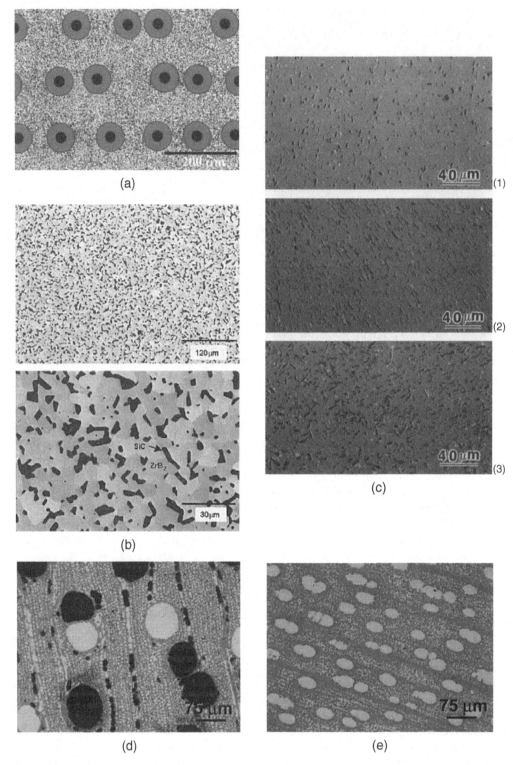

**Figure 9.1-17** (a) Microstructure of a polished section of an ultra-high temperature ceramic-matrix composite of $ZrB_2$-20 v/o SiC+ SCS-9a fibers showing representative fiber distribution. (S. R. Levine et al., Evaluation of Ultra-High Temperature Ceramics for Aeropropulsion, *Journal of the European Ceramics Society*, **22**, 2002, 2757–2767, Elsevier). (b) Microstructure of an ultrahigh temperature ceramic matrix composite of $ZrB_2$-20 v/o SiC particles showing particle distribution. (S. R. Levine et al., Evaluation of Ultra-High Temperature Ceramics for Aeropropulsion, *Journal of the European Ceramics Society*, **22**, 2002, p. 2757–2767, Elsevier). Reprinted with permission from Elsevier. Photo Courtesy of M. Singh, QSS Group, NASA Glenn Research Center, Cleveland OH. (c) SEM micrograph showing crack deflection and bridging in 3Y-TZP-$Al_2O_3$ composite containing (1) 5, (2) 10, and (3) 15 vol% of 10–15 μm platelets showing homogeneous distribution. (I. K. Cherian and W. M. Kriven, *American Ceramics Society Bulletin*, 80(12), December

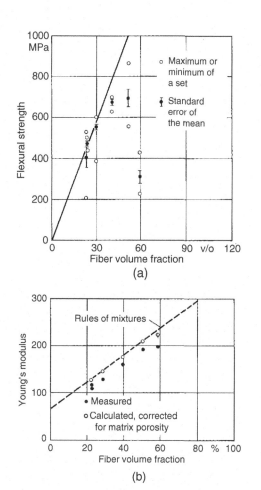

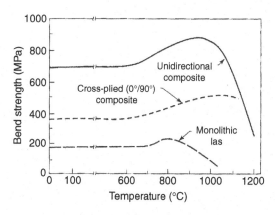

**Figure 9.1-19** Temperature dependence of bending strength of SiC-reinforced lithium aluminosilicate (LAS) ceramic-matrix composite. (R. W. Davidge and J. J. R. Davies, in B. F. Dyson, R. D. Lohr, and R. Morrel, eds., *Mechanical Testing of Engineering Ceramics at High Temperatures*, p. 264, 1989, Elsevier Science Publishers). Reprinted with permission from Elsevier.

**Figure 9.1-18** (a) Flexural strength as a function of fiber volume fraction for a borosilicate glass matrix composite containing continuous aligned carbon fibers. (D. C. Phillips, R. A. J. Sambell, and D. H. Bowen, *Journal of Materials Science*, **7**, 1972, p. 1454). (b) Young's modulus as a function of fiber volume fraction for a borosilicate glass matrix composite containing continuous aligned carbon fibers. (D. C. Phillips, R. A. J. Sambell, and D. H. Bowen, *Journal of Materials Science*, **7**, 1972, p. 1454).

Many applications require joining C-C composites to metals, and advanced joining techniques have been developed for these composites.

Carbon matrix in C-C composites binds the carbon fibes, maintains their orientation, and prevents fiber-to-fiber contact. Carbon-carbon composite can exist in a range of structures, from amorphous to fully graphitic. These composites are made by chemical vapor infiltration (CVI), and impregnation and pyrolysis using resin- or pitch-based precursors. Chemcial vapor infiltration involves thermal decomposition of a hydrocarbon vapor (e.g., $CH_4$ or $C_2H_2$, etc.) over a hot substrate on which carbon is deposited. Several different CVI methods have been developed to produce the C-C composites. In the isothermal CVI process, a porous carbon substrate is placed in a constant temperature zone, and reactant gases are passed over it. To avoid early pore closure, the surface reaction rate should be slower than the diffusion rate into the pores. The slow deposition rate is achieved by using low pressures (10–100 mbar) and low temperatures ($\sim 1100°C$). Eventually, pore blockage occurs even before full densification, and it is necessary to remove the part from the reactor, machine the surface, and reinfiltrate. The process is very slow, and several hundred hours are needed to achieve full densification. To reduce the composite processing times, thermal gradient-CVI (TG-CVI) and pressure gradient-CVI (PG-CVI) techniques were developed. In TG-CVI, the preform is supported on an inductively heated mandrel, which keeps the inside surface hot. The outer surface is cooler as it is close to water-cooled coils. A thermal gradient is, therefore, established across the preform, provided thermal conduction is slow. Carbon first deposits on the cooler outer surface. As density builds up, the substrate couples inductively and heats up, and the reaction front gradually moves through the component. The process is performed at atmospheric

**Figure 9.1-17 (Cont.)** 2001, 57–67. (d) Optical micrograph of as-fabricated SiC made from mahogany. View is perpendicular to the growth direction (white, Si; gray, SiC; black, pores). (M. Singh and J. A. Salem, Mechanical properties and microstructure of biomorphic silicon carbide ceramics fabricated from wood precursors, *Journal of the European Ceramic Society*, **22**, 2002, 2709–2717, Elsevier). (e) Optical micrograph showing the microstructure of silicon carbide made from maple wood. (M. Singh and J. A. Salem, Mechanical properties and microstructure of biomorphic silicon carbide ceramics fabricated from wood precursors, *Journal of the European Ceramic Society*, **22**, 2002, 2709–2717, Elsevier). Reprinted with permission from Elsevier. Photo Courtesy of M. Singh, QSS Group, Inc., NASA Glenn Research Center, Cleveland, OH.

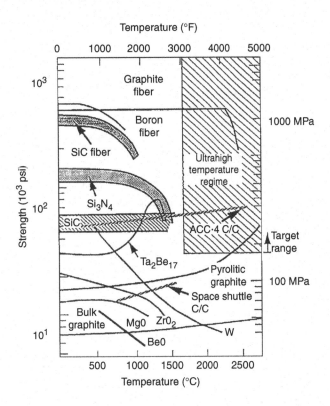

**Figure 9.1-20** Elevated-temperature strength of carbon-carbon composite and other high-temperature composites. (K. Upadhya, J. Yang, and W. P. Hoffman, Materials for Ultra high Temperature Structural Applications, American Ceramics Society Bulletin, December 1997, 51).

pressure, but the susceptor geometry should match the part geometry for best coupling. In the pressure-gradient (PG-CVI) technique, the preform is sealed off from the furnace chamber. The reactant gases are introduced into the inside surface of the preform at a pressure higher than that in the furnace; as a result, reactant gases flow through the porous carbon substrate under a pressure differential.

Carbon-carbon composites exhibit high propensity toward oxidation in air, especially above 500 °C. The oxidation reaction is $2C(s) + O_2(g) \rightarrow 2CO(g)$. The matrix in the C-C is more reactive that the fibers, and the fibers oxidize at a slower rate. The oxidation is, however, most severe at the fiber/matrix interface which presents a large surface area. The rate of oxidation increases with increasing temperature. In order to protect the composite from severe oxidation, protective coatings (e.g., in-situ grown oxide films) are applied, which must isolate the composite from the surroundings. For oxidation resistance, the protective coating must prevent oxygen diffusion into the composite and carbon diffusion to outer surface. The thermomechanical compatibility and adhesion between the coating and the composite are important. Usually the CTE of the carbon matrix is much less than that of the coating; as a result, cracks could form during cooling. A relatively successful approach of thermal protection is to use a glass- or glass-forming compound in the coating, which can flow into and seal the crack. Most protective coatings consist of three basic constituents: bond layer, functional layer, and erosion- and oxidation-resistant layer, such as a layer of SiC or $Si_3N_4$, which upon oxidation forms a $SiO_2$ skin. Figure 9.1-20 shows the experimental oxidation behavior and high-temperature strength of C-C and some advanced ceramic composites at various temperatures.

## Chemical vapor infiltration

CVI is used to fabricate ceramic-matrix composites, and involves infiltration of reactant vapors into a porous (e.g., fibrous) preform, and the deposition of the solid product phase (matrix) within the pores. For example, $Al_2O_3$ and TiC can form via vapor-phase reactions among $H_2$, $AlCl_3$, $TiCl_4$, $CH_4$, and $CO_2$. The reactions may be represented by

$$H_2(g) + CO_2(g) \rightarrow H_2O(g) + CO(g)$$
$$2AlCl_3(g) + 3H_2O(g) \rightarrow Al_2O_3(s) + 6HCl(g)$$
$$TiCl_4(g) + CH_4(g) \rightarrow TiC(s) + 4HCl(g)$$

In the case of deposition of $Al_2O_3$ within a porous SiC fiber bundle, the first two reactions apply, and the third reaction is replaced with

$$2AlCl_3(g) + 3H_2(g) + 3CO_2(g) \rightarrow Al_2O_3(s)$$
$$+ 3CO(g) + 6HCl(g)$$

An analysis of the deposition of alumina matrix in a SiC fiber bundle in an isothermal hot-wall CVI reactor has been given by Tai and Chou.[1] The deposition of $Al_2O_3$ matrix within cylindrical pores of a unidirectionally aligned SiC bundle occurs by the diffusion and reaction of $CO_2$ and $H_2$ at fiber surface within the pore (Figure 9.1-21). Because of the small pore size, gaseous transport through the pores via diffusion is the dominant process. It is assumed that the vapor concentration changes only along the radial ($r$) and axial ($z$) directions within the pore (variation along the $\theta$ coordinate is ignored). Let $N$ be the molar flux (moles/m².s), $C$ be the moles of the total vapor per unit volume (moles/m³), and $x$ be the mole fraction of a given species, then,

$$N_r = -CD\frac{\partial x}{\partial r}$$

---

[1] Tai N. and T.W. Chou, "Theoretical analysis of chemical vapor infiltration in ceramic/ceramic composites", MRS Symposium Proceedings, Vol. 120, 1998, 185-192, Materials Research Society, Boston.

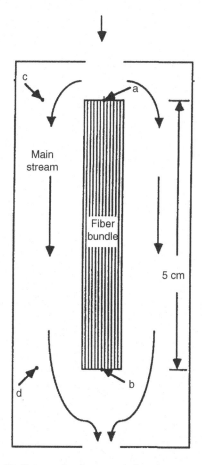

**Figure 9.1-21** Schematic diagram of chemical vapor infiltration of a fiber preform.

and

$$N_z = -CD\frac{\partial x}{\partial z},$$

where $D$ is the diffusion coefficient of a gaseous species. Under steady state, the diffusion equation can be written in terms of the mole fraction of the $i$th species as

$$\frac{\partial}{\partial r}\left(r\frac{\partial x_i}{\partial r}\right) + r\frac{\partial^2 x_i}{\partial z^2} = 0 \quad (9.1.7)$$

Because a homogenous composition of gaseous species is maintained throughout the reactor, including at the ends of the fiber bundle of length $L$, one boundary condition is $x_i = x_{i0}$ at $z = 0$ and $z = L$. Furthermore, at the deposition surface within the pore of radius $R$, the mole fraction of a gaseous species, $x_i$, can be expressed in terms of a reaction rate constant, $k$, and the rate, $r_1$, leading to the second boundary condition: $x_i = r_1/(Ck)$ at $r = R$. In terms of an excess mole fraction $X_i$ defined by $X_i = x_i - x_{i0}$, the boundary conditions can be written as $X_i = 0$ at $z = 0$ and $z = L$, and, $X_i = r_1/(Ck) - x_{i0}$ at $r = R$.

The governing diffusion equation (Equation 9.1.7) is transformed into

$$\frac{\partial}{\partial r}\left(r\frac{\partial X_i}{\partial r}\right) + r\frac{\partial^2 X_i}{\partial z^2} = 0 \quad (9.1.8)$$

This equation can be solved by the method of separation of variables. Define $X_i(r, z) = \phi(r)\psi(z)$, and substitute in Equation 9.1.8 above to yield the following two equations:

$$\frac{\phi''(r)}{\phi} + \frac{1}{r}\frac{\phi'(r)}{\phi(r)} = -\frac{\psi''(z)}{\psi(z)} = \lambda^2 \quad (9.1.9)$$

which can be rearranged to yield

$$r^2\phi''(r) + r\phi'(r) - \lambda^2 r^2 \phi(r) = 0 \quad (9.1.10)$$

and

$$\psi''(z) + \lambda^2 \psi(z) = 0 \quad (9.1.11)$$

The solutions to Equations 9.1.10 and 9.1.11 for $\lambda \neq 0$ are

$$\phi(r) = A'I_0(\lambda r) + B'K_0(\lambda r) \quad (9.1.12)$$

$$\psi(z) = E'\sin(\lambda z) + F'\cos(\lambda z) \quad (9.1.13)$$

The solution to the governing equation becomes

$$X_i(r, z) = I_0(\lambda r)[E\sin(\lambda z) + F\cos(\lambda z)] \quad (9.1.14)$$

where $I_0$ is the modified Bessel function. The boundary condition $X_i = 0$ at $z = 0$ and $z = L$ yields $F = 0$ and $\lambda = (n\pi/L)$, and the solution becomes

$$X_i(r, z) = \sum_{n=0}^{\infty} E_n I_0\left(\frac{n\pi r}{L}\right)\sin\left(\frac{n\pi z}{L}\right) \quad (9.1.15)$$

Substituting Equation 9.1.15 in

$$N_r = -CD\frac{\partial x}{\partial r} \quad \text{and} \quad N_z = -CD\frac{\partial x}{\partial z},$$

and using the second boundary condition,

$$X_i = r_1/(Ck) - x_{i0} \quad \text{at} \quad r = R,$$

yields

$$N_i(r, z) = -CD\sum_{n=0}^{\infty} E_n\left(\frac{n\pi}{L}\right)I_1\left(\frac{n\pi R}{L}\right)\sin\left(\frac{n\pi z}{L}\right) \quad (9.1.16)$$

where $I_1$ is the Bessel function of the first order, and

$$X_i(R,z) = -\frac{D}{k}\sum_{n=0}^{\infty} E_n\left(\frac{n\pi}{L}\right) I_1\left(\frac{n\pi R}{L}\right) \sin\left(\frac{n\pi z}{L}\right) - x_{i0} \quad (9.1.17)$$

The coefficient $E_n$ is obtained by equating the expression for $X_i$ at $r = R$ from Equations 9.1.15 and 9.1.17. This yields $E_n = 0$ when $n$ is even and the following expression when $n$ is odd,

$$E_n = \frac{-4x_{i0}}{n\pi\left[I_0\left(\frac{n\pi R}{L}\right) + \frac{D}{k}\left(\frac{n\pi}{L}\right) I_1\left(\frac{n\pi R}{L}\right)\right]} \quad (9.1.18)$$

The diffusion coefficient, $D$, in CVI depends on temperature, gas pressure, and pore size. For a given set of processing conditions (temperature and pressure), $D$ depends only on the pore diameter. Different diffusion regimes are possible in a CVI reactor depending on the pore size relative to the mean free path of gas molecules. In a dilute gaseous atmosphere (low concentration) and small pore sizes, intermolecular collisions are infrequent, and frequency of collision of gas molecules with the pore wall is high (Knudsen diffusion). At very high vapor concentrations where the frequency of intermolecular collisions is large (and molecules frequently change their direction), molecular diffusion dominates. At intermediate vapor concentrations, frequency of intermolecular collisions becomes comparable to the molecular collisions with the pore wall, and $D$ depends on both Knudsen diffusion and molecular diffusion processes. Expressions for $D$ for these regimes can be written in terms of temperature and molecular weights. The reaction rate constant, $k$, can be taken to be a constant at a fixed temperature; if the temperature is changed, $k$ can be expressed by the Arrhenius equation in terms of an activation energy, $Q$, by $k = k_0 \exp(-Q/R_g T)$, where $R_g$ is the gas constant and $k_0$ is a temperature-independent term. This equation transforms to a linear form: $\ln k = \ln k_0 - (Q/R_g T)$; thus, a plot of natural log of the deposition rate as a function of the reciprocal of absolute temperature will yield a straight line. Figure 9.1-22 shows such a plot for the deposition of $B_4C$ at three different total pressures of reactant gases, $BCl_3$, $CH_4$, and $H_2$. Figure 9.1-23 shows the calculated results for the effect of CVI temperature and pressure on the densification kinetics of porous preform. The densification is faster at higher temperatures and pressures in the CVI reactor, consistent with a diffusion-limited infiltration mechanism. The microstructures of a C/SiC composite synthesized using the isothermal CVI process is shown in Figure 9.1-24; fiber distribution, residual porosity, and some matrix cracking can be noted.

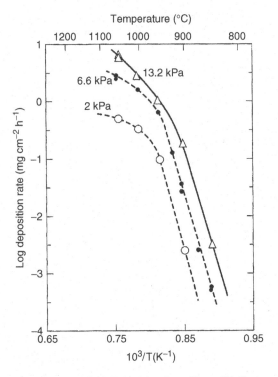

**Figure 9.1-22** Arrhenius plot for the deposition of $B_4C$ at various total pressures by the CVI process. (R. Naslain, CVI Composites, in R. Warren, ed., *Ceramic-Matrix Composites* pp. 199–244, Chapman and Hall, London, 1992).

## Metal-matrix composites

### Heat-resistant composites

#### Dispersion-strengthened composites

For elevated temperature use, the composite matrix must have a high melting point, be resistant to creep deformation, oxidation and hot corrosion, and be thermally stable and light. Advanced Ni-base superalloys, usually in single crystal form, have been used for such applications. To stabilize the matrix structure of these alloys for high-temperature use, the intermetallic phases in the matrix (primarily the $\gamma'$-phase in the $\gamma$-matrix) are stabilized by alloying with heavy metals, such as Re, W, Ta, etc. Whereas this improves the creep-resistance, it decreases the solidus temperature and increases the alloy density, thus degrading the specific properties (i.e., value of the property divided by the density). Complex thermal barrier coatings and increasingly more elaborate cooling systems have been used to increase the thermal efficiency of engines.

The development of oxide dispersion-strengthened (ODS) composites (e.g., $W-ThO_2$, $Ni-Al_2O_3$, $Cu-SiO_2$, $Cu-Al_2O_3$, $NiCrAlTi-Y_2O_3$, $Ni-ThO_2$, $Ni-HfO_2$, etc.) has led to considerable improvements in the elevated-temperature strength and creep-resistance of the matrix

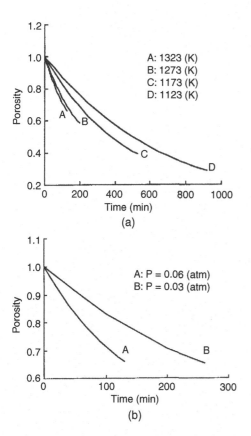

**Figure 9.1-23** (a) Theoretical projections of the effect of CVI temperature on the densification kinetics of porous preforms. The densification is faster at higher temperatures, consistent with a diffusion-limited infiltration mechanism. (N. Tai and T. W. Chou, Theoretical Analysis of Chemical Vapor Infiltration in Ceramic/Ceramic Composites. High Temperature/High Performance Composites, in Materials Research Society Symposium, vol. 120, 1988, 185). (b) Theoretical projections of the effect of pressure in the CVI reactor on the densification kinetics of porous preform. The densification is faster at higher pressures, consistent with a diffusion-limited infiltration mechanism. (N. Tai and T. W. Chou, Theoretical Analysis of Chemical Vapor Infiltration in Ceramic/Ceramic Composites. High Temperature/High Performance Composites, in Materials Research Society Symposium, vol. 120, 1988, 185). Reprinted with permission from Materials Research Society, Warrendale, PA.

alloys. The ODS composites contain small (<15%) amounts of fine second-phase particles, which resist recrystallization and grain coarsening, and act as a barrier to the motion of dislocations. Small quantities of fine oxide ceramics improve the strength without degrading the valuable matrix properties. Examples of ODS composites include yttria- or thoria-dispersed Ni-20Cr, and Ni-Cr alloys containing both yttria and $\gamma'$-phase (that forms due to additions of Al, Ti, or Ta). These composites have excellent resistance to oxidation, creep, carburization, and sulfidation, and in some cases, excellent resistance to molten glass. They are produced mainly by powder metallurgy (hot pressing) techniques which utilize powders of matrix alloy and the ceramic dispersoid as the feed materials. Besides oxide dispersions, carbides, borides and other refractory particulates have been used in dispersion-strengthened (DS) composites.

The hot-pressed oxide dispersion-strengthened (ODS) Ni-base superalloys have been used in aircraft gas turbine engine parts such as vane airfoils, blades, nozzles, and combustor assemblies. In these materials, the dispersed second phase primarily serves as a barrier to the motion of dislocations rather than as the load-bearing constituent (as in an engineered metal-base composite). The thermal stability of these materials is superior to precipitation-hardened alloys, in which the second phase softens and degrades at high temperatures. Besides the Ni-base superalloys, several other families of alloys have been dispersion-strengthened. Figure 9.1-25 shows a view of a dispersion-strengthened silver composite containing uniformly distributed nanometer-size $Al_2O_3$ particles. The composite was created by an internal oxidation method in which an Ag-Al alloy powder is oxidized in a controlled fashion to create a dispersion of fine, evenly distributed aluminum oxide particles.

In dispersion-strengthened composites, the fine dispersions of oxide, carbide, or boride particles impede the motion of dislocations so that the matrix is strengthened in proportion to the effectiveness of the dispersions as a barrier to the motion of dislocations. For a dislocation to pass through a dispersion of fine particles, the applied stress must be sufficiently large to bend the dislocation line into a semicircular loop. Calculations show that interparticle separation for effective dispersion hardening should be between 0.01 and 0.3 $\mu m$. To achieve this range of spacing between the dispersoid, the particle diameters should be less than 0.1 $\mu m$ at volume fractions below 15%. In the DS material, it is usually desirable to preserve as many of the properties (ductility, electrical and thermal conductivity, and impact strength, etc.) of the matrix material as possible. This requires that the dispersoid volume fraction be kept small because large-volume fractions decrease the fracture toughness, thermal conductivity, and other valuable matrix properties. Consider, for example, the room-temperature hardness data on Cu-alumina dispersion-strengthened composites containing only 3.5% submicron-size alumina. For this material, the hardness remains unchanged when the DS composite is annealed at different temperatures ranging from room temperature all the way to just a few degrees below the melting point of Cu. This remarkable hardness retention of the DS $Cu/Al_2O_3$ composites is achieved with minimum penalty on the other useful properties of Cu because the alumina volume fraction is so low.

### In situ composites

Another approach to developing heat-resistant composites is based on the idea of directionally solidifying

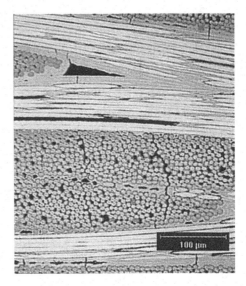

**Figure 9.1-24** Microstructure of a C-SiC composite synthesized using the isothermal CVI process showing the fiber distribution, residual porosity, and matrix cracking. (M. R. Effinger, G. G. Genge, and J. D. Kiser, *Advanced Materials and Processes*, June 2000, pp. 69-73). Reprinted with permission from ASM International, Materials Park, OH (www.asminternational.org).

two-phase alloys (e.g., Ni- or Co-base eutectic superalloys) to create composites containing in situ grown reinforcement (e.g., carbide whiskers). The structural stability of the reinforcement in a temperature gradient, and low crystallization rates needed to design the microstructure are major considerations in these composites.

**Figure 9.1-25** Longitudinal microstructure of an extruded bar of oxide dispersion-strengthened Ag composite containing 0.5 wt% $Al_2O_3$ dispersions. (J. Troxell, A. Nadkarni, and J. Abrams, in *Advanced Materials and Processes*, January 2001, 75–77). Reprinted with permission from ASM International, Materials Park, OH (www.asminternational.org).

Besides superalloys, Ni-Al inter-metallics ($Ni_3Al$ and NiAl) have been considered for the growth of in situ composites. Both $Ni_3Al$ and NiAl have relatively high melting points (higher for NiAl: $T_{m, NiAl} = 1950\,°K$, $T_{m, Ni_3Al} = 1638\,°K$), excellent oxidation resistance (further improved by alloying Zr or Hf), high strength and relatively low density (lower for NiAl, about 5860 kg/cu·m). NiAl has been used as a coating material in aircraft engines, but in a bulk polycrystalline form, the material is extremely brittle, and its strength above about 773 °K is very low. For example, the room-temperature yield strength (Y.S.) of NiAl is in the range of 120–300 MPa, but at 1300 °K, the Y.S. is only about 50 MPa. The ductility of polycrystalline NiAl is nearly zero below about 500 °K. However, as NiAl exists over a wide range of stoichiometries, it is possible to use suitable alloying additions to improve its mechanical properties. For example, Ti, Nb, Ta, Mn, Cr, Co, Hf, W, Zr, and B have been added to NiAl for increased resistance to compressive creep, with Nb and Ta being especially effective. Small additions of tungsten to NiAl refine the grain size and inhibit grain growth during hot consolidation, resulting in improvement in crack resistance. For toughness and strength gains, continuous fibers of W, Mo, and alumina have been incorporated in NiAl, and results with Mo and alumina have been promising (W tends to embrittle during processing, especially when solid-state diffusion bonding is used for composite fabrication). Thus, ductility of Ni-Al intermetallics has been increased by suitable alloying, and improvement in high-temperature creep resistance has been achieved mainly through fiber- and particulate reinforcements.

A variety of high-temperature in situ Ni-base composites consisting of dual-phase eutectic-type

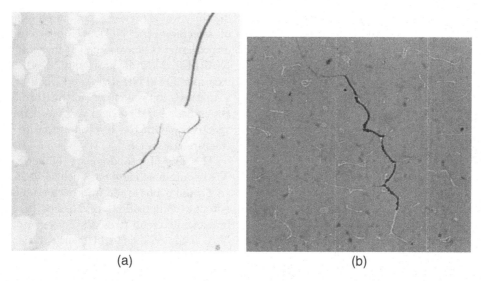

**Figure 9.1-26** (a) Crack deflection by a W particle in a compression-tested NiAl(W) alloy. (b) Crack propagation through the intercellular NiAl-Cr eutectic colonies in an extruded NiAl(Cr) bar. (R. Tiwari, S. N. Tewari, R. Asthana and A. Garg, *Materials Science & Engineering*, A 192/193. 1995,356–363).

microstructures in which one of the phases is sandwiched between the other have been developed. Toughening in in situ composites occurs due to inhibition of crack nucleation in the secondary ($\beta$) phase, inhibition of crack growth due to plastic bridging by the ductile phase (crack bridging), and blunting of the crack by the ductile phase. Figure 9.1-26 shows how the dispersed second phase in directionally solidified off-eutectic pseudobinary alloys of NiAl(W) and NiAl(Cr) deflect a propagating crack, thus improving the toughness of NiAl. For in situ composites, the interphase interfaces are clean and mutually compatible because the constitutent phases crystallize in situ rather than combined from separate sources. Directional solidification as well as deformation processing (extrusion and forging) are used to create the dual-phase microstructures having improved toughness, ductility, and creep strength. Directional solidification (DS) of pseudobinary NiAl-X(X could be Mo, W, Cr and Fe etc.) eutectic and off-eutectic alloys has been done to create a matrix containing an aligned $\beta$-phase sandwiched between a ductile second phase. The DS of NiAl-rich (hypoeutectic) compositions leads to aligned cells of $\gamma$-phase surrounded by the in situ composite (eutectic) microstructure created by the solidification of the intercellular eutectic liquid. The in situ composites of pseudobinary eutectic (and off-eutectic) compositions improve the room-temperature ductility and toughness, as well as the strength as in the case of directionally solidified ternary eutectics of NiAl with Mo, Cr, and Nb. The creep resistance and room temperature ductility of $\beta$-NiAl and Ni$_3$Al depends on the cell size, interlamellar spacing, and the second-phase morphology, all of which are controlled by varying the growth speed (e.g., the DS of Ni$_3$Al at 25 mm h$^{-1}$ resulted in columnar-grained single-phase Ni3Al with ~60% tensile ductility at room temperature. However, the same material grown above 50 mm h$^{-1}$ yielded lower ductility than the columnar-grained Ni$_3$Al grown at the lower speed). Figure 9.1-27 shows the microstructure of an in situ Ni-base eutectic composite.

The in situ grown composites can be reinforced with ceramic fibers to further enhance the high-temperature strength and creep-resistance. An example of this is single crystal sapphire fiber-reinforced off-eutectic alloys of NiAl with Cr or W. This approach can permit realization of toughening of the matrix resulting from the dual-phase aligned microstructures together with strengthening of the matrix from ceramic fibers. The

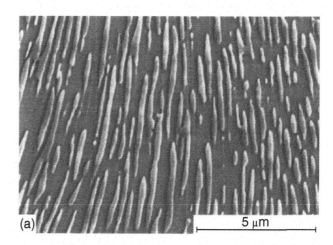

**Figure 9.1-27** Microstructure of a directionally solidified in situ Ni-base composite. (S. V. Raj and I. E. Locci, *Intermetallics*, 9, 2001, 217, Elsevier). Reprinted with permission from Elsevier. Photo Courtesy of S. V. Raj, NASA Glenn, Research Center, Cleveland, OH.

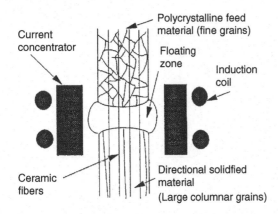

**Figure 9.1-28** Floating-zone directional solidification of a fiber-reinforced composite.

directionally solidified sapphire-NiAl(X) composites, where X is Cr or W, have high interfacial shear strength and promise of high-temperature strength as well as room-temperature toughness. These composites have been grown by various directional solidification techniques, most notably a containerless floating-zone directional solidification process (Figure 9.1-28). This technique allows growth of large, columnar NiAl grains with sapphire fibers residing in the eutectic colonies.

Besides directional solidification, deformation processes such as extrusion, rolling, and forging, in conjunction with suitable heat treatment, are used for ductile-phase toughening of $\beta$-NiAl. For example, heat treatment of an extruded Ni-30Al-20Co alloy and a Ni-36Al alloy produces equiaxed $\beta$-NiAl grains containing a necklace of continuous $\gamma'$ phase at the grain boundaries. This microstructure results in a slight improvement in the ductility (0.5%) over almost zero ductility of the alloy without the $\gamma'$ phase. Likewise, forging or rolling of other Ni-Al-base alloys such as Ni-20Al-20Cr, Ni-25Al-18Fe, Ni-15Al-65Fe, and Ni-26Al-50Co alloys leads to a uniform distribution of equiaxed $\beta$ and $\gamma$ grains with about 2–6% ductility at room temperature. Extrusion of cast Ni-20Al-30Fe alloys produces a fine equiaxed $\beta$-phase distributed in a $\beta + \gamma'$ eutectic, and depending on the fineness of the $\gamma'$ phase, this alloy exhibits a rather remarkable ductility (8 to 22%) at room temperature.

The first evidence of ductile-phase toughening of $\beta$-NiAl was provided in the directionally solidified Ni-34Fe-9.9Cr-18.2Al (atom%) alloy whose microstructure consisted of alternating lamellae of Ni-rich $\gamma$-phase and about 40% $\beta$-phase. This in situ composite exhibited 17% tensile elongation at room temperature with fracture occurring by cleavage of the $\beta$-phase that was sandwiched between ductile necked regions of $\gamma$-phase. Likewise, DS of Ni-30Al alloys led to aligned rodlike $\gamma'$ (Ni$_3$Al) precipitates in a $\beta$-phase matrix, and DS of Ni-30Fe-20Al alloys yielded aligned $\beta$-NiAl and $\gamma/\gamma'$ precipitates. In both these cases, about 10%

ductility was achieved as compared to near-zero ductility of stoichiometric NiAl at room temperature. Similar improvements have been achieved through DS in NiAl(Mo), NiAl(Cr,Mo), and Ni-40Fe-18Al alloys. For example, DS of NiAl(Mo) and NiAl(Cr,Mo) alloys led to a marked improvement in fracture resistance; these alloys had a rodlike and a layered microstructure, respectively, with the layered structure yielding superior fracture resistance.

The DS of pseudobinary eutectic compositions of NiAl-containing elements such as Re, Cr, and Mo yields an aligned $\alpha$-phase and improved strength. Similarly, off-eutectic compositions of NiAl with Cr are known to improve its creep strength. For example, rupture life of NiAl single crystals (<110> orientation) increases when chromium additions are made. A significant amount of strengthening in the NiAl(Cr) is provided by a high volume fraction of the fine $\alpha$-Cr particles that precipitate during a solution–reprecipitation treatment. Dislocation networks form on the precipitate boundaries in both extruded and DS NiAl(Cr) alloys, and contribute to the strengthening.

Ni-Al intermetallics utilizing the ductile-phase toughening concepts and containing relatively large quantities of solutes possess relatively low melting point, high density, and inferior oxidation resistance, thus defeating the purpose for which these materials are considered attractive. To avoid such problems, directional solidification (or extrusion) of ternary NiAl alloys containing the smallest quantities of solutes (Cr and W), consistent with respective pseudobinary eutectics-phase diagrams, has been attempted. Aligned dual-phase microstructure of Ni-43Al-9.7Cr and Ni-48.3Al-1W alloys has led to improvements in both 0.2% compressive yield strength and ductility as compared to the single-phase NiAl, with tungsten alloying yielding the greatest strength improvements at room temperature. All such DS alloys exhibit greater than near-zero ductility of polycrystalline $\beta$-NiAl at room temperature, with Cr yielding the largest gain in fracture strain, followed by W. Figure 9.1-29 shows the microstructure of a NiAl(W) in situ composite in which W fiber pullout contributed to toughening.

## Particulate and fiber-reinforced high-temperature alloys

For advanced gas turbine engine applications, particulate reinforcement and high-aspect ratio fibers also have been incorporated in Ni-base matrices especially Ni-base superalloys and Ni-base intermetallics (e.g., NiAl and Ni$_3$Al). Particulate reinforcements include AlN, Al$_2$O$_3$, HfC, TiC, TiB$_2$, and B$_4$C; a uniform distribution of the particulate reinforcement is necessary to achieve strengthening. From the standpoint of fiber-matrix

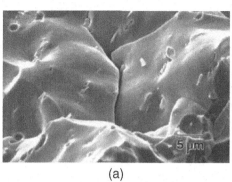
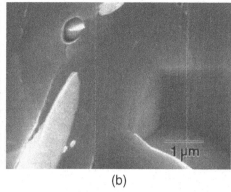

(a)              (b)

**Figure 9.1-29** (a) Scanning electron micrograph showing toughening of NiAl containing a small amount of W; toughening is provided by the deformation and pullout of the in situ-grown W fibers. (b) A higher magnification SEM view of micrograph of Figure 9.1-29a showing W fiber deformation and pullout. (R. Tiwari, S. N. Tewari, R. Asthana and A. Garg, *Journal of Materials Science*, 30, 1995, 4861–4870).

compatibility during fabrication and property enhancement, alumina-base ceramics appear to be an excellent choice as a reinforcement for Ni-base matrices, including superalloys and structural intermetallics such as NiAl and $Ni_3Al$. A variety of commercial alumina fibers (e.g., Du Pont's polycrystalline fiber FP and PRD166, Saphikon's sapphire, etc.) have been incorporated in NiAl and $Ni_3Al$ matrices, and alumina fiber-reinforced NiAl has been considered for applications in high-speed civil transport.

In the case of superalloy matrices, early studies of alumina reinforced Ni-base superalloys led to encouraging results in that single-crystal and polycrystalline alumina did not undergo significant chemical attack in composites formed by vacuum infiltration followed by prolonged heat treatment (up to 300 h at 1200°C). A very thin interaction zone formed at the fiber–matrix interface regardless of the type of alumina and duration of heat treatment. Owing to the chemical stability of alumina in Ni-base matrices, sapphire-reinforced superalloy composites attracted attention over 30 years ago. Tables 9.1-4 and 9.1-5 summarize the work done on Ni-base composites over the last several decades.

Other common reinforcement materials such as SiC, $Si_3N_4$, and refractory metal fibers showed less promise than alumina because of their chemical and physical instability in Ni-base matrices. Studies on the interactions between Ni-alloys and SiC indicated strong fiber–matrix chemical interaction that led to depletion of Si content of the fiber and excess of Ni and Cr in the reaction zone. Ni-base superalloys such as Nimocast 713 C, Nimocast 75, and Nimocast 258 were used with continuous fibers and whiskers of SiC. Both hot-pressing and vacuum infiltration were employed for composite fabrication, and processing conditions were found to strongly influence the chemical interaction. For example, mold temperature and heat treatment of the composite affected the interaction; both preheated molds and heat treatment led to considerably greater chemical attack than molds at room temperature and composites in the as-cast conditions, respectively.

Another reinforcement that is unstable in Ni-alloys is $Si_3N_4$, which exhibits an even more severe degradation than SiC. Nimocast 258 alloy matrix shows complete dissolution of the whiskers in the vacuum cast $Si_3N_4$ whisker-reinforced composites. However, hot-pressing of Ni-coated $Si_3N_4$ whiskers has been found to lead to minor whisker damage, although subsequent heat treatment leads to complete dissolution of whiskers.

Fibers of refractory metals such as W, W-Rh, Mo, and Nb show different levels of reactivity with Ni alloys depending on the processing conditions. For example, vacuum-infiltrated W- and W-Rh/Nimocast 713 C composites show a small interaction zone after heat treatment for 500 h at 1000 °C, and a considerably thicker reaction zone at 1100 °C after 600 h. On the downside, the high density and high oxidizing tendency of W fibers has been a problem, although in a composite, a dense (pore-free) oxidation-resistant superalloy matrix (or a similar pore-free fiber coating) could protect the fibers from oxidation, except perhaps near the exposed fiber ends. On the positive side, however, W fibers strengthened by doping them with HfC, Rh, and C have excellent resistance to thermal fatigue and hydrogen embrittlement, and have been used in superalloy matrices.

Both Mo and Nb fibers severely react with Ni-base alloys, but vapor-deposited W and alumina coatings provide excellent protection against penetration and reaction by Ni and Cr even after prolonged heat treatment (300 h at 1100°C). In the case of niobium fibers, chemical attack can be limited through oxide barrier layers prepared by controlled oxidation of Nb fibers themselves. Protective coatings and barrier layers are effective but must have long-term stability and physical integrity, both of which could be impaired by pores and microcracks that would cause increased metal penetration and chemical attack.

## Table 9.1-4 Early studies on Ni-base composites

| Fiber | Matrix | Comments (Source) |
|---|---|---|
| Ni$_3$Al | Ni-2 to 10% Al | PM. Ni$_3$Al fibers form in situ. Oxidation resistant, high-temperature strength (Cabot Corp.) |
| Ni$_3$Ta | Ni-Ta alloy | DS (Euratom) |
| Ni-Cr-Al-Y | Ni alloy | PM. Sealing elements in turbines and compressors (Brunswick Co.) |
| Stainless steel, Mo, Ti, Nb | Ni, and Ni-base superalloys | Electroforming, PM (NASA, U.S. Army, Imperial Metals) |
| W, Mo | Ni | PM |
| Alumina | Ni and Ni alloys | Infiltration, electrodeposition plus hot-pressing |
| B, W, and B-W | Ni, Nimocast 713C, MARM322E | Electrodeposition, hot-pressing, casting |
| Fiberfrax | Ni-Sn | Vacuum infiltration |
| SiC | Ni-Cr | PM, hot extrusion, chemical reaction (Tohoku University) |
| Graphite (Ni-coated), C-coated with carbides | Ni aluminide | Hot-pressing of fiber and Ni base powders (United Technologies, Union Carbide) |
| Carbides of Nb, Ta, W | Ni-Co alloys | DS (General Electric) |
| NiBe fibers | Ni-Cr alloy | Normal solidification |
| Mo fibers, Cr(Mo) plates | NiAl, Ni$_3$Al | Normal solidification |
| Cr-rich fibers, Cr and Mo fibers, Ni3Al fibers, Mo2NiB2 fibers | Ni-base alloys | Normal or directional solidification (United Technologies, General Electric) |
| Ni$_3$Al | Ni$_3$Ta | Normal solidification |
| Cr$_3$C$_2$ fibers | Ni alloy | Normal solidification |
| TaC, VC, orTaC+VC fibers | Ni base alloy | Normal solidification (GE) |
| Carbides of Ti, V, Cb, Hf, Zr | Ni-base alloy | Normal solidification |

## Fabrication of metal-matrix composites

The three generic methods of manufacturing metal-matrix composites are the solid-state, the liquid-state, and the vapor-phase methods. Solid-state methods include the various powder metallurgy and diffusion-bonding techniques of combining the matrix and the reinforcement, liquid-state methods include infiltration, in situ growth, spray-forming and mixing, and vapor-phase methods include reactive (liquid–gas) spray-forming and vapor-phase deposition techniques. The choice of fabrication technique is usually dictated by the type of composite selected, production cost, process efficiency, and the quality of the product. Thus, high-performance fiber-reinforced composites are usually fabricated to near-net shapes using the squeeze-casting technique in order to achieve superior product quality and to minimize or eliminate the need for secondary processing. However, high pressures tend to damage the preform and often limit the ability to make thin-walled shapes. In contrast, particulate-reinforced composites can be synthesized using relatively less expensive mixing techniques but the quality of the product is partly compromised; secondary processing may be necessary to improve the product quality. The solid-state techniques are relatively expensive but are more suited for reactive systems in which higher processing temperatures of liquid-phase fabrication will impair the properties of the composite. The liquid-state methods are preferred from the standpoint of economy, which is dictated to a large extent by the low viscosity of liquid metals. The "in situ composites" are limited to systems where chemical reactions lead to formation of reinforcement within the matrix phase, but the reinforcing phase is usually

**Table 9.1-5** Examples of Ni-base heat-resistant composites and fabrication method

| System | Fabrication Route |
|---|---|
| $Ni_3Al$-$SiC_p$ | VIM and stirring |
| $Ni_3Al$-$SiC_p$ | VIM/infiltration |
| $NiAl_3$-Ni-Al-$Al_2O_3$ | Reactive infiltration |
| $Ni_3Al$-$TiB_2$ | Plasma-spraying |
| INCONEL 718-$Al_2O_3$ (single crystal) | Pressure infiltration |
| $Ni_3Al$-$Al_2O_3$ FP (IC-15, IC-218), NiAl-PRD-166 | Pressure infiltration |
| NiAl-AlN | Extrusion or HIP |
| $Ni_3Al$-TiCp | PM or stir-casting |
| $Al_2O_3$-NiAl | Powder injection-modeling plus HIP |
| $Al_2O_3$-NiAl, $ThO_2$-NiAl, $Y_2O_3$-NiAl | Ball-milling |
| NiAl-AlN-$Y_2O_3$ | Reaction milling in liquid N2 |
| $TiB_2$-NiAl | XD—in situ growth, milling, and HIP |
| NiAl-W, NiAl-$Al_2O_3$ | Powder cloth |
| NiA1-Cr, NiAl-Mo, NiAlCr(Mo), NiFe-Al, Ni-Cr-Mo, NiAlVZr, NiAlCrNb, NiAlV, NiAlNb | Directional solidification (DS) |
| NiAl-$Al_2O_{3p}$ | Sedimentation technology (FG) |
| NiAl/aligned or chopped FP ($Al_2O_3$) | Hot-pressing |
| NiAl-$Al_2O_3$, SiC-coated $Al_2O_3$ | PM |
| NiAl-$ZrO_{2p}$ | — |
| NiA1, Ni-35Al-20Fe, Haynes (A214)-$Al_2O_{3f}$ | Powder cloth |
| NiAl-304SS-$B_4C$ | Blending, extrusion, or forging |
| Sapphire GS-32 or VKNA-4U, YAG-GS-32 or VKNA-4U | Internal crystallization method (ICM) |
| Sapphire-YAG-GS-32 or VKNA-4U | |
| Sapphire-NiAl(Me), Me = W, Cr, or Yb, sapphire-Hastealloy | DS or cast or powder-cloth (PC) preforms |

Note: GS-32 and VKNA-4U are Russian Ni-base alloys.

monocrystalline, and the reinforcement-matrix interfaces are very clean and thermodynamically stable in these materials. Also, the in situ composite growth techniques may be more economical, as the reinforcing phase is grown during the process of forming the composite and not manufactured separately. The vapor-phase fabrication processes such as chemical vapor infiltration (CVI) are slow and may take up to several hundred hours for completion, but have net-shape potential and reduced processing temperatures. The fastest processes for composite fabrication are the self-propagating high-temperature syntheses (SHS) in which near-explosive rates of reaction take place; however, secondary processing such as infiltration or solid-state consolidation are usually required because of a relatively high defect (porosity) content in the material.

## Solid-state fabrication

A wide variety of solid-state composite fabrication processes have been developed for metal-matrix composites. The spray-forming techniques combine rapid solidification of a fine dispersion of liquid droplets followed by consolidation of solidified droplets (powders) to produce the final component shape. The incoming material in the form of powder, wire, or rod is melted using plasma, arc or combustion sources. The melted material is then sprayed into fine droplets that hit a target at very high velocities in molten or partially solidified state and deform into splats that cool rapidly. The high energy of impact assists in powder consolidation and densification, although a high degree of porosity usually remains in the as-deposited material. The particulate-reinforced composites are generally made by injecting the solid reinforcement particles directly into a spray of the matrix alloy, followed by deposition of the mixture onto a substrate. The co-deposition on a substrate of discrete droplets of molten matrix material and the discontinuous reinforcement yields a product that usually has some very attractive properties, because the solidification process is very rapid and particle engulfment by the solid is highly likely due to high solidification rates and large shear forces from impact. This results in a fine-grained matrix structure and a homogeneous distribution of particles. Figure 9.1-30 show the basic concept of spraying and particle engulfment in the solidified matrix. In another variation of the spray process, the reactive spray-forming by gas-liquid reactions, spray nozzle is extended to include a reaction zone where reactive gases are fed into the liquid stream. The gases react with the molten particles (droplets) to form the reinforcement. Carbide, oxide, silicide, and nitride reinforcement in metal matrices have been produced in this manner.

The processing times in spray-forming are on the order of a few milliseconds, and the reinforcement is in contact with the liquid matrix for no more than a few tens of milliseconds. Hence, even for small particles,

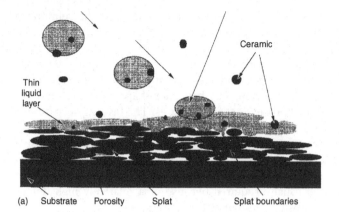

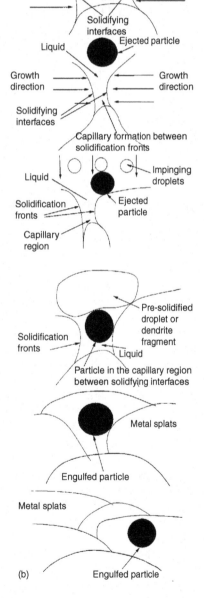

**Figure 9.1-30** (a) Sketch showing co-deposition of a metallic matrix and ceramic particulates via a thermal spray process. (b) Possible mechanisms of particle engulfment by solidifying droplets on impact during thermal spray co-deposition.

chemical interactions between the reinforcement and the matrix will not alter the composition and properties of the interface interphase, although thin interfacial layers of reaction compounds (e.g., intermetallics) may form by diffusional interactions that may in fact improve the interfacial bond strength and hence the composite properties.

For spray-forming of continuous fiber-reinforced composites, the fibers are wrapped around a mandrel with controlled interfiber spacing, and the matrix metal is sprayed onto the fibers. A composite monotape is thus obtained; bulk composites are formed by hot-pressing of composite monotapes. Fiber volume fraction and distribution is controlled by adjusting the fiber spacing and the number of fiber layers. With continuous ceramic fibers, however, the large thermal expansion mismatch between the ceramic fiber and the metal matrix, and the thermal shock and mechanical stresses arising from the initial exposure to plasma jet can result in displacement and fracture of fibers, especially because the brittle ceramic fibers are already bent to large curvatures when wrapped on a mandrel.

The spray-forming route of manufacturing composite materials offers several advantages over many other processes of fabricating composites. The spray-formed composite is relatively inexpensive, with a cost that is usually intermediate between melt stirring and powder metallurgy processes. The method produces a fine, nonsegregated structure with low levels (<2%) of porosity due to fast solidification rates, and imparts good mechanical properties to the material. The cooling rates in spray-forming are typically in the range $10^3$ to $10^6$ °K.s$^{-1}$. The porosity levels in the spray-formed material depend on the thermal conditions, impact velocity, and spray density or mass flux. If the spray density (rate of deposition per unit area) is controlled carefully with respect to the rate of heat extraction from the deposit by the substrate, and by radiation and convection to the gas impinging on the deposit, then only very thin (typically 20 μm to 1000 μm) molten or partially molten layers of metal remain exposed on the deposit surface at any instant during processing. Under these conditions, the liquid or semiliquid alloy droplets in the spray meet a thin, liquid layer at the deposit surface; the liquids flow together and eliminate what would otherwise be splat boundaries. If, in contrast, the spray density is reduced with the same total cooling, the result will be liquid droplets impinging on a solid deposit, in which case the solidification rate will be higher but the structure will retain splat boundaries and will have a higher porosity. A large increase in the spray density with the same thermal surroundings leads to an accumulation of liquid on the surface, causing a reduction in the solidification rate, a marked increase in the grain size and an undesirable shredding of liquid layer by the atomizing gas.

In the case of advanced Ni- and Ti-base high-temperature matrices, special techniques have been developed to fabricate the composite materials. Thus, for fabricating intermetallic-matrix composites, foil-fiber-foil technique, tape-casting, thermal spray, high-energy ball-milling, reactive consolidation, injection-molding, powder-cloth, exothermic dispersion, and deformation processing (rolling, forging, extrusion) are used.

The foil-fiber-foil technique (Figure 9.1-31), used mainly for Ti-base intermetallics (and also for fiber-reinforced Al composites) requires consolidation of alternating layers of woven fiber mats and matrix foil. The composite prepregs are then consolidated into final component shape by hot pressing. Nickel-base intermetallics have rather poor ductility, and are not amenable to this technique. They are generally fabricated using powder-based techniques. For example, tape-casting uses liquid polymer binders to create a slurry of matrix powders that is combined with the fibers and hot-consolidated. The major drawback is interface- and matrix contamination from the binder residues. Thermal spray combines atomization and spraying of matrix droplets over fibers to form monotapes that are laid up and consolidated. Particulate $TiB_2$- and $Al_2O_3$ fiber-reinforced $Ni_3Al$ and NiAl-matrix composites have been fabricated using this technique. Fiber damage from the mechanical impact of impinging droplets, powder surface contamination from oxides, and a somewhat inhomogeneous structure (unmelted grains, splat boundaries, voids) are some of the major limitations of spray forming techniques.

In reactive consolidation, fibers or particulates are combined with elemental or prealloyed matrix powders, followed by sintering and densification. Full-densification requires hot isostatic pressing (HIP). Figure 9.1-32 shows vacuum hot-pressed SiC-Ti and sapphire-NiAl composites. In SiC-Ti composites, SiC fibers were sputter-coated with a $\beta$-Ti alloy matrix and hot-pressed, and in sapphire-NiAl composites, plasma-sprayed NiAl powders were hot-consolidated with the fibers using NASA's patented "powder-cloth" process. The powder-cloth technique has also been used to fabricate NiAl-W and other advanced composites. The method first creates a cloth or sheet of atomized matrix powders by mixing them with Teflon binders and a solvent to a doughlike consistency. Fiber mats of specified thickness are produced by filament winding and application of a PMMA coating. The matrix cloth and fiber mats are then stacked in layers, and vacuum hot-pressed; the orientation of individual plies and stacking sequence provide flexibility in designing the composite properties. Complete binder removal is extremely important because small quantities of binder residues (carbon and fluorine) appreciably degrade the fiber–matrix interface strength and bulk composite properties.

During consolidation under high pressures, fiber damage can be a problem. This is overcome with powder injection-molding, although only the discontinuous reinforcement (particulate, chopped fibers, or whiskers) has been amenable to this approach. In powder injection-molding, up to 40% organic binders is combined with the matrix powders and the reinforcement, and the mixture

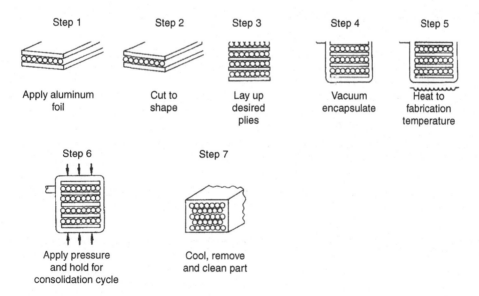

**Figure 9.1-31** Solid-state fabrication of fiber-reinforced Al composites by hot-pressing of Al foils and fibers. (K. K. Chawla, *Composite Materials: Science and Engineering*, Springer, New York, NY, 1987, p. 104).

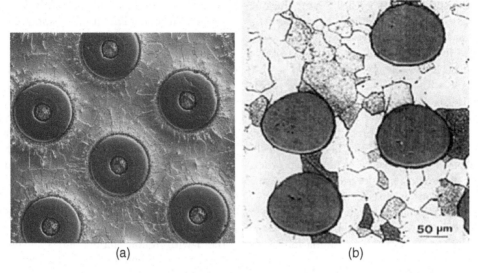

Figure 9.1-32 (a) A titanium-matrix composite consisting of SiC fibers sputter-coated with a β-Ti alloy and vacuum hot-pressed. (Photographed in polarized light by J. Baughman, NASA Langley Research Center, Langley, VA, reprinted from *Advanced Materials and Processes*, December 1998, pp. 19–24, ASM International, Materials Park, OH). (b) Single-crystal sapphire fiber-reinforced NiAl matrix composite produced by plasma spray and hot-consolidation techniques. (S. N. Tewari, R. Asthana, R. Tiwari, R. Bowman and J. Smith, *Metall. Mater. Trans.* 26A, 1995, 477–491).

is heated (above the glass transition of the binders), extruded through a die, and injected into a mold to create a "green" part. This is followed by debinding (removal of the major binders) using a catalyst or heat, and sintering. Both reactive consolidation and injection-molding have been used to make $TiB_2$- and $Al_2O_3$-reinforced NiAl matrix composites. The interface between final debinding and onset of sintering must be controlled, and binder formulations carefully chosen. This is because in the "green" part, powders are held together through binders rather than through a frictional bond (as in a press-and-sinter process), and complete removal of the binder prior to initiation of interparticle bond will result in failure.

High-energy ball milling is used for mechanical alloying of intermetallics and for fabrication of NiAl and $Ni_3Al$ composites containing $Al_2O_3$, $Y_2O_3$, $ThO_2$, and AlN particulates. The mechanically milled powder mixture is canned and extruded to create bulk composite specimens. Aluminum nitride (AlN) dispersions are produced by milling NiAl with finely divided $Y_2O_3$ in liquid nitrogen, a process often referred to as cryomilling or reaction milling because AlN reinforcement forms through a chemical reaction of Al with nitrogen during milling. One of the challenges in mechanical alloying is the need to separate the attrited powders from the unattrited powders to obtain a homogeneous material. Furthermore, canning materials must be ductile and nonreactive to the powder mixtures contained in the can.

Another powder-based technique is Martin-Marietta's exothermic dispersion (XD) process. It uses chemical reactions (e.g., between a gas bubbled through a melt) to create very fine discontinuous reinforcement. The XD feedstock (i.e., matrix and reinforcement mixture) is then milled and hot-pressed; the feedstock can be further processed via directional solidification to redesign the structure for properties (e.g., to preferentially orient the whiskers). The technique is likely to remain limited to discontinuous composites and to systems in which formation of the reinforcement via a chemical reaction is both thermodynamically and kinetically feasible.

## Liquid-state fabrication

### Infiltration

Infiltration processes use either a vacuum or a positive pressure to drive a liquid matrix into the pores between the reinforcement. Infiltration is accomplished with the assistance of a vacuum, mechanical pressure using hydraulic rams or with inert gas pressure, although centrifugal infiltration and electromagnetic field (Lorentz force)-driven infiltration also have been done. Pressure overcomes the capillary forces opposing liquid ingress and viscous drag on fluid. In the case of metal composites, if pressurization is continued through solidification, grains are refined, porosity is eliminated, interface bonding is improved, and better feeding of shrinkage occurs. High pressures can, however, cause fiber fracture, preform distortion, and uneven distribution. Metal coatings such as Cu, Ni, and Ag on the reinforcement surface, and suitable alloying additions to the matrix metal (e.g., Mg in Al alloys) improve the wetting, reduce the threshold pressure, and increase the infiltration rate. Premixed suspensions of chopped fibers, whiskers, or particulates in

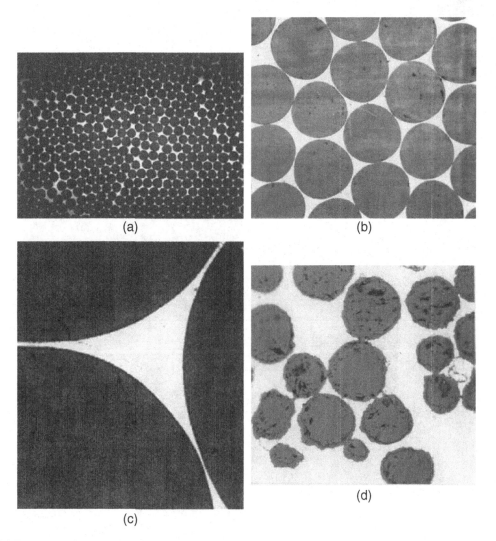

**Figure 9.1-33** (a) Transverse view showing fiber distribution in a sapphire-Hastealloy composite produced by a pressure-infiltration technique (Hastealloy, 47.5Ni-21.5Cr-17.8Fe-8.3Mo-1.7Co-0.3Mn-0.4Si-0.2Al-0.1Cu-0.4Nb, in wt%). (b) A higher magnification view of a region of the sapphire-Hastealloy composite of Figure 9.1-33a. (c) Scanning electron micrograph showing interfiber channels containing the solidified Hastealloy matrix. (d) Degradation of single-crystal sapphire fibers in the Hastealloy matrix on prolonged contact at high temperatures. (R. Asthana, S. N. Tewari, and S. L. Draper, *Metall. Mater. Trans.*, 29A, 1998, 1527–30).

the matrix melt can also be solidified under a large hydrostatic pressure, as in squeeze casting, with virtually no movement of the liquid metal. Figure 9.1-33 shows the microstructures of various continuous and discontinuous metal-matrix composites produced using the infiltration technique; the reinforcing phase includes sapphire and carbon fibers, SiC platelet, and fly ash microsphere. Figure 9.1-34 presents the data on infiltration kinetics (length versus time) for several metal-matrix composites produced using pressure infiltration techniques; depending on the system and processing conditions, both linear and parabolic kinetics have been observed. Rate of pressurization, magnitude of pressure, and cooling rate are important variables. The solidification path in infiltration must be guided (e.g., through use of chills) to feed the shrinkage. Infiltration is aided by alloying additions that promote wetting; for example, in the case of Ni-base matrices, Ti and Y are found to be effective. NiAl-, $Ni_3Al$-, and superalloy (Hastealloy and Inconel 718)-based matrices have been reinforced with various alumina fibers (single-crystal sapphire, polycrystalline fiber FP, and PRD-166). Some fiber-to-fiber contact may be unavoidable at large fiber fractions. Such regions may be difficult to fill with liquid metal even under large pressures, and become crack initiation sites.

A method to synthesize heat-resistant continuous fibers and their composites by pressure infiltration is the internal crystallization method (ICM), developed at the Russian Academy of Sciences. The method allows bundles of monocrystalline or eutectic fibers to actually crystallize in continuous channels of a matrix, usually Mo. The process is similar to the methods that are used

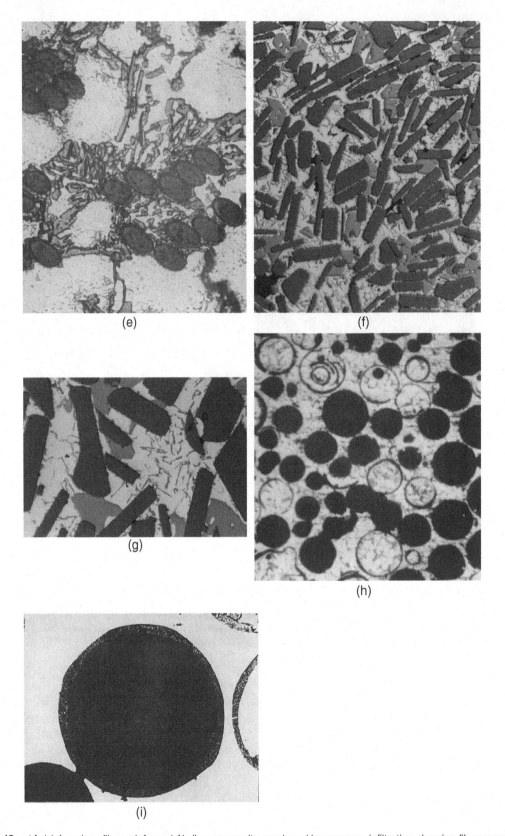

**Figure 9.1-33 (Cont.)** (e) A carbon fiber-reinforced Al alloy composite produced by pressure infiltration showing fiber segregation in the eutectic colonies. (R. Asthana and P. K. Rohatgi, *J. Mater. Sci. Lett.* 11, 1993, 442–445). (f) Pressure-infiltrated single-crystal SiC platelet-reinforced Al alloy composite showing platelet distribution in the solidified matrix. (R. Asthana and P. K. Rohatgi, *Composites Manufacturing*, 3(2), 1992, 119–123). (g) A higher magnification view of the sample of Figure 9.1-33f showing secondary-phase precipitation on the platelet surface. (h) Photomicrograph of a pressure-infiltrated Al alloy composite containing fly ash particles. (Photo Courtesy of P. K. Rohatgi, UW-Milwaukee, WI). (i) A higher magnification view of a Al-fly ash composite of Figure 9.1-33h showing the reaction layer around the fly ash particle (Photo Courtesy of P. Rohatgi, UW-Milwaukee, WI).

# Composite materials  CHAPTER 9.1

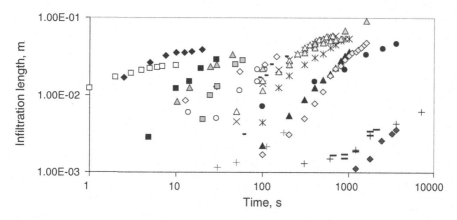

**Figure 9.1-34** Literature data on infiltration length as a function of time for various metal-matrix composites produced by melt infiltration of ceramics. (R. Asthana, M. Singh and N. Sobczak, J. *Korean Ceramic Society*, Nov. 2005).

to grow bulk oxide crystals and has been used to crystallize sapphire, mullite, yttrium-aluminum garnet (YAG), and alumina-YAG eutectic fibers. These fibers have been incorporated in Ni-base intermetallics and superalloys using pressure-casting, yielding considerable gains in creep resistance. Figure 9.1-35 shows typical microstructures of pressure-cast Ni-base composites containing internally crystallized fibers of sapphire,

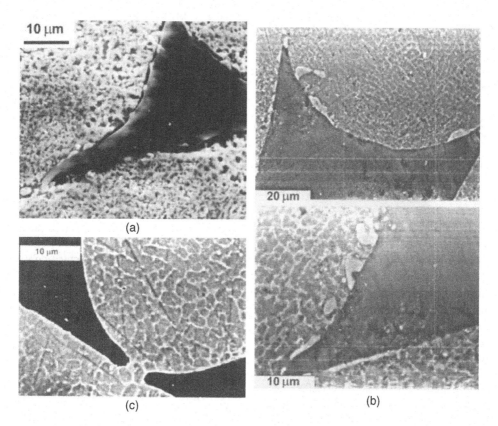

**Figure 9.1-35** (a) Scanning electron micrographs of a sapphire-YAG fiber-Ni-base superalloy matrix composite produced by pressure infiltration. The matrix composition is Ni-6Al-5Cr-1Mo-8.3W-4Ta-9Co-1.5Nb-4Re. (S. T. Mileiko et al., Oxide-fibre/nickel-based matrix composites-part I: fabrication and microstructure, *Composites Science and Technology*, 62, 2002, 167–179, Elsevier). (b) Scanning electron micrographs of the interface in a YAG fiber-Ni-base composite produced by pressure infiltration. The matrix composition is Ni-8.5Al-4.8Cr-2.2Mo-3.9W-4.4Co-1.1Ti). (S. T. Mileiko et al., Oxide-fibre/nickel-based matrix composites-part I: fabrication and microstructure, *Composites Science and Technology*, 62, 2002, 167–179, Elsevier). (c) Scanning electron micrographs of a sapphire-Ni-base composite produced by pressure infiltration. The matrix composition is Ni-8.5Al-4.8Cr-2.2Mo-3.9W-4.4Co-1.1Ti). (S. T. Mileiko et al., Oxide-fibre/nickel-based matrix composites-part I: fabrication and microstructure, *Composites Science and Technology*, 62, 2002, 167–179, Elsevier). Reprinted with permission from Elsevier. Photo Courtesy of S. T. Mileiko, Solid State Physics Institute, Russian Academy of Sciences, Moscow.

yttrium aluminum garnet (YAG), and YAG-alumina eutectic. These composites offer excellent creep resistance at elevated temperatures.

### Reactive infiltration

In reactive infiltration, the reinforcing phase forms via chemical reactions during infiltration. The reaction is usually exothermic (e.g., infiltration of silicon in porous carbon and infiltration of aluminum in $TiO_2$, mullite (3 $Al_2O_3$ .2$SiO_2$) and Ni-coated alumina) and may initiate explosively. The infiltration conditions can be controlled to achieve the desired level of conversion and structure. Clearly, the requirement of the in situ chemical formation of the reinforcement limits the choice of the systems to those in which large thermodynamic driving force exists for reaction at the processing temperatures, and the reaction kinetics are fast, to permit a substantial quantity of the product to be formed during a short period. The reactive infiltration of Si in porous C forms reacted SiC phase and unreacted C in a Si matrix. Preexisting SiC grains are employed as inert filler in the porous C preform to permit heterogeneous nucleation and bonding of SiC. The porous preform is made by pyrolyzing a high-char polymer precursor material, and some control on pore size, volume fraction, and morphology is possible. Reactive infiltration can produce two or more phases via chemical with a suitable choice of the precursor materials. For example, both $TiB_2$ and AlN phases form by reactive spontaneous infiltration of TiN, $TiC_xN_{1-x}$, and B powders with AlMg alloys. Low values of $x$ yield a greater quantity of product phases. The infiltration rate in this system is controlled by the reaction: TiN + 2B + Al → AlN + $TiB_2$. This is a highly exothermic reaction, and the heat of reaction depends on the magnitude of $x$; low values of $x$ lead to higher heat of reaction and faster infiltration rates whereas high values of $x$ lead to low heat of reaction and slower infiltration. The highest infiltration rates are achieved when TiN, B, and AlMg are used as starting materials and the lowest rates are achieved when the starting materials are $TiC_{0.7}N_{0.3}$, B, and AlMg. The presence of Mg is necessary for self-infiltration in this system; infiltration is equally good under both inert and reducing atmospheres.

Models of reactive infiltration consider pore size evolution during infiltration, and invoke the concept of a variable preform permeability $K$, which varies with both time $t$ and spatial coordinate $l$. The specific functional dependence of $K$ on $l$ and $t$ is determined by both initial pore structure, and the microscopic reaction kinetics and deposition morphologies (e.g., formation of a continuous reaction layer versus discontinuous deposition of product phase). For the simpler case of a permeability that is a function of time alone, the problem has been solved in analytic form by Messner and Chiang.[2] If the pores of radius $r$ shrink with either linear or parabolic time dependence, as may be appropriate for interface-controlled and diffusion-controlled reaction kinetics, respectively, then

$$r(t) = r_0 - kt$$

and

$$r(t) = r_0 - k_1\sqrt{t},$$

where $k$ and $k_1$ are linear and parabolic rate constants, respectively. The corresponding infiltration-distance-versus-time relationships are obtained from integration of an appropriate fluid-flow model. For the simplest capillary rise model that assumes parallel capillaries of identical diameter (Hagen-Poiseuille model), the permeability $K$ is given from the relationship $K = Pr^2/8$, where $P$ is the pore volume fraction. For more complex geometries, the concept of hydraulic radius is used; this changes only the numerical constant in the expression for permeability but not the parabolic dependence on pore radius ($r^2$ dependence). Assuming that geometry of the pore does not change during infiltration and reaction, the time-dependent permeability for the reactive infiltration is obtained from the temporal evolution of pore radius as a result of reaction. Thus, for the Hagen-Poiseuille model,

$$\frac{K(t)}{W_0} = \frac{r^4(t)}{8r_0^2} = \beta r^4(t),$$

where $W_0$ is the initial porosity. Here $r(t)$ is given from linear (interface-controlled) or parabolic (diffusion-controlled) rate expressions. Using these expressions in Darcy's equation for flow through porous media and integrating the resulting equation yields the solutions for infiltration kinetics for interface-controlled and diffusion-controlled cases, respectively, as

$$L^2(t) = \frac{2\beta\Delta P}{\mu}\left[r_0^4 t - 2r_0^3 k t^2 + 2r_0^2 k^2 t^3 - r_0 k^3 t^4 + 0.2 k^4 t^5\right] \quad (9.1.19)$$

$$L^2(t) = \frac{2\beta\Delta P}{\mu}\left[r_0^4 t - \frac{8}{3}r_0^3 k_1 t^{1.5} + 3r_0^2 k_1^2 t^2 - \frac{8}{5}r_0 k_1^3 t^{2.5} + 0.33 k_1^4 t^3\right] \quad (9.1.20)$$

where $\beta$ is a geometric factor given from $\beta = (8r_0)^2$. Of central importance to the final dimensions of the component to be processed is the final infiltration length

---

[2] Messner R.P. and Y.M. Chiang, Journal of the American Ceramic Society, 73(5), 1990, p. 1193.

$L_\text{f}$, at which reaction choking (analogous to freeze-choking during infiltration of a cold preform) would occur. This limiting length for interface-controlled and diffusion-controlled cases is given from

$$L_\text{f} = \sqrt{\frac{2\beta\Delta P r_0^5}{5\mu k}}$$

and

$$L_\text{f} = \sqrt{\frac{2\beta\Delta P r_0^6}{15\mu k_1^2}}.$$

These equations have been qualitatively verified in experiments on Si infiltration of porous C preforms. The model predicts the fast infiltration rates observed experimentally and shows that limiting infiltration depth is achieved literally in seconds. For fine-scaled carbon microstructure with fine pore size, the time to complete the reaction is only on the order of seconds for the appropriate values of reaction rate constants $k$ and $k_1$.

The limiting infiltration depth, $l_\text{f}$, can be increased by applying external pressure, increasing the pore size, or decreasing the reaction rate. The process parameters for preform fabrication, which typically involves pyrolysis of polymer precursor material, can be judiciously chosen to exercise control on the pore size, pore volume fraction, and pore morphology. The size of the pore determines the concentration gradients of solutes released from chemical reactions. The solutes released at the reaction interface should be able to diffuse away into the liquid metal in the capillary. A high transient concentration of solutes can build up around the liquid meniscus and limit further rejection of the solute into the melt to a rate at which the solute released can diffuse away from the interface. A high concentration of solutes released from reactions will also tend to retard the reaction because of a reduced thermodynamic driving force for the forward reaction. If the pore size is increased, solute diffusion becomes easier, because the metal sink becomes larger in size and can accommodate a larger quantity of solute; as a result, a larger concentration gradient is created. If, however, chemical reaction rather than solute diffusion is the driving force controlling infiltration, then as the pore size is increased the effective surface area of the ceramic phase decreases; this in turn limits the supply of the reactant species in the solid; e.g., in the $TiO_2$-Al system, an increase in the pore size reduces the available oxygen. As a result, chemical reaction rather than solute transport becomes the rate controlling step for infiltration. The limiting infiltration depth can also be increased by decreasing the reaction rate constants. However, controlling the reaction rate constants may be more difficult than controlling the pore size because the rate constants are strongly temperature dependent and will be altered appreciably if the reaction is either exothermic or endothermic. That this is usually the case in reactive systems is well established. Attempts to increase the limiting infiltration length may be hindered by reaction choking, product spallation, and pore closure. The volume changes that accompany chemical reactions in many systems because of different specific volumes of reactant and product phases, give rise to stresses and spalling-off of the reaction products, and to distortion and crack formation in the preform.

Ceramic- and metal-matrix composites are also fabricated via spontaneous infiltration without an external pressure. Self-infiltration can occur if the melt and preform compositions, temperatures, and gas atmosphere are judiciously selected to yield good wetting for wicking of the metal through a porous preform. In most cases, a critical level of Mg is needed for self-infiltration. The composite forms by growth of multiphase layers at the reaction front that comprise of a top MgO layer, a dense layer of magnesium spinel, and below this layer, another layer of spinel with microchannels above an aluminum reservoir. The formation of microchannels in the material allows "wicking" of Al alloy to the free surface and continuation of composite growth. Figure 9.1-36 shows the different stages in the process of self-infiltration of a porous ceramic preform by molten Al alloy. The rate of composite growth is determined by the supply of metal at the reaction front, and the transport of oxygen through grain boundaries, microcracks, and pores in the external spinel-MgO layer on the metal. If all the metal is used up, porous microstructures form. The self-infiltration process is slow and can take several hours to form the composite.

Fibers are made into preforms prior to infiltration. Continuous fibers may be woven into three-dimensional fabrics or ropes using weaving, braiding, and filament winding. The fiber volume fraction is controlled by controlling the fiber spacing and number of fiber layers. For stiff fibers, tapes, or laminates of fibers are made using a fugitive binder that burns off during composite fabrication. The amount of binder must be controlled; too little binder imparts little strength, whereas excessive binder could impede metal flow. Preforms of discontinuous reinforcement (e.g., short fibers) are made using the ceramic green bond-forming methods such as pressing, slip-casting, and injection-molding. Short fibers can be oriented by applying an electric field to fibers suspended in a nonconducting fluid between two electrodes. The electric field polarizes the reinforcement and orients it along the lines of force in the fluid medium. Preforms are made to net shape because of difficulty in machining ceramics. In hybrid preforms, particulates and whiskers are distributed between continuous fibers; this minimizes fiber-to-fiber contact and eliminates tiny cavities between touching fibers that are difficult to fill even under large pressures.

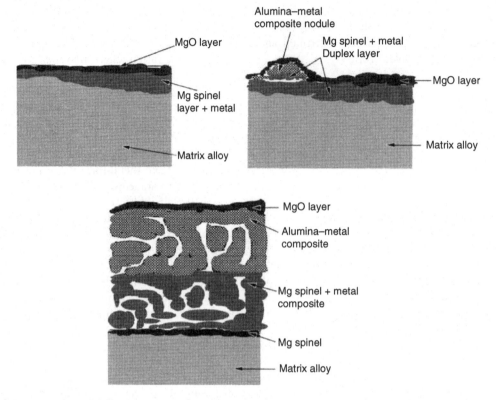

**Figure 9.1-36** Schematic of a self-infiltration process (Lanxide process) to grow metal-matrix composites.

At large applied pressures during infiltration, deformation and fracture of preform may take place and result in an unevenly reinforced casting and difficulty in continued infiltration due to collapse of relatively large interstices. The capillary pressure begins to increase for uninfiltrated regions as the compressive deformation begins, because fiber volume fraction begins to increase. Preform compliance and deformation depend strongly on the binder content in the preform; binders are added to the fiber preform to impart strength and rigidity during handling and fabrication. The deformation behavior of melt-infiltrated preforms involves further complexity as opposed to the deformation behavior of a dry preform. In pressure-casting, the elastic compression of the preform during initial hydrodynamic pressurization will increase the fiber volume fraction in the compressed zone, thereby making infiltration progressively more difficult. The subsequent elastic recovery of the preform will take place at the conclusion of hydrodynamic pressurization (i.e., after the passage of the infiltration front) and equalization of the pressure field. The localized relaxation of the preform at the conclusion of pressurization will remain incomplete if appreciable fiber fracture has occurred or if solidification has begun.

### Stirring techniques

Mechanical stirring using impellers or electromagnetic fields creates a vortexing flow in the liquid or semisolid alloy, and aids in the transfer and dispersion of the reinforcement. The solid–liquid slurry is then solidified to obtain the composite. Figure 9.1-37 is a microstructure of a SiC-Al alloy composite produced by stir-casting technique, where the SiC particles are distributed within the grain structure of the Al alloy matrix. Stirring under vacuum or inert gas shroud minimizes gas dissolution, and various impeller designs have been developed to provide high-shear mixing with minimum surface agitation. Reactive additives such as Mg or Ce improve particle dispersion in metals. Impeller erosion and dissolution due to direct contact with the melt can be minimized by employing an electromagnetic (EM) field. Special magnetohydrodynamic (MHD) stirrers have been developed to obtain various combinations of rotating and traveling magnetic fields in order to generate complex flow patterns. Ultrasonic vibrations aid the infiltration of fibers by a liquid metal, and the dispersion of fine particulates in the melt. Metal is poured over fibers (or preform) packed in a mold, and vibration is transmitted via an acoustic probe. Alternatively, the fiber bundle is passed through the melt while the latter is being irradiated. The vibrations, transmitted into the melt via a titanium horn with a $TiB_2$-coated tip, generate large accelerations ($10^3$g to $10^5$g) pressures that exceed the threshold pressure for infiltration. Reactive wetting additives (e.g., Ti in alumina fibers) further lower the vibration energy needed for complete infiltration.

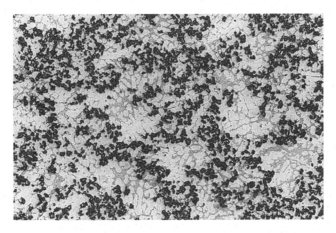

**Figure 9.1-37** Photomicrograph showing SiC particle distribution in a SiC-A359 Al alloy composite produced by a stir-casting technique. (R. Asthana, *J. Mater. Sci. Lett.*, 19, 2000, 2259–62).

Radiation also facilitates transfer and mixing of ceramic particulates, deflocculation of agglomerates, and refinement of the matrix structure. The propagation of the sonic wave causes "cavitation," which is the creation and catastrophic collapse (implosion) of tiny gas bubbles. This gives rise to enormous local accelerations that deflocculate the clusters by enabling the metal to penetrate the gas bridges between clustered particles. Vibration also maintains a uniform suspension, because particles are acoustically levitated and prevented from settling. When the vibration is continued through solidification, cavitation causes partial remelting of dendrites, grain refinement, and removal of gas porosity.

Many process variables must be controlled to achieve acceptable quality castings by the stir-casting techniques. For a fixed reinforcement content added to the melt via stirring, the process yield (i.e., % particles recovered in casting) varies with particle size. Generally, coarse particles exhibit better recovery than fine particles when properly heat treated; however, coarse particles lead to large-scale segregation due to settling or floating. In contrast, fine particles display considerable agglomeration and entrapped gases. Melt temperature is important, and an optimum temperature gives the highest yield. Very high temperatures cause severe erosion of the impeller, excessive chemical attack of the reinforcement, and increased gas-pick-up (although the wettability and fluidity may improve). In contrast, low temperatures hinder particle dispersion, and result in low fluidity, and low reinforcement content. High rates of particle addition to stirred metal cause severe agglomeration and low recovery; fast rates lead to covering of the melt by the powders, but do not appreciably increase the rate of assimilation in the melt.

For shape casting, the composite ingot must be remeltable, possess adequate fluidity, and the melt temperature and holding times should be low in order to inhibit reinforcement degradation. Excessive reinforcement degradation is prevented by limiting the temperature and holding time, or by reducing the activity of reactive solutes through composition adjustments (e.g., by increasing %Si in SiC-AlSi composite to inhibit the formation of brittle and hygroscopic $Al_4C_3$). The carbide dissolution reaction also influences the recyclability of composite scrap; if the scrap is free of $Al_4C_3$, it can be readily used. The amount of porosity in the scrap may also be a factor; due to increased viscosity, the composite does not vent air as readily as the low-viscosity base metal, and special gating and risering procedures are needed to minimize the entrained air.

Other considerations in remelting and shape casting are settling and fluidity. Heavy particles settle out and displace oxide inclusions and other solid impurities toward the top, which enables their removal. However, settling leads to an inhomogeneous particle distribution in the composite casting and forms particle-depleted and particle-enriched zones. Settling is retarded upon prolonged holding because of hindered settling and crowding of particles. Fine particles settle more slowly as compared to coarse particles, as do plate-like particles compared to blocky particles which experience lower fluid drag due to a smaller projected surface area. However, as already mentioned, fine particles severely agglomerate and cause defects. Another consideration is the fluidity which decreases due to the reinforcement, but may be adequate for casting purpose. However, chemical reactions with the melt may change the specific volume of the solid residing in the melt, which may adversely affect the fluidity (e.g., in Al-SiC slurries above 800° C, $Al_4C_3$ formation effectively increases the solid loading in the slurry, and makes shape casting difficult because of reduced fluidity). Industry has successfully explored use of virtually every casting method for particle-reinforced composites, such as green sand, bonded sand, permanent mold, plaster mold, investment, lost-foam, centrifugal casting, die casting and others. Wherever feasible and desirable, on-line determination of matrix chemistry and particle concentrations is performed during production to assure quality of the finished product.

The mixing techniques are technically and commercially viable at the industrial scale as demonstrated by their current use to make production quantities of SiC- and $Al_2O_3$-reinforced Al- and Mg-composites by the U.S. companies such as MC-21 Inc., Duralcan Co., and ECK Industries for applications in pistons, brake rotors, driveshaft tubes, bicycle frames and snow tire studs.

## Low-cost composites by casting

With the need to conserve precious and scarce materials and lower the material cost without sacrificing the essential engineering properties, it is necessary to identify and select fillers that preserve, and even enhance, the composite properties. Low-cost composites containing industrial wastes, industrial by-products, and inexpensive mineral fillers such as recycled glass, fly ash, waste mica, shell-char, rice husk ash, talc, and sand have been synthesized and evaluated for properties. Most of these fillers are lightweight materials so there is no penalty on component weight. An added benefit of using industrial wastes and byproducts as fillers is the elimination of disposal cost.

Several examples of waste byproducts can be identified, such as fly ash, mica, glass, foundry sand, slag etc. Fly ash is a waste byproduct of the electric utility industry that is generated in large quantities (e.g., 80 million tons annually in the U.S.) during the combustion of coal by thermal power plants. Such large quantities of fly ash present ecological problems associated with its storage and disposal. Currently, only 25 pct of fly ash is used in construction and other applications; the remainder goes to landfill and surface impoundments. Attempts to recover Al, Fe and other materials from fly ash are technically feasible but not commercially successful. Another use has been to synthesize mullite ceramics from waste fly ash.

Structurally, fly ash is a heterogeneous mixture of two types of solid particles: solid precipitator ash (1–150 μm, density: 1.6–2.6 g/cc) and hollow cenosphere ash (10–250 μm, density: 0.4–0.6 g/cc). Chemically, oxides such as $SiO_2$ (33–65%), $Al_2O_3$, $Fe_2O_3$, CaO are its major constituents. Another type of ash that has been experimented with in making low-cost composites is volcanic ash (e.g., Shirasu balloon in Japan). Light-weight Al composites containing ash have better specific compression strength and wear resistance relative to un-reinforced Al. Pressure casting using hydraulic rams or pressurized gas, and stir-casting have been used to make these composites. Even with 5.8% waste ash in Al, the tensile strength and hardness are comparable to commercial Al-SiC composites containing 20 vol% SiC particulates. Similarly, the thermal fatigue resistance of the ash composite is superior to that of the commercial Al-SiC composite.

Mica powder produced by grinding and milling waste mica sheets has been dispersed in Al, Cu, Ag and Ni by casting or powder metallurgy to produce self-lubricating composite bearings. Improvements in wetting by Al with the use of Mg alloying and Cu-coatings on mica improves the yield. Mica is a solid lubricant with a layered crystal structure and is more resistant to oxidation than graphite. The wear resistance of Al-mica composites is satisfactory under both lubricated and dry wear conditions. In addition, mica improves the vibration damping capacity of Al, and on a weight basis, the specific damping capacity is better than that of flake graphite cast iron, which is an excellent damping material. There is, however, some loss of strength and ductility of Al due to mica additions. Secondary processing (e.g., hot extrusion) of the composite can offset strength loss without impairing the composite's wear resistance. Another attractive solid lubricant filler is talc ($3MgO.4SiO_2.H_2O$) that is chemically inert and belongs to a family of layered silica minerals in which one layer of $Mg(OH)_2$ (also called brucite) is sandwiched between two sheets of Si-O tetrahedron. The adjacent crystal planes of talc are held together by weak Van der Waals forces, which allow the planes to easily shear and slide under an external load. Talc is also the softest material on Mohs hardness scale, and can be readily ground to fine size. Unfortunately, talc in Al alloys decreases the tensile strength and hardness but the composite responds to heat treatment, and some improvement in the strength and hardness is possible. In contrast, talc significantly improves the wear resistance of Al.

Dispersion of waste silica sand from foundries in molten Al enhances the hardness and abrasion resistance of Al. A strong chemical reaction partially converts silica to alumina according to: $3SiO_2(s) + 4Al(l) \rightarrow 3Si(s) + 2Al_2O_3(s)$. Mg as an alloying element reacts with silica according to $SiO_2(s) + 2Mg(l) \rightarrow Si(s) + 2MgO(s)$, and forms fine MgO crystals on silica while enriching the Al with Si, which improves the wear resistance and hardness (Si release in Al also compensates for the reduced fluidity of Al because of silica dispersion).

Rice husk and shell-char are agricultural wastes available in large quantities. Rice husk, which essentially protects the rice grain during growth, contains ~94% $SiO_2$, which points to possible improvements in hardness and wear resistance of light metals such as Al. The silica-rich ash from burnt rice husk has been incorporated in Al by stirring and casting. In addition, Si-rich rice husk ash from controlled burning of rice husk has been used in cement. The addition of shell-char to Al increases the hardness and wear resistance but decreases the strength.

Heat-resistant Si/SiC composites can be produced by infiltrating porous charcoal by Si, which will convert charcoal to SiC with improvements in strength of the Si matrix.

Glass is a common solid waste and can be combined with recycled metal to make the composite. It is a relatively low-energy, low-cost material with density close to Al, and may be added to Al for low-temperature use. The reclamation of waste glass involves color classification and crushing and melting with fluxes and raw materials. Colored glass obtained from different sources can be crushed and milled to achieve the fine particulate size required for effective compositing. Crushed and milled glass is preheated to remove volatile contaminants and combined with Al using either casting or powder metallurgy. Glass additions to Al improve the resistance to adhesive wear (possibly due to melting and smearing on the sliding surface), hardness, strength and vibration damping capacity of Al. Likewise, slag particles from solid waste dispersed in Al also reduce the wear of Al and increase the hardness and strength although the ductility decreases.

Rising energy costs favor recycling of metals for use in low-cost composites. Recycling of used beverage containers has become a large business; in 1992, 62.8 billion aluminum cans were collected which accounted for 68% of total produced in the U.S. Scrap metal is classified according to composition, cleanliness, and size. Impurities represent un-reclaimable mass and hinder metal recovery; therefore, some prior treatment is necessary. Because the reclaimed metal must compete with the primary metal in the user market, it must be of acceptable quality. This renders the entire reclamation enterprise capital- and energy-intensive, and rather elaborate. In view of this, judicious combination of fresh metal stock and recycled scrap may be used to make the low-cost composites. In the case of Al, the most difficult-to-process scrap goes through a reclamation plant before being melted in a reverberatory furnace (with or without salt additions, depending upon scrap cleanliness). The furnace produces a black dross containing salt, oxides, aluminum and impurities that is further processed to isolate the metal and the salt. Salt recovery plants crush the dross, dissolve the salt in water, and then evaporate the brine.

## Other liquid-phase techniques

Fine dispersions of thermodynamically stable refractory compounds can be produced in a matrix via reactions in the liquid–gas, liquid–solid, and liquid–liquid systems and salt mixtures. To grow TiC-Al composites, Ti and C are added to molten Al, and these react to form TiC particles. Titanium is prealloyed in Al whereas C is produced via decomposition of methane, which is bubbled through the AlTi alloy melt. Likewise, $TiB_2$ is incorporated in Al and Ti aluminides. To form $TiB_2$-Al composites, Ti, B, and Al in the form of either elemental powders or as prealloyed powders of Al-B and Al-Ti, are mixed in controlled amounts, and heated to melt Al in which Ti and B diffuse and chemically combine to form a fine dispersion of $TiB_2$ particles. In situ $TiB_2$-Al composites are also grown using reactions between mixed salts of Ti and B that react in molten Al to form $TiB_2$ dispersions. The dispersed phase is very fine, typically 0.1 to 3.0 $\mu$m, often monocrystalline, and uniformly distributed in the matrix. The process is flexible and permits growth of both hard and soft phases of various sizes and morphologies such as whiskers, particulates, and platelets. The growth rate is controlled by the interfacial reactions and the diffusion through the reacted layer. After an initial incubation, the early-stage growth is controlled by interfacial reactions, but once a product layer has formed the growth rate diminishes and is limited by the atomic diffusion through the reacted layer.

Many in situ composite-growth techniques can be grouped as "combustion synthesis" or self-propagating high-temperature synthesis (SHS). The process makes use of the ability of highly exothermic reactions in certain systems to be self-sustaining and energy efficient. The exothermic reaction to form a composite is initiated at the ignition temperature of the system. The reaction generates heat that is manifested in a maximum of combustion temperature that can exceed 3000 °K. In SHS, reactant powder mixture is placed in a special reactor under an argon blanket. The mixture is then ignited using laser beams or other high-energy sources to initiate the reaction. The reaction continues autocatalytically by the intense heat released from the reaction until all the feed material has combusted and reacted. High levels of open porosity (<50%) remain in the final product, and full densification for high-performance applications almost always requires secondary processing such as hot consolidation and infiltration.

In some systems, a liquid phase forms via reaction and infiltrates all open porosity. For example, excess liquid Al formed in the following reaction infiltrates and densifies the porous ceramic phase,

$$3TiO_2 + 3C + (4 + x)Al = 3TiC + 2Al_2O_3 + xAl.$$

The formation of a liquid phase during combustion synthesis is beneficial also because it increases the contact area between reactants and allows faster diffusion and reaction. Thus, in the case of combustion in the Ti-C system, ignition is controlled by the rate of surface reaction between Ti and C, which in turn is determined by the contact surface area between the two. The formation (or deliberate addition) of a low-melting-point phase (e.g., Al) in the Ti-C system further increases the surface area, and the rates of reaction and mass transfer. Besides infiltration, solid-state consolidation is also used to produce high-density composites.

### Wettability and bonding

Wetting of the fibers by molten metals is necessary for liquid-phase fabrication of composites. Prior surface treatment of the reinforcement (e.g., heat treatment, surface coatings) and alloying the matrix with a wetting promoter (e.g., Mg) are critical to high process yield. It may be difficult to predict the wettability of the filler by the matrix without conducting actual wettability tests. This is because many fillers are a complex solid solution of various constituents (e.g., oxides), and it is not always clear if the wettability of the filler will be a cumulative effect of all constituents acting in isolation, or of their mutual interactions, or if the wettability will be dominated by the major constituent. For example, the principal chemical constituent of glass, a filler for Al, is silica ($\sim$54–67%); other constituents include CaO, $Al_2O_3$, $Na_2O$, $B_2O_3$, BaO, $K_2O$ and MgO. It is known that the contact angles of Al on $SiO_2$ at 700 C and 950 C are 150° (non-wetting) and 80° (wetting), respectively. On the other hand, CaO is poorly wet by Al whereas MgO is adequately wet by Al. It may be conjectured that BaO and $B_2O_3$ will probably be wet by Al because barium wets and impregnates Al, and in the Al-B system a contact angle of 33° is attained at 1100°C. The wetting of glass shall probably be dominated by silica and, therefore, poor wetting at low temperatures (700°C) is anticipated. However, as silica in glass is in solid solution, a more complex behavior may actually occur. Extensive reaction of silica with Al forms a complex interfacial reaction zone rich in alumina, and reaction-induced wettability might aid particulate dispersion in the metal. In any case, wetting improvements are almost always necessary, and reactive additions such as Mg are quite effective.

### Ni-base composites

Experimental data on contact angles measured between various Ni-base alloys and oxides ($Al_2O_3$, $ZrO_2$), carbides (SiC, TiC, $Cr_3C_2$), borides ($TiB_2$, $TiB_2Cr$), and nitrides (TiN, AlN) are presented in Figure 9.1-38 as a function of test temperature, time, and alloying. These data cover both reactive and nonreactive systems. During the wettability test (sessile drop), alumina dissolves in the Ni droplet, and oxygen and Al released by the dissolution process diffuses into the drop and alters its chemistry. The dissolution process, however, does not appear to significantly lower the contact angle, and large obtuse contact angles are reported in $NiAl_2O_3$ (e.g., at 1773 °K, the values of $\theta$ are 128°, 133°, and 141° in vacuum, $H_2$, and He atmospheres, respectively). A progressive decrease in the $Ni-Al_2O_3$ bond strength (71–111 MPa) has been observed in joints made from liquid–solid contact with increasing contact angle. In contrast, the solid-state diffusion bonded $Ni-Al_2O_3$ couples (at 50 MPa pressure, 0.5 h, vacuum) had even lower bond strength, typically 16–60 MPa when hot-pressing was done in the temperature range of 1223 °K–1323 °K. As a generalization, therefore, liquid-phase techniques appear to provide better bonding than do solid-state fabrication techniques.

Even though $Al_2O_3$ is poorly wet by pure Ni, alloying Ni with chromium improves the wettability in $Ni-Al_2O_3$. Small quantities of Cr are sufficient to achieve good wetting and bonding in $Ni-Al_2O_3$; contact angle is close to 90° on both sapphire and polycrystalline alumina at 10% Cr in Ni, and acute angles close to $\sim$75° at 20% Cr. The solid–solid $Ni-Al_2O_3$ interfacial energy ($\sigma_{ss}$) is in the range 2.0–2.7 $J \cdot m^{-2}$, whereas the solid–liquid interfacial energy ($\sigma_{sl}$) is $\sim$2.44 $J \cdot m^{-2}$. In contrast, the solid–liquid interfacial energy in Ni-Cr-alumina couples is in the range of 1.3–1.8 $J \cdot m^{-2}$ for Cr contents in the range 0.001% to 10%. From Young's equation, $\cos \theta = (\sigma_{sv} - \sigma_{ls})/\sigma_{lv}$, it is noted that a decrease in $\sigma_{ls}$ by Cr alloying will decrease the contact angle, $\theta$, consistent with the experimental observations. However, both the contact angle and solid–liquid interfacial energy reach their respective maxima at $\sim$1% Cr, as does the bond strength. Thus, increasing the concentration of Cr in Ni at first strengthens, but later weakens the $Ni/Al_2O_3$ interface.

## Properties

### Interface strength

In composites, a high-bond strength is required for effective load transfer to the fiber, which is the primary load-bearing constituent. Normally, a matrix-matrix bond is achieved as a result of chemical interactions at the interface. For toughening of brittle matrix composites, however, weaker interfaces are preferred, to enable frictional sliding (pullout) of debonded fibers to contribute to toughening. Fiber pushout, fiber pullout, bend test, and other techniques have been used to characterize the strength of the fiber-matrix interface. In the fiber pushout test (Figure 9.1-39), interface strength is derived from the measurements of compressive load on a single fiber (residing in a matrix wafer) required to debond and displace the fiber. The test has been applied to SiC-Ti, glass-TiAl, sapphire-Nb, sapphire-NiAl, SiC-$Si_3N_4$ and other ceramic- and metal-based composites. In the conventional pushout test, thin (100- to 1000-μm) composite wafers, ground and polished to reveal the fiber ends and the surrounding matrix structure, are mounted on a support block containing grooves that permit unobstructed sliding of individual fibers through the wafer

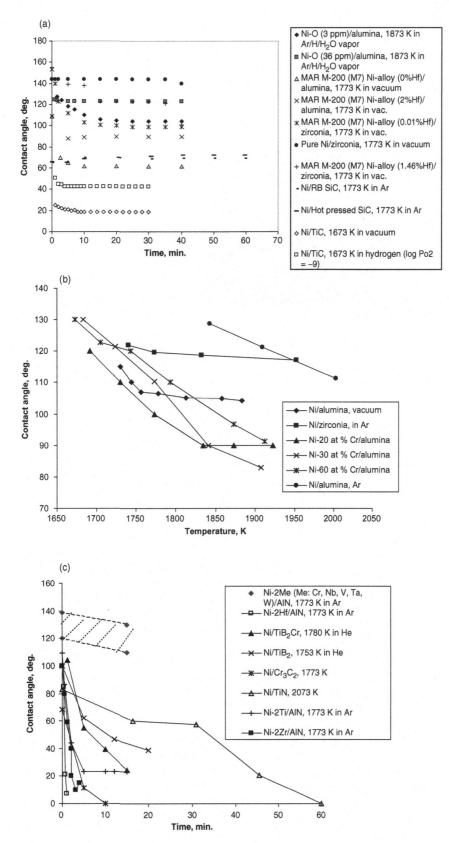

**Figure 9.1-38** (a) Contact angle versus time data for various oxide and carbide ceramics in contact with Ni-base matrices. (b) Contact angle of molten Ni and Ni-Cr alloys on alumina and zirconia as a function of temperature. (c) Contact angle of Ni and Ni-base alloys on various nitride, boride, and carbide substrates as a function of the time of contact.

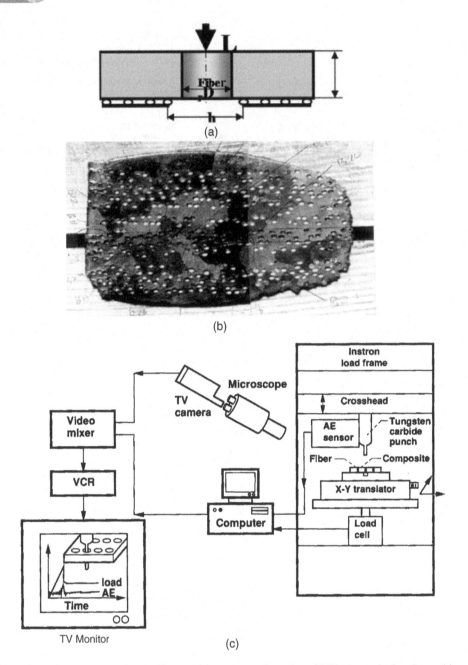

**Figure 9.1-39** (a) Schematic diagram showing the fiber-pushout test configuration. (b) Photograph showing a thin composite wafer mounted on a support block with a groove in it for fiber pushout test. The matrix grain structure and fiber ends are visible (tested fibers can be identified relative to their position at grain boundaries and grain interior). (c) Schematic diagram showing the fiber pushout test setup. (Courtesy of J. I. Eldridge, NASA Glenn Research Center, Cleveland, OH).

with the help of a flat-bottomed microindenter attached to an INSTRON frame. The load displacement data (and acoustic emission signal) are recorded, and interface strength $\tau^*$ is obtained from the measured debond load Q and surface area

$$\tau^* = \frac{Q}{\pi D t},$$

where $\tau$ is the shear stress, $D$ is fiber diameter, and $t$ is the wafer thickness. Figure 9.1-40 shows typical load displacement profiles from the fiber pushout test on sapphire-reinforced NiAl matrix composites. The interface strength (debond stress) corresponds to the maximum shear stress on the curve (other stress transitions, such as $\tau_p$, proportional stress, and $\tau_f$, frictional stress provide information on the crack initiation and propagation behaviors). The load displacement (or stress-strain) data are examined in light of the fractographic observations of debonded interfaces to interpret the results and obtain an estimate of the debond stress. In addition, interrupted

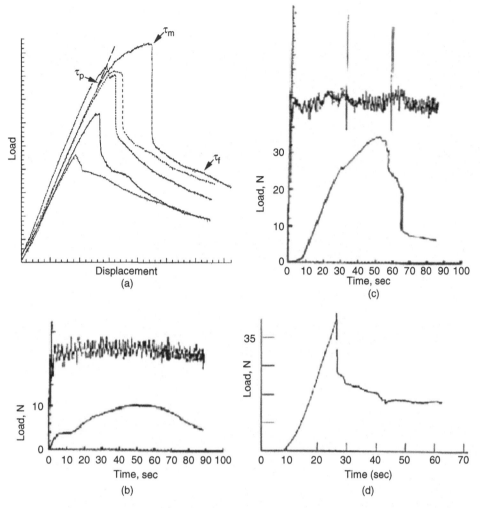

**Figure 9.1-40** (a) Load-displacement response in the fiber pushout test. Each curve represents the behavior of a single fiber. Three load transitions are identified for each fiber, $\tau_p$ (proportional load corresponding to the onset of deviation from the initial linear profile), $\tau_m$ (maximum debond load), and $\tau f$ (frictional load for sliding of the debonded fiber). (b) Load-time and acoustic emission profiles for a sapphire fiber in a NiAl matrix during fiber pushout. Data are on a thin (<300 μm) composite wafer. Load transitions $\tau_p$, $\tau_m$, and $\tau_f$ are indistinct. (c) Load-time and acoustic emission profiles for a sapphire fiber in a NiAl matrix during fiber pushout. Data are on a thick (>300 μm) composite wafer. Load transitions $\tau_p$, $\tau_m$, and $\tau_f$ are distinct in thick wafers. (d) Load-time profile during fiber pushout for a sapphire fiber in a NiAl matrix containing Cr as an alloying element. Data are on a thick (>300 μm) composite wafer. (R. Asthana, S. N. Tewari and R. Bowman, *Metallurgical & Materials Transactions*, 26A, 1995, 209–223).

pushout tests have been done in conjunction with fractography to identify the failure mode and to determine the stress corresponding to the initiation of disband. Figure 9.1-41 depicts some results from the interrupted pushout test on sapphire-NiAl composites. The results are sensitive to matrix plasticity and test configuration (e.g., to $t/h$ ratio, where $t$ is disk thickness and $h$ is support span). For fibers of variable cross-section and shape, it is difficult to load the fiber exactly at its center of gravity, leading to fiber fracture or unreliable strength measurements. A method developed at the Russian Academy of Sciences overcomes this problem by employing a metal ball indenter positioned between a flat-bottomed microindenter and the fiber; the ball indenter distributes the applied stress over a larger area.

Processing conditions, alloying, binders, and coatings all influence the measured strength. For example, in hot-pressed powder-cloth NiAl-alumina composites, polymethylmetacrylate (PMMA) is used as binder for fibers and polytetrafluoroethyline (Teflon) is used as a base binder for the NiAl matrix powders. The composites containing PMMA and Teflon binders have a purely frictional bond and low interface strength (50–150 MPa), whereas the composites free of binders had significantly higher strength (>280 MPa). Interface contamination from binder residues during hot-pressing interfered with interface development (carbon residue served as fracture initiation site). Processing temperature is another important variable that influences the bond strength. Figure 9.1-42 shows the effect of casting temperature on

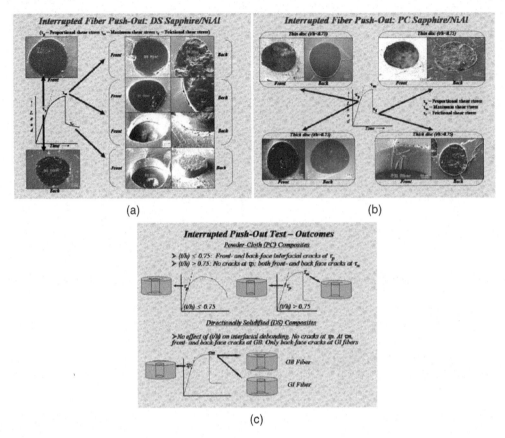

**Figure 9.1-41** (a) Fiber debond behavior as revealed in interrupted pushout test on directionally solidified sapphire-NiAl composites. Photographs show SEM views of the pushed fiber at the front and back faces of the composite wafer in the vicinity of the load transitions corresponding to $\tau_p$, $\tau_m$, and $\tau_f$. (b) Fiber debond behavior as revealed in interrupted pushout test on sapphire-NiAl composites fabricated using plasma spray and hot consolidation techniques. Photographs show SEM views of the pushed fiber at the front and back faces of the composite wafer in the vicinity of the load transitions corresponding to $\tau_p$, $\tau_m$, and $\tau_f$. (c) Schematic illustration of the fiber debond behavior as revealed in interrupted pushout test on sapphire-NiAl composites. The pushout behavior is shown at loads corresponding to $\tau_p$, $\tau_m$, and $\tau_f$ for both thin ($t/h \leq 0.75$) and thick ($t/h > 0.75$) wafers, where $t$ is the wafer thickness and $h$ is the width of the groove on the block that supports the wafer during loading. GB and GI fibers are the fibers positioned at the NiAl matrix grain boundaries, and the fibers engulfed in the grain interior, respectively.

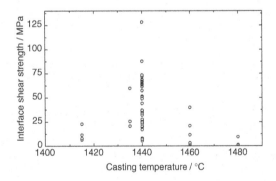

**Figure 9.1-42** Interfacial shear strength in a pressure-cast sapphire-Ni-base superalloy matrix composite as a function of the casting temperature (matrix composition is Ni-6Al-5Cr-1Mo-8.3W-4Ta-9Co-1.5Nb-4Re). (S. T. Mileiko et al., Oxide-fibre/nickel-based matrix composites-part I: fabrication and microstructure, *Composites Science and Technology*, 62, 2002, 167-179, Elsevier). Reprinted with permission from Elsevier.
Figure Courtesy of S. T. Mileiko, Solid State Physics Institute, Russian Academy of Sciences, Moscow.

the interfacial bond strength (from fiber pushout test) of a sapphire-Ni-base superalloy matrix composites synthesized using pressure-casting; despite the scatter in the data, it appears that both high and low temperatures decrease the interface strength in this system.

Fiber coatings and alloying additions to the matrix are commonly employed in the manufacture of composites for better compatibility with the matrix. Frequently, however, differences in the wetting and bonding behaviors are noted when the same chemical element is used as an alloying addition to the matrix and as an interfacial film on the ceramic. For example, Ti thin films on $Al_2O_3$ improve the wetting and bonding in Al-$Al_2O_3$ couples, whereas Ti alloying of Al matrix does not. In contrast, Cr, either as an alloying addition to Ni or as a surface film on both single-crystal and polycrystalline $Al_2O_3$, improves Ni-$Al_2O_3$ wetting and bond strength. Chromium has good thermoelastic compatibility with both Ni and $Al_2O_3$, and it can serve as a stress-absorbing compliant layer in Ni-base composites such as Ni-Al$_2$O$_3$ and NiAl-Al$_2$O$_3$. The mismatch of the coefficients of thermal expansion (CTE, $\alpha$) between Ni (or NiAl) and Al$_2$O$_3$ is rather large ($\alpha_{Ni} = 13.3 \times 10^{-6}$ °K, $\alpha_{sapphire} = 9.5 \times 10^{-6}$ °K$^{-1}$ and $\alpha_{NiAl} = 15 \times 10^{-6}$ °K$^{-1}$]. This results in thermal clamping stresses (compressive on the fiber and tensile in the matrix), which can cause matrix cracking in the vicinity of the fiber (the compressive stresses from radial clamping over the fiber length also increase the fiber debond stress). Because chromium has a CTE intermediate between Ni (or NiAl) and alumina, an interlayer of Cr could serve as a stress-absorbing compliant layer ($\alpha_{Cr} = 6.2 \times 10^{-6}$ °K$^{-1}$).

Ti coatings on alumina fibers lead to better bonding (interface strength > 150 MPa) in pressure-cast alumina-reinforced NiAl and Ni$_3$Al composites as compared to composites made from uncoated fibers. Zr and Y in Ni$_3$Al improve the bond strength, although Ti as an alloying addition to NiAl and Ni$_3$Al does not have a beneficial effect and low bond strength results. During composite fabrication, Ti atoms diffuse into the fiber, causing alumina grain growth. Titanium-sputter-coated Al$_2$O$_3$ fibers in Al$_2$O$_3$-NiAl show radial matrix cracks, suggesting that Ti layers bond well to the fiber and the matrix. Likewise, Pt-coated alumina also exhibits better bonding with NiAl than uncoated fibers. In a manner similar to coatings, matrix alloying with Zr, Cr, Yb, or W leads to improved NiAl-Al$_2$O$_3$ bonding. Alloying additions such as Zr also strengthen the NiAl matrix (however, Zr raises the ductile-to-brittle-transition temperature). Alloying NiAl with Cr, W, and especially the rare earth ytterbium has been found to markedly enhance the fiber-matrix shear strength. With Cr and W, the interface strength is in excess of 155 MPa. However, alloying NiAl with Yb led to interface strength of 205 MPa with sapphire, in contrast to 50–150 MPa in the absence of such alloying. A high frictional shear stress (53 to 126 MPa) and large load undulations characterize the sliding of debonded fibers, as seen in the load displacement response of this composite during fiber pushout presented in Figure 9.1-43. Extensive fiber surface reconstruction occurs due to a strong interfacial reaction between sapphire and NiAl(Yb) matrix. The ytterbium oxide (Yb$_2$O$_3$) is more stable than Al$_2$O$_3$, and sapphire is reduced by ytterbium (the free energies of formation of Yb$_2$O$_3$ and Al$_2$O$_3$ are $-1727.5$ kJ·mol$^{-1}$ and $-1583.1$ kJ·mol$^{-1}$, respectively). A complex multilayer interface consisting of Yb$_2$O$_3$ and Yb$_3$Al$_5$O$_{12}$ forms in sapphire-NiAl(Yb) composites containing only about 0.36 atom%. The microstructure of the fiber–matrix interface in this composite under different processing conditions is shown in Figure 9.1-44. The chemical makeup of the interface in this composite is schematically shown in Figure 9.1-45. Unlike Yb, Cr in NiAl only moderately attacks the sapphire; Cr preferentially precipitates on

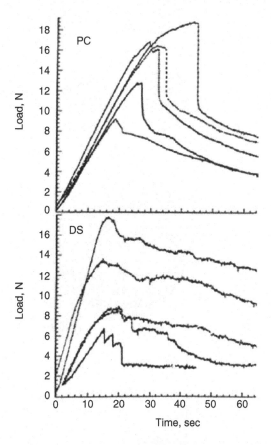

**Figure 9.1-43** Load-time profiles during fiber pushout in sapphire-NiAl(Yb) composites. PC, powder-cloth composite, DS, directionally solidified composite. The DS composite shows increased undulations in the frictional load beyond the (maximum) debond load because of increased fiber surface roughening caused by chemical reactions between the sapphire and the rare-earth ytterbium. (R. Asthana and S. N. Tewari, *Advanced Composite Materials*, 9(4), 2000, 265–307).

and bonds to the fibers without forming a visible reaction layer, as shown in Figure 9.1-46. Bonding in Ni-Cr-Al$_2$O$_3$ probably results from beneficial changes in the interfacial energy due to adsorption of Cr-O clusters on alumina, although thin interfacial films of Ni-spinel, NiAl$_2$O$_4$, have been noted in diffusion-bonded couples. The segregation of NiAl-Cr eutectic at the fiber–matrix interface results in a high frictional shear stress and formation of wear tracks due to matrix abrasion by hard Cr particles during sliding, as shown in the photomicrographs of Figure 9.1-47. In contrast, the unalloyed NiAl and W-alloyed NiAl matrices do not show interfacial segregation effects as shown in the photomicrographs of Figure 9.1-48. A high bond strength (>150 MPa) is also achieved in pressure-cast sapphire-INCONEL-718 superalloy composites in which the composite matrix contains reactive elements such as Cr.

The zirconia-toughened alumina (ZTA) fibers (PRD-166) in a pressure cast Ni-45Al-1Ti matrix degrade by

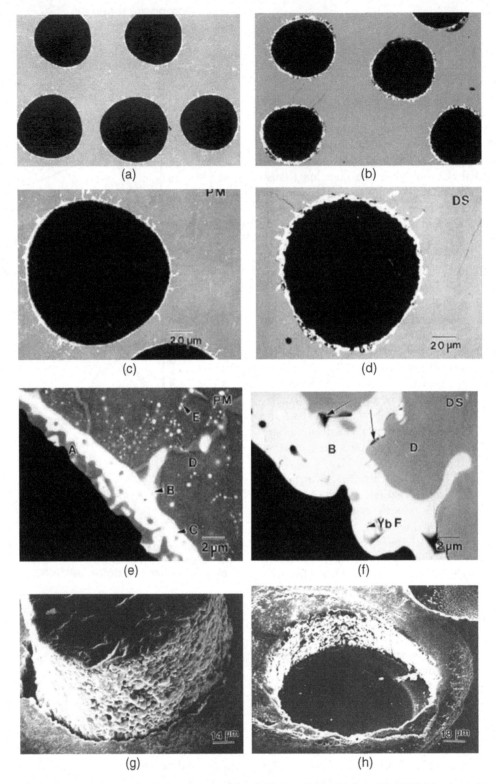

**Figure 9.1-44** (a) Powder-cloth sapphire-NiAl composite containing 0.36 atom% Yb. (b) Sapphire-NiAl(Yb) composite containing 0.19 atom% Yb directionally solidified from a powder-cloth feedstock. (c) Reaction zone around a sapphire fiber in the powder-cloth sapphire-NiAl(Yb) composite. (d) Reaction zone around a sapphire fiber in the directionally solidified sapphire-NiAl(Yb) composite. (S. N. Tewari, R. Asthana, R. Tiwari, R. Bowman and J. Smith, *Metall. Mater. Trans.*, 26A, 1995, 477–491). (e) Interfacial region in a powder-cloth sapphire-NiAl(Yb) composite showing compound layer formation due to reactions. A, oxygen-rich NiAl; B, $Yb_3Al_5O_{12}$; C, $Yb_2O_3$; and D, NiAl (the compound $Yb_3Al_5O_{12}$ is a spinel compound with formula $3Yb_2O_3 \cdot 5Al_2O_3$). (f) The fluoride compound (YbF) marked on the figure formed due to reaction with fluoride-based binders that were used in the powder-cloth feedstock specimen. (g) The back face of a pushed fiber in a wafer of sapphire-NiAl(Yb) composite showing extensive fiber surface reconstruction. (h) The front face of a pushed fiber in a wafer of sapphire-NiAl(Yb) composite showing extensive fiber surface reconstruction. (S. N. Tewari, R. Asthana, R. Tiwari, R. Bowman and J. Smith, *Metall. Mater. Trans.*, 26A, 1995, 477–491).

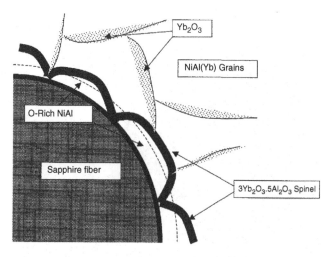

**Figure 9.1-45** Schematic representation of the interfacial compound layer formed in the sapphire-NiAl(Yb) composite.

grain coarsening and dissolution (Zr released from dissolution reduces and degrades the fibers). The PRD-166 fibers in a pressure-cast Ni-base alloy containing Cr, Ti, and Zr (Ni-16.8Al-7.9Cr-1.2Ti-0.5Zr-0.1B, in atom%) do not, however, form reaction products, although fiber dissolution leads to coarse $ZrO_2$ particles.

The short interaction times in pressure-casting minimize the interfacial reaction. For example, interfaces free of reaction products have been observed in pressure-cast sapphire-$Ni_3Al$-based $\gamma'/\gamma$ alloy containing Cr, Mo, W, and Co. However, fibers of yttrium aluminum garnet (YAG) and YAG-alumina eutectic show some reactivity with the $Ni_3Al$-based $\gamma'/\gamma$ alloy, and new interphases. Likewise, YAG fibers in a Ni-base superalloy containing (Re, Ta, W, etc.) form interfacial carbides (pure alumina does not show any significant amount of the new interphase). The fibers strongly bond to the matrix, which suggests the positive influence of yttrium of the fiber in promoting the interfacial bonding.

### Fiber strength

Whereas the fiber–matrix interfacial reactions could potentially enhance the interface strength, they also tend to degrade the fiber. Notches and grooves on the reconstructed fiber surface act as strength-limiting flaws in brittle ceramic fibers. Thus, a high bond strength may be achieved at the expense of fiber quality. Figure 9.1-49 shows the extent of surface degradation in sapphire fibers extracted from various NiAl-base composites synthesized using different processing conditions. The manufacturing process must be designed to limit the fiber degradation while providing a strong interfacial bond for load transfer. Usually it is necessary to characterize the strength of extracted fibers before process control and prediction are possible. This is also necessary because the behavior of separate fibers extracted from the matrix may be different from the behavior of the fibers residing in the matrix. The matrix (Hastealloy) composition is 47.5Ni, 21.5Cr, 17.8Fe, 8.3Mo, 1.7Co, 0.3 Mn, 0.4Si, 0.2Al, 0.1Cu, 0.4Nb, 0.06Ti, 0.08C, and 0.02 O, in wt%. The fiber strength data are represented using the two-parameter Weibull function

$$F(\sigma) = 1 - \exp\left(\frac{\sigma}{\sigma_o}\right)^\beta \quad (9.1.21)$$

where $F(\sigma)$ is the probability of failure of a fiber at a stress of $\sigma$, $\alpha$ is the scale parameter, and $\beta$ is the Weibull modulus. Figure 3.43a also displays the strength distribution of virgin (as-received) fibers and fibers extracted from a powder-cloth sapphire-NiAl composite. For the sapphire-Hastealloy composite, the values of $\alpha$ and $\beta$ are $5.561 \times 10^{-6}$ and 1.7479 respectively, and the mean strength, $\sigma^*$ standard deviation, $s$, and the coefficient of variation, $CV$, are 904 MPa, 562 MPa, and 62.2%, respectively. There is on average a 66% loss in fiber strength after casting; the as-received fiber strength is about 2.7 GPa.

Similar degradation of fiber strength due to extensive chemical attack has been observed in polycrystalline and single-crystal alumina fibers in pressure-cast as well as hot-pressed Fe- and Ni- base intermetallics. The strength loss in single-crystal sapphire in "powder-cloth" FeCrAlY, FeCrAl, Cr, FeAl, and NiAl matrix composites is in the range of 45 to 60%, and strength loss in pressure cast $Ni_3Al$-Ti matrix composites is about 67%. The initial fiber strength of 2.5–3.0 GPa drops to about 1.2–1.8 GPa after elevated-temperature contact between sapphire and FeCrAlY, FeCrAl, Cr, FeAl, and NiAl. In the case of FeCrAlY alloys, Y and Cr form brittle reaction products that weaken the fiber. The strength loss also seems to depend on the atmosphere because of the latter's influence on chemical reactions. For example, the strength of alumina fibers in contact with FeCrAlY alloys degrades under Argon but not under Ar+H2 atmosphere. Much of the deleterious effects of chemical attack of the fibers can be overcome through judicious selection of matrix alloy chemistry, use of reaction barrier layers, and control of process parameters (pressure and temperature).

### Stiffness, strength, and ductility

Generally, the tensile and compressive strengths, elastic modulus, and flexural strength of both continuous and discontinuous composites increase, whereas the ductility decreases with increasing volume fraction of the reinforcing phase (soft particulates such as graphite and mica decrease rather than increase the strength). Table 9.1-6

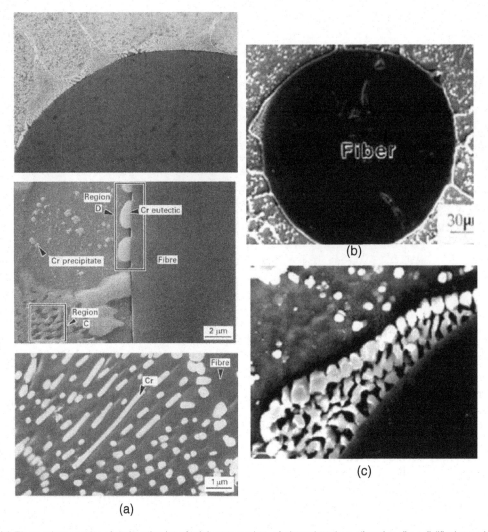

**Figure 9.1-46** (a) Photomicrographs showing the interfacial segregation of chromium in a directionally solidified sapphire-NiAl(Cr) composite. The matrix composition is 60.2Ni-28.2Al-11.4Cr (in wt%) The bottom figure shows eutectic chromium bonded to the sapphire after chemically dissolving the matrix. (b) Interfacial segregation of Cr in a vacuum-induction-melted and chill-cast sapphire-NiAl(Cr) composite. (c) A higher magnification view of the interfacial region in the sample of Figure 9.1-46b. (R. Asthana, R. Tiwari, and S. N. Tewari, *Metall. Mater. Trans.*, 26A, 1995, 2175–84).

lists the elastic modulus and strength of some metal-matrix composites (data on polymer composites is also shown for comparison). The composite's strength is strongly structure sensitive; structural defects, brittle reaction layers, stress concentration, dislocation pileups, grain structure, and texture influence the strength. The strength also depends on the spacing between and the size and distribution of the reinforcement; inhomogeneous distribution and clustering (with fiber-to-fiber contact) lower the strength. Clustering becomes severe at fine size, especially at large-volume fractions. The composite's strength is anisotropic, and with continuous fibers and high-aspect ratio whiskers, chopped fibers, or platelets, strength decreases, with increasing misorientation relative to the stress axis. Preform architecture also affects the composite's strength; with two-dimensional, planar random orientation of short fibers, the strength is better parallel to fiber array than normal to it (even though the reinforcement is randomly oriented in a plane). However, even in a direction normal to fiber axis, the composite's strength exceeds the matrix strength. In contrast, ductility markedly decreases with increasing reinforcement loading, especially at small size.

The composite modulus is anisotropic but relatively insensitive to the reinforcement distribution and structural defects; the modulus is less sensitive to fiber packing and clustering than strength, but is different along the fiber axis and transverse to it. Figure 9.1-50 shows the axial (parallel to fiber) and transverse elastic modulus and ultimate tensile strength of an alumina fiber (FP)-reinforced Al-Li composite as a function of fiber volume fraction. The longitudinal modulus of fiber-reinforced metals agrees well with the rule-of-mixture (ROM) values, but the transverse modulus is generally

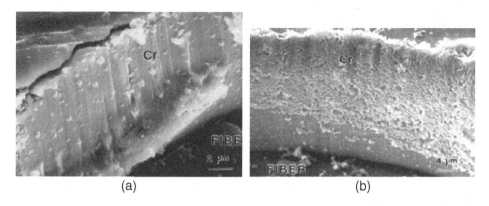

**Figure 9.1-47** (a) Wear tracks and primary Cr particles in the region of displaced fiber in a directionally solidified sapphire-NiAl(Cr) composite after the fiber pushout test. This is a front-face SEM view. (b) Eutectic Cr particles decorating the interfacial region around a displaced sapphire fiber in a directionally solidified sapphire-NiAl(Cr) composite after the fiber pushout test. This is a front-face SEM view. (R. Asthana, R. Tiwari, and S. N. Tewari, *Metall. Mater. Trans.*, 26A, 1995, 2175–84).

lower than ROM predictions. For the discontinuous reinforcement, modulus is independent of the distribution. Particulate composites also show an increase in the modulus with increasing additions of the reinforcement; the increase is, however, smaller than ROM predictions, which, in this case, are an upper bound.

A reinforcement with greater strength and stiffness than the matrix and well bonded to it will carry higher

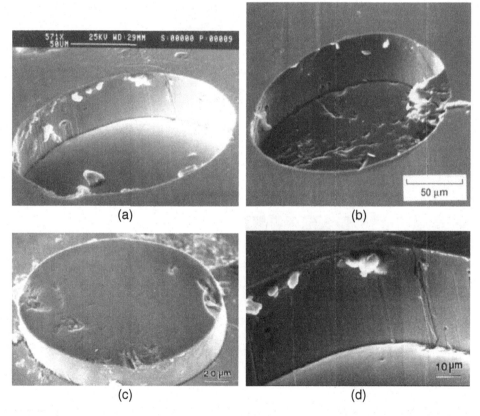

**Figure 9.1-48** (a) Front-face SEM view of a pushed fiber in a wafer of a directionally solidified sapphire-NiAl(W) composite specimen. No preferential phase segregation or reaction layers are seen (the interfacial deposit visible in the figure is an artifact). The matrix composition is 67.4Ni-31.4Al-1.5W (in wt%). (b) Front-face SEM view of a pushed fiber in a wafer of directionally solidified sapphire-NiAl composite specimen (equiatomic fractions of Ni and Al). No preferential-phase segregation or reaction layers are seen. (c) Back-face SEM view of a pushed fiber in a directionally solidified sapphire-NiAl(W) composite wafer. (d) A higher magnification view of the interfacial region of sapphire-NiAl(W) composite specimen of Figure 6-48a showing a clean fiber–matrix interface. (R. Asthana, R. Tiwari and S. N. Tewari, *Metall. Mater. Trans.*, 26 A, 1995, 2175–84). Figure (b) is from S. N. Tewari, R. Asthana and R. D. Noebe, *Metall. Mater. Trans.*, 24A, 1993, 2119–25.

## CHAPTER 9.1  Composite materials

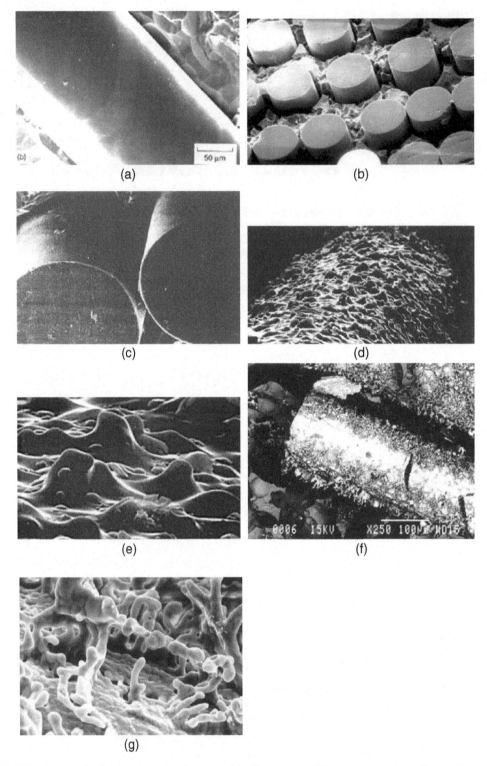

**Figure 9.1-49** (a) SEM micrograph of the surface of a single sapphire fiber extracted from a vacuum-induction-melted and chill-cast NiAl matrix. (b) Etched out fibers in a powder-cloth sapphire-NiAl composite. (c) A higher magnification view of the sample of Figure 9.1-49b. (d) SEM view of a sapphire fiber extracted from a powder-cloth sapphire-NiAl(Yb) composite specimen showing extensive roughening and degradation of the fiber. (e) A higher magnification view of the sample of Figure 9.1-49d. (f) SEM view of a sapphire fiber extracted from a directionally solidified sapphire-NiAl(Yb) composite. (R. Asthana and S. N. Tewari, *Advanced Composite Materials*, 9(4), 2000, 265–307). (g) A higher magnification view of the fiber of Figure 9.1-49f, showing extensive fiber degradation due to processing. (R. Asthana and S. N. Tewari, *Advanced Composite Materials*, 9(4), 2000, 265–307).

## Composite materials — CHAPTER 9.1

**Table 9.1-6** Mechanical properties of selected metal and polymer composites

| Composite | E, GPa | Strength, MPa |
|---|---|---|
| B-Al (50% fiber) | 210$^\parallel$ <br> 150$^\perp$ | 1500$^\parallel$ <br> 140$^\perp$ |
| SiC-Al (50% fiber) | 310$^\parallel$ <br> — | 250$^\parallel$ <br> 105$^\perp$ |
| Fiber FP-Al-Li (60% fiber) | 262$^\parallel$ <br> 152$^\perp$ | 690$^\parallel$ <br> 172–207$^\perp$ |
| C-Al (30% fiber) | 160 | 690 |
| E-Glass-epoxy (60% fiber) | 45$^\parallel$ <br> 12$^\perp$ | 1020$^\parallel$ <br> 40$^\perp$ |
| C-Epoxy (60% HS fiber) | 145$^\parallel$ <br> 10$^\perp$ | 1240$^\parallel$ <br> 41$^\perp$ |
| Aramid (Kevlar 49)-epoxy (60% fiber) | 76$^\parallel$ <br> 5.5$^\perp$ | 1380$^\parallel$ <br> 30$^\perp$ |

Note: $^\parallel$, value parallel to the fiber axis. $^\perp$, value perpendicular to the fiber axis.

load than the matrix. With perfectly bonded interfaces, ROM applies to the volume fraction-dependence of strength, i.e., $\sigma_c = V_f \sigma_f + V_m \sigma_m$, where $\sigma_m$ and $\sigma_f$ are the strength of the matrix and reinforcement, and $V_f$ and $V_m$ are the fiber- and matrix volume fractions, respectively. The rule of mixtures presupposes that the length of the fiber is above a minimum value to prevent pullout from the matrix under tensile loading. If the fiber is below a critical length (Equation 9.1.1), fiber pullout will precede fiber fracture, and no strengthening will be achieved because the load will not be effectively transferred from the matrix to the fiber. Strength predictions for particulate composites are more involved because the dislocations due to CTE mismatch contribute to the strength. Sharp ends lead to severe stress concentration and localized plastic flow at stresses well below the matrix yield stress. Large plastic strain causes void nucleation and growth, leading to premature failure. High-aspect-ratio discontinuous reinforcement provides greater improvement in the composite's modulus and strength at high-volume fractions; for example, platelets and whiskers are better than spheroidal particles. However, whiskers and platelets should be aligned along the direction of applied stress, because the effective composite modulus and strength fall rapidly with increasing misorientation. High-aspect-ratio platelets are less expensive than whiskers and pose fewer health hazards. However, the high aspect ratio of platelets needs to be achieved at relatively fine size if particle fracture during composite fabrication is to be avoided. A fine size usually leads to a greater tendency for clustering, and mechanical working may be necessary to decluster the platelets and preferentially orient them along the deformation axis. High-aspect-ratio platelets yield better gains in strength, modulus, and ductility in Al composites than low-aspect-ratio particulate

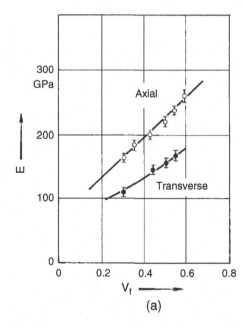
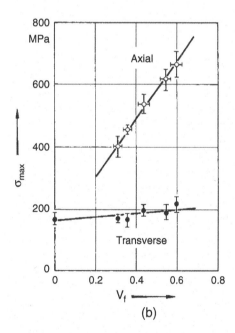

**Figure 9.1-50** (a) Axial and transverse Young's modulus of $Al_2O_3$ (FP)-Al-Li composite as a function of fiber volume fraction, and (b) axial and transverse ultimate tensile strength of the composite as a function of the fiber volume fraction. (A. R. Champion, W. H. Krueger, H. S. Hartman and A. K. Dhingra in Proceedings of the 1978 International Conference on Composite Materials [ICCM-2], TMS-AIME, New York, 1978, p. 883). Reprinted with permission from The Minerals, Metals & Materials Society, Warrendale, PA (www.tms.org).

reinforcement. Thermomechanical treatments such as forging, extrusion, and rolling improve the reinforcement distribution, decluster agglomerates, eliminate porosity, improve the interfacial bonding, and yield a fine matrix subgrain structure. These structural modifications markedly improve the strength and ductility of the composite.

In composites based on age-hardenable matrices, growth of second-phase precipitates is enhanced by the short-circuit diffusion paths that result from increased dislocation density from CTE mismatch, leading to faster hardening. Quenching at the conclusion of solutionizing forms a zone of plastically deformed matrix with a high dislocation density around each fiber or particle (prior mechanical deformation also accelerates the aging kinetics). The dislocations serve as sites for heterogeneous nucleation. The time to peak hardening decreases with increasing reinforcement volume fraction in well-bonded composites. However, interfacial reactions often lead to solute depletion and impair the aging response; in addition, thick reacted layers tend to impair the strength.

## Fatigue and fracture toughness

Continuous fibers, short fibers, whiskers, platelets, and particulates all enhance the composite's fatigue resistance. Fatigue properties are anisotropic and fiber orientation is important; improvement in the fatigue strength is greatest when the fibers are directionally aligned parallel to the stress axis. Fatigue crack propagation is affected by crack closure, crack bridging, and crack deflection, all of which are modulated by the dispersed particles or fibers. The fatigue life is very sensitive to processing conditions, and process improvements (e.g., in particulate composites, the control of particle distribution, particle clustering, slag and dross inclusions, and gas and shrinkage porosities) increase the fatigue strength. A high-particulate-volume fraction usually leads to a higher fatigue strength. This is because of lower elastic and plastic strains of composites due to increased modulus and work-hardening. For a fixed volume fraction, the fatigue strength usually increases as the particle size decreases. The size effect is related to increased propensity for cracking of large particles as well as a decrease in the slip distance at fine particle sizes. As particles serve as barriers to slip, a small interparticle distance will reduce the slip distance. Large particles fracture more easily than fine ones during fatigue testing and lower the fatigue strength by causing premature crack initiation. Particle fracture also leads to cycling softening of the matrix, and therefore an increase in the plastic strain. Fatigue cracks usually originate at defect sites such as particle clusters and microvoids in the matrix or at interfaces. Also because particle cracking lowers the fatigue strength, one method of enhancing the fatigue strength in a given system is to use single-crystal particulates with high fracture strength.

Both strength and toughness of composites depend on the interfacial shear strength, but their requirements are different. Weak interfaces improve the fracture toughness but lower the strength. For example, thermal cycling of fiber-reinforced metals leads to interfacial voids that impair the strength but it increases the work of fracture (toughness) in the fiber direction. Large-diameter fibers and large particulates yield higher toughness at a fixed loading because of the large interfiber distance that increases the effective amount of plastic matrix zones between fibers. The effect of reinforcement volume fraction and size on mode I fracture toughness ($K_{IC}$) for a SiC-Al alloy composite is shown in Figure 9.1-51. In low-toughness matrices (e.g., NiAl), the dissipation of the energy for crack propagation will be limited, with the result that cracks originating in the fiber can actually propagate across the interface and through the surrounding matrix, leading to catastrophic failure. Fracture toughness and fatigue strength both increase with decreasing particle size.

Cyclic changes in temperature lead to thermal fatigue, which is due to a mismatch of coefficient of thermal expansion (CTE). Cyclic temperature excursions impair the integrity of the interfacial bond, although void formation in the matrix and grain boundary sliding (as in W-Cu composites) may also play a role. The elastic thermal stress is $E\Delta\alpha\Delta T$, where $\Delta\alpha$ is the CTE mismatch, $\Delta T$ is the amplitude of temperature change, and $E$ is the elastic modulus of the matrix. The thermal fatigue damage can be reduced by employing a ductile matrix with a high yield strength, or by use of stress-absorbing compliant interfacial coatings.

## Other properties

### Creep

The high-temperature creep of in situ composites such as directional solidified eutectics and monotectics, and metal-matrix composites reinforced with metallic wires as well as ceramic fibers, whiskers, and particulates (e.g., W-Ag, $Al_2O_3$-Al, SiC-Al) is reduced relative to the unreinforced matrix alloy. High-volume fractions and large aspect ratios of the reinforcement yield better creep resistance in cast and extruded composites. The extrusion of composites aligns the whiskers and short fibers along the direction of deformation and greatly reduces the fraction of misoriented whiskers. This also contributes to improved creep resistance.

In Ni-base intermetallics, improvements in creep strength are achieved by mechanical alloying of with oxides, nitrides, and carbides. For example, mechanically alloyed NiAl containing AlN particulates (and a small amount of $Y_2O_3$) exhibits better strength and creep resistance in the temperature range 1200–1400 °K than does monolithic NiAl. At very high temperatures (1873 °K),

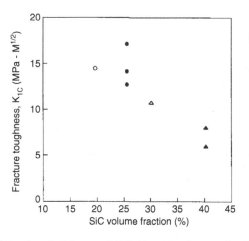 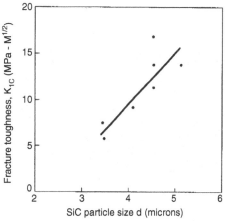

**Figure 9.1-51** (a) Fracture toughness of SiC-Al composite as a function of SiC particle volume fraction. (M. Taya and R. J. Arsenault, *Metal Matrix Composites: Thermomechanical Behavior*, Pergamon Press, 1989, p. 98). (b) Fracture toughness of SiC-Al composite as a function of the SiC particle size. The SiC volume fraction is constant at 20%. (M. Taya and R. J. Arsenault, *Metal Matrix Composites: Thermomechanical Behavior*, Pergamon Press, 1989, p. 99). Reprinted with permission from Elsevier.

creep strength deteriorates because of appreciable coarsening of AlN. Strongly bonded NiAl-alumina fiber composites also have creep resistance superior to monolithic NiAl. As a general rule, a strong interfacial bond is necessary to achieve high creep strength.

### Vibration damping

Many applications require high vibration-damping property in the use materials. In large-space structures, the dynamic response of structural components to space maneuvers needs to be controlled. Similarly, vibration damping is important in automotive engine components and electromechanical machinery. In multiphase materials, vibration energy is dissipated at the interphase boundary, and in the deformation of the matrix and the dispersed phase. Al composites have good damping properties, and the damping capacity increases with increasing additions of the second phase. Imperfectly bonded interfaces impart better damping than perfectly bonded interfaces because interfacial disbonds and microvoids provide sites for vibrational energy dissipation through frictional losses even at relatively low levels of applied stress intensity. The dislocations piled up at the fiber–matrix interface tend to become mobile under an applied stress; this provides an additional source of dissipation of vibrational energy in composites. However, a high dislocation density at the interface usually requires a good interfacial bonding, so that damping contributions from interfacial dislocations and from interfacial disbonds may not be equally effective in a given material.

### Wear and friction

Aluminum alloys are extensively used as matrix materials in light-weight composites. Aluminum is very susceptible to seizure during sliding, particularly under boundary lubrication, and even in fluid film lubrication, where localized breakdown of the lubricating oil film may occur. Graphite dispersions improve the seizure resistance of Al, and Al-graphite bearings successfully run under boundary lubrication. The sheared layers of graphite form solid lubricant films at the mating surface. Graphite, however, loses its lubricity in vacuum and dry environments, because absence of adsorbed vapors makes it difficult to shear its layers. The wear and friction data presented in Figure 9.1-52 and Figure 9.1-53 show that graphite dispersions reduce the wear volume and friction coefficient of Al. Transition metal dichalcogenides ($MoS_2$

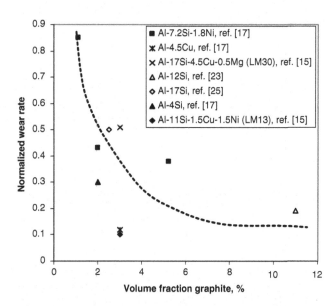

**Figure 9.1-52** Normalized wear rate of Al-graphite composites as a function of graphite volume fraction. Data on composites are normalized with respect to the wear rate of the base alloy. (Data are from studies cited in Prasad and Asthana, *Tribology Letters*, 17(3), 2004, 441–449).

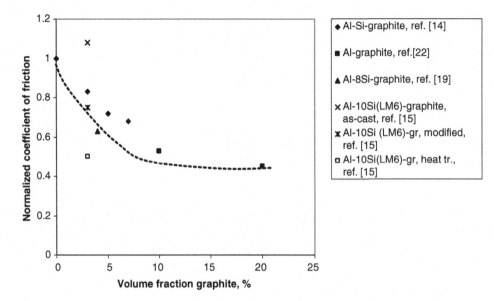

**Figure 9.1-53** Coefficient of friction of Al-graphite composite normalized with respect to the value for the base alloy as a function of the graphite volume fraction. (Data are from studies cited in Prasad and Asthana, *Tribology Letters*, 17(3), 2004, 441–449).

and $WS_2$, etc.) have lubricating properties superior to graphite in vacuum or dry environments but are thermally unstable in and react with molten Al. Low-temperature powder metallurgy and isostatic pressing techniques have been used to make Al composites containing $WS_2$ with self-lubricating behavior and low friction coefficients (0.05 to 0.10).

The dispersion of hard particles such as SiC, TiC, $Al_2O_3$, $Si_3N_4$, $SiO_2$, $B_4C$, $TiB_2$, $ZrSiO_4$, glass, fly ash, solid waste slag, and short fibers or whiskers in Al increases the resistance to abrasive wear (Figure 9.1-54) and the coefficient of friction (Figure 9.1-55) when mated against steel or cast iron. Abrasion is the removal of material from a relatively soft surface by the plowing or cutting action of hard grit particles. Test parameters (load, speed, and abrasive type, size, and shape), and whether the abrasive is free to roll or is fixed to a sheet, determine the wear mechanism and abrasion resistance. The normalized abrasion rates of metals decrease as the volume fraction of the hard phase increases, as shown in Figure 9.1-54, and significant improvements in wear resistance occur above ~20% hard phase. The coefficient of friction of Al composites decreases with increasing additions of hard particulates (Figure 9.1-55). Large

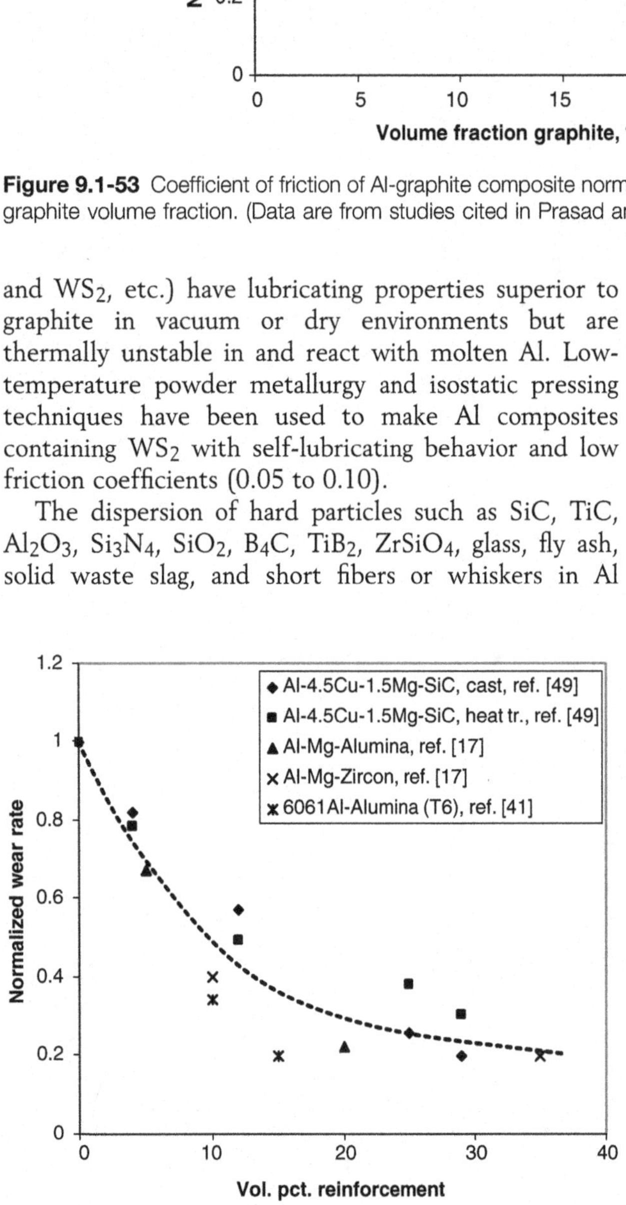

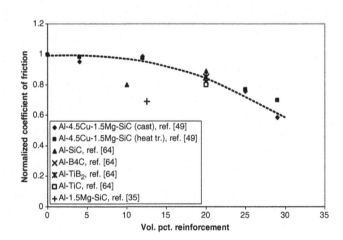

**Figure 9.1-54** Normalized wear rate of discontinuously reinforced Al composite containing hard particles as a function of the particle volume fraction (Data are from studies cited in Prasad and Asthana, *Tribology Letters*, 17(3), 2004, 441–449).

**Figure 9.1-55** Normalized coefficient of friction of Al composites containing hard particles as a function of the volume fraction of particles. (Data are from studies cited in Prasad and Asthana, *Tribology Letters*, 17(3), 2004, 441–449).

particles provide better wear resistance to Al than do fine particles. During abrasion, the ductile Al matrix is worn away by the cutting or plowing of the abrasive (or the asperities on the hard counterface) leaving the protrusions of the hard particles or fibers. Above a critical volume fraction (or interparticle spacing in relation to abrasive size), the hard-phase protrusions completely protect the matrix. The ductile matrix provides support to the hard phase, and insufficient support leads to the unsupported hard-phase edges becoming susceptible to fragmentation or pullout.

Table 9.1-7 lists some proven applications of metal-matrix composites, chiefly in the automotive industry. Diesel engine pistons containing Saffil ($Al_2O_3$) fibers are used by Toyota. Composite liners have better scuffing characteristics than conventional cast iron liners of the engine block. Al composites have superior thermal conductivity and lower density than cast iron, and this has been profitably used in disc brake rotors. Al composites save up to 60% weight when compared to cast iron. The abrasive wear rate is reduced 55–90% compared to Al, and at ~20% SiC, Al composite brake rotors have lower wear rate than cast iron.

### Thermal properties

The thermal properties such as coefficient of thermal expansion (CTE), thermal conductivity, and heat capacity are important in electronic packaging, high heat flux applications (e.g., rocket nozzles), and automotive engines. The CTE can be tailored by judicious choice of the material type and reinforcement content. For example, the addition of TiC to Al is most effective in reducing the CTE followed by alumina and SiC. Likewise, the longitudinal CTE of P-100 graphite fiber–reinforced copper composite decreases with increasing fiber content (the CTE of most commercial fibers is anisotropic, with different CTE values in the radial and axial directions). The thermal conductivity of aligned fiber composites is anisotropic and depends on both the fiber and matrix conductivities, reinforcement volume fraction, and the fiber orientation. Predictive models have been developed for both composite CTE and composite conductivity; these models generally assume perfectly bonded interfaces and a reinforcement that is evenly distributed in a defect-free matrix. The thermal conductivity along the axis of carbon fibers (grown with c-axis parallel to length) is very high, and unidirectional C-fiber–reinforced Al composites have thermal conductivity greater than Cu along fiber length over a wide range of temperatures. Likewise, in unidirectionally aligned graphite fiber–reinforced Cu, the longitudinal thermal conductivity is superior to Cu; the specific conductivity (conductivity-to-density ratio) at fiber volume fractions over 35%, is greater than both Cu and Be. The transverse conductivity is, however, less than for pure Cu and decreases with increasing fiber volume fraction. High thermal conductivity is needed to dissipate heat from the leading edges of the wings of high-speed airplanes, and from

**Table 9.1-7** Selected cast composite components with proven applications

| Manufacturer | Component and Composite |
|---|---|
| Duralcan, Martin Marietta, Lanxide | Pistons, Al-SiC$_p$ |
| Duralcan, Lanxide | Brake rotors, calipers, liners, Al-SiC$_p$ |
| GKN, Duralcan | Propeller shaft, Al-SiC$_p$ |
| Nissan | Connecting rod, Al-SiC$_w$ |
| Dow Chemical | Sprockets, pulleys, covers, Mg-SiC$_p$ |
| Toyota | Piston rings, Al-Al$_2$O$_3$ (Saffil), and Al-Boria$_w$ |
| Dont, Chrysler | Connecting rods, Al-Al$_2$O$_3$ |
| Hitachi | Current collectors, Cu-graphite |
| Associated Engineering, Inc. | Cylinders, pistons, Al-graphite |
| Martin Marietta | Pistons, connecting rods, Al-TiC$_p$ |
| Zollner | Pistons, Al-fiberfrax |
| Honda | Engine blocks, Al-Al$_2$O$_3$ – C$_f$ |
| Lotus Elite, Volkswagen | Brake rotors, Al-SiC$_p$ |
| Chrysler | Brake rotors, Al-SiC$_p$ |
| GM | Rear brake drum for EV-1, driveshaft, engine cradle, Al-SiC$_p$ |
| MC-21, Dia-Compe, Manitou | Bicycle fork brace and disk brake rotors, Al-SiC$_p$ |
| 3M | Missile fins, aircraft electrical access door, Al-Nextel$_f$ |
| Knorr-Bremse; Kobenhavn | Brake disc on ICE bogies, SiC-Al |
| Alcoa Innometalx | Multichip electronic module, Al-SiC$_p$ |
| Lanxide | PCB Heat sinks, Al-SiC$_p$ |
| Cercast | Electronic packages, Al-graphite foam |
| Textron Specialty Materials | PCB heat sinks, Al-B |

*p*: particle, *w*: whisker, *f*: fiber

high-density, high-speed integrated circuit packages for computers and in base plates of electronic equipment. Electronic packaging use requires tolerance for a small CTE mismatch to reduce the thermal stresses that arise from temperature variations. In addition, a high thermal conductivity of the materials helps in reducing temperature buildup. Table 9.1-8 summarizes the conductivity ($k$) and CTE of some metal-matrix composites and their constitutents.

The composite thermal conductivity depends on the conductivity of the constituents, and on the volume fraction, size, and the distribution of the reinforcing phase, as well as on any interfacial layers that could serve as a thermal discontinuity. In applications that require both strength and good thermal conductivity, particulate size needs to be optimized because there are conflicting requirements for strength and thermal conductivity enhancement. The strength decreases with increasing particle size, whereas the thermal conductivity in a given composite system increases with increasing particulate size. Similarly, the CTE (Figure 9.1-56) also increases with increasing particle size.

### Thermal fatigue

A composite's response to cyclic thermal load depends critically on the strength and integrity of the fiber–matrix interface. In strongly bonded (bond strength >280 MPa)

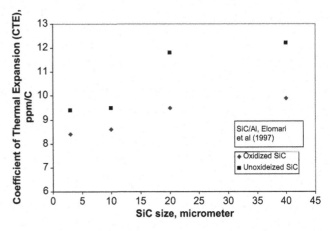

**Figure 9.1-56** The thermal expansion of SiC-Al composites containing oxidized and unoxidized SiC particles as a function of the particle size.

NiAl-sapphire composites fabricated using a "powder-cloth" process, matrix and fiber cracking after thermal cycling led to poor thermal fatigue resistance. In contrast, no matrix cracking occurred in weakly bonded composites (40–100 MPa); instead, failure occurred through interfacial debonding after thermal cycling. Ductile interfacial compliant layers can blunt cracks by localized matrix deformation. These layers must, however, bond well to both the fiber and the matrix in order to achieve fiber strengthening via load transfer to the fiber at elevated temperatures. One potential interlayer material for Ni-base composites is Mo, which is ductile and compatible with both alumina and Ni. Mo coatings (5–10 $\mu$m thick) improve the bonding in weakly bonded sapphire-NiAl composites compared to those without Mo coatings. However, during thermal cycling the interface cracks, suggesting the need for enhancing the fracture strength of Mo coating via alloying. One major drawback of Mo interlayers is their relatively poor oxidation resistance at high temperatures, which necessitates evaluation of alternative coating materials.

Composite fabrication conditions, matrix composition, and the type and volume fraction of the reinforcement influence the thermal fatigue resistance of composites. Figure 9.1-57 shows the thermal fatigue resistance of a variety of discontinuously reinforced Al-base composites produced using casting techniques. In this figure, the total length of all cracks formed on a particular surface of the test specimen is plotted as a function of the number of thermal cycles of ∼300-degree amplitude. The samples with the smallest total crack length at a fixed number of thermal cycles will be expected to have the best thermal fatigue resistance; squeeze-cast Al alloys containing $Al_2O_3$ and fly ash (a waste by-product of the electric utility industry) exhibit greatest resistance to thermal shock among the composites shown in Figure 9.1-57.

**Table 9.1-8** Thermal properties of selected composites and their constituents

| Material | $k$, W.m$^{-1}\cdot$°K$^{-1}$ | CTE, $10^{-6}\cdot$K$^{-1}$ |
|---|---|---|
| B-6061Al | | 0.65–1.87$^{\parallel}$ |
| | | 2.78–7.37$^{\perp}$ |
| BORSIC-Al | | 0.81–1.85$^{\parallel}$ |
| | | 2.68–6.08$^{\perp}$ |
| 6061Al-T6 | 166 | 3.7–8.39 |
| BORSIC | 38 | 5.0 |
| Graphite (36 vol%)/ Al6061-T6 | 246$^{\parallel}$ 76–92$^{\perp}$ | |
| Graphite (37%)/Mg AZ91C | 183$^{\parallel}$ 31–35$^{\perp}$ | |
| AZ91C | 70.9 | |
| Graphite | 396$^{\parallel}$ 0.35$^{\perp}$ | |

Note: $k$, thermal conductivity, CTE, coefficient of thermal expansion.
$^{\parallel}$, value parallel to the fiber axis.
$^{\perp}$, value perpendicular to the fiber axis.

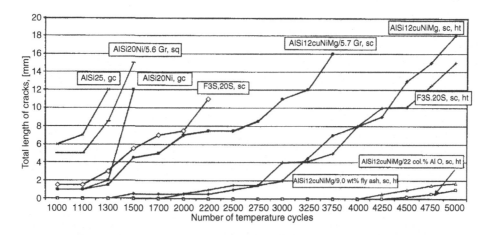

**Figure 9.1-57** The thermal fatigue resistance of some discontinuously reinforced cast Al composites as a function of the number of thermal cycles. The thermal fatigue resistance was characterized in terms of the total length of cracks on a prespecified area of the thermally cycled samples. (J. Sobczak, Z. Slawinski, N. Sobczak, P. Darlak, R. Asthana and P. K. Rohatgi, J. *Mater. Eng. and Performance*, 11(6), 2002, 595–602).

*Oxidation resistance*

The interface also influences the oxidation response of composites. In weakly bonded, powder-cloth sapphire-NiAl composites, Mo coatings slightly improve the oxidation resistance compared to uncoated composites, which have very poor oxidation resistance. In weakly bonded composites, internal oxidation occurs along the fiber–matrix interface, which in turn degrades the composite's oxidation resistance. Composites containing binders often leave residues (e.g., graphitic carbon) that contaminate and weaken the interface, and render the composite more susceptible to oxidation. Fiber-matrix reactions that form brittle interfacial compounds could weaken the interface and degrade the composite's oxidation resistance. For example, $ZrO_2$-toughened $Al_2O_3$ fibers (PRD-166) in pressure-cast $Ni_3Al$ matrices impair the oxidation resistance of the composite when the latter is annealed either in air or under partial vacuum. Whereas the as-cast interfaces are free of reaction products, subsequent vacuum-annealing has been observed to lead to precipitation of Cr-rich particles at the interface and ZrB particles in the matrix. In contrast, air-annealing leads to $ZrO_2$ particles at the interface and a thin layer of $\alpha$-$Al_2O_3$ around both the fiber and the $ZrO_2$ particles. Continued oxidation leads to $NiAl_2O_4$ formation around $ZrO_2$ particles at the interface together with a thin $Cr_2O_3$ layer. These brittle compounds deteriorate the interface and the overall oxidation resistance of the composite.

Suitable matrix alloying improves the oxidation resistance of the composite. For example, alloying NiAl with Zr or Hf significantly improves its static- as well as cyclic oxidation resistance far above conventional superalloys. In systems in which a protective oxide scale forms upon oxidation, the integrity of the scale is of importance. Thermal stresses and mechanical deformation can lead to fracture and spalling off of the scale, resulting in continued oxidation. In the case of mechanically alloyed NiAl-AlN particulate composites containing a small amount of $Y_2O_3$, oxidation resistance is good until 1400 °K, but at temperatures in excess of 1500 °K, rapid oxidation occurs in composites deformed at high strain rates; fracture of protective scale under rapid deformation is responsible for this behavior.

# References

Allison, J. E., and J. W. Jones. In S. Suresh, A. Mortensen, and A. Needlemann, eds., *Fundamentals of Metal-Matrix Composites*. London: Butterworth, 1993.

Ashby, M. F. *Acta. Metall. Mater.*, 41(5), 1993, 1313.

Asthana, R. *Solidification Processing of Reinforced Metals*. Key Engineering Materials, Trans Tech, 1998.

Chawla, K. K. In *Mater. Sci. Tech.*, vol. 13. Anheim: VCH Publ., 1993, 121.

Chawla, K. K. *Composite Materials—Science and Engineering*. New York: Springer, 1987.

Clyne, T. W., and P. J. Withers. *An Introduction to Metal-Matrix Composites*. Cambridge, UK: Cambridge University Press, 1993.

Frommeyer, G. In R. W. Cahn, and P. Haasen, eds., *Physical Metallurgy*, 3rd ed., Elsevier Sci. Publ., BV, 1983, p. 1854.

Nair, S. V., J. K. Tien, and R. C. Bates. *Int. Metals Revs.*; 30(6), 1985, 275.

Taya, M., and R. J. Arsenault. *Metal Matrix Composites—Thermomechanical Behavior*. New York: Pergamon Press, 1989.

# Section Ten

## Magnetic materials

# Chapter 10.1

# Magnetic materials

## 10.1.1 Introduction

In this chapter we study the principles and properties of magnetic materials. Our study begins with a brief review of the basic units of magnetism. This is followed by a summary of the major classifications of magnetic materials. Next, we consider the basic mechanisms of magnetism. We start at the atomic level with an analysis of an isolated single electron atom. Our analysis shows that the electron's magnetic moment is due to both its orbital motion and spin. The same analysis is then applied to an isolated multielectron atom where the spin and orbital moments of the constituent electrons couple to give a net atomic moment.

After discussing single atoms we consider a collection of atoms. Specifically, we apply classical statistical mechanics to determine the bulk magnetization of an ensemble on noninteracting atomic moments. The ensemble exhibits a bulk paramagnetic behavior. A similar analysis is applied to an ensemble of interacting atomic moments. In this case, there is a cooperative alignment of the moments (ferromagnetism) below the Curie temperature. Following this, we introduce the concepts of magnetostatic energy, magnetic anisotropy and domains, and then use these to explain the B-H (hysteresis) curves of bulk magnetization.

After describing bulk magnetization, we discuss soft and hard magnetic materials. For soft materials, we give a brief description of the most common materials and list their key physical properties. This is followed by a survey of commercially available hard materials (permanent magnets). The chapter concludes with a discussion of the magnetization and stability of permanent magnets. The topics covered in this chapter are summarized in Fig. 10.1-1.

## 10.1.2 Units

There are three systems of units that are commonly used in the magnetism. These are the CGS or Gaussian system, and two MKS or SI systems that are referred to as the Kennelly and Sommerfeld conventions, respectively. Throughout this chapter we use SI units in the Sommerfeld convention [1,2].

In the Sommerfeld convention the magnetic flux $\Phi$ is in webers (Wb), the flux density **B** is in teslas (T) or Wb/m$^2$, and both the field strength **H** and magnetization **M** are in A/m. The units for the SI (Sommerfeld) and CGS systems are as follows:

| Symbol | Description | SI | CGS |
|---|---|---|---|
| **H** | Magnetic Field Strength | A/m | Oe |
| **B** | Flux Density | Tesla | Gauss |
| **M** | Magnetization | A/m | emu/cm$^3$ |
| $\Phi$ | Flux | Webers | Maxwells |

(10.1.1)

The conversion factors for these systems are

$$\begin{aligned} 1 \text{ Oe} &= 1000/4\pi \text{ A/m} \\ 1 \text{ Gauss} &= 10^{-4} \text{ T} \\ 1 \text{ emu/cm}^3 &= 1000 \text{ A/m} \\ 1 \text{ Maxwell} &= 10^{-8} \text{ Webers} \end{aligned}$$

(10.1.2)

We can obtain an intuitive feeling for the magnitude of the fields (units) described in the foregoing text by considering the following physical examples:

**1.** Field strength **H**: Consider a straight, infinitely long wire carrying a current $i = 2\pi$A. The conductor generates a tangential field strength $H = 1$ A/m at

*Engineering Materials and Processes Desk Reference*; ISBN: 9781856175869
Copyright © 2009 Elsevier Ltd; All rights of reproduction, in any form, reserved.

# CHAPTER 10.1 Magnetic materials

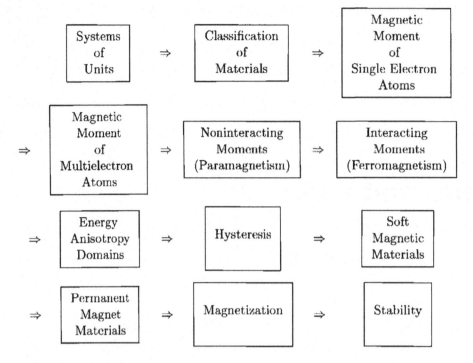

Figure 10.1-1 Organization of chapter topics.

a radial distance $r = 1$ m from its center. As another example, consider a long solenoid that has $n$ turns per meter and carries a current of $1/n$ A. It generates a field strength $H = 1$ A/m along its axis.

2. **Flux density B**: Consider an infinitely long conductor that carries a current $i = 1$ A perpendicular to an external **B**-field. A force of 1 newton will be imparted to each meter of the conductor when $B = 1$ T.

3. **Flux Φ**: Consider a single turn coil with 1 Wb of magnetic flux passing through it. One volt will be induced in the coil when the flux is uniformly reduced to zero in one second.

The fundamental element in magnetism is the magnetic dipole. This can be thought of as a pair of closely spaced magnetic poles, or equivalently as a small current loop (Fig. 10.1-2). A magnetic dipole has a magnetic dipole moment **m**. In the Sommerfeld convention **m** is measured in A · m². In the Kennelly and CGS systems it is measured in Wb· m and emu, respectively (1 emu = $4\pi \times 10^{-10}$ Wb · m). Magnetization **M** is a measure of the net magnetic dipole moment per unit volume. Specifically, it is given by

$$\mathbf{M} = \lim_{\Delta V \to 0} \frac{\sum_i \mathbf{m}_i}{\Delta V},$$

where $\sum_i \mathbf{m}_i$ is a vector sum of the dipole moments contained in the elemental volume $\Delta V$.

If a magnetic dipole is subjected to an external **B**-field, it acquires an energy

$$E = -\mathbf{m} \cdot \mathbf{B} \tag{10.1.3}$$

and experiences a torque

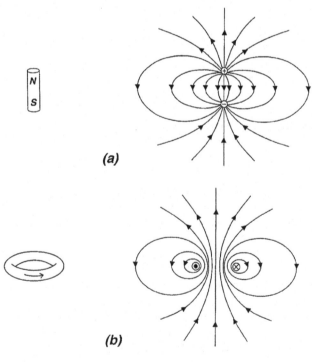

Figure 10.1-2 Magnetic dipole: (a) magnetic charge model and H-field; and (b) current loop and B-field.

$$\mathbf{T} = \mathbf{m} \times \mathbf{B}. \qquad (10.1.4)$$

In both the Kennelly and CGS systems these are given by $E = -\mathbf{m} \cdot \mathbf{H}$ and $\mathbf{T} = \mathbf{m} \times \mathbf{H}$.

In the Sommerfeld convention the fields are related by the constitutive relation

$$\mathbf{B} = \mu_0(\mathbf{H} + \mathbf{M}) \quad \text{(Sommerfeld convention)}, \qquad (10.1.5)$$

where $\mu_0 = 4\pi \times 10^{-7}$ T m/A is the permeability of free space. In the Kennelly convention, the constitutive relation is

$$\mathbf{B} = \mu_0 \mathbf{H} + \mathbf{J} \quad \text{(Kennelly convention)},$$

where $\mathbf{J}$ is called the magnetic polarization and is measured in tesla. Notice that $\mathbf{J} = \mu_0 \mathbf{M}$. The constitutive relation (10.1.5) simplifies for linear, homogeneous and isotropic media. For such materials, both $\mathbf{B}$ and $\mathbf{M}$ are proportional to $\mathbf{H}$. Specifically,

$$\mathbf{B} = \mu \mathbf{H}, \qquad (10.1.6)$$

and

$$\mathbf{M} = \chi_m \mathbf{H}, \qquad (10.1.7)$$

where $\mu$ and $\chi_m$ are the permeability and susceptibility of the material, respectively. These coefficients are related to one another. From Eqs (10.1.5), (10.1.6), and (10.1.7) we find that

$$\mu = \mu_0(\chi_m + 1), \qquad (10.1.8)$$

or

$$\chi_m = \frac{\mu}{\mu_0} - 1. \qquad (10.1.9)$$

The constitutive relations (10.1.6) and (10.1.7) need to be modified for nonlinear, inhomogeneous or anisotropic materials. A material is magnetically nonlinear if $\mu$ depends on $\mathbf{H}$, otherwise it is linear. For nonlinear materials, Eqs (10.1.6) and (10.1.7) become

$$\mathbf{B} = \mu(H)\mathbf{H},$$

and

$$\mathbf{M} = \chi_m(H)\mathbf{H},$$

respectively. A material is inhomogeneous if $\mu$ is a function of position, otherwise it is homogeneous. In inhomogeneous materials the permeability and susceptibility are functions of the coordinate variables, $\mu = \mu(x, y, z)$, $\chi_m = \chi_m(x, y, z)$. Last, a material is said to be anisotropic if $\mu$ depends on direction, otherwise it is isotropic. In anisotropic materials Eq. (10.1.6) generalizes to

$$B_x = \mu_{11}H_x + \mu_{12}H_y + \mu_{13}H_z,$$
$$B_y = \mu_{21}H_x + \mu_{22}H_y + \mu_{23}H_z,$$
$$B_z = \mu_{31}H_x + \mu_{32}H_y + \mu_{33}H_z.$$

A similar set of equations hold for Eq. (10.1.7).

## 10.1.3 Classification of materials

Magnetic materials fall into one of the following categories: diamagnetic, paramagnetic, ferromagnetic, antiferromagnetic, and ferrimagnetic (Fig. 10.1-3). Diamagnetic materials have no net atomic or molecular magnetic moment. When these materials are subjected to an applied field, atomic currents are generated that give rise to a bulk magnetization that opposes the field. Bismuth (Bi) is an example of a diamagnetic material.

Paramagnetic materials have a net magnetic moment at the atomic level, but the coupling between neighboring moments is weak. These moments tend to align with an applied field, but the degree of alignment decreases at higher temperatures due to the randomizing effects of thermal agitation.

Ferromagnetic materials have a net magnetic moment at the atomic level, but unlike paramagnetic materials there is a strong coupling between neighboring moments. This coupling gives rise to a spontaneous alignment of the moments over macroscopic regions called domains. The domains undergo further alignment when the material is subjected to an applied field.

Finally, antiferromagnetic and ferrimagnetic materials have oriented atomic moments with neighboring moments antiparallel to one another. In antiferromagnetic materials the neighboring moments are equal, and there is no net magnetic moment. In ferrimagnetic materials the neighboring moments are unequal, and there is a net magnetic moment.

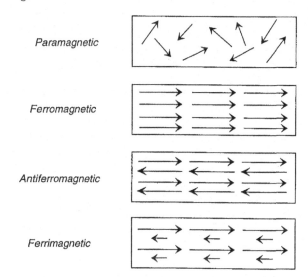

**Figure 10.1-3** Classifications of magnetic materials [16].

Rigorous treatments of these materials require a full quantum mechanical analysis that is beyond the scope of this text. We are primarily interested in ferromagnetic materials, and for our purposes it is sufficient to review some of the key principles and results. The interested reader can find more detailed presentations in numerous texts [1–7].

## 10.1.4 Atomic magnetic moments

In this section we study the magnetic moment of an isolated single electron atom. We start with a classical analysis. Consider an isolated atom with an electron of mass $m_e$ and charge $e$ moving in a circular orbit of radius $r$ with a linear velocity $u$ (angular velocity $\omega = u/r$) as shown in Fig. 10.1-4. The electron completes an orbit in $\tau = 2\pi/\omega$ s and has an orbital angular momentum $\mathbf{L}$ given by

$$\begin{aligned}\mathbf{L} &= m_e \mathbf{r} \times \mathbf{u} \\ &= m_e r^2 \boldsymbol{\omega}.\end{aligned} \quad (10.1.10)$$

The orientation of $\mathbf{L}$ is given by a right-hand rule in which the fingers of the right hand curl in the direction of the particle motion, and the thumb points in the direction of $\mathbf{L}$ (Fig. 10.1-4).

The orbiting electron defines a closed loop of current $i$ where

$$i = \frac{e}{\tau}.$$

The current $i$ defines the flow of positive charge and therefore circulates in a direction opposite to the motion of the electron (Fig. 10.1-4). The current loop gives rise to a magnetic dipole moment (Fig. 10.1-2b). In general, if a current $i$ circulates around an enclosed area $d\mathbf{s}$, it gives rise to a magnetic dipole moment $\mathbf{m}$,

$$\mathbf{m} = i d\mathbf{s} \quad \text{(magnetic moment)}. \quad (10.1.11)$$

The vector $d\mathbf{s}$ defines the orientation of $\mathbf{m}$ relative to the circulation of $i$. This is given by the right-hand rule, which states that if the fingers of the right hand follow the direction of current, then the thumb points in the direction of $d\mathbf{s}$. The magnitude of the moment is

$$m = ids, \quad (10.1.12)$$

and is measured in units of $\text{A} \cdot \text{m}^2$.

Let $\mathbf{m}_L$ denote the magnetic moment of the electron due to its orbital motion. The magnitude of this moment is given by Eq. (10.1.12)

$$\begin{aligned}m_L &= i\pi r^2 \\ &= \frac{e}{\tau}\pi r^2 \\ &= \frac{e}{2}\omega r^2.\end{aligned}$$

This can be expressed in terms of $\mathbf{L}$. Specifically, from Eq. (10.1.10), and the rules for determining the orientations of $\mathbf{m}_L$ and $\mathbf{L}$, we find that

$$\mathbf{m}_L = -\frac{e}{2m_e}\mathbf{L}. \quad (10.1.13)$$

Notice that $\mathbf{m}_L$ is antiparallel to $\mathbf{L}$ as shown in Fig. 10.1-4. Amazingly, Eq. (10.1.13) holds even at the atomic level when $\mathbf{L}$ is replaced by its quantized expression.

If the orbiting electron is subjected to an external $\mathbf{B}$-field, its moment $\mathbf{m}_L$ acquires an energy

$$\begin{aligned}E &= -\mathbf{m}_L \cdot \mathbf{B} \\ &= \frac{e}{2m_e}\mathbf{L} \cdot \mathbf{B},\end{aligned}$$

and experiences a torque

$$\begin{aligned}\mathbf{T} &= \mathbf{m}_L \times \mathbf{B} \\ &= -\frac{e}{2m_e}\mathbf{L} \times \mathbf{B}.\end{aligned}$$

As shown in Fig. 10.1-5, the torque causes $\mathbf{L}$ to precess around $\mathbf{B}$. In classical mechanics, $\mathbf{L}$ can take on a continuum of orientations relative to $\mathbf{B}$. However, when quantum effects are taken into account, the orientation of $\mathbf{L}$ is restricted to a discrete set of values as we shall see.

### 10.1.4.1 Single electron atoms

Atomic systems are governed by quantum theory with electronic states specified in terms of wavefunctions [8, 9]. The wavefunctions $\Psi$ are obtained by solving Schrödinger's equation

$$\left[-\frac{\hbar}{2m}\nabla^2 + V\right]\Psi = E\Psi,$$

where $m$ is the mass of the orbiting particle, $V$ is the potential energy, and $E$ is the total energy. The

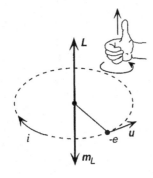

**Figure 10.1-4** Atomic system.

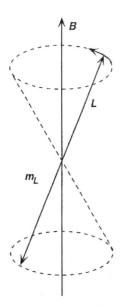

**Figure 10.1-5** Precession of a magnetic moment about an applied field.

wave-functions are indexed by a discrete set of quantum numbers ($n, l, m_l, m_s$) that distinguish the allowable energy levels, orbital configurations, etc. The principle quantum number is $n$ and this specifies the energy of a given orbit or shell. Observables such as energy and angular momentum are represented by operators that act on $\Psi$. The observed value $L_{obs}$ of an observable $\mathbf{L}$ for a given state $\Psi(n, l, m_l, m_s)$ is obtained by computing the expectation value of its operator $\hat{L}$,

$$L_{obs} = \int_{\text{all space}} \Psi^*(n, l, m_l, m_s) \hat{L} \Psi(n, l, m_l, m_s) dv,$$

(10.1.14)

where $\Psi^*$ is the complex conjugate of $\Psi$. It follows from Eq. (10.1.14) that observables are also quantized and indexed.

We now discuss the angular momentum of an atomic electron. Consider a hydrogenlike single electron atom. There are two contributions to the angular momentum: one due to orbital motion and the other due to spin. From quantum mechanics we know that for a given value of $n$ the electron's orbital angular momentum $\mathbf{L}$ has a magnitude

$$L = [l(l+1)]^{1/2}\hbar.$$

(10.1.15)

In Eq. (10.1.15), $l$ is the orbital angular momentum quantum number with allowable values $l = 0, 1, 2, \ldots, (n-1)$, and $\hbar = h/2\pi$ where $h$ is Planck's constant ($h = 6.6260755 \times 10^{-34}$ J s) [3]. Electrons with $l = 0, 1, 2, 3, \ldots$ are referred to as $s, p, d, f, \ldots$ electrons, respectively.

For example, the $n = 1$ shell has an $s$ electron, the $n = 2$ shell has $s$ and $p$ electrons, etc.

Notice that the magnitude of $\mathbf{L}$ is restricted to a discrete set of values. This is in contrast to classical theory, which allows for a continuum of values. The orientation of $\mathbf{L}$ is also restricted. Specifically, its projection onto a given axis (say, the $z$-axis) is given by

$$L_z = m_l \hbar,$$

(10.1.16)

where $m_l$ is called the magnetic quantum number, and is restricted to the values $m_l = l, (l-1), \ldots, 0, \ldots, -(l-1), -l$. Thus, for a given value of $l$ there are $2l + 1$ possible orientations of $\mathbf{L}$ relative to the axis (Fig. 10.1-6a).

An electron also has a spin angular momentum. The concept of spin was originally proposed to explain the multiplet structure of atomic spectra. Initially, it was thought that spin could be explained in terms of the electron spinning about an internal axis. However, predictions based on this model do not yield the correct magnetic moment. Therefore, spin is viewed as a purely quantum mechanical phenomenon.

The spin angular momentum $\mathbf{S}$ is quantized and has a magnitude

$$S = [s(s+1)]^{1/2}\hbar,$$

where $s = 1/2$. The projection of $\mathbf{S}$ along a given axis is also quantized with

$$S_z = m_s \hbar,$$

(10.1.17)

where $m_s = \pm 1/2$ (Fig. 10.1-6b).

The total angular momentum of the electron, which is denoted by $\mathbf{J}$, is the sum of its orbital and spin angular momenta $\mathbf{J} = \mathbf{L} + \mathbf{S}$. It has a magnitude

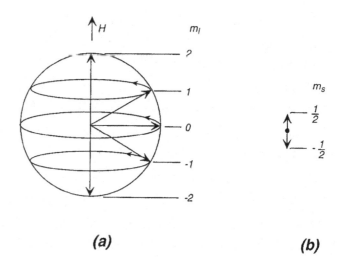

**(a)** **(b)**

**Figure 10.1-6** Spatial quantization of angular momentum: (a) orbital angular momentum ($l = 2$); and (b) spin angular momentum [4].

$$|\mathbf{J}| = [j(j+1)]^{1/2}\hbar,$$

where $j = l \pm 1/2$. The total angular momentum depends on the magnitude and relative orientations of the orbital and spin angular momenta.

We now consider the electron's magnetic moment. There are two contributions to the moment: one due to the orbital motion and the other due to spin. Recall from the classical result of Eq. (10.1.13) that the orbital moment $\mathbf{m}_L$ is proportional to $\mathbf{L}$. This same relation holds at the atomic level,

$$\mathbf{m}_L = -\frac{e}{2m_e}\mathbf{L},$$

but now the magnitude of $\mathbf{m}_L$ is quantized,

$$\begin{aligned} m_L &= -\frac{e}{2m_e}[l(l+1)]^{1/2}\hbar \\ &= -\mu_B [l(l+1)]^{1/2}, \end{aligned} \quad (10.1.18)$$

where $\mu_B = e\hbar/2m_e$ is the Bohr magneton. Notice that $m_L$ is restricted by the allowable values of $l$. If the electron is subjected to an external H-field, its orbital magnetic moment couples to $\mathbf{H}$, and the electron experiences a torque that causes $\mathbf{m}_L$ to precess around $\mathbf{H}$. However, the orientation of $\mathbf{m}_L$ relative to $\mathbf{H}$ is restricted to a discrete set of values as described in the preceding. Specifically, the projection of $m_L$ along $\mathbf{H}$ (denoted $m_{LH}$) is

$$m_{LH} = -\mu_B m_l,$$

where $m_l$ is as specified in Eq. (10.1.16). Thus, for a given value of $l$ there are $2l + 1$ possible orientations of $\mathbf{m}_L$ relative to $\mathbf{H}$ (Fig. 10.1-6a).

The magnetic moment $\mathbf{m}_s$ due to spin is given by

$$\mathbf{m}_s = -\frac{e}{m_e}\mathbf{S}.$$

It has a magnitude

$$m_s = -2\mu_B [s(s+1)]^{1/2}. \quad (10.1.19)$$

The spin also couples to an external field, and has a projection

$$m_{SH} = -2\mu_B m_s.$$

The electron's total magnetic dipole moment $\mathbf{m}_{tot}$ is the sum of its orbital and spin moments (Fig. 10.1-7):

$$\begin{aligned} \mathbf{m}_{tot} &= \mathbf{m}_L + \mathbf{m}_s \\ &= -\frac{e}{2m_e}(\mathbf{L} + 2\mathbf{S}) \end{aligned} \quad (10.1.20)$$

As $\mathbf{J} = \mathbf{L} + \mathbf{S}$, this can also be written as

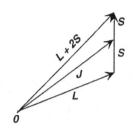

**Figure 10.1-7** Addition of angular momenta.

$$\mathbf{m}_{tot} = -\frac{e}{2m_e}(\mathbf{J} + \mathbf{S}). \quad (10.1.21)$$

Notice that $\mathbf{m}_{tot}$ is not antiparallel to $\mathbf{J}$.

From Eq. (10.1.20) we see that $\mathbf{m}_{tot}$ depends on both the magnitude and relative orientations of $\mathbf{L}$ and $\mathbf{S}$. However, $\mathbf{L}$ and $\mathbf{S}$ are neither independent of one another nor constant in time. Rather, they are coupled via a spin-orbit interaction. This interaction can be understood by considering a reference frame attached to the electron. In this frame, the nucleus appears to revolve around the electron. Therefore, the electron is subjected to a magnetic field $\mathbf{B}_n$ due to the orbiting charged nucleus. Moreover, $\mathbf{B}_n$ couples to the spin magnetic moment $\mathbf{m}_s$ with an energy $E_{so} \propto \mathbf{m}_s \cdot \mathbf{B}_n$. As $\mathbf{B}_n$ is parallel to $\mathbf{L}$ and $\mathbf{m}_s$ is antiparallel to $\mathbf{S}$, we have

$$E_{so} = \gamma \mathbf{L} \cdot \mathbf{S},$$

where $\gamma$ is the spin orbit parameter. Notice that $E_{so}$ depends on the angle between $\mathbf{S}$ and $\mathbf{L}$. Consequently, a torque is imparted perpendicular to both $\mathbf{L}$ and $\mathbf{S}$ causing them to precess. We may visualize this as a precession of $\mathbf{L}$ and $\mathbf{S}$ about the total angular momentum $\mathbf{J}$, which is constant (absent any external influences). It follows from Eq. (10.1.20) that $\mathbf{m}_{tot}$ also precesses around $\mathbf{J}$. Therefore, it is time dependent and cannot be specified as a fixed vector. However, we can define a constant average-value vector $\mathbf{m}_{ave}$. Specifically, as $\mathbf{J}$ is constant, we can specify $\mathbf{m}_{ave}$ by taking the time-averaged value of $\mathbf{m}_{tot}$ in the direction of $\mathbf{J}$ (Fig. 10.1-8). This is given by

$$\mathbf{m}_{ave} = \left(\mathbf{m}_{tot} \cdot \frac{\mathbf{J}}{|\mathbf{J}|}\right)\frac{\mathbf{J}}{|\mathbf{J}|}. \quad (10.1.22)$$

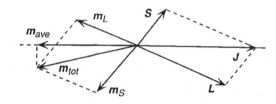

**Figure 10.1-8** Spin-orbit coupling and the orientation of $\mathbf{m}_{ave}$ [9].

Substituting Eq. (10.1.21) into Eq. (10.1.22), and making use of the fact that $\mathbf{J} = \mathbf{L} + \mathbf{S} \Rightarrow \mathbf{S} \cdot \mathbf{J} = 1/2(\mathbf{J} \cdot \mathbf{J} + \mathbf{S} \cdot \mathbf{S} - \mathbf{L} \cdot \mathbf{L})$, we get

$$\mathbf{m}_{ave} = -\left(\frac{e}{2m_e}\right) g \mathbf{J} \quad (10.1.23)$$

where

$$g = 1 + \frac{j(j+1) + s(s+1) - l(l+1)}{2j(j+1)}. \quad (10.1.24)$$

This is called the Landé $g$ factor. From Eq. (10.1.23) we see that the magnetic moment of an electron can be determined given its angular momentum. This generalizes to the case of multielectron atoms, which we study next.

### 10.1.4.2 Multielectron atoms

In this section we consider the magnetic moment of an isolated multielectron atom. When multiple electrons are present, each must occupy a different orbit as defined by the four quantum numbers $n$, $l$, $m_l$, and $m_s$. This is the Pauli exclusion principle which states that no two electrons may have the same set of quantum numbers.

The magnetic moment of a multielectron atom can be determined from the total angular momentum $\mathbf{J}_{tot}$ using a relation similar to Eq. (10.1.23). Since the atom is isolated, $\mathbf{J}_{tot}$ is a constant with a magnitude

$$\left|\mathbf{J}_{tot}\right| = [J(J+1)]^{1/2}\hbar.$$

The projection of $\mathbf{J}_{tot}$ onto a given axis is

$$J_{tot,z} = M_J \hbar,$$

where $M_J = J, J-1, J-2, \ldots, -(J-2), -(J-1), -J$. To determine $|\mathbf{J}_{tot}|$ or $J_{tot,z}$ we need to know the allowed values of $J$. These depend on the orbital and spin configurations of the constituent electrons, and the way in which their respective moments couple. There are two different modes of coupling. These are known as Russell-Saunders and spin-orbit coupling,

$$\mathbf{J}_{tot} = \begin{cases} \mathbf{L}_{tot} + \mathbf{S}_{tot} & \text{(Russell-Saunders)} \\ \text{or} \\ \sum_i (\mathbf{L}_i + \mathbf{S}_i) & \text{(spin-orbit)}. \end{cases}$$

$$(10.1.25)$$

The rule for evaluating Eq. (10.1.25) is that vectors with the strongest coupling are summed first. Russell-Saunders coupling applies when the orbital-orbital ($\mathbf{L}_i$–$\mathbf{L}_j$) and spin-spin ($\mathbf{S}_i$–$\mathbf{S}_j$) couplings between different electrons are stronger than the spin-orbit ($\mathbf{S}_i$–$\mathbf{L}_i$) coupling of each separate electron. This applies to all but the heaviest atoms. On the other hand, spin-orbit coupling is used when the spin-orbit coupling of each separate electron dominates the couplings between different electrons.

Russell-Saunders coupling applies to most magnetic atoms. In this coupling the total orbital angular momentum is the vector sum of orbital momenta of the individual electrons

$$\mathbf{L}_{tot} = \sum_i \mathbf{L}_i.$$

It has a magnitude

$$\left|\mathbf{L}_{tot}\right| = [L(L+1)]^{1/2}\hbar,$$

and its projection along a given direction (say, the $z$-axis) is

$$L_z = M_L \hbar,$$

where $M_L = L, (L-1), \ldots, 0, \ldots, -(L-1), -L$, and $M_L = \sum_i m_{Li}$.

Similarly, the total spin angular momentum is given by

$$\mathbf{S}_{tot} = \sum_i \mathbf{S}_i.$$

This has a magnitude

$$\left|\mathbf{S}_{tot}\right| = [S(S+1)]^{1/2}\hbar,$$

and a projection

$$S_z = M_s \hbar,$$

where $M_s = S, (S-1), \ldots, 0, \ldots, -(S-1), -S$, and $M_s = \sum_i m_{si}$. The net angular momentum of the electrons in a closed shell is zero,

$$\sum_{\text{closed shell}} \mathbf{L}_i = 0,$$

and

$$\sum_{\text{closed shell}} \mathbf{S}_i = 0,$$

Therefore, only the partially filled shells contribute to $\mathbf{J}_{tot}$. An example of Russell-Saunders coupling is shown in Fig. 10.1-9.

Recall that the total angular momentum is $\mathbf{J}_{tot} = \mathbf{L}_{tot} + \mathbf{S}_{tot}$. This has a magnitude

$$\left|\mathbf{J}_{tot}\right| = [J(J+1)]^{1/2}\hbar.$$

In Russell-Saunders coupling $J$ is restricted to the following values:

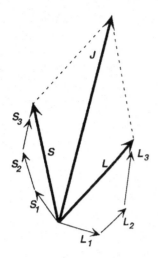

**Figure 10.1-9** Russell-Saunders coupling [9].

$$J = L+S, L+S-1, \ldots, |L-S|$$

To evaluate $J$ we need to know the ground state orbital and spin configurations of the electrons. These are determined using Hund's rules, which state that: (i) the spins $S_i$ combine to give the maximum value of S consistent with the Pauli exclusion principle; (ii) the orbital vectors $L_i$ combine to give the maximum value of $L$ compatible with the Pauli principle and condition (i); and (iii) the resultant $L_{tot}$ and $S_{tot}$ combine to form $J_{tot}$, with $J = L - S$ if the shell is less than half filled, and $J = L + S$ if it is more than half filled. When the shell is exactly half full, $L = 0$ and $J = S$

Finally, for an isolated atom $J_{tot}$ is constant. We determine an average magnetic moment of the atom $\mathbf{m}_{ave}$ by taking the projection of its magnetic moment $\mathbf{m}$ along the direction of $\mathbf{J}_{tot}$. Following an analysis similar to that for Eq. (10.1.22) we obtain

$$\left|\mathbf{m}_{ave}\right| = g\mu_B \sqrt{J(J+1)},$$

where

$$g = 1 + \frac{J(J+1) + S(S+1) - L(L+1)}{2J(J+1)}.$$

Notice that if the net orbital angular momentum is zero ($L = 0$), then $J = S$ and $g = 2$. On the other hand, if the net spin angular momentum is zero ($S = 0$), then $J = L$ and $g = 1$. The projection of $\mathbf{m}_{ave}$ along the direction of an applied H-field is denoted by $\mathbf{m}_H$

$$|\mathbf{m}_H| = g\mu_B M_J, \qquad (10.1.26)$$

where $M_J = J, J-1, J-2, \ldots, -(J-2), -(J-1), -J$. Thus, $\mathbf{m}_H$ is constrained to a discrete set of orientations relative to $\mathbf{H}$.

This completes our study of isolated atoms. In summary, we have found that the magnetic moment of the atom is due principally to the magnetic moments of its constituent electrons. These, in turn, are due to both orbital motion and spin. The magnitude of the atomic moment can be determined from the total angular momentum, and this is obtained by summing the contributions of the individual electrons using Russell-Saunders coupling, subject to Hund's rules.

## 10.1.5 Paramagnetism

Recall that paramagnetic materials consist of atoms that have a net magnetic moment, but negligible coupling between neighboring moments. In this section we study the response of such materials to an applied field. This serves as a prerequisite to our subsequent discussion of ferromagnetism.

Consider a paramagnetic specimen with $N$ noninteracting atomic magnetic moments per unit volume. We analyze its response to an applied field $\mathbf{H}_a$. Initially, we treat the moments classically assuming that they can take on a continuum of orientations. Later, we impose quantum constraints.

Let $\mathbf{m}$ denote the magnetic moment of a single atom. We seek an expression for the magnetization $\mathbf{M}$, which is the total magnetic moment per unit volume in the direction of $\mathbf{H}_a$. Each moment $\mathbf{m}$ couples to $\mathbf{H}_a$ and acquires a magnetostatic energy

$$\begin{aligned} E &= -\mu_0 \mathbf{m} \cdot \mathbf{H}_a \\ &= -\mu_0 m H_a \cos(\theta), \end{aligned} \qquad (10.1.27)$$

where $\theta$ is the angle between $\mathbf{m}$ and $\mathbf{H}_a$. We model the material as an ensemble of noninteracting atomic moments [10]. Specifically, we use Boltzmann statistics to describe the probability $p(E)$ of a moment having an energy $E$ at a temperature $T$

$$\begin{aligned} p(E) &= \exp\left(\frac{-E}{k_B T}\right) \\ &= \exp\left(\frac{\mu_0 m H_a \cos(\theta)}{k_B T}\right), \end{aligned} \qquad (10.1.27)$$

where $k_B$ is Boltzmann's constant. Physically, the alignment of $\mathbf{m}$ is opposed by thermal agitation and Eq. (10.1.28) gives the probability of $\mathbf{m}$ obtaining an angular orientation $\theta$ relative to $\mathbf{H}_a$ at temperature $T$. The magnetization is given by

$$M = \int_0^\pi m\cos(\theta) n(\theta) d\theta, \qquad (10.1.29)$$

where $m\cos(\theta)$ is the projection of $\mathbf{m}$ (with orientation $\theta$ relative to $\mathbf{H}_a$) onto the field direction and $n(\theta)\, d\theta$ is the number of atomic moments per unit volume with angular orientations between $\theta$ and $\theta + d\theta$. Specifically,

$$n(\theta)d\theta = 2\pi C \exp\left(\frac{\mu_0 m H_a \cos(\theta)}{k_B T}\right) \sin(\theta)d\theta, \tag{10.1.30}$$

where $C$ is a normalization constant determined from $\int_0^\pi n(\theta)d\theta = N$. An evaluation of Eq. (10.1.29) yields

$$M = Nm\mathcal{L}\left(\frac{\mu_0 m H_a}{k_B T}\right), \tag{10.1.31}$$

where $\mathcal{L}(\gamma)$ is the Langevin function

$$\mathcal{L}(\gamma) = \coth(\gamma) - \frac{1}{\gamma},$$

(see Fig. 10.1-10). Notice that $-1 < \mathcal{L}(\gamma) < 1$ and that

$$\lim_{\gamma \to \infty} \mathcal{L}(\gamma) = 1. \tag{10.1.32}$$

From Eqs. (10.1.31) and (10.1.32) we find that for low $T$ and high $H_a$, $M$ approaches its maximum value $Nm$, which means that the atomic moments tend towards perfect alignment in this limit.

In most cases $\gamma \ll 1$, and $\mathcal{L}(\gamma)$ can be represented as an infinite power series of the form,

$$\mathcal{L}(\gamma) = \frac{\gamma}{3} - \frac{\gamma^2}{45} + \dots .$$

For small $\gamma$ we ignore all but the first term and obtain

$$M = \frac{\mu_0 N m^2 H_a}{3 k_B T}.$$

This gives a parametric susceptibility of the form

$$\chi = \frac{M}{H_a} = \frac{\mu_0 N m^2}{3 k_B T}.$$

This is known as Curie's law.

The expression (10.1.31) follows from a classical analysis that allows for arbitrary orientations of the moments. However, from Eq. (10.1.26) we know that $\mathbf{m}$ is constrained to a discrete set of orientations relative to $\mathbf{H}_a$. We modify Eq. (10.1.29) taking this into account and obtain

$$M = N \frac{\sum_{m_J=-J}^{J} m_J g \mu_B \exp(\mu_0 m_J g \mu_B H_a / k_B T)}{\sum_{m_J=-J}^{J} \exp(\mu_0 m_J g \mu_B H_a / k_B T)}$$

This can be written as

$$M = M_0 \mathcal{B}_J\left(\frac{\mu_0 g J \mu_B H_a}{k_B T}\right), \tag{10.1.33}$$

where $M_0 = N g J \mu_B$ is the maximum possible moment in the direction of $\mathbf{H}_a$, and

$$\mathcal{B}_J(\gamma) = \frac{2J+1}{2J} \coth\left(\frac{(2J+1)\gamma}{2J}\right) - \frac{1}{2J} \coth\left(\frac{\gamma}{2J}\right) \tag{10.1.34}$$

is the Brillouin function. In the limiting case when $J \to \infty$ the allowed values and orientations of the atomic moments approach a continuum, and we recover the classical Langevin result of Eq. (10.1.31).

### 10.1.6 Ferromagnetism

Ferromagnetic materials consist of atoms with a net magnetic moment, and there is substantial coupling between neighboring moments. These materials exhibit a magnetic ordering that extends over a substantial volume and entails the cooperative orientation of a multitude of atomic moments. An early attempt to explain this phenomenon was made by Weiss, who postulated the existence of a "molecular field" $\mathbf{H}_m$ that is proportional to the magnetization

$$\mathbf{H}_m = \alpha \mathbf{M}. \tag{10.1.35}$$

In Eq. (10.1.35) $\alpha$ is the mean field constant [11].

Consider a material characterized by Eq. (10.1.35). We seek an expression for $\mathbf{M}$ and follow the same analysis as in the previous section, but now $\mathbf{H}_a = \mathbf{H}_m$. Specifically, the magnetostatic energy is given by.

$$\begin{aligned} E &= -\mu_0 \mathbf{m} \cdot \mathbf{H}_m \\ &= -\mu_0 m \alpha M \cos(\theta), \end{aligned} \tag{10.1.36}$$

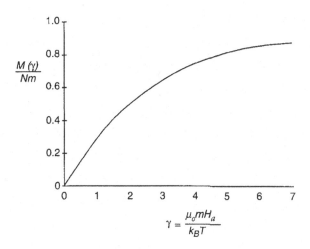

**Figure 10.1-10** Langevin function.

where $\theta$ is the angle between the local moment **m** and the bulk magnetization **M**. Substitute $H_a = \alpha M$ into Eq. (10.1.31) and obtain

$$M = Nm\mathcal{L}\left(\frac{\mu_0 m \alpha M}{k_B T}\right). \qquad (10.1.37)$$

Notice that Eq. (10.1.37) does not give $M$ explicitly in terms of $T$. Instead, graphical or self-consistent methods are used to determine $M$ for a given $T$. A careful analysis of Eq. (10.1.37) shows that spontaneous magnetization occurs at low temperatures, and the material becomes saturated with $M = M_0 = Nm$. However, as $T$ increases, the magnetization gradually decreases until $T$ approaches the Curie temperature,

$$\boxed{T_c = \frac{\mu_0 N \alpha m^2}{3 k_B}} \quad \text{(Curie temperature)}.$$

Near the Curie temperature, the magnetization drops dramatically to zero, and as shown in Fig. 10.1-11 the material becomes paramagnetic. Below the Curie temperature, the material divides into numerous distinct localized regions called domains. Each domain is uniformly magnetized to its saturation value. However, the bulk magnetization can be zero if the domains are randomly oriented. Domains are discussed in more detail in Section 10.1.10.

If we subject a ferromagnetic material to an applied field $H_a$, the same analysis as in the preceding applies, but now $H_m \to H_a + H_m$. The magnetostatic energy is given by

$$E = -\mu_0 \mathbf{m} \cdot (\mathbf{H}_a + \mathbf{H}_m)$$
$$= -\mu_0 m (H_a + \alpha M) \cos(\theta)$$

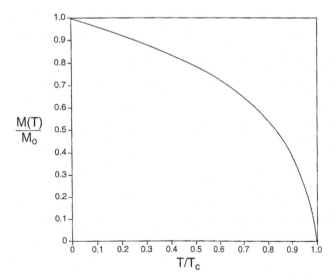

**Figure 10.1-11** Spontaneous magnetization vs temperature.

and we obtain

$$M = Nm\mathcal{L}\left(\frac{\mu_0 m (H_a + \alpha M)}{k_B T}\right). \qquad (10.1.38)$$

In ferromagnetic materials $\alpha M \gg H_a$ and, therefore, the applied field does not alter the local magnetization appreciably. That is, $H_a$ does not alter the magnetization within a domain. However, an applied field does alter the bulk magnetization of a specimen by altering the orientation and structure of individual domains. This is discussed in Section 10.1.11

Expression (10.1.38) is based on classical theory in which the magnetic moments can take on a full range of orientations. If we apply quantum theory, the orientations of the moments are restricted, and we obtain an expression similar to Eq. (10.1.33) with $H_a \to H_a + \alpha M_s$,

$$M_s = M_0 \mathcal{B}_J\left(\frac{\mu_0 g J \mu_B (H_a + \alpha M)}{k_B T}\right), \qquad (10.1.39)$$

where $M_0 = NgJ\mu_B$, and $\mathcal{B}_J(\gamma)$ is the Brillouin function (10.1.34). Because $\alpha M \gg H_a$, Eq. (10.1.39) reduces to

$$M_s = M_0 \mathcal{B}_J\left(\frac{\mu_0 g J \mu_B \alpha M_s}{k_B T}\right). \qquad (10.1.40)$$

As $J \to \infty$ the allowed values and orientations of the atomic moments approach a continuum, and Eq. (10.1.40) reduces to the classical result seen in Eq. (10.1.37).

Although the Weiss mean field theory explained some key features of ferromagnetism such as spontaneous magnetization below the Curie temperature, it nevertheless lacked a rigorous foundation. This would come later with the discovery of quantum mechanics [12]. Specifically, Heisenberg postulated the existence of an exchange interaction between spin angular momenta. This is characterized by a potential energy (exchange energy) of the form,

$$w_{ij} = -2J_{ex}\mathbf{S}_i \cdot \mathbf{S}_j$$
$$= -2J_{ex}S_i S_j \cos(\phi) \quad \text{(exchange energy)}, \qquad (10.1.41)$$

where $J_{ex}$ is called the exchange integral, $\mathbf{S}_i$ and $\mathbf{S}_j$ are the coupled spin angular momenta, and $\phi$ is the angle between $\mathbf{S}_i$ and $\mathbf{S}_j$. Notice that if $J_{ex} > 0$, the exchange energy is minimum when the spins are parallel ($\phi = 0$). However, if $J_{ex} < 0$, the exchange energy is a minimum when the spins are antiparallel ($\phi = \pi$). Therefore $J_{ex} > 0$ gives rise to ferromagnetic behavior, and $J_{ex} < 0$ gives rise to antiferromagnetic behavior. The exchange force falls off rapidly and is rarely effective beyond the nearest neighbors.

The forementioned analysis describes the local ordering (spontaneous alignment) of atomic moments and the temperature dependence of this ordering. However, it does not describe the hysteretic behavior of bulk materials. For this, we need to understand magnetic domains and their behavior in an external field. In the next few sections we introduce the concepts of magnetostatic energy, demagnetizing field, and magnetic anisotropy. These are key to understanding domains. Following this, we study domains specifically and then use them to describe hysteresis.

## 10.1.7 Magnetostatic energy

In this section we discuss the magnetostatic self-energy of a magnetized specimen, and the energy it acquires in an external field. Consider a specimen of volume $V$ with a fixed magnetization $\mathbf{M}$. It possesses a magnetostatic self-energy that is given by

$$W_s = -\frac{\mu_0}{2}\int_v \mathbf{M}\cdot \mathbf{H}_M\, dv \text{ (self-energy)}, \quad (10.1.42)$$

where $\mathbf{H}_M$ is the field in the specimen due to $\mathbf{M}$ (i.e., due to the continuum of dipole moments of which the specimen is composed [3]). The self-energy equation (10.1.42) can be derived by considering the energy required to assemble a continuum of dipole moments in the absence of an applied field [6]. Notice that the energy density associated with $W_s$ is

$$w_s = -\frac{\mu_0}{2}\mathbf{M}\cdot\mathbf{H}_M. \quad (10.1.43)$$

The self-energy equation (10.1.42) plays an important role in the formation of magnetic domains. We discuss this more in Section 10.1.10

If a magnetized specimen is subjected to an applied field $\mathbf{H}_a$, it acquires a potential energy

$$W_a = -\mu_0\int_v \mathbf{M}\cdot\mathbf{H}_a dv. \quad (10.1.44)$$

This can be viewed as the work required to move the specimen from an environment with zero field to a region permeated by $\mathbf{H}_a$. The expression (10.1.44) can be used to determine the force or torque imparted to a magnetized body by an external field. From Eq. (10.1.44) we see that the energy density due to the coupling of $\mathbf{M}$ to $\mathbf{H}_a$ is given by

$$w_a = -\mu_0\mathbf{M}\cdot\mathbf{H}_a$$
$$= -\mu_0 M H_a \cos(\theta),$$

where $\theta$ is the angle between $\mathbf{M}$ and $\mathbf{H}_a$.

## 10.1.8 Demagnetization field

When a specimen is magnetized, a self-field develops within it that opposes the magnetizing field. This is called the demagnetization field. We study this in some detail because it plays an important role in the magnetization process.

Consider a uniformly magnetized specimen with a volume $V$ and a surface $S$. Its magnetization $\mathbf{M}$ gives rise to surface poles. These, in turn, give rise to a demagnetization field $\mathbf{H}_d$ within the specimen. The field $\mathbf{H}_d$ is proportional to $\mathbf{M}$, but in the opposite direction. For example, along the $x$-axis we have

$$H_{dx} = -N_x M_x, \quad (10.1.46)$$

where $H_{dx}$ and $M_x$ are the $x$-components of $\mathbf{H}_d$ and $\mathbf{M}$, respectively, and $N_x$ is the demagnetizing factor along the axis. If the specimen is subjected to an applied field $\mathbf{H}_a$, then $\mathbf{H}$ inside the specimen (denoted $\mathbf{H}_{in}$) is the vector sum of $\mathbf{H}_a$ and $\mathbf{H}_d$,

$$\mathbf{H}_{in} = \mathbf{H}_a + \mathbf{H}_d. \quad (10.1.47)$$

For example, consider the bar magnet shown in Fig. 10.1-12. Outside the magnet $\mathbf{H}$ and $\mathbf{B}$ are in the same direction (Fig. 10.1-12a,b). However, inside the magnet $\mathbf{H} = \mathbf{H}_d$, which opposes both $\mathbf{M}$ and $\mathbf{B}$ (Fig. 10.1-12c)

Demagnetizing factors depend on the permeability and shape of a specimen and are difficult to obtain in closed-form. However, analytical expressions have been derived for ellipsoidal shapes. Ellipsoids have three demagnetizing factors $N_a$, $N_b$, and $N_c$ corresponding to the three orthogonal axes with lengths $a$, $b$, and $c$, respectively (Fig. 10.1-13a). These factors satisfy the following relation,

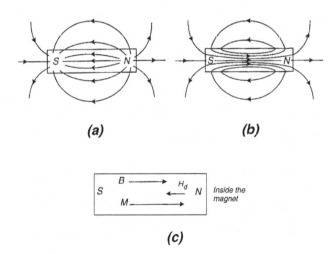

**Figure 10.1-12** Field distribution of a bar magnet: (a) H-field; (b) B-field; and (c) B, H, and M inside the magnet.

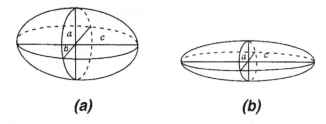

**Figure 10.1-13** Ellipsoidal shapes: (a) general ellipsoid; and (b) prolate ellipsoid.

$$N_a + N_b + N_c = 1 \quad \text{(ellipsoid)}. \tag{10.1.48}$$

For the special case of a sphere where $a = b = c$ we have $N_a = N_b = N_c = 1/3$ and

$$\mathbf{H}_d = -\frac{1}{3}\mathbf{M} \quad \text{(sphere)}. \tag{10.1.49}$$

Another shape of interest is the prolate spheroid with a semimajor axis $c$ and semiminor axis $a$ (i.e., $a = b \ll c$) (Fig. 10.1-13b). The demagnetization factor along the $c$-axis is

$$N_c = \frac{1}{k^2 - 1}\left[\frac{k}{\sqrt{k^2 - 1}} \ln\left(k + \sqrt{k^2 - 1}\right) - 1\right],$$

where $k = c/a$ [2]. For a long cylinder that is magnetized along its axis, the demagnetizing factors can be approximated by $N_c = 0$, and $N_a = N_b = 1/2$, where the $c$-direction is along the axis of the cylinder.

Finally, consider a magnetized specimen with $\mathbf{H}_d = -N\mathbf{M}$. Its self-energy is given by Eq. (10.1.42) with $\mathbf{H}_M = \mathbf{H}_d$. Specifically, we have

$$W_s = -\frac{\mu_0}{2}\int_v \mathbf{M} \cdot \mathbf{H}_d \, dv$$

$$= \frac{\mu_0 N}{2}\int_v M^2 \, dv. \tag{10.1.50}$$

If $M$ is constant throughout $V$, then Eq. (10.1.50) reduces to

$$W_s = \frac{\mu_0 N M^2 V}{2}.$$

We demonstrate some of these principles in the following example.

**Example 10.1.8.1** *Determine the magnetization and dipole moment of a sphere of linear isotropic material in free space that is subjected in an external field $\mathbf{H}_0$ (Fig. 10.1-14).*

**Solution 10.1.8.1** *Let $\mu$ and $\chi_m$ denote the permeability and susceptibility of the sphere, and let $R$ denote its radius. The field inside the sphere is*

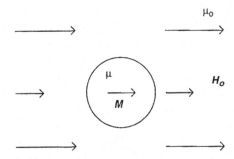

**Figure 10.1-14** Sphere with linear permeability in an external field.

$$\mathbf{H}_{in} = \mathbf{H}_0 + \mathbf{H}_d$$
$$= \mathbf{H}_0 - \frac{1}{3}\mathbf{M}.$$

*Notice that we have used Eq. (10.1.49). Recall from Eq. (10.1.7) that*

$$\mathbf{M} = \chi_m \mathbf{H}_{in}. \tag{10.1.51}$$

*Therefore,*

$$\mathbf{H}_{in} = \frac{\mathbf{H}_0}{1 + \chi_m/3}.$$

*As $\chi_m = \mu/\mu_0 - 1$ we have*

$$\mathbf{H}_{in} = \frac{3\mathbf{H}_0}{2 + \mu/\mu_0}. \tag{10.1.52}$$

*The magnetization inside the sphere is obtained from Eq. (10.1.51),*

$$\mathbf{M} = \frac{3(\mu - \mu_0)}{(\mu + 2\mu_0)}\mathbf{H}_0. \tag{10.1.53}$$

*The dipole moment $\mathbf{m}$ is*

$$\mathbf{m} = \text{Volume} \times \mathbf{M}$$
$$= \frac{4}{3}\pi R^3 \mathbf{M}.$$

*Therefore, we find that*

$$\mathbf{m} = \frac{4\pi R^3(\mu - \mu_0)}{(\mu + 2\mu_0)}\mathbf{H}_0. \tag{10.1.54}$$

## 10.1.9 Anisotropy

A material with magnetic anisotropy is one in which the magnetic properties are different in different directions. The principal classifications of anisotropy are as follows:

1. magnetocrystalline anisotropy;
2. shape anisotropy;
3. stress anisotropy; and
4. exchange anisotropy.

A thorough discussion of these can be found in Cullity [5]. In this section we review magnetocrystalline and shape anisotropy. These play an important role in the magnetization process.

### 10.1.9.1 Magnetocrystalline anisotropy

Magnetocrystalline anisotropy is an intrinsic property of a material in which the magnetization favors preferred directions (easy axes). For example, iron (Fe) has a body-centered cubic lattice structure, and there are six easy, equally preferred directions of magnetization: [0,0,1]; [0,1,0]; [1,0,0]; [0,0,−1]; [0,−1,0]; and [−1,0,0] (Miller indices) (Fig. 10.1-15). The preferred alignment within a material constitutes a lower energy state and is primarily due to spin-orbit coupling. Specifically, atomic electrons couple to the crystal lattice in such a way that their orbital moments tend to align along the crystallographic axes, and because of spin-orbit coupling the spin moments tend to align as well.

The most elementary form of magnetocrystalline anisotropy is uniaxial anisotropy. If a specimen with uniaxial anisotropy is polarized at an angle $\theta$ relative to its easy axis, the anisotropy energy density is

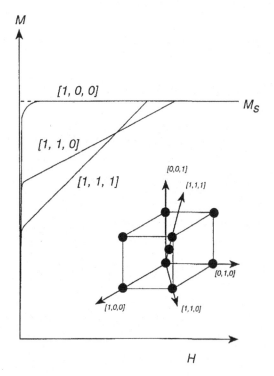

**Figure 10.1-15** Crystal structure of iron and magnetization curves along various axes.

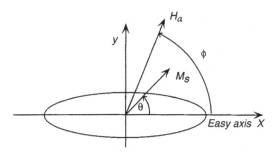

**Figure 10.1-16** Specimen with uniaxial anisotropy.

$$w_{ma} \approx K \sin^2(\theta) \quad (J/m^3). \tag{10.1.55}$$

Notice that energy minima occur when $\theta = 0°$ and $180°$.

Consider the response of a single domain specimen with uniaxial anisotropy when it is subjected to an applied field $\mathbf{H}_a$. Assume that $\mathbf{H}_a$ is directed at an angle $\phi$ with respect to the easy axis (Fig. 10.1-16). We analyze the behavior of this specimen following the presentation of Stoner and Wohlfarth [13]. Assume that the specimen is magnetized to saturation with $\mathbf{M}$ at an angle $\theta$ with respect to the easy axis. Further assume that demagnetization field $\mathbf{H}_d$ is negligible. Therefore, inside the specimen $\mathbf{H}_{in} = \mathbf{H}_a$ in accordance with Eq. (10.1.47). As the specimen constitutes a single domain, its energy density $w(\theta)$ consists of two parts: the magnetostatic energy density due to the coupling of $\mathbf{M}$ to $\mathbf{H}_a$ as given by Eq. (10.1.45) and the magnetocrystalline anisotropy energy density $w_{ma}$ as given by Eq. (10.1.55). The total energy density is

$$\begin{aligned} w(\theta) &= K \sin^2(\theta) - \mu_0 M_s H_a \cos(\phi - \theta) \\ &= K \sin^2(\theta) - \mu_0 M_s [H_x \cos(\theta) + H_y \cos(\theta)], \end{aligned} \tag{10.1.56}$$

where $H_x = H_a \cos(\phi)$ and $H_y = H_a \sin(\phi)$. We identify the values of $\theta$ that represent energy minima using the second derivative test where $\partial w(\theta)/\partial \theta = 0$ identifies the critical points, and $\partial^2 w(\theta)/\partial \theta^2 > 0$ confirms a minimum. We consider the case where the field is applied along the easy axis ($H_y = 0$) and obtain

$$\frac{\partial w(\theta)}{\partial \theta} = 2K \sin(\theta) \cos(\theta) + \mu_0 M_s H_x \sin(\theta) = 0, \tag{10.1.57}$$

and

$$\frac{\partial^2 w(\theta)}{\partial \theta^2} = 2K \cos(2\theta) + \mu_0 M_s H_x \cos(\theta) > 0. \tag{10.1.58}$$

Notice that minima occur when $\theta = 0$ or $\pi$. When $\theta = 0$ a minimum occurs when

$$-\frac{2K}{\mu_0 M_s} < H_x, \quad (10.1.59)$$

and we have $M_x = M_s \cos(\theta) = M_s$. When $\theta = \pi$ a minimum occurs when

$$H_x < \frac{2K}{\mu_0 M_s} \quad (10.1.60)$$

and we have $M_x = -M_s$. The conditions (10.1.59) and (10.1.60) identify the intrinsic coercivity $H_{ci}$ (or switching field) of the specimen. Specifically, $H_{ci}$ is given by

$$H_{ci} \equiv \frac{2K}{\mu_0 M_s}.$$

The intrinsic coercivity is an important property of permanent magnet materials. It gives a measure of the field required to magnetize and demagnetize a specimen (Section 10.1.19).

We apply these results to the behavior of a single domain specimen with uniaxial anisotropy. Assume that the specimen is initially magnetized along its easy axis with $\theta = 0$ ($\mathbf{M} = M_s \hat{\mathbf{x}}$) (Fig. 10.1-17a). We apply a reversal field in the direction $\theta = \pi$ ($H_x$ negative). As we gradually increase the field (in a negative sense), the specimen remains polarized along its easy axis (at an energy minimum) as long as Eq. (10.1.59) is satisfied (i.e., as long as $-H_{ci} < H_x < 0$). When the field increases to the point where $H_x = -H_{ci}$ then Eq. (10.1.59) is violated and $\mathbf{M}$ flips orientation to $\theta = \pi$ ($\mathbf{M} = -M_s \hat{\mathbf{x}}$). When this happens Eq. (10.1.60) is satisfied and the specimen is once again in a state of minimum energy. The specimen stays in this state as $H_x$ becomes more negative.

Next, we gradually decrease $H_x$ (making it less negative). The specimen stays with $\mathbf{M} = -M_s \hat{\mathbf{x}}$ for all negative field values. Eventually, $H_x$ reaches zero. We then gradually increase $H_x$ from zero until we reach $H_x = H_{ci}$. At this point, Eq. (10.1.60) is violated and $\mathbf{M}$ flips orientation to $\theta = 0$ ($\mathbf{M} = M_s \hat{\mathbf{x}}$) in order to satisfy Eq. (10.1.59). Once this occurs the specimen is again at an energy minimum. The specimen remains in this state for any further increase $H_x$.

This entire sequence of events can be summarized in graphical form by plotting $M$ vs $H$ (Fig. 10.1-17a). Notice that this plot has the form of a square loop. In particular, the specimen starts out with $M = M_s$. When $H$ is reversed and reaches $H = -H_{ci}$, the magnetization abruptly reverses direction to $M = -M_s$. When $H$ is reversed again and increases to $H = H_{ci}$, the magnetization abruptly reverses back to its original value $M = M_s$. Thus, except for the two points $H = \pm H_{ci}$, the magnetization has a constant value $M = \pm M_s$.

It is instructive to plot $B$ vs $H$ for this specimen (Fig. 10.1-17b). We obtain $B$ from the constitutive relation

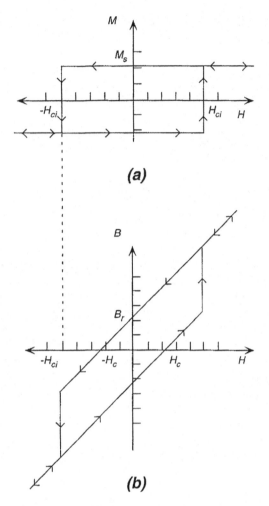

**Figure 10.1-17** Response of a specimen with uniaxial anisotropy: (a) $M$ vs $H$; and (b) $B$ vs $H$.

$$B = \mu_0 H \pm \mu_0 M_s, \quad (10.1.61)$$

where the $\pm$ sign takes into account the two possible orientations of $M$. Because $M_s$ is a constant, the plot of $B$ vs $H$ will be a straight line with a slope $\mu_0$ and with discontinuities at the points $H = \pm H_{ci}$. We set $H = 0$ in Eq. (10.1.61) and find that the curve crosses the $B$-axis at $B_r = \pm \mu_0 M_s$. The value $B_r = \mu_0 M_s$ is called remanent flux density or *remanence*. Similarly, we call the value of $H$ for which $B = 0$ the coercive force or *coercivity*. This is the value of $H$ required to reduce $B$ to zero and is denoted by $H_c$. If $H_{ci} > M_s$, then from Eq. (10.1.61) we find that $H_c = M_s$. When this is the case, the demagnetization curve is linear throughout the second quadrant as shown in Fig. 10.1-17b. This means that a weaker field is needed to reduce $B$ to zero than to reverse $M$, that is, $H_c < H_{ci}$. On the other hand, if $H_{ci} < M_s$, then $H_c = H_{ci}$ and the demagnetization curve is linear over only a portion of the second quadrant. This occurs when shape anisotropy is dominant.

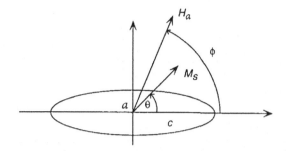

**Figure 10.1-18** Specimen with shape anisotropy.

## 10.1.9.2 Shape anisotropy

Shape anisotropy is not an intrinsic property of a material. Rather, it is due to a geometrically induced directional dependency of the demagnetization field $\mathbf{H}_d$ within the specimen. For example, consider a specimen in the shape of a prolate spheroid with a semimajor axis $c$ and semiminor axis $a$ (Fig. 10.1.18). Assume that the specimen is subjected to an applied field $\mathbf{H}_a$ at an angle $\theta$ with respect to the $c$-axis. Further assume that the specimen is uniformly polarized to a level $M_s$ with $\mathbf{M}$ at an angle $\theta$ with respect to the $c$-axis. The field inside the specimen is $\mathbf{H}_{in} = \mathbf{H}_a + \mathbf{H}_d$. The energy density $\omega(\theta)$ consists of the self-energy density $\omega_s$ as given by Eq. (10.1.43) with $\mathbf{H}_M = \mathbf{H}_d$, and the magnetostatic energy density due to the coupling of $\mathbf{M}$ to $\mathbf{H}_a$ as given by Eq. (10.1.45). Thus, the total energy density is

$$\omega(\theta) = -\frac{1}{2}\mu_0 \mathbf{M} \cdot \mathbf{H}_d - \mu_0 \mathbf{M} \cdot \mathbf{H}_a$$
$$= \frac{\mu_0}{2} M_s^2 [N_a \sin^2(\theta) + N_c \cos^2(\theta)]$$
$$\quad - \mu_0 M_s H_a \cos(\phi - \theta)$$
$$= \frac{\mu_0}{2} M_s^2 N_c + K_s \sin^2(\theta)$$
$$\quad - \mu_0 M_s [H_x \cos(\theta) + H_y \cos(\theta)], \quad (10.1.62)$$

where

$$K_s = \frac{\mu_0 M_s^2}{2}(N_a - N_c). \quad (10.1.63)$$

Notice that except for the leading constant term, Eq. (10.1.62) is formally similar to Eq. (10.1.56) of Section 10.1.9.1. We follow the same energy minimization analysis presented there except that here the $c$-axis plays the role of the easy axis. We find that the intrinsic coercivity (switching field) is given by

$$H_{ci} \equiv \frac{2K_s}{\mu_0 M_s}$$
$$= M_s(N_a - N_c).$$

From Eq. (10.1.48), and the fact that $N_a = N_b$ for a prolate spheroid, we have $2N_a + N_c = 1$. Therefore,

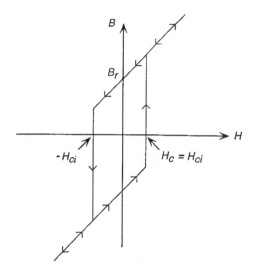

**Figure 10.1-19** The $B$-$H$ loop for a specimen with shape anisotropy.

$$|H_{ci}| \leq M_s \quad \text{(shape anisotropy)}. \quad (10.1.65)$$

From the discussion following Eq. (10.1.61) we know that in this case, $H_c = H_{ci}$, and the demagnetization curve is linear over only a portion of the second quadrant as shown in Fig. 10.1-19.

## 10.1.10 Domains

We have seen that the coupling of atomic moments in ferromagnetic materials gives rise to a spontaneous alignment of the moments below the Curie temperature (Section 10.1.6). Thus, one would expect ferromagnetic materials to be magnetically saturated at room temperature. However, we observe that these materials are often unmagnetized at a macroscopic level. This can be explained by the concept of magnetic domains [1,11].

Domains typically contain $10^{12}$–$10^{15}$ atoms. The atomic moments within a domain are subject to a magnetocrystalline anisotropy that favors spontaneous alignment along certain preferred crystallographic axes (Section 10.1.9). Thus, the moments within a domain align parallel to one another and the domain represents a local region of magnetic saturation.

A macroscopic or bulk sample of material consists of a multitude of domains that vary in size, shape, and orientation. The magnetization of the sample is defined by the collective structure and orientation of all its domains. For example, it is unmagnetized if the domains are randomly oriented, and saturated if the domains are uniformly aligned. This explains the apparent contradiction between spontaneous ferromagnetic alignment of atomic moments and the existence of macroscopically unmagnetized ferromagnetic materials.

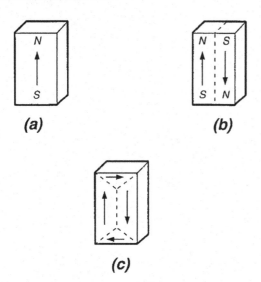

**Figure 10.1-20** Magnetized material: (a) high energy state; (b) intermediate energy state; and (c) low energy state

In homogeneous materials, domains form in such a way as to minimize the total energy of the specimen. For example, consider the block of material shown in Fig. 10.1-20. The magnetostatic energy of this block is highest in configuration a, and lowest in configuration c. In inhomogeneous materials the domain structure is affected by inhomogeneities such as inclusions, dislocations, crystal boundaries, internal stress, etc. Inclusions are regions with different magnetic properties than the surrounding material, and dislocations are line imperfections in the crystal lattice.

Neighboring domains are separated by a transition layer called a domain wall (Fig. 10.1-21a). These are also called Bloch walls since Bloch was the first to study their properties [14]. A domain wall has a finite thickness and an energy associated with it. The wall thickness and energy follow from an analysis of two competing effects: magnetocrystalline anisotropy that favors spin alignment along a preferred axis (thinner wall) and exchange interactions that favor parallel alignment of neighboring spins (thicker wall). The orientation of the spins within a wall varies gradually across its width so that the spins at either side of the wall are aligned with the respective adjacent domains. For example, if two neighboring domains are polarized in opposite directions, the spins in the wall between them rotate 180° across the wall from one side to the other (Fig. 10.1-21a).

Domain walls can move in response to an applied field. For example, if an external field is applied parallel to the magnetization in a domain, the spins within the wall experience a torque that tends to rotate them toward the field direction. As the spins rotate, the wall effectively moves with the aligned domain expanding and antialigned domain contracting (Fig. 10.1-21b).

The wall motion can be impeded by the forementioned listed inhomogeneities. For example, an inclusion lowers the energy of a domain wall when they intersect. Thus, a domain wall will tend to "stick" to an inclusion when it contacts it. Similarly, localized regions of inhomogeneous strain (which can be caused by defects and dislocations) give rise to local energy barriers that impede the motion of a domain wall. The energy barriers are caused by the interaction of the local magnetostriction with the magnetic moments of the wall. The various impediments to the movement of a domain wall tend to make its motion jerky and discontinuous. This effect is named after Barkhausen who was the first to observe it. Domain wall motion plays an important role in the magnetization process. We discuss this next.

## 10.1.11 Hysteresis

For the analysis and design of permanent magnet applications we need the macroscopic or bulk magnetic properties of a material. These are expressed in graphical form by plotting the magnetic induction **B** or the magnetization **M** as a function of **H**. Either of these plots can be obtained from the other using the constitutive relation $\mathbf{B} = \mu_0(\mathbf{H} + \mathbf{M})$.

A typical $B$ vs $H$ plot is shown in Fig. 10.1-22. This plot is nonlinear and multivalued, reflecting the fact that the response of a material depends on its prior state of magnetization. The $B$-$H$ plot is called a hysteresis loop. The word hysteresis is derived from the Greek word meaning "to lag". The energy expended in traversing a hysteresis loop is equal to the area of the loop. Detailed discussions of this phenomenon have been given by Jiles

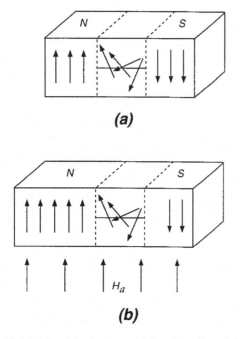

**Figure 10.1-21** Domain structure: (a) domain wall; and (b) domain wall movement in an applied field.

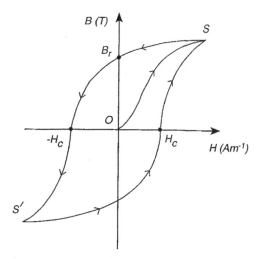

**Figure 10.1-22** Hysteresis loop.

[1] and Mayergoyz [15]. For our purposes, it suffices to review the magnetization cycle plotted in Fig. 10.1-22.

Consider a specimen with an isotropic distribution of preferred axes orientations (magnetocrystalline anisotropy). Initially, it is unmagnetized with its domains isotropically oriented as shown at point O in Fig. 10.1-23. We apply an **H**-field, and as $H$ increases from zero in the plus direction (weak field), the domains aligned with $H$ grow, and the antialigned domains shrink due to domain wall motion. As $H$ increases further (moderate field), the magnetic moments in the remaining unaligned domains rotate (or flip) into alignment along preferred axes that are in the general direction of $H$. For sufficiently high $H$, the magnetic moments along the preferred axes in domains that are not strictly aligned with $H$ rotate away from these axes and align with $H$ (point $S$) (Figs 10.1-22 and 10.1-23). At this point the material is saturated, and has achieved its saturation induction $B_s$. The segment $OS$ is called the initial magnetization curve.

As $H$ decreases from its saturation value, the moments along preferred axes strictly aligned with $H$ retain their orientation. However, the moments in domains with easy axes offset from $H$, rotate back towards the nearest easy axis, away from $H$. This results in a distribution of domain alignments about the plus direction. Even when $H$ is reduced to zero, there is still a net alignment in the plus direction, and the specimen exhibits a remanent (residual) induction $B_r$ (Fig. 10.1-23).

Now, as $H$ is reversed to the negative direction, the moments in domains oriented along the plus direction are first to rotate (or flip) into alignment along the negative direction. As $H$ increases further (in the negative direction), a sufficient number of domains obtain a predominant negative orientation, thereby offsetting the magnetization of the remaining positively oriented domains so that the net induction in the specimen is zero, $B = 0$. The value $H$ at which this occurs is called the coercivity and is denoted $H_c$, $H = -H_c$. Notice that although $B = 0$ at this point, the distribution of domain orientations is not isotropic as it was in the initial demagnetized state (Fig. 10.1-23).

As $H$ increases beyond $-H_c$ (in the negative direction), a progressively larger fraction of the domains obtain a predominant negative orientation until saturation is reached (point $S'$). From this point, $H$ is decreased to zero and a fraction of the domains reverse their orientation, resulting in a remanent induction $-B_r$. As $H$ increases from zero in a positive sense, a sufficient number of domains obtain a predominant positive orientation, thereby offsetting the magnetization of the remaining negatively oriented domains so that the net induction in the specimen is zero. This occurs when $H = H_c$. As $H$ increases further, the domains once again align themselves with $H$ until saturation is achieved. Figure 10.1-22 is called the major hysteresis loop. There are also minor hysteresis loops as shown in Fig. 10.1-24. These are formed by cycling $H$ within the range $-H_c < H < H_c$.

This section completes our study of the principles of magnetism. In the remaining sections we discuss various magnetic materials.

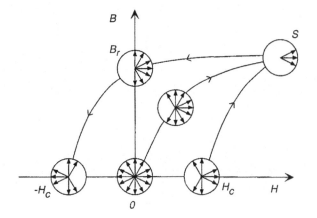

**Figure 10.1-23** Distribution of domain orientations at various points along a hysteresis loop [4].

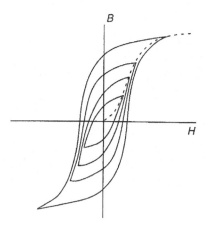

**Figure 10.1-24** Minor hysteresis loops.

## 10.1.12 Soft magnetic materials

Ferromagnetic materials are classified as either soft or hard depending on their coercivity $H_c$. Soft materials are characterized by a high permeability and a low coercivity ($H_c < 1000$ A/m), which makes them easy to magnetize and demagnetize. Hard materials have a relatively low permeability and a high coercivity ($H_c > 10{,}000$ A/m), which makes them more difficult to magnetize and demagnetize. The difference between these two materials is best illustrated by comparing their hysteresis loops. These are shown in Fig. 10.1-25. In this section we give a brief survey of common soft magnetic materials. More detailed treatments can be found in the texts by Jiles and Bozorth [1,7].

Soft magnetic materials are used as flux conduits to confine and direct flux, as flux amplifiers to amplify the flux density across a region, and as magnetic shields to shield a region from an external field. The most commonly used soft magnet materials are soft iron, alloys of iron-silicon, nickel-iron, and soft ferrites. They can be found in numerous devices including transformers, relays, motors, inductors, and electromagnets. When selecting a soft material, the properties that are weighed are its permeability, saturation magnetization, resistance, and coercivity. A higher permeability and saturation magnetization are desired for flux confinement and focusing. The resistance and coercivity are important for high-frequency applications. A higher resistance reduces eddy currents, whereas a lower coercivity reduces hysteresis loss.

Soft materials are magnetically linear over limited portions of their B-H curve where the permeability $\mu$ is constant (independent of $H$). However, they are nonlinear over other portions where $\mu$ is a function of $H$, $\mu = \mu(H)$. The B-H curves for various soft materials are shown in Figs 10.1.26 and 10.1.27.

Various measures of permeability are used. Specifically, there is the absolute permeability

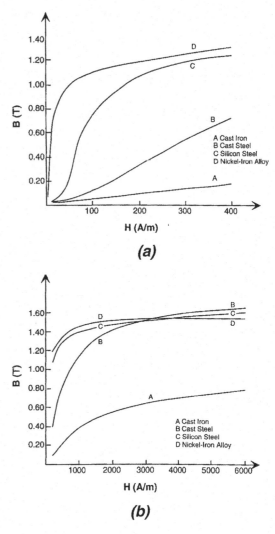

**Figure 10.1-26** The B-H curves for various soft magnetic materials: (a) $H < 400$ A/m; and (b) $H > 400$ A/m.

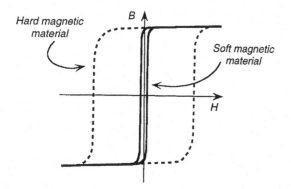

**Figure 10.1-25** The B-H loops for soft and hard magnetic materials.

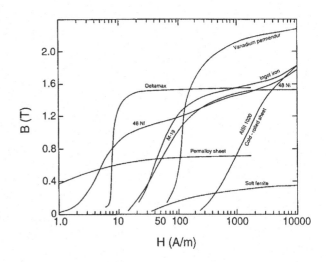

**Figure 10.1-27** The B-H curves of various soft magnetic materials.

$$\mu = \frac{B}{H} \quad \text{(absolute permeability),}$$

that changes along the B-H curve. There is also the relative permeability

$$\mu_r = \frac{\mu}{\mu_0} \quad \text{(relative permeability),}$$

that is normalized to the free space value $\mu_0 = 4\pi \times 10^{-7}$ Wb/A·m. A plot of $\mu_r$ vs H for silicon steel is shown in Fig. 10.1-28. This plot shows the nonlinear behavior of the material.

There is also the differential permeability

$$\mu_{\text{dif}} = \frac{dB}{dH} \quad \text{(differential permeability),}$$

which is the slope of the tangent line to the B-H curve. The initial permeability $\mu_i$ is $\mu_{\text{dif}}$ evaluated at the origin of the initial magnetization curve (segment OS in Fig. 10.1-22),

$$\mu_i = \frac{dB}{dH}\bigg|_{B=0, H=0}.$$

Other measures of permeability include the initial and maximum relative permeability $\mu_{i,r}$ and $\mu_{\text{max}r}$, respectively.

Similar definitions apply to the susceptibility of soft materials. Specifically, there are the absolute, differential and initial susceptibilities that are given by

$$\chi_m = \frac{M}{H} \quad \text{(susceptibility),}$$

$$\chi_{m,\text{dif}} = \frac{dM}{dH} \quad \text{(differential susceptibility),}$$

and

$$\chi_{m,i} = \frac{dM}{dH}\bigg|_{M=0, H=0},$$

respectively.

Soft iron was the original material of choice for electromagnetic applications, and is still widely used as a core material for dc electromagnets. However, for ac applications it has been supplanted by materials with a higher resistivity because these have lower eddy current losses. Commercially available soft iron typically contains low levels of impurities such as carbon (C) (0.02%), manganese (Mn) (0.035%), sulfur (S) (0.015%), phosphor (P) (0.002%), and silicon (Si). Iron with such impurities has a coercivity on the order of 80 A/m (1 Oe), a saturation magnetization of $1.7 \times 10^6$ A/m, and a maximum relative permeability of 10,000. However, these properties can be improved by annealing the iron in hydrogen, which removes the impurities. Such treatment can lower the coercivity to 4 A/m (0.05 Oe), and increase the maximum relative permeability to 100,000. The properties of soft iron also degrade when it is subjected to mechanical stress, such as when it is flexed or bent. However, these properties can also be restored by annealing, which relieves the internal stress.

Iron-silicon alloys typically contain 2–4% silicon. The addition of silicon has the beneficial effects of increasing the resistivity and reducing both magnetostriction and magnetic anisotropy. For example, an alloy with 3% silicon has a resistivity $\rho \approx 14\ \mu\Omega$ cm, which is approximately four times the resistivity of pure iron. Consequently, these alloys are superior to soft iron for low-frequency applications such as transformer cores. Their increased resistivity reduces eddy current loss. Iron-silicon alloys have an additional advantage in that their reduced magnetostriction reduces acoustic noise (transformer hum). Furthermore, their reduced anisotropy gives rise to increased permeability in nonoriented alloys. Alloys with 1–3% Si are used in rotating machines, whereas alloys with 3–4% Si are used in transformers. Iron-silicon alloys are inferior to soft iron in two respects. Specifically, they have a reduced saturation induction $B_s$, and they are more brittle. This latter property limits the level of silicon to 4% for most applications. Iron-silicon alloys have coercivities in the range of 5–120 A/m, and a maximum relative permeability in the range of 4000 to 20,000, depending on the percentage content of the silicon [1].

Nickel-iron alloys are used for a variety of applications such as inductors, magnetic amplifiers, and audio frequency transformer cores. Commercially available Ni-Fe alloys contain 50–80% Ni, and are characterized by a very high permeability. These include Permalloy (78% Ni, 22% Fe) with $\mu_{i,r} = 8000$ and $\mu_{\text{max},r} = 100,000$; Supermalloy (79% Ni, 16% Fe, 5% Mo) with $\mu_{i,r} = 100,000$ and $\mu_{\text{max},r} = 1,000,000$; and Mumetal (77% Ni, 16% Fe, 5% Cu, 2% Cr) with $\mu_{i,r} = 20,000$ and $\mu_{\text{max},r} = 100,000$ ($\mu_{i,r}$ and

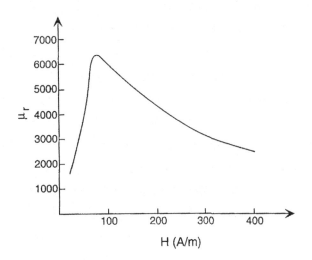

**Figure 10.1-28** Relative permeability $\mu_r$ for silicon steel vs H.

$\mu_{max,r}$ are the initial and maximum relative permeability, respectively). The resistivity of Ni-Fe alloys decreases from a peak value of 72 μΩ cm for 30% Ni to 15 μΩ cm for 80% Ni [1]. Permalloy has a coercivity $H_c = 4.0$ A/m (0.05 Oe), and a saturation magnetization $M_s = 0.86 \times 10^6$ A/m. The corresponding values for Supermalloy are $H_c = 0.16$ A/m and $M_s = 0.63 \times 10^6$ A/m.

Finally, there are the soft ferrites. These have a cubic crystalline structure with a chemical formula of the form $MO \cdot Fe_2O_3$ where M is a transition metal such as iron, nickel (Ni), magnesium (Mg), manganese, or zinc (Zn). Most soft ferrites have a relatively high resistivity (up to $10^{10}$ Ωm), a relatively low permeability (initial relative permeability in the range of 10 to 10,000), and a relatively low saturation magnetization ($M_s = 4 \times 10^5$ A/m) as compared to the soft magnetic alloys. Their low permeability and saturation magnetization render these materials inferior to soft magnetic alloys for dc and low-frequency applications. However, their high resistivity makes them the material of choice for high-frequency applications. Soft ferrites are sometimes categorized as either nonmicrowave ferrites (for use with audio frequencies to 500 MHz), or microwave ferrites (for use from 100 MHz to 500 GHz). Many soft ferrites have a permeability that is relatively constant up to frequencies of between 10–100 MHz. This makes them useful for broadband telecommunications applications such as specialized transformers and inductors [1,4].

## 10.1.13 Hard magnetic materials

Hard magnetic materials are characterized by a low permeability and high coercivity, typically >10,000 A/m. This latter property makes them difficult to magnetize and demagnetize. Such materials are referred to as permanent magnets because once magnetized they tend to remain magnetized. Permanent magnets are used as field source components in a wide range of products including consumer electronic equipment, computers, data storage devices, electromechanical devices, telecommunications equipment, and biomedical apparatus [2,16–18].

The properties of primary importance in the selection of a magnet are those that define the magnitude and stability of the field that it can provide. These include the coercivity $H_c$, saturation magnetization $M_s$, and remanence $B_r$, as well as the behavior of the hysteresis loop in the second quadrant (Fig. 10.1-29a). This portion of the hysteresis loop is called the demagnetization curve.

The points $(B, H)$ on the demagnetization curve define an energy product $BH$ that obtains a maximum $(BH)_{max}$ for some point in the interval $-H_c < H < 0$ as shown in Fig. 10.1-29b. When a magnet is used as a field source it becomes biased at an operating point $(B_m, H_m)$ on its demagnetization curve. The operating point

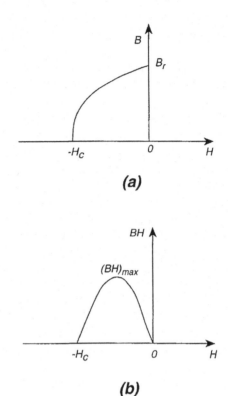

**Figure 10.1-29** Second quadrant B-H curve: (a) demagnetization curve; and (b) $B \cdot H$ vs $H$.

depends on the circuit in which it is used. It can be determined from the load line of the circuit. This intersects the demagnetization curve at the operating point $(B_m, H_m)$ as shown in Fig. 10.1-30.

We show there that it is desirable to bias a magnet at the point of maximum energy $(BH)_{max}$. This minimizes the volume of the magnet and reduces its cost. The energy product for a given material and circuit can be determined from second quadrant constant energy contours (Fig. 10.1-31). These are usually plotted on the demagnetization curves supplied by permanent magnet vendors.

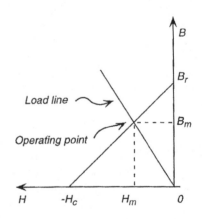

**Figure 10.1-30** Demagnetization curve and load line.

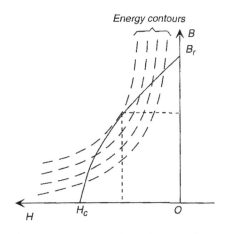

**Figure 10.1-31** Second quadrant constant energy contours.

In the following sections we give a brief survey of various commercially available hard materials. These include ferrites, alnico, samarium-cobalt (Sm-Co), neodymium-iron-boron (Nd-Fe-B) and bonded magnets. A more extensive treatment of hard materials is given by McCaig and Clegg, Parker, and Campbell [2, 16, 17].

## 10.1.14 Ferrites

Hard ferrites are the least expensive and most commonly used permanent magnet material. The development of ferrite magnets dates back to the 1950s [19]. These magnets are made from fine particle powders with compositions of the form $XO \cdot 6(Fe_2O_3)$ where X is either barium (Ba), strontium (Sr), or lead (Pb). They are fabricated using powder metallurgical methods and are commonly referred to as ceramic magnets. The fabrication cycle starts by either wet or dry mixing the correct proportions of iron oxide ($Fe_2O_3$) with the carbonate of either Ba, Sr, or Pb. The mixture is then calcined at a temperature of between 1000 and 1350 °C. The calcined material is then crushed and milled into a fine powder. Isotropic magnets are formed by drying and pressing the powder into a desired shape, with subsequent sintering at a temperature in the range of 1100—1300 °C.

Anisotropic ferrite magnets are fabricated using a fine grain powder with the grain size on the order of a single domain ($\sim 1$ $\mu$m). The powder is mixed with water to form a slurry, and the slurry is pressed and then sintered. An alignment field is applied during the compression of the powder, and the water acts as a lubricant to facilitate the alignment process. A shrinkage of 15% is common during sintering. Finished magnets are produced by grinding down the sintered material.

Ferrite magnets are characterized by a high coercivity and a nearly linear second quadrant $B$ vs $H$ curve. This is due to the dominant magnetocrystalline anisotropy of the constituent particles. Ceramic magnets are produced with a wide range of properties. These depend on various process variables such as grain size, milling sequence, and heating schedule in both the calcination and sintering phases of production. Demagnetization curves for different grades of ceramic magnets are given in Section 10.1.20. Various magnetic properties are listed in Table 10.1-1. Iron oxides are also used for various applications. The two most commonly used oxides are ferric oxide ($Fe_3O_4$), also known as magnetite or lodestone, and gamma ferric oxide

| Table 10.1-1 Properties of ceramic magnets | | | | | |
|---|---|---|---|---|---|
| Ceramic $\Rightarrow$ | 1 | 5 | 7 | 8 | 10 |
| $B_r$ | | | | | |
| (T) | 0.23 | 0.38 | 0.34 | 0.385 | 0.41 |
| (G) | 2300 | 3800 | 3400 | 3850 | 4100 |
| $H_c$ | | | | | |
| (kAm$^{-1}$) | 147 | 191 | 259 | 235 | 223 |
| (Oe) | 1850 | 2400 | 3250 | 2950 | 2800 |
| $H_{ci}$ | | | | | |
| (kAm$^{-1}$) | 259 | 203 | 318 | 243 | 231 |
| (Oe) | 3250 | 2550 | 4000 | 3050 | 2900 |
| $(BH)_{max}$ | | | | | |
| (kJm$^{-3}$) | 8.36 | 27.0 | 21.9 | 27.8 | 31.8 |
| (MG Oe) | 1.05 | 3.40 | 2.75 | 3.50 | 4.00 |
| $\mu_r$ | 1.1 | 1.1 | 1.1 | 1.1 | 1.1 |
| Density (kg/m$^3$) | 4982 | 4706 | 4706 | 4706 | 4706 |
| Curie temp (°C) | 450 | 450 | 450 | 450 | 450 |

($\gamma Fe_2O_3$). Ferric oxide has the following properties: $H_c = 7.16 \times 10^3$ A/m (90 Oe); $M_s = 4.8 \times 10^5$ A/m; density $\rho = 5000$ kg/m$^3$; and $T_c = 585$ °C. Some of the corresponding properties for $\gamma Fe_2O_3$ are $H_c = 21 \times 10^3 - 32 \times 10^3$ A/m (270–400 Oe), $\rho = 4700$ kg/m$^3$, and $T_c = 590$ °C.

## 10.1.15 Alnico

Alnico alloys date back to the early 1930s. Their principal components are iron, cobalt, nickel, and aluminum (Al) along with lesser amounts of other metals such as copper (Cu). Alnicos are hard as well as too brittle for cold working. Production methods are limited to casting the liquid alloy, or pressing and sintering the metal powders [2]. Once formed, the alloy is subjected to controlled heat treatment so as to precipitate a dispersion of fine magnetic particles throughout the Al-Ni-Fe-Co matrix. These particles are elongated and rod shaped with a dominant shape anisotropy that is manifest in the relatively high coercivity of the finished magnet.

Alnico magnets can be either isotropic or anisotropic, depending on whether or not the magnetic particles are oriented during their formation. These magnets exhibit a wide range of properties depending on the nature of the heat treatment, and whether the original alloy was prepared by casting or sintering. Demagnetization curves for various grades of Alnico are shown in Fig. 10.1-32a,b. The properties of Alnico magnets are listed in Tables 10.1-2 and 10.1–3.

From these tables we see that Alnico magnets have impressive magnetic properties. Unfortunately, they also have poor physical properties. As already noted, they are extremely hard and brittle. Thus, production of finished magnets with tight tolerances requires tedious and costly machining.

## 10.1.16 Samarium-cobalt

The development of samarium-cobalt (SmCo) magnets began in the 1960s [20]. Samarium is one of the rare-earth elements that form a transition group with atomic numbers ranging from 58 (Ce) to 71 (Lu). The development of SmCo resulted from research that was directed toward the formation of alloys of the rare-earth elements with the transition series ferromagnets: iron, cobalt, and nickel. The two most common SmCo materials are $SmCo_5$ and $Sm_2Co_{17}$.

The first samarium-cobalt magnets were made by bonding $SmCo_5$ powder in a resin. Subsequent production entailed the formation of the base alloy using either a reduction/melt or reduction/diffusion process. In the reduction/melt process Sm and Co are mixed and induction melted to form the alloy. The cast alloy is brittle, and readily ground into a fine grain powder.

In the reduction/diffusion process samarium oxide ($Sm_2O_3$) and cobalt powder are reacted with calcium (Ca) at approximately 1150 °C to form a compound:

$$10Co + Sm_2O_3 + 3Ca \rightarrow 2SmCo_5 + 3CaO.$$

The CaO is separated from the compound via a sequence of steps starting with a reaction with water, followed by a gravimetric separation of the hydroxide, followed by an acid rinse, and then drying. Once in powder form, powder metallurgical methods are used to form the desired magnet shape.

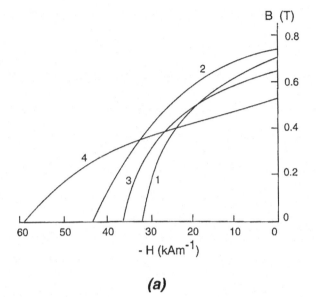

(a)

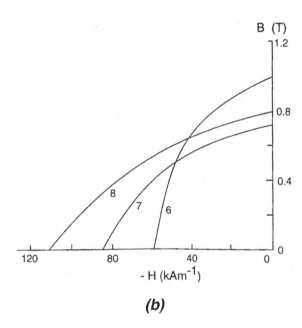

(b)

Figure 10.1-32 Demagnetization curves for Alnico magnets: (a) Alnico 1, 2, 3, and 4; (b) Alnico 6, 7, and 8 [1].

## Table 10.1-2 Properties of sintered alnico magnets

| Alnico (sintered) ⇒ | 2 | 5 | 6 | 8 |
|---|---|---|---|---|
| $B_r$ | | | | |
| (T) | 0.71 | 1.05 | 0.94 | 0.76 |
| (G) | 7100 | 10,500 | 9400 | 7600 |
| $H_c$ | | | | |
| (kAm$^{-1}$) | 43.8 | 47.8 | 62.9 | 119 |
| (Oe) | 550 | 600 | 790 | 1500 |
| $H_{ci}$ | | | | |
| (kAm$^{-1}$) | 45.3 | 49.3 | 65.2 | 134 |
| (Oe) | 570 | 620 | 820 | 1690 |
| $(BH)_{max}$ | | | | |
| (kJm$^{-3}$) | 11.9 | 24 | 23 | 36 |
| (MG Oe) | 1.5 | 3.0 | 2.9 | 4.5 |
| $\mu_r$ | 6.4 | 4.0 | 4.5 | 2.1 |
| Density (kg/m$^3$) | 6837 | 7000 | 6892 | 6975 |
| Curie temp. (°C) | 810 | 900 | 860 | 860 |

## Table 10.1-4 Properties of samarium-cobalt magnets

| | SmCo$_5$ | Sm$_2$Co$_{17}$ |
|---|---|---|
| $B_r$ | | |
| (T) | 0.83 | 1.0 |
| (G) | 8300 | 10,000 |
| $H_c$ | | |
| (kAm$^{-1}$) | 600 | 480 |
| (Oe) | 7500 | 6000 |
| $H_{ci}$ | | |
| (kAm$^{-1}$) | 1440 | 558 |
| (Oe) | 18,100 | 7000 |
| $(BH)_{max}$ | | |
| (kJm$^{-3}$) | 128 | 192 |
| (MG Oe) | 16 | 24 |
| $\mu_r$ | 1.05–1.1 | 1.05 |
| Density (kg/m$^3$) | 8200 | 8100 |
| Curie temp. (°C) | 700 | 750 |

Samarium-cobalt magnets are characterized by high coercivities and nearly linear second quadrant demagnetization curves. This suggests that the dominant magnetic mechanism for this material is magnetocrystalline anisotropy (Section 10.1.9). However, the grains in SmCo powders are typically 5–10 μm in diameter, which is approximately an order of magnitude larger than single domain size. This relatively large grain size makes it energetically favorable for the formation of domain walls within each grain. Thus, each grain may comprise multiple domains. In SmCo$_5$, the domain walls move with relative ease within each grain. The high coercivity of these magnets can be explained by the limited nucleation sites for domain walls, and the pinning of domain walls at the grain boundaries that hinders the growth of domains from grain to grain. In Sm$_2$Co$_{17}$, the high coercivity is attributable to the pinning of domain walls.

The magnetic properties of SmCo$_5$ and Sm$_2$Co$_{17}$ are given in Table 10.1.4. Demagnetization curves for these materials can be found in Section 10.1.20.

## 10.1.17 Neodymium-iron-boron

The development of neodymium-iron-boron (NdFeB) followed that of SmCo and dates back to the early 1980s [21, 22]. The initial motivation for its development was a desire for a cost effective alternative to SmCo. This was based on concern over the cost and availability of cobalt. Also, neodymium is much more abundant than samarium.

The NdFeB powders are produced using various methods including a reduction/diffusion process, and a rapid quenching (melt spinning) method. The finished magnets can be either isotropic or anisotropic [17]. The reduction/diffusion process is similar to that of SmCo$_5$ (Section 10.1.16). The finished magnets are produced from the powder via the steps of pressing and sintering, followed by heat treatment, and then grinding into the desired shape. The powder can be aligned with a bias field as it is pressed. The finished magnets have a high coercivity that is attributed to domain wall pinning at the grain boundaries.

## Table 10.1-3 Properties of cast alnico magnets

| Alnico (cast) ⇒ | 2 | 5 | 6 | 8 |
|---|---|---|---|---|
| $B_r$ | | | | |
| (T) | 0.75 | 1.24 | 1.5 | 0.82 |
| (G) | 7500 | 12,400 | 10,500 | 8200 |
| $H_c$ | | | | |
| (kAm$^{-1}$) | 44.6 | 50.9 | 62.1 | 131.3 |
| (Oe) | 560 | 640 | 780 | 1650 |
| $H_{ci}$ | | | | |
| (kAm$^{-1}$) | 46.1 | 54.1 | 63.7 | 148 |
| (Oe) | 580 | 680 | 800 | 1860 |
| $(BH)_{max}$ | | | | |
| (kJm$^{-3}$) | 13.5 | 43.8 | 31.0 | 42.2 |
| (MG Oe) | 1.7 | 5.5 | 3.9 | 5.3 |
| $\mu_r$ | 6.4 | 4.3 | 5.3 | 2.0 |
| Density (kg/m$^3$) | 7086 | 7308 | 7418 | 7252 |
| Curie temp. (°C) | 810 | 900 | 860 | 860 |

The rapid quenching or melt spinning method is commonly referred to as Magnequench [23, 24]. In this process the NdFeB alloy is melted and forced under argon (Ar) pressure through a small orifice onto the surface of a water-cooled revolving metal wheel. This yields a rapidly quenched thin ribbon of the alloy $\approx 35$ $\mu$m thick and 1–3 mm wide. This material has a very fine microcrystalline structure consisting of randomly oriented single domain NdFeB grains with a diameter on the order of 0.030 $\mu$m. The ribbon is milled into thin platelets $\approx 200$ $\mu$m across and 35 $\mu$m thick.

There are three grades of Magnequench known as MQ I, MQ II, and MQ III. In the MQ I process the milled platelets are annealed and then blended with an epoxy resin. The compound is pressed into a desired shape and then oven cured. The finished magnets are isotropic due to the random orientation of the NdFeB grains within the platelets. In the MQ II process the magnets are formed by hot-pressing the platelets at $\approx 700°$C. This results in a nearly fully densified magnet that is also isotropic.

In the MQ III process magnets are formed in a two-step sequence. First, the milled platelets are hot-pressed as in the MQ II process. Next, the hot-pressed material is subjected to a second hot-pressing in a die with a larger cross section. This process compresses the height by close to 50%, and is known as die-upsetting [25]. As the material thins, internal shear stresses develop at the granular level, and these tend to align the preferred magnetic axes parallel to the pressing direction. Thus, the finished magnet is anisotropic. The properties of the Magnequench materials are given in Table 10.1.5.

## 10.1.18 Bonded magnets

Permanent magnets made from fully dense materials tend to be hard and brittle. Consequently, substantial machining is required to form these materials into finished magnets with precession tolerances. A cost-effective solution to this problem is the use of bonded magnets. These magnets are fabricated by blending a magnetic powder with a binder material to form a compound, and then molding or extruding the compound into a desired shape. Typical binder materials include rubber, resins, and plastics. Bonded magnets are formed using compression-molding, injection-molding, and extrusion. Compression-molded magnets are made using a thermosetting binder such as epoxy resin, whereas injection-molded magnets are formed using a thermoplastic binder such as nylon. Extruded magnets are produced using an elastomer such as rubber. Bonded magnets can be made rigid or flexible depending on the volume fraction of powder used. Typical volume fractions are 80–85% for compression-molding, 60–65% for injection-molding, and 55–60% for extrusion [17].

**Table 10.1-5** Properties of Magnequench

|  | MQ I | MQ II | MQ III |
|---|---|---|---|
| $B_r$ |  |  |  |
| (T) | 0.61 | 0.80 | 1.18 |
| (G) | 6100 | 8000 | 11,800 |
| $H_c$ |  |  |  |
| (kAm$^{-1}$) | 424 | 520 | 840 |
| (kOe) | 5.3 | 6.5 | 10.5 |
| $H_{ci}$ |  |  |  |
| (kAm$^{-1}$) | 1200 | 1280 | 1040 |
| (kOe) | 15.0 | 16.0 | 13.0 |
| $(BH)_{max}$ |  |  |  |
| (kJm$^{-3}$) | 64 | 104 | 256 |
| (MG Oe) | 8 | 13 | 32 |
| Density (kg/m$^3$) | 6000 | 7500 | 7500 |
| Curie temp. (°C) | 312 | 312 | 312 |
| Maximum operat. temp. (°C) | 125 | 150 | 150 |
| $\mu_r$ | 1.15 | 1.15 | 1.05 |

The most common bonded magnets are ferrite based. These are usually fabricated using a fine powder (1-$\mu$m particle diameter) made from sintered $BaFe_{12}O_{19}$. The particles in this powder form as thin platelets with their planes perpendicular to the magnetic easy axis. Isotropic grades of these magnets can be produced by injection-molding the powder/binder compound. Typical energy products for these magnets are on the order of $(BH)_{max} = 4$ kJ/m$^3$ (0.5 MG Oe).

Anisotropic magnets can be produced using an injection-molding or extrusion process with an alignment field applied during the formation of the magnet. In the injection molding process, the alignment field is applied across the mold cavity while the powder/binder compound is still fluid. The alignment field can be provided by coils embedded in the mold itself. The energy product of anisotropic injection-molded ferrite magnets is typically $(BH)_{max} = 14$ kJ/m$^3$ (1.76 MG Oe).

A lesser degree of alignment can be achieved by applying mechanical pressure to the powder/binder compound during the formation of the magnet. The pressure tends to orient the platelets so that they face the pressing direction. Such mechanical alignment is sometimes implemented in the extrusion process. Magnetic properties of various grades of bonded ferrite magnets are listed in Table 10.1.6 [16].

Bonded magnets are also made using samarium-cobalt materials $SmCo_5$ and $Sm_2Co_{17}$. Such magnets have reduced performance relative to their sintered counterparts, but they are far less brittle and more easily

Magnetic materials CHAPTER 10.1

Table 10.1-6 Properties of bonded ferrite magnets (isotropic (i), anisotropic (a))

| Bonded ferrite ⇒ | Flexible (i) | Flexible (a) | Rigid (i) | Rigid (a) |
|---|---|---|---|---|
| $B_r$ | | | | |
| (T) | 0.17 | 0.25 | 0.14 | 0.30 |
| (G) | 1700 | 2500 | 1400 | 3000 |
| $H_c$ | | | | |
| (kAm$^{-1}$) | 127 | 175 | 84 | 191 |
| (Oe) | 1600 | 2200 | 1050 | 2400 |
| $H_{ci}$ | | | | |
| (kAm$^{-1}$) | 239 | 239 | — | 223 |
| (Oe) | 3000 | 3000 | — | 2800 |
| $(BH)_{max}$ | | | | |
| (kJm$^{-3}$) | 5.57 | 11.9 | 3.18 | 15.9 |
| (MG Oe) | 0.7 | 1.5 | 0.4 | 2.0 |
| $\mu_r$ | 1.05 | 1.05 | 1.05 | 1.05 |

This section completes our survey of permanent magnet materials. In the remaining sections we discuss the magnetization and stability of magnets.

## 10.1.19 Magnetization

Permanent magnet materials are of little use until they are magnetized. In this section we discuss the magnetization process. It is widely held that the field $H_s$ required to saturate a magnet is 3–5 times its intrinsic coercivity $H_{ci}$. Here, $H_s$ is the field inside the material. The applied field $H_a$ must be greater than $H_s$ to compensate for the internal demagnetization field $H_d$ (Section 10.1.8). Specifically,

$$H_s = H_a + H_d,$$

or

$$H_a = H_s - H_d.$$

Recall that $H_d = -NM$ where $N$ is the (geometry dependent) demagnetization factor and $M$ is the magnetization. For a saturated magnet $H_d = -NM_S$. Therefore, the (external) field required to saturate the magnet is

$$H_a = H_s + NM_s.$$

The values of $H_s$ for various materials are listed in Table 10.1-8.

Conventional materials such as Alnico and hard ferrites have lower values of $H_{ci}$ and can be magnetized with fields $H_s < 8.0 \times 10^5$ A/m. Such fields can be obtained by passing direct currents through conventional solenoids (Fig. 10.1-33). Magnetizing fixtures are also used to magnetize conventional materials. A typical fixture consists of a coil wrapped around a soft magnetic core that has a gap in which the magnet is placed (Fig. 10.1-34). To

machined. SmCo$_5$ powder lends itself to the injection-molding and extrusion processes where gate sizes limit the particle sizes to 5–10 μm. Energy products on the order of $(BH)_{max} = 60$ kJ/m$^3$ (7.5 MG Oe) are achieved with these magnets. The Sm$_2$Co$_{17}$ particles suffer a reduction in magnetic properties if their diameter is reduced much below 40 μm [26]. This material is best suited for compression molding where energy products on the order of $(BH)_{max} = 130$ kJ/m$^3$ (16 MG Oe) are achieved. Additional properties of bonded SmCo magnets are given in Table 10.1.7 [2]. Bonded magnets can also be made using NdFeB powder. The Magnequench MQ I material is one example of this (Section 10.1.17).

Table 10.1-7 Properties of bonded samarium-cobalt magnets

| Bonded SmCo ⇒ | SmCo$_5$ | Sm$_2$Co$_{17}$ |
|---|---|---|
| $B_r$ | | |
| (T) | 0.65 | 0.86 |
| (G) | 6500 | 8600 |
| $H_c$ | | |
| (kAm$^{-1}$) | 460 | 497 |
| (Oe) | 5750 | 6250 |
| $H_{ci}$ | | |
| (kAm$^{-1}$) | 620 | 800 |
| (Oe) | 7790 | 10,050 |
| $(BH)_{max}$ | | |
| (kJm$^{-3}$) | 80 | 130 |
| (MG Oe) | 10 | 16.2 |
| $\mu_r$ | 1.1 | 1.1 |

Table 10.1-8 Magnetizing fields of various permanent magnet materials

| Magnetizing field ⇒ | $H_s$ | |
|---|---|---|
| | kA/m | Oe |
| Ceramic ferrite | 796 | 10,000 |
| Alnico (2,5,6) | 199–318 | 2500–4000 |
| Alnico (8,9) | 636 | 8000 |
| SmCo$_5$ | 1193 | 15,000 |
| Sm$_2$Co$_{17}$ | 1989 | 25,000 |
| NdFeB | | |
| MQ I—MQ II | 2784 | 35,000 |
| MQ III | 1989 | 25,000 |

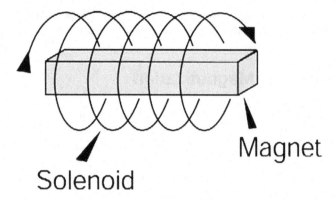

Figure 10.1-33 Magnetization of a bar magnet using a solenoid

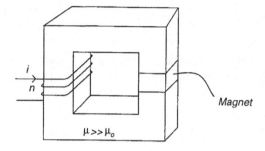

Figure 10.1-34 Conventional magnetizing fixture.

polarize the magnet, a direct current is applied to the coil and this generates a field through the core. The core guides and focuses the field across the magnet to produce the desired polarization pattern. High-grade core materials are available that are capable of carrying flux densities of up to 2.0 T, which is sufficient for the magnetization of low coercivity materials (i.e., $B = 2T \rightarrow H = B/\mu_0 = 16 \times 10^5$ A/m $> H_s$).

The design of such fixtures entails the following steps: (a) estimate the demagnetization field $H_d$ in the magnet assuming it is saturated; (b) compute the applied field $H_a = H_s - H_d$ required to saturate the magnet; and (c) determine the current required to generate $H_a$ assuming the magnet is an air gap. We demonstrate this procedure using the fixture shown in Fig. 10.1-34. This circuit (without the coil) is analyzed in Example 3.5.3. If the circuit is ideal ($\mu \approx \infty$), and there is no leakage, then the demagnetization field in the magnet is essentially zero, $H_d = 0$. Therefore, to saturate the magnet the applied field must equal the saturation field $H_a = H_s$. To determine $H_a$ we assume that the permeability of the magnet is $\mu_0$. Then, the magnet itself represents an air gap, and the field through it is given by

$$B = \frac{\mu_0 ni}{l_m},$$

or

$$H = \frac{ni}{l_m}, \quad (10.1.66)$$

where $n$ is the number of turns, $i$ is the current, and $l_m$ is the length of the magnet (height of the gap) (Example 3.5.1). Thus, to saturate the magnet we choose $ni = l_m H_s$ so that $H = H_s$.

While conventional magnetization methods are appropriate for low energy magnets, they are impractical for rare-earth materials. These require magnetizing fields on the order of $H_s = 2 \times 10^6$ A/m (25,000 Oe), and the current required to generate such fields cannot be sustained without overheating and damaging the conductors. The nominal limit of current density for continual operation is generally taken to be 1.5 to 2 A/mm². Fortunately, the magnetizing field needs to be applied only momentarily. Therefore, rare-earth magnets can be magnetized using impulse magnetizers that provide a high current pulse of short duration. Such magnetizers are relatively inexpensive and can operate from standard power lines.

A typical impulse magnetizing system consists of a charging circuit, a capacitor bank for energy storage, a thyristor or ignitron tube for switching the current, a step-down transformer, and a magnetizing fixture (Fig. 10.1-35). An equivalent electrical circuit for a basic magnetizing system (without the transformer) is shown in Fig. 10.1-36. The charging circuit consists of a voltage source $V_{dc}$ and a resistor $R_0$. The ignitron tube is represented by a toggle switch, and the capacitor bank is represented by capacitance C. The magnetizing fixture is represented as an RL load.

The magnetization process entails two distinct steps. In the first step, the capacitor bank is slowly charged to a predefined voltage. This corresponds to the switch being set to position (1) in Fig. 10.1-36. The switch is held in this position until the capacitor bank C is charged to the voltage $V_{dc}$. The duration of the charging phase is determined by the resistance $R_0$. In the second step, the ignitron tube is activated, which in effect toggles the switch to position (2). The current from the capacitor is

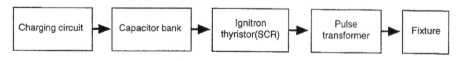

Figure 10.1-35 Components of an impulse magnetizing system.

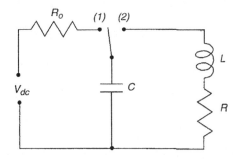

**Figure 10.1-36** Equivalent circuit of an impulse magnetizing system.

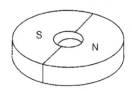

**Figure 10.1-37** Bipolar disk magnet.

discharged into the fixture in the form of a single high-current pulse. This produces a peak field of short duration across the magnet that is sufficient to magnetize it.

Impulse magnetizing fixtures typically consist of a coil wound on a nonmagnetic support structure that houses the magnet. There is usually little benefit in using a soft magnetic core because it would saturate at a level below the required magnetizing field. Moreover, such cores are conductive and can therefore generate eddy currents that oppose the magnetizing field. Without such a core, a fixture is magnetically linear and the magnetizing field that it generates scales linearly with the applied current.

It is instructive to study the electrical response of an impulse magnetizing system. Consider the circuit of Fig. 10.1-36. After the capacitor is charged to the voltage $V_{dc}$, the switch is moved to position (2), which sends current to the fixture. The current is governed by the equation

$$\frac{d^2 i}{dt^2} + \frac{R}{L}\frac{di}{dt} + \frac{i}{LC} = 0, \quad (10.1.67)$$

with initial conditions $i(0) = 0$ and $di/dt|_{t=0} = V_{dc}/L$. The solution to Eq. (10.1.67) is given by

$$i(t) = -\frac{V_{dc}}{2L\gamma}\left[e^{-t/\tau_1} - e^{-t/\tau_2}\right] \quad (10.1.68)$$

where

$$\tau_1 = \frac{1}{R/2L + \gamma}, \quad (10.1.69)$$

$$\tau_2 = \frac{1}{R/2L - \gamma}, \quad (10.1.70)$$

and $\gamma = \sqrt{(R/2L)^2 - 1/LC}$. Notice that $\tau_1 < \tau_2$. Initially, the current surges through the fixture and rises with a time constant $\tau_1$. After reaching a peak value, the current decays with a time constant $\tau_2$. The system is critically damped when $\gamma = 0$, and overdamped when $\gamma > 0$.

Thus, the system is overdamped when

$$R > 2\left(\frac{L}{C}\right)^{1/2}$$

and the current through the fixture will be unidirectional without oscillations. This is desired for magnetization.

A fixture needs to be designed with appropriate values of $R$ and $L$ so that sufficient energy with be transferred from the capacitor bank to the magnetizing field. To saturate a magnet, the fixture must provide a field $H_s$ throughout the magnet. Let $E_{mag}$ denote the magnetostatic energy required to saturate the magnet. We estimate $E_{mag}$ using Eq. (10.1.66),

$$E_{mag} = \frac{1}{2}\mu H_a^2 \times (\text{volume of the magnet}),$$

where $H_a > 3H_{ci}$, and $\mu$ is the permeability of the unmagnetized material ($\mu \approx \mu_0$ for rare-earth materials). The energy stored in an impulse magnetizer is $E_0 = 1/2CV_{dc}^2$ when its capacitor bank is fully charged. The energy dissipated through the resistor $R$ is $E_{loss} = 1/2R \int i^2\, dt$, where $i$ is given by Eq. (10.1.68). Thus $V_{dc}$, $C$, $R$, and $L$ are chosen so that $E_0 > E_{mag} + E_{loss}$. Moreover, the fixture needs to be designed to render the desired magnetization pattern throughout the magnet.

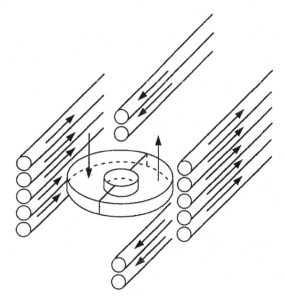

**Figure 10.1-38** Wire configuration for magnetizing a bipolar disk magnet.

## Magnetic materials

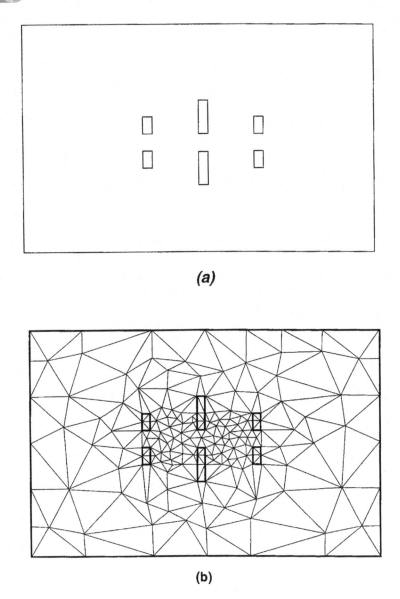

**Figure 10.1-39** Fixture analysis: (a) cross section of the wire configuration; and (b) FEA mesh.

The design of an impulse magnetizing fixture is straightforward [27]. First, a coil structure is chosen that will render the desired magnetization pattern within the magnet. Next, an initial wire gauge is selected for the coil. Third, an analysis is performed to determine the resistance $R$ and inductance $L$ of the coil as well as the average $B$-field that it generates across the magnet per ampere of applied current. This latter value is referred to as the tesla per amp ratio, or TPA. This ratio is used in a subsequent analysis to determine the level of current required to saturate the magnet, $B_{sat} = \mu_0 H_s$.

The coil resistance $R$ is easy to determine given the length and gauge of the wire. However, the inductance and TPA calculations are more complicated. The topology of the coil often precludes a simple analytical evaluation of these values. However, they can be determined using numerical methods such as finite-element analysis (FEA).

The FEA for fixture design is essentially a steady-state magnetostatic analysis. Briefly, the finite-element method amounts to dividing the fixture geometry and surrounding region into small elements (collectively known as a mesh) and assigning to each element a polynomial that will approximate the behavior of the magnetic potential in that element. The coefficients of these polynomials are arranged in a matrix, and their values are obtained via the solution of a matrix equation that contains all the information of the problem, including the location of each element relative to its neighbors, the underlying field equations, and the boundary conditions.

Once $R$, $L$, and TPA are known, a circuit analysis is performed to determine the current through the fixture and the field that develops across the magnet. A thermal analysis is also performed to determine the temperature rise in the fixture. The temperature $T$ is predicted using the formula

$$T(t) = \frac{1}{\rho c_v} \int_0^t \frac{J^2}{\sigma} dt,$$

where $\rho$, $c_v$, and $\sigma$ are the density, constant-volume specific heat, and conductivity of the conductor, and $J$ is the current density through it. For copper at room temperature, $\rho = 8950$ kg/m$^3$, $c_v = 383$ J/(kg°C), and $\sigma = 5.8 \times 10^7$ mho/m. The calculations are performed for a series of wire gauges and an optimum wire gauge is selected based on achieving the appropriate field strength, the sharpest pole transitions, and a tolerable fixture temperature (<220°C).

We demonstrate the design procedure via a simple example. Consider the bipolar disk magnet shown in Fig. 10.1-37. Such magnets are typically used for clamping or holding applications. Assume that the magnet is made from NdFeB MQ I material that requires a magnetizing field $H_s = 2785$ k Am$^{-1}$. This corresponds to a saturation flux density $B_{sat} = 3.5$ T (Table 10.1-8). The design starts with a wire configuration design such as the one shown in Fig. 10.1-38. The wires are arranged so as to induce the desired bipolar magnetization across the magnet. They are connected at their terminals to form a continuous circuit. Several magnets are to be magnetized at once and therefore the length of the fixture is much greater than its width. We perform a two-dimensional (2D) FEA of the cross-sectional geometry of the wires. This cross section is shown in Fig. 10.1-39a. The first step is the generation of the mesh (Fig. 10.1-39b). Once the geometry is meshed, a nominal value of current is defined for the wire regions and the appropriate boundary conditions are imposed. In this case the field is set to zero far from the fixture (Dirichlet condition). The field problem is then solved, and plots of the flux lines and the $B$-field are obtained (Fig. 10.1-40a,b). A flux line is a line that indicates the direction of the $B$-field at any point along its length. The inductance $L$ and TPA are obtained from this analysis. The TPA is computed by dividing the field generated at the center of a pole by the number of amps of applied current.

At this point $V_{dc}$, $R_0$, $R$, $L$, and TPA are known. A circuit analysis can now be performed to determine the optimum wire gauge. A sample output of an analysis is shown in Fig. 10.1-41. The vertical axis has been divided into 10 parts, and the top of the scale corresponds to the values $B_{max} = 5.0$ T, $I_{max} = 50,000$ A, and TEMP$_{max} = 200$°C. The negative swing in current is prevented in practice and therefore only the positive half of the cycle is shown. The analysis shows that a peak field of $\approx 4.5$ T is achieved with a maximum current of 25,000 A. The

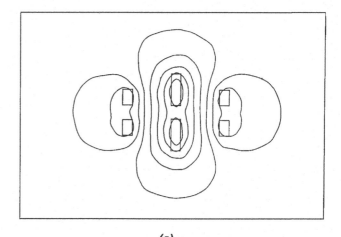

(a)

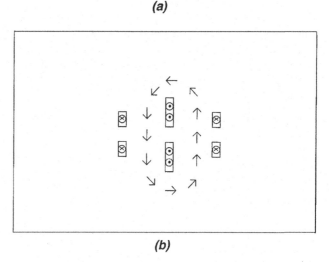

(b)

**Figure 10.1-40** Fixture analysis: (a) flux lines; and (b) **B**-field.

operating temperature of the coil reaches $\approx 140$°C, which is acceptable since the maximum working temperature for common grades of insulation is typically 200°C.

As a further example, consider the multipole disk magnet shown in Fig. 10.1-42. Eight north-south

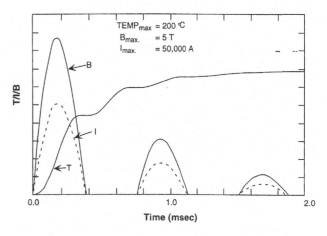

**Figure 10.1-41** Analysis of the magnetization process.

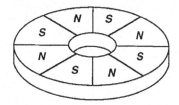

**Figure 10.1-42** Multipole disk magnet.

alternating magnetic poles are arranged in pie-shaped zones with vertical magnetization over each zone. This type of magnet is used in axial-field permanent magnet motors. A fixture for this magnet is shown in Fig.10.1-43. The coil is supported in a nonmagnetic structure and is wound to provide the desired magnetization. The resistance of the coil is easily calculated once an initial wire gauge is specified. The inductance and TPA are calculated using FEA. Once these have been determined, parametric circuit calculations are performed to determine the optimum wire gauge.

## 10.1.20 Stability

The operating point of a magnet depends on its load line as well as its demagnetization curve. These, in turn, depend on the operating environment. A good design must account for variations in the environment so that the performance of the magnet remains within an acceptable range throughout its working life. This is the issue of stability.

A magnet's load line is a function of the magnetic circuit used. Thus, load line stability needs to be considered on a case-by-case basis. However, the stability of the demagnetization curve depends on factors that are common to all applications. In this section we briefly review and summarize these factors. A thorough discussion of this subject is given by McCaig and Clegg [2].

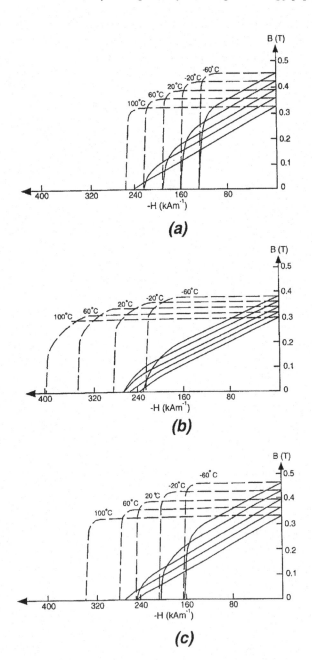

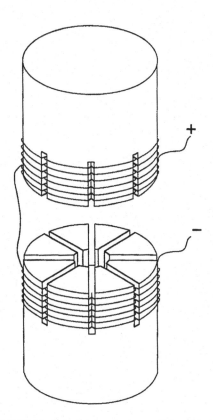

**Figure 10.1-43** Exploded view of a magnetizing fixture for an eight-pole disk magnet.

**Figure 10.1-44** Demagnetization curves of ceramic magnets (dashed lines = $\mu_0 M$): (a) ceramic 5; (b) ceramic 7; and (c) ceramic 8 [16].

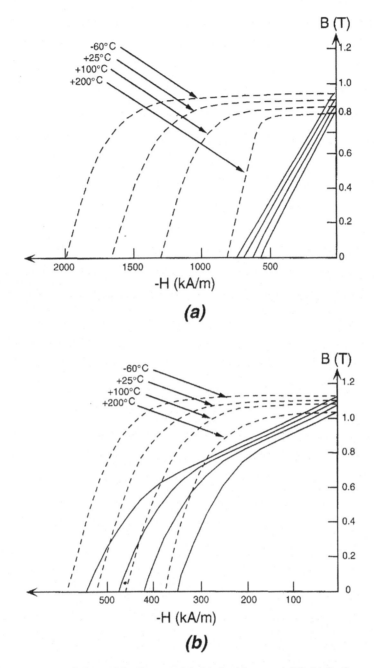

**Figure 10.1-45** Demagnetization curves of sintered SmCo materials (dashed lines = $\mu_0 M$): (a) SmCo$_5$; and (b) Sm$_2$Co$_{17}$ [17].

The magnetic properties and hence demagnetization curve of a magnet depend on many variables including temperature, pressure, and the applied field. The magnetic properties change due to variations in these variables, and such changes can be classified as reversible, irreversible, or structural.

Reversible changes occur when the magnetic properties change in response to a variation in a variable, but return to their original state when the variable returns to its initial value. For example, a magnet's hysteresis loop changes with temperature, and for a limited range of temperatures these changes are reversible and essentially linear. Such changes can be described by reversible temperature coefficients for the remanence $B_r$, coercivity $H_c$, and intrinsic coercivity $H_{ci}$. The coefficients are expressed as a percentage change per °C. All magnets suffer a reduction in $B_r$ as temperature increases. However, the change in $H_c$ can be positive or negative depending on the dominant anisotropic mechanism of the material. Most magnets exhibit a decrease in $H_c$ with temperature (a negative coefficient for $H_c$). However, ceramic magnets have a positive coefficient of $\approx 0.02\%/°C$. The designer must account for these changes to ensure that a magnet's operating point remains within an acceptable tolerance over

Table 10.1-9 Temperature coefficients of various permanent magnet materials

|  | Alnico 5 | Ferrite | SmCo$_5$ | NdFeB |
|---|---|---|---|---|
| Maximum service temp. (°C) | 520 | 400 | 250 | 150 |
| Curie temp. (°C) | 720 | 450 | 725 | 310 |
| $B_r$ Reversible coefficient (%/°C) | −0.02 | −0.20 | −0.04 | −0.12 |
| $H_{ci}$ Reversible coefficient (%/°C) | −0.03 | + 0.40 | −0.30 | −0.60 |

the range of operating temperatures. The demagnetization curves for ceramic and samarium cobalt materials at various temperatures are shown in Figs. 10.1-44 and 10.1-45, respectively. The temperature coefficients for various materials are listed in Table 10.1-9 [16].

In an irreversible change in magnetization, the magnetic properties change in response to a variation in a variable, and remain changed even when the variable returns to its initial value. To restore the original properties the magnet must be remagnetized. For example, a ferromagnetic specimen undergoes an irreversible change when its temperature is elevated above the Curie temperature (Section 10.1.6). Once above the Curie temperature, the specimen becomes paramagnetic. Its domains retain a random orientation as it cools, which renders it unmagnetized. The specimen can be remagnetized to its original state provided that its metallurgical structure has not been altered. Elevated temperatures can also cause changes in domain structure. In fact, immediately after magnetization a specimen undergoes a relatively slow process in which its unstable domains relax to a lower energy state via a variety of mechanisms including domain boundary movement. Such irreversible effects are collectively known as *magnetic viscosity*. This accelerates with temperature. Another example of an irreversible change is when an external demagnetizing field $H$ is applied that exceeds the intrinsic coercivity $H_{ci}$ of the specimen. The magnetization irreversibly flips to the opposite direction. Again, the specimen can be remagnetized to its original state.

Structural changes entail a permanent change in the metallurgical state of a magnet. Such changes cannot be undone by remagnetization. Corrosion and oxidation are examples of structural changes. Samarium cobalt, ceramic and Alnico magnets are resistant to oxidation, but rare-earth magnets are not. Oxidation is especially severe in sintered Nd$_2$Fe$_{14}$B. Oxygen diffuses into this material at elevated temperatures, causing an oxidized layer to form. The layer thickness $d$ grows according to the relation

$$d = f(T)\sqrt{t},$$

where $f(T)$ represents a nonlinear temperature dependence. The oxidized layer has a lower intrinsic coercivity than the body of the magnet, and is more easily demagnetized by the magnet's internal field. This layer degrades performance because: (a) it reduces the working volume of the magnet; and (b) it shunts the field from the magnet's interior. Oxidation can be controlled by varying the composition of the material, and/or coating the finished magnet. Common coating materials include nickel, zinc (Zn), and aluminum (Al) with thicknesses of 10–20 μm, or epoxy with a thickness of 20–30 μm.

# References

Jiles D. (1991). *Introduction to Magnetism and Magnetic Materials*, London: Chapman and Hall.

McCaig M. and Clegg A.G. (1987). *Permanent Magnets in Theory and Practice*, 2nd ed., New York: John Wiley and Sons.

Craik D. (1995). *Magnetism: Principles and Applications*, New York: John Wiley and Sons.

Chikazumi S. and Charap S. (1964). *Physics of Magnetism*, New York: John Wiley and Sons.

Cullity B.D. (1972). *Introduction to Magnetic Materials*, Reading, MA: Addison-Wesley.

Morrish A.H. (1983). *The Physical Principles of Magnetism*, Malabar, FL: R. Krieger Publishing Co.

Bozorth R.M. (1993). *Ferromagnetism*, New York: IEEE Press.

Messiah A. (1961). *Quantum Mechanics*, Amsterdam: North Holland.

Alonzo M. and Finn E.J. (1968). *Fundamentals of University Physics*, vol. III: *Quantum and Statistical Physics*, Reading, MA: Addison-Wesley.

Langevin P. (1905). *Annales de Chem. et Phys.* 5:70.

Weiss P. (1907). L'hypothèse du champ moléculaire et la propriété ferromagnétique. *J. de Phys.* 6:661.

Heisenberg W. (1928). Zur theorie des ferromagnetismus, Z. *Phys.* **49**:619.

Stoner E.C. and Wohlfarth E.P. (1948). A mechanism of magnetic hysteresis in heterogeneous alloys. *Phil. Trans. Roy. Soc.* **A240**:599.

Bloch F. (1932). Zur theorie des austauschproblems und der remanenzer-scheinung der ferromagnetika. *Z. Phys.* **74**:295.

Mayergoyz I.D. (1991). *Mathematical Models of Hysteresis*, New York: Springer-Verlag.

Parker R. (1990). *Advances in Permanent Magnetism*, 2nd ed., New York: John Wiley and Sons.

Campbell P. (1994). *Permanent Magnet Materials and their Application*, Cambridge: Cambridge Univ. Press.

Watson J.K. (1980). *Applications of Magnetism*, New York: John Wiley and Sons.

Went J.J., Rathenau G.W., Gorter E.W. and Van Oosterhout G.W. (1952). Ferroxdure, a class of new permanent magnet materials. *Philips Tech. Rev.* **13**: 194.

Strnat K., Hoffer G., Olson J., Ostertag W. and Becker J.J. (1967). A family of new cobalt based permanent magnetic materials. *J. Appl. Phys.* **38**: 1001.

Sagawa M., Fujimura S., Togawa N., Yamamoto H. and Matsuura Y. (1984). New material for permanent magnets on a base of Nd and Fe. *J. Appl. Phys.* **55**:2083.

Sagawa M., Horosawa S., Yamamoto H., Matsuura Y., Fujimura S., Tokuhara H. and Hiraga K. (1986). Magnetic properties of the bcc phase at grain boundaries in the Nd-Fe-B permanent magnet. *IEEE Trans. Magn.* **22**:910.

Croat J., Herbst J.F., Lee R.W. and Pinkerton F.E. (1984). Pr-Fe and Nd-Fe-based materials: a new class of high-performance permanent magnets. *J. Appl. Phys.* **55**:2078.

Lee, R.W. (1985). Hot-pressed neodymium-iron-boron magnets. *Appl. Phys. Lett.* **46**: 790.

Lee R.W., Brewer E.G. and Schaffel N.A. (1985). Processing of neodymium-iron-boron melt spun ribbons to fully dense magnets. *IEEE Trans. Magn.* **21** 1958.

Satoh K., Oka K., Ishii J. and Satoh T. (1985). Thermoplastic resin-bonded Sm-Co magnet. *IEEE Trans. Magn.* **21** 1979.

Lee J.K. and Furlani E.P. (1990). The optimization of multipole magnetizing fixtures for high-energy magnets. *J. Appl. Phys.* **67**(3).

# Section Eleven

## Nanomaterials

# Chapter 11.1

# Nanomaterials

## Introduction

Nanomaterials are an emerging family of novel materials that could be designed for specific properties. These materials will probably bring about significant shifts in the manner we design, develop, and use materials. For example, nanomaterials that are 1000 times stronger than steel, and 10 times lighter than paper, are cited as a possibility. The following properties can presumably be tailored: resistance to deformation and fracture, ductility, stiffness, strength, wear, friction, corrosion resistance, thermal and chemical stability, and electrical properties. With the emergence of novel fabrication and characterization technologies, new combinations of nanomaterials, or nanocomposites are beginning to be synthesized and characterized. Examples include nanoparticle-reinforced polymers for replacing structural metallic components in the auto industry for reduced fuel consumption and carbon dioxide emissions, and polymer composites containing nano-size inorganic clays as replacement for carbon black in tires for production of environmentally friendly, wear-resistant tires. Nanoparticles and fibers are too small to have substantial defects, and can be made stronger and used to develop ultra–high-strength composite materials.

The unique properties of nanomaterials result from the extremely large surface and interface area per unit volume (e.g., grain boundary area) and from the confinement effects at the nanoscale. These special attributes enable design of nanostructured materials that are harder and stronger but less brittle than comparable bulk materials with the same composition. The properties of *isolated* nanostructure units do not reflect the behavior of bulk nanomaterials (coatings, composites, etc.) that contain nanostructure networks in which the impact of interactions between nanostructured units modifies the properties. Nanoporous materials with large surface area such as aerogels and zeolites already offer improved chemical synthesis (faster reaction rates), cleanup (adsorbents), and separation (membranes, nanofilters) methods for chemical and biomedical sectors. In aeronautics and space exploration, lighter, stronger, and thermally stable nanostructured materials will permit fuel-efficient lifting of payloads into orbit, and reduced dependence on solar power for extended periods for travel away from the sun.

Nanotechnology is important for several reasons. Patterning matter at nanoscale will make it possible to control the fundamental properties of materials without changing their bulk composition (e.g., nanoparticles of different sizes emitting light at different frequencies and color, and nanoparticles of sizes comparable to magnetic domains for improved magnetic devices).

The ability to synthesize nanoscale building blocks with precisely controlled size and composition and then to assemble them into larger structures is believed to enable lighter, stronger, programmable, and self-healing materials to be synthesized. It will reduce life cycle costs through lower failure rates and permit molecular/cluster manufacturing (nanoscale manipulation and assembly of molecules). Improved printing ability with the use of nanoparticles (nanolithography), nanocoatings for cutting tools, and electronic and chemical applications are other potential benefits. Structural carbon and ceramic materials significantly stronger than steel, better heat-resistant polymeric materials stronger than the present generation of polymers, and nanofilters capable of removing finest contaminants from water and air are other potential benefits.

In the realm of nanoelectronics and computer technology, continued improvements in miniaturization, speed and power reduction in information-processing

*Engineering Materials and Processes Desk Reference*; ISBN: 9781856175869
Copyright © 2009 Elsevier Ltd; All rights of reproduction, in any form, reserved.

devices will be possible. Potential breakthroughs include nanostructured microprocessor devices, communication systems with higher transmission frequencies, small storage devices with capacities at multiterabit capacity, and integrated nanosensors capable of collecting, processing, and communicating massive amounts of data with minimal size, weight, and power consumption. Other benefits could include more sophisticated virtual reality systems for education, national defense, and entertainment, nanorobotics for nuclear waste management, chemical, biological, and nuclear sensing, and nano- and micromechanical devices for control of nuclear defense systems.

Prior to the emergence of nanoscience and nanotechnology as a discipline, several industrial sectors (chemical, aerospace) had developed novel technologies using the power of nanostructuring but without the aid of nanoscale analytical capabilities. For example, the aerospace industry had developed heat-resistant superalloys by dispersing 1 to 100 nanometer oxide particles in nickel superalloys with vastly improved elevated temperature strength and thermal stability, and the chemical industry had developed nanoporous catalysts with pore size of ~1 nm (their use is now the basis of a $30 billion/year industry).

Nanoporous ceramics for catalysis and filtration have been developed by the chemical industry. For example, several years ago Mobil Oil Company discovered a new class of zeolites, with pore size in the range of 0.45 to 0.6 nm, that is now widely used in hydrocarbon-cracking processes. The company also developed a porous aluminosilicate with 10-nm-size cylindrical pores; this development has been applied to both catalysis and filtration of fine dispersants in the environment. The discovery of the nanoporous material MCM-41 by the oil industry led to innovations in purification technologies (e.g., removal of ultrafine, 10–100 nm, contaminants from liquids and gases).

## Nanotubes, nanoparticles, and nanowires

Carbon nanotubes (CNT) are relatively new materials, discovered by Iijima in 1991, and were first observed as a minor by-product of the carbon arc process that is used to synthesize fullerenes. They present exciting possibilities for research and use. They have some remarkable properties, such as better electrical conductivity than copper, exceptional mechanical strength, and very high flexibility (with futuristic potential for use even in earthquake-resistant buildings and crash-resistant cars). There is already considerable interest in industry in the potential use of CNTs in chemical sensors, field emission elements, electronic interconnects in integrated circuits, hydrogen storage devices, and temperature sensors (Dai, 2003; Saito and Uemura, 2000; Wong and Lee, 2000).

Recently, record-breaking single-wall carbon nanotubes have been grown via catalytic chemical vapor deposition to a length of 40 mm by Zheng, O'Connell, Doorn, Liao, Zhao, Akhadov, Hoffbauer, Roop, Jia, Dye, Peterson, Huang, Liu, and Zhu (2004).

Since the discovery of CNTs, similar nanostructures were formed in other layered compounds such as BN, BCN, WS2, etc. (Bachtold, Hadley, Nakanishi, and Dekker, 2001). These different nanotubular materials offer different physical and engineering properties. For example, whereas CNTs are either metallic or semiconducting (depending on the shell helicity and diameter), BN nanotubes are insulating and could possibly serve as nanoshield for nanoconductors. BN nanotubes are also thermally more stable in oxidizing atmospheres than CNTs and have comparable modulus. The strength of nanotubular materials can be increased by assembling them in the form of ropes of 20–30 nm diameter and several micrometers in length. This has been done with CNT and BN nanotubes, with ropes made from SWCNTs being the strongest known material. The spacing between the individual nanotube strands in such a rope is in the subnanometer size range, e.g., ~0.34 nm in ropes made from multiwall BN nanotubes, which is on the order of the (0001) lattice spacing in the hexagonal BN cell. Due to their exceptional properties, there has been interest in incorporating nanotubes in polymers, ceramics, and metals.

Metal nanowires and nanoparticles are commonly formed by employing a template. For example, Au or Pt nanowires are first fabricated by flash evaporation and deposition of carbon onto glass. The coated glass is then thermally stressed by repeated dipping into liquid nitrogen, leading to cracks in the carbon film. Subsequent sputtering of Pt or Au and removal of surplus material results in nanowires. For certain special applications, metal powders in the micrometer-size range will be too coarse. For example, the catalytic property of gold particles occurs only at particle diameters of less than 3–5 nm.

Crystalline SiC nanowires have been created starting with thermal annealing at 1000 °C of polycrystalline Cu substrate and Si in a furnace. This produces grooves along grain boundaries in Cu. This Cu (copper) is used as a template for the subsequent growth of SiC nanowires. Methane gas is introduced in the furnace containing Cu template and Si, and 20 to 50 nm nanocrystals of SiC nucleate and grow to several 10s of micrometers along the Cu grain boundaries. These nanocrystals coalesce along the grain boundaries to form nanowires after sintering. The vapor–solid nucleation is suggested as the mechanism.

Pt nanowires can be mass-produced by reducing $H_2PtCl_6$ or $K_2PtCl_6$ by ethylene glycol (EG) at 110 °C in the presence of PVP (polyvinyl pyrrolidone) in air. This produces 5 nm diameter Pt nanoparticles. The Pt particles tend to agglomerate into spheres and larger structures. The nanowires grow at the surface of the agglomerates to which they are loosely attached, and are readily removed by sonic vibration. The Pt nanowires are separated from the agglomerates by centrifugation.

Nanoparticles are viewed by many as fundamental building blocks of nanotechnology. They are the starting point for many bottom-up approaches for preparing nanostructured materials and devices. As such, their synthesis is an important component of rapidly growing research efforts in nanoscale science and engineering. Nanoparticles of a wide range of materials can be prepared by a variety of methods including gas-phase synthesis for electronics-related applications. Nanoparticles are finding a myriad of uses, ranging from traditional applications, such as coloring agents (in stained-glass windows) and catalysts, to novel applications, such as magnetic drug delivery, hypothermic cancer therapy, contrast agents in magnetic resonance imaging, magnetic and fluorescent tags in biology, solar photovoltaics, nano bar codes, and emission control in diesel vehicles. This field is, in certain ways, reaching maturity, and to go to the next step it is becoming important to develop methods of scaling up the synthesis of these materials.

In vapor-phase synthesis of nanoparticles, conditions are created where the vapor-phase mixture is thermodynamically unstable relative to formation of the solid material to be prepared in nanoparticulate form. This includes the usual situation of a supersaturated vapor. It also includes "chemical supersaturation," in which it is thermodynamically favorable for the vapor-phase molecules to react chemically to form a condensed phase. If the degree of supersaturation is sufficient, and the reaction/condensation kinetics permit, particles will nucleate homogenously. Once nucleation occurs, remaining supersaturation can be relieved by condensation or reaction of the vapor-phase molecules on the resulting particles and particle growth will occur, rather than further nucleation. Therefore, to prepare small particles one wants to create a high degree of supersaturation, thereby inducing a high nucleation density, and then immediately quench the system, either by removing the source of supersaturation or slowing the kinetics, so that the particles do not grow.

Once particles form in the gas phase, they coagulate at a rate that is proportional to the square of their number concentration and that is only weakly dependent on particle size. At sufficiently high temperature, particles coalesce (sinter) faster than they coagulate, and spherical particles are produced. At lower temperatures, where coalescence is negligibly slow, loose agglomerates with quite open structures are formed. At intermediate conditions, partially sintered nonspherical particles are produced. If individual, nonagglomerated nanoparticles are desired, control of coagulation and coalescence is crucial. In contrast to the liquid phase, where a dispersion of nanoparticles can be stabilized indefinitely by capping the particles with appropriate ligands, nanoparticles in the gas phase will always agglomerate. So, by "nonagglomeration of nanoparticles," we usually mean particles agglomerated loosely enough that they can be redispersed without herculean effort, as compared to hard (partially sintered) agglomerates that cannot be fully redispersed. Figure 11.1-1 illustrates typical degrees of agglomeration

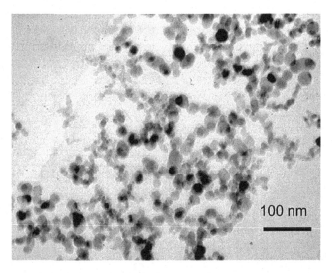

**Figure 11.1-1** *TEM image of agglomerated nanoparticles typical of those produced in many vapor-phase processes. These particular particles are silicon produced by laser pyrolysis of silane, but the degree of polydispersity and agglomeration is typical of many vapor-phase processes in which no special efforts have been made to avoid agglomeration or narrow the size distribution of the primary particles.* (Mark T. Swihart, Current Opinion in Colloid and Interface Science, 8, 2003, 127). Reprinted with permission from Elsevier. Photo courtesy of M. T. Swihart, State University of New York.

and polydispersity obtained in gas-phase processes when no special efforts have been made to control agglomeration or narrow the particle size distribution. In the cases of carbon black, fumed silica, and pigmentary titania, such particles are produced commercially in huge quantities.

In the following paragraphs, recent examples of and advances in gas-phase methods for preparing nanoparticles are reviewed. One useful way of classifying such methods is by the phase of precursor and the source of energy used to achieve a state of supersaturation. This part is structured around such a classification.

## Methods using solid precursors

One general class of methods of achieving the supersaturation necessary to induce homogeneous nucleation involves vaporizing the material into a background gas and cooling the gas.

### Inert gas condensation

The most straightforward method of achieving supersaturation is to heat a solid to evaporate it into a background gas, then mix the vapor with a cold gas to reduce the temperature. This method is well suited for production of metal nanoparticles, because many metals evaporate at reasonable rates at attainable temperatures. By including a reactive gas, such as oxygen, in the cold gas stream, oxides or other compounds of the evaporated material can be prepared. Detailed, systematic modeling and experimental study of this method, as applied to preparation of bismuth nanoparticles, including both visualization and computational fluid dynamics simulation of the flow fields in their reactor, have been presented in the literature. It has been shown that the particle size distribution can be controlled by controlling the flow field dynamics and the mixing of the cold gas with hot gas carrying the evaporated metal. Other advances of this method have been in preparing composite nanoparticles and in controlling the morphology of single-component nanoparticles by controlled sintering after particle formation. Maisels, Kruis, Fissan, and Rellinghaus (2000) prepared composite nanoparticles of PbS with Ag by separate evaporation/condensation of the two materials followed by coagulation of oppositely charged PbS and Ag particles selected by size and charge. Ohno et al. (2002) prepared Si-In, Ge-In, Al-In, and Al-Pb composite nanoparticles by condensation of In or Pb onto Si, Ge, or Al particles prepared by inert gas condensation and brought directly into a second condensation reactor.

### Pulsed laser ablation

Another method for the gas-phase synthesis of nanoparticles of various materials is based on the pulsed-laser vaporization of metals in a chamber filled with a known amount of a reagent gas followed by controlled condensation of nanoparticles onto the support. A schematic view of the installation for synthesizing nanoparticles is shown in Figure 11.1-2. As metal atoms diffuse from the target to the support, they interact with the gas to form the desired compound (for instance, oxide in the case of oxygen, nitride for nitrogen or ammonia, and carbide for methane). The pulsed-laser vaporization of metals in the chamber makes it possible to prepare nanoparticles of mixed molecular composition, such as mixed oxides-nitrides or mixtures of oxides of different metals. Along with the reagent gas, the chamber contains an inert gas, such as He or Ar, at a pressure of $10^{-21}$ Torr, which favors the establishment of steady convection between the heated bottom plate and cooled top plate. In a typical experiment with single pulse of a Nd:YAG laser (532 nm, 15–30 mJ/pulse, $10-9$ s pulse duration), over $10^{14}$ metal atoms are vaporized. A new compound is formed due to the reaction between the "hot" metal atoms and the gas molecules, which is accompanied by the energy loss of the molecules formed by collisions with the inert gas atoms. The metal atoms that did not enter into

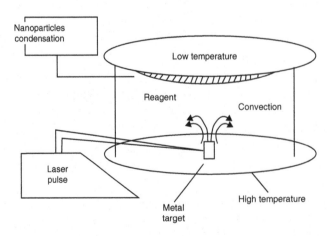

**Figure 11.1-2** *Schematic view of the installation for the synthesis of nanoparticles by laser vaporization of metals.*

reaction and the molecules of the new compound are carried by convection to the nucleation zone on the cooled top plate.

By changing the composition of the inert gas and the reagent gas in the chamber and by varying the temperature gradient and laser pulse power, it is possible to control the elemental composition and size of nanoparticles that are obtained. Marine et al. (2000) presented a recent analysis of this method in which they also reviewed its development. Nakata et al. (2002) used a combination of laser-spectroscopic imaging techniques to image the plume of Si atoms and clusters formed during synthesis of Si nanoparticles. They investigated the dependence of the particle formation dynamics on the background gas and found that it was substantial. Some other recent examples include the preparation of magnetic oxide nanoparticles by Shinde et al. (2000), titania nanoparticles by Harano et al. (2002) and hydrogenated-silicon nanoparticles by Makimura et al. (2002).

## Ion sputtering

A final means of vaporizing a solid is via sputtering with a beam of inert gas ions. Urban et al. (2002) recently demonstrated formation of nanoparticles of a dozen different metals using magnetron sputtering of metal targets. They formed collimated beams of the nanoparticles and deposited them as nanostructured films on silicon substrates. This process must be carried out at relatively low pressures ($\sim 1$ mTorr), which makes further processing of the nanoparticles in aerosol form difficult.

## Methods using liquid or vapor precursor

An alternate means of achieving the supersaturation required to induce homogenous nucleation of particles is chemical reaction. Chemical precursors are heated and/or mixed to induce gas-phase reactions that produce a state of supersaturation in the gas phase.

## Chemical vapor synthesis

In this approach, vapor-phase precursors are brought into a hot-wall reactor under conditions that favor nucleation of particles in the vapor phase rather than deposition of a film on the wall. Chemical vapor synthesis or chemical vapor deposition (CVD) are the processes used to deposit thin solid films on surfaces. This method has tremendous flexibility in producing a wide range of materials and can take advantage of the huge database of precursor chemistries that have been developed for CVD processes. The precursors can be solid, liquid, or gas at ambient conditions.

There are many good examples of the application of this method in the recent literature. Ostraat et al. (2001) have demonstrated a two-stage reactor for producing oxide-coated silicon nanoparticles that have been incorporated into high-density nonvolatile memory devices. By reducing the silane precursor composition to as low as 10 parts per billion, they were able to produce nonagglomerated single-crystal spherical particles with mean diameter below 8 nm. This is one of relatively few examples of a working microelectronic device in which vapor-phase–synthesized nanoparticles perform an active function. In other recent examples of this approach, Magnusson et al. (2000) produced tungsten nanoparticles by decomposition of tungsten hexacarbonyl, and Nasibulin et al. (2002) produced copper acetylacetonate.

## Laser pyrolysis

An alternate means of heating the precursors to induce reaction and homogenous nucleation is absorption of laser energy. Compared to heating the gases in a furnace, laser pyrolysis allows highly localized heating and rapid cooling, because only the gas (or a portion of the gas) is heated and its heat capacity is small. Heating is generally done using an infrared ($CO_2$) laser, whose energy is either absorbed by one of the precursors or by an inert photosensitizer such as sulfur hexafluoride. The silicon particles shown in Figure 11.1-1 were prepared by laser pyrolysis of silane. Nanoparticles of many materials have been made using this method. A few recent examples are $MoS_2$ nanoparticles prepared by Borsella et al. (2001), SiC nanoparticles produced by Kamlag et al. (2001), and Si nanoparticles prepared by Ledoux et al. (2002). Ledoux et al. used a pulsed $CO_2$ laser, thereby shortening the reaction time and allowing preparation of even smaller particles.

## Synthesis of nanoparticles by chemical methods

A number of different chemical methods can be used to make nanoparticles of metals and semiconductors. Several types of reducing agents can be used to produce nanoparticles such as $NaBEt_3H$, $LiBEt_3H$, and $NaBH_4$, where Et denotes the ethyl ($\cdot C_2H_5$) radical. For example, nanoparticles of molybdenum (Mo) can be reduced in toluene solution with $NaBEt_3H$ at room temperature, providing a high yield of Mo nanoparticles having dimensions of $1-5$ nm. The equation for the reaction is

$$MoCl_3 + 3NaBEt_3H \Rightarrow Mo + 3NaCl + 3BEt_3 + (3/2)H_2$$

Nanoparticles of aluminum have been made by decomposing $Me_2EtNAlH_3$ in toluene and heating the solution to 105 °C for 2 h (Me is methyl, $\cdot CH_3$). Titanium isopropoxide is added to the solution. The titanium acts as a catalyst for the reaction. The choice of catalyst determines the size of the particles produced. For instance, 80-nm particles have been made using titanium. A surfactant such as oleic acid can be added to the solution to coat the particles and prevent aggregation.

Nanoparticles of metal sulfides are usually synthesized by a reaction of a water-soluble metal salt and $H_2S$ or $Na_2S$ in the presence of an appropriate stabilizer, such as sodium metaphosphate. For example, the CdS nanoparticles can be synthesized by mixing $Cd(ClO_4)_2$ and $Na_2S$ solutions:

$$Cd(ClO_4)_2 + Na_2S = CdS(1-10\ nm) + 2NaClO_4$$

The growth of the CdS nanoparticles in the course of reaction is arrested by an abrupt increase in pH of the solution. Very recently Peng and Peng (2001) reproduced rice-shape CdSe nanocrystals (shown in Figure 11.1-3) by using CdO as precursor.

## Nanoparticles: Biomedical applications

In the past nanoparticles were studied because of their size-dependent physical and chemical properties (Murry et al., 2000). At present they have entered a commercial exploration period. In this section, biomedical applications of nanoparticles are considered.

Living organisms are built of cells that are typically 10 μm across. However, the cell parts are much smaller and are in the submicron-size domain. Even smaller are the proteins, with a typical diameter of just 5 nm, which is comparable with the dimensions of smallest manmade nanoparticles. This simple size comparison gives an idea of using nanoparticles as very small probes that would allow us to look at cellular machinery without introducing too much interference. Nowadays nanoparticles have many applications in biology and medicine; for example, (1) drug and gene delivery, (2) biodetection of pathogens, (3) fluorescent biological labels, (4) detection of proteins, (5) probing DNA structure, (6) tissue engineering, (7) tumor destruction via heating (hyperthermia), (8) separation and purification of biological molecules and cells, (9) MRI contrast enhancement, and (10) phagokinetic studies.

As mentioned above, the fact that nanoparticles exist in the same size domain as proteins makes nanomaterials suitable for biotagging or labeling. However, size is just one of many characteristics of nanoparticles, which by itself is rarely sufficient if one is to use

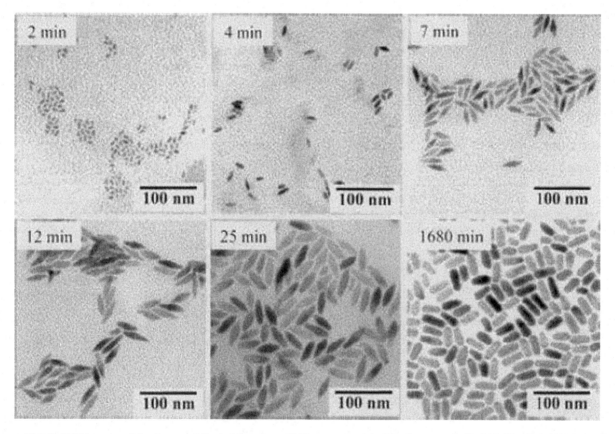

**Figure 11.1-3** *TEM images of the time evolution of rice-shape CdSe nanocrystals. The times are indicated.* (Reprinted by the order of Z. A. Peng and X. Peng, Formation of high-quality CdTe, CdSe, and CdS nanocrystals using CdO as precursor, Journal of the American Chemical Society, 123, 2001, 183). Reprinted with permission from the American Chemical Society, 1155 16th St. NW, Washington, DC 20036. Photo Courtesy of X. Peng, University of Arkansas.

nanoparticles as biological tags. To interact with a biological target, a biological or molecular coating or layer acting as a bioinorganic interface should be attached to the nanoparticle. Examples of biological coatings may include antibodies, biopolymers such as collagen (Sinani et al., 2003), or mono-layers of small molecules that make the nanoparticles biocompatible (Zhang et al., 2002). In addition, as optical detection techniques are widespread in biological research, nanoparticles should either fluoresce or change their optical properties. The approaches used in constructing nanobiomaterials are schematically presented in Figure 11.1-4. Nanoparticles usually form the core of a nanobiomaterial. The material can be used as a convenient surface for molecular assembly, and may be composed of inorganic or polymeric materials. It can also be in the form of a nanovesicle surrounded by a membrane or a layer. The shape is more often spherical, but cylindrical, platelike, and other shapes are possible. The size and size distribution might be important in some cases; for example, if penetration through a pore structure of a cellular membrane is required. The size and size distribution become extremely critical when quantum-size effects are used to control material properties. A tight control of the average particle size and a narrow distribution of sizes allow the creation of very efficient fluorescent probes that emit narrow light in a very wide range of wavelengths. This helps with creating biomarkers with many and well-distinguished colors. The core itself may have several layers and be multifunctional. For example, by combining magnetic and luminescent layers one can both detect and manipulate the particles. The core particle is often protected by several monolayers of inert material, such as silica. Organic molecules that are adsorbed or chemisorbed on the surface of the particle are also used for this purpose. The same layer may act as a biocompatible material. However, more often an additional layer of linker molecules is required to proceed with further functionalization. This linear linker molecule has reactive groups at both ends. One group is aimed at attaching the linker to the nanoparticle surface, and the other is used to bind various moieties such as biocompatibles. Recent developments using nanoparticles in biomedical applications are described later in more detail.

### Tissue engineering

Natural bone surface quite often contains features that are about 100 nm across. If the surface of an artificial

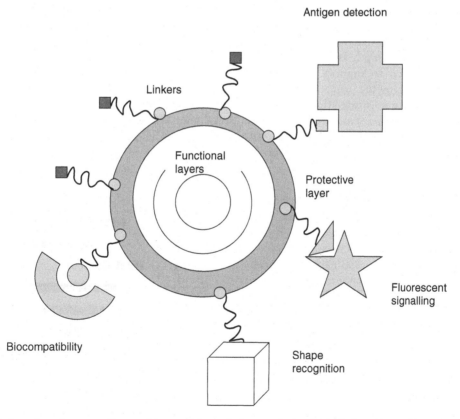

**Figure 11.1-4** *Typical configurations used in nanobiomaterials applied to medical or biological problems.* (O. V. Salata et al., Journal of Nanobiotechnology, 2, 2004, 3). Reprinted with permission from O. V. Salata.

bone implant were left smooth, the body would try to reject it. Because of that, the smooth surface is likely to cause production of a fibrous tissue covering the surface of the implant. This layer reduces the bone–implant contact, which may result in loosening of the implant and further inflammation. It has been demonstrated that by creating nano-sized features on the surface of the hip or knee prosthesis, one can reduce the chances of rejection of a hip or knee prosthesis and stimulate the production of osteoblasts, which are the cells responsible for the growth of the bone matrix and which are found on the advancing surface of the developing bone. The effect has been demonstrated with polymeric, ceramic, and, more recently, metal materials. For example, in one study more than 90% of the human bone cells formed a suspension that adhered to the nano-structured metal surface but only 50% in the control sample did so. In the end, this finding will allow design of a more durable and longer lasting hip or knee replacement and reduce the chances of the implant getting loose.

Titanium is a well-known bone-repairing material widely used in orthopedics and dentistry. It has high fracture resistance, ductility, and strength-to-weight ratio. Unfortunately, it suffers from the lack of bioactivity, as it does not support cell adhesion and growth. Apatite coatings are known to be bioactive and to bond to the bone. Hence, several techniques were used in the past to produce an apatite coating on titanium. Those coatings suffer from thickness nonuniformity, poor adhesion, and low mechanical strength. In addition, a stable porous structure is required to support the nutrients' transport through the cell growth.

It has been shown that using the biomimetic approach—a slow growth of nanostructured apatite film from the simulated body fluid—results in the formation of a strongly adherent and uniform nanoporous layer. The layer is built of nanometric crystallites and has a stable nanoporous structure and bioactivity.

A real bone is a nanocomposite material, composed of hydroxyapatite crystallites in the organic matrix, which mainly consists of collagen. Therefore, the bone is mechanically both tough and plastic, so it can recover from mechanical damage. The actual nanoscale mechanism leading to this useful combination of properties is still debated. Recently, an artificial hybrid material was prepared from 15- to 18-nm ceramic nanoparticles and poly (methyl methacrylate) copolymer. This hybrid material, deposited as a coating on the tooth surface, improved scratch resistance and exhibited a healing behavior similar to that of the tooth.

## Manipulation of cells and biomolecules

Functionalized magnetic nanoparticles have found many biological and medical applications (Chung et al., 2004).

The size of the particles can range from a few nanometers to several microns (shown in Figure 11.1-5) and thus is compatible with biological entities ranging from proteins (a few nm) to cells and bacteria (several µm). Generally the magnetic particles are coated with a suitable ligand, which allows chemical binding of the particles to different biological environments. Conversely, absence of ferromagnetism in most biological systems, which typically have only diamagnetism or paramagnetism, means that in a biological environment the magnetic moment from the ferromagnetic particles can be detected with little noise.

Based on these ideas, a variety of applications has emerged. A direct application is to bind magnetic particles to the biological system of interest, which then allows manipulating the biological material via magnetic field gradients. This can be used for high-gradient magnetic field separation, which has already been applied to several problems, such as separating red blood cells from blood, cancer cells from bone marrow, and radioactive isotopes from food products. In contrast, one can use magnetic nanoparticles for targeted drug delivery. In this case, a drug is bound to a magnetic particle and either DC magnetic fields are used to confine the drug in a specific location of the body, or AC magnetic fields are used to trigger the release of the drugs. A related application uses magnetic nanoparticles for hypothermal treatment, wherein heating the diseased tissue destroys the cancerous cells. This can be achieved by heating magnetic nanoparticles with AC magnetic fields to about 42 °C for at least 30 minutes. Besides these therapeutic applications, magnetic nanoparticles can also be used for diagnostics. One common application is the enhancement of contrast in magnetic resonance imaging, where the local stray field of the magnetic nanoparticles can modify the magnetic relaxation of the surrounding tissue. Last but not least, magnetic nanoparticles are also used for

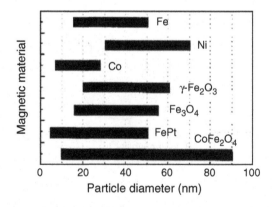

**Figure 11.1-5** *Particle diameters for stable single-domain magnetic nanoparticles.* (Reprinted by the order of S. H. Chung, A. Hoffmann, S. D. Bader, C. Liu, B. Kay, and L. Chen, Applied Physics Letters, 85, 2004, 2971).

biomagnetic sensing, where a target of interest is typically tagged with the magnetic particles, so that stray fields of the nanoparticles can be used for signal transduction.

## Protein detection

Proteins are an important part of the cell's language, machinery, and structure, and understanding their functionalities is extremely important for further progress in human well-being. Gold nanoparticles are widely used in immunohistochemistry to identify protein–protein interaction. However, the multiple simultaneous detection capabilities of this approach are fairly limited. Surface-enhanced Raman scattering spectroscopy is a well-established technique for detection and identification of single dye molecules. By combining both methods in single nanoparticle probe, one can drastically improve the multiplexing capabilities of protein probes. Mirkin et al. (2003) designed a sophisticated multifunctional probe that was built around 13-nm gold nanoparticles. The nanoparticles are coated with hydrophilic oligonucleotides containing Raman dye at one end and terminally capped with a small molecule-recognition element (e.g., biotin). Moreover, this molecule is catalytically active and is coated with silver in the solution of Ag(I) and hydroquinone. After the probe is attached to a small molecule or an antigen it is designed to detect, the substrate is exposed to silver and hydroquinone solution. A silver-plating occurs close to the Raman dye, which allows for dye signature detection with a standard Raman microscope. Apart from being able to recognize small molecules, this probe can be modified to contain antibodies on the surface to recognize proteins. When tested in the protein array format against both small molecules and proteins, the probe shows no cross-reactivity.

## Cancer therapy

Photodynamic cancer therapy is based on the destruction of the cancer cells by laser-generated atomic oxygen, which is cytotoxic. A greater quantity of special dye that is used to generate the atomic oxygen is taken in by the cancer cells when compared with a healthy tissue. Hence, only the cancer cells are destroyed when exposed to a laser radiation. Unfortunately, the remaining dye molecules migrate to the skin and the eyes and make the patient very sensitive to daylight exposure. This effect can last for up to six weeks. To avoid this side effect, a hydrophobic version of the dye molecule was enclosed inside a porous nanoparticle (Cao et al., 2003). The dye stayed trapped inside the Ormosil nanoparticle and did not spread to the other parts of the body. At the same time, its oxygen-generating ability had not been affected, and the pore size of about 1 nm freely allowed the oxygen to diffuse out.

## Semiconductor nanowires

Semiconductor nanowires exhibit novel electronic and optical properties owing to their unique structural one-dimensionality and possible quantum confinement effects in two dimensions. With a broad selection of compositions and band structures, these one-dimensional semiconducting nanostructures are considered the critical components in a wide range of potential nanoscale device applications. The understanding of general nanocrystal growth mechanisms serves as the foundation for the rational synthesis of semiconductor heterostructures in one dimension. Availability of these high-quality semiconductor nanostructures allows systematic structure-property correlation investigations, particularly the effects of size and dimensionality. Novel properties including nanowire microcavity lasing, phonon transport, interfacial stability, and chemical sensing are surveyed. This section is divided into three subsections. The first section explores the advances in gas-phase production methods, especially the vapor–liquid–solid (VLS) and vapor–solid (VS) processes with which most one-dimensional heterostructures and ordered arrays are now grown. Several approaches for fabricating one-dimensional nanostructures in solution, focusing especially on those that use a selective capping mechanism are discussed. In the second section, focus is on interesting fundamental properties exhibited by rods, wires, belts, and tubes. In the third section, progress in the assembly of one-dimensional nanostructures into useful architectures is addressed and illustrates the construction of novel devices based on such schemes.

### General synthetic strategies

A novel growth mechanism should satisfy three conditions: it must (1) explain how one-dimensional growth occurs, (2) provide a kinetic and thermodynamic rationale, and (3) be predictable and applicable to a wide variety of systems. Growth of many one-dimensional systems has been experimentally achieved without satisfactory elucidation of the underlying mechanism, as is the case for oxide nanoribbons. Nevertheless, understanding the growth mechanism is an important aspect for developing a synthetic method for generating one-dimensional nanostructures of desired material, size, and morphology.

In general, high-quality, single crystal nanowire materials are synthesized by promoting the crystallization of solid-state structures along one direction. The actual mechanisms of coaxing this type of crystal growth include (1) growth of an intrinsically anisotropic crystallographic structure, (2) the use of various templates with one-dimensional nanostructures, (3) the introduction of a liquid–solid interface to reduce the symmetry of a seed, (4) use of an appropriate capping reagent to kinetically control the growth rates of various facets of a seed, and (5) the self-assembly of zero-dimensional nanostructures.

The ability to form heterostructures through carefully controlled doping and interfacing is responsible for the success of semiconductor integrated-circuit technology, and the two-dimensional semiconductor interface is ubiquitous in optoelectronic devices such as light-emitting diodes (LEDs), laser diodes, quantum cascade lasers, and transistors (Weisbuch and Vinter, 1991). Therefore, the synthesis of one-dimensional heterostructures is equally important for potential future applications, including efficient light-emitting sources and thermoelectric devices. This type of one-dimensional nanoscale heterostructure can be rationally prepared once the fundamental one-dimensional nanostructure growth mechanisms are understood.

In general two types of one-dimensional heterostructures can be formed: longitudinal heterostructures and coaxial heterostructures. The term "longitudinal heterostructures" refers to nanowires composed of different stoichiometries along the length of the nanowire, and "coaxial heterostructures" refers to nanowire materials having different core and shell compositions. Various approaches to fabricate heterostructure and inorganic nanotube materials derived from three-dimensional bulk crystal structures are discussed here.

Among all vapor-based methods, the VLS mechanism seems the most successful for generating a large quantity of nanowires with single-crystal structures. This process was originally developed in the 1960s by Wagner and Ellis to produce micrometer-sized whiskers (Wagner and Ellis, 1964), later justified thermodynamically and kinetically (Givargizov, 1975), and recently reexamined by other researchers to generate nanowires and nanotubes from a rich variety of inorganic materials (Wu and Yang, 2000; Zhang et al., 2001; Westwater et al., 1997; Wu and Wang, 2001; Gudiksen and Lieber, 2000; Wu et al., 2002; Duan and Lieber, 2000; Chen et al., 2001; Zhang et al., 2001; He et al., 2001; Shi et al., 2001).

Figure 11.1-6 shows image from in situ transmission electron microscopy (TEM) technique to monitor the VLS growth mechanism in real time (Wu and Wang, 2001). A typical VLS process starts with the dissolution of gaseous reactants into nanosized liquid droplets of a catalyst metal, followed by nucleation and growth of single crystal rods and then wires. The one-dimensional growth is induced and dictated by liquid droplets, whose sizes remain essentially unchanged during the entire process of wire growth. Each liquid droplet serves as a virtual template to strictly limit the lateral growth of an individual wire. The major stages of the VLS process can be seen in the example of Figure 11.1-6, where the growth of a Ge nanowire as observed by in situ TEM is shown. Based on the Ge-Au binary-phase diagram, Ge and Au form liquid alloys when the temperature is raised above the eutectic point (361 °C). Once the liquid droplet is supersaturated with Ge, nanowire growth will start at the solid-liquid

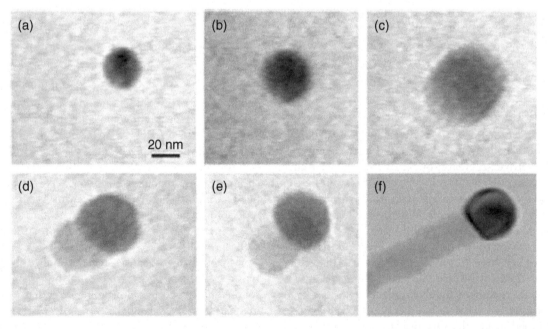

**Figure 11.1-6** In situ TEM images recorded during the process of nanowire growth. (a) Au nanoclusters in solid state at 500 °C; (b) alloying initiates at 800 °C; at this stage, Au exists mostly in solid-state; (c) liquid Au-Ge alloy; (d) the nucleation of Ge nanocrystals on the alloy surface; (e) Ge nanocrystal elongates with further Ge condensation; and (f) eventually forms a wire. (Reprinted with permission of Y. Wu and P. Yang, Journal of the American Chemical Society, 123, 1999, 3165–3166). Reprinted with permission from the American Chemical Society, 1155 16th St. NW, Washington DC 20036. Photo Courtesy of P. Yang, University of California, Berkeley.

interface. The establishment of the symmetry-breaking solid–liquid interface is the key step for the one-dimensional nanocrystal growth in this process, whereas the stoichiometry and lattice symmetry of the semiconductor material systems are less relevant. The growth process can be controlled in various ways. Because the diameter of each nanowire is largely determined by the size of the catalyst particle, smaller catalyst islands yield thinner nanowires or tubes. The VLS process has now become a widely used method for generating one-dimensional nanostructures from a rich variety of pure and doped inorganic materials that include elemental semiconductors (Si, Ge) (Wu and Yang, 2000; Zhang et al., 2001; Westwater et al., 1997), III–V semiconductors (GaN, GaAs, GaP, InP, InAs) (Gudiksen and Lieber, 2000; Wu et al., 2002; Duan and Lieber, 2000; Chen et al., 2001; Zhang et al., 2001; He et al., 2001; Shi et al., 2001; Chen and Yeh, 2000; Shimada et al., 1998; Hiruma et al., 1995; Yazawa et al., 1993; Kuykendall et al., 2003; Zhong et al., 2003), II–VI semiconductors (ZnS, ZnSe, CdS, CdSe) (Wang et al., 2002; Wang et al., 2002; Lopez-Lopez et al., 1998), oxides (indium-tin oxide, ZnO, MgO, SiO2, CdO) (Wang et al., 2002; Lopez-Lopez et al., 1998; Peng et al., 2002; Yang and Lieber, 1996; Wu et al., 2001; Huang et al., 2001; Liu et al., 2003; Naguyen et al., 2003), carbides (SiC, B$_4$C) (Ma and Bando, 2002; Kim et al., 2003), and nitrides (Si$_3$N$_4$) (Kim et al., 2003). The nanowires produced using the VLS approach are remarkable for their uniformity in diameter, which is usually on the order of 10 nm over a length scale of >1 μm.

### Fabrication of semiconductor nanowires

High-crystalline silicon nanowires are ingredients for electronic devices, light-emitting devices, field emission sources, and sensors. Thermal vapor growth from solid precursors, usually in a high-temperature furnace, is the most common way to achieve bulk production of nanowires.

Typically, quartz or alumina boats filled with suitable precursor powders are placed in a furnace tube and heated to high temperatures, while a lower temperature substrate is placed at one end of the tube, in the downstream direction of inert gas flow (Ar or N$_2$). Schematic diagrams of experimental setup for the synthesis of semiconductor nanostructures are shown in Figure 11.1-7. In the case of silicon, different strategies are possible for the growth of nanowires. One strategy is to anneal Si-SiO$_2$ powders well above 1000 °C to evaporate them. Silicon nanowires then grow on the substrate either by the vapor-liquid-solid method, by placing gold as the catalyst, or by the so-called oxide-assisted method, which does not require a catalyst particle and is triggered by the self-condensation of the vapor in a low-temperature region of the furnace. Wang et al., (2005)

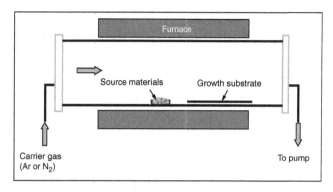

**Figure 11.1-7** *Schematic of experimental setup for the synthesis of semiconductor nanostructures.* (Zhong Lin Wang, Journal of Materials Chemistry, 15, 2005, 1021). Reprinted with permission from The Royal Society of Chemistry, Thomas Graham House, Science Park, Cambridge, U.K. Photo Courtesy of Z. L. Wang, Georgia Institute of Technology.

synthesized a wide range of polar-surface–dominated nanostructures of ZnO under controlled conditions by thermal evaporation of solid powders at high yield (Figure 11.1-8).

Recently, Xu et al. (2005) grew silicon nanowires by thermal chemical vapor deposition (Figure 11.1-9). This method is based on the VLS idea, in which gold acts like a catalyst. VLS mechanism and high-resolution transmission electron microscopy (HRTEM) of Si nanowires are presented in Figures 11.1-10 and 11.1-11, respectively. In another example of the growth of GaN nanowires prepared by using metal organic chemical vapor deposition (MOCVD), the SEM, TEM, and HRTEM images of a sample are shown in Figure 11.1-12.

A few of the major disadvantages of high-temperature approaches to nanowires synthesis include the high cost of fabrication and scaleup, and the inability to produce metallic wires. Recent progress using solution-phase techniques has resulted in the creation of one-dimensional nanostructures in high yields (gram scales) via selective capping mechanisms. It is believed that molecular capping agents play a significant role in the kinetic control of the nanocrystal growth by preferentially adsorbing to specific crystal faces, thus inhibiting growth of that surface (although defects could also induce such one-dimensional crystal growth). The growth of semiconductor nanowires has also been realized using a synthetic mechanism. Microrods of ZnO have been produced via the hydrolysis of zinc salts in the presence of amines (Vayssieres et al., 2001). Hexamethylenetetramine as a structural director has been used to produce dense arrays of ZnO nanowires in aqueous solution (Figure 11.1-13) having controllable diameter of 30-100 nm and lengths of 2-10 μm (Greene et al., 2003). Most significantly, these oriented nanowires can be prepared on any substrate. The growth process ensures that a majority of the nanowires in the array are in

# CHAPTER 11.1 Nanomaterials

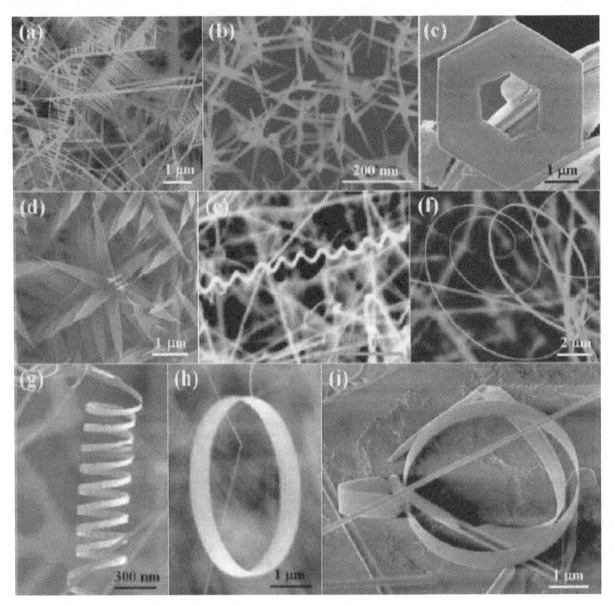

Figure 11.1-8 A collection of polar-surface induced/dominated nanostructures of ZnO, synthesized under controlled conditions by thermal evaporation of solid powders unless notified otherwise: (a) nanocombs induced by asymmetric growth on the Zn-(0001) surface, (b) tetraleg structure due to catalytically active Zn-(0001) surfaces, (c) hexagonal disks/rings synthesized by solution-based chemical synthesis, (d) nanopropellers created by fast growth along the c-axis, (e) deformation-free nanohelices as a result of block-by-block self-assembly, (f) spiral of a nanobelt with increased thickness along the length, (g) nanosprings, (h) single-crystal seamless nanoring formed by loop-by-loop coiling of a polar nanobelt, and (i) a nanoarchitecture composed of a nanorod, nanobow, and nanoring. (Z. W. Wang et al., Journal of Materials Chemistry 15, 2005, 1021). Reprinted with permission from The Royal Society of Chemistry, Thomas Graham House, Science Park, Cambridge U.K. Photo Courtesy of Z. L. Wang, Georgia Institute of Technology.

direct contact with the substrate and provide a continuous pathway for carrier transport, an important feature for future electronic devices based on these materials.

## Fabrication of metal nanowires

Metal nanowires are very attractive materials because their unique properties may lead to a variety of applications. Examples include interconnects for nanoelectronics, magnetic devices, chemical and biological sensors, and biological labels. Metal nanowires are also attractive because they can be readily fabricated with various techniques. Various methods for fabricating metal nanowires are discussed next.

The diameters of metal nanowires range from a single atom to a few hundreds of nm. The lengths vary over an even greater range: from a few atoms to many micrometers. Because of the large variation in the aspect ratio

# Nanomaterials  CHAPTER 11.1

**Figure 11.1-9** *(a) SEM image of aligned amorphous silicon nanowires grown on silicon wafers at 1300 °C by thermal CVD method. (b) High-magnification image illustrates that the diameter of silicon nanowires ranges between 80 and 100 nm. (c) TEM image of silicon nanowires and inset shows that the silicon wires are pure amorphous.* (Y. Xu, C. Cao, B. Zhang, and H. Zhu. Preparation of aligned amorphous silica nanowires. Chemistry Letters, 34, 2005, 414). Reprinted with permission from The Chemical Society of Japan, Chiyoda-Ku, Tokyo 101-8307, Japan.

(length to diameter ratio), different names have been used in the literature to describe the wires; those with large aspect ratios (e.g., >20) are called nanowires, whereas those with small aspect ratios are called nanorods. When short "wires" are bridged between two larger electrodes, they are often referred to as nanocontacts. In terms of electron transport properties, metal wires have been described as classical wires and quantum wires. The electron transport in a classical wire obeys the classical relation

$$G = \sigma \frac{A}{L}$$

where $G$ is the wire conductance, and $L$ and $A$ are the length and the cross-sectional area of the wire, respectively; $\sigma$ is the conductivity, which depends on the material of the wire.

## Electrochemical fabrication of metal nanowires

A widely used approach to fabricating metal nanowires is based on various templates, which include negative, positive, and surface step templates. Each approach is discussed below.

### Negative template methods

Negative template methods use prefabricated cylindrical nanopores in a solid material as templates. Depositing metals into the nanopores fabricates nanowires with a diameter predetermined by the diameter of the nanopores. There are several ways to form nanowires, but the electrochemical method is a general and versatile method. If one dissolves away the host solid material, free-standing nanowires are obtained. This method may

## CHAPTER 11.1 Nanomaterials

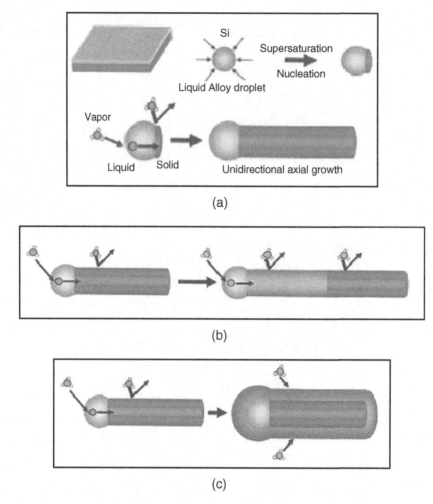

**Figure 11.1-10** *Illustration of nanowire synthesis process, the main concept. (a) Vapor-liquid-solid (VLS) mechanism for nanowire growth. (b) Synthesis method to form nanowire axial heterostructure. (c) Synthesis method for nanowire core-shell heterostructure formation. (Reprinted by D. C. Bell et al., Microscopy Research and Technique, 64, 2004, 373). Reprinted with permission from Z. L. Wang, Georgia Institute of Technology.*

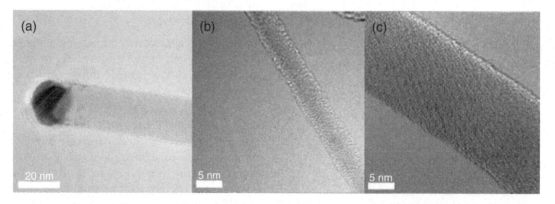

**Figure 11.1-11** *HRTEM images of silicon nanowires. (a) Silicon nanowire showing attachment of gold catalyst particle on the end. (b) Thin ~5-nm silicon nanowire showing lack of contrast along edges, but atomic structure information indicated along the length of the wire. (c) Silicon nanowire ~17 nm wide; micrograph shows clear atomic structure detail. (D. C. Bell et al., Microscopy Research and Technique, 64, 2004, 373). Reprinted with permission from Z. L. Wang, Georgia Institute of Technology.*

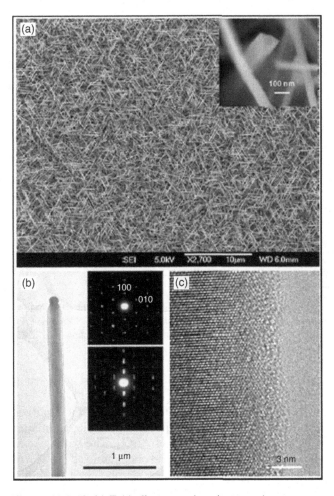

**Figure 11.1-12** (a) Field-effect scanning electron microscope (FESEM) image of the GaN nanowires grown on a gold-coated -plane sapphire substrate. Inset shows a nanowire with its triangular cross-section. (b) TEM image of a GaN nanowire with a gold metal alloy droplet on its tip. Insets are electron diffraction patterns taken along the [001] zone axis. The lower inset is the same electron diffraction pattern but purposely defocused to reveal the wire growth direction. (c) Lattice-resolved TEM image of the nanowire. (Kuykendall T, Pauzauskie P, Lee S, Zhang Y, Goldberger J, Yang P. Metallorganic chemical vapor deposition route to GaN nanowires with triangular cross sections. Nano Lett. 3:1063 1066, 2003). Reprinted with permission from the American Chemical Society, 1155 16th St. N.W., Washington D. C. 20036. Photo Courtesy of P. Yang, University of California Berkeley.

be regarded as a "brute force" method, because the diameter of the nanowires is determined by the geometric constraint of the pores rather than by elegant chemical principles (Foss, 2002). However, it is one of the most successful methods for fabricating various nanowires that are difficult to form by the conventional lithographic process.

There are a number of methods for fabricating various negative templates. Examples include porous alumina membranes, polycarbonate membranes, mica sheets, and diblock polymer materials. These materials contain a large number of straight, cylindrical nanopores with a narrow distribution in the diameters of the nanopores.

### Anodic porous alumina

Anodic porous alumina is a commonly used negative template. Figure 11.1-14a shows a schematic drawing of an anodic porous alumina template. The nanopores in the template are formed by anodizing aluminum films in an acidic electrolyte. The individual nanopores in the alumina can be ordered into a close-packed honeycomb structure (Figure 11.1-14b). The diameter of each pore and the separation between two adjacent pores can be controlled by changing the anodization conditions.

To achieve highly ordered pores, high-purity (99.999% pure) aluminum films are used. In addition, they are first preannealed to remove mechanical stress and enhance grain size. Subsequently, the films are electropolished in a 4:4:2 (by weight) mixture of $H_3PO_4$, $H_2SO_4$, and $H_2O$ to create homogenous surfaces. Without the preannealing and electropolishing steps it is hard to form well-ordered pores (Jessensky et al., 1998). The order of the pores depends also on other anodization conditions, such as anodization voltage and electrolyte. Control of anodization voltage can produce an almost ideal honeycomb structure over an area of several µm (Jessensky et al., 1998; Masuda and Fukuda, 1995). The optimal voltage depends on the electrolyte used for anodization (Jessensky et al., 1998; Masuda and Fukuda, 1995; Masuda et al., 1998; Masuda and Hasegawa, 1997). For example, the optimal voltage for long-range ordering is 25 V in sulfuric acid, 40 V in oxalic acid, and 195 V in phosphoric acid electrolyte, respectively (Jessensky et al., 1998; Masuda and Fukuda, 1995; Masuda et al., 1998; Masuda and Hasegawa, 1997; Shingubara et al., 1997; Li et al., 1998).

The diameter and depth of each pore, as well as the spacing between adjacent pores, can be controlled by the anodizing conditions. Both the pore diameter and the pore spacing are proportional to the anodizing voltage, with proportional constants of 1.29 nm $V^{-1}$ and 2.5 nm $V^{-1}$, respectively. The dependence of the diameter and the spacing on the voltage is not sensitive to the electrolyte, which is quite different from the optimal voltage for ordered distribution of the pores. By properly controlling the anodization voltage and choosing the electrolyte, one can make highly ordered nanopores in alumina with desired pore diameter and spacing.

The order of the pores achieved by anodizing an aluminum film over a long period is often limited to a domain of several µm. The individual ordered domains are separated by regions of defects. Recently, a novel approach has been reported to produce a nearly ideal hexagonal nanopore array that can extend over several

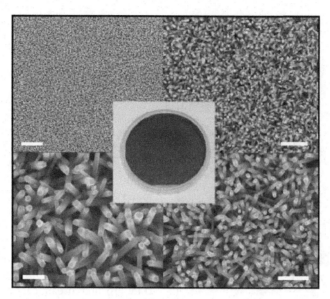

**Figure 11.1-13** *ZnO nanowires array on a 4-inch silicon wafer. Centered is a photograph of a coated wafer, surrounded by SEM images of the array at different locations and magnifications. These images are representative of the entire surface. Scale bars, clockwise from upper left, correspond to 2 μm, 1 μm, 500 nm, and 200 nm.* (L. E Greene, et al., Angewandte Chemie International Edition, 42, 2003, 3031-3034). Reprinted with permission from P. Yang, University of California Berkeley.

millimeters (Masuda et al., 1997; Asoh et al., 2001). The approach uses a pretexturing process of Al in which an array of shallow, concave features is initially formed on Al by indentation. The pore spacing can be controlled by the pretexured pattern and the applied voltage. Another widely used method for creating highly ordered nanopore arrays is a two-step anodization method (Masuda et al., 1997; Asoh et al., 2001; Masuda and Satoh, 1996; Li et al., 1999; Li et al., 2000; Foss et al., 1992, 1994). The first step involves a long-period anodization of high-purity aluminum to form a porous alumina layer. Subsequent dissolution of the porous alumina layer leads to a patterned aluminum substrate with an ordered array of concaves that serve as the initial sites to form a highly ordered nanopore array in a second anodization step.

Acidic anodization of Al normally results in a porous alumina structure that is separated from the aluminum substrate by a barrier layer of $Al_2O_3$. The barrier layer and aluminum substrate can be removed to form a free-standing porous alumina membrane. The aluminum can be removed with saturated $HgCl_2$ and the barrier layer of $Al_2O_3$ with a saturated solution of KOH in ethylene glycol. An alternative strategy to separate the porous alumina from the substrate is to take advantage of the dependence of pore diameter on anodization voltage. By repeatedly decreasing the anodization voltage several times at 5% increments, the

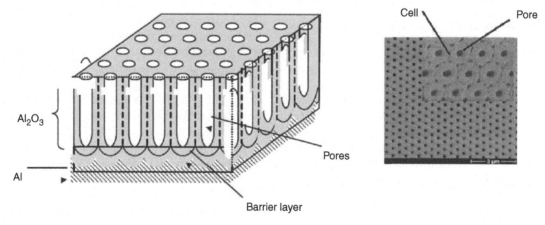

**Figure 11.1-14** *(a) Schematic drawing of an anodic porous alumina template. (b) Surface study of anodic porous alumina (inset shows cell and pore).* (Reprinted by permission of K. H. Lee, H. Y. Lee, and W. Y. Jeung. Magnetic properties and crystal structures of self-ordered ferromagnetic nanowires by ac electroforming. Journal of Applied Physics, 91, 2002, 8513).

barrier layer becomes a tree-root-like network with fine pores.

## Fabrication of metal nanowires

Using the membrane templates previously described, nanowires of various metals, semiconductors (Lakshmi et al., 1997), and conducting polymers (Van Dyke and Martin, 1990; Wu and Bein, 1994) have been fabricated. These nanostructures can be deposited into the pores by either electrochemical deposition or other methods, such as chemical vapor deposition (CVD) (Che et al., 1998), chemical polarization (Martin et al., 1993; Parthasarathy and Martin, 1994; Sapp et al., 1999), electroless deposition (Wirtz et al., 2002), or sol-gel chemistry (Lakshmi et al., 1997). Electrodeposition is one of the most widely used methods to fill conducting materials into the nanopores to form continuous nanowires with large aspect ratios. One of the great advantages of the electrodeposition method is the ability to create highly conductive nanowires. This is because electrodeposition relies on electron transfer, which is the fastest along the highest conductive path. Structural analysis shows that the electrodeposited nanowires tend to be dense, continuous, and highly crystalline in contrast to nanowires deposited using other deposition methods, such as CVD. Yi and Schwarzacher demonstrated that the crystallinity of superconducting Pb nanowires can be controlled by applying a potential pulse with appropriate parameters (Yi and Schwarzacher, 1999). The electrodeposition method is not limited to nanowires of pure elements. It can fabricate nanowires of metal alloys with good control over stoichiometry. For example, by adjusting the current density and solution composition, Huang et al. (2002) controlled the compositions of the CoPt and FePt nanowires to 50:50 to obtain the highly anisotropic face-centered tetragonal phases (Yang et al., 2002). Similar strategies have been used in other magnetic nanowires (Yang et al., 2002; Qin et al., 2002) and in thermoelectronic nanowires (Sapp et al., 1999; Sander et al., 2002). Another important advantage of the electrodeposition method is the ability to control the aspect ratio of the metal nanowires by monitoring the total amount of passed charge. This is important for many applications. For example, the optical properties of nanowires are critically dependent on the aspect ratio (Foss et al., 1992, 1994; Preston and Moskovits, 1993). Nanowires with multiple segments of different metals in a controlled sequence can also be fabricated by controlling the potential in a solution containing different metal ions (Liu et al., 1995).

Electrodeposition often requires deposition of a metal film on one side of the freestanding membrane to serve as a working electrode on which electrodeposition takes place. In the case of large pore sizes, the metal film has to be rather thick to completely seal the pores on one side. The opposite side of the membrane is exposed to an electrodeposition solution, which fills up the pores and allows metal ions to reach the metal film. However, one can avoid using the metal film on the backside by using anodic alumina templates with the natural supporting Al substrate. The use of the supported templates also prevents a worker from breaking the fragile membrane during handling. However, it requires the use of AC electrodeposition (Lee et al., 2002; Caboni, 1936). This is because of the rather thick barrier layer between the nanopore membrane and the Al substrate.

Detailed studies of the electrochemical fabrication process of nanowires have been carried out by a number of groups (Schonenberger et al., 1997; Whitney et al., 1993). The time dependence of the current curves recorded during the electrodeposition process reveals three typical stages. Stage I corresponds to the electrodeposition of metal into the pores until they are filled up to the top surface of the membrane. In this stage, the steady-state current at a fixed potential is directly proportional to the metal film area that is in contact with the solution, as found in the electrodeposition on bulk electrodes. However, the electrodeposition is confined within the narrow pores, which has a profound effect on the diffusion process of the metal ions from the bulk solution into the pores before reaching the metal film. The concentration profiles of Co ions in the nanopores of polycarbonate membranes during electrodeposition of Co have been studied by Valizadeh et al. (2001). After the pores are filled up with deposited metal, metal grows out of the pores and forms hemispherical caps on the membrane surface. This region is called stage II. Because the effective electrode area increases rapidly during this stage, the electrochemical current increases rapidly. When the hemispherical caps coalesce into a continuous film, stage III starts, which is characterized by a constant value of the current. By stopping the electrodeposition process before stage I ends, an array of nanowires filled in pores is formed. When freely standing nanowires are desired, one has to remove the template hosts after forming the nanowires in the templates. This task is usually accomplished by dissolving away the template materials in a suitable solvent. Methylene chloride can readily dissolve away track-etched polycarbonate film and $0.1\,M$ NaOH removes anodic alumina effectively. If one wants to also separate the nanowires from the metal films on which the nanowires are grown, a common method is to first deposit sacrificial metal. For example, to fabricate freely standing Au nanowires, one can deposit a thin layer of Ag onto the metal film coated on one side of the template membrane before filling the pores with Au. The Ag layer can be etched away later in concentrated nitric acid, which separates the Au nanowires from the metal film.

Although DC electrodeposition can produce high-quality nanowires, it is challenging to obtain an ordered nanowire array using this method. Normally only 10–20% of the pores in the membrane are filled up completely using the simple DC method (Prieto et al., 2001). Using AC electrodeposition with appropriate parameters (Lee et al., 2002; Schonenberger et al., 1997), a high filling ratio can be obtained using a sawtooth wave (Yin et al., 2001). Furthermore, the researchers found that the filling ratio increases with the AC frequency. A possible reason is that nuclei formed at higher frequencies are more crystalline, which makes the metal deposition easier in the pores and promotes homogenous growth of nanowires. Nielsch et al. (2000) and Sauer et al. (2002) developed a pulsed electrodeposition method. After each potential pulse, a relatively long delay follows before application of the next pulse. The rationale is that the long delay after each pulse allows ions to diffuse into the region where ions are depleted during the deposition (pulse). These researchers demonstrated that the pulsed electrodeposition is well suited for a uniform deposition in the pores of porous alumina with a nearly 100% filling rate.

*Positive template method*
The positive template method is used to make wirelike nanostructures. In this case DNA and carbon nanotubes act like templates, and nanowires form on the outer surface of the templates. Unlike negative templates, the diameters of the nanowires are not restricted by the template sizes and can be controlled by adjusting the amount of materials deposited on the templates. By removing the templates after deposition, wirelike and tubelike structures can be formed.

*Carbon nanotube template*
Fullam et al. (2000) have demonstrated a method to fabricate Au nanowires using carbon nanotubes as positive templates. The first step is to self-assemble Au nanocrystals along carbon nanotubes. After thermal treatment, the nanocrystal assemblies are transformed into continuous polycrystalline Au nanowires of several microns. Carbon nanotubes have also been used as templates to fabricate Mo-Ge superconducting nanowires (Bezryadin et al. 2000) and other metal nanowires (Zhang and Dai, 2000; Yun et al., 2000). Choi et al. (2002) reported highly selective electrodeposition of metal nanoparticles on single-wall carbon nanotubes (SWNTs). Because $HAuCl_4$ ($Au^{3+}$) or $Na_2PtCl^4$ ($Pt^{2+}$) have much higher reduction potentials than SWNTs, they are reduced spontaneously and form Au or Pt nanoparticles on the side walls of SWNTs (shown in Figure 11.1-15). This is different from traditional electroless deposition because no reducing agents or catalysts

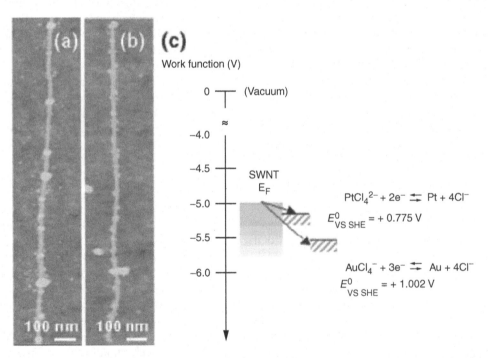

**Figure 11.1-15** *AFM images of metal nanoparticles formed on a SWNT template. (a) Au nanoparticles spontaneously and selectively formed on an individual SWNT after immersion in a $Au^{3+}$ solution for 3 min. (b) Pt nanoparticles formed on a SWNT after 3-min exposure to a $Pt^{2+}$ solution. (c) Diagram showing the Fermi energy ($E_F$) of a SWNT, and the reduction potentials of $Au^{3+}$ and $Pt^{2+}$ versus SHE, respectively. The reduction potentials of most other metal ions lie above $E_F$, except for $Ag^+$. (H. C. Choi et al., Journal of the American Chemical Society, 124, 2002, 9058). Reprinted with permission from the American Chemical Society, 1155 16th St. N.W., Washington DC 20036. Photo Courtesy of H. Dai, Stanford University.*

are required. Charge transfer during the reaction is probed electrically, because it causes significant changes in the electrical conductance of the nanotubes by hole doping. The nanoparticles deposited on the nanotube can coalesce and cover the entire surface of the nanotube. By removing the nanotube via heating, a Au tubelike structure with a outer diameter <10 nm can be fabricated.

### DNA template

DNA is another excellent choice as a template to fabricate nanowires because its diameter is ~2 nm and its length and sequence can be precisely controlled (Mbindyo et al., 2001). Coffer and co-workers (1996) synthesized micrometer-scale CdS rings using a plasmid DNA as a template. They reported a two-step electrodeposition process to fabricate Pd nanowires using DNA as templates. The first step is to treat DNA in a Pd acetate solution. The second step is to add a reducing agent, typically dimethylamine borane, which reduces Pd ions into Pd along the DNA chains. If the reduction time is short, it leads to individual isolated Pd clusters with a diameter of 3–5 nm. With increasing reduction time, Pd clusters aggregate and form a quasi-continuous Pd nanowire.

This metallization method has been applied to both DNA in solution and DNA immobilized on a solid surface. Braun et al. (117) fabricated a Ag nanowire of ~100 nm in diameter and ~15 μm in length using a linear DNA template.

The procedure used by Braun et al. (1998) to form nanowires using DNA templates is illustrated in Figure 11.1-16. The first step is to fix a DNA strand between two electrical contacts. The DNA is then exposed to a solution containing Ag+ ions. The Ag+ ions bind to DNA and are then reduced by basic hydroquinone solution to form Ag nanoparticles decorating along the DNA chain. In the last step, the nanoparticles are further "developed" into a nanowire using a standard photographic enhancement technique. The nanowires are highly resistive because they are composed of individual Ag clusters of ~50 nm in diameter. Recently, researchers developed a DNA sequence-specific molecular lithography to fabricate metal nanowires with a predesigned insulating gap (Keren et al., 2002). The approach uses homologous recombination process and the molecular recognition capability of DNA. Homologous recombination is a protein-mediated reaction by which two DNA molecules having some sequence homology crossover at equivalent sites. In the lithography process, RecA proteins are polymerized on a single-strand DNA (ssDNA) probe to form a nucleoprotein filament. Then the nucleoprotein filament binds to an aldehyde-derivatized double-strand DNA (dsDNA) substrate at a homologous sequence. Incubation of the formed complex in $AgNO_3$ solution results in the formation of Ag aggregates along the substrate DNA molecules at regions unprotected by RecA. The Ag aggregates serve as catalysts for specific Au deposition, converting the unprotected regions to conductive Au wires. Thus, a Au nanowire with an insulating gap is formed. The position and size of the insulating gap can be tailored by choosing the template DNA with the special sequence and length.

### Polymer templates

Like DNA, many other polymer chains can also be excellent choices as positive templates for nanowire fabrications. For example, single synthetic flexible polyelectrolyte molecules, poly2-vinylpyridene ($P_2VP$) are used as templates to fabricate nanowires. Because these polymers are thinner than DNA, it is possible to fabricate thinner nanowires. Under appropriate conditions, the polymer chains are stretched into wormlike coils by the electrostatic repulsion between randomly distributed positive charges along the chain. This stretched conformation is frozen when the polymer is attached to a solid substrate. Exposing the polymer to palladium acetate acidic aqueous solution causes $Pd^{2+}$ to coordinate to the polymer template via an ion exchange reaction. In the following step, $Pd^{2+}$ is reduced by dimethylamine borane. The procedure results in metal nanoparticles of 2–5 nm in diameter, which deposit along the template into a wirelike structure.

Djalali et al. (2002) used the core-shell cylindrical polymer brushes as templates to synthesize metal cluster arrays and wires (shown in Figure 11.1-17). The starting material of the templates is methacryloyl end-functionalized block copolymers, consisting of styrene and vinyl-2-pyridine, which are polymerized to poly(block comacromonomer)s. The formed poly(block comacromonomer)s exhibit an amphipolar core-shell cylindrical brush structure with a core of vinylpyridine and a shell of polystyrene. The vinylpyridine cores of the cylindrical brushes are loaded with $HAuCl_4$ in toluene or methylene chloride, followed by reduction of the Au salt by the electron beam, UV light, or chemical-reducing agents. Depending on the amount of $AuCl_4$−ions loaded in the cores and the reduction conditions, either a linear array of Au clusters or a continuous Au nanowire is formed within the core of the cylindrical brushes. The resulting Au metal nanowires are much longer than the individual core-shell macromolecules, which are caused by a yet unexplained specific end-to-end aggregation of the cylindrical polymers on loading with $HAuCl_4$. Since the metal formation occurs within the cores of the polymers, the polystyrene shells may serve as the electrically insulating layers.

## Applications of the nanowires

Metal nanowires are promising materials for many novel applications, ranging from chemical and biological sensors to optical and electronic devices. This is not only

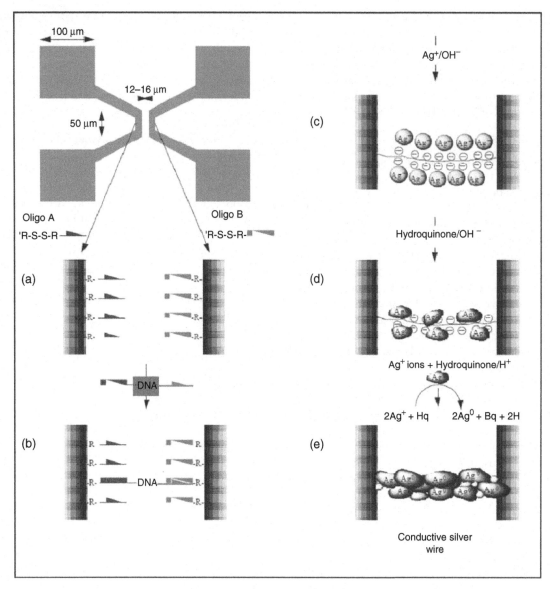

**Figure 11.1-16** *Construction of an Ag wire connecting two gold electrodes using DNA as template. The top left image shows the electrode pattern used in the experiments. The two 50-μm-long parallel electrodes are connected to four (100- × 100-μm) bonding pads. (a) Oligonucleotides with two different sequences attached to the electrodes. (b) λ-DNA bridge connecting the two electrodes. (c) Ag-ion-loaded DNA bridge. (d) Metallic Ag aggregates bound to the DNA skeleton. (e) Fully developed Ag wire. (Reprinted by the permission of E. Braun etal., Nature, 391, 1998, 775).*

because of their unique geometry, but also because they have many unique physical properties, including electrical, magnetic, optical, as well as mechanical properties. Some of the applications are discussed below.

### Magnetic materials and devices

The electrodeposition methods described above have been used to fabricate magnetic nanowires of a single metal (Whitney et al., 1993), multiple metals in segments (Piraux et al., 1994), as well as alloys (Dubois et al., 1997). Magnetic nanowires with relatively large aspect ratios (e.g., >50), exhibit an easy axis along the wires. An important parameter that describes magnetic properties of materials is the remanence ratio, which measures the remanence magnetization after switching off the external magnetic field. The remanence ratios of the Fe, Co, and Ni nanowires can be larger than 0.9 along the wires and much smaller in the perpendicular direction of the wires. This finding clearly shows that the shape anisotropy plays an important role in the magnetism of the nanowires. Another important parameter that describes the magnetic properties is coercivity, which is the coercive field required to demagnetize the magnet after full magnetization. The magnetic nanowires exhibit

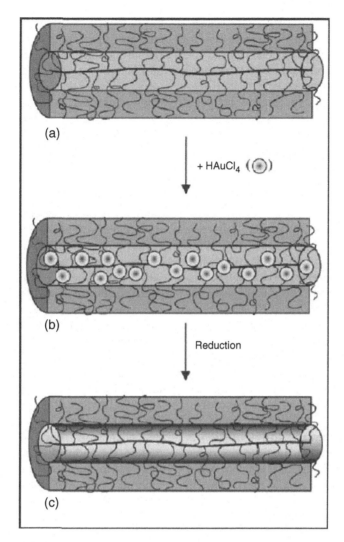

**Figure 11.1-17** *Fabrication of nanowires with a polymer template. (a) Core-shell cylindrical brushes with a PVP core and PS shell. (b) Loading the core with HAuCl$_4$. (c) Subsequent reduction of HAuCl$_4$ yields a one-dimensional Au wire within the macromolecular brush. (D. Djalali et al., Macromolecules, 35, 2002, 4282). Reprinted with permission from the American Chemical Society, 1155 16$^{th}$ St. NW, Washington, DC 20036.*

greatly enhanced magnetic coercivity (Chien, 1991). In addition, the coercivity depends on the wire diameter and the aspect ratio, which shows that it is possible to control the magnetic properties of the nanowires by controlling the fabrication parameters. The diameter dependence of the coercivity reflects a change of the magnetization reversal mechanism from localized quasi-coherent nucleation for small diameters to a localized curling like nucleation as the diameter exceeds a critical value (Thurn-Albrecht et al., 2000).

Another novel property of magnetic nanowires is giant magnetoresistance (GMR) (Liu et al.; Evans, et al., 2000). For example, Evans et al. have studied Co-Ni-Cu/Cu multilayered nanowires and found a magnetoresistance ratio of 55% at room temperature and 115% at 77 K for current perpendicular to the plane (along the direction of the wires). Giant magnetoresistance has also been observed in semimetallic Bi nanowires fabricated by electrodeposition (Liu et al., 1998; Hong et al., 1999; Lin et al., 2000). Hong et al. (1999) have studied GMR of Bi with diameters between 200 nm and 2 μm in magnetic fields up to 55 T and found that the magnetoresistance ratio is between 600 and 800% for magnetic fields perpendicular to the wires and ~200% for the fields parallel to the wires. The novel properties and small dimensions have potential applications in the miniaturization of magnetic sensors and the high-density magnetic storage devices.

The alignment of magnetic nanowires in an applied magnetic field can be used to assemble the individual nanowires (Lin et al., 2000). Tanase et al. studied the response of Ni nanowires in response to magnetic field (Tanase et al., 2001). The nanowires are fabricated by electrode-position using alumina templates and functionalized with luminescent porphyrins so that they can be visualized with a video microscope. In viscous solvents, magnetic fields can be used to orient the nanowires. In mobile solvents, the nanowires form chains in a head-to-tail configuration when a small magnetic field is applied. In addition, Tanase et al. demonstrated that three-segment Pt-Ni-Pt nanowires can be trapped between lithographically patterned magnetic microelectrodes (Tanase et al., 2002). The technique has a potential application in the fabrication and measurement of nano-scale magnetic devices.

## Optical applications

Dickson and Lyon studied surface plasmon (collective excitation of conduction electrons) propagation along 20-nm-diameter Au, Ag, and bimetallic Au-Ag nanowires with a sharp Au-Ag heterojunction over a distance of tens of μm (Dickson and Lyon, 2000). The plasmons are excited by focusing a laser with a high-numerical-aperture microscope objective, which propagate along a nanowire and reemerge as light at the other end of the nanowire via plasmon scattering. The propagation depends strongly on the wavelength of the incident laser light and the composition of the nanowire. At the wavelength of 820 nm, the plasmon can propagate in both Au and Ag nanowires, although the efficiency in Ag is much higher that in Au. In the case of bimetallic nanowire, light emission is clearly observed from the Ag end of the nanowire when the Au end is illuminated at 820 nm. In sharp contrast, if the same bimetallic rod is excited at 820 nm via the Ag end, no light is emitted from the distal Au end. The observations suggest that the plasmon mode excited at 820 nm is able to couple from the Au portion into the Ag portion with high efficiency, but not from the Ag portion into Au. The unidirectional propagation has been explained using a simple two-level potential model. Since surface plasmons propagate much more efficiently in Ag than in Au, the Au → Ag boundary is largely transmissive, thus enabling efficient plasmon propagation in this direction from Ag to Au and a much steeper potential wall, which allow less optical energy to couple through to the distal end. The experiments suggest that one can initiate and control the flow of optically encoded information with nanometer-scale accuracy over distances of many microns, which may find applications in future high-density optical computing.

## Biological assays

We have already mentioned that by sequentially depositing different metals into the nanopores, multisegment or striped metal nanowires can be fabricated (Martin et al., 1999). The length of each segment can be controlled by the charge passed in each plating step, and the sequence of the multiple segments is determined by the sequence of the plating steps. Due to the different chemical reactivities of the "stripe" metals, these strips can be modified with appropriate molecules. For example, Au binds strongly to thiols and Pt has high affinity to isocyanides. Interactions between complementary molecules on specific strips of the nanowires allow different nanowires to bind to each other and form patterns on planar surfaces. Using this strategy, nanowires could assemble into cross- or T-shaped pairs, or into more complex shapes (Reiss et al., 2001). It is also possible to use specific interactions between selectively functionalized segments of these nanowires to direct the assembly of nanowire dimers and oligomers, to prepare a two-dimensional assembly of nanowire-substrate epitaxy, and to prepare three-dimensional colloidal crystals from nanowire-shaped objects (Yu et al., 2000). As an example, single-strand DNA can be exclusively modified at the tip or any desired location of a nanowire, with the rest of the wire covered by an organic passivation monolayer. This opens the possibility for site-specific DNA assembly (Martin et al., 1999).

Nicewarner-Pena et al. showed that the controlled sequence of multisegment nanowires can be used as "bar codes" in biological assays (Nicewarner-Pena et al., 2001). The typical dimension of the nanowire is $\sim 200$ nm thick and $\sim 10$ μm long. Because the wavelength dependence of reflectance is different for different metals, the individual segments are easily observed as "stripes" under an optical microscope with unpolarized white-light illumination. Different metal stripes within a single nanowire selectively adsorb different molecules, such as DNA oligomers, which can be used to detect different biological molecules simultaneously. These multisegment nanowires have been used like metallic bar codes in DNA and protein bioassays.

The optical scattering efficiency of the multisegment nanowires can be significantly enhanced by reducing the dimensions of the segment, such that excitation of the surface plasmon occurs. Mock et al. (2002) have studied the optical scattering of multisegment nanowires of Ag, Au, and Ni that have diameters of $\sim 30$ nm and length up to $\sim 7$ μm. The optical scattering is dominated by the polarization-dependent plasmon resonance of Ag and Au segments. This is different from the case of the thicker nanowires used by Nicewarner-Pena et al., where the reflectance properties of bulk metals determine the contrast of the optical images (Nicewarner-Pena et al., 2001). Because of the large enhancement by the surface plasmon resonance, very narrow ($\sim$30-nm-diameter) nanowires can be readily observed under white-light illumination, and the optical spectra of the individual segments are easily distinguishable (Mock et al., 2002). The multisegment nanowires can host a large number of segment sequences over a rather small spatial range, which promises unique applications.

## Chemical sensors

Penner, Handley, and Dagani et al. exploited hydrogen sensor applications using arrays of Pd nanowires (Walter et al., 2002). Unlike the traditional Pd-based hydrogen sensor that detects a drop in the conductivity of Pd on exposure to hydrogen, the Pd-nanowire sensor measures an increase in the conductivity (Figure 11.1-18). This happens because the Pd wire consists of a string of Pd particles separated with nanometer-scale gaps. That these gaps close to form a conductive path in the presence of hydrogen molecules as Pd particles expand is well known; this closure is due to the disassociation of hydrogen molecules into hydrogen atoms that penetrate into the Pd lattice and expand the lattice. Although macroscopic Pd-based hydrogen sensors are readily available, they have the following two major drawbacks. First, their response time is between 0.5 s to several minutes, which is too slow to monitor gas flow in real time. Second, they are prone to the contamination by a number of gas molecules, such as methane, oxygen, and carbon monoxide, which adsorb onto the sensor surfaces and block the adsorption sites for hydrogen molecules. The Pd nanowires offer remedies to the above problems. Pd nanowires have a large surface-to-volume ratio, which makes the nanowire sensor less prone to contamination by common substances.

## Carbon nanotubes

Carbon nanotubes (CNTs) are very interesting nanostructures with a wide range of potential applications. CNTs were first discovered by Iijima in 1991 (Iijima, 1991); since then, great progress has been made toward many applications, including, for example, the following:

**Materials** Chemical and biological separation, purification, and catalysis. Energy storage such as hydrogen storage, fuel cells, and the lithium battery. Composites for coating, filling, and use as structural materials

**Devices** Probes, sensors, and actuators for molecular imaging, sensing, and manipulation Transistors, memories, logic devices, and other nanoelectronic devices. Field emission devices for x-ray instruments, flat-panel display, and other vacuum nanoelctronic applications. These applications and advantages can be understood by the unique structure and properties of nanotubes, as outlined below:

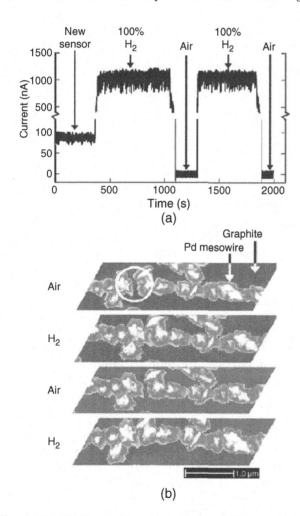

Figure 11.1-18 Chemical sensor application of Pd nanowires. (a) Plot of sensor current versus time for the first exposure of a Pd nanowire sensor to hydrogen and one subsequent $H_2$-air cycle. (b) AFM image of a Pd nanowire on a graphite surface. These images were acquired either in air or in a stream of $H_2$ gas, as indicated. A hydrogen-actuated break junction is highlighted (circle). (E. C. Walter et al., Analytical Chemistry, 74, 2002, 1546). Reprinted with permission from the American Chemical Society, 1155 16th St. N.W., Washington, DC 20036. Photo Courtesy of M. Penner, University of California, Irvine.

## Structures

*Bonding* The $sp2$ hybrid orbital allows carbon atoms to form hexagons and occasionally pentagon units by in-plane $\sigma$ bonding and out-of-plane $\pi$ bonding.

*Defect-free nanotubes* These are tubular structures of hexagonal network with a diameter as small as 0.4 nm. Tube curvature results in $\sigma-\pi$ rehybridization or mixing.

*Defective nanotubes* Occasionally pentagons and heptagons are incorporated into a hexagonal network to form bent, branched, helical, or capped nanotubes.

## Properties

*Electrical* Electron confinement along the tube circumference makes a defect-free nanotube either semiconducting or metallic with quantized conductance, whereas pentagons and heptagons generate localized states.

*Optical and optoelectronic* Direct band gap and one-dimensional band structure make nanotubes ideal for optical applications with wavelength ranging possibly from 300 to 3000 nm.

*Mechanical and electrochemical* The rehybridization gives nanotubes the highest Young's modulus (over 1 Tpa), tensile strength of over 100 GPa, and remarkable electronic response to strain and metal-insulator transition.

*Magnetic and electromagnetic* Electron orbits circulating around a nanotube give rise to many interesting phenomena such as quantum oscillation and metal-insulator transition.

*Chemical and electrochemical* High specific surface and rehybridization facilitate molecular adsorption, doping, and charge transfer on nanotubes, which, in turn, modulates electronic properties.

*Thermal and thermoelectric* Possessing a property inherited from graphite, nanotubes display the highest thermal conductivity, whereas the quantum effect shows up at low temperature.

## Structure of the carbon nanotube

This section describes the structure of various types of carbon nanotubes (CNT). A CNT can be viewed as a hollow cylinder formed by rolling graphite sheets. Bonding in nanotubes is essentially *sp2*. However, the circular curvature will cause quantum confinement and $\sigma-\pi$ rehybridization in which three bonds are slightly out of plane; for compensation, the orbital is more delocalized outside the tube. This makes nanotubes mechanically stronger, electrically and thermally more conductive, and chemically and biologically more active than graphite. In addition, they allow topological defects such as pentagons and heptagons to be incorporated into the hexagonal network to form capped, bent, toroidal, and helical nanotubes, whereas electrons will be localized in pentagons and heptagons because of redistribution of electrons. A nanotube is called defect free if it is of only hexagonal network and defective if it also contains topological defects such as pentagonal and heptagonal or other chemical and structural defects. A large amount of work has been done in studying defect-free nanotubes, including single- or multiwall nanotubes (SWNTs or MWNTs). A SWNT is a hollow cylinder of a graphite sheet, whereas a MWNT is a group of coaxial SWNTs. SWNTs were discovered in 1993 (Iijima and Ichihashi, 1993), two years after the discovery of MWNTs (Bethune et al., 1993). They are often seen as straight or elastic bending structures individually or in ropes (Thess et al., 1996) by transmission electron microscopy (TEM), scanning electron microscopy (SEM), atomic force microscopy (AFM), and scanning tunneling microscopy (STM). In addition, electron diffraction (EDR), x-ray diffraction (XRD), Raman, and other optical spectroscopy can be also used to study structural features of nanotubes.

A SWNT can be visualized as a hollow cylinder, formed by rolling over a graphite sheet. It can be uniquely characterized by a vector $C_h$ where $C_h$ denotes chiral vector in terms of a set of two integers $(n, m)$ corresponding to graphite vectors $a_1$ and $a_2$, (Figure 11.1-19; Dresselhaus, Dresselhaus, and Eklund, 1996).

$$C_h = na_1 + ma_2 \qquad (11.1.1)$$

Thus, the SWNT is constructed by rolling up the sheet such that the two end-points of the vector $C_h$ are superimposed. This tube is denoted as $(n, m)$ tube with diameter given by

$$D = |C_h|/\pi = a(n^2 + nm + m^2)^{1/2}/\pi \qquad (11.1.2)$$

where $a = |a_1| = |a_2|$ is the lattice constant of graphite. The tubes with $m = n$ are commonly referred to as armchair tubes, and $m = 0$, as zigzag tubes. Others are called chiral tubes in general, with the chiral angle, $\theta$, defined as that between the vector $C_h$ and the zigzag direction $a_1$,

$$\theta = \tan^{-1}[3^{1/2}m/(m+2n)] \qquad (11.1.3)$$

$\theta$ ranges from 0 for zigzag $(m = 0)$ and 30° for armchair $(m = n)$ tubes.

The lattice constant and intertube spacing are required to generate a SWNT, SWNT bundle, and MWNT. These two parameters vary with tube diameter or in radial direction. Most experimental measurements and theoretical calculations agree that, on average, the $c - c$ bond length $d_{cc} = 0.142$ nm or $a = |a_1| = |a_2| = 0.246$ nm, and intertube spacing $d_{tt} = 0.34$ nm (Dresselhaus et al., 1996). Thus, Equations 11.1.1 to 11.1.3 can be used to model various tube structures and interpret experimental observation. Figure 11.1-20 illustrates examples of nanotube models.

## Synthesis of carbon nanotubes

Carbon nanotubes can be made by laser evaporation, carbon arc methods, and chemical vapor deposition. Figure 11.1-21 illustrates the apparatus for making carbon nanotubes by laser evaporation. A quartz tube containing argon gas and a graphite target are heated to 1200 °C (Smalley et al., 1997). Contained in the tube, but somewhat outside the furnace, is a water-cooled copper collector. The graphite target contains small amounts of Fe, Co, and Ni that act as seeds for the growth of carbon nanotubes. An intense, pulsed laser beam is incident on the target, evaporating carbon from the graphite. The argon then sweeps the carbon atoms from the high-temperature zone to the colder

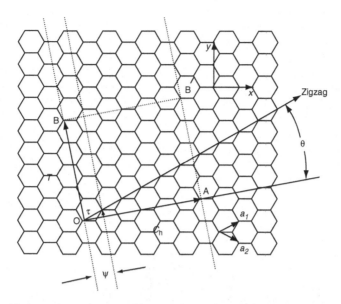

**Figure 11.1-19** A nanotube (n,m) is formed by rolling a graphite sheet along the chiral vector $C_h = na_1 + ma_2$ on the graphite where $a_1$ and $a_2$ are graphite lattice vectors. (M. S. Dresselhaus, G. Dresselhaus, and P. C. Eklund, Science and Fullerenes and Carbon Nanotubes, Academic Press, New York, 1996, Chapter 21). Reprinted with permission from Academic Press, an imprint of Elsevier.

collector, on which they condense into nanotubes. Tubes 10–20 nm in diameter and 100 μm long can be made by this method.

Nanotubes can also be synthesized using a carbon arc method (Figure 11.1-22). An electrical potential of 20–25 V and a DC electric current of 50–120 A flowing between the electrodes of 5- to 20-mm diameter and separated by ~1 mm at 500 Torr pressure of flowing helium are used (Saito et al., 1996). As the carbon nanotubes form, the length of the positive electrode decreases, and a carbon deposit forms on the negative electrode. To produce single-wall nanotubes, a small amount of Fe, Co, and Ni is incorporated as a catalyst in the central region of the positive electrode. These catalyst act like seeds for the growth of single and multiwalled carbon nanotubes.

For the large-scale production of carbon nanotubes, thermal CVD is the most favorable method (Figure 11.1-22). The thermal CVD apparatus is very simple for the growth of carbon nanotubes (Cassell et al., 1999). It consists of a quartz tube (1- to 2-in. diameter) inserted into a tubular furnace capable of maintaining a temperature of 1 °C over a 25-cm zone. Thus, it is a hot-wall system at primarily atmospheric pressure (CVD) and hence does not require any pumping systems. In thermal CVD, either CO or some hydrocarbon such as methane, ethane, ethylene, acetylene, or other higher hydrocarbon is used without dilution. The reactor is first filled with argon or some other inert gas until the reactor reaches the desired growth temperature. Then the gas flow is switched to the feedstock for the specified growth period. At the end, the gas flow is switched back to the inert gas while the reactor cools down to 300 °C or lower before exposing the nanotubes to air. Exposure to air at elevated temperatures can cause damage to the CNTs. Typical growth rates range from a few nm/min to 2 to 5 μm/min. Hongjie et al. (1999) reported a patterned growth of MWNTs by thermal CVD method. Figure 11.1-23 shows MWNTs grown by CVD at 700 °C in a 2-in. tube furnace under an ethylene flow of 1000 sccm for 15–20 min on a porous silicon substrate.

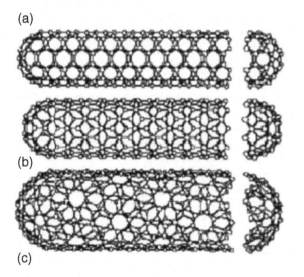

**Figure 11.1-20** Illustration of some possible structures of carbon nanotubes, depending on how graphite sheets are rolled: (a) armchair, (b) zigzag, and (c) chiral structures.

# CHAPTER 11.1 — Nanomaterials

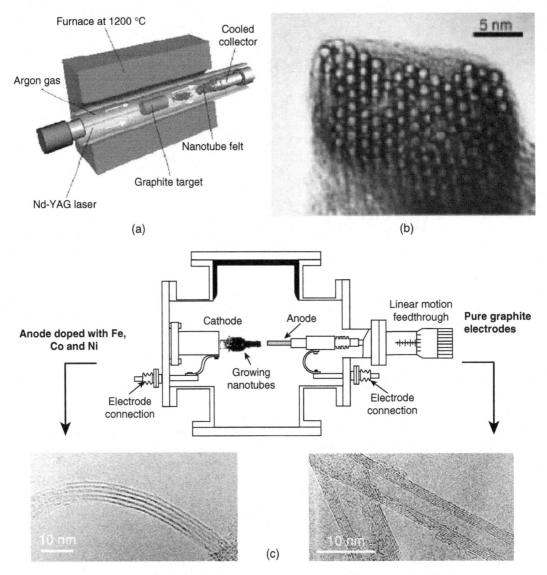

**Figure 11.1-21** *(a) Schematic of laser ablation apparatus. (b) TEM image of SWNT grown by laser ablation technique.* (R. E. Smalley et al., American Scientist, 85, 1997, 324). Reprinted with permission from Hazel Cole, exec. Asst to late Prof. Smalley, Rice University. *(c) Schematic diagram of carbon arc apparatus for the production of carbon nanotubes.* (Saito et al., Journal of Applied Physics, 80(5), 1996, 3062-3067). Reprinted with permission from Y. Saito, Nagoya University, Japan.

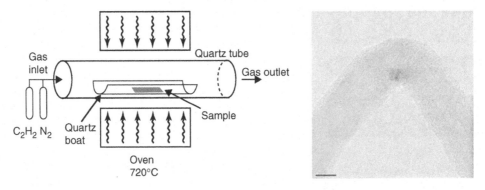

**Figure 11.1-22** *Schematic of a thermal CVD apparatus and TEM image of multiwall carbon nanotube.*

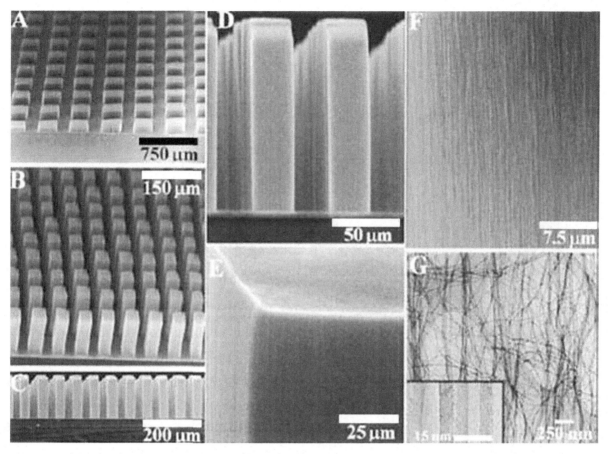

**Figure 11.1-23** Electron micrographs of self-oriented nanotubes synthesized on porous silicon substrates. (a) SEM image of nanotube blocks synthesized on 250- by 250-mm catalyst patterns. The nanotubes are 80 μm long and oriented perpendicular to the substrate. (b) SEM image of nanotube towers synthesized on 38-μm catalyst patterns. The nanotubes are 130 μm long. (c) Side view of the nanotube towers in (b). The nanotube self-assembly such that the edges of the towers are perfectly perpendicular to the substrate. (d) Nanotube "twin towers", a zoom-in view of (c). (e) SEM image showing sharp edges and corners at the top of a nanotube tower. (f) SEM image showing that nanotubes in a block are well aligned to the direction perpendicular to the substrate surface. (g) TEM image of pure multiwall nanotubes in several nanotube blocks grown on a porous silicon substrate. The inset is a high-resolution TEM image that shows two nanotubes bundling together. (Hongjie Dai et al., Science, 283, 1999, 512). Reprinted with permission from H. Dai, Stanford University, CA.

### Growth mechanisms of carbon nanotubes

The growth mechanism of nanotubes may vary depending on which method is used; with the arc-discharge and laser-ablation method MWNT can be grown without a metal catalyst, contrary to carbon nanotubes synthesized with CVD method, where metal particles are necessary. For growing SWNTs, in contrast, metals are necessary for all three methods mentioned previously.

The growth mechanism of nanotubes is not well understood; different models exist, but some of them cannot unambiguously explain the mechanism. The metal or carbide particles seem necessary for the growth because they are often found at the tip inside the nanotube or also somewhere in the middle of the tube. In 1972, Baker et al. developed a model for the growth of carbon fibers, which is shown in Figure 11.1-24 on the right side. It is supposed that acetylene decomposes at 600 °C on the top of nickel cluster on the support. The dissolved carbon diffuses through the cluster due to a thermal gradient formed by the heat release of the exothermic decomposition of acetylene. The activation energies for filament growth were in agreement with those for diffusion of carbon through the corresponding metal (Fe, Co, Cr) (Baker et al., 1973). Whether the metal cluster moves away from the substrate (tip growth) or stays on the substrate (base growth) is explained by a weaker or stronger metal–support interaction, respectively. As stated by Baker et al., the model has a number of shortcomings. It cannot explain the formation of fibers produced from the metal catalyst decomposition of methane, which is an endothermic process.

Oberlin et al. (1976) proposed a variation of this model shown in Figure 11.1-25. The fiber is formed by

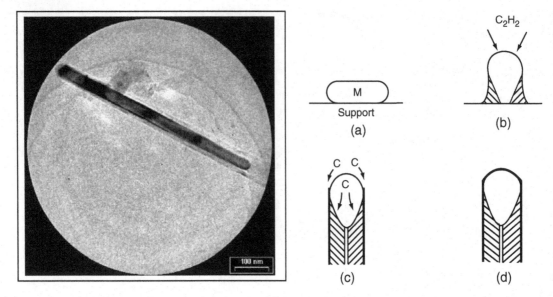

**Figure 11.1-24** Growth model of vapor-grown carbon fibers (right), according to Baker et al. (Carbon, 27, 1989, 315); and TEM image of a MWNT with metal nanorod inside the tip (left). (M. K. Singh et al., Journal of Nanomaterials and Nanotechnology, 3(3), 2003, 165). Reprinted with permission from Elsevier.

a catalytic process involving the surface diffusion of carbon around the metal particle, rather than by bulk diffusion of carbon through the catalytic cluster. In this model the cluster corresponds to a seed for the fiber nucleation. Amelinckx et al. (1994) adapted the growth model of Baker et al. (1973) to explain the growth of carbon nanotubes.

For the synthesis of single wall-carbon nanotubes, the metal clusters have to be present in the form of nanosized particles. Furthermore, it is supposed that the metal cluster can have two roles: first, it acts as a catalyst for the dissociation of the carbon-bearing gas species. Second, carbon diffuses on the surface of the metal cluster or through the metal to form a nanotube. The most active metals are Fe, Co, and Ni, which are good solvents for carbon (Kim et al., 1991).

## Carbon nanotube composite materials

Composite materials containing carbon nanotubes are a new class of materials. These materials will have a useful role in engineering applications. Many nano-enthusiasts are working in the area of polymer composites; efforts in metal and ceramic matrix composite are also of interest.

New conducting polymers, multifunctional polymer composites, conducting metal matrix composites, and

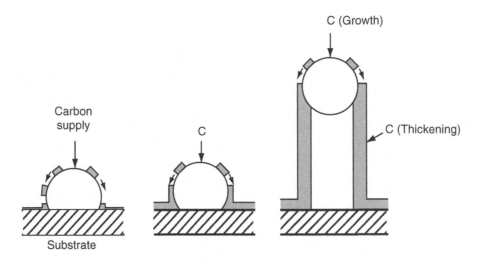

**Figure 11.1-25** *Alternative growth model of vapor-grown carbon fibers where the metal cluster acts as a seed for the growth.* (Oberlin et al., Journal of Crystal Growth, 32, 1976, 335). Reprinted with permission from Elsevier.

higher fracture-strength ceramics are just a few of the new materials being processed that make them near-term opportunities. Moreover, high conductors that are multifunctional (electrical and structural), highly anisotropic insulators, and high-strength, porous ceramics, are other examples of new materials that can come from nanotubes.

Most researchers who are developing new type of composite materials with nanotubes work with nanotube concentrations below 10 wt% because of the limited availability of nanotubes. Collective studies show broader promise for composite materials with concentrations as high as 40 and 50 wt% that may be limited only by our ability to create a complete matrix and fiber registry with high-surface-area nanofibers. In the following sections processing, properties and numerous potential applications and the application range of a wide variety of nanocomposites are discussed.

## Polymer nanocomposites

Plastics are compounded with inorganic fillers to improve processability, durability, and thermal stability. The mechanical properties of the composite are related to the volume fraction and aspect ratio of the filler. There is interest in incorporating nanoscale fillers in plastics. For example, the fire retardancy of polymers such as polypropylene has been increased by dispersing very small (<1%) amount of nanoscale copper particles. The thermoplastic PVC (polyvinyl chloride) has been compounded with inorganic fillers to improve its processability and thermal stability. Similarly, polymers containing nanoscale montmorillonite clay exhibit high strength, high modulus, good heat distortion temperature, and enhanced flame-retarding properties. The strength, modulus, and fracture toughness of PVC has been improved by dispersing nanosize calcium carbonate in PVC. *In-situ* polymerization of vinyl chloride in the presence of $CaCO_3$ nanoparticles was used to produce the composite. The in situ polymerization reduces the agglomeration of nanopowders, which is a major problem with adding nanoparticles to preformed polymers, thus permitting homogeneous distribution to be achieved. In a manner similar to the classic polymer composites, the polymer nanocomposites also exhibit shear thinning and power law behaviors. For example, for PVC-$CaCO_3$ at low shear rates, the viscosity was found to be higher than that of pristine PVC.

The incorporation of CNTs in polymers has been found to improve a variety of engineering properties. Because of their metallic or semiconducting character, CNT incorporation in polymer matrices permits attainment of an electrical conductivity sufficient to provide an electrostatic discharge at low CNT concentrations. The addition of CNT to polymers also results in significant weight savings. Surface treatments such as oxidation of CNTs to $CO_2$, or their graphitization, are found to improve the tensile strength of polypropylene-CNT composites. In addition to particles and nanotubes, nano plates (e.g., layered silicates, exfoliated graphite) have been used as fillers in polymers.

There are, however, some key differences between polymeric nanocomposites and classic filled polymers or composites. These include: low percolation threshold (i.e., network-forming tendency, $\sim 0.1$ to 2 vol% for nanodispersions), large numerical density of nanoparticles per unit volume ($10^6$ to $10^8$ particles per $\mu m^3$), extensive interfacial area per unit volume (1 km to 10 km per mL), and short distances between particles (10 to 50 nm at 1 to 8 vol% nanoparticles).

The aspect ratio (length-to-diameter ratio) of nano fillers is extreme. The size of the interfacial zone in a nanocomposite is comparable to the spacing between the filler, and the ratio of the volume of the interface to the volume of the bulk (matrix) increases dramatically as the nanoregime is approached. This affects many interface-sensitive properties such as vibration-damping capacity and strength. The mechanical behavior of polymeric nanocomposites is likely to be different from micrometer-size polymer composites. This is because the total interfacial area becomes the critical characteristic rather than the volume fraction, as is the case with micrometer-size composites. This, in turn, impacts the engineering issues related to irreversible agglomeration, nanoparticle network formation, and ultralong times for relaxation (glasslike behavior) in the case of nanocomposites.

The classic view of reinforced composites suggests that a strong fiber–matrix interface leads to high composite stiffness and strength but also low toughness, because of the brittle nature of the fiber and because of a lack of crack deflection at a strong interface, which is needed for strengthening. With nanocomposites having strong interfaces, it may be possible to attain very high strength and stiffness, and also high toughness, because of the nanotube's ability to considerably deform before fracture. In contrast, with weak interfaces toughness would be possible in a nanocomposite via the mechanism of crack deflection that also operates in classic polymeric composites.

Nanotubes are considered to have varying degrees of defects depending on whether they are multiwall nanotubes or single-wall tubes. Initially, nanotube functionalization was thought to occur at the various defect sites (MWNTs being highly defective compared with SWNTs), but functionalizations both at nanotube ends and along the side walls without disruption or degradation of the tubes have been demonstrated. A variety of functionalized nanotubes are being developed for composite applications, including fluoronanotubes (f-SWNTs); carboxyl-nanotubes with various end functionalization; and numerous covalently bonded SWNTs such as amino-SWNTs, vinyl-SWNTs, epoxy-SWNTs, and many others, to provide for matrix bonding, cross-linking, and initiation of polymerization. Wrapped nanotubes (w-SWNTs) with noncovalent bonding are also another variety of nanotubes and find particular use when the electrical properties of the nanotubes need to be preserved. Figure 11.1-26a. shows nanotube side walls cross-linked in a polymer matrix. The insert, Figure 11.1-26b, shows integration of the functionalized SWNTs into epoxy resin. Nanotube chemistry is not only for polymeric systems but will also be evolved for metals (particularly for Al, Cu, and Ti) and ceramics that are carbide-, oxide-, and nitride-based, as well as many other varieties. Nanotube functionalization in metals and ceramics will likely play a role in nanotube stabilization, defect refinement, and overcoating methodologies. If ends are considered defects, then methodologies for welding nanotubes may well include other species that would be delivered to the nanotubes from a metal or ceramic matrix. These opportunities coupled to a variety of processing modes may well produce next-generation metal hybrids and porous, high-conducting, high-toughness ceramic structures (Barrera, 2000; Mickelson et al., 1998; Zhu, Jiang, et al., 2003).

### Ceramic nanocomposites

Nanoceramics and ceramic nanocomposites are novel and technologically ascendant materials. Examples include tough ceramics that contain dispersion of

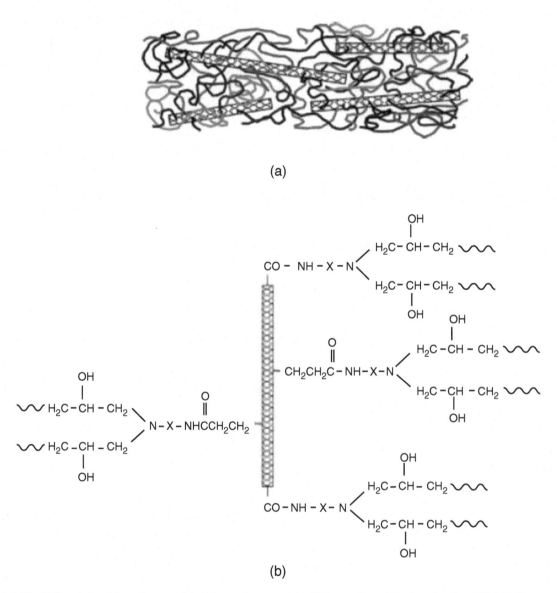

**Figure 11.1-26** (a) Nanotube side walls cross-linked in a polymer matrix. (b) Integration of the functionalized SWNTs into epoxy. (Reprinted by permission of E.V. Barrera et al., Advanced Functional Materials, 14, 2004, 643).

nanoscale metals (Co, Ni), and semiconductor or metal nanoparticles in glass matrices for optoelectronic devices. These materials have been processed in small quantities using high-energy ball milling, internal oxidation, hot consolidation, crystallization of amorphous solid, sol-gel processing, and vapor-phase and vapor-liquid-solid deposition.

Many nanoceramics have outstanding physical and mechanical properties. For example, CNTs have excellent thermal conductivity, and ceramic-CNT nanocomposites can be produced with tailored thermal conductivity for thermal management applications. Although the primary motivation to incorporate CNT in ceramics is to toughen the ceramic, the high thermal conductance of CNTs suggests that their incorporation in ceramics will facilitate thermal transport and thus improve the thermal shock resistance of the ceramic. Toughening is achieved via weak fiber–matrix interfaces that permit debonding and sliding of the fiber within the matrix. The closing forces exerted by the fibers on matrix cracks that propagate around the fibers constrain crack growth, and the work required to pull broken fibers out against sliding friction at the interface imparts toughness.

Many ceramic nanocomposites possessing improved fracture toughness have been synthesized. These include hot-pressed powder composites of CNT-SiC, $Al_2O_3$-SiC, MgO-SiC, W-$Al_2O_3$, $Al_2O_3$-Co, $ZrO_2$-Ni, and $Si_3N_4$-SiC. The powders are first synthesized using chemical reactions and precipitation. For example, oxide ceramic nanocomposites such as $Al_2O_3$-Co and $ZrO_2$-Ni are made from nickel-nitrate or cobalt-nitrate solutions. These solutions are mixed with $Al_2O_3$ or $ZrO_2$ powders and decomposed in air above 400 °C, causing heterogeneous nucleation and precipitation of metal oxide nanoparticles on the surface of ceramic particles. The resulting material is then reduced by hydrogen and hot press sintered at high temperatures and pressures (typically, 1400–1600 °C under 20–40 MPa) to produce Ni- or Co-dispersed ceramic-matrix composites. Considerable improvements in mechanical strength, hardness, and fracture toughness are achieved.

The improvement in the fracture toughness of brittle ceramics through nanoscale metal dispersions, which is especially attractive, is achieved because of the extremely high concentration of internal interfaces that become available for energy dissipation. Nanocomposites of fine particles dispersed in a dielectric matrix have been developed; for example, semiconductor or metal nanoparticles dispersed in a glass matrix have been synthesized for optoelectronic devices, catalysts, and for magnetic, and solar energy conversion devices. CNT incorporation in ceramics such as SiC and alumina by hot-pressing, increases the bending strength and fracture toughness, and lowers the material density, thus creating strong and tough lightweight materials.

To obtain fully dense nanocrystalline ceramics and avoid damaging nanotube reinforcements during sintering, spark-plasma sintering (SPS) is considered the leading technique for composite consolidation because it is a rapid sintering process (Sun et al., 2002; Zhan et al., 2003). This technique is a pressure-assisted fast sintering method based on a high-temperature plasma (spark plasma) that is momentarily generated in the gaps between powder materials by electrical discharge during on/off DC pulsing. It has been suggested that the DC pulse could generate several effects: Spark-plasma sintering, SPS, can rapidly consolidate powders to near-theoretical density through the combined effects of rapid heating, pressure, and powder surface cleaning. Fully dense SWNT-$Al_2O_3$ nanocomposites were fabricated using SPS. Figure 11.1-27 shows a transmission electron micrograph of the nanocomposite, in which the SWNTs are found between the grain boundaries.

### Ceramic nanotube composite systems

There is a lot of interest in developing nanoceramic composites (NCCs) to enhance the mechanical properties of brittle ceramics. The conventional ceramic-processing techniques were first employed to develop MWNT-reinforced SiC ceramic composites where a 20% and 10% increase in strength and fracture toughness, respectively, were measured on bulk composite samples over that of the monolith ceramic (Ma et al., 1998). These increases are believed to be due to the introduction of high-modulus MWNTs into the SiC matrix, which contribute to crack deflection and nanotube debonding. Furthermore, good interfacial bonding is required to achieve adequate load transfer across the MWNT–matrix interface, which is a condition necessary for further improving the mechanical properties in all NCC systems. Therefore, nonconventional processing techniques such as colloidal processing and in situ chemical methods are important processing methods for NCC systems. These novel techniques allow for the control of the interface developed after sintering through manipulation of the surface properties of the composite materials during green processing. Carbide and nitride ceramic materials have been widely used as matrix materials due to the extreme processing conditions required to consolidate and sinter the matrix material. Average sintering temperatures for most nitride and carbide ceramic materials are well above 1800 °C. The following sections deal with specific novel processing routes for NCCs and their related applications.

### Ceramic-coated MWNTs and SWNTs

Multiwall nanotubes coated with $Al_2O_3$ ceramic particles via simple colloidal processing methods provide a significant enhancement to the mechanical properties of the

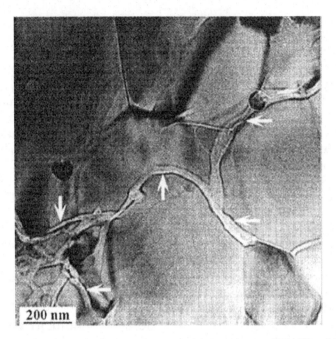

**Figure 11.1-27** *TEM image of a 5.7 vol% SWNT/$Al_2O_3$ nanocomposite fabricated using SPS. The white arrows indicate SWNTs within $Al_2O_3$ grains. (Reprinted by permission of G.-D Zhan et al., Applied Physics Letters, 83(6), 2003, 1228).*

monolithic ceramic (Sun and Gao, 2003; Sun et al., 2002). Coating the surface of MWNTs with alumina improves the homogenous distribution of MWNTs in the ceramic matrix and enables tighter binding between two phases after sintering. MWNTs are treated in $NH_3$ at 600 °C for 3 hours to change their surface properties. Treated MWNTs are put into a solution containing polyethyleneamine (PEI), a cationic polymer used as a dispersant. Alumina is dispersed into deionized water, and polyacrylic acid (PAA) is added into this very dilute alumina suspension. Sodium hydroxide is used to adjust the pH. The prepared dilute alumina suspension with PAA is added into the as-prepared carbon nanotube suspension with PEI, and the suspensions are ultrasonicated. The coated carbon nanotubes collected from the mixed suspension are subsequently added into the concentrated alumina suspension in about 50 wt% ethanol. Finally, the content of MWNTs is only 0.1 wt% of alumina amount. Further drying and grinding result in MWNT-alumina composite powder that is sintered by SPS in a graphite die at 1300 °C with a pressure of 50 MPa for 5 minutes in an Ar atmosphere.

A colloidal processing route is an effective way to improve the mechanical properties of MWNT-alumina composites. Adjusting the surface properties of the alumina powder and those of MWNTs helps bind them together with attractive electrostatic forces, producing strong cohesion between two phases after sintering. During the sintering processing, the growth of alumina particles is believed to wrap MWNTs, which should increase the effectiveness of the reinforcement. The addition of only 0.1 wt% MWNTs in alumina composites increases the fracture toughness from 3.7 to 4.9 MPa $\sqrt{m}$, an improvement of 32% compared with that of the single-phase alumina.

Individual single-wall carbon nanotubes coated with $SiO_2$ can be used to develop a highly sensitive sensing device structure because of the unique properties of SWNTs (Whitsitt and Barron, 2003). The selective etching of silica-coated SWNTs either as small ropes or as individual tubes provides a route to site-selective chemical functionalization as well as the spontaneous generation of tube-to-tube interconnects. The individual SWNTs may be coated in solution and isolated as a solid mat. Either the end or the center of the SiO-SWNT may be etched and exposed. The exposure of the central section of SWNTs has potential for sensor and device structures. Processing of the coated SWNTs takes place during an in situ chemical reaction between fumed silica and SWNT-surfactants structures in suspension. The choice of surfactant type is important in determining whether individual SWNTs or small ropes are coated. It has been found that anionic surfactants result in the formation of coated ropes, whereas cationic surfactants result in individual coated nanotube ropes. It has been proposed that this effect is a consequence of the pH stability of the surfactant−SWNT interaction.

### Conductive ceramics

SWNTs can be used to convert insulating nanoceramics to metal-like conductive composites. Using nonconventional consolidation techniques, spark-plasma sintering (SPS), and conventional powder-processing techniques, dense SWNT-$Al_2O_3$ nanocomposites can be synthesized. These nanocomposites show increasing electrical conductivity with increasing SWNT content (Zhan et al., 2003). The conductivity of these nanocomposites (15 vol% SWNT-$Al_2O_3$), at room temperature, is a 13-fold increase in magnitude over pure alumina. In addition, these nanocomposites show a significant enhancement to fracture toughness over the pure alumina: a 194% increase in fracture toughness over pure alumina, up to 9.7 MPa $\sqrt{m}$ in the 10 vol% SWNT-$Al_2O_3$ nanocomposite has been achieved (Zhan et al., 2003).

## Nanostructured metals and metal composites

A variety of metal-nonmetal nanocomposites have been synthesized and characterized. These nanocomposites have been produced to improve the physical, electrical, and mechanical properties (strength, hardness, fracture toughness, etc.). Metal-matrix nanotube composites have been proposed as future materials for advanced space propulsion structures as well as for radiators and heat pipes, because of their high conductivity and light weight.

Metal-nanocomposites in particular, and nanocrystalline metals in general, are produced in the form of thin ribbons or splats, and powders. Ductile metal-metal nanocomposites such as Nb-Cu have been produced by severe plastic deformation (e.g., cold-drawing Cu and Nb filaments). Other methods to synthesize metal nanocomposites include hot consolidation, electrodeposition, and liquid- and vapor-phase processing.

High-modulus (1–4 TPa) carbon nanotubes (CNT) have been incorporated in metals to obtain modulus greater than that of advanced C fiber-reinforced Al at reduced density (the density of Al-CNT is lower than C-fiber-Al composite, because of the hollow spaces within the CNT). Al-CNT nanocomposites have been synthesized using hot-pressing of powders; however, brittle aluminum carbide phases forms during processing, and the electrical resistivity of the composite increases slightly with CNT volume fraction at room temperature.

Comparison of mechanical properties of various materials for aerospace vehicles, including aluminum, foam, composites, and CNT-Al composites suggests that CNT-Al composite is likely to have better mechanical properties than other enabling materials. As with conventional metal composites, the mechanical properties of CNT-Al composites will strongly depend on the distribution and volume fraction of the CNT and CNT type. Multiwall CNTs generally have better mechanical properties than SWCNTs.

Ceramic-metal nanocomposites essentially belong to the family of heat-resistant, dispersion-strengthened (DS) composites such as W-$ThO_2$, Ni-$Al_2O_3$, Cu-$SiO_2$, Cu-$Al_2O_3$, NiCrAlTi-$Y_2O_3$, etc., that were developed in the 1960s. These DS composites contain 10–15% particles of diameter 10–100 nm. The second phase acts as a barrier to the motion of dislocations rather than as the load-bearing constituent (as in an engineered composite). The dispersoid size must be less than 100 nm for the mean free path in the matrix to be in the range (10–300 nm) needed for dislocation strengthening. The DS composites are designed to retain high yield strength, creep resistance, and oxidation resistance at elevated temperatures rather than to enhance the room-temperature yield strength. The thermal stability of the DS composites is usually superior to precipitation-hardened alloys. As it is desirable to preserve as many of the properties of the matrix metal as possible (e.g., ductility, oxidation resistance, thermal conductivity, impact strength), the volume fraction of the nanodispersoid is kept small (< 15%). In the emerging field of nanocomposites, this limitation is sought to be alleviated, although it is not clear from the current literature how anticipated strength gains will be achieved without tremendous losses in the ductility and toughness at high solid loadings.

### Solid-state powder-based processing

The two solid-state fabrication techniques most widely used for the synthesis of nanocomposites have been high-energy ball milling in conjunction with hot-pressing and deformation processing. For example, Al, Mg, Cu, Ti, and W matrix composites reinforced with nanoscale SiC, $Al_2O_3$, and CNT have been synthesized using hot-press consolidation techniques. In contrast, Nb filament–reinforced Cu matrix composites have been fabricated using the deformation processing (cold-drawing); Cu-Nb nanocomposites have excellent strength and electrical conductivity. For Cu-Nb nanocomposites, continuous multifilaments of Nb are plastically drawn in a Cu matrix to form the nanocomposite. Initially, a Nb rod is drawn in a Cu jacket to a hexagonal shape. Several such hexagonal segments are stacked in a Cu jacket and cold-drawn. The process is repeated several times to create 100- to 250-nm-size whisker-like Nb filaments in the Cu matrix containing a high density of dislocations. Alternatively, extensive plastic deformation of cast Cu-Nb alloys transforms Nb dendrites into nanosize Nb ribbons and stringers. The composite is able to reach high strains without fracture.

Powder metallurgy (PM) -based Al and Mg nanocomposites are fabricated using high-energy ball-milling followed by hot consolidation. Creep-resistant, lightweight Mg-SiC nanocomposites have been prepared using milling and hot extrusion. The nanoscale SiC may be formed by $CO_2$ laser-induced reaction of Si and a carbon-bearing gas such as acetylene, and the micrometer-scale Mg powder is formed by argon gas atomization of molten Mg. It has been found that the ultimate tensile strength (UTS) of the hot extruded Mg-SiC composite doubled as compared to unreinforced Mg. Significant improvement in the hardness and a lowered creep rate were noted even at low-volume fractions of nanoscale SiC (3 vol%). Similarly, CNT-Cu nanocomposites have been synthesized using a press-and-sinter PM technique. The CNT was formed by thermal decomposition of acetylene, and CNT and Cu powders were mixed to a 1:6 ratio in a ball mill. The mixture was pressed at 350 MPa for 5min., and then sintered at 850 °C for 2h. The CNT-Cu composites exhibited reduced coefficient of friction and lower wear loss with increasing CNT content in Cu. Additionally, the CNT-Cu nanocomposites exhibited lower wear loss with increasing load as compared to the Cu-C fiber composites.

Internal oxidation and hot deformation has been used to create fine dispersions of $Al_2O_3$ particles in a nanocrystalline Cu matrix; the fine reinforcement and matrix grain sizes (80 nm for Cu and 20 nm for $Al_2O_3$) led to high strength, hardness, creep resistance and good ductility and electrical conductivity in the nanocomposite. Conversely, $Al_2O_3$ matrix composites containing Cu nanoparticles have been fabricated using hot-pressing of $Al_2O_3$ and CuO powder mixtures to achieve enhanced toughness compared to monolithic $Al_2O_3$, although melting of Cu dispersions during hot-pressing (sintering) can occur.

Internal oxidation is limited to production of composites containing a low-volume fraction of ceramic particles such as oxide dispersion–strengthened composites. Other metal nanoparticle–dispersed ceramics such as $Al_2O_3$-W, $Al_2O_3$-Mo, and $Al_2O_3$-Ni have been synthesized for improved fracture toughness and fracture strength; these composites have nanoscale reinforcements with melting temperatures above the conventional hot-pressing temperatures.

The major drawback of PM to fabricate nanocomposites is the surface oxidation and contamination due to powder exposure to processing environment at high temperatures. For example, in one study fully dense, nanocrystalline iron compacts made by PM exhibited high hardness, but the hardness increased and fracture stress and elongation to failure decreased significantly with decreasing grain size from 33 to 8 nm. Microscopic examination confirmed that processing defects such as micropores, minute quantities of oxide contaminants on powders, and residual binders had masked the benefits of size effects in the nanomaterial.

## Liquid-phase processing

Conventional liquid-phase methods to synthesize metal-matrix nanocomposites have been employed with some success. Agglomerate-free Al-SiC nanocomposite castings containing coarse nanometer-size SiC have been synthesized by stirring nano-SiC in semisolid alloys. Surface pretreatment and wetting agents are usually necessary to improve the wetting and facilitate dispersion. For example, acid treatment with $HNO_3$ and HF has been used to eliminate the agglomeration, and wetting agents (e.g., Ca or Mg) have been added to improve the wetting and to facilitate particle transfer. A potential problem facing successful implementation of stirring techniques at large nanophase loading will be a dramatic increase in the suspension viscosity, which would adversely affect the ability to shape-cast engineering components. This is because the significantly greater surface area of nanoparticles (compared to micrometer-size particles) will greatly increase the viscosity even at moderate loadings.

Metal infiltration has been used to synthesize metal nanocomposites. Studies at the High Pressure Research Institute of Polish Academy of Sciences on metal infiltration of nanoparticles have reported the synthesis of a variety of metal-matrix nanocomposites. Extremely large (7.7 GPa) pressures were used to infiltrate $Al_2O_3$, SiC, and diamond nanoparticles by Mg, Sn, Zn, Al, Ag, Cu, and Ti. The matrix crystallized within nanopores to form grains with size in the range 10–50 nm. The infiltration method used was somewhat unconventional: It consisted of placing pellets of nanoparticles and metal in a die, applying the pressure, and then heating. Pressurization prior to melting the metal permitted cleaner interfaces to be created. X-ray diffraction patterns were recorded during heating (and cooling after infiltration) to determine the inception of melting. For ductile metals, infiltration commenced even in the solid state prior to melting. At 20 kbar pressure, Al melted at 670 °C (instead of 660 °C), and solidified at 600 °C. The infiltration permitted up to 80% ceramic content in the composite. In addition carbon nanotubes have been infiltrated with molten lead.

Capillary infiltration of nanotubes and nanoparticles has been more extensively investigated with nonmetallic liquids than with metals. For example, CNTs have been infiltrated with vanadium oxide and silver nitrate. In the case of silver nitrate, the open-ended CNTs were penetrated by the solution under capillary forces, and the solution-filled nanotubes were irradiated with an electron beam to create CNTs containing chains of tiny silver beads, with beads separated by gas pockets under as high pressures as 1300 atmospheres ($\sim 132 MPa$). Only CNTs of relatively large (>4 nm) diameter could be infiltrated with silver nitrate solution. Even though in a bulk form carbon is wetted by silver nitrate, in the nanoscale regime strong surface forces seem to affect the polarizability, which in turn affects the capillary forces driving the liquid in the tube. In addition, CNTs with internal diameter of $\sim 1$ nm have been filled with halides such as $CdCl_2$, CsCl, and KI.

The magnitude of capillary pressure to infiltrate nanoreinforcement with molten metals could be very large. A simple calculation using the Young-Laplace equation shows that the pressure required to infiltrate CNTs by molten Al will be in the gigapascals range. Although special techniques can be used to achieve such extreme pressures (as was done in the Polish study cited earlier), the ability of common foundry tools and techniques to achieve such pressures is doubtful. Use of ultrasound to disperse nanoscale reinforcement in molten metals could be a viable method to make nanocomposites. The extremely large pressure fluctuations due to "cavitation" will cause the nanophase to be incorporated in the melt.

Prior studies suggest the feasibility of using ultrasound to disperse micrometer-size particles in liquids (to make composites such as silica-Al, glass-Al, C-Al, alumina-Sn, WC-Pb, SiC-Al). Sonic radiation also permits pressureless infiltration of fibers, dispersion and deagglomeration of clusters, and structural refinement of the matrix. The enormous accelerations ($10^3$ to $10^5$ g) due to cavitation aid in deflocculating the agglomerates, rupturing gas bridges, and disintegrating oxide layers, thus allowing physical contact and improved wetting. Sonic radiation also stabilizes the suspension by acoustically levitating particles against gravity, thus enabling a homogeneous particle distribution in the matrix to be created. In the case of solidifying alloys, partial remelting of dendrites by the radiation leads to refinement of structure and removal of air pockets around the particles.

Intriguing possibilities can be envisioned in the application of external pressure (or ultrasound) to synthesize nanocomposites reinforced with ropes made out of nanotubes. The strength of CNT and BN nanotubes has been increased by assembling them in the form of ropes of 20–30 nm in diameter. Combining these ropes with metals presents opportunity to produce ultra-high-strength metals. The spacing between the nanotube strands in a rope is small (e.g., ~0.34 nm in BN nanotube rope); such a small spacing will require extremely large local pressures for infiltration, possibly achievable through intense sonic radiation.

Possible effects of solidification of metal-nanotube composites have been discussed in the literature. These include depression of the solidification temperature due to capillary forces inside the nanotube, and large undercoolings required for heterogeneous nucleation, solidification of the liquid as a glassy (amorphous) rather than a crystalline phase, and the deformation (buckling) of the nanotubes due to stress generated from the volume changes accompanying the phase transformation during solidification and cooling.

Many material defects caused by the solid-state PM processes may be eliminated by liquid-phase processing, which also permits greater flexibility in designing the matrix and interface microstructure. Rapid solidification is already used to produce amorphous or semicrystalline micro- and nanostructured materials. In liquid-phase processed composites, featureless and segregation-free nanostructured matrix may be created because of solidification in nanosize interfiber regions. This is likely to enhance the properties of the matrix, thereby adding to the property benefits achieved due to the nanophase (nanostructured metals have been shown to have better mechanical properties than micrograin metals). Another benefit of combining nanoscale reinforcement with a nanocrystalline matrix is that the latter will be partially stabilized against grain growth by the dispersed nanophase. The nanometer-size matrix grains will increase the total grain boundary area, which will serve as an effective barrier for the dislocation slip and will inhibit creep. Solidification could also permit self-assembly of nanoparticles and nanotubes in interdendritic boundaries and create functionally gradient microstructure.

Despite the successful (albeit limited) demonstration of liquid-phase processing of metal-matrix nanocomposites, several processing challenges can be visualized. These include (1) difficulty in homogeneously dispersing nanoscale reinforcement; (2) extreme flocculation tendency of nanoscale reinforcement; (3) increased viscosity (poor or inadequate fluidity) even at small loadings, due to large surface-to-volume ratio; (4) prohibitively large external pressures needed to initiate flow in nanoscale pores; and (5) possible rejection of dispersed nanophase by solidifying interfaces. The distribution of particles in a solidified matrix depends on whether the nanophase is pushed or engulfed by the growing solid. The dynamics of interactions has been studied theoretically and experimentally. Most experiments have used coarse (>700 nm) particles; only a few studies have used nano-size (30–100 nm) particles (e.g., Ni, Co, $Al_2O_3$) mainly in low-temperature organics (e.g., succinonitrile). However, measurements of critical solidification front velocity to engulf the nanophase are non-existent.

In the case of micrometer-size particles, experiments show that above a critical velocity of the growing solid, the dispersed particles are engulfed, resulting in a uniform distribution, but severe agglomeration and segregation occur at low growth velocities, because the particles are pushed ahead of the growing solid. For pushing to occur, a liquid film must occupy the gap between the particle and the front; hence, the stability of thin liquid films supported between two solids is important. The critical velocity for engulfment increases with decreasing particle size ($V_{cr} \alpha R^{-n}$, $R$ = particle radius) and with decreasing viscosity (which is a strong function of particle concentration). Particles often pile up ahead of the solidification front due to pushing, and the entire pileups could be pushed for various distances. The measured critical velocity for micrometer-size particles varies widely depending on the system; for liquid metals it is in the range 13,000–16,000 μm/s. For nano-SiC (<100nm) in Al, the calculated critical velocity is $>10^7$ μm/s, which would be impossible to achieve by conventional solidification techniques unless ultrasonic irradiation accompanied the solidification. Sonic radiation (and Brownian motion) could induce interparticle collisions and collisions with the growth front, leading to mechanical entrapment. Collisions with the front could induce shape perturbations in the growing solid that might favor capture (e.g., localized pitting or concavity due to altered undercooling via the Gibbs-Thompson effect). However, nanoparticles will likely be ineffective in obstructing the thermal and diffusion fields, and the latter's effect on interactions with growing crystals will probably be negligible. Similarly, fluid convection will be less important than for micrometer-size particles because nanoparticles will likely be smaller than the hydrodynamic boundary layer's (thickness ∼10–1000 μm). For nanosize particles, the critical velocity for the engulfment will likely be much higher than that for the micron-size particles, provided the fundamental nature of the interactions is not altered.

## Surface, interface, nucleation, and reactivity

Surface and interface phenomena achieve overriding importance at the nanoscale, and cause departure from the bulk behavior. For example, even though in bulk form carbon is wetted by many liquids, and therefore, a narrow carbon tube will create a strong attraction for the liquid, at the nanoscale strong surface forces and the curvature of the rolled C sheet adversely affect the capillary forces and cause CNT to become hydrophobic and develop strong attractive forces, leading to self-organizing structures (clusters). Structurally, CNTs are rolled-up sheets of graphite, and the significant curvature of the CNT influences their surface properties. Wetting tests have been done on CNT in liquid epoxy, polypropylene, polyethylene glycol, and other polymers using the Wilhelmy force balance method and transmission electron microscopy (TEM). It seems certain that the wetting behavior is influenced by the nanotube geometry, such as wall thickness and helicity, and by the physical chemistry of the external graphene surface. Likewise, the flow of water and stabilization of water droplets within nanotubes have been studied using molecular dynamic simulation, and contact angle and density of the water droplets within CNTs have been determined.

For nanocomposites, surface and interface considerations lead to two important questions: First, is the nanoreinforcement (e.g., CNT) wetted by the liquid matrix (e.g., liquid polymer)? This is important because wetting is a necessary (though not sufficient) condition for adhesion. And, second, is there evidence of stress transfer across the interface? What experimental techniques can be used at the nanoscale to measure the interfacial strength and adhesion?

From a practical viewpoint, the issue of stress transfer between the matrix and the nanoreinforcement is of major interest. In reality, the physics of the interface, in particular, measurement of the extent and efficiency of load transfer between nanoreinforcement and the matrix is quite difficult, even though the large interfacial area implies that high composite strengths are possible. Raman spectroscopy is the main tool that has been used to evaluate whether stress transfer took place. This is done by monitoring the extent of peak shift under strain. In addition, molecular dynamic simulations have been done to simulate the pullout tests, and to study the effect of interfacial chemical cross-linking on the shear strength in CNT and amorphous or crystalline polymers.

The first direct quantitative tests for interface strength of a single nanofiber in a polymer matrix were reported in 2002. The tests used samples in which nanotubes bridged voids in a polymer matrix. Once these voids were located using the TEM, the bridging nanotubes were dragged out of the polymer using an atomic force microscope (AFM) tip. The lateral force acting on the tip was monitored as the nanotube was progressively extracted from the polymer. An average interfacial shear strength of 150 MPa was obtained to drag a single multiwall CNT out of epoxy. In another technique, a CNT was directly attached to the tip of the AFM probe stylus and was pushed into a liquid polymer. The semicrystalline polymer was then cooled around the CNT. The single nanotube was then pulled out using the AFM, with the forces acting on the nanotubes recorded from the deflection of the AFM tip cantilever. The interfacial strengths far exceeded the polymer tensile strength; thus failure will be expected to take place in the polymer rather than at the interface. Residual stresses due to thermal expansion mismatch and their effect on interface behavior in nanocomposites are still unexplored.

Surface phenomena are also manifested in nucleation of phases. For example, grain refinement in metals is achieved via heterogeneous nucleation as a result of adding inoculants (e.g., $TiB_2$ and $TiAl_3$ in Al). As nucleation is often masked by subsequent growth of crystals, observations of nucleation are difficult. However, many glass-forming alloys serve as slow-motion models of undercooled liquids, and permit high-resolution electron microscopic (HREM) observations of nucleation. Studies show that $TiB_2$ particles dispersed in amorphous Al alloys form an adsorbed layer of $TiAl_3$ over which nanometer-size Al crystals nucleate. The atomic matching at the interfacing planes of $TiB_2$-$TiAl_3$-Al nucleus gives rise to considerable strains in the nucleated crystals. The crystallographic anisotropy of surface energies is important in nucleation; the nucleating substrate must be oriented in a manner so as to present a low-energy interface to the crystallizing liquid. For example yttrium is more effective as a nucleating agent when its prismatic plane (rather than the basal plane) is exposed to the solid. Similarly, the basal and prismatic planes of graphite have different surface energies, and this is expected to influence the crystal nucleation on CNT. The addition to the matrix of a reactive element, which promotes the formation of a low-energy transition layer (even a monolayer) at the interface via adsorption, reaction, or solute segregation has been known to aid nucleation. The atomic disregistry between the particle and the crystal must be small for nucleation to occur. The catalytic potency of a solid for nucleation increases with decreasing atomic disregistry at the interfacing atomic planes. Direct observations of disregistry have been made only in a few systems, and this area of study remains unexplored.

The chemical stability of nanoreinforcement in the liquid matrix at the processing temperature is an important consideration, especially because the nanoreinforcement has a much higher surface-to-volume ratio than does micrometer-size reinforcement. This is seldom a concern when dealing with polymeric liquids but could become important in ceramic and metal matrices. With CNT in Al, for example, aluminum carbide ($Al_4C_3$) formation could embrittle the composite. In the current family of advanced carbon fiber-Al composites, $Al_4C_3$ grows epitaxially on the prismatic planes. The surface energy of the prismatic plane ($\sim 4.8\ J\cdot m^{-2}$) is greater than that of the basal plane ($\sim 0.15\ J\cdot m^{-2}$), and there is anisotropy of chemical reactivity of graphite crystals. Because carbon nanotubes consist of rolled-up (less reactive) graphite basal planes oriented along their axis, they should be chemically stable in molten Al although direct verification of this is lacking. In any case, once formed, the relatively large quantities of the reaction product (e.g., $Al_4C_3$) in the nanocomposite will increase the specific volume of the solid and the viscosity, thus making shape-casting a challenge.

Special precautions may be needed to minimize the extent of deleterious matrix-reinforcement reactions in nanocomposites. For example, rapid densification under very large pressures (1.5–5 GPa) has been used to fabricate well-bonded, nonreacted high-density SiC-Ti nanocomposites (even though SiC and Ti show significant reactivity). Similarly, SiC-Al nanocomposites have been produced using high-energy ball-milling at room temperature followed by hot consolidation using plasma-activated sintering at 823 K (below the melting temperature of Al). The composites had clean interfaces and did not exhibit any strength-limiting reaction products.

## Agglomeration, dispersion, and sedimentation

A dispersed nanophase will likely experience pronounced clustering, which must be resisted to realize the benefits of compositing. Very fine nanoparticles undergo thermal (Brownian) collisions even in a monodisperse suspension, leading to singlets, doublets, triplets, and higher-order aggregates. To deagglomerate the cluster, liquid must flow into the gap, but the thinner the liquid film, the greater is the hydrodynamic resistance to its thinning (disjoining pressure), and the greater will be the energy needed for deagglomeration. Agglomeration depends on both the collision frequency (diffusion) and sticking probability (interparticle forces). Classical theories of flocculation kinetics and floc stability in the surface science literature are probably applicable to nanodispersions (as they are to colloidal suspensions).

The dispersion properties of nanopowders in liquids are important in slurry-based manufacturing processes. The frequency of agglomeration is increased, because of large surface-to-volume ratio of nanopowders. In the case of nonmetallic liquids, the dispersant molecules that are adsorbed on the particle surface generate strong surface charge and electrostatic forces that counter the physical (van der Waals) attraction. This permits stabilization of the suspension. If there are enough ionizable surface groups, electrostatic stabilization is also achieved with proper pH control. In addition, surfactant molecules of uncharged polymers are used to stabilize fine suspensions (steric stabilization). These adsorbed molecules have extended loops and tails, so particle surfaces with such adsorbed molecules begin to show repulsion forces that tend to stabilize the suspension. An intermediate group of dispersants (polyelectrolytes) combine both steric and electrostatic repulsion to stabilize the suspension. The rheological behavior of nanoslurries containing these different dispersants will likely depart from the behavior shown by micrometer-size slurries. As a result, solid loading and viscosity of nanoslurries will be limited by the size, and this will influence the slurry behavior in injection-molding and slip-casting.

High-intensity ultrasound could be a viable tool for declustering nanoagglomerates. However, propagation of ultrasound through a suspension can also induce relative motion between differently sized particles. Finer particles respond better to the inducing frequency and vibrate with greater amplitude than larger particles; as a result, differently sized particles will collide during propagation of sonic waves. This may cause some agglomeration. A narrow size distribution of nanoparticles should facilitate deagglomeration through sonication.

Gravity plays an important role in flocculation. Because particles of different size or density settle in a fluid at different rates, the relative motion between them could cause particle collisions and agglomeration. Because of their extremely fine size, nanoparticles will likely form neutrally buoyant suspensions in liquids (including molten metals). Segregation because of floatation (or sedimentation) may not be severe, due to fine size, as long as agglomeration can be inhibited. Experiments on the settling of micrometer-size SiC in Al and the computed velocity for SiC particles (10 nm SiC, 10 vol%) indicate that the settling rate of nanosize SiC will be several orders of magnitude smaller than micrometer-size SiC (the settling rate is calculated from the hindered-settling equation developed by Richardson and Zaki: $u = u_0(1 - \phi)^{4.65}$, where $u_0$ is the Stokes velocity of a sphere of radius $R$, $u$ is the hindered settling velocity, and $\phi$ is the particulate volume fraction). The settling velocity decreases with particle size at a fixed volume fraction, with nanosize particles yielding settling rates over five orders of magnitude smaller than those of micrometer-size particles, which have settling rates on the order of $1 \times 10^{-5}$ to $1 \times 10^{-4}$ m/s. In the case of CNT, the mass of a single CNT (diameter 100 nm, length 1000 nm) will be about $2.13 \times 10^{-14}$ grams. Such a small mass will experience little buoyancy in a liquid and will have a negligibly small settling or floating rate (velocity varies as the square of the particle radius). Even under centrifugation (acceleration of 3 g to 5 g at a few hundred rpm), segregation effects are likely to be small.

## Properties

Nanostructured materials exhibit property improvements over conventional coarse-grained materials; for example, nanostructured aluminum alloys can be designed to have higher strength than low-carbon steel. In the case of composites with micrometer-size reinforcement, fracture toughness and fatigue strength increase with decreasing particle size (although thermal expansion, thermal conductivity, and wear resistance decrease). At nanometric grain sizes, the proportion of the disordered interfacial area becomes large when compared to a characteristic physical length (e.g., Frank–Read loop size for dislocation slip). Interfacial defects such as grain boundaries, triple points, and segregation begin to make an increasing contribution to the physical and mechanical properties. Deviations from the classical behavior (e.g., Hall–Petch relationship) could occur, and properties of micrometer-size composites may not be reliably extrapolated down to the nanometer size (this is especially true of CNT that has distorted electronic structure due to cylindrical C layers). In monolithic nanomaterials, grain size effects on properties show considerable departure from the behavior expected for micrometer-size grains.

### Strength and modulus

The composite modulus is generally estimated from a rule-of-mixtures relationship. Using the literature values of relevant parameters ($E_{CNT} = 1.81$ TPa, $E_{C-fiber} = 800$ GPa, and $E_{Al} = 70$ GPa), it is found that at a volume fraction of 0.45, Al-CNT composites will attain a composite modulus higher than the conventional vapor-grown C fiber (the specific modulus of Al-CNT, i.e., modulus-to-density ratio, can be even higher). However, a dispersion in the strength measurements on CNT, and the variability introduced by processing could cause deviations from the predictions.

The strength of Al-CNT composite can be estimated from the Kelly–Tyson equation, according to which the composite strength $\sigma_c$ is given from $\sigma_c = \sigma_f \cdot V_f \cdot [1 - (l_c/2l)] + \sigma'_m \cdot (1 - V_f)$, where $\sigma_c$ = composite strength, $\sigma_f$ = tensile strength of the nanotube, $\sigma'_m$ = stress of Al matrix at the failure strain of the composite, $l_c$ = critical length of a nanotube in Al, $l$ = average length of a nanotube. For Al-CNT, the literature data show that $\sigma_{CNT} = 3$ GPa, $\sigma'_m = 40$ MPa, $l_c = 0.85$ μm, and $l = 2$ μm. Calculations using these values confirm that significantly improved strength will be expected to result, and very high composite strength could be achieved at a relatively small-volume fraction of the CNT (in comparison to conventional C fibers). A nonhomogeneous distribution of CNT, and processing defects could, however, substantially lower the strength. The strength predictions are based on classical composite mechanics that do not take into account the hollow nature of the CNT.

### Nanotribology

Nanoscale materials present interesting opportunities for tribological performance. Ultra-low-friction states have been observed at the nanoscale using AFM capable of measuring lateral forces in the piconewton range ($10^{-12}$ Newtons). The energy dissipation in atomic friction is measured as an AFM stylus tip is dragged over a surface. Normally, the tip sticks to an atomic position on the surface, and then, when the force is sufficient, slips to the next atomic position, and so on, resulting in a "sawtooth" modulation of the lateral force and stick-slip-type motion. In one experiment, the tip of the AFM stylus was coated with a tiny flake of graphite, which slid over an extended graphite surface. When the atomic lattice of the flake was aligned with that of the surface, stick-slip sliding was observed, but at all other orientations the friction was near zero. This indicates that when the two surfaces are not in atomic-scale registry, there is a high degree of force cancellation. Similar observations have been made on NaCl crystals with a Si tip.

Significantly greater hardness of nanocomposites than microscale composites leads to improved wear resistance. Thus, CNT-Ni-P composites tested under lubricated conditions led to low friction and wear relative to similar composites containing micrometer-size SiC or graphite, and to virgin Ni-P itself. Rapidly solidified nanopowders have been used to fabricate nanoscale Al alloys and composites possessing improved wear resistance and low friction. Al alloys for bearing applications contain either layered dispersoids (graphite, mica, talc, etc.) or dispersoids of soft metals such as Pb, Bi, and In. Studies show that dispersoids in the nanosize range provide better antifriction properties than those in the micron-size range. The material transfer is significantly less with nanodispersions providing a more uniform thin layer of soft material at the mating interface than do the coarser, micron-size dispersoids.

## Carbon nanotubes: biosensor applications

Biomolecules such as nucleic acids and proteins carry important information of biological processes. The ability to measure extremely small amounts of specific biomarkers at molecular levels is highly desirable in biomedical research and health care. Current technologies rely on well-equipped central laboratories for molecular diagnosis, which is expensive and time consuming, often causing delay in medical treatments. There is a strong need for smaller, faster, cheaper, and simpler biosensors for molecular analysis. The recent advance in carbon nanotube (CNT) nanotechnology has shown great potential in providing viable solutions. CNTs with well-defined nanoscale dimension and unique molecular structure can be used as bridges linking biomolecules to macro/micro-solid-state devices so that the information of bioevents can be transduced into measurable signals. Exciting new biosensing concepts and devices with extremely high sensitivities have been demonstrated using CNTs.

As the size of the materials reach the nanometer regime, approaching the size of biomolecules, they directly interact with individual biomolecules, in contrast to conventional macro- and microdevices, which deal with assembly of relatively large amount of samples. Nanomaterials exhibit novel electronic, optical, and mechanical properties inherent with the nanoscale dimension. Such properties are more sensitive to the environment and target molecules in the samples. Although a big portion of nanomaterials are isotropic nanoparticles or thin films, high-aspect-ratio one-dimensional nanomaterials such as CNTs and various inorganic nanowires (NWs) are more attractive as building blocks for device fabrication. The potential of CNTs and NWs as sensing elements and tools for biomolecular analysis as well as sensors for the gases and small molecules have been recently recognized (Li and Ng, 2003). Promising results in improving sensitivity, lowering detection limit, reducing sample amount, and increasing detection speed have been reported using such nanosensors (Kong et al., 2000; Cui et al., 2001; Li et al., 2003). CNTs integrated with biological functionalities are expected to have great potential in future biomedical applications.

The electronic properties of CNTs are very sensitive to molecular adsorption. Particularly, in a semiconducting single-wall CNT (SWNT), all carbon atoms are exposed at the surface so that a small partial charge induced by chemisorption of gas molecules is enough to deplete the local charge carrier and cause dramatic conductance change (Collins et al., 2000). Because biomolecules typically carry many ions, they are expected to affect CNT sensing elements and transducers more dramatically than are simple gases and small molecules. Sensing devices have been fabricated for various applications using single CNTs, single semiconducting SWNT field-effect-transistors (Star et al., 2003; Besteman et al., 2003; Li et al., 2002), vertically aligned nanoelectrode arrays (Snow et al., 2003), and random networks or arrays.

### FET-based biosensors

The extreme high sensitivity and potential for fabricating high-density sensor array make nanoscale field-effect-transistors (FETs) very attractive for biosensing, particularly because biomolecules such as DNA and proteins are heavily charged under normal conditions. SWNT FETs are expected to be more sensitive to the binding of such charged species than are chemisorbed gas molecules. However, the wet chemical environment with the presence of various ions and of other biomolecules makes it much more complicated than gas sensors. Besteman et al. (2003) successfully demonstrated that the enzyme-coated SWNT FETs can be used as single-molecule biosensors. As shown in Figure 11.1-28, the redox enzyme glucose oxidase (GOx) is immobilized on SWNT using a linking molecule, which on one side binds to the SWNT through van der Waals coupling with a pyrene group and on the other side covalently binds the enzyme through an amide bond, as developed by Chen et al. (2001). The FET preserves the p-type characteristic but shows much lower conductance on GOx immobilization, which is likely the result of the decrease in the capacitance of the tube caused by GOx immobilization, because GOx blocks the liquid from access to the SWNT surface.

The GOx-coated SWNT showed a strong pH dependence as well as high sensitivity to glucose. Figure 11.1-28 shows the real-time measurements where the conductance of a GOx-coated SWNT FET has been recorded as a function of time in milli-Q water. No significant change in conductance is observed when more milli-Q water is added, as indicated by the first arrow. However, when $0.1\ M$ glucose is added, the conductance increases by 10%, as indicated by the first arrow. Both studies on liquid-gated FETs use very low ionic strength with 10 m$M$ NaCl (Besteman et al., 2003) and 0.1 m$M$ KCl (Rosenblatt et al. 2002), respectively. The salt concentrations are more than 10 times lower than physiology buffers.

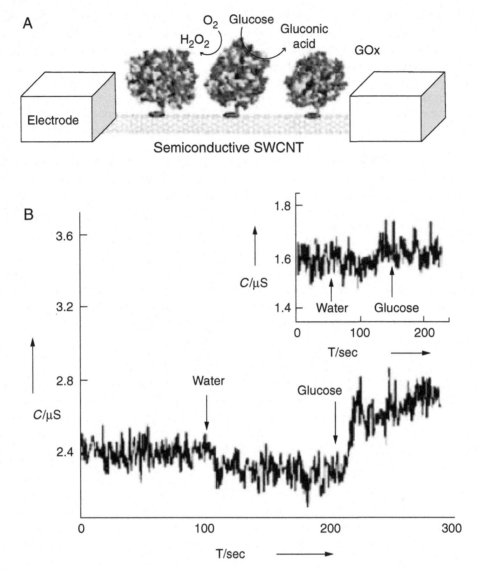

**Figure 11.1-28** (A) A GOx-functionalized CNFET for glucose analysis. (B) Electronic response of the GOx-CNFET to glucose. The source-drain and liquid-gate voltages were kept constant at 9.1 mV and -500 mV, respectively. Arrows show the addition of water and a glucose sample (2 μm, 0.1 mM) to the system. Inset shows the same measurement on a control CNFET without GOx. (K. Besteman et al., Nano Lett., 3 (6), 2003, 727-730). Reprinted with permission from the American Chemical Society, 1155 16th St. NW, Washington, DC 20036. Photo Courtesy of C. Dekker, Delft University of Technology, The Netherlands.

### Aligned nanoelectrode array-based electronic chips

Researchers at NASA Ames Research Center have developed the concept of making DNA chips that are more sensitive than current electrochemical biosensors. They cover the surface of a chip with millions of vertically mounted carbon nanotubes 30–50 nm in diameter (Figure 11.1-29). When the DNA molecules attached to the ends of the nanotubes are placed in a liquid containing DNA molecules of interest, the DNA on the chip attaches to the target and increases its electrical conductivity. This technique, expected to reach the sensitivity of fluorescence-based detection systems, may find application in the development of a portable sensor.

An embedded array minimizes the background from the side walls, whereas the well-defined graphitic chemistry at the exposed open ends allows the selective functionalization of −COOH groups with primary amine-terminated oligonucleotide probes through amide bonds. The wide electropotential window of carbon makes it possible to directly measure the oxidation signal of guanine bases immobilized at the electrode surface. Such a nanoelec-trode array can be used as ultrasensitive DNA sensors based on an electrochemical platform (Koehne et al., 2003, 2004). As shown in Figure 11.1-30, oligonucleotide probes of 18 bases with a sequence of [Cy3]5-CTIIATTTCICAIITCCT-3[AmC7-Q] are covalently attached to the open end of MWNT exposed at

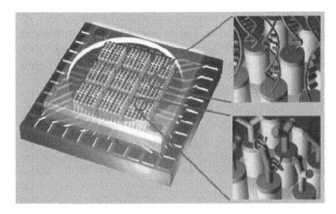

**Figure 11.1-29** Vertical carbon nanotubes are grown on a silicon chip. DNA and antigen molecules attached at the ends of the tubes detect specific types of DNA (top) and antigen detection (bottom) in an analyte. (R. E. Smalley "Chip Senses Trace DNA," Technology Research News, 6(22), 2003). Reprinted with permission from Hazel Cole, exec. asst. to late Prof. Smalley, Rice University.

the $SiO_2$ surface. This sequence is related to the wild-type allele (Arg1443stop) of *BRCA 1* gene (Miki et al., 1994). The guanine bases in the probe molecules are replaced with nonelectroactive inosine bases, which have the same base-pairing properties as guanine bases. The oligonucleotide target molecule has a complimentary sequence [Cy5]5-AGGAC-CTGCGAAATCCAGGGG GGGGGGG-3, including a 10 mer polyG as the signal moieties. Hybridization was carried out at 40 °C for about 1 hour in ~100 n$M$ target solution in 3 × SSC buffer. Rigorous washing—in three steps using 3 × SSC, 2 × SSC with 0.1% SDS, and 1 × SSC respectively at 40 °C for 15 minutes after each probe functionalization and target hybridization process—was applied to get rid of nonspecifically bound DNA molecules, which is critical for getting reliable electrochemical data.

Such solid-state nanoelectrode arrays have great advantages in stability and processing reliability over other electrochemical DNA sensors based on mixed self-assembled monolayers of small organic molecules. The density of nanoelectrodes can be controlled precisely using lithographic techniques, which in turn define the number of probe molecules. The detection limit can be optimized by lowering the nanoelectrode density. However, the electrochemical signal is defined by the number of electrons that can be transferred between the electrode and the analytes. Particularly, the guanine oxidation occurs at rather high potential (~1.05 V versus saturated calomel [SCE]) at which a high background is produced by carbon oxidation and water electrolysis. This problem can be solved by introducing $Ru(bpy)_3^{2+}$ mediators to amplify the signal based on an electrocatalytic mechanism

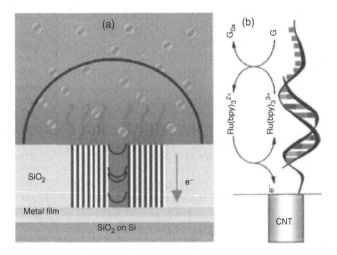

**Figure 11.1-30** Schematic of the MWNT nanoelectrode array combined with $Ru(bpy)_3^{2+}$ mediated guanine oxidation for ultrasensitive DNA detection. (J. Koehne et al., Journal of Material Chemistry, 14(4), 2004, 676). Reprinted with permission from the Royal Society of Chemistry, Thomas Graham House, Science Park, Cambridge, U.K, Photo Courtesy of J. Li, NASA Ames Research Center, Moffet Field, CA.

(Miki et al., 1994). Combining the MWNT nanoelectrode array with Ru(bpy)$_3^{2+}$-mediated guanine oxidation (as schematically shown in Figure 11.1-30), the hybridization of fewer than ~1000 oligonucleotide targets can be detected with a 20- × 20-µm$^2$ electrode, with orders of magnitude improvement in sensitivity compared with previous EC-based DNA detectors (Sistare, Holmberg, and Thorp, 1999; Koehne et al., 2003).

# References

Amelinckx S., X. B. Zhang, D. Bernaerts, X. F. Zhang, V. Ivanov, and J. B. Nagy. *Science 265*, 1994, 635.

Asoh H., N. Kazuyuki, M. Nakao, T. Tamamura, and H. Masuda. *Journal of the Electrochemical Society 148*, 2001, B152.

Baker R. T. K., M. A. Barber, P. S. Harris, F. S. Feates, and R. J. Waite. *J. Catal. 26*, 1972, 51.

Baker R. T. K., P. S. Harris, R. B. Thomas, and R. J. Waite. *J. Catal. 30*, 1973, 86.

Barrera E. V. *J. Mater. 52*, 2000, 38.

Besteman K., et al. *Nano Letters 3*(6), 2003, 727.

Bethune D. S., et al. *Nature 363*, 1993, 605.

Bezryadin A., C. N. Lau, and M. Tinkham. *Nature 404*, 2000, 971.

Borsella E., S. Botti, M. C. Cesile, S. Martelli, A. Nesterenko, and P. G. Zappelli. MoS$_2$ nanoparticles produced by laser induced synthesis from gaseous precursors. *Journal of Mater Science Lett 20*, 2001, 187.

Cao Y. C., R. Jin, and C. A. Mirkin. Raman dye-labeled nanoparticle probes for proteins. *JACS 125*, 2003, 14677.

Cassell A. M., J. A. Raymakers, J. King, and H. Dai. *J. Phys. B. Chem. 103*, 1999, 6484.

Che G., B. B. Lakshmi, C. R. Martin, and E. R. Fisher. *Chem. Mater. 10*, 1998, 260.

Chen C. C., and C. C. Yeh. Large-scale catalytic synthesis of crystalline gallium nitride nanowires. *Adv. Mater. 12*, 2000, 738.

Chen C. C., C. C. Yeh, C. H. Chen, M. Y Yu, and H. L. Liu, et al. Catalytic growth and characterization of gallium nitride nanowires. *Journal of Am. Chem. Soc. 123*, 2001, 2791.

Chen R. J., et al. *J. Am. Chem. Soc. 123*, 2001, 3838.

Chien C. L. *Journal of Applied Physics 69*, 1991, 5267.

Choi H. C., M. Shim, S. Bangsaruntip, and H. J. Dai. *Journal of the American Chemical Society 124*, 2002, 9058.

Chung S. H., A. Hoffmann, S. D. Bader, C. Liu, B. Kay, and L. Chen. *Applied Physics Letters 85*, 2004, 2971.

Coffer J. L., S. R. Bigham, X. Li, R. F. Pinizzotto, Y. G. Rho, R. M. Pirtle, and I. L. Pirtle. *Applied Physics Letters 69*, 1996, 3851.

Collins P. G., et al. *Science 287*, 2000, 1801.

Crouse D., Y.-H. Lo, A. E. Miller, and M. Crouse. *Applied Physics Letters 76*, 2000, 49.

Cui Y., et al. *Science 293*, 2001, 1289.

Dai H., et al. *Science 283*, 1999, 512.

Despic A. R. *Journal of Electroanal. Interfacial Electrochem. 191*, 1985, 417.

Dickson R. M., and L. A. Lyon. *J. Phys. Chem., B 104*, 2000, 6095.

Djalali D., S. Y Li, and M. Schmidt. *Macromolecules 35*, 2002, 4282.

Dresselhaus M., G. Dresselhaus, and P. Eklund. *Science of Fullerenes and Carbon Nanotubes*. San Diego: Academic Press, 1996.

Duan X., and C. M. Lieber. General synthesis of compound semiconductor nanowirese. *Adv. Mater. 12*, 2000, 298.

Dubois S., C. Marchal, and J. L. Maurice. *Applied Physics Letters 70*, 1997, 396.

Evans P. R., et al. *Applied Physics Letters 76*, 2000, 481.

Forrer P., F. Schlottig, H. Siegenthaler, and M. Textor. *Journal of Appl. Electrochem. 30*, 2000, 533.

Foss C. A. J. *Metal Nanoparticles Synthesis Characterization and Applications*. Dekker, 2002.

Foss C. A., G. L. Hornyak, J. A. Stocket, and C. R. Martin. *Journal of Phys Chem 96*, 1992, 7497.

Foss C. A., G. L. Hornyak, J. A. Stocket, and C. R. Martin. *Journal of Phys. Chem. 98*, 1994, 2963.

Fullam S., D. Cottell, H. Rensmo, and D. Fitzmaurice. *Adv. Mater. 12*, 2000, 1430.

Furneaux R. C., W. R. Rigby, and A. P. Davidson. *Nature 337*, 1989, 147.

Givargizov E. I. Fundamental aspects of VLS growth. *Journal of Cryst. Growth 31*, 1975, 20.

Greene L. E., M. Law, J. Goldberger, F. Kim, and J. C. Johnson, et al. Low-temperature wafer-scale production of ZnO nanowire arrays. *Angew. Chem. Int. Ed. 42*, 2003, 3031–3034.

Grieve K., P. Mulvaney, and F. Grieser. Synthesis and electronic properties of semiconductor nanoparticles/quantum dots. *Curr Opin Colloid Interf Sci, 5*, 2000, 168.

Gudiksen M. S., and C. M. Lieber. Diameter-selective synthesis. *J. Am. Chem. Soc. 122*, 2000, 8801.

Dai H.. *Nano Lett. 3*, 2003, 347.

Han H. Gas phase synthesis of nanocrystalline materials. *Nanostruct Mater 9*, 1997, 3.

Harano A., K. Shimada, T. Okubo, and M. Sadakata. Crystal phases of TiO$_2$ ultrafine particles prepared by laser ablation solid rods. *Journal of Nanoparticles Research 4*, 2002, 215.

He M., P. Zhou, S. N. Mohammad, G. L. Harris, and J. B. Halpern, et al. Growth of GaN nanowires by direct reaction of Ga with NH$_3$. *Journal of Crystal Growth 231*, 2001, 357.

Hiruma K., M. Yazawa, T. Katsuyama, K. Ogawa, and K. K. Haraguchi, et al. Growth and optical properties of nanometer-scale GaAs and InAs whiskers. *J. Appl. Physics 77*, 1995, 447.

Hong K., F. Y. Yang, K. Liu, and C. L. Chien, et al. *J. Applied Physics 85*, 1999, 6184.

Huang M. H., Y. Wu, H. Feick, N. Tran, E. Weber, and P. Wang. Catalytic growth of zinc oxide nanowires by vapor transport. *Adv. Mater. 13*, 2001, 113.

Huang Y. H., H. Okumura, and G. C. Hadjipanayis. *Journal of Applied Physics 91*, 2002, 6869.

Huixin He and Nongjian J. Tao. "Electrochemical Fabrication of Metal Nanowires" *Encyclopedia of Nanoscience and Nanotechnology*, Edited by H.S. Nalwa.

Iijima S.. *Nature 354*, 1991, 56.

Iijima S., and T. Ichihashi. *Nature 363*, 1993, 605.

Jessensky O., F. Muller, and U. Gosele. *Applied Physics Letters 72*, 1998, 1173.

Kamlag Y., A. Goossens, I. Colbeck, and J. Schoonman. Laser CVD of cubic SiC nanocrystals. *Appl. Surf. Science* 184, 2001, 118.

Keren K., M. Krueger, R. Gilad, U. Sivan, and E. Braun. *Science* 297, 2002, 72.

Kim H. Y., J. Park, and H. Yang. Direct synthesis of aligned carbide nanowires from the silicon substrates. *Chem. Commun.* 2, 2003, 256–257.

Kim H. Y., J. Park, and H. Yang. Synthesis of silicon nitride nanowires directly from the silicon substrates. *Chem. Phys. Letters* 372, 2003, 269.

Kim M. S., N. M. Rodriguez, and R. T. K. Baker. *J. Catal.* 131, 1991, 60.

Koehne J., et al. *J. Mater. Chem.* 14(4), 2004, 676.

Koehne J., et al. *Nanotechnology* 14, 2003, 1239.

Kong J., et al. *Science* 287, 2000, 622.

Kruis F. E., H. Fissan, and A. Peled. Synthesis of nanoparticles in the gas phase for electronic, optical and magnetic applications—a review. *Journal of Aerosol Sci.* 29, 1998, 511.

Kuykendall T., P. Pauzauskie, S. Lee, Y. Zhang, J. Goldberger, and P. Yang. Metallorganic chemical vapor deposition route to GaN nanowires with triangular cross sections. *Nano Letters* 3, 2003, 1063.

Gutwein L. G., and T. J. Webster. Osteoblast and Chondrocyte Proliferation in the Presence of Alumina and Titania Nanoparticles. *J. Nanoparticle Res.* 4, 2002, 231–238.

Lakshmi B. B., P. K. Dorhout, and C. R. Martin. *Chem. Mater.* 9, 1997, 857.

Ledoux G., J. Gong, F. Huisken, O. Guillois, and C. Reynaud. Photoluminescence of size-separated silicon nanocrystals: confirmation of quantum confinement. *Applied Physics Letters* 80, 2002, 4834.

Lee K. H., H. Y Lee, W. Y. Jeung, and W. Y Lee. *Journal of Applied Physics* 91, 2002, 8513.

Li A.-P., F. Muller, A. Birner, K. Nielsch, and U. Gosele. *Adv. Mater.* 11, 1999, 483.

Li A.-P., F. Muller, A. Birner, K. Nielsch, and U. Gosele. *Journal of Vacuum Science Technology, A* 17, 1999, 1428.

Li A.-P., F. Muller, and U. Gosele. *Electrochem. Solid-State Letters* 3, 2000, 131.

Li F., L. Zhang, and R. M. Metzger. *Chem. Mater.* 10, 1998, 2470.

Li J., and H. T. Ng. Carbon nanotube sensors. In: Nalwa H. S. (ed), *Encyclopedia of Nanoscience and Nanotechnology*. American Scientific Publishers, Santa Barbara, CA.

Li J., et al. *J. Phys. Chem. B* 106, 2002, 9299.

Li J., et al. *Nano Letters* 3(5), 2003, 597.

Lin Y. M., and S. B. Cronin, et al. *Applied Physics Letters* 76, 2000, 3944.

Liu K., and C. L. Chien, et al. *Phys. Rev., B* 51, 1995, 7381.

Liu K., C. L. Chien, P. C. Searson, and Y. J. Kui. *Applied Physics Letters* 73, 1998, 1436.

Liu K., K. Nagodawithana, P. C. Searson, and C. L. Chien. *Phys. Rev., B* 51, 1995, 7381.

Liu X., C. Li, S. Han, J. Han, and C. Zhou. Synthesis and electronic transport studies of CdO nanoneedles. *Applied Physics Letters* 82, 2003, 1950.

Lopez-Lopez M., A. Guillen-Cervantes, Z. Rivera-Alvarez, and I. Hernandez-Calderon. Hillock formation during the molecular beam epitaxial growth of ZnSe on GaAs substrates. *J. Cryst. Growth* 193, 1998, 528.

Ma R. Z., et al. *Journal of Material Science* 33, 1998, 5243.

Ma R., and Y. Bando. Investigation of the growth of boron carbide nanowires. *Chem. Mater.* 14, 2002, 4403.

Magnusson M. H., K. Deppert, and J.-O. Malm. Single-crystalline tungsten nanoparticles produced by thermal decomposition of tungsten hexacarbonyl. *Journal of Materials Research* 15, 2000, 1564.

Maisels A., F. E. Kruis, H. Fissan, and B. Rellinghaus. Synthesis of tailored composite nanoparticles in the gas phase. *Applied Physics Letters* 77, 2000, 4431.

Makimura T., T. Mizuta, and K. Murakami. Laser ablation synthesis of hydrogenated silicon nanoparticles with green photoluminescence in the gas phase. *Japan Journal of Applied Physics* 41, 2002, L144.

Marine W., L. Patrone, and M. Sentis. Strategy of nanocluster and nanostructure synthesis by conventional pulsed laser ablation. *Appl. Surf. Sci.* 345, 2000, 154–155.

Martin B. R., et al. *Adv. Mater.* 11, 1999, 1021.

Martin C. R., R. Parthasarathy, and V. Menon. *Synth. Met.* 55–57, 1993, 1165.

Masuda H., and F. Hasegwa. *Journal of the Electrochemical Society* 144, 1997, L127.

Masuda H., and K. Fukuda. *Science* 268, 1995, 1466.

Masuda H., and M. Satoh. *Jpn Journal of Applied Physics* 35, 1996, L126.

Masuda H., H. Yamada, M. Satoh, H. Asoh, M. Nakao, and T. Tamamura. *Applied Physics Letters* 71, 1997, 2770.

Masuda H., K. Yaka, and A. Osaka. *Jpn. Journal of Applied Physics, Part 2* 37, 1998, L1340.

Mbindyo J. K. N., B. D. Reiss, B. R. Martin, C. D. Keating, M. J. Natan, and T. E. Mallouk. *Adv. Mater.* 13, 2001, 249.

Mickelson E. T., et al. *Chem. Phys. Letters* 296, 1998, 188.

Miki Y., et al. *Science* 266, 1994, 66.

Mock J. J., et al. *Nano Letters* 2, 2002, 465.

Murray C. B., C. R. Kagan, and M. G. Bawendi. Synthesis and characterization of monodisperse nanocrystals and close-packed nanocrystal assemblies. *Annu Rev Mater Sci.* 30, 2000, 545.

Naguyen P., H. T. Ng, J. Kong, A. M. Cassell, and R. Quinn, et al. Epitaxial directional growth of indium-doped tin oxide nanowire arrays. *Nano Letters* 3, 2003, 925.

Nakata Y., J. Muramoto, T. Okada, and M. Maeda. Particle dynamics during nanoparticle synthesis by laser ablation in a background gas. *Journal of Applied Physics* 91, 2002, 1640.

Nasibulin A. G., O. Richard, E. I. Kauppinen, D. P. Brown, J. K. Jokiniemi, and I. S. Altman. Nanoparticle synthesis by copper(II) acetylacetonate vapor decomposition in the presence of oxygen. *Aerosol Science Technol* 36, 2002, 899.

Nicewarner-Pena S. R., and etal. *Science* 294, 2001, 137.

Nielsch K., F. Muller, A.-P. Li, and U. Gosele. *Adv. Mater.* 12, 2000, 582.

Oberlin A., et al. *J. Crystal Growth* 32, 1976, 335.

Ohno T. Morphology of composite nanoparticles of immiscible binary system prepared by gas-evaporation technique and subsequent vapor condensation. *Journal of Nanoparticle Research* 4, 2002, 255.

Ostraat M. L., J. W. De Blauwe, M. L. Green, L. D. Bell, H. A. Atwater, and R. C. Flagan. Ultraclean two-stage aerosol reactor for production of oxide-passivated silicon nanoparticles for novel memory devices. *Journal of the Electrochemical Society* 148, 2001, G265.

Park S. J., T. A. Taton, and C. A. Mirkin. *Science* 295, 2002, 1503.

Parthasarathy R., and C. R. Martin. *Nature* 369, 1994, 298.

Peng X. S., G. W. Meng, X. F. Wang, Y. W. Wang, and J. Zhang, et al. Synthesis of oxygen-deficient indium-tin-oxide (ITO nanofibers). *Chem. Mater. 14*, 2002, 4490.

Peng Z. A., and X. Peng. Formation of high-quality CdTe, CdSe, and CdS nanocrystals using CdO as precursor. *Journal of Am. Chem. Soc. 123*, 2001, 183.

Piraux L., et al. *Applied Physics Letters 65*, 1994, 2484.

Preston C. K., and M. J. Moskovits. *Journal of Phys Chem. 97*, 1993, 8495.

Prieto A. L., M. S. Sander, M. S. Martin-Gonzalez, R. Gronsky, T. Sands, and A. M. Stacy. *Journal of the American Chemical Society 123*, 2001, 7160.

Qin D. H., C. W. Wang, Q. Y. Sun, and H. L. Li. *Applied Physics A-Mater. 74*, 2002, 761.

Reiss B.D., et al., *Mater. Res. Symp. Proc.*, in press, 2001.

Richter J., R. Seidel, R. Kirsch, and H. K. Schackert. *Adv. Mater. 12*, 2000, 507.

Rodriguez-Ramos N. M.. *J. Mat. Research 8*, 1993, 3233.

Rosenblatt S., et al. *Nano Letters 2*(8), 2002, 869.

Roy I., T. Y Ohulchanskyy, H. E. Pudavar, E. J. Bergey, A. R. Oseroff, J. Morgan, T. J. Dougherty, and P. N. Prasad. Ceramic-based nanoparticles entrapping water-insoluble photosensitizing anticancer drugs: a novel drug-carrier system for photodynamic therapy. *Journal of Am. Chem. Soc. 125*, 2003, 7860.

Iijima S.. *Nature 354*, 1991, 56.

Saito, et al. *Journal of Applied Physics 80*(5), 1996, 3062–3067.

Sander M. S., A. L. Prieto, R. Gronsky, T. Sands, and A. M. Stacy. *Adv. Mater. 14*, 2002, 665.

Sapp S. A., B. B. Lakshmi, and C. R. Martin. *Adv. Mater. 11*, 1999, 402.

Sauer G., G. Breshm, S. Schneider, K. Nielsch, R. B. Wehrspohn, J. Choi, H. Hofmeister, and U. Cosele. *Journal of Applied Physics 91*, 2002.

Schonenberger C., B. M. I. van der Zande, L. G. J. Fokkink, M. Henny, C. Schmid, M. Kruger, A. Bachtold, R. Huber, H. Birk, and U. Staufer. *Journal Phys Chem.*, B 101, 1997, 5497.

Shi W. S., Y. F. Zheng, N. Wang, C. S. Lee, and S. T. Lee. Synthesis and microstructure of gallium phosphide nanowires. *Journal of Vacuum Science Technology B19*, 2001, 1115.

Shimada T., K. Hiruma, M. Shirai, M. Yazawa, and K. Haraguchi, et al. Size, position and direction control on GaAs and InAs nanowhisker growth. *Superlattice Microstr.*, 24, 1998, 453.

Shinde S. R., S. D. Kulkarni, A. G Banpurkar, and S. B. Ogale. Magnetic properties of nanosized powders of magnetic oxides synthesized by pulsed laser ablation. *Journal of Applied Physics 88*, 2000, 1566.

Shingubara S., O. Okino, Y. Sayama, H. Sakaue, and T. Takahagi. *Jpn. Journal of Applied Physics 36*, 1997, 7791.

Sinani V. A., D. S. Koktysh, B. G. Yun, R. L. Matts, T. C. Pappas, M. Motamedi, S. N. Thomas, and N. A. Kotov. Collagen coating promotes biocompatibility of semiconductor nanoparticles in stratified LBL films. *Nano Letters 3*, 2003, 1177.

Sistare M. F., R. C. Holmberg, and H. H. Thorp. *J. Phys. Chem.*, B 103, 1999, 10718.

Smalley R. E., et al. *American Scientist 85*, 1997, 324.

Snow E. S., et al. *Applied Phys. Letters 82*(13), 2003, 2145.

Star A., et al. *Nano Letters 3*(4), 2003, 459.

Sun J., and L. Gao. *Carbon 41*, 2003, 1063.

Sun J., L. Gao, and W. Li. *Chem. Mater. 14*, 2002, 5169.

Tanase M., D. M. Silevitch, A. Hultgren, L. A. Bauer, P. C. Searson, G. L. Meyer, and D. H. Reich. *Nano Letters 1*, 2001, 155.

Tanase M., et al. *Journal of Applied Physics 91*, 2002, 8549.

Thess A., et al. *Science 273*, 1996, 483.

Thurn-Albrecht T., and J. Schotter, et al. *Science 290*, 2000, 2126.

Ttrindade T., P. O'Brien, and N. L. Pickett. Nanocrystalline semiconductors: synthesis, properties, and perspectives. *Chem Mater 13*, 2001, 3843.

Urban F. K., A. Hosseini-Tehrani, P. Griffiths, A. Khabari, Y.-W. Kim, and I. Petrov. Nanophase films deposited from a high rate, nanoparticle beam. *Journal of Vacuum Science Technology*, B 20, 2002, 995.

V.T.S Wong, W.J. Li. Micro Electro Mechanical Systems, 2003. MEMS-03 Kyoto. IEEE The Sixteenth Annual International Conference, 2003

Valizadeh S., J. M. George, P. Leisner, and L. Hultman. *Electrochim Acta 47*, 2001, 865.

Van Dyke L. S., and C. R. Martin. *Langmuir 6*, 1990, 1123.

Vayssieres L., K. Keis, S.-E. Lindquist, and A. Hagfeldt. Purpose-built anisotropic metal oxide material: three-dimensional highly oriented microrod array of ZnO. *Journal of Phys. Chem.*, B 105, 2001, 3350.

Wagner R. S., and W. C. Ellis. Vapor-liquid-solid mechanism of single crystal growth. *Applied Physics Letters 4*, 1964, 89.

Walter E. C., et al. *Anal. Chem. 74*, 2002, 1546.

Wang Y., G. Meng, L. Zhang, C. Liang, and J. Zhang. Catalytic growth of large scale single-crystal CdS nanowires by physical evaporation and their photoluminescence. *Chem. Mater. 14*, 2002, 1773.

Wang Y., L. Zhang, C. Liang, G. Wang, and X. Peng. Catalytic growth and photoluminescence properties of semiconductor single-crystal ZnS nanowires. *Chem. Phys. Letters 357*, 2002, 314.

Wang, and Zhong Lin. Self-assembled nanoarchitectures of polar nanobelts/nanowires. *Journal of Mater. Chem. 15*, 2005, 1021.

Wegner K., B. Walker, S. Tsantilis, and S. E. Pratsinis. Design of metal nanoparticle synthesis by vapor flow condensation. *Chem. Eng. Sci. 57*, 2002, 1753.

Weisbuch C., and B. Vinter (eds.), 1991 *Quantum Semiconductor StructuResearch*. Boston:Academic, Boston.

Westwater J., D. P. Gossain, S. Tomiya, S. Usui, and H. Ruda. Growth of silicon nanowires via gold/silane vapor–liquid–solid reaction. *Journal of Vacuum Science Technology*, B 15, 1997, 554.

Whitney T. M., J. S. Jiang, P. C. Searson, and C. L. Chien. *Science 261*, 1993, 1316.

Whitsitt E. A., and A. R. Barron. *Nano Lett. 3*, 2003, 775.

Wirtz M., M. Parker, Y. Kobayashi, and C. R. Martin. *Chem. Eur. J. 8*, 2002, 353.

Wu C.-G., and T. Bein. *Science 264*, 1994, 1757.

Wu X. C., W. H. Song, B. Zhao, Y. P. Sun, and J. J. Du. Preparation and photoluminescence properties of crystalline $GeO_2$ nanowires. *Chem. Phys. Letters 349*, 2001, 210.

Wu Y., and P. Wang. Direct observation of vapor–liquid–solid nanowire growth. *Journal of Am. Chem. Soc. 123*, 2001, 3165.

Wu Y., and P. Yang. Germanium nanowire growth via simple vapor transport. *Chem. Mater. 12*, 2000, 605.

Wu Y., H. Yan, M. Huang, B. Messer, J. H. Song, and P. Yang. Inorganic semiconductor nanowires: rational growth, assembly, and novel properties. *Chem. Eur. J. 8*, 2002, 1260.

Xu Y., C. Cao, B. Zhang, and H. Zhu. Preparation of aligned amorphous silica nanowires. *Chemistry Letters 34*, 2005, 414.

Saito Y., and S. Uemura. *Carbon D 38*(169), 2000, 738.

Yang, C. Z., G.W. Meng, Q. Q. Fang, X. S. Peng, Y. W. Wang, Q. Fang, and L. D. Zhang. *Journal of Phys. D 35*, 2002, 738.

Yang P. C., and M. Lieber. Nanorodsuperconductor composites: a pathway to materials with high critical current density. *Science 273*, 1996, 1836.

Yazawa M., M. Koguchi, A. Muto, and K. Hiruma. Semiconductor nanowhiskers. *Adv. Mater. 5*, 1993, 577.

Yi G., and W. Schwarzacher. *Applied Physics Letters 74*, 1999, 1746.

Yin A. J., J. Li, W. Jian, A. J. Bennett, and J. M. Xu. *Applied Physics Letters 79*, 2001, 1039.

Yu J. S., et al. *Chem. Commun. 24*, 2000, 2445.

Yun W. S., J. Kim, K. H. Park, J. S. Ha, Y. J. Ko, K. Park, S. K. Kim, Y. J. Doh, H. J. Lee, J. P. Salvetat, and L. Forro. *Journal of Vac. Science. & Technol., A 18*, 2000, 1329.

Zeng H., R. Skomki, L. Menon, Y. Liu, S. Bandyopadhyay, and D. J. Sellmyer. *Phys. Rev., B 65*, 2002, 13426.

Zhan G.-D., et al. *Applied Phys. Letters 83*(6), 2003, 1228.

Zhan G.-D., et al. *Nat. Mater. 2*, 2003, 38.

Zhang J., X. S. Peng, X. F. Wang, Y. W. Wang, and L. D. Zhang. Micro-Raman investigation of GaN nanowires by direct reaction of Ga with $NH_3$. *Chem. Phys. Letters 345*, 2001, 372.

Zhang Y. J., Q. Zhang, N. L. Wang, Y. J. Zhou, and J. Zhu. Synthesis of thin Si whiskers (nanowires using $SiCl_4$). *Journal of Cryst. Growth 226*, 2001, 185.

Zhang Y., and H. J. Dai. *Applied Physics Letters 77*, 2000, 3015.

Zhang Y., N. Kohler, and M. Zhang. Surface modification of superparamagnetic magnetite nanoparticles and their intracellular uptake. *Biomaterials 23*, 2002, 1553.

Zhong Z., F. Qian, D. Wang, and C. M. Lieber. Synthesis of p-type gallium nitride nanowires for electronic and photonic nanodevices. *Nano Letters 3*, 2003, 343.

Zhu Jiang, et al. *Nano Letters 3*(8), 2003, 1107.

## General References

1. National Nanotechnology Initiative: Leading to the Next Industrial Revolution (a report by the Interagency working group on nanoscience, engineering and technology), Committee on Technology, National Science & Technology Council, February 2000, Washington, DC.

2. Curtin W. A., and B. W Sheldon. CNT–reinforced ceramics and metals. *Materials Today*, November 2004: 44–49.

# Section Twelve

## Joining materials

# Chapter 12.1

# Joining materials

## 12.1.1 Joining defined

From the dawn of humankind (in fact, maybe even before, if Figure 12.1-1 is any more than a fanciful anthropomorphism), the ability to join similar or dissimilar materials has been central to the creation of useful tools, the manufacture of products, and the erection of structures. Joining was undoubtedly one of the first, if not the first, manufacturing technology. It began when a naturally shaped or broken stone was first joined to a naturally forked or split stick; first wedging the stone into the fork or split, and later, as the first "engineering improvement" took place, lashing the stone into place with a vine or piece of animal sinew to produce a hammer, ax, or spear. This earliest creation of functional tools by assembling simple components surely must have triggered a whole rash of increasingly more complex, useful, and efficient tools, as well as an entirely new approach to building shelters from Nature's elements and from enemies. It also must have quickly advanced—or degenerated—into creative ways of producing efficient defensive and offensive weapons for war: longbows and longboats, crossbows and castles, swords and siege machines.

With the passage of time, the need for and benefits of joining have not abated; they have grown. More diverse materials were fabricated into more sophisticated

**Figure 12.1-1** An artist's concept that joining, as an important process in manufacturing, began with—or maybe even before—the dawn of humankind, making it one of the oldest of all processes. (Courtesy of Victoria Messler-Kaufman, with permission.)

*Engineering Materials and Processes Desk Reference*; ISBN: 9781856175869
Copyright © 2009 Elsevier Ltd; All rights of reproduction, in any form, reserved.

components, and these components were joined in more diverse and effective ways to produce more sophisticated assemblies. Today, from a Wheatstone resistance bridge to the Whitestone suspension bridge,[1] from missiles to MEMS,[2] joining is a critically important consideration in both design and manufacture. In fact, we as a species and joining as a process are at the dawn of a new era—one in which joining changes from simply a pragmatic process of the past to an enabling technology for the future, to be practiced as much by physicists and physicians as by hard-hatted riveters and helmeted welders.

In the most general sense, *joining* is the act or process of putting or bringing things together to make them continuous or to form a unit. As it applies to fabrication, joining is the process of attaching one component, structural element, detail, or part to create an assembly, where the assembly of component parts or elements is required to perform some function or combination of functions that are needed or desired and that cannot be achieved by a simple component or element alone. At the most basic level, it is the joining (of materials into components, devices, parts, or structural elements, and then, at a higher level, the joining of these components into devices, devices into packages, parts into assemblies, and structural elements into structures) that is of interest here.

An *assembly* is a collection of manufactured parts, brought together by joining to perform one or more than one primary function. These primary functions can be broadly divided into the following three categories: (1) structural, (2) mechanical, and (3) electrical. In *structural assemblies*, the primary function is to carry loads—static, dynamic, or both. Examples are buildings, bridges, dams, the chassis of automobiles, or the airframes of aircraft or spacecraft. In *mechanical assemblies*, the primary function, while often seeming to be (and, in fact, also having to be) structural, is really to create, enable, or permit some desired motion or series of motions through the interaction of properly positioned, aligned, and oriented components. Examples are engines, gear trains, linkages, actuators, and so on. Without question, such assemblies must be capable of carrying loads and, therefore, must be structurally sound, but load carrying is incidental to creating or permitting motion. Finally, in *electrical assemblies*, the primary purpose is to create, transmit, process, or store some electromagnetic signal or state to perform some desired function or set of functions. The most noteworthy examples are microelectronic packages and printed circuit boards but also include motors, generators, and power transformers. Here, too, there is also often a need to provide structural integrity, but only to allow the primary electromagnetic-based function(s) to occur.

Usually, assemblies must perform multiple functions, albeit with one function generally being primary and the others being secondary. Thus, the joints in assemblies must also support multiple functions. For example, soldered joints in an electronic device have the primary function of providing connectivity—for the conduction of both electricity and heat—but they must also be able to handle mechanical forces applied to or generated within the system. They must also hold the assembly of electrical components together in the proper arrangement under applied stresses, acceleration, motion, vibration, or differential thermal expansion and contraction. Regardless of the primary or secondary functions of an assembly and its component joints, joints are an extremely important and often critical aspect of any assembly or structure, and they are found in almost every structure. In fact, joints make complex structures, machines, and devices possible, so joining is a critically important and pervasive process.

At some level, joining anything comes down to joining materials, with the inherent microscopic structure and macroscopic properties of the material(s) thus dictating how joining must be accomplished to be possible, no less successful. After all, everything and anything one might need or wish to join is made of materials. Nevertheless, there surely are issues and considerations associated with joining structures that go beyond material issues and considerations.

## 12.1.2 Reasons for joining materials and structures

For many structures, and certainly for static structures,[3] an ideal design would seemingly be one containing no joints, since joints are generally a source of local weakness or excess weight, or both. However, in practice, there are actually many reasons why a structure might need or be wanted to contain joints, sometimes by necessity and sometimes by preference.

There are four generally accepted goals of any design (Ashby, 1999; Charles et al., 1997): (1) functionality, (2) manufacturability, (3) cost, and (4) aesthetics. While one could argue about the order of relative importance

---

[1] The Whitestone Bridge links the boroughs of Queens and the Bronx outside of the borough of Manhattan in New York city.
[2] MEMS are "micro-electromechanical systems" — machines on a microscopic scale.
[3] *Static structures* are structures that are not required or intended to move. In fact, when such structures are required or intended to move, at least on any gross scale (beyond normal elastic deflection, for example), it is usually considered a failure of the structure. *Dynamic structures*, on the other hand, are intended to move from place to place or within themselves.

Joining materials CHAPTER 12.1

**Figure 12.1-2** The use and importance of joining pervades our world and our lives, as shown in this depiction; it enables the creation of structures from beneath the seas to the outermost regions of space, and everything in between. (Courtesy of Avram Kaufman, with permission.)

of the latter three, there is no arguing about the primal importance of the first (i.e., functionality—at least if the designer is putting things in proper perspective!). Without functionality, whether something can be manufactured at all or at a low cost while looking good or even be pleasing to look at is of little, if any, consequence. It should thus come as no surprise that the reasons for joining materials or structures composed of materials are directly related to achieving one or more of these four goals. Let us look at these reasons goal by goal.

If one thinks about structures (on any size scale), there are two fundamental types: (1) those that are not required, intended, or wanted to move, either from place to place or within themselves, to function, or both, and (2) those that are. The former can be referred to as *static structures*, while the latter can be referred to as *dynamic structures*. Achieving functionality in both types requires that the structural entity[4] be able to carry loads, whether applied from the outside (i.e., external loads) or generated from within (i.e., internal loads). In both types of structures, functionality depends on any and all parts responsible for some aspect of the overall function of the structure or assembly to be held in proper arrangement, proximity, and orientation. In dynamic structures, however, there is the added requirement that these component parts must be capable of needed motion relative to one another while still having the ability to carry any and all loads generated by and/or imposed on the assembly. It is immediately obvious that a dynamic structure must contain joints. If it did not, implying it was made from one piece, there could not possibly be any relative motion between parts. Hence, joining is essential for allowing relative motion between parts in a dynamic structure. Less obvious is the fact that static structures usually (albeit not always) require joints, too, and thus require joining. If a static structure is very large, however, the likelihood that it can be created from one piece decreases as the size increases.

Hence, joining is needed in large structures since such large structures (or even components of very large structures) cannot be produced by any primary fabrication process, whether these structures are static or dynamic. There is, in fact, a limitation on the size—and also the shape complexity—for any and every primary fabrication process, such as casting; molding; deformation processing by forging, rolling, or extrusion; powder processing; or lay-up and other special fabrication processes for composites. Once this primary process limitation is exceeded, joining, as a secondary fabrication process, is necessary. Figure 12.1-3 shows an example of the need for joining to produce large-scale structures.

Sometimes special functionality is required of a structure that necessitates joining. An example is the desire or need to see through or into or out of a sealed structure. One could make the entire structure from a transparent material, such as glass, but doing so could seriously compromise the structure's integrity for other functions, such as resisting impact loads or tolerating flexing. Hence, joining is necessary to achieve special functionality achievable only by mixing fundamentally different materials (e.g., metals and glasses in an automobile's windows). Figure 12.1-4 shows an example of joining for this reason.

For some products or structures, it is necessary for them to be portable (e.g., to bring them to a site for short-term use, and then be removed, often for use elsewhere). Clearly, for something to be portable it either has to be small or has to be able to be disassembled and re-assembled. Examples range from temporary modular buildings for providing shelter or security to climbing cranes used in erecting skyscrapers, to huge tunneling machines such as those used to build the "Chunnel" under the English Channel between France and England. In all cases, joining—by some means that is preferably, but not necessarily, easy to reverse—is needed.

Finally, there are situations where service loads threaten a structure's integrity due to the propagation of internal damage (e.g., a crack). The tolerance of a structure to ultimate failure from a propagating flaw can be dealt with in two ways: (1) by making the structure from a material with inherent tolerance for damage (in the form of high fracture toughness, for example); and/or (2) by building crack-arresting elements into the structure (often, if not always, in the form of joints). Hence, joining can be used to impart structural damage tolerance, beyond inherent material damage tolerance. Figure 12.1-5 shows the superb example of built-up riveted structure in a metal aircraft airframe structure for damage tolerance.

Most of the time, the second most important goal of design is manufacturability. If a functional design cannot be manufactured at any cost, it will never have a chance to function. Joining plays a key role in achieving manufacturability in several ways. First and foremost is the use of joining to achieve structural efficiency, which clearly relates closely to functionality. Structural efficiency means providing required structural integrity (e.g., static strength, fatigue strength and/or life, impact strength or toughness, creep strength, etc.) at minimum structural weight. As an example, a fighter aircraft's wing plank or

---

[4] *Structural entity* refers to a device composed of materials (e.g., a p-n-p transistor), a package composed of devices (e.g., a logic chip), an assembly of parts or packages, or a collection of structural elements used to produce a structure (e.g., a bridge).

Joining materials   CHAPTER 12.1

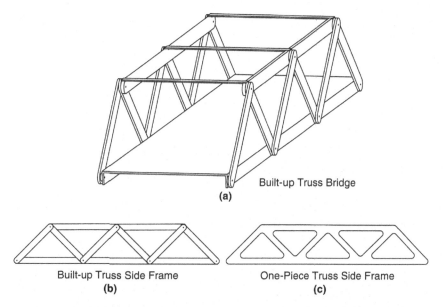

**Figure 12.1-3** An important reason for joining is to enable the construction of objects or structures that are simply too large to fabricate in one piece by any means. Here a truss bridge (a) is assembled from pinned, riveted, or bolted elements (b), since creating the bridge from one piece (c) would be impractical, if not impossible.

**Figure 12.1-4** Joining allows the use of fundamentally dissimilar materials to achieve special function. Here a glass windshield consisting of glass mounted in a metal frame and sealed with a polymer is being robotically assembled into a modern automobile constructed of metal, plastic, or reinforced plastic. (Courtesy of KUKA Schweissanlagen GmbH, Augsburg, Germany, with permission.)

**Figure 12.1-5** Joining one part to another can result in enhanced damage tolerance in a structural assembly over that inherent in the materials used to create the individual parts of the assembly, nowhere more apparent or important than in the riveted, built-up Al-alloy structure of an airplane. Here, the riveted fuselage of a T38 trainer is shown. (Courtesy of Northrop Grumman Corporation, El Segundo, CA, with permission.)

cargo aircraft's floor plank in a conventional Al-alloy design can be made lightweight while still providing required structural stiffness by creating "pockets" in thick plates by machining or by creating built-up stiffeners (e.g., ribs and frames) by riveting. Both end up using only as much metal as is absolutely needed to carry the loads. However, building up small pieces into a structurally efficient assembly trades off increased assembly labor against wasted material (i.e., scrap) and machining time, and, as a byproduct, favorably impacts structural damage tolerance. These two approaches, both of which seek to maximize structural efficiency, are shown schematically in Figure 12.1-6. Obviously, the riveted assembly offers the added advantage of optimized (i.e., maximized) material utilization known as "buy-to-fly ratio" in the aerospace industry. Figure 12.1-7 shows a comparison between the main landing gear door of an E2C aircraft fabricated from all composite details by adhesive bonding versus all Al-alloy details by riveting, to reduce part count, virtually eliminate fasteners, dramatically reduce assembly labor (required to drill holes and install rivets), and save weight. Thus, joining offers structural efficiency and an opportunity for optimized material utilization.

Related to optimized utilization of material is optimized selection of material. Optimum functionality sometimes requires a material of construction to satisfy two opposing requirements. For example, while it is often desirable for a portion of a structure (such as the ground-engaging edge of a bulldozer blade) to resist wear by being hard, making the entire structure from a hard, wear-resistant material would compromise the structure's toughness under expected impact (e.g., with

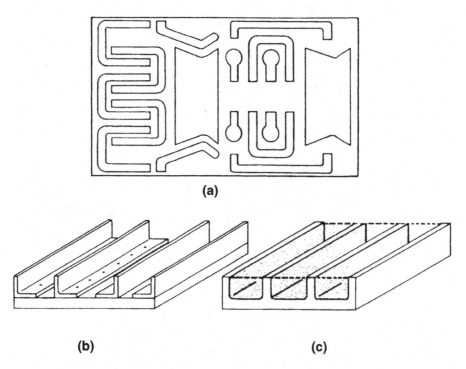

**Figure 12.1-6** Schematic illustration showing how joining, here by fastening (b), can be used as effectively as machining (c) to achieve structural efficiency; the former by building up details, the latter by removing material (say by machining) to minimize weight and carry service loads. The need to nest parts to optimize material utilization is shown in (a) (Reprinted from *Joining of Advanced Materials*, Robert W. Messler, Jr., Butterworth-Heinemann, Stoneham, MA, 1993, Fig. 1.4, page 8, with permission of Elsevier Science, Burlington, MA.)

Joining materials  CHAPTER 12.1

**Figure 12.1-7** An adhesively bonded composite main landing gear door for an E2C (left) dramatically reduces part count, assembly labor, and weight compared to a conventional built-up, riveted Al-alloy door (right) for the same aircraft. (Courtesy of Northrop Grumman Corporation, El Segundo, CA, with permission.)

boulders). It would also make fabrication of the large and complex shaped blade, in the example of a bulldozer, terribly difficult. Using joining, it is possible to mix two different materials to achieve both goals (e.g., a wear-resistant material at the blade's ground-engaging edge and a tough material in the blade body). So, joining allows optimum material selection (i.e., the right material to be used in the right place). This could also allow an inherently damage-tolerant material to be mixed with a less damage-tolerant material using joining to achieve the aforementioned structural damage tolerance.

As mentioned earlier, large size and/or complex shape can pose a problem for certain fabrication processes and certain materials. As examples, casting allows complex shapes to be produced at relatively low cost (using simple mold-making techniques for small-run castings or using more elaborate mold-making techniques for large-run castings), but has greater limitations on size than a forging process with its inherently more limited shape complexity capability. These can also be considered manufacturability issues, both of which can be overcome by joining.

Finally, there are many structures (e.g., all civil- or built-infrastructure structures) that must be erected, if not fully fabricated, on site. In either case (i.e., prefabricated parts shipped to the site or parts fabricated on site), joining is essential. Figure 12.1-8 shows a bridge that obviously had to be erected on site using prefabricated detailed parts.

Cost is often a key consideration, even if not the driver, for a manufactured product or structure. Joining allows cost to be minimized by (1) allowing optimal material selection (versus forcing compromise); (2) allowing optimal material utilization (versus forcing scrap losses); (3) keeping the weight of materials needed to a minimum (i.e., maximizing structural efficiency); (4) achieving functionality through large size and/or complex shape (without pressing primary processing limits); and (5) sometimes (depending on the process) allowing automated assembly

**Figure 12.1-8** Sometimes joining is necessary not only because it allows something too large or too complicated to be created from one piece to be made, but also because the structure has to be constructed or erected on site, as is clearly the case for the bolted bridge shown. In modern bridge building, pre-fabricated details are joined on site after being pre-fitted in the more controlled environment of a fabrication shop (Courtesy of the American Bridge Company, Coraopolis, PA, with permission.)

503

**Figure 12.1-9** Modern manufacturing often benefits when labor-intensive, quality-critical assembly is automated, as exemplified in the automobile industry by robotic welding. (Courtesy of DaimlerChrysler AG, Stuttgart, Germany, with permission.)

(to reduce labor cost and improve product consistency). Figure 12.1-9 shows how joining can be automated, thereby lowering the cost of a product's manufacture.

Cost-effectiveness means more than low cost of manufacture, however. It also means low cost of maintenance, service, repair, and upgrade, all of which are made practical, beyond feasible, by joining. Finally, joining facilitates responsible disposal, whether by recycling or other means.

How a finished product looks and how it makes the user feel (aesthetics) can be enabled by joining also. From the adhesive bonding application of expensive wood veneers, to less expensive wood furniture or plastic veneers that simulate wood, to the thermal spray application of protective and/or decorative coatings, to the application of attractive architectural façades, joining is often an enabler of improved aesthetics. By allowing more complex shapes to be produced cost-effectively, joining may further contribute to aesthetics through form beyond appearance.

Table 12.1-1 summarizes the reasons for joining structures and the materials that comprise them.

## 12.1.3 Challenges for joining materials

When one thinks about it, joining always comes down to joining materials. Whether one is erecting a concrete block wall by cementing block to block, constructing a ship by welding steel plates to one another, or implanting a titanium-alloy artificial hip joint into a sufferer of chronic and crippling rheumatoid arthritis, what is being joined is one material to another more fundamentally than one structure to another. This is most obvious in the case of implantations, where biocompatibility of the implant material is the key to successful implantation. Hence, the real challenges of joining (for any of the reasons described in the previous section) are usually directly the challenges of joining materials and usually indirectly the challenges of joining structural shapes (i.e., structures).

It is fairly safe to say that fewer parts, simpler shapes, and less-sophisticated, lower-performance materials require less elaborate joining processes and procedures. Not surprisingly, the corollary is also true (i.e., more-sophisticated, higher-performance—so-called "advanced"—materials require special attention and more elaborate joining processes and procedures). In every case, however, the general rules are: (1) select a joining process that minimally alters or disrupts the material's inherent microstructure (including chemistry), while still achieving required or desired functionality; and (2) consider the effect of the process of joining on the resulting properties of the final material and structure.

The challenges to joining posed by materials are growing as (1) the sheer diversity of materials continues to grow (e.g., the challenge of joining ceramic-matrix composites to monolithic ceramics in advanced-concept engines); (2) the degree of "engineered" microstructure and properties of materials increases (e.g., directionally solidified eutectic superalloy gas turbine blades to

**Table 12.1-1** Reasons for joining structures and materials (by design goals)

Goal 1: Achieve Functionality
- To carry or transfer loads in an array of parts needing to act together without moving (i.e., a static structure)
- To carry and transfer loads in an array of parts needing to act together by moving (i.e., a dynamic structure)
- To achieve size and/or shape complexity beyond the limits of primary fabrication processes (e.g., casting, molding, forging, forming, powder processing, etc.)
- To enable specific functionality demanding mixed materials
- To allow structures to be portable (i.e., able to be moved to or from sites)
- To allow disassembly for ultimate disposal
- To impact damage tolerance in the structure beyond that inherent in the materials of construction (i.e., structural damage tolerance)

Goal 2: Facilitate Manufacturability
- To obtain structural efficiency through the use of built-up details and materials
- To optimize choice and use of just the right materials in just the right place
- To optimize material utilization (i.e., minimize scrap losses)
- To overcome limitations on size and shape complexity from primary fabrication processes
- To allow on-site erection or assembly of prefabricated details

Goal 3: Minimize Costs
- To allow optimal material selection and use (versus forcing compromise)
- To maximize material utilization and minimize scrap losses
- To keep the total weight of materials to a minimum (through structural efficiency)
- To provide more cost-effective manufacturing alternatives (versus forcing a primary fabrication process to its limit)
- To facilitate automation of assembly, for some methods
- To allow maintenance, service, repair, or upgrade; all of which reduce life-cycle costs
- To facilitate responsible disposal

Goal 4: Provide Aesthetics
- To enable application of veneers, facades, etc., different from the underlying structure
- To allow complex shapes to be formed

monolithic superalloy rotors); and (3) designers and users demand and modern, sophisticated analysis techniques allow higher operating stresses, permit combined or complex loading, and enable combined properties for severe environments all at minimum weight, minimum cost, minimum environmental impact, and maximum flexibility. Often to meet these demands, designers combine diverse materials in individual functional elements to create hybrid structures that truly do optimize overall function, performance, and cost. An example is shown in Figure 12.1-10.

Clearly, the pressure on processes for joining materials is growing.

## 12.1.4 Challenges for joining structures

We live in a world where we are being pushed to—and are thus moving toward—new and extreme conditions. Bigger, faster aircraft, deeper-water offshore drilling platforms (Figure 12.1-11), smaller machines and microelectromechanical systems (or MEMS) (Figure 12.1-12), longer and more comfortable stays in space (Figure 12.1-13), greater need to extend the operating life of nuclear power plants (Figure 12.1-14)—all of these and more pose new challenges to our ability to join structures beyond joining materials. Bigger supertankers and petrochemical refineries demand larger and thicker-section structures be joined and be leak-tight. Offshore drilling platforms demand erection, anchoring, and periodic repair to occur underwater. The intriguing possibilities of MEMS demand micro- (if not nano-) joining. Ventures into space and the need to make repairs on radioactive nuclear reactors demand automation of joining processes heretofore operated manually. And, past successes in limb reattachment and the promise of tissue engineering make new demands that pragmatic manufacturing processes like joining become enabling technologies for biotechnology. Past lessons learned in manufacturing suggest joining must adapt and evolve to meet new demands and realize new possibilities.

Let us take a look at how joining is already changing and how it must change in the future.

## 12.1.5 How joining is changing or must change

Until quite recently and, for most applications even now, joining has been a "secondary" fabrication process when classified with all other generic fabrication processes in manufacturing (Charles et al., 1997). Not secondary in

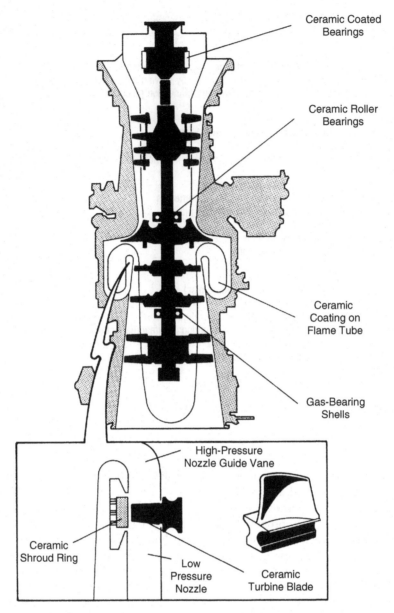

**Figure 12.1-10** Joining makes possible the use of just the right material, in just the right amount, in just the right places to create "hybrid structures," as exemplified by the schematic of an advanced ceramic engine for a helicopter. (Reprinted from *Ceramic Joining*, Mel M. Schwartz, Figure 7.1, page 167, ASM International, Materials Park, OH, with permission.)

the sense that it is of lesser importance (although that, too, is often the thinking!), but in the sense that it occurs after parts, components, or structural elements have been fabricated by other means. Five generic process categories are usually considered primary: (1) casting; (2) molding; (3) deformation processing (using mill processes like rolling or extrusion, or other processes like forging or sheet-metal forming); (4) powder processing; and the catch-all (5) special processing, which is epitomized by processes used for fabricating items from polymer-matrix composites (e.g., broad-good and tape lay-up, filament winding, weaving, etc.). Being primary, these processes either create the starting stock (e.g., rough casting, rolled plate, forged billet or rough forging, powder preform, etc.), or they produce a part to near-net shape (e.g., investment casting, injection molding, precision forging, pressed-and-sintered part, etc.). Most of the time, joining is one of the later, if not the last, steps in a product's manufacture. And, worse yet, it is often an afterthought; examples include alloys that are not designed to be welded being used in products or structures needing welding, using add-on screws to back up integrally snap-fit plastic assemblies to prevent accidental disassembly, and applying a bolted reinforcement (or "band-aid") over a weld repair on a cast-iron machine frame.

**Figure 12.1-11** Larger and larger drilling platforms for use in deeper water require extensive welding during their construction, erection, and maintenance and repair. (Courtesy of Bechtel Corporations, San Francisco, CA, with permission.)

This is beginning to change and must continue to change for joining to achieve its full potential and to have its full impact. The best examples are in microelectronics, where semiconductor devices (e.g., MOSFETs) are created by synthesizing the p- and n-type extrinsic semiconductor materials as integral device elements in a single device. Material synthesis, device or part synthesis, and assembly or system synthesis occur in an integrated, even if not simultaneous, fashion. This trend must—and will—continue, making joining an integral part of primary processing.[5]

One only has to look at news releases or technical briefs in present-day materials or manufacturing journals to see terms like "self-forming joints", "self-limiting joining", "self-healing materials", and "self-assembling structures" to sense the change of joining from a secondary to a primary process. Self-forming joints can be found in microelectronics when lean Cu-Al or Cu-Mg alloys are sputtered onto $SiO_2$-coated Si substrates and heat-treated to create a bond-forming $Al_2O_3$ or MgO joint layer. In this same process, such joint formation can be made self-limiting by carefully controlling the composition and amount (i.e., thickness) of the sputtered alloy. The quest in the military aerospace industry for self-healing or self-repairing of damaged skins or understructures now reveals a technical and practical reality using nanotechnology. Encapsulated resins and catalysts in the form of nanoparticles can be embedded in thermosetting polymer-matrix composites to affect automatic "healing" of any flaws that develop and rupture the encapsulants in the process of the flaw's propagation. And, finally, self-assembly of microscale (or eventually nanoscale) components into MEMS (or eventually NEMS) is being employed by carefully designing the shapes of the components to enable and assure that they can be "shaken" into proper arrangement and orientations.

A second major change that is occurring, and must continue to occur, is accepting joining as a value-adding, not a value-detracting, process. While it surely is accepted in some instances, it is not in far too many other instances. Designers and process engineers must accept high-cost joining (often arising from high labor intensity and/or high-priced labor) for high-value applications and highly valued benefits. The best example might be heeded by those charged with joining continuous, unidirectional-reinforced composites for demanding service by watching surgeons reattach a severed limb. First, patience, time, and precision are accepted costs for the high value to be gained. Second, joining begins with the critical internal structure (analogous to the reinforcements in composites) and ends with the less critical external structure (analogous to the matrix of composites). Bones and blood vessels (as essential

---

[5] By the way, there are already examples where welding is being used to produce finished or near-net shapes.

**Figure 12.1-12** Special joining techniques and methods are needed to enable micro-electromechanical systems (MEMS) to be assembled. In this cut-away sample, various micron-scale details have been joined to create an accelerometer through the use of a silver-filled glass to bond the die to the ceramic package base, ultrasonic aluminum wire bonds between aluminum bond pads on the die and Alloy 42 lead frame, and use of a glass frit to seal the package lid to the package base. (Courtesy of Analog Devices, Inc., Cambridge, MA, with permission.)

structural elements) are joined, followed by muscles and tendons and ligaments (as actuators), followed by nerves (as sensors). After all these critical elements have been joined, the surrounding tissue and skin are joined (as analogues to the matrix of the composite). Think of this when the joining of composites is discussed.

Figure 12.1-15 shows how precision microjoining is accepted in microelectronics manufacturing to obtain highly valued hermeticity.

Finally, joining must continue to change from a pragmatic process in fabrication, as much an afterthought and a necessary evil as a value-adding step in manufacturing, to an enabling technology. Microelectronics could not have achieved what is has without joining as a technology enabling solid-state devices. The future of information technology will be enabled by microelectronics and nanoelectronics, optoelectronics, photonics, and molecular electronics (called "moletronics"), and joining will enable these to act as a technology, not simply as a process. Likewise, much of the tremendous promise of biotechnology (e.g., gene splicing, tissue engineering, and the like) will also depend on joining as a technology more than as a pragmatic process.

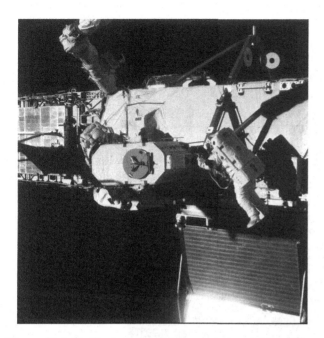

**Figure 12.1-13** Allowing humans to work, learn, and live at the edge of outer space is made possible in the Orbiting International Space Station by many types of joining, including mechanical fastening, snap-fit assembly, and welding. (Courtesy of the National Aeronautics and Space Administration, Washington, DC, with permission.)

## 12.1.6 Joining options

### 12.1.6.1 Fundamental forces involved in joining

Joining is made possible by the following three—and only three—fundamental forces: (1) mechanical forces, (2) chemical forces, and (3) physical forces, which have their origin in electromagnetic forces. Not coincidentally, these three fundamental forces are, in turn, responsible for the three fundamental methods or processes by which materials and structures can be joined: (1) mechanical joining, (2) adhesive bonding, and (3) welding.

Mechanical forces arise from interlocking and resulting interference between parts, without any need for chemical or physical (electromagnetic) interaction. As shown in Figure 12.1-16, such interlocking and interference can (and to some extent always does) arise at the microscopic level with surface asperities[6] giving rise to friction or, at macroscopic levels, using macroscopic features of the parts being joined. Chemical forces arise from chemical reactions between materials. Such reactions can take place entirely in the solid state of the materials involved or can take place (often much more rapidly, uniformly, and completely) between a liquid and a solid phase of the materials involved, relying on wetting of the solid by the liquid. Finally, the naturally occurring attraction between atoms, oppositely charged ions, and molecules leads to bond formation and joining due to physical forces in what is generally referred to as welding. Brazing and soldering are special subclassifications of welding, that find their origin and effectiveness in the combined effects of chemical and mechanical forces (albeit with the strength of the ultimate joints, in both sub-classifications, arising from the physical forces of atomic bonding). Unlike adhesive bonding, neither brazing nor soldering, nor welding for that matter, is dependent upon chemical forces to produce joint strength. They depend just on physical forces.

Table 12.1-2 summarizes how different fundamental forces give rise to the different joining options.

Let us look at each of these major joining options.

### 12.1.6.2 Mechanical fastening and integral attachment: Using mechanical forces

Mechanical fastening and integral mechanical attachment are the two ways in which mechanical forces can be used to join structures, rather than materials. Together, mechanical fastening and integral mechanical attachment constitute what is properly known as *mechanical joining*. In both methods, joining or attachment is achieved completely through mechanical forces, arising from interlocking—at some scale—and resulting physical interference between or among parts. At the macroscopic level, interlocking and interference arise from designed-in or processed-in (or, in nature, from naturally occurring) geometric features. In *mechanical fastening*, these features are the result of the parts being joined and a supplemental part or device known as a "fastener." The role of the fastener is to cause the interference and interlocking between the parts, which, by themselves, would not interlock. In integral mechanical attachment, on the other hand, these interlocking features occur naturally in, are designed in, or are processed into the mating parts being joined.

Figure 12.1-17 shows a typical mechanically fastened structure, while Figure 12.1-18 shows a typical integrally attached structure using snap-fit features.

In both mechanical fastening and integral mechanical attachment, interlocking and interference also arise at the microscopic level in the form of friction. Friction has its origin in the microscopic asperities—or "peaks and valleys"—present on all real surfaces, regardless of the effort to make these surfaces smooth. Not only do these

---

[6] *Asperities* are the "peaks and valleys" found on all real surfaces, regardless of efforts to make the surface smooth.

CHAPTER 12.1  Joining materials

**Figure 12.1-14** Joining is essential to the routine scheduled maintenance and unscheduled emergency repair, not only the construction, of nuclear power plant components; it sometimes demands that welding, for example, be done using either mechanized systems operated by welders outside of radioactive areas or by remotely controlled robots within such areas. (Courtesy of Bechtel Corporation, San Francisco, CA, with permission.)

asperities interfere and interlock with one another mechanically, but also, under the right circumstances (e.g., adhesive wear or abrasion) with the right materials (e.g., metals), atomic bonding actually can and does occur. Localized "welding" of asperities by these naturally occurring physical forces causes metal transfer manifested as "seizing."

Common examples of mechanical fasteners are nails, bolts (with or without nuts), rivets, pins, and screws. Less well recognized, but still common, mechanical fasteners are paper clips, zippers, buttons, and snaps (actually "eyelets and grommets"). Special forms of mechanical fasteners are staples, stitches, and snap-fit fasteners. Common examples of designed-in integral mechanical attachments are dovetails and grooves, tongues-and-grooves, and flanges, while common examples of processed-in attachments are crimps, hems, and punchmarks or "stakes." A common use of friction for

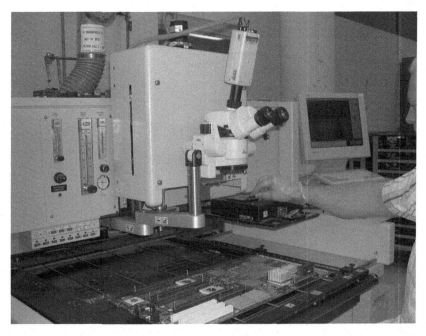

**Figure 12.1-15** Joining has already become a more integrated part of the synthesis of materials, devices, and systems in microelectronics, where microjoining is used to hermetically seal critical electronic packages. (Courtesy of International Business Corporation, Poughkeepsie, NY, with permission.)

mechanical joining is roughened or "knurled" mating (or faying) surfaces, as in Morse tapers.

The principal advantage of all mechanical joining (with the sole exception of some processed-in features) is that it uniquely allows intentional relative motion (i.e., intentional movement) between mating parts. It also rather uniquely allows intentional disassembly without damaging the parts involved. Regrettably, this major advantage can also be a major disadvantage (i.e., the ability to intentionally disassemble can lead to unintentional or accidental disassembly if special care is not exercised). More will be said about the relative advantages and disadvantages of mechanical joining processes.

Mechanical fastening and, to a lesser extent, integral mechanical attachment can be used with any material, but is best with metals and, to a lesser extent, with composites. Problems arise in materials that are susceptible to damage through easy (especially "cold") deformation (such as certain polymers under high point loads) or fracture by stress concentration at points of mechanical interference due to poor inherent damage tolerance (such as brittle ceramics and glasses). Problems also arise in materials that are susceptible to severe reductions in strength or damage tolerance in certain directions due to anisotropy (such as in continuous, unidirectionally reinforced composite laminates through their thickness). Another great advantage of all mechanical joining is that, since it involves neither chemical nor physical forces, it causes no change in the part's or material's microstructure and/or composition. This makes it possible to join inherently different materials mechanically.

Specific problems associated with mechanical joining of specific materials will be discussed.

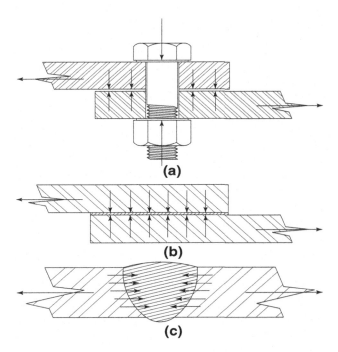

**Figure 12.1-16** A schematic illustration of the various forces used in joining materials and structures: (a) mechanical forces for fastening, (b) chemical forces for adhesive bonding, or (c) physical forces for welding.

# CHAPTER 12.1 — Joining materials

**Table 12.1-2** Fundamental forces used in different joining processes, sub-processes, variants, and hybrids

|  | Primary | Secondary |
|---|---|---|
| **Mechanical Joining** | | |
| Mechanical Fastening | Mechanical | - |
| Integral Attachment | Mechanical | - |
| **Adhesive Bonding** | | |
| Using Adhesives | Chemical | Mechanical/Physical |
| Solvent Cementing | Chemical | Physical |
| Cementing/Mortaring | Chemical | Mechanical |
| **Welding** | | |
| Fusion Welding | Physical | - |
| Non-fusion Welding | Physical | - |
| Brazing | Physical | Chemical (Reaction) |
| Soldering | Physical | Chemical (Reaction)/ Mechanical |
| **Variant Processes** | | |
| Braze Welding | Physical | Chemical (Reaction) |
| Thermal Spraying—Metals/Ceramics | Physical* | Mechanical/Chemical (Reaction) |
| Thermal Spraying—Polymers | Chemical* | Mechanical |
| **Hybrid Processes** | | |
| Rivet-Bonding | Mechanical/Chemical | - |
| Weld-Bonding | Physical/Chemical | - |
| Weld-Brazing | Physical | Chemical (Reaction) |

\* If done correctly!

**Figure 12.1-17** Floor trusses are typically mechanically fastened to the vertical structure of modern skyscrapers using high-strength bolts and nuts, such as those shown here in the Quaker Tower, Chicago, IL. (Courtesy of Bechtel Corporation, San Francisco, CA, with permission.)

## 12.1.6.3 Adhesive bonding: Using chemical forces

In adhesive bonding, materials and the structures they comprise are joined one to the other with the aid of a substance capable of holding those materials together by surface attraction forces arising principally (but not usually solely) from chemical origins. The bonding agent, called an "adhesive," must be chemically compatible with and chemically bondable to each substrate of what are called "adherends." Sometimes actual chemical reactions take place that give rise to the bonding and "adhesion," while more often no actual reaction is involved, just the development of surface bonding forces from other sources such as adsorption or diffusion. In such cases, adhesion arises from chemical bond formation, usually (but not always) of a secondary type (e.g., van der Waal's, hydrogen, or Loudon bonding). Occasionally, chemical bonding is aided and abetted by contributions from mechanical interlocking (i.e., mechanical forces) and/or physical forces (e.g., electrostatic forces).

Depending on the nature of the adhesive chosen and the adherends involved, adhesive bonding usually causes little or no disruption of the microscopic structure of the parts

Joining materials  CHAPTER 12.1

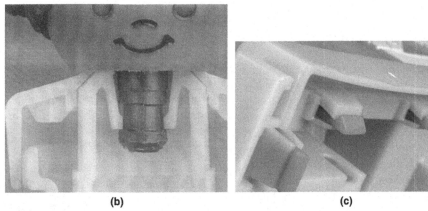

**Figure 12.1-18** Children's toys are commonly assembled from molded plastic parts using integral "snap-fit" attachment features to avoid using screws and other small objects that children can put in their mouths and choke on. In this Little People Farm™ set (a), cantilever hooks and catches (b) and annular post snaps (c) are used. (Courtesy of Fisher-Price Corporation, East Aurora, NY, with permission.)

involved, but it may cause varying (but usually minor) degrees of chemical alteration or disruption. Because attachment forces arise and occur over the surfaces of the parts being joined, loads that must be carried and transferred by the joint are spread out or distributed so that no stress concentrations (like those found at the points of actual fastening or attachment in mechanical joining) arise. The greatest shortcoming of adhesive bonding is the susceptibility of adhesives, particularly those that are organic as opposed to inorganic in nature, to environmental degradation. More will be said about the relative advantages and disadvantages of this joining process later.

Metals, ceramics, glasses, polymers, and composites of virtually all types, as well as dissimilar combinations of these, can be successfully adhesive-bonded. Disassembly can occasionally be accomplished, but never without difficulty and seldom without causing some damage to the parts involved.

Figure 12.1-19 shows the use of adhesive bonding in the airframe structure of a modern aircraft.

**Figure 12.1-19** Adhesive bonding is used in the assembly of the airframes of modern aircraft, especially when thermoplastic or thermosetting polymer-matrix composites are employed as they are here on the hybrid thermosetting epoxy–graphite/epoxy–boron and titanium-alloy horizontal stabilizer of the F14 fighter. (Courtesy of Northrop Grumman Corporation, El Segundo, CA, with permission.)

## 12.1.6.4 Welding: Using physical forces

Welding is in many respects the most natural of all joining processes. It has its origin in the natural tendency of virtually all atoms (except those of the inert gases), all oppositely charged ions, and all molecules to form bonds to achieve stable electron configurations, thereby lowering their energy states. In practice, *welding* is the process of uniting two or more materials (and, thereby, the parts or structures made from those materials) through the application of heat or pressure or both to allow the aforementioned bonding to occur. Figure 12.1-20 shows a typical application of welding employing an electric arc as a heat source to construct a ship from pre-fabricated (also welded) modules.

The terms "welding," "welding processes," and "welds" commonly pertain to metallic materials. But it is possible—and it is the practice—to also produce welds in certain polymers (i.e., thermoplastics) and glasses and, to a lesser extent, in some ceramics. Welding of composite materials can be accomplished to the degree that it is possible and acceptable to join only the matrix, as the process is performed today. By definition for a process that must form primary bonds to accomplish joining, welds cannot be produced between fundamentally different types or classes of materials (e.g., metallic-bonded metals to ionic- or covalent-bonded ceramics).

The relative amount of heat or pressure or both required to produce a weld can vary greatly. This is, in fact, one of the great advantages of this joining process—versatility through a vast variety of process embodiments. There can be enough heat to cause melting of two abutting base materials to form a weld with very little pressure beyond what is needed to simply hold these materials in contact. When this is the case, the process is known as

**Figure 12.1-20** Ships of all kinds are fabricated by welding small parts into large parts, large parts into structural modules, and modules to one another to construct the hull and superstructure. Here, a larger pre-fabricated modular section of the hull of the carrier USS Reagan is shown being lifted into place for welding to the rest of the hull. (Courtesy of Northrop Grumman Corporation's Newport News Shipbuilding, Newport News, VA, with permission.)

# Joining materials  CHAPTER 12.1

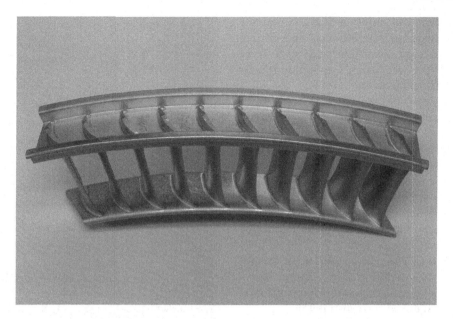

**Figure 12.1-21** Brazing is used to assemble various superalloy components of a gas turbine engine, such as this vane section. (Courtesy of The General Electric Company's Aircraft Engine Division, Evansdale, OH, with permission.)

"fusion welding." Alternatively, there might be little or no conscious or intentional heating, but with enough pressure to cause some degree of plastic deformation (commonly called "upsetting" if it occurs on a macroscopic scale and friction or creep if it occurs on a microscopic scale), welds can be produced. In any case, melting or fusion is not required to establish primary bonds; only pressure is required to cause large numbers of atoms (or ions or molecules) to come into intimate contact. Such processes are known as "solid-phase welding" or "non-fusion welding" or, if the pressure is significant, "pressure welding."

Not surprisingly, because primary bonds are formed during joining, welding results in extremely strong joints per unit area, so it is often the process of choice for particularly demanding high-load/high-stress applications.

There are two subclassifications of welding in which the base materials are heated but *not* melted, a filler material is added and melted, and little or no pressure is applied; the molten filler spreads to fill the joint by capillary attraction forces. These two, known as *brazing* and *soldering*, are described next.

## 12.1.6.5 Brazing: A subclassification of welding

*Brazing* is a subdivision or subclassification of welding in which the materials comprising the joint are heated to a suitable temperature in the presence of a filler material having a liquidus temperature[7] above 450°C (840°F) and below the solidus temperature(s) of the base material(s). This allows flow of the molten filler under the action of wetting and capillary attraction forces. Bonding is accomplished without melting and mixing the substrate materials, making possible the joining of dissimilar base materials, so long as each can form primary bonds with the filler. The filler material (usually a metal, but possibly a ceramic or glass) is caused to distribute between close-fitting, intentionally gapped joint element faying surfaces. Bonding occurs by the formation of primary bonds—metallic in metals and ionic or covalent or mixed in ceramics. In brazing, joint strength tends to also depend fairly significantly on interdiffusion between the filler and the substrate(s).

Figure 12.1-21 shows a typical brazed assembly.

## 12.1.6.6 Soldering: A subset of brazing

Like brazing, *soldering* is a subdivision or subclassification of welding. Also like brazing, soldering requires a filler material that melts and substrates that do not melt. It is distinguished from brazing by the fact that the filler's liquidus temperature is below (not above) 450°C (840°F). As in brazing, the filler material (which is almost always a metal but can be a glass for some joining applications), or "solder," is distributed using surface

---

[7] The *liquidus temperature* is the temperature at which an alloy, which melts over a range as opposed to at a discrete temperature, becomes completely (100%) liquid on heating. The *solidus temperature* is the temperature at which an alloy just begins to melt to form liquid on heating. On cooling, the liquidus is where the first solid appears, while the solidus is where the last liquid disappears to leave it 100% solid.

wetting, capillary action, and surface tension, sometimes causing the molten solder to flow between close-fitting, intentionally gapped joint elements and sometimes simply letting the solder wet a joint element and "self-form" a smooth transitioning joint.

In soldering, because of the lower temperatures involved, the joining can be the result of some combination of primary (e.g., metallic or covalent) bonds and mechanical interlocking. The mechanical interlocking is sometimes itself the combination of interlocking at a macroscopic and a microscopic scale. It is macroscopic when "pigtail" leads are folded under circuit boards or behind terminal strips after the lead is passed through a hole or "via". Microscopic interlocking, of course, results from the solder interacting with the substrate's surface asperities.

As in brazing (although less than 20 years ago it was not recognized), successful soldering also requires some degree of dissolution of the substrate(s) and interdiffusion between the molten filler and the substrate(s).

Figure 12.1-22 shows typical mass-soldered joints in a microelectronic assembly or printed circuit or wire board.

### 12.1.6.7 Variant and hybrid joining processes

While mechanical joining, adhesive bonding, and welding are the fundamental processes for joining materials and structures, and brazing and soldering are sub-classifications of welding, there are some processes that are either variants or hybrids of these.

Two variants of welding are (1) *braze welding* and (2) *thermal spraying*. Braze welding uses a low-melting, braze-like filler material to fill a pre-prepared joint without relying on capillary action but still relying on wetting and dissolution of the substrate(s) without their melting. Hence, the process looks something like welding and something like brazing but is neither hide nor hair! Thermal spraying, often considered a variant of welding, has some characteristics of adhesive bonding in some applications. It is a special means of applying solid or softened (but rarely fully molten) material to an always solid substrate. It is often not used to join parts, but rather to simply join the material as a coating to the substrate. There are applications, however, where the process is actually used to join parts. Figure 12.1-23 shows thermal spraying being employed to join a coating to a substrate.

There are three examples of what are really "hybrid" joining processes in which two fundamental joining processes are combined (i.e., used together) to create some synergistic benefit(s). These three are as follows: (1) *rivet-bonding*, (2) *weld-bonding*, and (3) *weld-brazing*. In addition to these hybrids between fundamental processes,

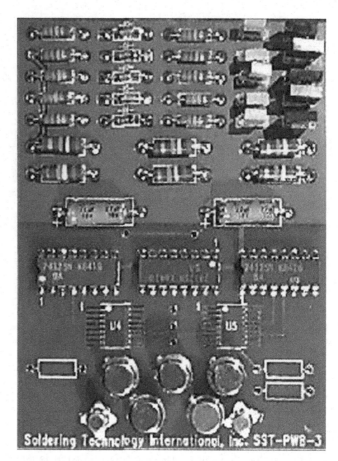

**Figure 12.1-22** Soldering is the process of choice for producing self-shaping soldered joints en masse in the microelectronics industry. Here, rather conventional through-hole and modern surface-mount technology soldered joints are shown on a printed circuit board. (Courtesy of Sandia National Laboratory, Albuquerque, NM, with permission.)

there are some hybrids between specific welding processes, to be described later.

### 12.1.7 Some key concepts relating to joints

#### 12.1.7.1 Joint loading or stress state

How a structure is loaded determines its *stress state*, and the stress state or the complexity of loading in a structure is critical to its performance. From the material's standpoint, the state of stress determines the point at which the material will yield (i.e., from Von Mises's or Treska's yield criteria, for example), and how able the material will be to respond in a ductile rather than brittle manner. The state of stress on a joint in a structure is also critically important in selecting an appropriate joining method or process.

Joining materials    CHAPTER 12.1

**Figure 12.1-23** Thermal spraying, which is a variant of welding in some manifestations and of adhesive bonding or brazing in others, can be used for creating shapes as well as for applying coatings (as shown here). (Courtesy of Foster-Wheeler Corporation, Perryville/Clinton, NJ, with permission.)

Figure 12.1-24 schematically illustrates the progressively more severe stress states of uniaxial stress, biaxial stress, and triaxial stress. It also shows how combined loading from tension or compression, bending, torsion, and internal or external pressure in a closed cylinder gives rise to a severe stress state in the cylinder's wall (biaxial if the wall is relatively thin, triaxial if the wall is relatively thick).

As a general rule, the more complex the loading, the more complex the stress state, all else (e.g., structural geometry) being equal. Hence, the greater the demands on a given joint, the poorer the performance to be expected. Biaxial loading is more severe than uniaxial loading, and triaxial loading is more severe than biaxial loading. However, the effect of stress state complexity is much greater for some forms of joining than for others. Stress state complexity and its effects are described in detail in any good reference on mechanical behavior of materials (Dieter, 1991).

For a welded joint, the stress state does not matter very much as long as the weld filler metal's solidified structure is reasonably nondirectional (thus exhibiting isotropic properties) and as long as the volume of weld metal is reasonable and fairly three-dimensional. For some alloys, welding processes, welding operating parameters, and joint configurations, however, one or the other or both of these conditions are not met, and anisotropic properties result. This can lead to serious

517

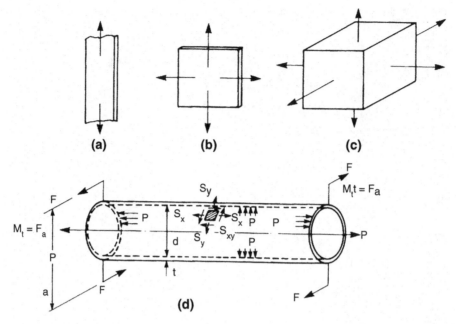

**Figure 12.1-24** A schematic illustration of the various states of stress that can arise in structural joints: (a) uniaxial tension, (b) biaxial tension, (c) triaxial tension, and (d) complex loading from combined internal pressure and external torsion. (Reprinted from *Joining of Advanced Materials*, Robert W. Messler, Jr., Butterworth-Heinemann, Stoneham, MA, 1993, Fig. 1.13, page 22, with permission of Elsevier Science, Burlington, MA.)

problems if loading and stress state are complex. For adhesive bonded joints, on the other hand, it is critically important to keep the state of stress as near to perfect shear as possible because the thin layer of adhesive usually performs badly under out-of-plane peel or cleavage loading. Brazed and soldered joints exhibit similar but less dramatic behavior, since these processes also employ thin layers of filler with little ability to tolerate strain through the filler's thickness. In mechanically fastened or integrally attached joints, how well the joint tolerates different stress states depends greatly on the particular fastener or attachment feature employed. Actually, the selection of a particular fastener, a fastener over an integral attachment, or a particular attachment feature depends greatly on the type of loading to be endured.

## 12.1.7.2 Joint load-carrying capacity versus joint efficiency

The first step in designing a joint in a structural assembly is to consider the magnitude as well as the complexity of the load(s) to be carried or transferred. The load that the joint must carry is the same as the load being carried by the structural elements on each side of the joint for simple joints but can be higher for more complex joints composed of more than two joint elements. After thinking about the load(s) to be carried, the designer considers the stress, or the load-per-unit-cross-sectional area, in each structural element to be sure that this stress does not exceed the allowable stress for the material used in each of the elements. At this point, the designer must also consider the stress in the joint.

*The joint stress*[8] is determined by dividing the load in the joint by the effective cross-sectional or load-bearing area of the joint. The effective joint area, in turn, depends on the type of joint or joint design (e.g., straight- versus scarf-butt joints or single- versus double-lap joints), the size or dimensions of the joint, and the joining method, since the method of joining directly determines how much of the joint is really carrying loads. For welded, brazed, soldered, or adhesively bonded joints, the effective joint area is almost always the same as the area of the faying surfaces (assuming continuous, full-penetration welding, or continuous full-area brazing, soldering, or adhesive

---

[8] The *joint stress* as defined here is different from the stress in the joint section. The stress in the joint section is simply the load carried by the joint divided by the effective area of the structure at the joint. For fastened joints, this is the area of the structural element minus the area of the fastener holes along some plane cutting through the joint. The *joint stress*, on the other hand, would be the load carried by the joint divided by the load-bearing cross-sectional area of fasteners or welds along some plane cutting through the joint.

Joining materials CHAPTER 12.1

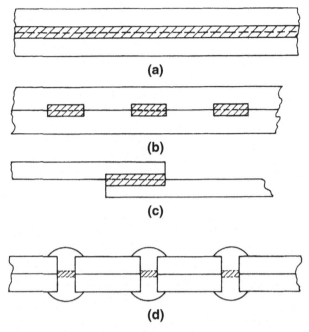

**Figure 12.1-25** A schematic illustration showing how different joining processes result in different "effective load-carrying areas" and, thus, efficiencies for: (a) continuous welding, (b) intermittent or skip welding, (c) adhesive bonded, brazed, or soldered, and (d) riveted joints. (Reprinted from *Joining of Advanced Materials*, Robert W. Messler, Jr., Butterworth-Heinemann, Stoneham, MA, 1993, Fig. 1.11, page 18, with permission of Elsevier Science, Burlington, MA.)

bonding). For mechanically fastened or integrally attached joints, the effective area is almost always much less than the area of the joint faying surfaces, and is given by the cross-sectional area of all of the fasteners or attachments used in making the joint. The actual points of joining or attachment are virtually never continuous. Brazing, soldering, and adhesive bonding, on the other hand, are almost always continuous, while welding can be continuous or discontinuous (or intermittent). Welding may be continuous when sealing against fluid leaks is required, and can be either continuous or discontinuous when welds are strictly to carry loads, with continuous welds being used when loads become larger.

Figure 12.1-25 schematically illustrates the effective area of various joints, including continuous and discontinuous welded joints, as well as brazed, soldered, and adhesively bonded joints, and various fastened and integrally attached joints.

## Illustrative example 12.1.1—calculation of joint stress

For the single-overlap joint shown in Figure IE 12.1-1, the use of two $\frac{1}{4}$-in. diameter rivets is compared to the use of a structural adhesive applied over the full $1\frac{1}{2}$-in. overlap. The actual stress (in tension) in the joint elements for an 1,800-lb. force at planes A or C is:

$$\sigma_{A(or\ C)} = 1{,}800\ lbs./(3.0\ in.)(0.125\ in.)$$
$$= 4{,}800\ lbs/in^2\ or\ psi.$$

The actual stress (in shear) in the rivets and, alternatively, in the adhesive are:

$$\tau_{rivets} = 1{,}800\ lbs./(2)(\pi)[(0.250\ in.)/2]^2$$
$$= 18{,}355\ psi$$

and

$$\tau_{adhesive} = 1{,}800\ lbs./(1.5\ in.)(3.0\ in.) = 400\ psi.$$

The much lower stress carried in shear in the adhesive joint than in the structural elements carried in tension is due to the much greater area of the adhesive bond than of the element's cross-section. Likewise, the much higher stress carried in shear in the rivets than in the structural elements carried in tension is due to the much smaller area of the rivets' combined cross-sections compared to the cross-section of the structural elements. Because of this effect of effective load-carrying area, lower strength adhesives can be compared favorably to the much higher strength fasteners in total load-carrying capacity.

It is critical that a joint be able to carry imposed loads successfully. However, there are ways to get a joint to be able to carry imposed loads that cause the joint to be unacceptably heavy. For example, the joint could be made heftier (e.g., by using a thicker section at the

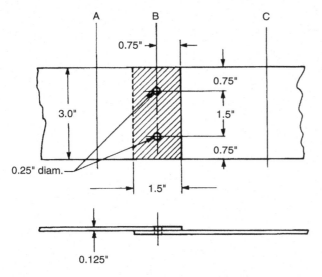

**IE 12.1-1** A schematic illustration of a fastened versus adhesively bonded single-overlap joint. Sections at A and C pass through the joint elements with their full cross-sectional area. Section B passes through the overlap area where, if fasteners requiring holes are needed, the cross-sectional area is reduced by the total area occupied by the holes. (Reprinted from *Joining of Advanced Materials*, Robert W. Messler, Jr., Butterworth-Heinemann, Stoneham, MA, 1993, Fig. 1.12, page 19, with permission of Elsevier Science, Burlington, MA.)

location of a weld or by doubling the depth of overlap of an adhesive bond). The true measure of a joint's structural effectiveness is thus whether it can safely carry the loads imposed, but the ultimate efficiency of the structure, in terms of its overall load-carrying capacity, its size, and its weight, is dependent on the efficiency of the joints making up the structural assembly. *Joint efficiency* is a measure of the effectiveness of the joint compared to the rest of the structure for carrying the design or service loads, and is defined by:

$$\text{Joint efficiency} = \frac{\text{Joint stress}}{\text{Stress in the structure} \times 100\%} \quad (12.1.1)$$

Joint efficiency varies widely depending on the joining process or method used, and can range from very low values (say 10%) to over 100%. Some examples will help to illustrate this point.

For two pieces of metal containing a continuous, full-penetration straight-butt weld whose composition is the same as the base metal, the joint efficiency is typically 100%. That is, the strength of and stress developed in the weld itself are typically equal to the strength of and stress developed in the base metal structural elements containing the weld. A 100% joint efficiency could reasonably be considered to be a characteristic of a perfect joint. But, it is possible to have joint efficiencies that are higher or lower than 100%. If a filler metal is used that is weaker than the base metal(s) adjacent to the weld, and the section thickness of the weld is the same as that of the adjacent base metal(s), the joint efficiency will have to be less than 100% if the allowable strength of the weld is not to be exceeded. If, on the other hand, a stronger filler metal is used, the joint efficiency could exceed 100%, but to load this stronger weld to its allowable strength limit would over-stress the adjacent base metal(s). In fact, the purpose of using a higher-strength filler is to keep the stress in the weld below 100% of the allowable stress for the filler when the joint elements are loaded to their design limit.

Obviously, the effective load-bearing area in a joint has an extremely strong effect on the joint's efficiency; with small effective areas leading to high efficiencies (as with mechanical fasteners, integral attachment features, and spot or intermittent or "skip" welds) and large effective areas leading to low efficiencies (as with adhesive bonded, brazed, or some continuously soldered joints).

Welded joints exhibit efficiencies of less than 100% for one of two major reasons: (1) the weld is over-designed (i.e., is longer in length or is caused to be larger in cross-section than the adjacent structural elements by creating positive reinforcement at the weld crown and root or by placing the weld in a locally thicker section or land), or (2) the weld filler metal is of lower inherent strength than the base metal(s) (i.e., it is said to be "under-matched"). Suffice it to say here, oversized welds are (or "over welding" is) done to provide a measure of added safety, to make a structure containing welded joints be more forgiving, while "undermatched" filler may be used to force the weld to fail before the rest of the structure, acting like a "safety valve." An example of the former is often the welding found on pressure vessels, while an example of the latter is found in the use of spot welds in automobiles to absorb energy of a crash to protect the vehicle's occupants.

Welded joint efficiencies typically range from approximately 50% to 100%, due to the degrading effects of the heat of fusion welding on either weld filler metal or immediately surrounding base metal properties, but they can be made to exceed 100%, as just explained. Brazed and soldered joints typically exhibit efficiencies lower than 100% (often much lower!) due to the typical use of lap (versus butt) joints and the use of lower melting (and, hence, inherently less strong) filler alloys. As service temperatures are increased, and brazed or soldered joints are loaded while at high fractions of their homologous temperatures,[9] joint efficiencies must be low at room

---

[9] *Homologous temperature* refers to temperature as a fraction of a material's absolute melting temperature, in degrees Kelvin.

**Table 12.1-3** Typical achievable joint efficiencies for various processes in various materials in terms of static strength (as a percentage)

| Joining Method | Metals | Ceramics | Glass | Polymers | Composites |
|---|---|---|---|---|---|
| Mechanical Fastening | 75–100+ | <50 | <50 | 50–100 | 50–100 |
| Mechanical Attachment | 75–100+ | near 100 | N/A | 100+ | 50–100+ |
| Adhesive Bonding | | | | | |
|   Organic adhesive | <20 | <20 | <20 | 40–100+ | 20–60 (100+)* |
|   Inorganic adhesive | N/A | 50–100+ | 20–50+ | N/A | 50+ |
|   Cement/Mortar | N/A | 50–100+ | N/A | N/A | 50–100+** |
| Welding | | | | | |
|   Fusion processes | 50–100+ | 30–80 | 100 | 100 | about 50 |
|   Non-fusion processes | 100 | 80–100 | N/A | 100 | about 50 |
| Brazing | 40–90 | 20–70 | N/A | N/A | about 50 |
| Soldering | 5–20 | <20 | 10–40 | N/A | UNK |
| Braze Welding | 50–75 | N/A | N/A | N/A | UNK |
| Thermal Spraying | | | | | |
|   Metals/Ceramics | 30–80 | 30–80 | N/A | N/A | N/A |
|   Polymers | <20 | <20 | N/A | 50–100+ | >50 |
| Rivet-Bonding | 75–100 | N/A | N/A | 100+ | UNK |
| Weld-Bonding | 25–75 | N/A | N/A | 100+ | UNK |
| Weld-Brazing | 50–80 | N/A | N/A | N/A | UNK |

N/A = Not applicable (generally)
UNK = Unknown for the process
* With wood
** With cement or concrete

temperature. Adhesive bonded joints can exhibit widely varying joint efficiencies, largely depending on the design intent in terms of the ultimate acceptable location of the failure (i.e., in the adhesive or in the adherends) and the adherend. For bonded polymers, joint efficiencies often approach 100%, while for bonded metals or ceramics, joint efficiencies are usually quite low (say 10–20%). What makes all of these low joint efficiencies tolerable is that the joints can still have high load-carrying capacity as a result of the large effective bonded area.

Mechanically fastened and some integrally attached joints also tend to exhibit high joint efficiencies, often exceeding 100%. In these joints, high efficiency is obtained by using fasteners made from higher strength materials than the joint elements are made of. This is done to compensate for the relatively low effective joint area associated with points versus areas of joining. Obviously, this approach (i.e., to use different materials) is impossible for integral attachments.

It should be recognized that all discussion on joint efficiency has been for statically loaded joints at room temperature. Obviously, dynamic loads (e.g., impact, fatigue), extreme temperature (e.g., well into the creep regime for materials involved), or aggressive corrosion environments must also be taken into account. Thus, one could and should assess joint efficiency in terms of other appropriate properties, like fatigue strength, fracture toughness, and creep or stress-rupture strength.

Table 12.1-3 summarizes the joint efficiencies in terms of static strength, typically obtained for various joining methods in various materials.

## Illustrative example 12.1.2—calculation of joint efficiency

The stress in the structural elements at the planes at A or C in Illustrative Example 12.1.1 is:

$$\sigma_{element} = Load/Area_{element}$$
$$= 1,800 \text{ lbs.}/(3.0 \text{ in.})(0.125 \text{ in.})$$
$$= 4,800 \text{ psi.}$$

The stress in the rivets, carried in shear, and in the adhesive, also carried in shear, were calculated in Illustrative Example 1.1 as $\tau_{rivets} = 18,355$ psi and $\tau_{adhesive} = 400$ psi.

Thus, the joint efficiencies for the riveted configuration versus the adhesive-bonded configuration are:

Joint efficiency for rivets
$= 18,355 \text{ psi}/4,800 \text{ psi} \times 100\% = 375\%$

while

Joint efficiency for adhesive
$= 400 \text{ psi}/4,800 \text{ psi} \times 100\% = 8.3\%.$

Obviously, the effective load-bearing area in a joint has a profound effect on the joint's efficiency, with small effective areas leading to high efficiencies and large effective areas leading to low efficiencies. Joints that are fastened with rivets or bolts tend to have high efficiencies, while joints that are bonded, brazed, or soldered tend to have low efficiencies. What matters is whether the joint, however it is made, can carry the required loads safely. Joint efficiency tends to affect the weight added to a structural assembly by the joint, with more weight added when low-efficiency joints are employed.

## Summary

The need for and the ability to join materials into components and structural elements and components and elements into assemblies and structures began with the dawn of humankind. From deep below the oceans to deep into space, from the information highway that is built upon microelectronics to the vehicles that ply the superhighways, joining pervades our world and our lives. The reasons for joining abound, with goals of achieving functionality, facilitating manufacturability, minimizing cost, and obtaining aesthetics, pretty much in that order. The ability to join allows products and structures to be created with sizes and shapes and performance unattainable in single pieces of a single material, and overcomes the limitations of primary fabrication processes (like casting, molding, forging, powder processing, and composite lay-up) and single material properties. First and foremost, joining allows static structures to remain static and dynamic structures to perform needed motions. Joining also allows both the choice of material and its utilization to be optimized. It imparts enhanced structural damage tolerance while improving structural efficiency. It allows service and maintenance and upgrading, and it facilitates ultimate disposal. But, none of this comes without challenges from the materials being joined as well as from the structures themselves. And all of it is forcing this age-old pragmatic process to also evolve to become an enabler for the future offered by information technology and biotechnology.

Using only three fundamental forces with their origin in mechanical interlocking and interference, chemical reactions, and atomic-level bonding, three fundamental options of mechanical joining (including fastening and integral attachment), adhesive bonding, and welding (including the subclasses of brazing and soldering) emerge. Each has its own advantages and disadvantages. Together, along with some variants (braze welding and thermal spraying) and hybrids (rivet-bonding, weld-bonding, and weld-brazing), these joining processes provide extraordinary diversity and capability. All require an understanding of the stress state to be tolerated by the resulting joint to provide structural integrity with structural efficiency. All require careful consideration of the materials being joined because, when all is said and done, joining of structures is the joining of materials.

## Questions and problems

1. Define the term *joining* in your own words.
2. What are the three major types of assemblies or structures, and what is the primary function of each? Give two examples of each type of assembly or structure in which there is essentially no other function than the primary function.
3. Give some examples of assemblies or structures with multiple functions. What are these functions? Which is (are) primary, and which is (are) secondary? Can an assembly or structure really have more than one primary function? Explain and give an example.
4. What are the four major goals of all design? Explain each goal and why it is important in manufacturing. Give an example of a design that places a preponderance of value on each of the four goals you identify.
5. For each design goal in Problem #4, give two reasons why joining is necessary or useful.
6. How can joining be used to render a structure that is more damage tolerant, even if the materials comprising the structural elements or components are not inherently damage tolerant? Give an example of this possibility using mechanical fastening, adhesive bonding, and welding.
7. What are the three fundamental forces that enable joining of materials and/or structures, and what is the origin of each force that allows joining to take place?
8. Modern manufacturing speaks of the desirability of "net-shape" or "near-net-shape" processing methods. The most common examples of such processing methods are casting, molding, certain deformation processing (e.g., forging), powder processing, and certain special processes particularly amenable to composites (e.g., tape lay-up). Explain how joining methods can be used for net-shape or near-net-shape

processing, giving at least one example each for mechanical fastening, integral (mechanical) attachment, adhesive bonding, and welding.

9. What are three special challenges associated with the joining of so-called "advanced" materials? Give an example of each.
10. What are some special challenges posed to joining by the structure itself independent of the materials involved? Give an example of each.
11. Explain why brazing and soldering are logically considered sub-classifications of the fundamental process of welding. Differentiate brazing from fusion welding, and soldering from brazing, in the most meaningful way.
12. Define what is meant by a "variant joining process," and name the two principal examples. Suggest a logical or known application of each.
13. Define what is meant by a "hybrid joining process," and name the three principal examples. Suggest a logical or known application of each.
14. Define what is meant by a "hybrid structure," and explain how joining is especially useful for producing such structures. Give a couple of modern examples from your experience.
15. Each specific joining process has relative advantages and shortcomings, if not disadvantages. What is the predominant advantage and what is the predominant shortcoming or disadvantage of each of the following?
    a. Mechanical fastening relative to all other processes
    b. Adhesive bonding relative to mechanical joining
    c. Adhesive bonding relative to fusion welding
    d. Fusion welding relative to brazing and soldering
    e. Brazing or soldering relative to fusion welding
    f. Brazing relative to adhesive bonding
    g. Soldering relative to brazing
    h. Brazing relative to soldering
    i. Braze welding relative to fusion welding and to brazing
    j. Weld-bonding relative to adhesive bonding and to spot welding.
16. Explain why a biaxial stress state is a more severe stress state than a uniaxial stress state, and why a triaxial stress state is the most severe of all. (Hint: consider the yield criteria for a ductile metal.)
17. Explain what limits the load-carrying capacity of a joint. Give two ways that loading capacity can be increased.
18. How would you calculate the joint efficiency of a joint required to function at elevated temperatures such as those found in operating gas turbines in aircraft? Based upon your answer, how would the efficiencies of joints produced by brazing compare to those produced by fusion welding using a filler alloy matching (i.e., of the same composition as) the base alloy?
19. A single-lap joint between two pieces of 50.0-mm-wide by 2.5-mm-thick aluminum alloy has a 25.0 mm overlap and contains four 2.5-mm-diameter aluminum alloy rivets arranged along a line, on 10.0 mm centers, with the first and last rivets located 10.0 mm from opposite edges of the joint pair. Calculate the shear stress in the rivets (i.e., the joint stress) for a unidirectional 8,800 N load applied along the longitudinal (i.e., long) centerline of the aluminum joint elements. What is the joint efficiency for this joint? Assume the following for the aluminum alloy: $\sigma_f$ = 470 MPa, $\tau_f$ = 320 MPa.
    Bonus: What is the net tensile stress in the aluminum joint elements along the line of rivets?
20. For the joint in Problem #19, calculate the stress in the joint if, instead of being riveted, the joint were brazed over the entire area of the overlap. What is the joint efficiency for this case? Assume the shear strength of the braze filler is 120 MPa and the tensile strength is 180 Mpa.
    Extra: What is the net tensile stress in the elements at the midpoint of the overlap?
21. What would be the stress in the brazed joint in Problem #19 if the joint configuration were a straight-butt rather than a single-overlap joint (i.e., the two pieces of aluminum were simply butted end to end)? What would be the joint efficiency for this case?
22. Of simple tension, simple compression, simple shear, bending, and torsion, which would cause the least problem with adhesive-bonded joints? Rank all of these loading types from least problematic to most problematic.
23. How do you think riveted joints might respond to the types of loading given in Problem #22? Would you change your answer for joints that were bolted using nuts and bolts?
24. Calculate the load in a pinned assembly of four structural elements (e.g., trusses) that come to a single pinning point (one from the left and one from the right on the same horizontal line, and one each coming into the pinning point at 45-degree downward angles from the left and the right) if a 10,000 lb. load is hung from the pinning point. How does the load in this case compare to the load in each structural element?

**25.** A 37.5-mm-wide pair of aluminum alloy AA5754 strips, 1.0 mm thick each, are to be overlapped and adhesive-bonded using a structural adhesive with a tensile shear strength of 24 MPa. How much overlap can there be if the bonded assembly is to fail just in tensile shear (not in tensile overload) in either of the aluminum alloy strips? Assume the following: Tensile strength of AA5754 = 240 MPa; WHERE IS THE REST?

## Cited References

Ashby M.F. *Materials Selection in Mechanical Design*, 2$^{nd}$ ed., Butterworth-Heinemann, Oxford, United Kingdom, pp. 1–7, 13–14, and 246–280, 1999.

Charles J.A., Crane F.A.A., and Furness J.A.G. *Selection and Use of Engineering Materials*, 3$^{rd}$ ed., Butterworth-Heinemann, London, pp. 3–31, 1997.

Dieter G.E. *Engineering Design: A Materials and Processes Approach*, 2$^{nd}$ ed., McGraw-Hill, New York. pp. 273–365, 1991.

## Bibliography

Ashby M.F., and Jones, D.R.H. *Engineering Materials 2: An Introduction to Microstructures, Processing and Design*. Pergamon Press, Oxford, 1992.

Brandon D., and Kaplan, W.D. *Joining Processes*. John Wiley & Sons, Inc., New York, 1997.

Charles J.A., Crane, F.A.A., and Furness J.A.G. *Selection and Use of Engineering Materials*, 3$^{rd}$ ed., Butterworth-Heinemann, London, 1997.

Datsko, J. *Materials Selection for Design and Manufacturing: Theory and Practice*. Marcel Dekker, New York, 1997.

Dieter, G.E. *Engineering Design: A Materials and Processes Approach*, 2$^{nd}$ ed. McGraw-Hill, New York, 1991.

Faupel, J.H. and Fisher F.E. *Engineering Design*, 2$^{nd}$ ed. John Wiley & Sons, Inc., New York, 1981.

Lindberg R.A. *Processes and Materials of Manufacture*, 4$^{th}$ ed. Allyn and Bacon, Needham Heights, MA, 1990.

Marganon P.L. *The Principles of Materials Selection for Engineering Design*. Upper Saddle River, NJ Prentice Hall. 1999.

Messler R.W. Jr. *Joining of Advanced Materials*. Butterworth-Heinemann. Stoneham, MA, 1993.

Parmley R.O. *Standard Handbook of Fastening & Joining*, 2$^{nd}$ ed., McGraw-Hill, New York, 1989.

Poli C. *Design for Manufacturing: A Structured Approach*. Butterworth-Heinemann, Boston, 2001.

Swift K.G., and Booker J.D. *Process Selection: From Design to Manufacture*. Arnold, London, 1997.

Todd R.H., Allen D.K., and Alting L. *Manufacturing Processes Reference Guide*. New York. Industrial Press, 1994.

# Index

AC electrodeposition, 466
Acidic anodization, 464
Adherends, 512
Adhesive, 512
   bonding, chemical forces, 512–14
Adiabatic process, 291
Agglomerated nanoparticles, 451
Alnico (Al–Ni–Co) alloys, 147, 434, 435
Aircraft braking systems, 367
Aligned nanoelectrode array-based electronic chips, 488–90
Alkyl silanes, 356
Allied Corporation's Spectra, 900, 355
Alloys, ordering in:
   detection of, 132–4
   influence on properties, 134–5
   long-range and short-range, 131–2
Alloy system, 115
Alumina fibers, commercial, 355
Aluminum nitride (AlN) dispersions, 382
Anelasticity and internal friction, metals, 129–31
Anisotropy:
   magnetocrystalline anisotropy, 425–7
   shape anisotropy, 427
Annealing, 21–2
   definition, 198
   effects of, 198
   grain growth, 202–4
   recovery process, 198–9
   recrystallization, 199–202
   recrystallization textures, 205–6
   twins, 204–5
Anodic porous alumina, 463–5
Anodization, acidic, 464
Antiferromagnetic materials, 415
Anti-ferromagnetism and ferrimagnetism, 148–50
APB-locking model, 174–5
Aramid, 355
Archard wear constant $(K_A)$, 11
Arrhenius equations, 129, 372
Artificial hip joints, metals for, 102–3
Atomic bonding, in materials, 15–16
Atomic magnetic moments:
   multielectron atoms, 419–20
   single electron atoms, 416–19
Atomic system, 416
Autoclave moulding, 336
Automatic processing processes:
   for reinforced thermosets, 338
Automotive handle:
   stages in gas injection moulding of, 318
AVCO Specialty Materials Company, 356
Avrami equation, 36

Ballistic particle manufacture (BPM), 77
Barkhausen effect, 146
Barrels, used in injection moulding, 309
Basquin's law, 215
$BaTiO_3$, 151
Beryllium, 355
B-field, 441
B-H curves:
   hard magnetic materials, 430
   second quadrant, 432
   soft magnetic materials, 430
B-H loop, 427
BHN, see Brinell hardness number
Biaxial orientation, in blow moulding, 302–3
Biomolecule manipulation, 456–7
Bipolar disk magnet:
   wire configuration for magnetizing, 439
Blister packs, 322
Block copolymer, 362
Blow moulding, extrusion, 300–3
   analysis of, 301–3
   stages in, 301
Bohr magneton, 418
Boltzmann's constant, 18
Boltzmann statistics, 420
Bonded magnets, 436–7
Bonding, wettability and, 392
Boron carbide ($B_4C$) fibers, 356
Boron fibers, 353, 356
Bragg's law, 133
Brass, 101
Braze welding, 516
Brazing, 515
Breaker plate with filter pack, extruder screw, 290–1
Brillouin function, 421, 422
Brillouin zone, 136–7
Brinell hardness number (BHN), 159
Brute force method, 463
Bulk modulus, 157
Bulk moulding compound (BMC), see Dough moulding compounds (DMC)

CAD solid-modeling software, 76
Calendaring, 364
Calendering, 326–8
Cancer therapy, 457
Capacitor bank, 438
Carbide whiskers, 374
Carbon-carbon (C-C) composites, 367–70
Carbon fibers, 353–5
   vapor-grown, 476
Carbonization, 354

*Engineering Materials and Processes Desk Reference*; ISBN: 9781856175869
Copyright © 2009 Elsevier Ltd; All rights of reproduction, in any form, reserved.

# Index

Carbon nanotubes (CNT), 354, 450–4, 471–81
  biosensor applications, 487–90
    aligned nanoelectrode array-based electronic chips, 488–90
    FET-based, 487
  composite materials, 476–81
    ceramic-coated nanotubes, 479–80
    ceramic nanocomposites, 478–9
    ceramic nanotube composite systems, 479
    conductive ceramics, 480–1
    polymer nanocomposites, 477–8
  growth mechanisms, 475–6
  properties, 471–2
  structure, 471, 472, 473
  synthesis, 472–3
Carbon nanotube template, for metal nanowires fabrication, 466–7
Casting, low-cost composites by, 390–1
Cell manipulation, 456–7
Cement, 344, 345
Centrifugal casting, 338
Ceramic-coated MWNT, 479–80
Ceramic-coated nanotubes, 479–80
Ceramic composites, 345
Ceramic fibers, 355–7
Ceramic magnets, properties, 433
Ceramic-matrix composites, 365–7
Ceramic nanocomposites, 478–9
Ceramics, 4
  data for, 345
  high-performance engineering, 344
  natural, 345
  properties of, 346
  vitreous, 344
Ceramic 'windows,', 154
C-glass, 352
CGS, *see* Gaussian system (CGS)
Charpy V-notch test, 32
Chemical methods, for nanoparticle synthesis, 453–4
Chemical vapor deposition (CVD), 453
Chemical vapor infiltration (CVI), 365, 367, 370–2, 379
Chopped fibres, 334
Chopped strand mat, 334
Clamping systems:
  used in injection moulding, 309–10
Coating processes, extrusion, 303–4
Coaxial heterostructures, 458
Co-base superalloys, 355
Coefficient of linear thermal expansion (CTE), 42–3
Coercive force, 426
Co-extrusion, 304–5
Coffin-Manson law, 215
Cold press moulding, 335, 336
Composite fabrication methods, 75–6
Composite materials, definition, 351
Compression-molded magnets, 436
Compression moulding, 331–2, 337
Compression zone, extruder screw, 288
Computer-aided material selection, 61–3, 91–4

Concrete, 344, 345
Conductive ceramics, 480–1
Continuous cooling transformation (CCT) diagram, 37–8, 40
Conversion factors:
  CGS, 413–15
  MKS, 413–15
Copper interconnects, for microelectronic packages, 14
Copper–nickel diagram, 118
Cost model, 89–91
Coulomb flow law, 7
Creep, 404–5
  deformation, 34–5
  design of creep-resistant alloy, 210–12
  fracture, 210
  grain boundary sliding, 209–10
  steady-state, 206–9
  tertiary, 210
  testing, 160
  transient, 206–9
Creep-resistant alloy, design, 210–12
Critical shear stress, law of, 164–5
Cross-linked thermosetting resins, 362
Crystalline SiC nanowires, 450
Crystalline solids:
  defects in, 17–21
  diffusion in, 22–5
  structure, 16–17
Crystal structure:
  of iron, 424
  yield points and, 172–4
Cunico (Cu–Ni–Co) alloys, 147
Cup-and-cone fracture, 159
Curie's law, 421
Curie temperature, 146, 422, 427, 445
CVD, *see* Chemical vapor deposition (CVD)

Darcy's equation for flow, 386
DC electrodeposition, 466
Debye's maximum frequency, 124
Defect-free nanotubes, 471
Defective nanotubes, 471
Deformation mechanism map, 212–13
Degree of polymerization (DP), 362
Demagnetization curve, 432
  alnico magnets, 434
  ceramic magnets, 442
  sintered SmCo materials, 443
Demagnetization field, 423–4
Dense random packing (drp), 108
Density, of metals, 121
Design for assembly (DFA), 80, 85, 94
Diamagnetic materials, 415
Diamagnetism and paramagnetism, metals and, 144–5
Die characteristics, extruder screw, 294–7
Dielectric constant, 10
Diffusion, metals:
  factors affecting, 129
  laws, 126–8
  mechanisms of, 128–9

Discontinuous yielding, characteristics:
  Lüders band formation, 171–2
  overstraining, 171
  strain-age hardening, 171
  yield point, 171
Dislocation:
  discontinuous yielding:
    characteristics, 171–2
    ordered material, 174–5
  edge, 167
  Frank-Read source, 170–1
  locking of, 177–8
  mobility of, 166–8
  screw, 169
  solute-dislocation interaction, 175–7
  temprature, 177–8
  yield points, 172–4
  yield stress variations, 168–9
Dispersion-hardened alloys, 191–2
Dispersion-strengthened composites, 372–3
Distribution tube system, 314
DNA template, for metal nanowires fabrication, 467
Domains, magnetic materials, 427–8
Domain walls, 428
Double-strand DNA (dsDNA), 467
Dough moulding compounds (DMC), 337
Drag flow, extruder, 291–2
Drape forming, 322
Draw-down, 299
Drinks cans, metals for, 102
Dry-spun yarn, 355
DsDNA, see Double-strand DNA (dsDNA)
Dual-sheet thermoforming, 323
Ductility, 158
Du Pont's aramid fiber Kevlar, 355
Dynamic structures, 500

Economic batch-size chart, 88–9
E-glass, 352
Einstein characteristic temperature, 124
Elastic deformation, 160–2
Elastic modulus, 6, 26, 157
Elastomers, 4
Electrical assemblies, 498
Electrical conductivity, of metals, 135–8
Electrical resistivity, 10
Electrochemical fabrication, metal nanowires, see Metal nanowires
Electrodeposition, in nanowires, 465
  AC, 466
  DC, 466
Electro-discharge machining (EDM), 78
Electronic chips, aligned nanoelectrode array-based, 488–90
Electron metallographic observations, 189
Electro-optic ceramics, 154
Ellipsoids, 423, 424
Endurance limit, 160
Environment, fatigue, 215–16
Extruded panel sections, 299

Extruded window profile, 299
Extruder screw, 288
  compression zone, 288
  development of, 294
  die characteristics, 294–7
  feed zone, 287–8
  general features of twin, 297
    versus single, 298
  mechanism of flow, 291
  metering zone, 288–91
  outputs for different plastics, 289
  processing methods, 297–306
  schematic view, 288
  venting zone, 289–90
Extrusion, 287–304
  analysis of flow, 291–4
  die characteristics, 294–7
  mechanism of flow of plastic, 291
  processing methods, 297–306
  recent technological developments, 304–6
  zones of extruder screw, 287–91

Fabrication, metal nanowires, see Metal nanowires
Fabrication of metal-matrix composites, 378–9
Fabrication process, manfacturing, 505–6
Fatigue, 32–3
Fatigue, metallic:
  corrosion, 215
  cracks, 218–20
  at elevated tempratures, 220
  engineering aspects of, 214–16
  failures of, 213, 218–20
  hardening, 216–18
Fatigue and fracture toughness, 404
Fatigue testing, 160
FEA, see Finite-element analysis (FEA)
Feed zone, extruder screw, 287–8
Fe-Fe$_3$C diagram, 36
Ferrimagnetic materials, 415
Ferrimagnetism:
  anti-ferromagnetism and, 148–50
Ferrites, 147, 433–4
Ferromagnetic materials, 415, 427
Ferromagnetism, 145–6, 421–3
FET, see Field-effect-transistors (FET)
FET-based biosensors, 487
Fiber FP, 355
Fiber(s), 351–2
  strength, 399
  strengthening, 359–60
Fibre reinforced thermoplastics, 333
Fibre reinforced thermosets, 333–8
  manufacturing methods, 334
Fick's equations, 23, 43
Fick's law of diffusion, 126–7
Field distrubution, bar magnet, 423
Field-effect-transistors (FET), 487

# Index

Filament, 333
   winding, 336, 337, 338
   woven strands of, 334
Film blowing, extrusion, 298–300
Finite-element analysis (FEA), 440
Fixture analysis, 440, 441
Flux density, 426
Flux lines, 441
Foil-fiber-foil technique, 381
Fracture mechanics, 30–1
   creep deformation, 34–5
   fatigue, 32–3
Fracture toughness parameter ($K_c$), 160
Free energy of transformation, of metals, 124–6
Frequency of stress cycle, fatigue, 214
Functionalized magnetic nanoparticles, 456
Fundamental forces, joining, 509
Fused deposition modeling (FDM), 77

Gas injection moulding, 317–18
Gates, 310
   used in injection moulding, 311
Gauss error function (erf (y)), 127
Gaussian system (CGS), 413
GFRP:
   pre-form moulding of, 335
Glasses, 4, 344, 352–3, 356
Glass fiber-reinforced plastics (GRPs), 353
Glass transition temperature, 359–60
Gold, 457
Grain boundary:
   influence on plasticity, 178–80
Grains and phases
   boundaries, 110–11
   shapes of, 111–12
   see also Metals
Granule production/compounding, extrusion, 297
Graphitic form of carbon, 353
Graphitization, 354
Grid structure, extrusion:
   highly oriented, 305–6
Griffith equation, 31

Hagen-Poiseuille model, 386
Hall effect, 140–1
Hall-Petch equation, 29, 177
Hand lay-up techniques:
   for processing reinforced thermosets, 334–5
Hard magnetic materials:
   alnico, 434
   bonded magnets, 436–7
   ferrites, 433–4
   neodymium-iron-boron, 435–6
   samarium-cobalt, 434–5
Hardness testing, 159
Heaters:
   used in injection moulding, 309
Heat sinks, for hot microchips, 54
Heusler alloy, 144, 147

Hexagonal metals, 173
High-energy ball milling, 382
High-performance engineering ceramics, 344
Homopolymer, 361
Hooke's law, 25–7, 157, 161, 362
Hot isostatic pressing (HIP), 381
Hot pressing, 373
Hot press mouldings, 336–7
Hot runner moulds, 313–14
Hume-Rothery rules, 18
Hybrid joining process, 516
Hybrids, 4–5
Hydrocarbon-cracking processes, 450
Hydrogen decrepitation (HD), 148–9
Hydrolysis, 309
Hypernik, 147
Hysteresis:
   loop, 428, 429, 430
   magnetic materials, 428–9

Ignitron tube, 438
Impact testing, 159–60
Impulse magnetizing fixtures, 439
   design, 440
Inert gas condensation, 452
Infiltration processes for liquid matrix, 382–6
Information-processing devices, 449
Inhomogeneity interaction, 178
Injection blow moulding, 319–20
Injection moulding, 306–21, 338
   cycle, 308
   stages during, 309
   temperature control, 312
   of thermosetting materials, 320–1
Injection orientation blow moulding, 320, 321
Injection stretch blow moulding, 303
In situ composites, 374–6
Insulated runner moulds, 314
Integral mechanical attachment, mechanical forces, 509–12
Interfaces, 357–9
Interface strength, 392–9
Internal crystallization method (ICM), 383
Ion sputtering, 453
Iron-carbon alloys, 36
Iron-silicon alloys, 431

Johnson-Mehl equation, 200
Joining, 509–16
   chemical forces, 512–14
   fundamental forces, 509
   materials, see Joining materials
   mechanical forces, 509–12
   physical forces, 514–16
Joining materials:
   challenges, 504–5
   change, 505–9
   defined, 497–8
   key concepts, 516–22
   reasons for, 498–504

value adding process, 507
Joining structures:
   challenges, 505
   reasons, 498–504
Joint efficiency, 520
   calculations of, 521–2
   vs joint load, 518–19
Joint load:
   calculations, 519–21
   vs joint efficiency, 518–19
Joint stress, see Joint load
Jominy end-quench test (ASTM Standard A 255), 39

Kê torsion pendulum, 130
Kevlar, 355, 356
Kinetics of strain ageing, 178
Kirkendall effect, 128–9
Knudsen diffusion, 372

L, see Orbital angular momentum (L)
Lamellaé, 361
Landé g factor, 419
Langevin function, 421
Laser pyrolysis, 453
Lasers, metals and, 153–4
Lattice spacing, in BN cell, 450
LDR, see Limiting draw ratios
Lead-tin phase diagram, 116–18
Leakage flow, extruder, 293–4
Limiting draw ratios (LDR), 195
Linear thermal expansion coefficient, 9
Lingo-cellulosic fibers, 357
Liquid matrix, infiltration processes for, 382–6
Liquid-phase processing, in metal nanocomposites, 482–4
Liquid-phase techniques, 391–2
Longitudinal heterostructures, 458
Lorentz force, 382
Low-cost composites by casting, 390–1

Macroscopic plasticity:
   stress–strain relationship, 197
   tresca yield criterion of, 196–7
   von Mises yield criterion of, 196–7
Magnequench (MQ), 435
Magnetic alloys, 146–8
Magnetic dipole, 414
Magnetic materials:
   anisotropy, 424–7
   classification of, 415–16
   domains, 427–8
   hard, 432–3
   hysteresis, 428–9
   soft, 430–2
   stability, 442–5
Magnetic nanowires, 469
Magnetic susceptibility, metals and, 144
Magnetic viscosity, 445
Magnetization, materials, 437–42
Magnetization curves, 424

Magnetizing system, 438
Magnetocrystalline anisotropy, 425–7
Magnetohydrodynamic (MHD) stirrers, 388
Magnetostatic energy, 423
Magnox (Mg), 220
Manual processing methods:
   for reinforced thermosets, 334–6
Mass bar-chart, 82–3
Matched die forming, 323
Material indices:
   heat sinks for hot microchips, 54–5
   for light, stiff beam, 57
   for light, strong tie-rod, 56–7
Materials, engineering:
   behavior:
      annealing, 21–2
      atomic bonding in, 15–16
      crystalline solids, see Crystalline solids
      crystal structure, 16–17
      deformation processing, 35
      dielectric and magnetic properties, 45–7
      electrical properties and, 43–5
      fracture mechanics, 30–5
      heat treatment, 35–41
      mechanical, 25–9
      optical properties, 47–8
      precipitation hardening, 41
      process innovation as driver of technological growth, 13–15
      strengthening of metals, 29–30
      thermal properties and, 41–3
   families of, 3–5
   process cost, 84–91
   processes classification, 70–2
      finishing, 80
      joining, 79–80
      shaping, 72–9
   properties:
      eco, 10
      electrical, 10
      environmental resistance, 11
      general, 5–6
      mechanical, 6–9
      optical, 10
      thermal, 9–10
   selection, 51
      attribute limits and material indices, 54–7
      computer-aided, 61–3, 91–4
      deriving indices, 57–8
      procedure for, 58–61
      strategies for, 52–4
      structural index, 63–4
   systematic process selection, 80–4
Materials science and engineering (MSE), 13
MCM-41, 450
Mean stress, fatigue, 214–15
Mechanical assemblies, 498
Mechanical fastening, mechanical forces, 509–12
Mechanical forces, joining, 509–12

# Index

Mechanical test:
  creep testing, 160
  fatigue testing, 160
  hardness testing, 159
  impact testing, 159–60
  tensile test, 157–9
Mechanical twinning:
  crystallography of, 182
  dislocation mechanism, 184
  effect of impurities on, 184
  effect of prestrain on, 184
  fracture and, 184–5
  growth of, 182–4
  nucleation of, 182–4
Meissner effect, 142
Melt infiltration method, 366–7
MEMS, see Microelectromechanical systems (MEMS)
Metallic fibers, 355
Metallization, in nanowires, 467
Metal nanocomposites, 481–7
  agglomeration, 485–6
  dispersion, 485–6
  interface phenomena, 484–5
  liquid-phase processing, 482–4
  nucleation, 484–5
  properties, 486–7
  reactivity, 484–5
  sedimentation, 485–6
  solid-state powder-based processing, 481–2
  surface phenomena, 484–5
Metal nanowires:
  electrochemical fabrication, 461–5
    anodic porous alumina, 463–5
    negative template methods, 461–3
  fabrication, 460–1, 465–7
    carbon nanotube template, 466–7
    DNA template, 467
    polymer templates, 467
    positive template method, 466
Metals, 3
  alloys, ordering in:
    detection of, 132–4
    influence on properties, 134–5
    long-range and short-range, 131–2
  alloy system, 115
  aluminium-based, 102
  anelasticity and internal friction, 129–31
  for artificial hip joints, 102–3
  copper-based, 101
  copper–nickel diagram, 118
  density, 121
  dielectric materials:
    capacitors and insulators, 151
    piezoelectric materials, 151
    polarization, 150–1
    pyroelectric and ferroelectric materials, 151–2
  diffusion:
    factors affecting, 129
    laws, 126–8
    mechanisms of, 128–9
  for drinks cans, 102
  electrical properties:
    electrical conductivity, 135–8
    Hall effect, 140–1
    oxide superconductors, 143–4
    semiconductors, 138–40
    superconductivity, 141–3
  incompletely defined constitutions, 118–19
  iron-based, 100
  lead-tin phase diagram, 116–18
  magnetic properties:
    anti-ferromagnetism and ferrimagnetism, 148–50
    diamagnetism and paramagnetism, 144–5
    ferromagnetism, 145–6
    magnetic alloys, 146–8
    magnetic susceptibility, 144
  for metals, 103–4
  for model traction engine, 99–102
  nickel-based, 101
  optical properties:
    ceramic 'windows,', 154
    electro-optic ceramics, 154
    lasers, 153–4
    optical fibers, 153
    reflection, absorption and transmission effects, 152–3
  structures:
    crystal and glass, 107–8
    grain and phase boundaries, 110–11
    phases, 110
    shapes of grains and phases, 111–12
    of solutions and compounds, 108–10
  thermal properties:
    free energy of transformation, 124–6
    specific heat capacity, 123–4
    specific heat curve and transformations, 124
    thermal expansion, 121–3
  titanium-based, 103
Metering zone, extruder screw, 288–91
Mg, see Magnox
Microelectromechanical systems (MEMS), 505
Miner's concept of cumulative damage, 215
Miniaturization, 449
MKS, see SI systems (MKS)
Mobil Oil Company, 450
Model traction engine, metals for, 99–102
Modulus of rupture, 345
Molecule-recognition element, 457
Morse tapers, 511
Mould clamping force, 314–16
Moulds, in injection moulding, 310–11
  hot runner, 313–14
  insulated runner, 314
  multi-daylight, 313
MQ, see Magnequench (MQ)
Multi-daylight moulds, 313
Multielectron atoms, 419–20
Multiple-chip modules (MCMs), 54
Multiple slip, 165–6

Multipole disk magnet, 442
Multiwall nanotubes (MWNT), 472, 474, 479–80
MWNT, see Multiwall nanotubes (MWNT)

Nanoceramic composites (NCC), 479
Nanoparticles, 450–4
   biomedical applications, 454–67
      biomolecule manipulation, 456–7
      cancer therapy, 457
      cell manipulation, 456–7
      protein detection, 457
      semiconductor nanowires, 457
      tissue engineering, 455–6
Nanoparticle synthesis, by chemical methods, 453–4
Nanoporous ceramics, 450
Nanostructured metals, see Metal nanocomposites
Nanostructured microprocessor devices, 450
Nanotechnology, 449
Nanotribology, 486
Nanotubes, see Carbon nanotubes (CNT)
Nanowires, 450–4
   aligned amorphous silicon, 461
   applications:
      biological assays, 470
      chemical sensors, 471
      magnetic devices, 468–70
      magnetic materials, 468–70
      optical applications, 470
   GaN, 463
   growth, 458
   metal nanowire electrochemical fabrication, 461–5
   metal nanowire fabrication, 460–1, 465–7
   semiconductor, 457
   semiconductor nanowire fabrication, 459–60
   silicon, 462
   synthesis, 462
   ZnO, 464
Natural ceramics, 345
NCC, see Nanoceramic composites (NCC)
Neck ring stretch blow moulding, 303
N(E)–E curve, 137
Negative forming, 322
Negative temperature coefficient (NTC), 140
Negative template methods, 461–3
Neodymium-iron-boron, 435–6
Neomax ($Nd_2Fe_{14}B$), 147
Neumann bands, 183
Nextel, 312, 355
Nextel fiber, 353
Ni-Al inter-metallics, 374–6
Ni-base composites, 392
Ni-base superalloys, 355
Nicalon fibers, 356
Nickel, 101–2
Nickel-base intermetallics, 381
Nickel-iron alloys, 431–2
Nimocast, 75, 377
Nimocast, 258, 377
Nimocast 713 C, 377

Nimonic alloys, 211
Niobium fibers, 377
NLP 101 fiber, 356
Nonagglomeration, of nanoparticles, 451
Nozzle:
   used in injection moulding, 309

Ohm's law, 43, 140
Optical computing, 470
Optical fibers, 153
Optical scattering, 470
Orbital angular momentum (L), 416
Organic fibers, 355
Organosilazane compound, 356
Orientation of crystals:
   crystallographic aspects, 193–4
   texture hardening, 194–6
Ormosil, 457
Oxidation resistance, 408
Oxide dispersion-strengthened (ODS) composites, 372–3
Oxide superconductors, 143–4

PAA, see Polyacrylic acid (PAA)
Paramagnetic materials, 415
Paramagnetism, 420–1
Particle diameters, for single-domain magnetic nanoparticles, 456
Particulate and fiber-reinforced high-temperature alloys, 376–7
Particulate reinforcements, 376–7
Pauli exclusion principle, 145, 419
PEI, see Polyethyleneamine (PEI)
Peierls–Nabarro (P-N) stress, 167
Permalloys, 147
PFZs, see Precipitate-free zones
Photodynamic cancer therapy, 457
Pipe extrusion, 300
Pitch-based raw materials, 354
Plain-strain fracture toughness, 31
Planck's constant, 47, 123
Plastic deformation:
   critical shear stress law, 164–5
   dislocation behavior, 166–82
   multiple slip, 165–6
   relation between work hardening and slip, 166
   relation of slip to crystal structure, 164
   resolved shear stress, 163–4
   slip and twinning process, 163
Plastics:
   processing of, 287–338
      extrusion, 287–304
   water sensitivity of, 310
Plug assisted forming, 322
Plunger type injection moulding machine, 307
P–n junction rectification, 139
P-N stress, see Peierls-Nabarro stress
Point defect hardening, 185–6
Polarization, metals and, 150–1
Polyacrylic acid (PAA), 480
Polyacrylonitrile (PAN), 353

# Index

Polycarbosilane, 356
Polycrystals, 191
  see also Crystal structure
Polyethylene, 362
Polyethyleneamine (PEI), 480
Polyethylene (PE), 151
Polymer brush, 467
Polymer composites:
  properties of, 363–5
Polymer-matrix composites, 360–2
  properties of, 362–3
Polymer nanocomposites, 477–8
Polymers, 4
Polymer templates, for metal nanowires fabrication, 467
Polymethylmetacrylate (PMMA), 362
Polyvinyl acetate, 352
Polyvinyl chloride (PVC), 151
Positive forming, 322
Positive temperature coefficient (PTC), 140
Positive template method, for metal nanowires fabrication, 466
Powder methods, 74–5
PRD16, 377
Precipitate-free zones (PFZs), 219
Pre-form moulding, 337
  of GFRP, 335
Prepregs, 364
Pressure flow, extruder, 292–3
Pressure forming, 322–3
Process–material matrix, 81–2, 84
Production energy, 10
Profile production, extrusion, 297–8
Protein detection, 457
Pulsed laser ablation, 452–3
Pultrusion, 338

Quantum confinement, 472
Quantum numbers, 417

Random copolymer, 361–2
Rapid prototyping systems (RPS), 76–8
Rare-earth materials, 438
Ratchet mechanism, 218
Rayon, 353
Reaction choking, 387
Reaction injection moulding, 319
Reactive extrusion, 306
Reactive infiltration, 386–8
Recrystallization, 199–202
  textures, 205–6
Refractive index, 10
Refractory metals, fibers of, 377
Relative permeability, silicon steel, 431
Resin injection process, 336, 338
Resolved shear stress, 163–4
Right-hand rule, 416
Rigidity modulus, see Shear modulus
River lines, 159
Rotational moulding, 328–31
Roving, 333

Rule of mixture (ROM), 359
Runners, 310
  used in injection moulding, 312
Russell-Saunders coupling, 419, 420

Samarium-cobalt, 434–5
Samarium–cobalt alloys ($SmCo_5$ and $Sm_2(Co, Fe, Cu, Zr)_{17}$), 147
Sandwich moulding, 317
SAP, see Sintered aluminum powder
Schrödinger's equation, 416
Screen packs, extruder screw, 290
Screws:
  used in injection moulding, 309
Section thickness bar-chart, 83
Seizing, 510
Selected laser sintering (SLS), 77
Self-propagating hightemperature syntheses (SHS), 379
Semi-automatic processing methods:
  for reinforced thermosets, 336–8
Semiconductor devices, 507
Semiconductor nanostructure synthesis, 459
Semiconductor nanowire fabrication, 459–60
Semiconductor nanowires, 457
Semiconductors, 138–40
S-glass, 352
Shape anisotropy, 427
Shear controlled orientation in injection moulding (SCORIM), 318–19
Shear direction, 182
Shear modulus, 157
Sheet extrusion, 299
Sheet moulding compounds:(SMC), manufacturing of, 336
SiC, 365
SiC-coated filament, 356
Silicon carbide fibers, 356
Single-crystal turbine blades, 13–14
Single electron atoms, 416–19
Single-strand DNA (ssDNA), 467
Single-wall nanotubes (SWNT), 472, 474, 479–80
Sintered aluminum powder (SAP), 202
SI systems (MKS), 413
Skin packaging, 322
Slip planes, 163
Slip process, 163
Slush moulding, 331
S–N diagram, 160, 162
Soft ferrites, 432
Soft magnetic materials, 430–2
Soldering, 515–16
Sol-gel-type chemical precipitation, 352–3
Solid ground curing (SGC), 77
Solid-state composite fabrication, 379–82
Solid-state powder-based processing, in metal nanocomposites, 481–2
Sommerfeld convention, 413
Spatial quantization, angular momentum, 417
Specific heat, of metals, 123–4
  curve and transformations, 124

Spin angular momentum, electron, 417
Spin-orbit coupling, 418, 419, 425
Spontaneous magnetization vs temperature, 422
Spray-forming route, 380
Spray-up techniques:
    for reinforced thermosets, 335, 336
Sprue, 310
    used in injection moulding, 312
ssDNA, see Single-strand DNA (ssDNA)
Stability, magnetic materials, 442–5
Stainless steel, 100–1, 108
    see also Metals
Static structures, 500
Steady-state creep, 206–9
Stereo-lithography (SLA), 77
Stiffness, strength, and ductility, 399–404
Stirring, mechanical, 388–90
Strain ageing, 171
Strain-hardening exponent, 28, 30
Strain tensor, 161
Strengthening/hardening mechanisms:
    orientation of crystals, 193–6
    point defect hardening, 185–6
    work hardening, 186–93
Stress–elongation curves, 157, 158
Stress state, 516–18
    effects, 517
    joint stress, calculation, 519–21
Stretch blow moulding, extrusion, 302–3
Stretcher strains, 172
Structural assemblies, 498
Structural foam injection moulding, 317
Superconductivity, of metals, 141–3
Superplasticity, 180–2
Supersaturation, 451
    methods:
        using liquid/vapor precursors, 453
        using solid precursors, 452–3
Surface preparation, fatigue, 214
SWNT, see Single-wall nanotubes (SWNT)
Symbol description:
    CGS system, 413
    MKS system, 413
Symmetry-breaking solid-liquid interface, 459

T, see Teslas (T)
Tailor-welded blanks (TWBs), 14–15
Temperature effects on, fatigue, 214
Tensar, 305, 306
Tensile strength (TS), 158
Tensile test, 157–9
Teslas (T), 413
Thermal conductivity, 9
Thermal fatigue, 408
Thermal properties, 407–8
    metals, 121–3
Thermal shock resistance, 10, 345
Thermal spraying, 516
Thermal state of extruder melt, 291

Thermal vapor growth, 459
Thermoforming, 321–6
    analysis of, 323–6
    dual-sheet thermoforming, 323
    matched die forming, 323
    pressure forming, 322–3
    vacuum forming, 321–2
Thermoplastics, 361, 362
    processing reinforced, 333
Thermosets:
    injection moulding of, 320–1
    processing reinforced, 333–8
Thermosetting:
    polymers, 361
    resins, cross-linked, 362
Tissue engineering, 455–6
Titanium, 456
Tolerance and surface-roughness bar-charts, 83
Tow, 333
Transfer moulding, 332–3
Transient creep, 206–9
Transmission electron microscopy (TEM), 19
Tresca yield criterion, 196–7
TS, see Tensile strength
Tungsten filament, for light bulbs, 14
Twin extruder screw, 297
    versus single screw, 298
Twinning:
    plane, 182
    process, 163
Twin-sheet forming, 323
Two-dimensional dislocation structures, 189

U see Linear velocity (u)
Units, magnetism, 413–15

Vacuum forming, 321–2
Vacuum injection, 336, 338
Vapor-grown carbon fibers, 476
Vapor-liquid-solid (VLS) process, 457
Vapor-phase synthesis, of nanoparticles, 451
Vapor-solid (VS) process, 457
Variant joining process, 516
Vegard's law, 121
Venting zone, extruder screw, 289–90
Vibration damping, 405
Vickers hardness number (VPN), 159
Vitreous ceramics, 344
VLS process, see Vapor-liquid-solid (VLS) process
von Mises yield criterion, 196–7
VPN, see Vickers hardness number
VS process, see Vapor-solid (VS) process
Vulcanized rubber, 361

Wavefunctions $\Psi$, 416
Wb, see Webers (Wb)
Wear and friction, 405–7
Webers (Wb), 413
Weiss mean field theory, 422

# Index

Welding, 510
  physical forces, 514–16
    brazing, 515
    hybrid joining process, 516
    soldering, 515–16
    variant joining process, 516
Wettability and bonding, 392
Wiedemann-Franz law, 43, 135
Wire covering die, 304
Work hardening:
  definition, 166
  dispersion-hardened alloys, 191–2
  in ordered alloys, 192–3
  in polycrystals, 191
  theoretical treatment, 186–8
  Three-stage, 188–91
Woven strands of filament, 334

Yarn, 333
Young's modulus, *see* Elastic modulus
Young's modulus (E), 6, 53
Yttria-stabilized zirconia (YSZ), 365–6
Yttrium aluminum garnet (YAG), 383

Zirconium diboride ($ZrB_2$) matrix, 367

Printed in the United States
By Bookmasters